T0313199

Basic Analysis I

Basic Analysis I: Functions of a Real Variable

The cephalopods wanted to learn advanced mathematics
and decided to contact Jim.

James K. Peterson
Department of Mathematical Sciences
Clemson University

 CRC Press
Taylor & Francis Group
Boca Raton London New York

CRC Press is an imprint of the
Taylor & Francis Group, an **informa** business

A CHAPMAN & HALL BOOK

First edition published 2020
by CRC Press
6000 Broken Sound Parkway NW, Suite 300, Boca Raton, FL 33487-2742

and by CRC Press
2 Park Square, Milton Park, Abingdon, Oxon, OX14 4RN

ISBN: 9781138055025 (hbk)
ISBN: 9781315166254 (ebk)
LCCN: 2019059882

Dedication We dedicate this work to all of our students who have been learning these ideas of analysis through our courses. We have learned as much from them as we hope they have from us. Each time we have returned to this material, we have honed our understanding of how to explain what we think is important in this area. Our hat is off to their patience and careful attention. We have been teaching this material for thirty some years in a variety of universities and courses and we have paid close attention to how our students respond to our efforts and tried to make our approach better each time.

We are a firm believer that all our students are capable of excellence and that the only path to excellence is through discipline and study. We have always been proud of our students for doing so well on this journey. We hope these notes in turn make you proud of our efforts.

Abstract We will study basic analysis in this text. This is the analysis a mathematics major or a student from another discipline who needs this background should learn in their early years. Learning how to think in this way changes you for life. You will always know how to look carefully at assumptions and parse their consequences. That is a set of tools you can use for your own profit in your future. In this text, you learn a lot about how functions which map numbers to numbers behave in the context of the usual ideas from calculus such as limits, continuity, differentiation and integration. However, these ideas are firmly rooted in the special properties of the numbers themselves and we take great pains in this set of notes to place these ideas into the wider world of mappings on sets of objects. This is done anecdotally as we cannot study those things in careful detail yet but we want you exposed to more general things. The text also covers ideas from sequences of functions, series of functions and Fourier Series.

Acknowledgments I also want to acknowledge the great debt I have to my wife, Pauli, for her patience in dealing with those vacant stares and the long hours spent in typing and thinking. You are the love of my life.

The cover for this book is an original painting by us done in July 2017. It shows when the cephalopods first made contact with us to begin their training in mathematics.

Table of Contents

Part I

Introduction

Chapter 1

Introduction

We believe that all students who are seriously interested in mathematics at the master's and doctoral level should have a passion for analysis even if it is not the primary focus of their own research interests. This is also true if you are a mathematics major who is finishing their undergraduate degree and moving on to either work or study in a different area. Learning the concepts in analysis changes the way you think and makes you excel at understanding the consequences of assumptions. A student at any level can take this training and use it to do well in any area. We use a blend of mathematics, computation and science in the things we pursue for our research and to do this well, a good understanding of the principles of mathematical analysis is essential. So you should all understand that our own passion for the subject will shine though in the notes that follow!

In our view, to become a good mathematical scientist, you should know a large amount of fundamental mathematics and the text you are starting now is a first step in this direction. If you want a Ph.D. in mathematics or something closely aligned to that field, you can't really do your own research without understanding a lot of core material. Also, in order to read research papers and follow that sort of densely organized reading you need a lot of training in abstraction and hard thinking. Hence, in your own work to come, you need to be able to read, assess and use very abstract material and to do that well, you need to have proper core training. You must learn the appropriate core concepts and those of us who teach these courses must provide satisfying educational experiences that let you learn these ideas efficiently. Of course, we expect you to work hard but if you want to be a mathematical scientist or use advanced mathematics to make a living, you have all the motivation you need! And, it goes without saying that we assume that you are all mature mathematically and eager and interested in the material! The analysis courses fit together into a very interesting web of ideas.

Eventually, if you are so inclined, you can work your way through a set of five basic discourses on analysis: these are in a sense primers but they are not simple at all. They will make you go through the reasoning behind a large amount of the tools we used to formulate and solve problems. These discussions can be found in (Peterson (13) 2020) (this set of notes), (Peterson (15) 2020) (more $\Re^n$ stuff), (Peterson (14) 2020) (what is called linear analysis: the first graduate level text here), (Peterson (12) 2020) (the extension of the ideas of Riemann Integration to a more general integration with respect to what is called a measure; where a measure is an extension of the idea of the length of an interval) and (Peterson (11) 2020) (a blend of topology and analysis).

We typically organize this kind of class as follows:

- Three exams and one final which is cumulative and all have to take it. The days for the exams and the final are already set and you can't change those so make your travel plans accordingly.

- Usually HW is assigned every lecture.

- Two projects which let you look at some topics in detail as well as train you how to write mathematics.

Some words about studying: Homework is essential. We give about 2 - 4 exercises per lecture so that is about 80 - 160 exercises in all. In addition, there are two projects. All are designed to make you learn and to see how to think about things on your own. This takes time so do the work and be patient. We give 3 sample exams on the web site. You should do this to prepare for an exam:

- Study hard and then take sample exam 1 as a timed test for 60 minutes. You will know what you didn't get right and so you can figure out what to study. We don't give answers to these tests as this is stuff you should know.

- After studying again, take the second sample exam as a 60 minute timed exam and see how you did. Then study again to fill in the gaps.

- Take the third sample exam as a timed 60 minute exam. You should be able to do well and you are now prepared for the real exam.

We use Octave (Eaton et al. (4) 2020), which is an open source GPL licensed (Free Software Foundation (5) 2020) clone of MATLAB®, as a computational engine and we are not afraid to use it as an adjunct to all the theory we go over. Of course, you can use MATLAB® (MATLAB (9) 2018 - 2020) also if your university or workplace has a site license or if you have a personal license. Get used to it: theory, computation and science go hand in hand! Well, buckle up and let's get started!

1.1   Table of Contents

This text gathers material from many other sets of notes that we have written over the years since our first senior analysis course taught in 1982. From all of our experiences, we have selected the following material to cover.

Part One: Introduction These are our beginning remarks you are reading now which are in Chapter 1.

Part Two: Understanding Smoothness Here we are concerned with developing continuity and the idea of differentiation for both calculus on the real line and calculus in $\Re^2$.

- We think learning abstraction is a hard thing, so we deliberately start slow. You have probably seen treatments of the real line and mathematical induction already, but in Chapter 2, we start at that basic level. We go over induction carefully and explain how to organize your proofs. We also work out the triangle inequality for numbers and introduce the important concept of the infimum and supremum of sets of numbers.

- In Chapter 3, we introduce sequences of real numbers and lay out a lot of notational machinery we will use later.

- In Chapter 4, we prove the important Bolzano - Weierstrass Theorem for bounded sequences and bounded sets and introduce sequential compactness. We also discuss carefully the limit inferior and superior of sequences in two equivalent ways.

- In Chapter 5, we go over more set theory and introduce topological compactness and finally prove the full equivalence that sequential and topological compactness on the real line are equivalent to the set being closed and bounded.

- In Chapter 6, we define limits of functions and explore limit inferior and limit superiors carefully.

- In Chapter 7, we talk about continuity.

- In Chapter 8, we go over the consequences of continuity of a function on an interval such as the existence of global extreme values when the domain is compact.

- In Chapter 9, we step back and talk about other ways to guarantee the maximum and minimum of a function without requiring continuity and compactness of the domain. We introduce convex functions and subdifferentials.

- In Chapter 10, we introduce the idea of a derivative and talk about this level of smoothness in general.

- The many properties of differentiation are discussed in Chapter 11. We also introduce the trigonometric functions.

- In Chapter 12, the consequences of differentiation are explored such as Taylor polynomials and remainders.

- In Chapter 13, we use our knowledge of sequences and limits to develop exponential and logarithmic functions. We develop e^x first and then use it to define $\ln(x)$ as the inverse function of e^x.

- In Chapter 14, we go over standard extremal theory for functions of one variable.

- In Chapter 15, we introduce calculus ideas in $\Re^2$. This includes a nice development of simple surface graphing tools to let us see tangent planes and so forth visually. We develop tools to explore tangent plane approximation error and discuss carefully the theorems that let us know when mixed partials match and when the local behavior of the first order partials guarantees the differentiability of the function.

- In Chapter 16, we develop standard extremal theory of functions to two variables. We also take a deeper look at symmetric matrices and their connection to extremal behavior.

Part Three: Integration and Sequences of Functions In this part, we discuss integration theory as well as sequence and series of functions. This includes going over Fourier series and applying these ideas to some ordinary and partial differential equations.

- In Chapter 17, we introduce uniform continuity and some of its consequences.

- In Chapter 18, we begin discussing the important idea of the completeness of a space by introducing the idea of Cauchy sequences.

- In Chapter 19, we can use the idea of Cauchy sequences to better understand the behavior of the special sequences of partial sums of a given sequence that give rise to the idea of a series. We discuss some of the standard tests for convergence and define the classic ℓ^p spaces.

- In Chapter 20, we explore series more by adding the ratio and root tests to our repertoire. We also make the study of the ℓ^p spaces more concrete by proving the Hölder's and Minkowski's Inequalities. This allows us to talk out norms and normed spaces a bit.

- In Chapter 21, we define the Riemann Integral and explore some basic properties.

- In Chapter 22, we add a discussion of Darboux Integration and prove the equivalence of it with Riemann Integration and what is called the Riemann Criterion. We also prove some more properties of integrable functions.

- In Chapter 23, we prove the most important theorem yet, the Fundamental Theorem of Calculus, which connects the idea of differentiation to the Riemann Integral. This leads to the Cauchy Fundamental Theorem of Calculus which is the one that allows us to evaluate many Riemann Integrals easily. We also use the Fundamental Theorem of Calculus to define $\ln(x)$ and then develop its inverse e^x. This is backwards from the order we used for this development in the first part. We also start to discuss when two functions have the same integral value.

- In Chapter 24, we begin our discussion of pointwise and uniform convergence of sequences of functions and interchange theorems. We also discuss the Weierstrass Approximation Theorem.

- In Chapter 25, we discuss the convergence of sequences of functions in the context of the special sequence of partial sums we get by constructing series of functions. In particular, we go over power series in detail and look at the interchange theorems for power series.

- In Chapter 26, we have enough tools to carefully study when two functions have the same Riemann Integral and what kinds of functions are Riemann Integrable.

- In Chapter 27, we discuss Fourier Series. This is important because we can't handle the convergence issues here with our usual ratio tests and so forth. So new ideas are needed such as sequences of orthonormal functions in a vector space. Hence, more abstraction but for a good cause! We also introduce a fair bit of code to help with the calculations.

- In Chapter 28, we apply Fourier Series ideas to both ordinary and partial differential equations. We also show how to apply power series to ordinary differential equations when the coefficients are simple polynomials.

Part Four: Summary In Chapter 29 we talk about the things you have learned here, and where you should go next to continue learning basic analysis at a more advanced level.

Jim Peterson
School of Mathematical and Statistical Sciences
Clemson University

Part II

Understanding Smoothness

Chapter 2

Proving Propositions

2.1 Mathematical Induction

We begin our study of analysis by looking at a powerful tool for proving certain types of propositions: the **Principle of Mathematical Induction**;

Theorem 2.1.1 The Principle of Mathematical Induction

> *For each natural number n, let $P(n)$ be a statement or proposition about the numbers n.*
>
> - *If $P(1)$ is true:* This is called the BASIS STEP
>
> - *If $P(k+1)$ is true when $P(k)$ is true:* This is called the INDUCTIVE STEP
>
> *then we can conclude $P(n)$ is true for all natural numbers n.*

A proof using the POMI is organized as follows:

Proof 2.1.1

State the Proposition Here

Proof:

BASIS

Verify $P(1)$ is true

INDUCTIVE

Assume $P(k)$ is true for arbitrary $k > 1$ and use that information to prove $P(k+1)$ is true.

We have verified the inductive step. Hence, by the POMI, $P(n)$ holds for all n.

$$\text{QED}$$

You must include this finishing statement as part of your proof and show the **QED** *as above. Here* **QED** *is an abbreviation for the Latin* Quod Erat Demonstratum *or that which was to be shown. We often use the symbol* ∎ *instead of* **QED**.

Note, the natural numbers or counting numbers are usually denoted by the symbol $\mathbb{N}$. The set of all integers, positive, negative and zero is denoted by $\mathbb{Z}$ and the real numbers is denoted by $\Re$ or $\mathbb{R}$. There are many alternative versions of this. One useful one is this:

Theorem 2.1.2 The Principle of Mathematical Induction II

For each natural number n, let $P(n)$ be a statement or proposition about the numbers n.
If

- *If there is a number n_0 so that $P(n_0)$ is true:*
  BASIS STEP

- *If $P(k+1)$ is true when $P(k)$ is true for all $k \geq n_0$:*
  INDUCTIVE STEP

then we can conclude $P(n)$ is true for all natural numbers $n \geq n_0$.

A proof using this version of the POMI is organized as follows:

Proof 2.1.2

State the Proposition Here
Proof:
BASIS
Verify $P(n_0)$ is true
INDUCTIVE
Assume $P(k)$ is true for arbitrary $k > n_0$ and use that
information to prove $P(k+1)$ is true.

We have verified the inductive step. Hence, by
the POMI, $P(n)$ holds for all n.

$$\textbf{QED } or \; \blacksquare$$

It helps to do a few examples. Here the symbol $\forall$ is read as "for all" or "for each" depending on the context.

Theorem 2.1.3 n **factorial** $\geq 2^{n-1}$ $\forall n \geq 1$

$n! \geq 2^{n-1}$ $\forall n \geq 1$

Proof 2.1.3

BASIS : *$P(1)$ is the statement $1! \geq 2^{1-1} = 2^0 = 1$ which is true. So the basis step is verified.*
INDUCTIVE : *We assume $P(k)$ is true for an arbitrary $k > 1$. We use $k > 1$ because in the basis step we found out the proposition holds for $k = 1$. Hence, we know $k! \geq 2^{k-1}$. Now look at $P(k+1)$. We note $(k+1)! = (k+1)\,k!$ but, by the induction assumption, we know $k! \geq 2^{k-1}$. Plugging this fact in, we have*

$$(k+1)! \;=\; (k+1)\,k! \geq (k+1)\,2^{k-1}.$$

To finish, we note since $k > 1$, $k + 1 > 2$. Thus, we have the final step

$$(k+1)! \;=\; (k+1)\,k! \geq (k+1)\,2^{k-1} \geq 2 \times 2^{k-1} = 2^k.$$

This is precisely the statement $P(k+1)$. Thus $P(k+1)$ is true and we have verified the inductive step. Hence, by the POMI, $P(n)$ holds for all n. $\blacksquare$

Let's change it a bit.

Theorem 2.1.4 n **factorial** $\geq 3^{n-1}$ $\quad \forall n \geq 5$

$$n! \geq 3^{n-1} \quad \forall n \geq 5$$

Proof 2.1.4

BASIS : *$P(5)$ is the statement $5! = 120 \geq 3^{5-1} = 3^4 = 81$ which is true. So the basis step is verified.*
INDUCTIVE : *We assume $P(k)$ is true for an arbitrary $k > 5$. We use $k > 5$ because in the basis step we found out the proposition holds for $k = 5$. Hence, we know*

$$k! \quad \geq \quad 3^{k-1}.$$

Now look at $P(k+1)$. We note

$$(k+1)! \quad = \quad (k+1)\, k!$$

But, by the induction assumption, we know $k! \geq 3^{k-1}$. Plugging this fact in, we have

$$(k+1)! \quad = \quad (k+1)\, k! \geq (k+1)\, 3^{k-1}.$$

To finish, we note since $k > 5$, $k+1 > 6 > 3$. Thus, we have the final step

$$(k+1)! \quad = \quad (k+1)\, k! \geq (k+1)\, 3^{k-1} \geq 3 \times 3^{k-1} = 3^k.$$

This is precisely the statement $P(k+1)$. Thus $P(k+1)$ is true and we have verified the inductive step. Hence, by the POMI, $P(n)$ holds for all $n \geq 5$. ∎

Comment 2.1.1 *Note we use the second form of POMI here.*

This one is a bit harder.

Theorem 2.1.5

$$1^2 - 2^2 + 3^2 - \cdots + (-1)^{n+1}n^2 = \tfrac{1}{2}(-1)^{n+1}\, n\,(n+1)$$

Proof 2.1.5

BASIS : *$P(1)$ is what we get when we plug in $n = 1$ here. This gives 1^2 on the left hand side and $\frac{1}{2}(-1)^2 1(2)$ on the right hand side. We can see $1 = 1$ here and so $P(1)$ is true.*
INDUCTIVE : *We assume $P(k)$ is true for an arbitrary $k > 1$. Hence, we know*

$$1^2 - 2^2 + 3^2 - \cdots + (-1)^{k+1}k^2 \quad = \quad \frac{1}{2}(-1)^{k+1}\, k\,(k+1)$$

Now look at the left hand side of $P(k+1)$. We note at $k+1$, we have the part that stops at $k+1$ and a new term that stops at $k+2$. We can write this as

$$\left(1^2 - 2^2 + 3^2 - \cdots + (-1)^{k+1}k^2\right) \quad + \quad (-1)^{k+2}(k+1)^2.$$

But the first part of this sum corresponds to the induction assumption or hypothesis. We plug this into the left hand side of $P(k+1)$ to get

$$\frac{1}{2}(-1)^{k+1}\,k\,(k+1)+(-1)^{k+2}(k+1)^2.$$

Now factor out the common $(-1)^{k+1}(k+1)$ to get

$$(-1)^{k+1}(k+1)\left\{\frac{1}{2}\,k-(k+1)\right\}\;=\;\frac{1}{2}(-1)^{k+1}(k+1)\left\{k-2k-2\right\}$$

The term $k-2k-2=-k-2$ and so bringing out another -1 and putting it into the $(-1)^{k+1}$, we have

$$(-1)^{k+1}(k+1)\left\{\frac{1}{2}\,k-(k+1)\right\}\;=\;\frac{1}{2}(-1)^{k+2}(k+1)(k+2)$$

This is precisely the statement $P(k+1)$. Thus $P(k+1)$ is true and we have verified the inductive step. Hence, by the POMI, $P(n)$ holds for all $n\geq 1$. ∎

This argument was a lot harder as we had to think hard about how to manipulate the left hand side. So remember, it can be tricky to see how to finish the induction part of the argument! Note we try to use *pleasing indentation* strategies and *white space* to improve readability and understanding in these proofs.

Now many of our proofs in this course will not really be so scripted, but we start by learning this kind of attack and as we learn more we can use the arguments that do have a template as a kind of skeleton on which to base other types of approaches.

2.1.1   Homework

Provide a careful proof of this proposition.

Exercise 2.1.1 *Prove $n!\geq 4^n$ $\forall n\geq 9$,*

Exercise 2.1.2 *Prove $n!\geq 3^n$ for suitable n.*

Exercise 2.1.3 *Prove $n!\geq 4^n$ for suitable n,*

Exercise 2.1.4 *Prove $n!\geq 5^n$ for suitable n,*

Exercise 2.1.5 *Prove $(n-2)!\geq 7^n$ for suitable n,*

2.2   More Examples

Let's work out some of these type arguments.

Theorem 2.2.1

$$1+2+3+\cdots+n=\tfrac{1}{2}n(n+1),\quad \forall n\geq 1$$

Proof 2.2.1

BASIS : *$P(1)$ is the statement $1=\frac{1}{2}(1)(2)=1$ which is true. So the basis step is verified.*

INDUCTIVE : *We assume $P(k)$ is true for an arbitrary $k > 1$. Hence, we know*

$$1 + 2 + 3 + \cdots + k = \frac{1}{2}k(k+1)$$

Now look at $P(k+1)$. We note

$$1 + 2 + 3 + \cdots + (k+1) = \{1 + 2 + 3 + \cdots + k\} + (k+1)$$

Now apply the induction hypothesis and let $1 + 2 + 3 + \cdots + k = \frac{1}{2}k(k+1)$ We find

$$1 + 2 + 3 + \cdots + (k+1) = \frac{1}{2}k(k+1) + (k+1) = (k+1)\left\{\frac{1}{2}k+1\right\} = \frac{1}{2}(k+1)(k+2)$$

This is precisely the statement $P(k+1)$. Thus $P(k+1)$ is true and we have verified the inductive step. Hence, by the POMI, $P(n)$ holds for all n. ∎

Recall when you first encountered Riemann integration, you probably looked at taking the limit of Riemann sums using right side partitions. So for example, for $f(x) = 2 + x$ on the interval $[0, 1]$ using a partition width of $\frac{1}{n}$, the Riemann sum is

$$\sum_{i=1}^{n} f\left(0 + \frac{i}{n}\right)\frac{1}{n} = \sum_{i=1}^{n}\left(2 + \frac{i}{n}\right)\frac{1}{n} = \frac{2}{n}\sum_{i=1}^{n} 1 + \frac{1}{n^2}\sum_{i=1}^{n} i.$$

The first sum, $\sum_{i=1}^{n} 1 = n$ and so the first term is $2\frac{n}{n} = 2$. To evaluate the second term, we use our formula from above: $\sum_{i=1}^{n} i = \frac{1}{2}n(n+1)$ and so the second term becomes $\frac{n(n+1)}{2\,n\,n}$ which simplifies to $\frac{1}{2}(1 + \frac{1}{n})$. So the Riemann sum here is $2 + \frac{1}{2}\left(1 + \frac{1}{n}\right)$ which as n gets large clearly approaches the value 2.5. The terms $2.5 + \frac{1}{2n}$ form what is called a **sequence** and the **limit** of this sequence is 2.5. We will talk about this a lot more later. From your earlier calculus courses, you know

$$\int_0^1 (2 + x)dx = \left(2x + \frac{1}{2}x^2\right)\Big|_0^1 = 2 + \frac{1}{2}.$$

which matches what we found with the Riemann sum limit. In later chapters, we discuss the theory of Riemann integration much more carefully, so consider this just a taste of that kind of theory!

Theorem 2.2.2

$$1^2 + 2^2 + 3^2 + \cdots + n^2 = \frac{1}{6}n(n+1)(2n+1) \ \forall n \geq 1$$

Proof 2.2.2

BASIS : *$P(1)$ is the statement $1 = \frac{1}{6}(1)(2)(3) = 1$ which is true. So the basis step is verified.*
INDUCTIVE : *We assume $P(k)$ is true for an arbitrary $k > 1$. Hence, we know*

$$1^2 + 2^2 + 3^2 + \cdots + k^2 = \frac{1}{6}k(k+1)(2k+1)$$

Now look at $P(k+1)$. We note

$$1^2 + 2^2 + 3^2 + \cdots + (k+1)^2 = \{1^2 + 2^2 + 3^2 + \cdots + k^2\} + (k+1)^2$$

Now apply the induction hypothesis and let $1^2 + 2^2 + 3^2 + \cdots + k^2 = \frac{1}{6}k(k+1)(2k+1)$ *We find*

$$
\begin{aligned}
1^2 + 2^2 + 3^2 + \cdots + (k+1)^2 &= \frac{1}{6}k(k+1)(2k+1) + (k+1)^2 \\
&= (k+1)\left\{\frac{1}{6}k(2k+1) + (k+1)\right\} \\
&= \frac{1}{6}(k+1)\left\{k(2k+1) + 6(k+1)\right\} = \frac{1}{6}(k+1)\left\{2k^2 + 7k + 6\right\} \\
&= \frac{1}{6}(k+1)(k+2)(2k+3)
\end{aligned}
$$

This is precisely the statement $P(k+1)$. *Thus* $P(k+1)$ *is true and we have verified the inductive step. Hence, by the POMI,* $P(n)$ *holds for all* n. ∎

Now look at $f(x) = 2 + x^2$ on the interval $[0,1]$. Using a partition width of $\frac{1}{n}$, the Riemann sum is

$$
\sum_{i=1}^{n} f\left(0 + \frac{i}{n}\right)\frac{1}{n} = \sum_{i=1}^{n}\left(2 + \left(\frac{i}{n}\right)^2\right)\frac{1}{n} = \frac{2}{n}\sum_{i=1}^{n}1 + \frac{1}{n^3}\sum_{i=1}^{n}i^2.
$$

The first sum, $\sum_{i=1}^{n} 1 = n$ and so the first term is again $2\frac{n}{n} = 2$. To evaluate the second term, we use our formula from above: $\sum_{i=1}^{n} i^2 = \frac{1}{6}n(n+1)(2n+1)$ and so the second term becomes $\frac{n(n+1)(2n+1)}{6\,n\,n\,n}$ which simplifies to $\frac{1}{6}\left(1 + \frac{1}{n}\right)\left(2 + \frac{1}{n}\right)$. So the Riemann sum here is $2 + \frac{1}{6}\left(1 + \frac{1}{n}\right)\left(2 + \frac{1}{n}\right)$ which as n gets large clearly approaches the value $2 + \frac{1}{3}$. The terms $2 + \frac{1}{6}\left(1 + \frac{1}{n}\right)\left(2 + \frac{1}{n}\right)$ form a **sequence** and the **limit** of this sequence is $7/3$. From your earlier calculus courses, you know

$$
\int_0^1 (2 + x^2)dx = \left(2x + \frac{1}{3}x^3\right)\Big|_0^1 = 2 + \frac{1}{3}.
$$

which matches what we found with the Riemann sum limit.

2.3　More Abstract Proofs and Even Trickier POMIs

The induction step can have a lot of algebraic manipulation so we need to look at some of those. Let's start with this one.

Theorem 2.3.1

$$\boxed{2^n \geq n^2, \quad \forall n \geq 4}$$

Proof 2.3.1

$\boxed{\text{BASIS}}$: $P(4)$ *is the statement* $2^4 = 16 \geq 4^2 = 16$ *which is true. So the basis step is verified.*
$\boxed{\text{INDUCTIVE}}$: *We assume* $P(k)$ *is true for an arbitrary* $k > 4$. *Hence, we know* $2^k \geq k^2$. *Now look at* $P(k+1)$. *We note*

$$
2^{k+1} = 2 \times 2^k \geq 2 \times k^2.
$$

We need to show $2^{k+1} \geq (k+1)^2$ so we must show $2 \times k^2 \geq (k+1)^2$. We can simplify this by multiplying both sides out to get

$$2k^2 \overset{?}{\geq} k^2 + 2k + 1 \quad \Rightarrow \quad k^2 \overset{?}{\geq} 2k + 1.$$

We can answer this question by doing another POMI inside this one or we can figure it out graphically. Draw the graph of x^2 and $2x + 1$ together and you can clearly see $k^2 > 2k + 1$ when $k > 3$.

Thus $P(k+1)$ is true and we have verified the inductive step. Hence, by the POMI, $P(n)$ holds for all $n \geq 4$. ∎

Here is another one that is quite different.

Theorem 2.3.2

$(1+x)^n \geq 1 + nx, \quad \forall n \in \mathbb{N}, \forall x \geq -1$

Proof 2.3.2

BASIS : *When $n = 1$, we are asking if $1 + x \geq 1 + x$ when $x \geq -1$ which is actually true for all x. So the basis step is verified.*
INDUCTIVE : *We assume the proposition is true for any $x \geq -1$ and for any $k > 1$. Thus, we assume $(1+x)^k \geq 1 + kx$. Now look at the proposition for $k + 1$. We have*

$$(1+x)^{k+1} \quad = \quad (1+x)(1+x)^k \geq (1+x)(1+kx).$$

We must show $(1+x)(1+kx) \geq (1+(k+1)x)$ for $x \geq -1$. We have to show

$$1 + (k+1)x + kx^2 \quad \overset{?}{\geq} \quad 1 + (k+1)x.$$

We can cancel the $1 + (k+1)x$ on both sides which tells us we must check if $kx^2 \geq 0$ when $x \geq -1$. This is true. Thus $P(k+1)$ is true and we have verified the inductive step. Hence, by the POMI, $P(n)$ holds for all n. ∎

A totally different kind of induction proof is the one below. We want to count how many subsets a set with a finite number of objects can have. We let the number of objects in a set S be denoted by $|S|$. We call this the **cardinality** of the set S. For example, the cardinality of the set $\{Jim, Pauli, Qait, Quinn\}$ is 4. Given a set S, S how many subsets does it have?

For example, $\{Jim, Pauli\}$ is a subset of the original S defined above. Since the original set has just 4 objects in it, there are 4 subsets with just one object, There are $\binom{4}{2}$ ways to choose subsets of 2 objects. Recall $\binom{4}{2}$ is $\frac{4!}{2!\,2!} = \frac{24}{4} = 6$. There are then $\binom{4}{3}$ ways to choose 3 objects which gives $\frac{4!}{3!\,1!} = \frac{24}{6} = 4$. Finally there is just one way to choose 4 objects. So the total number of subsets is $1 + 4 + 6 + 4 = 15$. We always also add in the empty set $\emptyset = \{\ \}$ to get the total number of subsets is 16. Note this is the same as 2^4. Hence, we might conjecture that if S had only a finite number of objects in it, the number of subsets of S is $2^{|S|}$. The collection of all subsets of a set S is denoted by 2^S for this reason and the cardinality of 2^S is thus $2^{|S|}$. We can prove this using an induction argument.

Theorem 2.3.3 The Number of Subsets of a Finite Set

$|2^S| = 2^{|S|}, \quad \forall |S| \geq 1$

Proof 2.3.3

$\boxed{\textbf{BASIS}}$: *When S is 1, there is only one element in S and so the number of subsets is just S itself and the empty set. So $2^{|S|}$ is $2^1 = 2$. This shows the proposition is true for $|S| = 1$. So the basis step is verified.*

$\boxed{\textbf{INDUCTIVE}}$: *We assume the proposition is true for an arbitrary $|S| = k$ for any $k > 1$. We can thus label the objects in S as $S = \{a_1, a_2, \ldots, a_k\}$. Now consider a new set which has one more object in it. Call this set S'. We see $S' = \{S, a_{k+1}\}$ and we can easily count or enumerate how many new subsets we can get. We can add or adjoin the object a_{k+1} to each of the $2^{|S|}$ subsets of S to get $2^{|S|}$ additional subsets. Thus, the total number of subsets is $2^{|S|} + 2^{|S|} = 2^{|S|+1}$. But $|S| + 1 = |S'|$. Thus $P(|S| = k + 1)$ is true and we have verified the inductive step. Hence, by the POMI, $P(|S| = n)$ holds for all n.* ∎

2.3.1 Homework

Use the POMI to prove the following propositions.

Exercise 2.3.1 $\frac{1}{1\cdot 2} + \frac{1}{2\cdot 3} + \cdots + \frac{1}{n\cdot(n+1)} = \frac{n}{n+1}$.

Exercise 2.3.2 $\frac{d}{dx}x^n = nx^{n-1}$, $\quad \forall x, \forall n \in \mathbb{N}$. *You can assume you know the powers $f(x) = x^n$ are differentiable and that you know the product rule: if f and g are differentiable, then $(fg)' = f'g + fg'$.*

Exercise 2.3.3 $1 + x + \cdots + x^n = \frac{1-x^{n+1}}{1-x}$, $\quad \forall x \neq 1, \forall n \in \mathbb{N}$.

Exercise 2.3.4 $\int x^n \, dx = \frac{1}{n+1} x^{n+1}$, $\quad \forall n \in \mathbb{N}$. *You can assume you know integration by parts. The basis step is $\int x\, dx = x^2/2$ which you can assume you know. After that it is integration by parts.*

2.4 Some Contradiction Proofs

Another type of proof is one that is done by contradiction.

Theorem 2.4.1 $\sqrt{2}$ is not a Rational Number

> *$\sqrt{2}$ is not a rational number.*

Proof 2.4.1
We will prove this technique using a technique called **contradiction**. *Let's assume we can find positive integers p and q so that $2 = (p/q)^2$ with p and q having no common factors. When this happens we say p and q are* **relatively prime**. *This tells us $p^2 = 2q^2$ which also tells us p^2 is divisible by 2. Thus, p^2 is even. Does this mean p itself is even? Well, if p was odd, we could write $p = 2\ell + 1$ for some integer ℓ. Then, we would know*

$$p^2 \;=\; (2\ell + 1)^2 = 4\ell^2 + 4\ell + 1.$$

The first two terms, $4\ell^2$ and 4ℓ are even, so this implies p^2 would be **odd**. *So we see p odd implies p^2 is* **odd**. *Thus, we see p must be even when p^2 is even. So we now know $p = 2k$ for some integer k as it is* **even**. *But since $p^2 = 2q^2$, we must have $4k^2 = 2q^2$. But this says q^2 must be* **even**.

The same reasoning we just used to show p **odd** *implies p^2 is* **odd**, *then tells us q* **odd** *implies q^2 is* **odd**. *Thus q is even too.*

Now here is the contradiction. We assumed p and q were relatively prime; i.e. they had no common factors. But if they are both **even**, *they share the factor 2. This is the contradiction we seek.*

Hence, our original assumption must be incorrect and we can not find positive integers p and q so that $2 = (p/q)^2$. ∎

Note the similarities in the argument below to the one just presented.

Theorem 2.4.2 $\sqrt{3}$ is not a Rational Number

> $\sqrt{3}$ *is not a rational number.*

Proof 2.4.2

Let's assume we can find positive integers p and q so that $3 = (p/q)^2$ with p and q being relatively prime. This tells us $p^2 = 3q^2$ which also tells us p^2 is divisible by 3. Does this mean p itself is divisible by 3? Well, if p was not divisible by 3, we could write $p = 3\ell + 1$ or $3\ell + 2$ for some integer ℓ. Then, we would know

$$p^2 = (3\ell + 1)^2 = 9\ell^2 + 6\ell + 1.$$

or

$$p^2 = (3\ell + 2)^2 = 9\ell^2 + 12\ell + 4.$$

The first two terms in both choices are divisible by 3 and the last terms are not. So we see p^2 is not divisible by 3 in both cases. Thus, we see p must be divisible by 3 when p^2 is divisible by 3.

So we now know $p = 3k$ for some integer k as it is divisible by 3. But since $p^2 = 3q^2$, we must have $9k^2 = 3q^2$. But this says q^2 must be divisible by 3. The same reasoning we just used to show p^2 divisible by 3 implies p is divisible by 3, then tells us q^2 divisible by 3 implies q is divisible by 3.

Now here is the contradiction. We assumed p and q were relatively prime; i.e. they had no common factors. But if they are both divisible by 3, they share the factor 3. This is the contradiction we seek. Hence, our original assumption must be incorrect and we can not find positive integers p and q so that $3 = (p/q)^2$. ∎

Let's introduce some notation: if p and q are relatively prime integers, we say $(p, q) = 1$ and if the integer k divides p, we say $k|p$. Now let's modify the two proofs we have seen to attack the more general problem of showing $\sqrt{n}$ is not rational if n is prime.

Theorem 2.4.3 $\sqrt{n}$ is not a Rational Number when n is a Prime Number

> $\sqrt{n}$ *is not a rational number when n is a prime number.*

Proof 2.4.3

Let's assume there are integers u and v with $(u, v) = 1$ so the $n = (u/v)^2$ which implies $nv^2 = u^2$. This tells us $n|u^2$. Now we will invoke a theorem from **number theory** *or* **abstract algebra** *which tells us every integer u has a* **prime factor decomposition***:*

$$u = (p_1)^{r_1} (p_2)^{r_2} \cdots (p_s)^{r_s}$$

for some prime numbers p_1 to p_s and positive integers r_1 to r_s. For example, here are two such prime decompositions.

$$66 = 2 \cdot 3 \cdot 11$$
$$80 = 2^4 \cdot 5$$

It is easy to see what the integers p_1 to p_s and r_1 to r_s are in each of these two examples and we leave that to you!

Thus, we can say

$$u^2 \;=\; (p_1)^{2r_1}\,(p_2)^{2r_2}\cdots(p_s)^{2r_s}$$

Since $n|u^2$, n must divide one of the terms in the prime factor decomposition: i.e. we can say n divides the term $p_i^{r_i}$. Now the term $p_i^{r_i}$ is a prime number to a positive integer power r_i. The only number that can divide into that evenly are appropriate powers of p_i. But, we know n is a prime number too, so n must divide p_i itself. Hence, we can conclude $n = p_i$. But this tells us immediately that n divides u too. ∎

2.5   Triangle Inequalities

Now let's study some of the properties of the real numbers.

Definition 2.5.1 Absolute Values

Let x be any real number. We define the **absolute value** *of x, denoted by $|x|$, by*

$$|x| \;=\; \begin{cases} x, & \text{if } x \geq 0 \\ -x, & \text{if } x < 0. \end{cases}$$

For example, $|-3| = 3$ and $|4| = 4$.

Using this definition of the absolute value of a number, we can prove a fundamental inequality called the **triangle inequality** which we will use frequently to do estimates.

Theorem 2.5.1 Triangle Inequality

Let x and y be any two real numbers. Then

$$|x+y| \;\leq\; |x| + |y|$$
$$|x-y| \;\leq\; |x| + |y|$$

and for any number z.

$$|x-y| \;\leq\; |x-z| + |z-y|$$

Proof 2.5.1
We know $(|x+y|)^2 = (x+y)^2$ which implies $(|x+y|)^2 = x^2 + 2xy + y^2$. But $2xy \leq 2|x||y|$ implying

$$(|x+y|)^2 \;\leq\; x^2 + 2|x|\,|y| + y^2 = |x|^2 + 2|x|\,|y| + |y|^2 = (|x|+|y|)^2.$$

Taking square roots, we find $|x+y| \leq |x| + |y|$. Of course, the argument for $x - y$ is similar as $x - y = x + (-y)$. To do the next part, we know $|a+b| \leq |a| + |b|$ for any a and b. Let $a = x - z$ and $b = z - y$. Then $|(x-z)+(z-y)| \leq |x-z| + |z-y|$. ∎

Comment 2.5.1 *The technique where we do $x - y = (x-z) + (z-y)$ is called the* **Add and Subtract Trick** *and we will use it a lot!*

Comment 2.5.2 *Also note $|x| \le c$ is the same as $-c \le x \le c$ and we use this other way of saying it a lot too.*

Theorem 2.5.2 Backwards Triangle Inequality

Let x and y be any two real numbers. Then

$$|x| - |y| \quad \le \quad |x - y|$$
$$|y| - |x| \quad \le \quad |x - y|$$
$$| \, |x| - |y| \, | \quad \le \quad |x - y|$$

Proof 2.5.2
Let x and y be any real numbers. Then by the triangle inequality

$$|x| \quad = \quad |(x - y) + y| \le |x - y| + |y| \Rightarrow |x| - |y| \le |x - y|$$

Similarly,

$$|y| \quad = \quad |(y - x) + x| \le |y - x| + |x| \Rightarrow |y| - |x| \le |x - y|$$

Combining these two cases we see

$$-|x - y| \quad \le \quad |x| - |y| \quad \le \quad |x - y|$$

But this is the same as saying $| \, |x| - |y| \, | \le |x - y|$. ∎

Another tool we use a lot is the following small theorem which we use to show some value is zero.

Lemma 2.5.3 Proving a Number x is Zero via Estimates

Let x be a real number that satisfies $|x| < \epsilon, \quad \forall \epsilon > 0$. Then, $x = 0$.

Proof 2.5.3
We will prove this by contradiction. Let's assume x is not zero. Then $|x| > 0$ and $|x|/2$ is a valid choice for ϵ. The assumption then tells us that $|x| < |x|/2$ or $|x|/2 < 0$ which is not possible. So our assumption is wrong and $x = 0$. ∎

We can extend the triangle inequality to finite sums using POMI!

Theorem 2.5.4 Extended Triangle Inequality

Let x_1 to x_n be a finite collection of real numbers with $n \ge 1$. Then

$$|x_1 + \cdots x_n| \quad \le \quad |x_1| + \cdots + |x_n|$$

or using summation notation $|\sum_{i=1}^{n} x_i| \le \sum_{i=1}^{n} |x_i|$.

Proof 2.5.4

BASIS : *$P(1)$ is the statement $|x_1| \le |x_1|$; so the basis step is verified.*

$\boxed{\text{INDUCTIVE}}$: *We assume $P(k)$ is true for an arbitrary $k > 1$. Hence, we know*

$$\left| \sum_{i=1}^{k} x_i \right| \leq \sum_{i=1}^{k} |x_i|.$$

Now look at $P(k+1)$. We note by the triangle inequality applied to $a = \sum_{i=1}^{k} x_i$ and $b = x_{k+1}$, we have $|a + b| \leq |a| + |b|$ or

$$\left| \sum_{i=1}^{k+1} x_i \right| \leq \left(\left| \sum_{i=1}^{k} x_i \right| \right) + |x_{k+1}|$$

Now apply the induction hypothesis to see

$$\left| \sum_{i=1}^{k+1} x_i \right| \leq \sum_{i=1}^{k} |x_i| + |x_{k+1}| = \sum_{i=1}^{k+1} |x_i|$$

This shows $P(k+1)$ is true and by the POMI, $P(n)$ is true for all $n \geq 1$.   ∎

2.5.1   A First Look at the Cauchy - Schwartz Inequality

We can use our new tools to provide a first look at something you probably have seen before: the Cauchy - Schwartz Inequality. The ℓ_2 norm of a vector x

$$x = \begin{bmatrix} x_1 \\ x_2 \\ \vdots \\ x_n \end{bmatrix}$$

is defined by $\|x\| = \sqrt{\sum_{i=1}^{n} x_i^2}$ and if we have two such vectors a and b, we can prove this fundamental inequality.

Theorem 2.5.5 ℓ_2 Cauchy - Schwartz Inequality

Let $\{a_1, \dots, a_n\}$ and $\{b_1, \dots, b_n\}$ be finite collections of real numbers with $n \geq 1$. Then

$$\left| \sum_{i=1}^{n} a_i b_i \right|^2 \leq \left(\sum_{i=1}^{n} a_i^2 \right) \left(\sum_{i=1}^{n} b_i^2 \right)$$

Proof 2.5.5

$\boxed{\text{BASIS}}$: *$P(1)$ is the statement $|a_1 b_1|^2 \leq a_1^2 b_1^2$; the basis step is true.*

$\boxed{\text{INDUCTIVE}}$: *We assume $P(k)$ is true for $k > 1$. Hence, we know*

$$\left| \sum_{i=1}^{k} a_i b_i \right|^2 \leq \left(\sum_{i=1}^{k} a_i^2 \right) \left(\sum_{i=1}^{k} b_i^2 \right)$$

Now look at $P(k + 1)$.

$$\left| \sum_{i=1}^{k+1} a_i b_i \right|^2 = \left| \sum_{i=1}^{k} a_i b_i + a_{k+1} b_{k+1} \right|^2$$

Let A denote the first piece; i.e. $A = \sum_{i=1}^{k} a_i b_i$. Then expanding the term $|A + a_{k+1} b_{k+1}|^2$, we have

$$\left| \sum_{i=1}^{k+1} a_i b_i \right|^2 = \left(\sum_{i=1}^{k+1} a_i b_i \right)^2 = A^2 + 2A a_{k+1} b_{k+1} + a_{k+1}^2 b_{k+1}^2$$

$$= \left| \sum_{i=1}^{k} a_i b_i \right|^2 + 2 \left(\sum_{i=1}^{k} a_i b_i \right) a_{k+1} b_{k+1} + a_{k+1}^2 b_{k+1}^2$$

or

$$\left| \sum_{i=1}^{k+1} a_i b_i \right|^2 \leq \left| \sum_{i=1}^{k} a_i b_i \right|^2 + 2 \left| \sum_{i=1}^{k} a_i b_i \right| a_{k+1} b_{k+1} + a_{k+1}^2 b_{k+1}^2$$

Now use the induction hypothesis to see

$$\left| \sum_{i=1}^{k+1} a_i b_i \right|^2 \leq \sum_{i=1}^{k} a_i^2 \sum_{i=1}^{k} b_i^2 + 2 \sqrt{\sum_{i=1}^{k} a_i^2} \sqrt{\sum_{i=1}^{k} b_i^2} a_{k+1} b_{k+1} \qquad (2.1)$$
$$+ a_{k+1}^2 b_{k+1}^2$$

Now let $\alpha = \sqrt{\sum_{i=1}^{k} a_i^2}\, b_{k+1}$ and $\beta = \sqrt{\sum_{i=1}^{k} b_i^2}\, a_{k+1}$. We know for any real numbers α and β that $(\alpha - \beta)^2 \geq 0$. Thus, $\alpha^2 + \beta^2 \geq 2\alpha\beta$. We can use this in our complicated sum above. We have

$$2\alpha\beta = 2 \sqrt{\sum_{i=1}^{k} a_i^2} \sqrt{\sum_{i=1}^{k} b_i^2}\, a_{k+1}\, b_{k+1}$$

$$\alpha^2 + \beta^2 = \left(\sum_{i=1}^{k} a_i^2 \right) b_{k+1}^2 + \left(\sum_{i=1}^{k} b_i^2 \right) a_{k+1}^2$$

Hence, the middle part of Equation 2.1 can be replaced by the $2\alpha\beta \leq \alpha^2 + \beta^2$ inequality above to get

$$\left| \sum_{i=1}^{k+1} a_i b_i \right|^2 \leq \sum_{i=1}^{k} a_i^2 \sum_{i=1}^{k} b_i^2 + \left(\sum_{i=1}^{k} a_i^2 \right) b_{k+1}^2 + \left(\sum_{i=1}^{k} b_i^2 \right) a_{k+1}^2 + a_{k+1}^2 b_{k+1}^2 \quad (2.2)$$

$$= \left(\sum_{i=1}^{k} a_i^2 + a_{k+1}^2 \right) \left(\sum_{i=1}^{k} b_i^2 + b_{k+1}^2 \right)$$

But this says

$$\left| \sum_{i=1}^{k+1} a_i b_i \right|^2 \leq \left(\sum_{i=1}^{k+1} a_i^2 \right) \left(\sum_{i=1}^{k+1} b_i^2 \right)$$

This shows $P(k+1)$ is true and by the POMI, $P(n)$ is true for all $n \geq 1$.   ■

Comment 2.5.3 *This is a famous type of theorem. For two vectors*

$$V = \begin{bmatrix} a_1 \\ a_2 \end{bmatrix} \quad and \quad W = \begin{bmatrix} b_1 \\ b_2 \end{bmatrix}$$

the inner product of V and W, $< V, W >$ is $a_1b_1 + a_2b_2$ and the norm or length of the vectors is the ℓ_2 norm

$$||V||_2 = \sqrt{a_1^2 + a_2^2} \quad and \quad ||W||_2 = \sqrt{b_1^2 + b_2^2}$$

The theorem above then says $< V, W > \leq | < V, W > | \leq ||V||_2 ||W||_2$. This is called the **Cauchy - Schwartz Inequality** *also. This works for three and n dimensional vectors too. Note this inequality then says*

$$-1 \leq \frac{< V, W >}{||V||_2 ||W||_2} \leq 1$$

which implies we can define the angle between the vectors V and W by

$$\cos(\theta) = \frac{< V, W >}{||V||_2 ||W||_2}.$$

2.5.2   Homework

Prove the following propositions.

Exercise 2.5.1 *$\sqrt{5}$ is not a rational number using the same sort of argument we used in the proof of $\sqrt{3}$ is not rational.*

Exercise 2.5.2 *$\sqrt{7}$ is not a rational number using the same sort of argument we used in the proof of $\sqrt{3}$ is not rational.*

Exercise 2.5.3 *$\sqrt{43}$ is not a rational number using the same sort of argument we used in the proof of $\sqrt{7}$ is not rational. But this time, argue more succinctly, that is, find a way to argue without all the cases!*

Exercise 2.5.4 *On the interval $[1, 10]$, use factoring and the triangle inequality to prove $|x - y| \leq 2\sqrt{10}\,|\sqrt{x} - \sqrt{y}|$.*

2.6   The Supremum and Infimum of a Set of Real Numbers

Let S be a nonempty set of real numbers. We need to make precise the idea of a set of real numbers being **bounded**.

2.6.1   Bounded Sets

First, we start with a set being bounded above.

Definition 2.6.1 Sets Bounded Above

> *We say a nonempty set S is bounded above if there is a number M so that $x \leq M$ for all x in S. We call M an **upper bound** of S or just an **u.b.***

Example 2.6.1 *If* $S = \{y : y = x^2 \text{ and } -1 \leq x \leq 2\}$, *there are many u.b.'s of S. Some choices are* $M = 5$, $M = 4.1$. *Note* $M = 1.9$ *is* **not** *an u.b. You should draw a graph of this to help you understand what is going on.*

Example 2.6.2 *If* $S = \{y : y = \tanh(x) \text{ and } x \in \Re\}$, *there are many u.b.'s of S. Some choices are* $M = 2$, $M = 2.1$. *Note* $M = 0$ *is* **not** *an u.b. Draw a picture of this graph too.*

We can also talk about a set being bounded below.

Definition 2.6.2 Sets Bounded Below

> *We say a set S is bounded below if there is a number m so that* $x \geq m$ *for all x in S. We call m a* **lower bound** *of S or just a l.b.*

Example 2.6.3 *If* $S = \{y : y = x^2 \text{ and } -1 \leq x \leq 2\}$, *there are many l.b.'s of S. Some choices are* $m = -2$, $m = -0.1$. *Note* $m = 0.3$ *is* **not** *a l.b.*

Example 2.6.4 *If* $S = \{y : y = \tanh(x) \text{ and } x \in \Re\}$, *there are many l.b.'s of S. Some choices are* $m = -1.1$, $m = -1.05$. *Note* $m = -0.87$ *is* **not** *a l.b. Draw a picture of this graph again.*

We can then combine these ideas into a definition of what it means for a set to be bounded.

Definition 2.6.3 Bounded Sets

> *We say a set S is bounded if S is bounded above and bounded below. That is, there are finite numbers m and M so that* $m \leq x \leq M$ *for all* $x \in S$. *We usually overestimate the bound even more and say S is bounded if we can find a number B so that* $|x| \leq B$ *for all* $x \in S$. *A good choice of such a B is to let* $B = \max(|m|, |M|)$ *for any choice of l.b. m and u.b. M.*

Example 2.6.5 *If* $S = \{y : y = x^2 \text{ and } -1 \leq x < 2\}$, *here* $S = [0, 4)$ *and so for* $m = -2$ *and* $M = 5$, *a choice of B is* $B = 5$. *Of course, there are many other choices of B. Another choice of m is* $m = -1.05$ *and with* $M = 2.1$, *we could use* $B = 2.1$.

Example 2.6.6 *If* $S = \{y : y = \tanh(x) \text{ and } x \in \Re\}$, *we have* $S = (-1, 1)$ *and for* $m = -1.1$ *and* $M = 1.2$, *a choice of B is* $B = 1.2$.

2.6.2 Least Upper Bounds and Greatest Lower Bounds

The next material is more abstract! We need to introduce the notion of **least upper bound** and **greatest lower bound**. We also call the **least upper bound** the **l.u.b.**. It is also called the **supremum** of the set S. We use the notation $\sup(S)$ as well. We also call the **greatest lower bound** the **g.l.b.**. It is also called the **infimum** of the set S. We use the notation $\inf(S)$ as well.

Definition 2.6.4 Least Upper Bound and Greatest Lower Bound

> *The* **least upper bound**, **l.u.b.** *or* sup *of the set S is a number U satisfying*
>
> 1. *U is an upper bound of S*
>
> 2. *If M is any other upper bound of S, then* $U \leq M$.
>
> *The* **greatest lower bound**, **g.l.b.** *or* inf *of the set S is a number u satisfying*
>
> 1. *u is a lower bound of S*
>
> 2. *If m is any other lower bound of S, then* $u \geq m$.

Example 2.6.7 *If $S = \{y : y = x^2$ and $-1 \leq x < 2\}$, here $S = [0, 4)$ and so $\inf(S) = 0$ and $\sup(S) = 4$.*

Example 2.6.8 *If $S = \{y : y = \tanh(x)$ and $x \in \Re\}$, we have $\inf(S) = -1$ and $\sup(S) = 1$. Note the* inf *and* sup *of a set S need* **NOT** *be in S!*

Example 2.6.9 *If $S = \{y : \cos(2n\pi/3), \quad \forall n \in \mathbb{N}\}$, The only possible values in S are $\cos(2\pi/3) = -1/2$, $\cos(4\pi/3) = -1/2$, and $\cos(6\pi/3) = 1$. There are no other values and these 2 values are endlessly repeated in a cycle. Here $\inf(S) = -1/2$ and $\sup(S) = 1$.*

Comment 2.6.1 *If a set S has no finite lower bound, we set $\inf(S) = -\infty$. If a set S has no finite upper bound, we set $\sup(S) = \infty$.*

Comment 2.6.2 *If the set $S = \emptyset$, we set $\inf(S) = \infty$ and $\sup(S) = -\infty$.*

These ideas then lead to the notion of the minimum and maximum of a set.

Definition 2.6.5 Maximum and Minimum of a Set

> *We say $Q \in S$ is a maximum of S if $\sup(S) = Q$. This is the same, of course, as saying $x \leq Q$ for all x in S which is the usual definition of an upper bound. But this is different as Q is in S. We call Q a* **maximizer** *or a* **maximum element** *of S.*
> *We say $q \in S$ is a minimum of S if $\inf(S) = q$. Again, this is the same as saying $x \geq q$ for all x in S which is the usual definition of a lower bound. But this is different as q is in S. We call q a* **minimizer** *or a* **minimal element** *of S.*

2.6.3   The Completeness Axiom and Consequences

There is a fundamental **axiom** about the behavior of the real numbers which is very important.

Axiom 1 The Completeness Axiom

> *Let S be a set of real numbers which is nonempty and bounded above. Then the supremum of S exists and is finite.*
> *Let S be a set of real numbers which is nonempty and bounded below. Then the infimum of S exists and is finite.*

Comment 2.6.3 *So nonempty bounded sets of real numbers always have a finite infimum and supremum. This does not say the set has a finite minimum and finite maximum. Another way of saying this is that we don't know if S has a minimizer and maximizer.*

We can prove some basic results about these things.

Theorem 2.6.1 A Set has a Maximum if and only if its Supremum is in the Set

> *Let S be a nonempty set of real numbers which is bounded above. Then $\sup(S)$ exists and is finite. Then S has a maximal element if and only if (IFF) $\sup(S) \in S$. We also use the symbol $\Longleftrightarrow$ to indicate IFF.*

Proof 2.6.1
($\Leftarrow$):
Assume $\sup(S)$ is in S. By definition, $\sup(S)$ is an upper bound of S and so must satisfy $x \leq \sup(S)$ for all x in S. This says $\sup(S)$ is a maximizer of S.
($\Rightarrow$):
Let Q denote a maximizer of S. Then by definition $x \leq Q$ for all x in S and is an upper bound. So by

the definition of a supremum, $\sup(S) \leq Q$. *Since* Q *is a maximizer,* Q *is in* S *and from the definition of upper bound, we have* $Q \leq \sup(S)$ *as well. This says* $\sup(S) \leq Q \leq \sup(S)$ *or* $\sup(S) = Q$. ∎

Theorem 2.6.2 A Set has a Minimum if and only if its Infimum is in the Set

Let S *be a nonempty set of real numbers which is bounded below. Then* $\inf(S)$ *exists and is finite. Then*
S has a minimal element $\Longleftrightarrow$ $\inf(S) \in S$.

Proof 2.6.2

($\Leftarrow$): *Assume* $\inf(S)$ *is in* S. *By definition,* $\inf(S)$ *is a lower bound of* S *and so must satisfy* $x \geq \inf(S)$ *for all* x *in* S. *This says* $\inf(S)$ *is a minimizer of* S.
($\Rightarrow$): *Let* q *denote a minimizer of* S. *Then by definition* $x \geq q$ *for all* x *in* S *and is a lower bound. So by the definition of an infimum,* $q \leq \inf(S)$. *Since* q *is a minimizer,* q *is in* S *and from the definition of lower bound, we have* $\inf(S) \leq q$ *as well. This says* $\inf(S) \leq q \leq \inf(S)$ *or* $\inf(S) = q$. ∎

2.6.4 Some Basic Results

Two results that we use all the time are the infimum and supremum tolerance lemma which are proven using a simple contradiction argument.

Lemma 2.6.3 Infimum Tolerance Lemma

Let S *be a nonempty set of real numbers that is bounded below. Let* $\epsilon > 0$ *be arbitrarily chosen. Then*

$$\exists\, y \in S \ni \inf(S) \leq y < \inf(S) + \epsilon$$

Proof 2.6.3

We do this by contradiction. Assume this is not true for some $\epsilon > 0$. *Then for all* y *in* S, *we must have* $y \geq \inf(S) + \epsilon$. *But this says* $\inf(S) + \epsilon$ *must be a lower bound of* S. *So by the definition of infimum, we must have* $\inf(S) \geq \inf(S) + \epsilon$ *for a positive epsilon which is impossible. Thus our assumption is wrong and we must be able to find at least one* y *in* S *that satisfies* $\inf(S) \leq y < \inf(S) + \epsilon$. ∎

Lemma 2.6.4 Supremum Tolerance Lemma

Supremum Tolerance Lemma: *Let* S *be a nonempty set of real numbers that is bounded above. Let* $\epsilon > 0$ *be arbitrarily chosen. Then*

$$\exists\, y \in S \ni \sup(S) - \epsilon < y \leq \sup(S)$$

Proof 2.6.4

We do this by contradiction. Assume this is not true for some $\epsilon > 0$. *Then for all* y *in* S, *we must have* $y \leq \sup(S) - \epsilon$. *But this says* $\sup(S) - \epsilon$ *must be an upper bound of* S. *So by the definition of supremum, we must have* $\sup(S) \leq \sup(S) - \epsilon$ *for a positive epsilon which is impossible. Thus our assumption is wrong and we must be able to find at least one* y *in* S *that satisfies* $\sup(S) - \epsilon < y \leq \sup(S)$. ∎

Example 2.6.10 *Let* $f(x, y) = x + 2y$ *and let* $S = [0, 1] \times [1, 3]$ *which is also* $S_x \times S_y$ *where* $S_x = \{x : 0 \leq x \leq 1\}$ *and* $S_y = \{y : 1 \leq y \leq 3\}$. *Note* $\inf_{(x,y) \in [0,1] \times [1,3]} f(x, y) = 0 + 2 = 2$

and $\sup_{(x,y)\in[0,1]\times[1,3]} f(x,y) = 1 + 6 = 7.$

$$\inf_{1\le y\le 3} f(x,y) = \inf_{1\le y\le 3} (x+2y) = x+2$$

$$\sup_{0\le x\le 1} f(x,y) = \sup_{0\le x\le 1} (x+2y) = 1+2y$$

$$\sup_{0\le x\le 1} \inf_{1\le y\le 3} (x+2y) = \sup_{0\le x\le 1} (x+2) = 3$$

$$\inf_{1\le y\le 3} \sup_{0\le x\le 1} (x+2y) = \inf_{1\le y\le 3} (1+2y) = 3$$

so in this example,

$$\inf_{y\in S_y} \sup_{x\in S_x} f(x,y) = \sup_{x\in S_x} \inf_{y\in S_y} f(x,y).$$

Example 2.6.11 *Let*

$$f(x,y) \;=\; \begin{cases} 0, & (x,y)\in(1/2,1]\times(1/2,1] \\ 2, & (x,y)\in(1/2,1]\times[0,1/2] \text{ and } [0,1/2]\times(1/2,1] \\ 1, & (x,y)\in[0,1/2]\times[0,1/2] \end{cases}$$

and let $S = [0,1]\times[0,1]$ *which is also* $S_x\times S_y$ *where* $S_x = \{x: 0\le x\le 1\}$ *and* $S_y = \{y: 0\le y\le 1\}$. *Note* $\inf_{(x,y)\in[0,1]\times[0,1]} f(x,y) = 0$ *and* $\sup_{(x,y)\in[0,1]\times[0,1]} f(x,y) = 2$. *Then, we also can find*

$$\inf_{0\le y\le 1} f(x,y) = \begin{cases} 1, & 0\le x\le 1/2 \\ 0, & 1/2 < x\le 1 \end{cases}$$

and

$$\sup_{0\le x\le 1} f(x,y) = \begin{cases} 2, & 0\le y\le 1/2 \\ 2, & 1/2 < y\le 1 \end{cases}$$

and

$$\sup_{0\le x\le 1} \inf_{0\le y\le 1} f(x,y) = \sup_{0\le x\le 1} \begin{cases} 1, & 0\le x\le 1/2 \\ 0, & 1/2 < x\le 1 \end{cases} = 1.$$

and

$$\inf_{0\le y\le 1} \sup_{0\le x\le 1} f(x,y) = \inf_{0\le y\le 1} \begin{cases} 2, & 0\le y\le 1/2 \\ 2, & 1/2 < y\le 1 \end{cases} = 2$$

so in this example

$$\inf_{y\in S_y} \sup_{x\in S_x} f(x,y) \ne \sup_{x\in S_x} \inf_{y\in S_y} f(x,y)$$

and in fact

$$\sup_{x\in S_x} \inf_{y\in S_y} f(x,y) < \inf_{y\in S_y} \sup_{x\in S_x} f(x,y)$$

The moral here is that **order** matters. For example, in an applied optimization problem, it is not always true that

$$\min_x \max_y f(x,y) = \max_y \min_x f(x,y)$$

where x and y come from some domain set S. So it is probably important to find out when the order does not matter because it might be easier to compute in one ordering than another.

Theorem 2.6.5 The Infimum and Supremum are Unique

> *Let S be a nonempty bounded set of real numbers. Then $\inf(S)$ and $\sup(S)$ are unique.*

Proof 2.6.5
By the completeness axiom, since S is bounded and nonempty, we know $\inf(S)$ and $\sup(S)$ are finite numbers. Let u_2 satisfy the definition of supremum also. Then, we know $u_2 \leq M$ for all upper bounds M of S and in particular since $\sup(S)$ is an upper bound too, we must have $u_2 \leq \sup(S)$. But since $\sup(S)$ is a supremum, by definition, we also know $\sup(S) \leq u_2$ as u_2 is an upper bound. Combining, we have $u_2 \leq \sup(S) \leq u_2$ which tells us $u_2 = \sup(S)$. A similar argument shows the $\inf(S)$ is also unique. ∎

Let's look at some basic results involving functions. What does it mean for a function to be bounded?

Definition 2.6.6 Bounded Functions

> *Let $f : D \longrightarrow \Re$ be a function whose domain D is a nonempty set. We say f is bounded on D if there is a number B so that $|f(x)| \leq B$ for all $x \in D$.*
> *If we let $f(D)$ be the set $\{y : y = f(x), \, x \in D\}$, note this is the same as saying $f(D)$ is a bounded set in the sense we have previously used.*

Using these we can prove some fundamental results.

Lemma 2.6.6 The Supremum of a Sum is Smaller than or Equal to the Supremum of the parts

> *Let $f, g : D \longrightarrow \Re$ be two functions whose domain D is a nonempty set. Then, if f and g are both bounded on D, we have*
> $$\sup_{x \in D} \{f(x) + g(x)\} \quad \leq \quad \sup_{x \in D} \{f(x)\} \; + \; \sup_{x \in D} \{g(x)\}$$

Proof 2.6.6
Since f is bounded on D, there is a number B_f so that $|f(x)| \leq B_f$. This tells us the set $f(D)$ is bounded and non empty and so by the completeness axiom, the set $f(D)$ has a finite supremum, $\sup_{x \in D} f(x)$. A similar argument shows that the set $g(D)$ has a finite supremum, $\sup_{x \in D} g(x)$. Then, by the definition of a supremum, we have
$$f(x) + g(x) \quad \leq \quad \sup_{x \in D} f(x) \; + \; \sup_{x \in D} g(x).$$

This implies the number $\sup_{x \in D} f(x) + \sup_{x \in D} g(x)$ is an upper bound for the set $\{f(x) + g(x) : x \in D\}$. Hence, by definition
$$\sup_{x \in D} (f(x) + g(x)) \quad \leq \quad \sup_{x \in D} f(x) \; + \; \sup_{x \in D} g(x).$$

∎

There is a similar result for infs.

Lemma 2.6.7 The Infimum of a Sum is Greater than or Equal to the Infimum of the Parts

Let f, $g : D \longrightarrow \Re$ be two functions whose domain D is a nonempty set. Then, if f and g are both bounded on D, we have

$$\inf_{x \in D} \{f(x) + g(x)\} \geq \inf_{x \in D} \{f(x)\} + \inf_{x \in D} \{g(x)\}$$

Proof 2.6.7

Since f and g are bounded on D, by the completeness axiom, the set $f(D)$ and $g(D)$ have finite infima, $\inf_{x \in D} f(x)$ and $\inf_{x \in D} f(x)$. Then, by the definition of an infimum, we have $f(x) + g(x) \geq \inf_{x \in D} f(x) + \inf_{x \in D} g(x)$. This implies the number $\inf_{x \in D} f(x) + \inf_{x \in D} g(x)$ is a lower bound for the set $\{f(x) + g(x) : x \in D\}$. Hence, by definition $\inf_{x \in D}(f(x) + g(x)) \geq \inf_{x \in D} f(x) + \inf_{x \in D} g(x)$. ∎

2.6.5   Homework

Exercise 2.6.1 *Let*

$$f(x,y) = \begin{cases} 3, & (x,y) \in (1/2,1] \times (1/2,1] \\ -2, & (x,y) \in (1/2,1] \times [0,1/2] \text{ and } [0,1/2] \times (1/2,1] \\ 4, & (x,y) \in [0,1/2] \times [0,1/2] \end{cases}$$

and let $S = [0,1] \times [0,1]$ which is also $S_x \times S_y$ where $S_x = \{x : 0 \leq x \leq 1\}$ and $S_y = \{y : 0 \leq y \leq 1\}$. Find

1. *$\inf_{(x,y) \in S} f(x,y)$, and $\sup_{(x,y) \in S} f(x,y)$,*

2. *$\inf_{y \in S_y} \sup_{x \in S_x} f(x,y)$ and $\sup_{x \in S_x} \inf_{y \in S_y} f(x,y)$.*

3. *$\inf_{x \in S_x} \sup_{y \in S_y} f(x,y)$ and $\sup_{y \in S_y} \inf_{x \in S_x} f(x,y)$.*

Exercise 2.6.2 *Let*

$$f(x,y) = \begin{cases} 6, & (x,y) \in (1/2,1] \times (1/2,1] \\ 1, & (x,y) \in (1/2,1] \times [0,1/2] \text{ and } [0,1/2] \times (1/2,1] \\ 14, & (x,y) \in [0,1/2] \times [0,1/2] \end{cases}$$

and let $S = [0,1] \times [0,1]$ which is also $S_x \times S_y$ where $S_x = \{x : 0 \leq x \leq 1\}$ and $S_y = \{y : 0 \leq y \leq 1\}$. Find

1. *$\inf_{(x,y) \in S} f(x,y)$, and $\sup_{(x,y) \in S} f(x,y)$,*

2. *$\inf_{y \in S_y} \sup_{x \in S_x} f(x,y)$ and $\sup_{x \in S_x} \inf_{y \in S_y} f(x,y)$.*

3. *$\inf_{x \in S_x} \sup_{y \in S_y} f(x,y)$ and $\sup_{y \in S_y} \inf_{x \in S_x} f(x,y)$.*

Exercise 2.6.3 *Let $S = \{z : z = e^{-x^2 - y^2} \text{ for } (x,y) \in \Re^2\}$. Find $\inf(S)$ and $\sup(S)$. Does the minimum and maximum of S exist and if so what are their values?*

Chapter 3

Sequences of Real Numbers

The next thing to look at is what are called sequences of real numbers. You probably have seen them in your earlier calculus courses, but now it is time to look at them very carefully.

3.1 Basic Definitions

To define sequences and subsequences of a sequence, we must have a way of stating precisely what kinds of subsets of the integers we want to focus on.

Definition 3.1.1 Right Increasing Subsets of Integers

*We know $\mathbb{Z} = \{\ldots, -2, -1, 0, 1, 2, 3, \ldots\}$ is the set of integers. Let $\mathbb{Z}_{\geq k}$ denote the set of all integers $\geq k$ Thus, $\mathbb{Z}_{\geq -3} = \{-3, -2, -1, 0, 1, 2, 3, \ldots\}$. Let T be a subset of $\mathbb{Z}$. We say T is a **Right Increasing Infinite Subset** or **RII** of $\mathbb{Z}$ if T is **not** bounded above and T **is** bounded below and the entries in T are always increasing.*

Example 3.1.1

$$\begin{aligned} T &= \{2, 4, 6, \ldots, 2n, \ldots\} = \{2k\}_{k=1}^{\infty} = (2k)_{k=1}^{\infty} = (2k) \\ T &= \{-17, -5, 7, 19, \ldots\} = \{-17 + 12k\}_{k=0}^{\infty} = (-17 + 12k)_{k=0}^{\infty} \end{aligned}$$

In general, a **RII** subset T of $\mathbb{Z}$ can be characterized by

$$T = \{n_0, n_1, n_2, \ldots, n_k, \ldots\}$$

where n_0 is the starting integer or index and

$$n_0 < n_1 < n_2 < \ldots < n_k < \ldots$$

with $n_k \to \infty$ as $k \to \infty$.

Note $T = \mathbb{Z}_{\geq 1} = \{1, 2, 3, \ldots\}$ is our usual set of counting numbers $\mathbb{N}$. We can use this ideas to define sequences and subsequences carefully.

3.2 The Definition of a Sequence

Definition 3.2.1 A Sequence of Real Numbers

> A sequence of real numbers is simply a real valued function whose domain is the set $\mathbb{Z}_{\geq k}$ for some k. A subset of the set $f(\mathbb{Z}_{\geq k})$ defined by $f(T)$ for any RII T of $\mathbb{Z}_{\geq k}$ is called a subsequence of the sequence.

To help you see what sequences are all about, you need to look at examples!

Example 3.2.1 *Consider* $f : \mathbb{N} \to \Re$ *defined by* $f(n) = a_n$ *where each* a_n *is a number. There are many notations for this sequence:*

$$
\begin{aligned}
\{f(n) : n \in \mathbb{N}\} &= \{f(n)\}_{n=1}^{\infty} = (f(n))_{n=1}^{\infty} = (f(n)) \\
&= \{f(1), f(2), \ldots, f(n), \ldots\} = \{a_1, a_2, \ldots, a_n, \ldots\}
\end{aligned}
$$

Example 3.2.2 *Consider* $f : \mathbb{N} \to \Re$ *defined by* $f(n) = \sin(n\pi/4)$. *Then*

$$
\begin{aligned}
\{f(n)\}_{n=1}^{\infty} &= \{\sin(\pi/4), \sin(2\pi/4), \sin(3\pi/4), \sin(4\pi/4), \\
&\quad \sin(5\pi/4), \sin(6\pi/4), \sin(7\pi/4), \sin(8\pi/4), \ldots\} \\
&= \{\sqrt{2}/2, 1, \sqrt{2}/2, 0, -\sqrt{2}/2, -1, -\sqrt{2}/2, 0, \ldots\}
\end{aligned}
$$

Let the numbers $\{\sqrt{2}/2, 1, \sqrt{2}/2, 0, -\sqrt{2}/2, -1, -\sqrt{2}/2, 0\}$ *be called block* B. *Then we see the values of this sequence consist of the infinitely repeating blocks* B *as follows:*

$$
(\sin(n\pi/4)) = \{B, B, \ldots, B, \ldots\}
$$

where each block B *consists of 8 numbers.*

So the range of this function, i.e. the range of this sequence, is the set $\{-1, -\sqrt{2}/2, 0, \sqrt{2}/2, 1\}$ *which is just 5 values.*

Now define RII sets as follows:

$$
\begin{aligned}
T_1 &= \{1 + 8k\}_{k \geq 0} = \{1, 9, 17, \ldots\} \\
T_2 &= \{2 + 8k\}_{k \geq 0} = \{2, 10, 18, \ldots\} \\
T_3 &= \{4 + 8k\}_{k \geq 0} = \{4, 12, 20, \ldots\} \\
T_4 &= \{5 + 8k\}_{k \geq 0} = \{5, 13, 21, \ldots\} \\
T_5 &= \{6 + 8k\}_{k \geq 0} = \{6, 14, 22, \ldots\}
\end{aligned}
$$

Define new functions $f_i : T_i \to \Re$ *by*

$$
\begin{aligned}
f_1(1 + 8k) &= f(1 + 8k) = \sin((1 + 8k)\pi/4) = \sin(\pi/4) = \sqrt{2}/2 \\
f_2(2 + 8k) &= f(2 + 8k) = \sin((2 + 8k)\pi/4) = \sin(2\pi/4) = 1 \\
f_3(4 + 8k) &= f(4 + 8k) = \sin((4 + 8k)\pi/4) = \sin(4\pi/4) = 0 \\
f_4(5 + 8k) &= f(5 + 8k) = \sin((5 + 8k)\pi/4) = \sin(5\pi/4) = -\sqrt{2}/2 \\
f_5(6 + 8k) &= f(6 + 8k) = \sin((6 + 8k)\pi/4) = \sin(6\pi/4) = -1
\end{aligned}
$$

Each of these functions **extracts** *a subset of the original set* $(f(n)) = (\sin(n\pi/4))$ *where we did not explicitly indicate the subscripts* $n \geq 1$ *as it is understood at this point. These five functions give five subsequences of the original sequence.*

We usually don't go through all this trouble to find subsequences. Instead of the f_i *notation, we define these subsequences by another notation. The original sequence is* $f(n) = \sin(n\pi/4)$ *so the values of the sequence are* $a_n = \sin(n\pi/4)$. *These five subsequences are then defined by*

$$
\begin{aligned}
a_{1+8k} &= f(1+8k) = \sin((1+8k)\pi/4) \\
a_{2+8k} &= f(2+8k) = \sin((2+8k)\pi/4) = \sin(2\pi/4) = 1 \\
a_{4+8k} &= f(4+8k) = \sin((4+8k)\pi/4) = \sin(4\pi/4) = 0 \\
a_{5+8k} &= f(5+8k) = \sin((5+8k)\pi/4) = \sin(5\pi/4) = -\sqrt{2}/2 \\
a_{6+8k} &= f(6+8k) = \sin((6+8k)\pi/4) = \sin(6\pi/4) = -1
\end{aligned}
$$

*There is a subsequence for each RII subset of $\mathbb{N}$ here but really these are the **interesting** ones. Consider this table:*

Sequence/ Function	All Values	Range
$(a_n), f$	$\{\sqrt{2}/2, 1, \sqrt{2}/2, 0, -\sqrt{2}/2, -1,$ $-\sqrt{2}/2, 0,$ *repeat* $\}$	$\{-1, -\sqrt{2}/2, 0, \sqrt{2}/2, 1\}$
a_{1+8k}, f_1	$\{\sqrt{2}/2,$ *repeat* $\}$	$\{\sqrt{2}/2\}$
a_{2+8k}, f_2	$\{1,$ *repeat* $\}$	$\{1\}$
a_{4+8k}, f_3	$\{0,$ *repeat* $\}$	$\{0\}$
a_{5+8k}, f_4	$\{-\sqrt{2}/2,$ *repeat* $\}$	$\{-\sqrt{2}/2\}$
a_{6+8k}, f_5	$\{-1,$ *repeat* $\}$	$\{-1\}$

Each T_i is contained in the domain of f and the range of each subsequence f_i is contained in the range of the original sequence. Note, t he set of all values of the sequence and the subsequences is different from their range. The set of values is a sequence and the range is the distinct values seen in the set of all sequence values.

Example 3.2.3 $f(n) = 1/n$ *for $n \in \mathbb{N}$. We often write this as $(1/n)$.*

- *The domain of the sequence is $\mathbb{N}$.*

- *The range of the sequence is $\{1, 1/2, 1/3, 1/4, \ldots, 1/n, \ldots\}$*

- *The set of values of the sequence is $\{1, 1/2, 1/3, 1/4, \ldots, 1/n, \ldots\}$; in this case the same as the range.*

- *$f(2n)$ is the subsequence corresponding to the RII set $T = \{2, 4, \ldots, 2n, \ldots\}$. We have $(a_{2n}) = (1/(2n))$.*

Example 3.2.4 *We can define a sequence recursively. This time we will use the notation (x_n) instead of (a_n) for the sequence to show you the choice of letter does not matter. Let $x_1 = 1$ and for $n \geq 1$, define $x_{n+1} = x_n + 2$.*

- *The domain of the sequence is $\mathbb{N}$.*

- *The range of the sequence is $\{1, 3, 5, 7, \ldots, 2n+1, \ldots\}$*

- *The set of values of the sequence is $\{1, 3, 5, 7, \ldots, 2n-1, \ldots\}$ for $n \geq 1$; in this case the same as the range.*

- *$f(2n)$ is the subsequence corresponding to the RII set $T = \{2, 4, \ldots, 2n, \ldots\}$. We have $(a_{2n}) = \{3, 9, 11, \ldots, 4n-1, \ldots\}$ for $n \geq 1$.*

Example 3.2.5 *Here is another recursive sequence. This time we will use the notation (x_n) instead of (a_n) for the sequence to show you the choice of letter does not matter. Let $x_1 = 1$, $x_2 = 1$ and for $n > 1$, define $x_{n+2} = x_{n+1} + 2x_n$. This gives $x_3 = 1 + 2 = 3$, $x_4 = x_3 + 2x_2 = 3 + 2 = 5$, $x_5 = x_4 + 2x_3 = 5 + 10 = 15$, $x_6 = x_5 + 2x_4 = 25$ etc.*

- *The domain of the sequence is $\mathbb{N}$.*

- *It is not clear what the pattern for the range is. So far we have $\{1, 1, 3, 5, 15, 25, \ldots\}$.*

- *The set of values of the sequence is again the same as the range.*

- *The subsequence corresponding to the RII set $T = \{2, 4, \ldots, 2n, \ldots\}$ is $(x_{2n}) = \{x_2, x_4, \ldots\}$ $= \{2, 5, \ldots\}$.*

Example 3.2.6 *Here is an example of a sequence that is defined recursively that occurs in the solution of a second order differential equation that is discussed in Section 19.2.1. The solution of*

$$(1+t)x''(t) + t^2 x'(t) + 2x(t) = 0$$
$$x(0) = 1$$
$$x'(0) = 3$$

can be written as a sum of terms of the form

$$a_0 + a_1 t + a_2 t^2 + \ldots$$

for values of t in a circle about $t = 0$. We find that $a_0 = 1$, $a_1 = 3$, $a_2 = -1$ and $a_3 = -2/3$ and the other terms satisfy the recursion

$$(n+2)(n+1)a_{n+2} + (n+1)(n)a_{n+1} + (n-1)a_{n-1} + 2a_n = 0, \quad \forall\, n \geq 2$$

and so we can calculate $a_4 = 1/4$ and so forth.

3.2.1   Homework

Exercise 3.2.1 *Examine the sequence $(\sin(3n\pi/5))$ as we have done for $(\sin(n\pi/4))$. Do this for $n \geq 1$. Find the blocks, the subsequences etc.*

Exercise 3.2.2 *Examine the sequence $(\sin(7n\pi/5))$ as we have done for $(\sin(n\pi/4))$. Do this for $n \geq 1$. Find the blocks, the subsequences etc.*

Exercise 3.2.3 *Examine the sequence $(\cos(7n\pi/5))$ as we have done for $(\sin(n\pi/4))$. Do this for $n \geq 1$. Find the blocks, the subsequences etc.*

Exercise 3.2.4 *Examine the sequence $(\cos(n\pi/6))$ as we have done for $(\sin(n\pi/4))$. Do this for $n \geq 1$. Find the blocks, the subsequences etc.*

Exercise 3.2.5 *Examine the sequence $(\sin(n\pi/4) + \cos(n\pi/3))$ as we have done for $(\sin(n\pi/4))$. Do this for $n \geq 1$. Find the blocks, the subsequences etc.*

Exercise 3.2.6 *Examine the sequence $(\sin(3n\pi/4) + \cos(2n\pi/3))$ as we have done for $(\sin(n\pi/4))$. Do this for $n \geq 1$. Find the blocks, the subsequences etc.*

3.3   The Convergence of a Sequence

What do we mean by the phrase **a sequence converges**? For a simple sequence like $(a_n = 1/n)$ it is easy to see that as n gets large the value of the sequence gets closer and closer to 0. It is also easy to see for sequences with finite blocks that repeat in their range, like $((-1)^n)$ or $(\sin(n\pi/3))$, the values of the sequences bounce around among the finite possibilities in the block. But we need a careful way to say when the sequence gets closer to some fixed value and a careful way to say the sequence can not do that. This language is handled in the definition below:

Definition 3.3.1 Sequence Convergence and Divergence

Let $(a_n)_{n \geq k}$ be a sequence of real numbers. Let a be a real number. We say the sequence a_n converges to a if

$$\forall \epsilon > 0, \exists N \ni n > N \Longrightarrow |a_n - a| < \epsilon$$

We usually just write $a_n \to a$ as $n \to \infty$ to indicate this convergence. We call a the limit of the sequence $(a_n)_{n \geq k}$.

*The sequence $(a_n)_{n \geq k}$ **does not converge** to a number a if we can find a positive number ϵ so that no matter what N we choose, there is always at least one $n > N$ with $|a_n - a| > \epsilon$. We write this using mathematical language as*

$$\exists \epsilon > 0, \ni \forall N \exists n > N \text{ with } |a_n - a| \geq \epsilon$$

*If a sequence does not converge, we say it **diverges**.*

Here are some specific examples to help these ideas sink in.

3.3.1 Proofs of Divergence

Theorem 3.3.1 $((-1)^n)$ Diverges

The sequence $((-1)^n)_{n \geq 1}$ does not converge.

Proof 3.3.1
The range of this sequence is the set $\{-1, 1\}$. We will show there is no number a so that $(-1)^n \to a$ for this sequence.
   Case 1:
Let $a = 1$. Note the difference between -1 and $+1$ is 2 which suggests we pick a tolerance $\epsilon < 2$ to show the sequence does not converge to 1 as we want to isolate which value we are trying to get close to. Let $\epsilon = 1$. Then

$$|(-1)^n - 1| = \begin{cases} |1 - 1| = 0, & \text{if } n \text{ is even.} \\ |-1 - 1| = 2, & \text{if } n \text{ is odd.} \end{cases}$$

Now pick an N. Then there is an odd integer larger than N, say $2N + 1$, for which $|(-1)^{2N+1} - 1| = 2 > \epsilon$. Since we can do this for all N, we see the sequence can not converge to 1.
   Case -1:
We can repeat this argument for this case. Let $\epsilon = 1$. Then

$$|(-1)^n - (-1)| = \begin{cases} |1 + 1| = 2, & \text{if } n \text{ is even.} \\ |-1 + 1| = 0, & \text{if } n \text{ is odd.} \end{cases}$$

Now pick an N. Then there is an even integer larger than N, say $2N$, for which $|(-1)^{2N} - (-1)| = 2 > \epsilon$. Since we can do this for all N, we see the sequence can not converge to -1.
   Case $a \neq 1, -1$:
If a is not -1 or 1, let d_1 be the distance for a to 1 which is $|a - 1|$ and let d_2 be the distance to -1 which is $|a - (-1)|$. Let d be the minimum of these two distances, $d = \min(d_1, d_2)$. Then, we have

$$|(-1)^n - a| = \begin{cases} |1 - a| = d_1, & \text{if } n \text{ is even.} \\ |-1 - a| = d_2, & \text{if } n \text{ is odd.} \end{cases}$$

If we pick $\epsilon = (1/2)d$, we see if both cases $|(-1)^n - a| > d$ for any value of n. Thus, $(-1)^n$ can not converge to this a either. Since all choices of a are now exhausted, we can say this sequence does not converge. ∎

Comment 3.3.1 *This argument works for any sequence with a finite range. For example, for $a_n = \sin(n\pi/4)$ for $n \geq 1$, the range is $\{-1, -\sqrt{2}/2, 0, \sqrt{2}/2, 1\}$. We can argue that the limit can not be any of the values in the range by choosing an ϵ less that $1/2$ of the distances between the values in the range. If we let the range values be $y_1, \ldots, y_5$, the minimum distance to a number a other than these values is is $\min_{1 \leq i \leq 5} |a - y_i|$. We use then use the last argument above choosing ϵ as before. Thus, no a can be the limit of this sequence.*

Comment 3.3.2 *No sequence with a finite range can have a limit unless there is only one value in its range.*

Theorem 3.3.2 $(\cos(n\pi/4))$ **Diverges**

> *The sequence $(\cos(n\pi/4))_{n \geq 1}$ does not converge.*

Proof 3.3.2
The block for this sequence is $B = \{\sqrt{2}/2, 0, -\sqrt{2}/2, -1, -\sqrt{2}/2, 0, \sqrt{2}/2, 1\}$ and the range of this sequence is the set $\{-1, -\sqrt{2}/2, 0, \sqrt{2}/2, 1\}$. We show there is no number a so that $\cos(n\pi/4) \to a$ for this sequence.
Case -1:
Let $a = -1$. Note the minimum distance between range values is $1 - \sqrt{2}/2 \approx .3$ which suggests we pick $\epsilon < .3$ to show the sequence does not converge to -1. Let $\epsilon = .1$. Then

$$|\cos(n\pi/4) - (-1)| = \begin{cases} |-1+1| = 0, \text{ if } n \text{ for range value } -1 \\ |-\sqrt{2}/2 + 1| = .303, \text{ for range value } -\sqrt{2}/2 \\ |0+1| = 1, \text{ for range value } 0 \\ |\sqrt{2}/2 + 1| = 1.707, \text{ for range value } \sqrt{2}/2 \\ |1+1| = 2.0, \text{ for range value } 1 \end{cases}$$

Now pick an N. Then any $n > N$ with $\cos(n\pi/4) \neq -1$ satisfies $|\cos(n\pi/4) - (-1)| > \epsilon = .1$. Since we can do this for all N, we see the sequence can not converge to -1.
     Case $-\sqrt{2}/2$:
Let $a = -\sqrt{2}/2$. Let $\epsilon = .1$ for the same reason as before. Then

$$|\cos(n\pi/4) - (-\sqrt{2}/2)| =$$
$$\begin{cases} |-1+\sqrt{2}/2| = .303, \text{ if } n \text{ for range value } -1 \\ |-\sqrt{2}/2 + \sqrt{2}/2| = 0, \text{ for range value } -\sqrt{2}/2 \\ |0 + \sqrt{2}/2| = .707, \text{ for range value } 0 \\ |\sqrt{2}/2 + \sqrt{2}/2| = 1.414, \text{ for range value } \sqrt{2}/2 \\ |1 + \sqrt{2}/2| = 1.707, \text{ for range value } 1 \end{cases}$$

Now pick an N. Then any $n > N$ whose sequence value $\cos(n\pi/4) \neq -\sqrt{2}/2$ satisfies $|\cos(n\pi/4) - (-\sqrt{2}/2)| > \epsilon = .1$. Since we can do this for all N, we see the sequence can not converge to $-\sqrt{2}/2$.
     We will leave it to you to mimic these arguments for the remaining cases.
Next, we will do the case where a is not equal to any of the range values.
     Case a is not a range value:
If a is not equal to a range value, let d_1 be the distance for a to -1, $|a - (-1)|$, d_2 the distance for a to $-\sqrt{2}/2$, $|a - (-\sqrt{2}/2)|$, d_3 the distance for a to 0, $|a - 0|$, d_4 the distance for a to $\sqrt{2}/2$,

$|a - \sqrt{2}/2|$ *and d_5 the distance for a to 1, $|a-1|$, and let d_2 be the distance to -1 which is $|a-(-1)|$.* *Let d be the minimum of these five distances, $d = \min\{d_1, \ldots, d_5\}$ and choose $\epsilon = d/2$. Then, we have*

$$| \cos(n\pi/4) - a | \quad = \quad \begin{cases} |-1-a| = d_1, & \textit{if n for range value } -1 \\ |-\sqrt{2}/2 - a| = d_2, & \textit{for range value } -\sqrt{2}/2 \\ |0-a| = d_3, & \textit{for range value } 0 \\ |\sqrt{2}/2 - a| = d_4, & \textit{for range value } \sqrt{2}/2 \\ |1-a| = d_5, & \textit{for range value } 1 \end{cases}$$

For $\epsilon = d/2$, in all cases $| \cos(n\pi/4) - a | > d/2$ for any value of n. Hence, for any N, any $n > N$ gives $| \cos(n\pi/4) - a | > \epsilon$. Thus $\cos(n\pi/4)$ can not converge to an a that is not a range value. Since all choices of a are now exhausted, we can say this sequence does not converge. ∎

3.3.2   Uniqueness of Limits and So Forth

First, we need to show the limit of a convergent sequence is unique.

Theorem 3.3.3 Convergent Sequences have Unique Limits

If the sequence (x_n) converges, then the limit is unique.

Proof 3.3.3
Assume $x_n \to a$ and $x_n \to b$. Then for an arbitrary $\epsilon > 0$, there is an N_1 and an N_2 so that

$$n > N_1 \quad \Rightarrow |x_n - a| < \epsilon/2$$
$$n > N_2 \quad \Rightarrow |x_n - b| < \epsilon/2$$

Now pick a $P > \max\{N_1, N_2\}$. Then, we have

$$|a - b| \quad = \quad |a - x_P + x_P - b| \le |a - x_P| + |x_P - b| < \epsilon/2 + \epsilon/2 = \epsilon.$$

Since $\epsilon > 0$ is arbitrary and $|a - b| < \epsilon$ for all $\epsilon > 0$, this shows $a = b$. ∎

Next we need to look at subsequences of convergent sequences.

Theorem 3.3.4 Subsequences of Convergent Sequences have the Same Limit

Assume (x_n) converges and (x_{n_k}) is a subsequence. Then (x_{n_k}) converges to the same limiting value.

Proof 3.3.4
Since $x_n \to a$ for some a, for arbitrary $\epsilon > 0$, there is an N so that $n > N \Rightarrow |x_n - a| < \epsilon$. In particular, $n_k > N \Rightarrow |x_{n_k} - a| < \epsilon$. This says $x_{n_k} \to a$ also. ∎

3.3.3   Proofs of Convergence

Theorem 3.3.5 $(1 + (4/n))$ Converges

The sequence $(1 + (4/n))_{n \ge 1}$ converges to 1.

Proof 3.3.5

The range of this sequence is not finite and indeed does not repeat any values. However, we can guess that as n gets large, the values in the sequence are closer and closer to 1. So let's assume $a = 1$ as our guess for the limit. Pick an arbitrary $\epsilon > 0$ and consider

$$|a_n - a| \quad = \quad |(1 + (4/n)) - 1| = |4/n| = 4|1/n|.$$

Pick any N so that $4/N < \epsilon$ or $N > 4/\epsilon$. Then for any $n > N$, $n > 4/\epsilon$ or $4/n < \epsilon$. So we have shown

$$n > N > 4/\epsilon \quad \Rightarrow \quad |(1 + (4/n)) - 1| < \epsilon$$

Since ϵ was chosen arbitrarily, we know the sequence converges to 1. ∎

The convergence proofs follows a nice template which we show next.

Theorem 3.3.6 $a_n \to a$ **Template**

> $a_n \to a.$

Proof 3.3.6

Step 1: *Identify what a is.*
Step 2: *Choose $\epsilon > 0$ arbitrarily.*
Step 3: *Now follow this argument:*

$$
\begin{aligned}
|\text{ original sequence - proposed limit }| \quad &= \quad |\text{ simplify using algebra etc to get a new expression}| \\
&\leq \quad |\text{ use triangle inequality etc to get a new expression again}| \\
&= \quad \text{call this last step the overestimate. We have now} \\
&= \quad |overestimate| < \epsilon
\end{aligned}
$$

Step 4: *Solve for n in terms of ϵ to give a simple equation.*
Step 5: *Choose N to satisfy the inequality you get from* **Step 4**.
Step 6: *Then for any $n > N$, $|overestimate| < \epsilon$ and we have $|(original) - (proposed\ limit)| < \epsilon$ proving $a_n \to a$.* ∎

Theorem 3.3.7 $\left(\frac{1+4n}{5+6n}\right)$ **Converges**

> $\left(\dfrac{1+4n}{5+6n}\right)_{n \geq 1}$ *converges.*

Proof 3.3.7

We can guess the value of the limit is $a = 2/3$ so pick $\epsilon > 0$ arbitrarily. Consider

$$\left| \frac{1+4n}{5+6n} - \frac{2}{3} \right| \quad = \quad \left| \frac{3(1+4n) - 2(5+6n)}{(3)(5+6n)} \right| = \left| \frac{-7}{(3)(5+6n)} \right|$$

The denominator can be underestimated: $15 + 18n \geq 18n$ which implies $1/(3(5+6n))| < 1/(18n)$. Thus, our original calculation gives

$$\left| \frac{1+4n}{5+6n} - \frac{2}{3} \right| < (7/18)(1/n)$$

Now set this overestimate less than ϵ and solve for n.

Now $7/(18n) < \epsilon \Rightarrow n > 7/(18\epsilon)$ and so choosing $N > 7/(18\epsilon)$ implies the following chain:

$$n \geq N \quad \Rightarrow \quad n > 7/(18\epsilon) \Longrightarrow |original - limit| < \epsilon$$

This shows $\frac{1+4n}{5+6n} \to \frac{2}{3}$. ■

Here is another one.

Theorem 3.3.8 ($\frac{1+4n}{5-6n}$) Converges

$\left(\frac{1+4n}{5-6n} \right)_{n \geq 1}$ *converges.*

Proof 3.3.8

We can guess the value of the limit is $a = -2/3$ so pick $\epsilon > 0$ arbitrarily. Consider

$$\left| \frac{1+4n}{5-6n} - -\frac{2}{3} \right| = \left| \frac{3(1+4n) + 2(5-6n)}{(3)(5-6n)} \right| = \left| \frac{13}{(3)(5-6n)} \right|$$

We want to underestimate $|15 - 18n| = |18n - 15|$. Now $18n - 15$ is positive for $n \geq 1$ so we don't need the absolute values here. Pick a number smaller than 18. We usually go for half or 9. Set $18n - 15 > 9n$ and figure out when this is true. A little thought tells us this works when $n \geq 2$. So $|18n - 15| > 9n$ when $n \geq 2$ implying $|13/(15 - 18n)| < 13/(9n)$ when $n \geq 2$. Thus,

$$\left| \frac{1+4n}{5-6n} + \frac{2}{3} \right| < (13/(9n)), \forall n \geq 2$$

Next, set this overestimate less than ϵ to get $13/(9n) < \epsilon$ or $n > 13/(9\epsilon)$. But we must also make sure $N \geq 2$ too, so if we set $N > \max(13/(9\epsilon), 2)$ implies the following chain:

$$n \geq N \quad \Rightarrow \quad n > 13/(9\epsilon) \Longrightarrow |original - limit| < \epsilon$$

This shows $\frac{1+4n}{5-6n} \to -\frac{2}{3}$. ■

Another fractional limit:

Theorem 3.3.9 ($\frac{11+4n+8n^2}{6+7n^2}$) Converges

$\left(\frac{11+4n+8n^2}{6+7n^2} \right)$ *converges.*

Proof 3.3.9

We guess the limit is $a = 8/7$. Consider

$$\left| \frac{11+4n+8n^2}{6+7n^2} - \frac{8}{7} \right| = \left| \frac{7(11+4n+8n^2) - 8(6+7n^2)}{7(6+7n^2)} \right|$$

$$= \left| \frac{29 + 28n}{7(6 + 7n^2)} \right|$$

Now, for the numerator, $29 + 28n \le 29n + 28n = 57n$ and for the denominator, $7(6 + 7n^2) > 49n^2$ and so

$$\left| \frac{11 + 4n + 8n^2}{6 + 7n^2} - \frac{8}{7} \right| < \frac{57n}{49n^2} = \frac{57}{49n}$$

Set $57/(49n) < \epsilon$ to get $n > 57/(49\epsilon)$. Now pick any $N > 57/(49\epsilon)$ and we see $n > N$ implies $|original - limit| < \epsilon$. We conclude $8/7$ is the limit of the sequence.   ■

Theorem 3.3.10 ($\frac{11+4n-8n^2}{6-n-2n^2}$) Converges

$(\frac{11+4n-8n^2}{6-n-2n^2})$ converges.

Proof 3.3.10

We guess the limit is $a = 4$. Consider

$$\left| \frac{11 + 4n - 8n^2}{6 - n - 2n^2} - 4 \right| = \left| \frac{11 + 4n - 8n^2 - 4(6 - n - 2n^2)}{6 - n - 2n^2} \right|$$

$$= \left| \frac{-13 + 8n}{6 - n - 2n^2} \right|$$

Now, for the numerator, $|-13 + 8n| \le 13 + 8n \le 13n + 8n = 21n$ by the triangle inequality and for the denominator, $|6 - n - 2n^2| = |2n^2 + n - 6|$. A good choice for underestimating here is n^2; i.e. use half of the coefficient of the n^2. For $n > 2$, the absolute values are not needed and so we see when $2n^2 + n - 6 > n^2$ implying $n^2 + n - 6 > 0$. This is true for all $n \ge 3$ so there are restrictions here on n. We have $|6 - n - 2n^2| \ge n^2$ if $n \ge 3$.

Thus, if $n \ge 3$

$$\left| \frac{11 + 4n - 8n^2}{6 - n - 2n^2} - 4 \right| < \frac{21n}{n^2} = 21\,(1/n)$$

Set $21/n < \epsilon$ to get $n > 3/\epsilon$. Now pick any $N > \max(21/\epsilon, 3)$ and we see $n > N$ implies $|original - limit| < \epsilon$. We conclude 4 is the limit of the sequence.   ■

Theorem 3.3.11 $(3 + \sin(2n + 5)/n)$ Converges

$(3 + \sin(2n + 5)/n)_{n \ge 1}$ converges.

Proof 3.3.11

We suspect the limit is 3. Hence, pick $\epsilon > 0$ arbitrary and consider

$$\left| 3 + \frac{\sin(2n + 5)}{n} - 3 \right| = \left| \frac{\sin(2n + 5)}{n} \right| \le 1/n$$

because $|\sin| \leq 1$ *no matter its argument. Set this overestimate less than ϵ to conclude if $N > 1/\epsilon$, then $n > N$ implies $|original - limit| < \epsilon$. Thus, we see this sequence converges to* 3. ∎

3.3.4 Homework

Exercise 3.3.1 *Prove the sequence* $(\sin(2n\pi/5))_{n\geq 1}$ *does not converge.*

Exercise 3.3.2 *Prove the sequence* $(\frac{3+5n}{6+8n})_{n\geq 1}$ *converges.*

Exercise 3.3.3 *Prove the sequence* $(\frac{4+5n}{6-7n})_{n\geq 1}$ *converges.*

Exercise 3.3.4 *Prove the sequence* $(\frac{4+5n^2}{6-3n-4n^2})_{n\geq 1}$ *converges.*

Exercise 3.3.5 *Prove the sequence* $(\sin(3n\pi/7))_{n\geq 1}$ *does not converge.*

Exercise 3.3.6 *Prove the sequence* $(\frac{30+2n}{6+9n})_{n\geq 1}$ *converges.*

Exercise 3.3.7 *Prove the sequence* $(\frac{8+15n}{60-17n})_{n\geq 1}$ *converges.*

Exercise 3.3.8 *Prove the sequence* $(\frac{4+6n-5n^2}{6-3n-4n^3})_{n\geq 1}$ *converges.*

3.4 Sequence Spaces

Let S denote the set of all sequences of real numbers. To make it easy to write them down, let's assume all these sequences start at the same integer, say k. So S is the set of all objects x where x is a sequence of the form $(a_n)_{n\geq k}$. Thus, $S = \{x : x = (a_n)_{n\geq k}\}$.

- We define **addition** operation $+$ like this: $x + y$ is the new sequence $(a_n + b_n)_{n\geq k}$ when $x = (a_n)_{n\geq k}$ and $y = (b_n)_{n\geq k}$. So if $x = (3+4/n+5/n^2)$ and $y = (-7+\sin(n+2)+3/n^3)$ the sequence $x + y = (-4 + 4/n + 5/n^2 + 3/n^3 + \sin(n + 2))$.

- We can do a similar thing with the **subtraction** operation, $-$.

- We can **scale** a sequence with any number α by defining the sequence $\alpha\, x$ to be $\alpha\, x = (\alpha\, a_n)_{n\geq k}$. Thus, the sequence $2(13 + 5/n^4) = (26 + 10/n^4)$.

- With these operations, S is a **vector space** over the Real numbers. This is an idea you probably heard about in your Linear Algebra course.

The set of all sequences that converge is also a vector space which we denote by c but we have to **prove** that the new sequence $\alpha(a_n) + \beta(b_n)$ also converges when we know (a_n) and (b_n) converge.

The set of all sequences that converge to zero is also a vector space which we denote by c_0 but we have to **prove** that the new sequence $\alpha(a_n) + \beta(b_n)$ also converges to 0 when we know (a_n) and (b_n) converge to 0.

3.4.1 Limit Theorems

There are many results we can prove about convergence sequences. Here are a few.

Theorem 3.4.1 A Convergent Sequence is Bounded

If a sequence (a_n) converges it must be bounded; i.e. $\exists D > 0 \ni |a_n| \leq D \ \forall n$. Further, if the sequence limit a is not zero, $\exists N \ni |a_n| > |a|/2 \ \forall n > N$.

Proof 3.4.1

Let (a_n) be a sequence which converges to a. Pick $\epsilon = 1$. Then there is an N so that $n > N \Rightarrow$ $|a_n - a| < 1$. Use the backwards triangle inequality to write $n > N \Rightarrow |a_n| - |a| < |a_n - a| < 1$. The first part of this is what we need: we have $|a_n| < 1 + |a|$ when $n > N$ (we ignore the middle part and only use the far left and the far right). Let $B = \max\{|a_1|, \ldots, |a_N|\}$. Then we see $|a_n| < \max\{B, 1 + |a|\} = D$ for all n. This shows the sequence is bounded.

For the next part, use $\epsilon = |a|/2 > 0$. Then $\exists N$ so that $n > N$ implies $|a_n - a| < |a_n|/2$. Using the backwards triangle inequality again, we have $|a| - |a_n| < |a_n - a| < |a|/2$. Now use the far left and far right of this to see $\Rightarrow |a_n| > |a|/2$ when $n > N$.   ■

A very important fact is called the squeezing lemma which we use a lot.

Lemma 3.4.2 The Squeezing Lemma

> **The Squeezing Lemma**
> *Let (a_n) and (b_n) be two sequences that converge to L. If (c_n) is another sequence satisfying there is an N so that $a_n \leq c_n \leq b_n$ for $n > N$, then (c_n) converges to L also.*

Proof 3.4.2

Let $\epsilon > 0$ be chosen arbitrarily. Since (a_n) converges to L, there is an N_1 so that $n > N_1 \Rightarrow -\epsilon < a_n - L < \epsilon$. Also, since (b_n) converges to L, there is an N_2 so that $n > N_2 \Rightarrow -\epsilon < b_n - L < \epsilon$. So for $N_3 > \max\{N_1, N_2, N\}$, we have $L - \epsilon < a_n \leq c_n \leq b_n < L + \epsilon$. This says $n > N_3 \Rightarrow |c_n - L| < \epsilon$ which tells us $c_n \to L$ as well.

Note this is a much shorter proof than the one in the notes. The one in the notes first shows the sequence $(a_n - b_n)$ converges to 0 and uses that to show $c_n \to L$. This is much more cumbersome!
■

Since there is so much linear and multiplicative structure with sequences that converge, we can prove what we call the **Algebra of Limits Theorem** easily.

Theorem 3.4.3 Algebra of Limits Theorem

> *Let $a_n \to a$ and $b_n \to b$. Then*
>
> - $c\, a_n \to c\, a$.
>
> - $a_n + b_n \to a + b$
>
> - $a_n\, b_n \to a\, b$
>
> - *if $\exists N$ so that $b_n \neq 0$ for all $n > N$ and $b \neq 0$, then $a_n/b_n \to a/b$ where we only use a_n/b_n for $n > N$.*

Proof 3.4.3

$c\, a_n \to c\, a$:
Let $\epsilon > 0$ be arbitrary. Then $a_n \to a$ implies $\exists N$ so that $|a_n - a| < \epsilon/(|c| + 1)$. We use $|c| + 1$ in the denominator as this argument will work even if $c = 0$! We have $|c\, a_n - c\, a| = |c|\, |a_n - a| < \epsilon\, |c|/(|c| + 1)$. But $|c|/(|c| + 1) < 1$ always. Thus, $n > N$ implies $|c\, a_n - c\, a| < \epsilon$; i.e. $c\, a_n \to c\, a$.

$a_n + b_n \to a + b$:
Let $\epsilon > 0$ be arbitrary. Then since $a_n \to a$, $\exists N_1$ so that $n > N_1 \Rightarrow |a_n - a| < \epsilon/2$. And since

$b_n \to b$, $\exists N_2$ so that $n > N_2 \Rightarrow |b_n - b| < \epsilon/2$. *Thus,* $n > \max\{N_1, N_2\}$, *we have*

$$
\begin{aligned}
|a_n + b_n - (a + b)| &= |(a_n - a) + (b_n - b)| \leq |a_n - a| + |b_n - b| \\
&< \epsilon/2 + \epsilon/2 = \epsilon
\end{aligned}
$$

which shows $a_n + b_n \to a + b$.

$a_n\, b_n \to a\, b$

First, note

$$
\begin{aligned}
|a_n\, b_n - a\, b| &= |a_n\, b_n - a_n\, b + b\, a_n - a\, b| = |a_n(b_n - b) + b(a_n - a)| \\
&\leq |a_n|\, |b_n - b| + |b|\, |a_n - a|
\end{aligned}
$$

Now we can see what to do. Since $a_n \to a$, (a_n) *is bounded by a positive number* A. *Thus,* $|a_n\, b_n - a\, b| \leq A\, |b_n - b| + |b|\, |a_n - a|$. *Now, for an arbitrary* $\epsilon > 0$, *since* $a_n \to a$, $\exists N_1$ *so that* $n > N_1 \Rightarrow |a_n - a| < \epsilon/(2(|b| + 1))$. *And since* $b_n \to b$, $\exists N_2$ *so that* $n > N_2 \Rightarrow |b_n - b| < \epsilon/(2A)$. *Hence, if* $n > N = \max\{N_1, N_2\}$, *we have*

$$
|a_n\, b_n - a\, b| < A\, \epsilon/(2A) + |b|\, \epsilon/(2(|b| + 1)) < \epsilon/2 + \epsilon/2 = \epsilon
$$

This tells us $a_n\, b_n \to a\, b$. $a_n/b_n \to a/b$ *under some restrictions:*
We have to assume $\exists N$ *so that* $b_n \neq 0$ *for all* $n > N$ *and* $b \neq 0$. *So, for* $n > N$, *the fraction* a_n/b_n *makes sense to write.*

Now if $n > N$, *we have*

$$
\begin{aligned}
\left| \frac{a_n}{b_n} - \frac{a}{b} \right| &= \left| \frac{a_n\, b - a\, b_n}{b_n\, b} \right| \\
&= \left| \frac{a_n\, b - ab + ab - a\, b_n}{b_n\, b} \right| \quad \text{*add and subtract trick*} \\
&= \frac{|b(a_n - a) + a(b - b_n)|}{|b_n\, b|} \\
&\leq \frac{1}{|b_n|\, |b|} \{|a_n - a|\, |b| + |b_n - b|\, |a|\}
\end{aligned}
$$

Since $b_n \to b$, (b_n) *is bounded below by* $|b|/2$ *for* $n > N_1$; *hence* $1/|b_n| < 2/|b|$. *Thus, using the triangle inequality,*

If $n > \max\{N, N_1\}$, *we have*

$$
\begin{aligned}
\left| \frac{a_n}{b_n} - \frac{a}{b} \right| &\leq \frac{2}{|b|^2} \{|a_n - a|\, |b| + |a|\, |b - b_n|\} \\
&\leq \frac{2}{|b|^2} |a_n - a||b| + \frac{2}{|b|^2} |b - b_n|\, |a| \\
&= \frac{2}{|b|} |a_n - a| + \frac{2\, |a|}{|b|^2} |b - b_n|
\end{aligned}
$$

Since $a_n \to a$, *let's pick the tolerance so that the first term above becomes* $\epsilon/2$. *This means we want* $\frac{2}{|b|} |a_n - a| < \frac{\epsilon}{2} \Rightarrow |a_n - a| < \frac{\epsilon |b|}{4}$. *Using this tolerance,* $\exists N_2$ *so that* $n > N_2 \Rightarrow |a_n - a| < \epsilon |b|/4$.

Similarly, since $b_n \to b$, *let's pick the tolerance so the second term becomes* $\epsilon/2$. *We want* $\frac{2\, |a|}{|b|^2} |b_n - b| < \frac{\epsilon}{2} \Rightarrow |b_n - b| < \frac{\epsilon |b|^2}{4(|a| + 1)}$. *Hence* $\exists N_3$ *so that* $n > N_3 \Rightarrow |b_n - b| < \epsilon |b|^2/(4(|a| + 1))$ *where we use* $(|a| + 1)$ *just in case* $a = 0$. *Then for* $n > \max\{N, N_1, N_2, N_3\}$, *we have*

$$\left| \frac{a_n}{b_n} - \frac{a}{b} \right| \quad < \quad \frac{2}{|b|} \frac{\epsilon |b|}{4} + \frac{2|a|}{|b|^2} \frac{\epsilon |b|^2}{4(|a| + 1)} < \epsilon/2 + \epsilon/2$$

as $|a|/(|a| + 1) < 1$ *as usual. This shows* $a_n/b_n \to a/b$. ∎

3.4.2   Some Examples

Now let's work out some examples.

Theorem 3.4.4 $a_n \to 3$ **and** $b_n \to -8$, **then** $a_n b_n \to -24$

> *If* $a_n \to 3$ *and* $b_n \to -8$, *then* $a_n b_n \to -24$.

Proof 3.4.4
Pick $\epsilon > 0$. *Then consider*

$$
\begin{aligned}
|a_n b_n - (-24)| &= |a_n b_n - 3b_n + 3b_n + 24| \\
&= |b_n (a_n - 3) + 3 (b_n - (-8))| \\
&\leq |b_n| \, |a_n - 3| + 3|b_n + 8|
\end{aligned}
$$

Since $b_n \to -8$, *for tolerance* $\xi = 1$, *there is a* N_1 *so that* $-1 < b_n - (-8) < 1$ *if* $n > N_1$. *Thus,* $-9 < b_n < -7$ *if* $n > N_1$. *Thus,* $|b_n| < 9$ *if* $n > N_1$.

So we have if $n > N_1$

$$|a_n b_n - (-24)| \quad \leq \quad |b_n| \, |a_n - 3| + 3|b_n + 8| < 9|a_n - 3| + 3|b_n + 8|$$

Since the two sequences converge, for our ϵ,

$$
\begin{aligned}
\exists \, N_2 \quad &\ni \quad n > N_2 \Longrightarrow |a_n - 3| < \epsilon/(2(9)) \\
\exists \, N_3 \quad &\ni \quad n > N_3 \Longrightarrow |b_n + 8| < \epsilon/(2(3))
\end{aligned}
$$

Thus, if $n > \max(N_1, N_2, N_3)$ *all conditions hold and we have*

$$n > N \quad \Longrightarrow \quad |a_n b_n - (-24)| < 9(\epsilon/18) + 3(\epsilon/6) = \epsilon.$$

∎

Theorem 3.4.5 $a_n \to 3$ **and** $b_n \to -8$, **then** $a_n/b_n \to -3/8$

> *Assume there is a* N *so that* $b_n \neq 0$ *when* $n > N$. *Then if* $a_n \to 3$ *and* $b_n \to -8$, *then* $a_n/b_n \to -3/8$.

Proof 3.4.5
Let $\epsilon > 0$ *be chosen arbitrarily. Consider*

$$
\begin{aligned}
\left| \frac{a_n}{b_n} - \frac{-3}{8} \right| &= \left| \frac{8a_n + 3b_n}{8b_n} \right| = \left| \frac{8a_n - 24 + 24 + 3b_n}{8b_n} \right| \\
&\leq \quad 8 \frac{|a_n - 3|}{|b_n|} + 3 \frac{|b_n - (-8)|}{8|b_n|}
\end{aligned}
$$

Since $b_n \to -8$ there is an N_1 so that $|b_n| > |-8|/2 = 4$ when $n > N_1$ by our bounded below result. Thus, we have $1/|b_n| < 1/4$ when $n > N_1$, So when $n > \max(N, N_1)$ both conditions hold and we have

$$\left| \frac{a_n}{b_n} - \frac{-3}{8} \right| \leq (1/4)|a_n - 3| + (3/32)|b_n + 8|$$

Now

$$\exists N_2 \ni n > N_2 \implies |a_n - 3| < \epsilon/(2(1/4)) = 2\epsilon$$
$$\exists N_3 \ni n > N_3 \implies |b_n + 8| < \epsilon/(2(3/32)) = (16/3)\epsilon$$

So if $n > \max(N, N_1, N_2, N_3)$ all conditions hold and

$$\left| \frac{a_n}{b_n} - \frac{-3}{8} \right| \leq (1/4)|a_n - 3| + (3/32)|b_n + 8|$$
$$< 2\,\epsilon/4 + (3/32)\,(16/3)\,\epsilon = \epsilon.$$

∎

Here is a nice result which we use frequently. It is proved differently from the way we have been attacking sequence convergence so far and because of that it is a good thing to see.

Theorem 3.4.6 The nth Root of a Positive Number goes to One

If c is a positive number, the $\lim_{n \to \infty}(c)^{1/n} = 1$.

Proof 3.4.6

$c \geq 1$:
Then, we can say $c = 1 + r$ for some $r \geq 0$. It is easy to see $c^{1/n} = (1 + r)^{1/n} \geq 1$ for all n. Let $y_n = (1 + r)^{1/n}$. Then $c^{1/n} = y_n$ with $y_n \geq 1$. Thus, $c^{1/n} - 1 = y_n - 1 \geq 0$. Let $x_n = y_n - 1 \geq 0$. It follows we have $c^{1/n} = 1 + x_n$ with $x_n \geq 0$.

Using a POMI argument, we can show $c = (1 + x_n)^n \geq 1 + nx_n$. We leave that to you!

Thus, $0 \leq (c)^{1/n} - 1 = x_n \leq (c-1)/n$. This show $(c)^{1/n} \to 1$.

$0 < c < 1$:
If $0 < c < 1$, $1/c \geq 1$ and so if we rewrite this just right we can use our first argument. We have $\lim_n (c)^{1/n} = \lim_n \frac{1}{(1/c)^{1/n}} = 1/1 = 1$. ∎

3.4.3 Homework

Exercise 3.4.1 *Prove the sequence $7a_n + 15b_n \to 7a + 15b$ if we know $a_n \to a$ and $b_n \to b$. This is an $\epsilon - N$ proof.*

Exercise 3.4.2 *Prove the sequence $-2a_n\, b_n \to -2a\, b$ if we know $a_n \to a$ and $b_n \to b$. This is an $\epsilon - N$ proof.*

Exercise 3.4.3 *Prove the sequence $55a_n/b_n \to 55a/b$ if we know $a_n \to a$ and $b_n \to b > 0$ and $b_n \neq 0$ for all n. This is an $\epsilon - N$ proof.*

Exercise 3.4.4 *Prove the sequence $2a_n + 5b_n \to 2a + 5b$ if we know $a_n \to 4$ and $b_n \to 6$. This is an $\epsilon - N$ proof.*

Exercise 3.4.5 *Prove the sequence $2a_n\, b_n \to 2a\, b$ if we know $a_n \to -2$ and $b_n \to 3$. This is an $\epsilon - N$ proof.*

Exercise 3.4.6 *Prove the sequence $5a_n/b_n \to 5a/b$ if we know $a_n \to 2$ and $b_n \to 7$ and $b_n \neq 0$ for all n. This is an $\epsilon - N$ proof.*

Chapter 4

Bolzano - Weierstrass Results

Let's study bounded sequences in greater detail.

4.1 Bounded Sequences with a Finite Range

We have already looked at sequences with finite ranges. Since their range is finite, they are bounded sequences. We also know they have subsequences that converge which we have explicitly calculated. If the range of the sequence is a single value, then we know the sequence will converge to that value and we now know how to prove convergence of a sequence. Let's formalize this into a theorem. But this time, we will argue more abstractly. Note how the argument is still essentially the same.

Theorem 4.1.1 A Sequence with a Finite Range Diverges Unless the Range is One Value

> Let the sequence (a_n) have a finite range $\{y_1, \ldots, y_p\}$ for some positive integer $p \geq 1$. If $p = 1$, the sequence converges to y_1 and if $p > 1$, the sequence does not converge but there is a subsequence $(a_{n_k^i})$ which converges to y_i for each y_i in the range of the sequence.

Proof 4.1.1

If the range of the sequence consists of just one point, then $a_n = y_1$ for all n and it is easy to see $a_n \to y_1$ as given $\epsilon > 0$, $|a_n - y_1| = |y_1 - y_1| = 0 < \epsilon$ for all n which shows convergence.

If the range has $p > 1$, let a be any number not in the range and calculate $d_i = |a - y_i|$, the distance from a to each point y_i in the range. Let $d = (1/2)\min\{d_1, \ldots, d_p\}$ and choose $\epsilon = d$. Then $|a_n - a|$ takes on p values, $|y_i - a| = d_i$ for all n. But $d_i > d$ for all i which shows us that $|a_n - a| > \epsilon$ for all n. A little thought then shows us this is precisely the definition of the sequence (a_n) not converging to a.

If a is one of the range values, say $a = y_i$, then the distances we defined above are positive except d_i which is zero. So $|a_n - y_i|$ is zero for all indices n which give range value y_i but positive for all other range values. Let $\epsilon = d = (1/2)\min_{j \neq i}|d_i - d_j|$. Then, for any index n with $a_n \neq y_i$, we have $|a_n - y_i| = |y_j - y_i| = d_j > d$ for some j. Thus, no matter what N we pick, we can always find $n > N$ giving $|a_n - y_i| > \epsilon$. Hence, the limit can not be y_i. Since this argument works for any range value y_i, we see the limit value can not be any range value.

To make this concrete, say there were 3 values in the range, $\{y_1, y_2, y_3\}$. If the limit was y_2, let $\epsilon = d = (1/2)\min\{|y_1 - y_2|, |y_3 - y_2|\}$. Then,

$$|a_n - y_2| \quad = \quad \begin{cases} |y_2 - y_1| > d = \epsilon, & a_n = y_1 \\ |y_2 - y_2| = 0, & a_n = y_2 \\ |y_3 - y_2| > d, = \epsilon & a_n = y_3 \end{cases}$$

Given any N we can choose $n > N$ so that $|a_n - y_2| > \epsilon$. Hence, the limit can not be y_2. **Note how this argument is much more abstract than our previous ones.** ■

Comment 4.1.1 *For convenience of exposition (cool phrase...) let's look at the range value y_1. The sequence has a block which repeats and inside that block are the different values of the range y_i. There is a first time y_1 is present in the first block. Call this index n_1. Let the block size by Q. Then the next time y_1 occurs in this position in the block is at index $n_1 + Q$. In fact, y_1 occurs in the sequence at indices $n_1 + jQ$ where $j \geq 1$. This defines the subsequence a_{n_1+jQ} which converges to y_1. The same sort of argument can be used for each of the remaining y_i.*

4.2 Sequences with an Infinite Range

We are now ready for our most abstract result so far.

Theorem 4.2.1 Bolzano - Weierstrass Theorem

Every bounded sequence has at least one convergent subsequence.

Proof 4.2.1

As discussed, we have already shown a sequence with a bounded finite range always has convergent subsequences. Now we prove the case where the range of the sequence of values $\{a_1, a_2 \ldots, \}$ has infinitely many distinct values. We assume the sequences start at $n = k$ and by assumption, there is a positive number B so that $-B \leq a_n \leq B$ for all $n \geq k$. Define the interval $J_0 = [\alpha_0, \beta_0]$ where $\alpha_0 = -B$ and $\beta_0 = B$. Thus at this starting step, $J_0 = [-B, B]$. Note the length of J_0, denoted by ℓ_0 is $2B$.

Let S be the range of the sequence which has infinitely many points and for convenience, we will let the phrase infinitely many points *be abbreviated to* **IMPs.**

Step 1:

Bisect $[\alpha_0, \beta_0]$ into two pieces u_0 and u_1. That is the interval J_0 is the union of the two sets u_0 and u_1 and $J_0 = u_0 \cup u_1$. Now at least one of the intervals u_0 and u_1 contains IMPs of S as otherwise each piece has only finitely many points and that contradicts our assumption that S has IMPS. Now both may contain IMPS so select one such interval containing IMPS and call it J_1. Label the endpoints of J_1 as α_1 and β_1; hence, $J_1 = [\alpha_1, \beta_1]$. Note $\ell_1 = \beta_1 - \alpha_1 = \frac{1}{2}\ell_0 = B$ We see $J_1 \subseteq J_0$ and

$$-B = \alpha_0 \leq \alpha_1 \leq \beta_1 \leq \beta_0 = B$$

Since J_1 contains IMPS, we can select a sequence value a_{n_1} from J_1.

Step 2:

Now bisect J_1 into subintervals u_0 and u_1 just as before so that $J_1 = u_0 \cup u_1$. At least one of u_0 and u_1 contain IMPS of S.

Choose one such interval and call it J_2. Label the endpoints of J_2 as α_2 and β_2; hence, $J_2 = [\alpha_2, \beta_2]$. Note $\ell_2 = \beta_2 - \alpha_2 = \frac{1}{2}\ell_1$ or $\ell_2 = (1/4)\ell_0 = (1/2^2)\ell_0 = (1/2)B$. We see $J_2 \subseteq J_1 \subseteq J_0$ and

$$-B = \alpha_0 \leq \alpha_1 \leq \alpha_2 \leq \beta_2 \leq \beta_1 \leq \beta_0 = B$$

Since J_2 contains IMPS, we can select a sequence value a_{n_2} from J_2. It is easy to see this value is different from a_{n_1}, our previous choice.

You should be able to see that we can continue this argument using induction.

Proposition:

$\forall p \geq 1, \exists$ an interval $J_p = [\alpha_p, \beta_p]$ with the length of J_p, $\ell_p = B/(2^{p-1})$ satisfying $J_p \subseteq J_{p-1}$, J_p contains IMPS of S and

$$\alpha_0 \leq \ldots \leq \alpha_{p-1} \leq \alpha_p \leq \beta_p \leq \beta_{p-1} \leq \ldots \leq \beta_0.$$

Finally, there is a sequence value a_{n_p} in J_p, different from $a_{n_1}, \ldots, a_{n_{p-1}}$.

Proof *We have already established the proposition is true for the basis step J_1 and indeed also for the next step J_2.*

Inductive: *We assume the interval J_q exists with all the desired properties. Since by assumption, J_q contains IMPs, bisect J_q into u_0 and u_1 like usual. At least one of these intervals contains IMPs of S. Call the interval J_{q+1} and label $J_{q+1} = [\alpha_{q+1}, \beta_{q+1}]$. We see immediately that*

$$\ell_{q+1} = (1/2)\ell_q = (1/2)(1/2^{q-1})B = (1/2^q)B$$

with $\ell_{q+1} = \beta_{q+1} - \alpha_{q+1}$ with

$$\alpha_q \leq \alpha_{q+1} \leq \beta_{q+1} \leq \beta_q.$$

This shows the nested inequality we want is satisfied.

Finally, since J_{q+1} contains IMPs, we can choose $a_{n_{q+1}}$ distinct from the other a_{n_i}'s. So the inductive step is satisfied and by the POMI, the proposition is true for all n. □

From our proposition, we have proven the existence of three sequences, $(\alpha_p)_{p \geq 0}$, $(\beta_p)_{p \geq 0}$ and $(\ell_p)_{p \geq 0}$ which have various properties.

The sequence ℓ_p satisfies $\ell_p = (1/2)\ell_{p-1}$ for all $p \geq 1$. Since $\ell_0 = 2B$, this means $\ell_1 = B$, $\ell_2 = (1/2)B$, $\ell_3 = (1/2^2)B$ leading to $\ell_p = (1/2^{p-1})B$ for $p \geq 1$.

$$
\begin{aligned}
-B = \alpha_0 \quad &\leq \quad \alpha_1 \leq \alpha_2 \leq \ldots \leq \alpha_p \\
&\leq \quad \ldots \leq \\
\beta_p \quad &\leq \quad \ldots \leq \beta_2 \leq \ldots \leq \beta_0 = B
\end{aligned}
$$

Note $(\alpha_p)_{p \geq 0}$ is bounded above by B and $(\beta_p)_{p \geq 0}$ is bounded below by $-B$. Hence, by the completeness axiom, $\inf (\beta_p)_{p \geq 0}$ exists and equals the finite number β; also $\sup (\alpha_p)_{p \geq 0}$ exists and is the finite number α.

So if we fix p, it should be clear the number β_p is an upper bound for all the α_p values (look at our inequality chain again and think about this). Thus β_p is an upper bound for $(\alpha_p)_{p \geq 0}$ and so by definition of a supremum, $\alpha \leq \beta_p$ for all p. Of course, we also know since α is a supremum, that $\alpha_p \leq \alpha$. Thus, $\alpha_p \leq \alpha \leq \beta_p$ for all p.

A similar argument shows if we fix p, the number α_p is a lower bound for all the β_p values and so by definition of an infimum, $\alpha_p \leq \beta \leq \beta_p$ for all the α_p values. This tells us α and β are in $[\alpha_p, \beta_p] = J_p$ for all p. Next we show $\alpha = \beta$.

Now let $\epsilon > 0$ be arbitrary. Since α and β are in J_p whose length is $\ell_p = (1/2^{p-1})B$, we have $|\alpha - \beta| \leq (1/2^{p-1})B$. Pick P so that $1/(2^{P-1}B) < \epsilon$. Then $|\alpha - \beta| < \epsilon$. But $\epsilon > 0$ is arbitrary.

Hence, by a previous proposition, $\alpha - \beta = 0$ implying $\alpha = \beta$.

We now must show $a_{n_k} \to \alpha = \beta$. This shows we have found a subsequence which converges to $\alpha = \beta$. We know $\alpha_p \le a_{n_p} \le \beta_p$ and $\alpha_p \le \alpha \le \beta_p$ for all p. Pick $\epsilon > 0$ arbitrarily. Given any p, we have

$$
\begin{aligned}
|a_{n_p} - \alpha| &= |a_{n_p} - \alpha_p + \alpha_p - \alpha|, & \text{add and subtract trick} \\
&\le |a_{n_p} - \alpha_p| + |\alpha_p - \alpha| & \text{triangle inequality} \\
&\le |\beta_p - \alpha_p| + |\alpha_p - \beta_p| & \text{definition of length} \\
&= 2|\beta_p - \alpha_p| = 2\,(1/2^{p-1})B.
\end{aligned}
$$

Choose P so that $(1/2^{P-1})B < \epsilon/2$. Then, $p > P$ implies $|a_{n_p} - \alpha| < 2\,\epsilon/2 = \epsilon$. Thus, $a_{n_k} \to \alpha$.
∎

4.2.1   Extensions to $\Re^2$

We can extend our arguments to bounded sequences in $\Re^2$. We haven't talked about it yet, but given a sequence in $\Re^2$, the elements of the sequence will be vectors. The sequence then will be made up of vectors.

$$
((x_n)) = \begin{bmatrix} x_{1,n} \\ x_{2,n} \end{bmatrix}
$$

We would say the sequence converges to a vector $\boldsymbol{x}$ is for all $\epsilon > 0$, there is an N so that

$$
n > N \implies ||\boldsymbol{x_n} - \boldsymbol{x}||_2 < \epsilon
$$

where we measure the distance between two vectors in $\Re^2$ using the standard Euclidean norm, here called $|| \cdot ||_2$ defined by

$$
||\boldsymbol{x_n} - \boldsymbol{x}||_2 = \sqrt{(x_{1,n} - x_1)^2 + (x_{2,n} - x_2)^2}
$$

where $\boldsymbol{x} = [x_1, x_2]'$. This sequence is bounded if there is a positive number B so that $||\boldsymbol{x_n}||_2 < B$ for all appropriate n. We can sketch the proof for the case where there are infinitely many vectors in this sequence which is bounded.

Theorem 4.2.2 The Bolzano - Weierstrass Theorem in $\Re^2$

> *Every bounded sequence of vectors in $\Re^2$ with an infinite range has at least one convergent subsequence.*

Proof 4.2.3
We will just sketch the argument. Since this sequence is bounded, there are positive numbers B_1 and B_2 so that

$$
-B_1 \le x_{1n} \le B_1 \quad \text{and} \quad -B_2 \le x_{2n} \le B_2
$$

The same argument we just used for the Bolzano - Weierstrass Theorem in $\Re$ works. We bisect both edges of the box to create 4 rectangles. At least one must contain IMPs and we choose one that does. Then at each step, continue this subdivision process always choosing a rectangle with IMPs. Here are a few of the details. We start by labeling the initial box by

$$J_0 = [-B_1, B_1] \times [-B_2, B_2] = [\alpha_0, \beta_0] \times [\delta_0, \gamma_0].$$

We pick a first vector x_{n_0} from the initial box. The area of this rectangle is $A_0 = (\beta_0 - \alpha_0)(\gamma_0 - \delta_0)$ and at the first step, the bisection of each edge leads to four rectangles of area $A_0/4$. At least one of these rectangles contains IMPs and after our choice, we label the new rectangle $J_1 = [\alpha_1, \beta_1] \times [\delta_1, \gamma_1]$ and pick a vector x_{n_1} different from the first one from J_0.

At this point, we have

$$\begin{aligned} -B_1 &= \alpha_0 \leq \alpha_1 \leq \beta_1 \leq \beta_0 = B_1 \\ -B_2 &= \delta_0 \leq \delta_1 \leq \gamma_1 \leq \gamma_0 = B_2 \end{aligned}$$

and after p steps, we have

$$\begin{aligned} -B_1 &= \alpha_0 \leq \alpha_1 \leq \ldots \alpha_p \leq \beta_p \leq \ldots \leq \beta_1 \leq \beta_0 = B_1 \\ -B_2 &= \delta_0 \leq \delta_1 \leq \ldots \delta_p \leq \gamma_p \leq \ldots \leq \gamma_1 \leq \gamma_0 = B_2 \end{aligned}$$

with the edge lengths $\beta_p - \alpha_p$ and $\gamma_p - \delta_p$ going to zero as p increases, As before, we pick a value x_{n_p} different from the ones at previous stages. Similar arguments show that $\alpha_p \to \alpha$ and $\beta_p \to \beta$ with $\alpha = \beta$ and $\delta_p \to \delta$ and $\gamma_p \to \gamma$ with $\delta = \gamma$. We find also $x_{n_p} \to [\alpha, \delta]'$ which gives us the result. The convergence arguments here are indeed a bit different as we have to measure distance between two vectors using $||\cdot||_2$ but it is not too difficult to figure it out. ∎

We can then extend our arguments to bounded sequences in $\Re^3$. Given a sequence in $\Re^3$, the elements of the sequence will be 3D vectors. The sequence then will be made up of vectors.

$$((x_n)) = \begin{bmatrix} x_{1,n} \\ x_{2,n} \\ x_{3,n} \end{bmatrix}$$

We would then say the sequence converges to a vector x is for all $\epsilon > 0$, there is an N so that

$$n > N \implies ||x_n - x||_2 < \epsilon$$

where we measure the distance between two vectors in $\Re^3$ using the standard Euclidean norm, here called $||\cdot||_2$ defined by

$$||x_n - x||_2 = \sqrt{(x_{1,n} - x_1)^2 + (x_{2,n} - x_2)^2 + (x_{3,n} - x_3)^2}$$

where $x = [x_1, x_2]'$. This sequence is bounded now means there is a positive number B so that $||x_n||_2 < B$ for all appropriate n; this means the sequence elements live in a cube in $\Re^3$. We can then sketch the proof for the case where there are infinitely many vectors in this sequence which is bounded.

Theorem 4.2.3 The Bolzano - Weierstrass Theorem in $\Re^3$

Every bounded sequence of vectors with an infinite range has at least one convergent sub-sequence.

Proof 4.2.4

We now bisect each edge of a cube and there are now 8 pieces at each step, at least one of which has IMPs. The vectors are now 3 dimensional but the argument is quite similar. ∎

The extension to $\Re^n$ is a standard induction argument: really!

Theorem 4.2.4 The Bolzano - Weierstrass Theorem in $\Re^n$

> *Every bounded sequence of vectors with an infinite range has at least one convergent subsequence.*

Proof 4.2.5

We now bisect each edge of what is called a n dimensional hypercube and there are now 2^n pieces at each step, at least one of which has IMPs. The vectors are now n dimensional but the argument is quite similar. ∎

4.3   Bounded Infinite Sets

A more general type of result can also be shown which deals with sets which are bounded and contain infinitely many elements.

Definition 4.3.1 Accumulation Point

> *Let S be a nonempty set. We say the real number a is an **accumulation** points of S if given any $r > 0$, the set*
> $$B_r(a) = \{x : |x - a| < r\}$$
> *contains at least one point of S different from a. The set $B_r(a)$ is called the **ball** or **circle** centered at a with radius r.*

Example 4.3.1 $S = (0, 1)$. *Then 0 is an accumulation point of S as the circle $B_r(0)$ always contains points greater than 0 which are in S, Note $B_r(0)$ also contains points less than 0. Note 1 is an accumulation point of S also. Note 0 and 1 are not in S so accumulation points don't have to be in the set. Also note all points in S are accumulation points too. Note the set of all accumulation points of S is the interval $[0, 1]$.*

Example 4.3.2 $S = ((1/n)_{n \geq 1})$. *Note 0 is an accumulation point of S because every circle $B_r(0)$ contains points of S different from 0. Also, if you pick a particular $1/n$ in S, the distance from $1/n$ to its neighbors is either $1/n - 1/(n+1)$ or $1/n - 1/(n-1)$. If you let r be half the minimum of these two distances, the circle $B_r(1/n)$ does not contain any other points of S. So no point of S is an accumulation point. So the set of accumulation points of S is just one point, $\{0\}$.*

4.3.1   Homework

Exercise 4.3.1 *Let $S = (2, 5)$. Show 2 and 5 are accumulation points of S.*

Exercise 4.3.2 *Let $S = (-12, 15)$. Show -12 and 15 are accumulation points of S.*

Exercise 4.3.3 *Let $S = (\cos(n\pi/4))_{n \geq 1}$. Show S has no accumulation points.*

Exercise 4.3.4 *Let $S = (\cos(3n\pi/5))_{n \geq 1}$. Show S has no accumulation points.*

Exercise 4.3.5 *Let $S = (\sin(n\pi/4))_{n \geq 1}$. Show S has no accumulation points.*

Exercise 4.3.6 *Let $S = (\sin(3n\pi/5))_{n \geq 1}$. Show S has no accumulation points.*

Exercise 4.3.7 *This one is a problem you have never seen. So it requires you look at it right! Let (a_n) be a bounded sequence and let (b_n) be a sequence that converges to 0. Then $a_n b_n \to 0$. This is an $\epsilon - N$ proof. Note this is **not** true if (b_n) converges to a nonzero number.*

Exercise 4.3.8 *If you know* $(a_n b_n)$ *converges does that imply both* (a_n) *and* (b_n) *converge?*

4.3.2 Bounded Infinite Sets Have at Least One Accumulation Point

We can extend the Bolzano - Weierstrass Theorem for sequences to bounded sets having an infinite number of points.

Theorem 4.3.1 Bolzano - Weierstrass Theorem for Sets: Accumulation Point Version

> *Every bounded infinite set of real numbers has at least one accumulation point.*

Proof 4.3.1
We let the bounded infinite set of real numbers be S. We know there is a positive number B so that $-B \leq x \leq B$ *for all x in S because S is bounded.*
Step 1:
By a process essentially the same as the **subdivision** *process in the proof of the Bolzano - Weierstrass Theorem for Sequences, we can find a sequence of sets*

$$J_0, J_1, \ldots, J_p, \ldots$$

These sets have many properties.

- $\ell_p = B/2^{p-1}$ *for* $p \geq 1$ *(remember B is the bound for S)*

- $J_p \subseteq J_{p-1}$,

- $J_p = [\alpha_p, \beta_p]$ *and* J_p *contains IMPs of S.*

- $-B = \alpha_0 \leq \ldots \leq \alpha_p < \beta_p \leq \ldots \leq \beta_0 = B$

Since J_0 *contains IMPs of S choose* x_0 *in S from* J_0. *Next, since* J_1 *contains IMPs of S, choose* x_1 *different from* x_0 *from S. Continuing (note we are using what we called our* **relaxed** *use of POMI here!) we see by an induction argument we can choose* x_p *in S, different from the previous ones, with* x_p *in* J_p.

We also know the sequence (α_p) *has a finite supremum* α *and the sequence* (β_p) *has a finite infimum* β. *Further, we know* $\alpha_p \leq \alpha, \beta \geq \beta_p$ *for all p. Since the lengths of the intervals* J_p *go to 0 as* $p \to \infty$, *we also know* $\alpha = \beta$ *and the sequence* (x_p) *converges to* α. *Thus, the ball* $B_r(\alpha) = (\alpha - r, \alpha + r)$ *contains* J_p *for* $p > P$. *So in particular,* $\alpha \in J_p$ *and* $x_p \in J_p = [\alpha_P, \beta_P]$ *for any* $p \geq P$. *It remains to show that* α *is an accumulation point of S. Choose any* $r > 0$. *Since* $\ell_p = B/2^{p-1}$, *we can find an integer P so that* $B/2^{P-1} < r$. *So we know also*

$$|x_p - \alpha| \quad \leq \quad |\beta_p - \alpha_p| \leq |\beta_P - \alpha_P| = \frac{B}{2^{P-1}} < r$$

Hence, x_p *is in* $B_r(\alpha)$ *for all* $p \geq P$ *and so* $B_r(\alpha)$ *contains IMPs of S distinct from* α. *Since our choice of* $r > 0$ *was arbitrary, this tells us* α *is an accumulation point of S.* ■

Comment 4.3.1 *Note the Bolzano - Weierstrass Theorem for sets shows the bounded infinite set S must have at least one accumulation point* α **and** *we found out much more:*

$$\exists \ a \ sequence \ (x_p) \subseteq S \ so \ that \ x_p \to \alpha$$

Note we don't know the value of $\alpha \in S$.

4.3.3   Cluster Points of a Set

The existence of a sequence converging to the accumulation point in the bounded infinite set above seems like a fairly commonplace occurrence. This leads to a new definition.

Definition 4.3.2 Cluster Points

> *Let S be a nonempty set of real numbers. We say the number a is a **cluster point** of S if there is a sequence $(x_p) \subseteq S$ with each $x_p \neq a$ so that $x_p \to a$. Note any accumulation point is also a cluster point and any cluster point is an accumulation point.*

We can therefore restate our theorem like this:

Theorem 4.3.2 Bolzano - Weierstrass Theorem for Sets: Cluster Point Version

> *Every bounded infinite set of real numbers has at least one cluster point.*

Proof 4.3.2
The proof is the same: we just note the accumulation point we have found is also a cluster point. ∎

Example 4.3.3 $S = \{2\}$. *This is a set of one element only. Note 2 is **not** an accumulation point of S as $B_{0.5}(2) = (1.5, 2.5)$ only contains 2 from S. Since no other points of S are in this circle, 2 can not be an accumulation point of S. It is also not a cluster point because we can't find a sequence (x_n) in S with each $x_n \neq 2$ that converges to 2. The only sequence that works would be the constant sequence $(x_n = 2)$ which we can't use.*

Example 4.3.4 $S = (0, 4)$. *If c is in S, the sequence $(x_n = c + 1/n)$ for n small enough will be in S and will converge to c. So every c in $(0, 4)$ is a cluster point.*

We note 0 is also a cluster point because the sequence $(x_n) = (1/n)$ is contained in S and converges to 0. We also see 4 is also a cluster point because the sequence $(x_n) = (4 - 1/n)$ is contained in S and converges to 4. The set of cluster points of S is thus $[0, 4]$.

Example 4.3.5 $S = \{1, 2, 3, 4\} \cup (6, 7] \cup \{10\}$. *What are the cluster points of S? You should be able to see that the cluster points of S are $[6, 7]$ and the accumulation points of S are also $[6, 7]$*

4.4   More Set and Sequence Related Topics

Now it is time to define more types of sets of numbers: **open sets**, **closed sets** and so on. These are useful concepts we will use a lot.

Definition 4.4.1 Open Sets

> *Let S be a set of real numbers. We say S is an open set for every point p in S. we can find a ball or radius r centered at p which is strictly contained in S. That is $\forall p \in S \; \exists r > 0 \ni B_r(p) \subset S$.*

Example 4.4.1 $S = \{2\}$ *is not an open set because every circle $B_r(2)$ contains lots of points not in S!*

Example 4.4.2 $S = (0, 4)$ *is an open set because for every p in S, we can find an $r > 0$ with $B_r(p) \subset (0, 4)$.*

Example 4.4.3 $S = (0, 5]$ *is **not** an open set because although for every p in $(0, 5)$, we can find an $r > 0$ with $B_r(p) \subset (0, 5]$, we can not do that for 5. Every circle centered at 5 **spills** outside of $(0, 5]$.*

We can define more types of sets.

Definition 4.4.2 The Complement of a Set

S is a set of real numbers. Note $\notin$ means **not in** *. Then, we define $S^C = \{x : x \notin S\}$, the* **complement** *of S. We say S is* **closed** *if its complement S^C is open.*

The points of an open set are *interior* to the set in a special way. We have a special name for such points:

Definition 4.4.3 Interior Points, Boundary Points and Set Closure

S is a set of real numbers. Then, we define p in S to be an **interior point** *of S if we can find a positive r so that the ball $B_r(p) \subset S$. Note from our definition of open set,* **every point in an open set is an interior point.** *Then we have*

$$\begin{aligned} Int\, S &= \{x : x \text{ is an interior point of } S\}, \quad \text{the interior of } S \\ \partial S &= \{x : \text{ all } B_r(x) \text{ contains points of both } S \text{ and } S^C\}, \\ &\quad \text{the boundary of } S \\ S' &= S \cup \partial S, \quad \text{the closure of } S \end{aligned}$$

Note if S is open, $Int(S) = S$. Also a point x which is in ∂S is called a **boundary point.** *It is ok if $B_r(x)$ is just s itself. Hence in the set $S = \{2, 3, 4\}$, 2, 3 and 4 are boundary points* **but** *they are not accumulation points as each $B_r(x)$ only contains x from the set.*

Note if $S = \{2\}$, 2 is not an accumulation point but the constant sequence $\{2, 2, 2, 2, \ldots\}$ from S does converge to 2. This leads to the idea of a **limit point** of a set.

Definition 4.4.4 Limit Points

Let S be a set of numbers. We say p is a limit point of S if there is a sequence of points in S, (x_n), so that $x_n \to p$.

Example 4.4.4 *If $S = (0, 2)$, 0 and 2 are accumulation points and they are also boundary points. If $S = [0, 2]$, 0 and 2 are accumulation points and they are also boundary points. If $S = (0, 2)$, the set of all limit points of S is $[0, 2]$ as there is sequence from inside S that converges to 0 and there is one that converge to 2.*

Example 4.4.5 *If $S = \{2\} \cup (3, 4)$,*
$Int\, S = (3, 4)$
$S^C = (-\infty, 2) \cup (2, 3] \cup [4, \infty)$,
$\partial S = \{2\} \cup \{3\} \cup \{4\}$
$S' = \{2\} \cup [3, 4]$,
$(S')^C = (-\infty, 2) \cup (2, 3) \cup (4, \infty)$ which is open and so S' is closed.
The set of limit points of S is $\{2\} \cup [3, 4]$.

Example 4.4.6 *Let $S = (2/n + 3)_{n \geq 1}$. Then $S = \{5, 4, 3\frac{2}{3}, 3\frac{2}{4}, \ldots\}$.*
$Int\, S = \emptyset$; there are no interior points.

$$\begin{aligned} S^C &= (-\infty, 3] \\ &\cup (4, 5) \cup (3\frac{2}{3}, 4) \cup (3\frac{2}{4}, 3\frac{2}{3}), \ldots \\ &(5, \infty) \end{aligned}$$

which is not an open set because 3 is a boundary point and so it is not an interior point.
$\partial S = \{5, 4, 3\frac{2}{3}, 3\frac{2}{4}, \ldots\}$ *and* $\{3\}$. *Note 3 is both an accumulation point and a boundary point. But each entry in the sequence, i.e.* $3\frac{2}{3}$, *is not an accumulation point although it is a boundary point.*
$S' = \{5, 4, 3\frac{2}{3}, 3\frac{2}{4}, \ldots\}$ *and* $\{3\}$.
Note the limit points of S is all of S plus the point 3.

4.4.1 Homework

Exercise 4.4.1 *Let* $S = \{2, 3, 6\}$. *Find* $Int(S)$, ∂S, S' *and the accumulation points of S.*

Exercise 4.4.2 *Let* $S = (2 + 1/n)_{n \geq 1}$. *Find* $Int(S)$, ∂S, S' *and the accumulation points of S.*

Exercise 4.4.3 *Let* $S = \{12, 13, 16\}$. *Find* $Int(S)$, ∂S, S' *and the accumulation points of S.*

Exercise 4.4.4 *Let* $S = (5 + 1/n + 3/n^2)_{n \geq 1}$. *Find* $Int(S)$, ∂S, S' *and the accumulation points of S.*

Exercise 4.4.5 *Let* $P(x) = b_0 + b_1 x + b_2 x^2 + \ldots + b_n x^n$ *with* $b_n \neq 0$ *be an arbitrary polynomial of degree n. If* $a_n \to a$, *prove* $P(a_n) \to P(a)$ *using the POMI. Here the basis is an arbitrary polynomial of degree 1, i.e.* $P(x) = b_0 + b_1 x$ *with* $b_1 \neq 0$ *and our algebra of sequence limits results show the proposition is true here. For the inductive step, assume it is true for an arbitrary polynomial of degree k and then note a polynomial of degree* $k + 1$ *looks a polynomial of degree* $\leq k$ *plus the last term* $b_{k+1} x^{k+1}$. *You should be able to see how to finish the argument from that.*

4.4.2 Some Set Properties

Recall a set S is closed if its complement is open. We can characterize closed sets using the closure too.

Theorem 4.4.1 S is Closed if and only if it equals its Closure

Let S be a set of real numbers. Then S is a closed set $\Leftrightarrow$ $S' = S$.

Proof 4.4.1
($\Rightarrow$): *We assume S is a closed set. Then* S^C *is open. We wish to show* $S' = S$. *Now* S' *contains S already, so we must show all boundary points of S are in S. Let's assume x is a boundary point of S which is in* S^C. *Then, since* S^C *is open, x is an interior point and there is an* $r > 0$ *so that* $B_r(x) \subseteq S^C$. *But* $B_r(x)$ *must contains points of S and* S^C *since x is a boundary point. This is impossible. So our assumption that x was in* S^C *is wrong. Hence,* $\partial S \subseteq S$. *This shows* $S' = S$.
    ($\Leftarrow$):
We assume $S' = S$ *and show S is closed. To do this, we show* S^C *is open. It is easy to see the complement of the complement gives you back the set you started with, so if* S^C *is open, by definition,* $(S^C)^C = S$ *is closed. We will show any x in* S^C *is an interior point. That will show* S^C *is open. Let's do this by contradiction. To show* S^C *is open, we show that for any x in* S^C, *we can find a positive r so that* $B_r(x) \subseteq S^C$. *So let's assume we can't find such an r. Then every* $B_r(x)$ *contains points of S and points of* S^C. *Hence, x must be a boundary point of S. Since* $S' = S$, *this means* $\partial S \subseteq S$ *which implies* $x \in S$. *But we assumed x was in* S^C. *So our assumption is wrong and we must be able to find an* $r > 0$ *with* $B_r(x) \subset S^C$. *Hence* S^C *is open which means S is closed.* ∎

Another way to check to see if a set is closed is this:

Theorem 4.4.2 S is Closed if and only if it Contains its Limit Points

A set S is closed if and only if it contains all its limit points.

Proof 4.4.2

$(\Longrightarrow)$:

*If S is closed, then its complement is open. Assume (x_n) is a sequence in **S** with limit point x and $x \notin S$, i.e. $x \in S^C$. Then x is an interior point of S^C and so there is a radius $r > 0$ so that $B_r(x) \subset S^C$. Since $x_n \to x$, given $\epsilon = r/2$, there is an N so that $|x - x_n| < r/2$ when $n > N$. This says $x_n \in B_r(x) \subset S^C$. But $x_n \in S$ so this is a contradiction. So each limit point of S must be in S.*

$(\Longleftarrow)$:

We assume all of S's limit points are in S. Let $x \in S^C$. If we could not find a radius $r > 0$ so that $B_r(x) \subset S^C$, then S_C would not be open and so S would not be closed. But if this were the case, for each $r_n = 1/n$, there is an $x_n \in S$ with $|x - x_n| < 1/n$. But this says $x_n \to x$ with $(x_n) \subset S$. So x is a limit point of S and by assumption must be in S. This is a contradiction and so S^C must be open implying S is closed. ∎

Here are some more set based proofs.

Theorem 4.4.3 If A and B are Open, so is $A \cup B$

If A and B are open, so is $A \cup B$.

Proof 4.4.3

Let $p \in A \cup B$. Then $p \in A$ or $p \in B$ or both. If $p \in A$, p is an interior point because A is open. So there is an $r > 0$ so that $B_r(p) \subset A$. This says $B_r(p) \subset A \cup B$ too. A similar argument works for the case $p \in B$.

Since p is arbitrary, all points in $A \cup B$ are interior points and we have shown $A \cup B$ is open. ∎

There is a great connection between open and complementation which is called DeMorgan's Laws. Here is a simple three set version of that.

Theorem 4.4.4 $(A \cup B)^C = A^C \cap B^C$ and $(A \cap B)^C = A^C \cup B^C$.

$(A \cup B)^C = A^C \cap B^C$ and $(A \cap B)^C = A^C \cup B^C$.

Proof 4.4.4

$(A \cup B)^C \subset A^C \cap B^C$:

If $p \in (A \cup B)^C$, $p \notin A \cup B$. Thus $p \notin A$ and $p \notin B$ implying $p \in A^C$ and $p \in B^C$; i.e. $p \in A^C \cap B^C$. Since p is arbitrary, this shows $(A \cup B)^C \subset A^C \cap B^C$

$A^C \cap B^C \subset (A \cup B)^C$:

If $p \in A^C \cap B^C$, $p \in A^C$ and $p \in B^C$. So $p \notin A$ and $p \notin B$. Thus $p \notin A \cup B$ telling us $p \in (A \cup B)^C$. Since p is arbitrary, this shows $A^C \cap B^C \subset (A \cup B)^C$.

The other one is left for you as a homework. ∎

We can also take unions of closed sets.

Theorem 4.4.5 If A and B are Closed, so is $A \cup B$.

If A and B are closed, so is $A \cup B$.

Proof 4.4.5

Let's show $A \cup B$ is closed by showing $(A \cup B)^C$ is open. Let $p \in (A \cup B)^C = A^C \cap B^C$. Then

$p \in A^C$ and $p \in B^C$. *So there is a radius r_1 so that $B_{r_1}(p) \subseteq A^C$ and there is a radius r_2 so that $B_{r_2}(p) \subseteq B^C$. So if $r = \min\{r_1, r_2\}$, $B_r(p) \subseteq A^C \cap B^C$. Thus, p is an interior point of $(A \cup B)^C$.*

Since p is arbitrary, all points in $(A \cup B)^C$ are interior points and we have shown $(A \cup B)^C$ is open. ■

4.5   The Notion of Sequential Compactness

The next big topic is that of sequential compactness which is strongly related to the Bolzano - Weierstrass result of a set.

Definition 4.5.1 Sequential Compactness

> *Let S be a set of real numbers. We say S is **sequentially compact** or simply **compact** if every sequence (x_n) in S has at least one subsequence which converges to an element of S. In other words Given $(x_n) \subseteq S$, $\exists (x_{n_k}) \subseteq (x_n)$ and an $x \in S$ so that $x_{n_k} \to x$.*

We can characterize sequentially compact subsets of the real line fairly easily.

Theorem 4.5.1 Sequentially Compact Subsets of $\Re$ if and only Closed and Bounded

> *A set S is sequentially compact $\iff S$ is closed and bounded.*

Proof 4.5.1

$(\Rightarrow)$:

We assume S is sequentially compact and we show S is both closed and bounded. First, we show S is closed. Let $x \in S'$. If we show $S' = S$, by the previous theorem, we know S is closed. Since $S \subseteq S'$, this means we have to show all $x \in \partial S$ are also in S.

Now if $x \in \partial S$, x is a boundary point of S. Hence, for all $B_{1/n}(x)$, there is a point $x_n \in S$ and a point $y_n \in S^C$. This gives a sequence (x_n) satisfying $|x_n - x| < 1/n$ for all n which tells us $x_n \to x$. If this sequence is the constant sequence, $x_n = x$, then we have $x \in S$.

But if (x_n) satisfies $x_n \neq x$ for all n, we have to argue differently. In this case, since S is sequentially compact, (x_n) has a subsequence (x_{n_k}) which converges to some $y \in S$. Since limits of a sequence are unique, this says $x = y$. Since y is in S, this shows x is in S too. Hence, since x was arbitrarily chosen, we have $S' \subseteq S$ and so $S' = S$ and S is closed.

Next, we show S is bounded by contradiction. Let's assume it is not bounded. Then given any positive integer n, there is an x_n in S so that $|x_n| > n$. This defines a subsequence (x_n) in S. Since S is sequentially compact, (x_n) has a convergent subsequence (x_{n_k}) which converges to an element y in S. But since this subsequence converges, this tells us (x_{n_k}) is bounded; i.e. there is a positive number B so that $|x_{n_k}| \leq B$ for all n_k. But we also know $|x_{n_k}| > n_k$ and so $n_k < |x_{n_k}| \leq B$ for all n_k. This is impossible as $n_k \to \infty$ and B is a finite number. Thus, our assumption that S was unbounded is wrong and so S must be bounded.

$(\Leftarrow)$:

We assume S is closed and bounded and we want to show S is sequentially compact. Let (x_n) be any sequence in S. Since S is a bounded set, by the Bolzano - Weierstrass Theorem for Sequences, since (x_n) is bounded, there is at least one subsequence (x_{n_k}) of (x_n) which converges to a value y. If this sequence is a constant sequence, then $x_{n_k} = y$ always which tells us y is in S too. On the other hand, if it is not a constant sequence, this tells us y is an accumulation point of S. Now if y was not in S, then y would be in S^C and then since $x_{n_k} \to y$, every $B_r(y)$ would have to contain points of S and points of S^C. This says y has to be a boundary point. But since S is closed, S contains all its *boundary points. Thus the assumption $y \in S^C$ can't be right and we know $y \in S$. This shows the arbitrary sequence (x_n) in S has a subsequence which converges to an element of S which shows S is sequentially compact.* ∎

Example 4.5.1 *Let $S = \{2\} \cup (3, 4)$. Note S is not closed as 3 and 4 are boundary points not in S. Hence, S is not sequentially compact either.*

Example 4.5.2 *Let $S = \{2\} \cup [5, 7]$. Then S is closed as it contains all of its boundary points and so S is also sequentially compact.*

Example 4.5.3 *Let $S = [1, \infty)$. Then S is not bounded so it is not sequentially compact.*

There are a number of consequences to sequential compactness.

Theorem 4.5.2 Convergent Sequences in a Sequentially Compact Set Have Limit in the Set

If S is sequentially compact, then any sequence (x_n) in S which converges, converges to a point in S.

Proof 4.5.2
Since S is sequentially compact, such a sequence does have a convergent subsequence which converges to a point in S. Since the limit of the subsequence must be the same as the limit of this convergent sequence, this tells us the limit of the sequence must be in S. ∎

4.5.1 Sequential Compactness and the Existence of Extrema

There is a strong connection between sequential compactness and the existence of extreme values. Let S be a nonempty and bounded set of numbers. Then $\alpha = \inf(S)$ and $\beta = \sup(S)$ are both finite numbers.

Given any $r_n = 1/n$, the Supremum Tolerance Lemma then tells us there is a $x_n \in S$ so that $\beta - 1/n < x_n \leq \beta$ for all n. Thus, rearranging this inequality, we have $-1/n < x_n - \beta \leq 0 < 1/n$ which says $|x_n - \beta| < 1/n$ for all n. This tells us $x_n \to \beta$. We can do a similar argument for the infimum α and so there is a sequence (y_n) in S so that $y_n \to \alpha$.

Now in general, we don't know if α or β are the minimum and maximum of S, respectively. But if we also knew S was sequentially compact, we can say more. By the Theorem 4.5.2, since $x_n \to \beta$ and $(x_n) \subset S$, we know $\beta \in S$. Hence β is a maximum. A similar argument shows α is a minimum. These types of sequences are important enough to have special names.

Definition 4.5.2 Minimizing and Maximizing Sequences

Let S be a nonempty and bounded set of numbers. Then $\alpha = \inf(S)$ and $\beta = \sup(S)$ are both finite numbers. A sequence $(y_n) \subset S$ which converges to $\inf(S)$ is called a **minimizing sequence** *and a sequence $(x_n) \subset S$ which converges to $\sup(S)$ is called a* **maximizing sequence***.*

What we want are conditions that force minimizing and maximizing sequences to converge to points inside S; i.e. make sure the minimum and maximum of S exist. One way to make this happen is to prove S is sequentially compact.

4.5.2 Homework

Exercise 4.5.1 *Let $S = \{3\} \cup (0, 3)$. Explain why S is or is not sequentially compact.*

Exercise 4.5.2 *Let $S = (-\infty, 3]$. Explain why S is or is not sequentially compact.*

Exercise 4.5.3 *Let* $S = \{23\} \cup (= (-10, 12]$. *Explain why* S *is or is not sequentially compact.*

Exercise 4.5.4 *Let* $S = (-\infty, 30]$. *Explain why* S *is or is not sequentially compact.*

Exercise 4.5.5 *Let* $S = (x_n) = (\sin(n))_{n \geq 1}$. *Since* $(x_n) \subset [-1, 1]$, *why do you know that* (x_n) *has a subsequence which converges to a point* x *in* $[-1, 1]$? *Can you see why it is hard to decide if* S *is a closed set? This question is just to show you deciding if a set is closed or open can be difficult.*

Exercise 4.5.6 *Let* $S = (x_n) = (\cos(n))_{n \geq 1}$. *Since* $(x_n) \subset [-1, 1]$, *why do you know that* (x_n) *has a subsequence which converges to a point* x *in* $[-1, 1]$? *Can you see why it is hard to decide if* S *is a closed set? This question is just to show you deciding if a set is closed or open can be difficult.*

Exercise 4.5.7 *If* A *and* B *are open, prove* $A \cap B$ *is open also.*

4.6   The Deeper Structure of Sequences

We now want to explore some deeper structure inherent in sequences. We need better tools to probe exactly how a sequence fails to converge. The notion of the **limit inferior** and **limit superior** of a sequence will enable us to do that.

4.6.1   The Limit Inferior and Limit Superior of a Sequence

Definition 4.6.1 The Limit Inferior and Limit Superior of a Sequence

> *Let* $(a_n)_{n \geq k}$ *be a sequence of real numbers which is bounded. Also let* $S = \{y : \exists (a_{n_p}) \subseteq (a_n) \ni a_{n_p} \to y\}$. *Since* S *is non empty by the Bolzano - Weierstrass Theorem for Sequences,* $\inf S$ *and* $\sup S$ *both exist and are finite. We define*
>
> $$\liminf(a_n) \quad = \quad = \underline{\lim}(a_n) = \text{ limit inferior } (a_n) = \inf S$$
> $$\limsup(a_n) \quad = \quad = \overline{\lim}(a_n) = \text{ limit superior } (a_n) = \sup S$$
>
> S *is called the set of subsequential limits of* (a_n).

Example 4.6.1 *For* $((-1)^n)$, $S = \{-1, 1\}$ *and* $\underline{\lim}((-1)^n) = -1$ *and* $\overline{\lim}((-1)^n) = 1$.

Example 4.6.2 *For* $(\cos(n\pi/3))$, $S = \{-1, -1/2, 1/2, 1\}$ *and* $\underline{\lim}(a_n) = -1$ *and* $\overline{\lim}(a_n) = 1$.

Example 4.6.3 *For* $(\cos(n\pi/4))$, $S = \{-1, -1/\sqrt{2}, 0, 1/\sqrt{2}, 1\}$ *and* $\underline{\lim}(a_n) = -1$ *and* $\overline{\lim}(a_n) = 1$.

Example 4.6.4 *For* $(5 + \cos(n\pi/4))$, $S = \{5-1, 5-1/\sqrt{2}, 5+0, 5+1/\sqrt{2}, 5+1\}$ *and* $\underline{\lim}(a_n) = 4$ *and* $\overline{\lim}(a_n) = 6$.

There are connections between the limit of a sequence and the limit inferior and superior values of the sequence.

Theorem 4.6.1 If a Sequence Converges then the Limit Inferior and Limit Superior Match the Limit Value

> *Let* $(a_n)_{n \geq k}$ *be a bounded sequence and let* a *be a real number. Then* $a_n \to a \iff \underline{\lim}(a_n) = \overline{\lim}(a_n) = a$.

Proof 4.6.1

Assume $a_n \to a$. *Then all subsequences* (a_{n_k}) *also converge to* a *and so* $S = \{a\}$. *Thus,* $\inf S = \sup S = a$. *Thus, by definition,* $\underline{\lim}(a_n) = \overline{\lim}(a_n)$.

($\Leftarrow$):
We assume $\underline{\lim}(a_n) = \overline{\lim}(a_n)$ and so we have $S = \{a\}$. Suppose $a_n \not\to a$. Then there is an ϵ_0 so that for all k, there is an n_k with $|a_{n_k} - a| \geq \epsilon_0$. Since (a_n) is bounded, (a_{n_k}) is also bounded. By the Bolzano - Weierstrass Theorem, there is a subsequence $(a_{n_{k_p}})$ which we will denote by $(a_{n_k}^1)$. We will let n_{k_p} be denoted by n_k^1 for convenience.

The superscript 1 plays the role of adding another level of subscripting which would be pretty ugly! This subsequence of the subsequence converges to a number y. So by definition, $y \in S$. But S is just one point, a, so we have $y = a$ and we have shown $a_{n_k}^1 \to a$ too. Now pick the tolerance ϵ_0 for this sub subsequence. Then there is an Q so that $|a_{n_k}^1 - a| < \epsilon_0$ when $n_k^1 > Q$. But for an index $n_k^1 > Q$, we have both $|a_{n_k}^1 - a| < \epsilon_0$ and $|a_{n_k}^1 - a| \geq \epsilon_0$. This is not possible. Hence, our assumption that $a_n \not\to a$ is wrong and we have $a_n \to a$. ∎

Example 4.6.5 *$(a_n) = (\sin(n\pi/6))_{n \geq 1}$ has $S = \{-1, -\sqrt{3}/2, -1/2, 0, 1/2, \sqrt{3}/2, 1\}$. So $\underline{\lim}(a_n) = -1$ and $\overline{\lim}(a_n) = 1$ which are not equal. This tells us immediately, $\lim(a_n)$ does not exist.*

4.6.2 Limit Inferior$_*$ Star and Limit Superior*

We can approach the inferior and superior limit another way.

Definition 4.6.2 Limit Inferior$_*$ and Limit Superior*

Let (a_n) be a bounded sequence. Define sequences (y_k) and (z_k) by $y_k = \inf\{a_k, a_{k+1}, a_{k+2}, \ldots\} = \inf_{n \geq k}(a_n)$ and $z_k = \sup\{a_k, a_{k+1}, a_{k+2}, \ldots\} = \sup_{n \geq k}(a_n)$. Then we have $y_1 \leq y_2 \leq \ldots \leq y_k \leq \ldots \leq B$ and $z_1 \geq z_2 \geq \ldots \geq z_k \geq \ldots \geq -B$ where B is the bound for the sequence. We see $y = \sup(y_k) = \lim_{k \to \infty} y_k$ and $z = \inf(z_k) = \lim_{k \to \infty} z_k$. We denote z by $\overline{\lim}^*(a_n)$ and y by $\underline{\lim}_*(a_n)$. Since $y_k \leq z_k$ for all k, we also know $\lim_k y_k = y \leq \lim_k z_k = z$.

We will show $\underline{\lim}_*(a_n) = \underline{\lim}(a_n)$ and $\overline{\lim}^*(a_n) = \overline{\lim}(a_n)$. Thus, we have two ways to characterize the limit inferior and limit superior of a sequence. Sometimes one is easier to use than the other!

Let's look more closely at the connections between subsequential limits and the $\underline{\lim}_*(a_n)$ and $\overline{\lim}^*(a_n)$.

Theorem 4.6.2 Limit Inferior$_*$ and Limit superior* are Subsequential Limits

There are subsequential limits that equal $\overline{\lim}^*(a_n)$ and $\underline{\lim}_*(a_n)$.

Proof 4.6.2
Let's look at the case for $z = \overline{\lim}_*(a_n)$. Pick any $\epsilon = 1/k$. Let $S_k = \{a_k, a_{k+1}, \ldots\}$. Since $z_k = \sup S_k$, applying the Supremum Tolerance Lemma to the set S_k, there are sequence values a_{n_k} with $n_k \geq k$ so that $z_k - 1/k < a_{n_k} \leq z_k$ for all k. Thus, $-1/k < a_{n_k} - z_k \leq 0 < 1/k$ or $|a_{n_k} - z_k| < 1/k$. Pick an arbitrary $\epsilon > 0$ and choose N_1 so that $1/N_1 < \epsilon/2$. Then, $k > N_1 \Rightarrow |a_{n_k} - z_k| < \epsilon/2$. We also know since $z_k \to z$ that there is an N_2 so that $k > N_2 \Rightarrow |z_k - z| < \epsilon/2$.

Now pick $k > \max\{N_1, N_2\}$ and consider

$$|a_{n_k} - z| = |a_{n_k} - z_k + z_k - z| \leq |a_{n_k} - z_k| + |z_k - z|$$
$$< \epsilon/2 + \epsilon/2 = \epsilon$$

This shows $a_{n_k} \to z$.
A very similar argument shows that we can find a subsequence (a_{n_k}') which converges to y. These arguments shows us y and z are in S, the set of all subsequential limits. ∎

To prove these two ways of looking at limit inferiors and limit superiors are equivalent is a bit difficult. The best way to do it is to prove a transitional result which gives us a better characterization of $\underline{\lim}_*(a_n)$ and $\overline{\lim}^*(a_n)$.

Theorem 4.6.3 A Characterization of Limit Inferior$_*$ and Limit superior*

$$
\begin{aligned}
y = \underline{\lim}_*(a_n) &\iff (c < y \Rightarrow a_n < c \text{ for only finitely many indices }) \\
&\text{and} \quad (y < c \to a_n < c \text{ for infinitely many indices }) \\
z = \overline{\lim}^*(a_n) &\iff (c < z \Rightarrow a_n > c \text{ for infinitely many indices }) \\
&\text{and} \quad (z < c \Rightarrow a_n > c \text{ for only finitely many indices })
\end{aligned}
$$

Proof 4.6.3

Let's look at the proposition for $\underline{\lim}_*(a_n) = y$. The proof of the other one is similar and is worth your time to figure out!

($\Rightarrow$):
We assume $y = \underline{\lim}_*(a_n)$. For $c < y = \sup\{y_k\}$, there has to be a $y_{k_0} > c$. Now $y_{k_0} = \inf\{a_{k_0}, a_{k_0+1}, \ldots\}$ and so we have $a_n \geq y_{k_0} > c$ for all $n \geq k_0$. Hence, the set of indices where the reverse inequality holds must be finite. That is, $\{n : a_n < c\}$ is a finite set. This shows the first part.

Next, assume $y < c$. Thus, $y_k < c$ for all k. Hence, $y_k = \inf\{a_k, a_{k+1}, \ldots\} < c$. Let $\epsilon = c - \inf\{a_k, a_{k+1}, \ldots\} = c - y_k > 0$. Note, $y_k + \epsilon = c$, Now, by the Infimum Tolerance Lemma, there is a_{n_k} so that $y_k \leq a_{n_k} < y_k + \epsilon = c$. This shows the set $\{n : a_n < c\}$ must be infinite.
($\Leftarrow$):
We will just show the first piece. We assume the number A satisfies the first condition on the left. We must show $A = y$. Since A satisfies both conditions, we have if $c < A$, $a_n < c$ for only finitely many indices and if $A < c$, $a_n < c$ for infinitely many indices.

So given $c < A$, since $a_n < c$ for only finitely many indices, we can find an index k_0 so that $a_n \geq c$ for all $k \geq k_0$. This tells us $y_{k_0} = \inf\{a_{k_0}, a_{k_0+1}, \ldots\} \geq c$ also. But then we have $y_k \geq y_{k_0} \geq c$ for all $k \geq k_0$ too. But this tells us $y = \sup y_k \geq c$. Now $y \geq c < A$ for all such c implies $y \geq A$ as well.

Now assume $A < c$. Then we have $a_n < c$ for infinitely many indices. Then we have $y_k = \inf\{a_k, a_{k+1}, \ldots\} < c$ also for all k. Since $y_k \to y$, we see $y \leq c$ too. Then since this is true for all $A < c$, we have $y \leq A$ also.

Combining, we have $y = A$ as desired. The argument for the other case is very similar. We will leave that to you and you should try to work it out as it is part of your growth in this way of thinking! ∎

We are now ready to prove the main result.

Theorem 4.6.4 The Limit of a Sequence Exists if and only the Limit Inferior$_*$ and Limit Superior* Match That Value

$$\lim(a_n) = a \iff \underline{\lim}_*(a_n) = \overline{\lim}^*(a_n) = a$$

Proof 4.6.4

$(\Rightarrow)$:

We assume $\lim(a_n) = a$. *We also have* $y = \underline{\lim}_*(a_n)$ *and* $z = \overline{\lim}^*(a_n)$. *Let's assume* $y < z$. *Then pick arbitrary numbers* c *and* d *so that* $y < d < c < z$. *Now use the previous Theorem. We have* $a_n < d$ *for infinitely many indices and* $a_n > c$ *for infinitely many indices also.*

The indices with $a_n > c$ *define a subsequence,* $(a_{n_k}) \subseteq (a_n)$. *Since this subsequence is bounded below by* c *and it is part of a bounded sequence, the Bolzano - Weierstrass Theorem tells us this subsequence has a convergent subsequence. Call this subsequence* $(a_{n_k}^1)$ *and let* $a_{n_k}^1 \to u$. *Then* $u \geq c$. *Further, since* $a_n \to a$, *we must have* $u = a \geq c$.

We can do the same sort of argument with the indices where $a_n < d$ *to find a subsequence* $(a_{m_k}^1)$ *of* (a_n) *which converges to a point* $v \leq d$. *But since* $a_n \to a$, *we must have* $v = a \leq d$.

This shows $a \leq d < c \leq a$ *which is impossible as* $d < c$. *Thus, our assumption that* $y < z$ *is wrong and we must have* $y = z$.

$(\Leftarrow)$:

Now we assume $y = z$. *Using what we know about* y, *given* $\epsilon > 0$, $y - \epsilon/2 < y$ *and so* $a_n < y - \epsilon/2$ *for only a finite number of indices. So there is an* N_1 *so that* $a_n \geq y - \epsilon/2$ *when* $n > N_1$. *This used the* y *part of the IFF characterization of* y *and* z. *But* $y = z$, *so we can also use the characterization of* z. *Since* $z = y < y + \epsilon/2$, $a_n > y + \epsilon/2$ *for only a finite number of indices. Thus, there is an* N_2 *so that* $a_n \leq y + \epsilon/2$ *for all* $n > N_2$. *We conclude if* $n > \max\{N_1, N_2\}$, *then* $y - \epsilon/2 \leq a_n \leq y + \epsilon/2$ *which implies* $|a_n - y| < \epsilon$. *We conclude* $a_n \to y$ *and so* $\lim(a_n) = \underline{\lim}_*(a_n) = \overline{\lim}^*(a_n)$. ∎

Now we can finally tie all the pieces together and show these two ways of finding inferior and superior limits are the same.

Theorem 4.6.5 Limit inferior is the Same as Limit Inferior$_*$ and the Limit Superior is the Same as the Limit Superior*

For the bounded sequence (a_n), $\underline{\lim}_*(a_n) = \underline{\lim}(a_n)$ *and* $\overline{\lim}^*(a_n) = \overline{\lim}(a_n)$.

Proof 4.6.5

Since we can find subsequences (a_{n_k}) *and* (a'_{n_k}) *so that* $\underline{\lim}_*(a_n) = \lim a_{n_k}$ *and* $\overline{\lim}^*(a_n) = \lim a'_{n_k}$, *we know* $\underline{\lim}_*(a_n)$ *and* $\overline{\lim}^*(a_n)$ *are subsequential limits. Thus, by definition,* $\underline{\lim}(a_n) \leq \underline{\lim}_*(a_n) \leq \overline{\lim}^*(a_n) \leq \overline{\lim}(a_n)$.

Now let c *be any subsequential limit. Then there is a subsequence* (a_{n_k}) *so that* $\lim_k a_{n_k} = c$. *Hence, we know* $\underline{\lim}_*(a_{n_k}) = \overline{\lim}^*(a_{n_k}) = c$ *also. We also know, from their definitions,* $\underline{\lim}_*(a_{n_k}) \geq \underline{\lim}_*(a_n)$ *and* $\overline{\lim}^*(a_{n_k}) \leq \overline{\lim}^*(a_n)$. *Thus,*

$$\underline{\lim}_*(a_n) \leq \underline{\lim}_*(a_{n_k}) = \overline{\lim}^*(a_{n_k}) = c \leq \overline{\lim}^*(a_n).$$

Now, since the subsequential limit value c *is arbitrary, we have* $\underline{\lim}_*(a_n)$ *is a lower bound of the set of subsequential limits,* S, *and so by definition* $\underline{\lim}_*(a_n) \leq \underline{\lim}(a_n)$ *as* $\underline{\lim}(a_n)$ *is* $\inf S$. *We also know* $\overline{\lim}^*(a_n)$ *is an upper bound for* S *and so* $\overline{\lim}(a_n) \leq \overline{\lim}^*(a_n)$.

Combining inequalities $\overline{\lim}(a_n) \leq \overline{\lim}^*(a_n) \leq \overline{\lim}(a_n)$ *and* $\underline{\lim}(a_n) \leq \underline{\lim}_*(a_n) \leq \underline{\lim}(a_n)$. *This shows us* $\underline{\lim}_*(a_n) = \underline{\lim}(a_n)$ *and* $\overline{\lim}(a_n) = \overline{\lim}(a_n)$. ∎

4.6.3   Homework

Exercise 4.6.1 *If $(a_n) \to a$ and $(b_n) \to b$ and we know $a_n \le b_n$ for all n, prove $a \le b$. This might seem hard, but pick $\epsilon > 0$ and write down the $\epsilon - N$ inequalities without using the absolute values. You should be able to see what to do from there.*

Exercise 4.6.2 *Prove there is a subsequence (a'_{n_k}) which converges to $\underline{\lim}_*(a_n)$. This is like the one we did, but uses the Infimum Tolerance Lemma.*

Exercise 4.6.3 *If $y \ge c$ for all such $c < A$, then $y \ge A$ as well. The way to attack this is to look at the sequence $c_n = A - 1/n$. You should see what to do from there.*

Chapter 5

Topological Compactness

We now want to explore some deeper questions about sequential compactness and its connection to open sets.

5.1 More Set Theory

There are more things we can talk about with sets. So let's get started. You'll see that the way we prove propositions in set theory is very different from the way we do this in other parts of this class. All part of your growth!

Theorem 5.1.1 The Union of Two Open Sets is Open

> *Let A and B be open sets. Then $A \cup B$ is open too.*

Proof 5.1.1
Remember $A \cup B = \{c : a \in A \text{ and } / \text{ or } b \in B\}$. To show this result, we must show an arbitrary c in $A \cup B$ is an interior point. Now for such a c it is clear c is in A only or it is in B only or it is in both. All that matters here is that c is in one or the other. If $c \in A$, then c is an interior point and there is a $B_r(c) \subset A \subseteq A \cup B$. And if $c \in B$, again c is an interior point and there is a $B_r(c) \subset B \subseteq A \cup B$. So in all cases, c is an interior point and so $A \cup B$ is open. ■

A **countable** union is one where the index set is an RII of $\mathbb{Z}$. We will talk about this more later but this is enough of a definition for now. Usually $\mathbb{N}$ is what we use, but it could be such a subset.

Theorem 5.1.2 Countable Unions of Open Sets are Open

> *The countable union of open sets is open.*

Proof 5.1.2
Let our RII set be $\mathbb{N}$ just for convenience of notation and let the individual open sets here be $\mathbb{O}_i$. We let $A = \cup_{n=1}^{\infty} \mathbb{O}_n = \cup_{n \in \mathbb{N}} \mathbb{O}_n$ denote the countable union. Let x be in the union. Then x is in at least one $\mathbb{O}_Q$. Since $\mathbb{O}_Q$ is open, x is an interior point of $\mathbb{O}_Q$. So there is a $B_r(x) \subset \mathbb{O}_Q$. But $\mathbb{O}_Q \subseteq \cup_{n \in \mathbb{N}} \mathbb{O}_n$. Hence, x is an interior point of the countable union. Since x was arbitrary, the countable union is open. ■

Comment 5.1.1 *For a countable union of sets, we usually just write $\cup_n \mathbb{O}_n$.*

Example 5.1.1 *Unions need not be over countable index sets. A good example is this. Let $S = [1,3)$ which has an infinite number of points. At each $x \in S$, pick the radius $r_x > 0$ and consider the set $A = \cup_{x \in S} B_{r_x}$. This index set is not an RII so indeed is an example of what is called a uncountable set. To be precise, a* **countable** *set Ω is one for which there is a map f which assigns each element in Ω one and only one element of $\mathbb{N}$. A RII is countable because n_k is assigned to k in $\mathbb{N}$ and this assignment is unique. Hence there is a $1 - 1$ and onto map from the RII to $\mathbb{N}$.*

Theorem 5.1.3 The Rational Numbers are Countable

The rational numbers are countable.

Proof 5.1.3

We will do the construction for positive fractions only but you should be able to see how to extend it to negative fractions also. We can set up the map like this:

Write down all the fraction in a table:

$$
\begin{array}{ccccccccc}
1/1 & 1/2 & 1/3 & 1/4 & 1/5 & 1/6 & 1/6 & \ldots & 1/n & \ldots \\
2/1 & 2/2 & 2/3 & 2/4 & 2/5 & 2/6 & 2/6 & \ldots & 2/n & \ldots \\
3/1 & 3/2 & 3/3 & 3/4 & 3/5 & 3/6 & 3/6 & \ldots & 3/n & \ldots \\
4/1 & 4/2 & 4/3 & 4/4 & 4/5 & 4/6 & 4/6 & \ldots & 4/n & \ldots \\
\vdots & \vdots & \vdots & \vdots & \vdots & \vdots & \vdots & \vdots & \vdots & \vdots \\
n/1 & n/2 & n/3 & n/4 & n/5 & n/6 & n/6 & \ldots & n/n & \ldots \\
\vdots & \vdots & \vdots & \vdots & \vdots & \vdots & \vdots & \vdots & \vdots & \vdots
\end{array}
$$

The mapping is then defined by the numbers in parenthesis chosen in diagonal sweeps:

$$
\begin{array}{ccccccccc}
1/1\,(1) & 1/2\,(2) & 1/3\,(6) & 1/4\,(7) & 1/5 & 1/6 & 1/6 & \ldots & 1/n & \ldots \\
2/1\,(3) & 2/2\,(5) & 2/3\,(8) & 2/4 & 2/5 & 2/6 & 2/6 & \ldots & 2/n & \ldots \\
3/1\,(4) & 3/2\,(9) & 3/3 & 3/4 & 3/5 & 3/6 & 3/6 & \ldots & 3/n & \ldots \\
4/1\,(10) & 4/2 & 4/3 & 4/4 & 4/5 & 4/6 & 4/6 & \ldots & 4/n & \ldots \\
\vdots & \vdots & \vdots & \vdots & \vdots & \vdots & \vdots & \vdots & \vdots & \vdots \\
n/1 & n/2 & n/3 & n/4 & n/5 & n/6 & n/6 & \ldots & n/n & \ldots \\
\vdots & \vdots & \vdots & \vdots & \vdots & \vdots & \vdots & \vdots & \vdots & \vdots
\end{array}
$$

eliminating the fractions, it is easier to see the pattern:

$$
\begin{array}{cccccc}
(1) & (2) & (6) & (7) & (15) & \ldots \\
(3) & (5) & (8) & (14) & \vdots & \ldots \\
(4) & (9) & (13) & \vdots & & \ldots \\
(10) & (12) & \vdots & & & \ldots \\
(11) & \vdots & & & & \ldots
\end{array}
$$

∎

So the rationals $\mathbb{Q}$ are countable too. We often call this countably infinite to distinguish it from the case where the index set is countably finite like $\{1,2,3,4,5\}$. Hence, $[1,5]$ would not be countable as it contains lots of numbers which are not fractions. Now let's look at uncountable unions of open sets.

Theorem 5.1.4 Uncountable Unions of Open Sets are Open

The uncountable union of open sets is open.

Proof 5.1.4
Let the uncountable index set be Ω. and let the individual open sets here be $\mathbb{O}_\alpha$ where α is a member of the index set Ω.

We let $A = \cup_{\alpha \in \Omega} \mathbb{O}_\alpha = \cup_\alpha \mathbb{O}_\alpha$ denote the uncountable union. Let x be in the union. Then x is in at least one $\mathbb{O}_\beta$. Since $\mathbb{O}_\beta$ is open, x is an interior point of $\mathbb{O}_\beta$. So there is a $B_r(x) \subset \mathbb{O}_\beta$.

But $\mathbb{O}_\beta \subseteq \cup_\alpha \mathbb{O}_\alpha$. Hence, x is an interior point of the uncountable union. Since x was arbitrary, the uncountable union is open. ∎

Now let's look at intersections of open sets.

Theorem 5.1.5 Finite Intersections of Open Sets are Open

The intersection of a finite number of open sets is open.

Proof 5.1.5
Let $A = \cap_{n=1}^N \mathbb{O}_n$ be a finite intersection of open sets. Let x be in this intersection. Then x is an interior point of all N sets. Hence, we have N circles: $B_{r_1}(x) \subset \mathbb{O}_1$ through $B_{r_N}(x) \subset \mathbb{O}_N$. If we let $r = \min\{r_1, \ldots, r_N\}$, then $B_r(x)$ is contained in each of the N sets. Hence A is open. ∎

However, countable intersections of open sets is another story. The *property of being open* can be lost.

Example 5.1.2 *The intersection of a countable number of open sets need not be open.*

Solution *Take the sets $(1 - 1/n, 4 + 1/n)$. We claim $\cap_n (1 - 1/n, 4 + 1/n) = [1, 4]$, a closed set. To prove these sets are equal, we first show $\cap_n (1 - 1/n, 4 + 1/n) \subset [1, 4]$ and then $[1, 4] \subseteq \cap_n (1 - 1/n, 4 + 1/n)$.*

$(\cap_n (1 - 1/n, 4 + 1/n) \subset [1, 4])$:
Let $x \in \cap_n(1 - 1/n, 4 + 1/n)$. Then, $x \in (1 - 1/n, 4 + 1/n)$ for all n. Thus, $1 - 1/n \leq x \leq 4 + 1/n$ for all n. Now just let $n \to \infty$ to show $x \in [1, 4]$.

$([1, 4] \subseteq \cap_n(1 - 1/n, 4 + 1/n))$:
Let $x \in [1, 4]$. Then $x \in (1 - 1/n, 4 + 1/n)$ for all n implying $x \in \cap_n(1 - 1/n, 4 + 1/n)$.

Comment 5.1.2 *So the property of being open can be violated if a countably infinite number of operations are used.*

We need lots of small results in this kind of thinking. This is one of De Morgan's Laws.

Theorem 5.1.6 The Complement of a Finite Intersection is the Union of the Individual Complements

$\left(\cap_{n=1}^N \mathbb{B}_n\right)^C = \cup_{n=1}^N \mathbb{B}_n^C$ *for any sets.*

Proof 5.1.6
$\left(\cap_{n=1}^N \mathbb{B}_n\right)^C \subseteq \cup_{n=1}^N \mathbb{B}_n^C$

If $x \in \left(\cap_{n=1}^{N} \mathbb{B}_n \right)^{C}$, then $x \notin \cap_{n=1}^{N} \mathbb{B}_n$. That tells us $x \notin \mathbb{B}_n$ for at least one k. Hence, $x \in \mathbb{B}_k^{C}$ implying $x \in \cup_{n=1}^{N} \mathbb{B}_n^{C}$.

Going the other way, $\cup_{n=1}^{N} \mathbb{B}_n^{C} \subseteq \left(\cap_{n=1}^{N} \mathbb{B}_n \right)^{C}$.
If $x \in \cup_{n=1}^{N} \mathbb{B}_n^{C}$, there is an index Q with $x \in \mathbb{B}_Q^{C}$. Thus $x \notin \mathbb{B}_Q$ which tells us $x \notin \cap_{n=1}^{N} \mathbb{B}_n$. Hence, $x \in \left(\cap_{n=1}^{N} \mathbb{B}_n \right)^{C}$. ∎

Using this result, we can look at the intersection of closed sets and see if the property of being closed in retained.

Theorem 5.1.7 Finite Intersections of Closed Sets Are Closed

> *The intersection of a finite number of closed sets is closed.*

Proof 5.1.7
Let $A = \cap_{n=1}^{N} \mathbb{D}_n$ be a finite intersection of closed sets. We now know $A^{C} = \left(\cap_{n=1}^{N} \mathbb{D}_n \right)^{C} = \cup_{n=1}^{N} \mathbb{D}_n^{C}$. Since each $\mathbb{D}_n$ is closed, $\mathbb{D}_n^{C}$ is open. We also know the finite union of open sets is open. Then A^{C} is open implying A is closed. ∎

Comment 5.1.3 *We can also prove $\left(\cap_{n=1}^{\infty} \mathbb{B}_n \right)^{C} = \cup_{n=1}^{\infty} \mathbb{B}_n^{C}$ for any sets B_n and prove the intersection of a countable number of closed sets is still closed. You should be able to do these!*

Example 5.1.3 *The union of a countable number of closed sets need not be closed.*

Solution *We claim $\cup_n [1 + 1/n, 4 - 1/n] = (1, 4)$, an open set. $(\cup_n [1 + 1/n, 4 - 1/n) \subset (1, 4]$):*
Let $x \in \cup_n [1 + 1/n, 4 - 1/n]$. Then, $x \in [1 + 1/Q, 4 - 1/Q]$ for some Q. Thus, the distance from x to 1 or 4 is positive. So $x \in (1, 4)$.

($(1, 4) \subseteq \cup_n [1 + 1/n, 4 - 1/n]$):
If $x \in (1, 4)$, x is an interior point and so there is $B_r(x) \subset (1, 4)$. That is, $1 < x - r < x < x + r < 4$. Let the distance from $x - r$ to 1 be d_1 and the distance from $x + r$ to 4 be d_2. Let $d = (1/2) \min\{d_1, d_2\}$. Choose N so that $1/N < d$. Then $x \in [1 + 1/N, 4 - 1/N]$ and so $x \in \cup_n [1 + 1/n, 4 - 1/n]$.

Comment 5.1.4 *The property of closed can be violated using countably infinite operations.*

5.2   The Notion of Topological Compactness

Now we can introduce another form of compactness called topological compactness which we will show is equivalent to sequential compactness although it will takes us awhile to do that.

Definition 5.2.1 Open Covers and Finite Subcovers

> *We say a collection of open sets $\mathscr{U} = \{\mathbb{O}_\alpha : \alpha \in \Omega\}$ indexed by an arbitrary set Ω (so it can be finite, countably infinite or uncountable) is a **open cover** of a set A if $A \subseteq \cup_{\alpha \in \Omega} \mathbb{O}_\alpha$. We say the collection $\{\mathbb{O}_{\alpha_1}, \ldots, \mathbb{O}_{\alpha_N}\}$ from $\mathscr{U}$, for some natural number N, is a **finite subcover** or fsc of A if $A \subseteq \cup_{n=1}^{N} \mathbb{O}_{\alpha_n}$.*

We are now ready for the **big new idea**!

Definition 5.2.2 Topological Compactness

> *We say a set A is **topologically compact** if every open cover of A has a finite subcover.*

We need to show this idea in our setting $\Re$ is equivalent to sequential compactness.

5.2.1 Finite Closed Intervals and Topological Compactness

We start by showing a topologically compact set is closed and bounded.

Theorem 5.2.1 Topologically Compact Implies Closed and Bounded

A is topologically compact $\Rightarrow$ A is closed and bounded.

Proof 5.2.1

Let's assume A is topologically compact. Consider the union of all $B_r(x)$ for a fixed radius r over all x in S. This is an open cover. Since A is topologically compact, this cover has a fsc which contains A. We can express the fsc as $\mathscr{V} = \{B_r(x_1), \ldots, B_r(x_N)\}$ for some N. Now let $x \in A$. Then $x \in B_r(x_j)$ for some $1 \leq j \leq N$. Thus, $|x - x_j| < r \Rightarrow |x| < r + |x_j|$ using the backwards triangle inequality. And in turn, $r + |x_j| \leq \max\{r + |x_1|, \ldots, r + |x_N|\} = B$. So $|x| \leq B$ for all $x \in A$ implying A is bounded.

To show A is closed, we will show A^C is open. Let $x \in A^C$. Given any $y \in A$, since $x \neq y$, the two points are separated by a distance d^{xy}. Thus, $B_{d^{xy}/2}(y)$ and $B_{d^{xy}/2}(x)$ have no points in common. The collection of all the circles $B_{d^{xy}/2}(y)$ for $y \in A$ gives an open cover $\mathscr{U}$ of A which must have a fsc, $\mathscr{V} = \{B_{d_1}(y_1), \ldots, B_{d_N}(y_N)\}$, so that $A \subseteq \cup_{n=1}^{N} B_{d_n}(y_n)$ where each $d_i = d^{xy_i}/2$. Let $\mathscr{W} = \cap_{n=1}^{N} B_{d_n}(x)$. The circles $B_{d_n}(x)$ are all separate (we say disjoint)from $B_{d_n}(y_i)$. Hence, points in the intersection are outside the union of the fsc $\mathscr{V}$. Then $\mathscr{W}$ is an open circle as it is a finite intersection of open circles and it is outside of the union over the fsc $\mathscr{V}$ which contains A. We see $\mathscr{W} \subset A^C$ implying x is an interior point of A^C. Since x was arbitrary, we see A^C is open. Hence A is closed. ∎

Now we go the other way! To get the other direction, we have to be indirect. We start by showing a finite interval that is closed is also topologically compact.

Theorem 5.2.2 $[a, b]$ is Topologically Compact

The closed and bounded finite interval $[a, b]$, $a < b$, is topologically compact.

Proof 5.2.2

This proof is similar to how we proved the Bolzano - Weierstrass Theorem (BWT). So look at the similarities!
Assume this is false and let $J_0 = [a, b]$. Label this interval as $J_0 = [\alpha_0, \beta_0]$; i.e. $\alpha_0 = a$ and $\beta_0 = b$. The length of J_0 is $b - a = B$ which we will call ℓ_0. Assume there is an open cover $\mathscr{U}$ which does not have a fsc.

Now divide J_0 into two equal pieces $\{x_0 = \alpha_0, x_1 = \alpha_0 + 1\frac{b-a}{2}, x_2 = a + 2\frac{b-a}{2} = b\}$. We have $[\alpha_0, \beta_0] = [x_0, x_1] \cup [x_1, x_2]$. At least one of these intervals can not be covered by a fsc as otherwise the entire interval $[\alpha_0, \beta_0]$ has a fsc which we have assumed is not true. Call this interval J_1 and note the length of J_1 is $\ell_1 = (1/2)\ell_0 = B/2$. Label this interval as $J_1 = [\alpha_1, \beta_1]$.

We can continue in this way by induction. Assume we have constructed the interval $J_n = [\alpha_n, \beta_n]$ in this fashion with length $\ell_n = (1/2)\ell_{n-1} = B/2^n$ and there is no fsc of J_n. Now divide J_n into 2 equal pieces as usual $\{x_0 = \alpha_n, x_1 = \alpha_n + 1\,\ell_n, x_2 = \alpha_n + 2\,\ell_n = \beta_n\}$. These points *determine* 2 *subintervals and at least one of them can not be covered by a fsc. Call this interval* $J_{n+1} = [\alpha_{n+1}, \beta_{n+1}]$ *which has length* $\ell_{n+1} = (1/2)\ell_n$ *like required.*

We can see each $J_{n+1} \subset J_n$ *by the way we have chosen the intervals and their lengths satisfy* $\ell_n \to 0$. *Using the same sort of arguments that we used in the proof of the BWT for Sequences, we see there is a unique point* z *which is the* $\sup(\alpha_n) = \inf(\beta_n)$. *Another way to say this in our new set language is that* $\cap_{n=1}^{\infty} [\alpha_n, \beta_n] = \{z\}$!

Since z *is in* J_1, *there is a* $\mathbb{O}_1$ *in* $\mathscr{U}$ *with* $z \in \mathbb{O}_1$ *because* $\mathscr{U}$ *covers A. Also, since* $\mathbb{O}_1$ *is open, there is a circle* $B_{r_1}(z) \subset \mathbb{O}_1$. *Now choose K so that* $\ell_K < \ell_1$. *Then,* J_k *is contained in* $\mathbb{O}_1$ *for all* $k > K$. *This says* $\mathbb{O}_1$ *is a fsc of* J_k *which contradicts our construction process. Hence, our assumption that there is an open cover with no fsc is wrong and we have* $[a, b]$ *is topologically compact.* ∎

5.2.2 Homework

Exercise 5.2.1 *Explain why the set* $[1, 4]$ *is or is not topologically compact.*

Exercise 5.2.2 *Explain why the set* $[-101, 45]$ *is or is not topologically compact.*

Exercise 5.2.3 *Find an open cover with no fsc for the set* $(1, 3)$. *Explain why this shows* $(1, 3)$ *is not topologically compact.*

Exercise 5.2.4 *Find an open cover with no fsc for the set* $(1, 3]$. *Explain why this shows* $(1, 3)$ *is not topologically compact.*

Exercise 5.2.5 *Find an open cover with no fsc for the set* $[-11, 3)$. *Explain why this shows* $(1, 3)$ *is not topologically compact.*

Exercise 5.2.6 *If* A_1, A_2 *and* A_3 *are nonempty sets, show all the details of the proof for* $(A_1 \cup A_2 \cup A_3)^C = A_1^C \cap A_2^C \cap A_3^C$.

Exercise 5.2.7 *If* A_1, A_2 *and* A_3 *are nonempty sets, show all the details of the proof for* $(A_1 \cap A_2 \cap A_3)^C = A_1^C \cup A_2^C \cup A_3^C$.

5.2.3 The Equivalence of Topological Compactness and Closed and Bounded

We are now ready to look at a general bounded and closed subset, not just a closed and bounded interval like $[a, b]$.

Theorem 5.2.3 Closed Subsets of Bounded Intervals are Topologically Compact

A closed subset B of $[a, b]$, $a < b$, *is topologically compact.*

Proof 5.2.3
Let $\mathscr{U}$ *be an open cover of B. Then the collection of open sets* $\mathbb{O}' = \mathscr{U} \cup B^C$ *is an open cover of* $[a, b]$ *and hence has a fsc* $\mathscr{V}' = \{\mathbb{O}_1, \ldots, \mathbb{O}_N, B^C\}$. *for some N with each* $\mathbb{O}_n \in \mathscr{U}$. *This shows B is topologically compact.* ∎

We can now handle a general closed and bounded set.

Theorem 5.2.4 A Closed and Bounded Set is Topologically Compact

A closed and bounded set A is topologically compact.

Proof 5.2.4
Since A is bounded, there is a positive number so that $A \subseteq [-B, B]$. Since A is a closed subset of $[-B, B]$, by the previous theorem, A is topologically compact. ∎

We have now done what we set out to do. We already knew a set A is sequentially compact IFF it is closed and bounded and now we know A is topologically compact IFF it is closed and bounded. Hence A **is sequentially compact IFF topologically compact IFF closed and bounded**. Hence, from now on we will just say a set is **compact** and understand we can use these different ways of looking at it as convenient to our arguments. Let's state this as a Theorem because it is very important!

Theorem 5.2.5 A set is Sequentially Compact if and only if it is Topologically Compact

> *Let S be a subset of real numbers. Then S is topologically compact if and only if it is sequentially compact.*

Proof 5.2.5

($\Longrightarrow$:)
If S is topologically compact, then it is closed and bounded which implies it is sequentially compact.
($\Longleftarrow$:)
If S is sequentially compact, then it is closed and bounded which implies it is topologically compact.
∎

5.2.4 Homework

Remember, we can just use the term **compact** now for a set.

Exercise 5.2.8 *Here are some questions about compactness.*

- *If a nonempty subset A is compact, is it true that any subset of A is also?*

- *If a nonempty subset A is sequentially compact, it is true that any closed and bounded subset of A is topologically compact?*

- *If a nonempty subset A is closed, is it true that any subset of A is also?*

- *If a nonempty subset A is open, is it true that any subset of A is also?*

Exercise 5.2.9 *Here are a few proofs about the union of two compact sets.*

- *Prove if A and B are topologically compact, then $A \cup B$ is also. Do this proof using topologically compact ideas only.*

- *Prove if A and B are sequentially compact, then $A \cup B$ is also. Do this proof using sequentially compact ideas only.*

As we have mentioned before, the Bolzano - Weierstrass Theorem extends nicely to $\Re^n$. So at this point, we have an obvious series of questions:

- Does the Bolzano - Weierstrass Theorem work in a general metric space?

- Are the ideas of topological compactness and sequential compactness always equivalent?

Chapter 6

Function Limits

We are now going to discuss the limits of functions.

6.1 The Limit Inferior and Limit Superior of Functions

Recall if a nonempty set S is unbounded below, we set $\inf S = -\infty$ and if it is unbounded above, we set $\sup S = \infty$. We all know what functions are: they are a way to assign each point of one set D to a point in another set R.

Definition 6.1.1 Functions

> *Let f be a mapping that takes each point in the **domain set** D into the **range set** R. We say f is a function if to each $x \in D$, the corresponding value $y \in R$ is unique. We use the notation $y = f(x)$ to denote this value in the range. Hence, $R = \{f(x) : x \in D\}$.*

Comment 6.1.1 *It is ok for two x values to go to one y value: for example $f(x) = x^2$ sends ± 1 to 1. But if you look at the relationship defined by $x = y^2$ which you know is a parabola going sideways rather than up, for $x = 1$, there are two possibilities: $y = 1$ and $y = -1$. This relationship does **not** define a function. There are two functions here: $y = \sqrt{x}$ and $y = -\sqrt{x}$.*

Comment 6.1.2 *We have seen a lot of functions already: every sequence is a function f from a RII domain. But we now want to look at functions whose domain is an* interesting *subset of $\Re$. These are functions are at least* locally *defined.*

We usually just need a function to defined **locally** at a point.

Definition 6.1.2 Locally Defined Functions

> *We say a function f is locally defined at the point p, if there is a circle $B_r(p)$ for some radius $r > 0$ so that $f(x)$ is defined as a range value for all x in $B_r(p)$. Of course, f may have a larger domain than this, but at p, f is defined at least on a circle about p.*
>
> *Let a be a real number. If f is locally defined at p except possibly p itself, we say a is a* **cluster point** *of f at p if there is a sequence $(x_n) \subset B_r(p)$ so that $x_n \to p$ and $f(x_n) \to a$. We let $S(p) = \{a : a$ is a cluster point of f at $p\}$. If $S(p)$ is nonempty, we define the* **limit inferior** *of f at p to be $\underline{\lim}_{x \to p}(f) = \inf S(p)$. We also define* **limit superior** *of f at p to be $\overline{\lim}_{x \to p}(f) = \sup S(p)$.*

6.1.1   A Poorly Behaved Function

Let's examine how badly behaved a function can be.

6.1.2   A Function with Only One Set of Cluster Points That Is One Value

Now let's begin a long extended example. The function below, although defined on the whole interval $[0,1]$ has only one point, $p = 0$, where the set of cluster points consists of just one point. This means at $p = 0$, $\underline{\lim}_{x \to 0} f(x) = \overline{\lim}_{x \to 0} f(x)$.

Example 6.1.1 *Define f by*

$$f(x) \;=\; \begin{cases} x, & x \in \mathbb{Q} \cap [0,1] \\ -x, & x \in \mathbb{IR} \cap [0,1] = \mathbb{Q}^C \cap [0,1] \end{cases}$$

Then

$$S(p) \;=\; \begin{cases} \{-p, p\}, & p \neq 0 \\ \{0\}, & p = 0 \end{cases}$$

and

$$\lim_{x \to p} f(x) \;=\; \begin{cases} -p, & p \neq 0 \\ 0, & p = 0 \end{cases} \quad and \quad \overline{\lim}_{x \to p} f(x) = \begin{cases} p, & p \neq 0 \\ 0, & p = 0 \end{cases}$$

Solution *where* $\mathbb{IR}$ *denotes the set of all irrational numbers.*
Case: $p = 0$
This point is locally defined because it is defined on any $B_r(0)$ for a positive r. Pick any such r. Let (x_n) be any sequence such that $x_n \to 0$. We suspect 0 is a cluster point of f at 0. Consider

$$|f(x_n) - 0| \;=\; \begin{cases} |x_n|, & x_n \in \mathbb{Q} \cap [0,1] \\ |-x_n|, & x_n \in \mathbb{IR} \cap [0,1] \end{cases} = |x_n|, \; \forall n.$$

Pick $\epsilon > 0$. then since $x_n \to 0$, there is a N so that $n > N \Rightarrow |x_n - 0| < \epsilon$. Thus $n > N \Rightarrow |f(x_n) - 0| < \epsilon$ also. We see $f(x_n) \to 0$ and so 0 is a cluster point of f at 0.

Can any other point $a \neq 0$ be a cluster point of f at 0? The set up we used above still works, except now we consider

$$|f(x_n) - a| \;=\; \begin{cases} |x_n - a|, & x_n \in \mathbb{Q} \cap [0,1] \\ |-x_n - a|, & x_n \in \mathbb{IR} \cap [0,1] \end{cases}$$

$$=\; \begin{cases} |x_n - a|, & x_n \in \mathbb{Q} \cap [0,1] \\ |x_n + a|, & x_n \in \mathbb{IR} \cap [0,1] \end{cases}$$

Since $x_n \to 0$, for $\epsilon = |a|/2$, there is an N so that $n > N$ implies $|x_n - 0| < |a|/2$. Let's do the case $a > 0$ for convenience. The other case is similar and is left to you. We have for $n > N$,

$$-a/2 - a < x_n - a < a/2 - a \quad and \quad -a/2 + a < x_n + a < a/2 + a$$

or

$$-3a/2 < x_n - a < -a/2 \quad and \quad a/2 < x_n + a < 3a/2$$

This says $|f(x_n) - a|$ can not be made small for n sufficiently large. Hence, $a \neq 0$ can not be a cluster point of f. Thus $S(0) = \{0\}$ telling us $\underline{\lim}_{x \to 0} f(x) = \inf S(0) = 0$ and $\overline{\lim}_{x \to 0} f(x) =$

$\sup S(0) = 0$.

Another way to show $S(0) = 0$ is to use this argument: Can any other point $a \neq 0$ be a cluster point of f at 0? The set up we used above still works, except now we consider

$$f(x_n) - a \; = \; \begin{cases} x_n - a, & x_n \in \mathbb{Q} \cap [0,1] \\ -x_n - a, & x_n \in \mathbb{IR} \cap [0,1] \end{cases}$$

(1) If $(x_n) \subset \mathbb{Q}$ except for finitely many values in $\mathbb{IR}$, then $f(x_n) - a \to 0 - a$.

(2) This is nonzero unless $a = 0$. So these sequences will converge to 0. If $(x_n) \subset \mathbb{IR}$ except for finitely many values in $\mathbb{IR}$, then $f(x_n) - a \to 0 - a$ and this is not zero unless $a = 0$ also. Thus these sequences converge to 0.

(3) If (x_n) has infinitely many values in both $\mathbb{Q}$ and $\mathbb{IR}$, then

$$f(x_n) - a \; \to \; \begin{cases} -a, & x_n \in \mathbb{Q} \cap [0,1] \\ -a, & x_n \in \mathbb{IR} \cap [0,1] \end{cases}$$

and these two limits are not zero unless $a = 0$. Thus again, the only cluster point for f at 0 is $a = 0$. So $S(0) = \{0\}$ telling us $\underline{\lim}_{x \to 0} f(x) = 0$ and $\overline{\lim}_{x \to 0} f(x) = 0$.
Case: $p \neq 0$:
The argument is very similar to what we have done. We now have $x_n \to p \neq 0$. We have

$$f(x_n) - a \; = \; \begin{cases} x_n - a, & x_n \in \mathbb{Q} \cap [0,1] \\ -x_n - a, & x_n \in \mathbb{IR} \cap [0,1] \end{cases}$$

(1): If $(x_n) \subset \mathbb{Q}$, except for finitely many irrationals, then $f(x_n) - a \to p - a$. This is nonzero unless $a = p$. So these sequences will converge to p.

(2): If $(x_n) \subset \mathbb{IR}$, except for finitely many rationals, then $f(x_n) - a \to -p - a$ and this is not zero unless $a = -p$. Thus these sequences converge to $-p$.

(3): If (x_n) has infinitely many values in both $\mathbb{Q}$ and $\mathbb{IR}$, then since $x_n \to p$, for any $\epsilon > 0$, $\exists N$ so that $n > N \implies p - \epsilon < x_n < p + \epsilon$. Thus, for $n > N$,

$$f(x_n) - a \; \in \; \begin{cases} (p - a - \epsilon, p - a + \epsilon), & x_n \in \mathbb{Q} \cap [0,1] \\ (-p - a - \epsilon, -p - a + \epsilon), & x_n \in \mathbb{IR} \cap [0,1] \end{cases}$$

This says $f(x_n) - a$ has values in two sets with no common values as long as $p - a \neq -p - a$ or $p \neq 0$. Hence if (x_n) has infinitely many values in both $\mathbb{Q}$ and $\mathbb{IR}$, then the limit of $f(x_n)$ can not exist.

Thus the only sequence limits are those for sequences with infinitely many rational values and finitely many irrational values which will have limit p and the other case for sequences with infinitely many irrational values and finitely many rational values which will have limit $-p$.
We have shown

$$S(p) \; = \; \begin{cases} \{-p, p\}, & p \neq 0 \\ \{0\}, & p = 0 \end{cases}$$

and

$$\underline{\lim}_{x \to p} f(x) \; = \; \begin{cases} -p, & p \neq 0 \\ 0, & p = 0 \end{cases} \quad \text{and} \quad \overline{\lim}_{x \to p} f(x) = \begin{cases} p, & p \neq 0 \\ 0, & p = 0 \end{cases}$$

In this example, note there is only one point where the limit inferior and limit superior values match. This occurs at 0 and $\underline{\lim}_{x \to 0} f(x) = \overline{\lim}_{x \to 0} f(x) = 0$. We might conjecture that the limit of f as $x \to p$ exists IFF $\underline{\lim}_{x \to p} f(x) = \overline{\lim}_{x \to p} f(x)$ and this common value is the value of

$\lim_{x \to p} f(x)$. We could also conjecture that what we call continuity for f at the point p should mean $\underline{\lim}_{x \to p} f(x) = \overline{\lim}_{x \to p} f(x) = a$ and $f(p) = a$. That is, $\lim_{x \to p} f(x)$ exists and equals a and this matches the function value $f(p)$. In short, we would say, f is continuous at p if $\lim_{x \to p} f(x) = f(p)$. **So our function in Example 6.1.1 here is continuous at only one point.** So far no ϵ's in how we define things. That comes later.

Here is another example of this type of function, but we will do the work faster.

Example 6.1.2

$$f(x) \;\; = \;\; \begin{cases} 3x, & x \in \mathbb{Q} \cap [0,1] \\ -2x, & x \in \mathbb{R} \cap [0,1] \end{cases}$$

Find $S(p)$ for all p.

Solution *Look at the point p. Let (x_n) be any sequence such that $x_n \to p$. Let a be any number and consider*

$$f(x_n) - a \;\; = \;\; \begin{cases} 3x_n - a, & x_n \in \mathbb{Q} \cap [0,1] \\ -2x_n - a & x_n \in \mathbb{R} \cap [0,1] \end{cases}$$

If $(x_n) \in \mathbb{Q}$ for infinitely many n and in $\mathbb{R}$ for only finitely many n, $f(x_n) - a \to 3p - a$ which is nonzero unless $a = 3p$. Hence these type of sequences $f(x_n)$ converges to $3p$ and so $3p$ is a cluster point of f at any p.
If $(x_n) \in \mathbb{R}$ for infinitely many n and in $\mathbb{Q}$ for only finitely many n, $f(x_n) - a \to -2p - a$ which is nonzero unless $a = -2p$. Hence these type of sequences $f(x_n)$ converges to $-2p$ and so $-2p$ is a cluster point of f at any p.
If (x_n) has infinitely many values in both $\mathbb{Q}$ and $\mathbb{R}$, then

$$f(x_n) - a \;\; \to \;\; \begin{cases} 3p - a, & x_n \in \mathbb{Q} \cap [0,1] \\ -2p - a & x_n \in \mathbb{R} \cap [0,1] \end{cases}$$

and these two values are distinct unless $3p - a = -2p - a$ or $p = 0$. Hence,

$$S(p) \;\; = \;\; \begin{cases} \{-2p, 3p\}, & p \neq 0 \\ \{0\}, & p = 0 \end{cases}$$

and

$$\lim_{x \to p} f(x) \;\; = \;\; \begin{cases} -3p, & p \neq 0 \\ 0, & p = 0 \end{cases} \quad \text{and} \quad \overline{\lim}_{x \to p} f(x) = \begin{cases} 3p, & p \neq 0 \\ 0, & p = 0 \end{cases}$$

In this example, again note there is only one point where the limit inferior and limit superior values match. This occurs at 0 and $\underline{\lim}_{x \to 0} f(x) = \overline{\lim}_{x \to 0} f(x) = 0$.
*So our function here is continuous at only **one** point.*

Homework

Exercise 6.1.1 *Define f by*

$$f(x) \;\; = \;\; \begin{cases} 5x, & x \in \mathbb{Q} \cap [0,1] \\ -8x, & x \in \mathbb{R} \cap [0,1] = \mathbb{Q}^C \cap [0,1] \end{cases}$$

- *Find $S(p)$ for all p.*

- *Find $\underline{\lim}_{x\to p} f(x)$ and $\overline{\lim}_{x\to p} f(x)$ for all p.*

- *Where do you think this function is continuous?*

Exercise 6.1.2 *Define f by*

$$f(x) \;=\; \begin{cases} -2x, & x \in \mathbb{Q} \cap [0,1] \\ -7x, & x \in \mathbb{R} \cap [0,1] = \mathbb{Q}^C \cap [0,1] \end{cases}$$

- *Find $S(p)$ for all p.*

- *Find $\underline{\lim}_{x\to p} f(x)$ and $\overline{\lim}_{x\to p} f(x)$ for all p.*

- *Where do you think this function is continuous?*

Exercise 6.1.3 *Define f by*

$$f(x) \;=\; \begin{cases} 15x, & x \in \mathbb{Q} \cap [0,1] \\ -81x, & x \in \mathbb{R} \cap [0,1] = \mathbb{Q}^C \cap [0,1] \end{cases}$$

- *Find $S(p)$ for all p.*

- *Find $\underline{\lim}_{x\to p} f(x)$ and $\overline{\lim}_{x\to p} f(x)$ for all p.*

- *Where do you think this function is continuous?*

Exercise 6.1.4 *Define f by*

$$f(x) \;=\; \begin{cases} 2x, & x \in \mathbb{Q} \cap [0,1] \\ 3x, & x \in \mathbb{R} \cap [0,1] = \mathbb{Q}^C \cap [0,1] \end{cases}$$

- *Find $S(p)$ for all p.*

- *Find $\underline{\lim}_{x\to p} f(x)$ and $\overline{\lim}_{x\to p} f(x)$ for all p.*

- *Where do you think this function is continuous?*

6.1.3 A Function with All Cluster Point Sets Having Two Values

Here is an example of a function where the cluster point set is always two values.

Example 6.1.3

$$f(x) \;=\; \begin{cases} 1, & x \in \mathbb{Q} \\ -1, & x \in \mathbb{R} \end{cases}$$

Find $S(p)$ for all p.

Solution
Let a be a real number. Let (x_n) be a sequence with $x_n \to p$. We don't assume p is rational or irrational.

$$f(x_n) - a \;=\; \begin{cases} 1 - a, & x_n \in \mathbb{Q} \\ -1 - a, & x_n \in \mathbb{R} \end{cases}$$

Let $(x_n) \in \mathbb{Q}$ for infinitely many n and in $\mathbb{R}$ for only finitely many n, For this sequence, even though $x_n \to p$, it is still possible that p is irrational. For example, we can easily find a sequence of rational numbers that converge to $\sqrt{2}$. $f(x_n) - a \to 1 - a$ which is nonzero unless $a = 1$. Hence for these *type of sequences $f(x_n)$ converges to 1 and so 1 is a cluster point of f at any p.*

If $(x_n) \in \mathbb{IR}$ for infinitely many n and in $\mathbb{Q}$ for only finitely many n, $f(x_n) - a \to -1 - a$ which is nonzero unless $a = -1$. Hence these type of sequences $f(x_n)$ converges to -1 and so -1 is a cluster point of f at any p.

If (x_n) is a sequence with infinitely many rational and irrational values, then

$$f(x_n) - a \quad \to \quad \begin{cases} 1 - a, & x_n \in \mathbb{Q} \\ -1 - a, & x_n \in \mathbb{IR} \end{cases}$$

Since these two values are distinct for all values of a, we see for none of these sequences does $f(x_n)$ converge. Thus the only sequence limits are those for sequences with infinitely many rational values and finitely many irrational values which will have limit 1 and the other case for sequences with infinitely many irrational values and finitely many rational values which will have limit -1.
We have $S(p) = \{-1, 1\}$ and $\underline{\lim}_{x \to p} f(x) = -1$ and $\overline{\lim}_{x \to p} f(x) = 1$
Note the limit inferior and limit superior never match. So we suspect this is a function which is continuous nowhere.

Example 6.1.4

$$f(x) \quad = \quad \begin{cases} -14, & x \in \mathbb{Q} \\ 10, & x \in \mathbb{IR} \end{cases}$$

Find $S(p)$ for all p.

Solution *Let a be real number. Let (x_n) be a sequence with $x_n \to p$.*

$$f(x_n) - a \quad = \quad \begin{cases} -14 - a, & x_n \in \mathbb{Q} \\ 10 - a, & x_n \in \mathbb{IR} \end{cases}$$

If $(x_n) \in \mathbb{Q}$ for infinitely many n and in $\mathbb{IR}$ for only finitely many n, $f(x_n) - a \to -14 - a$ which is nonzero unless $a = -14$. Hence these type of sequences $f(x_n)$ converges to -14 and so -14 is a cluster point of f at any p.

If $(x_n) \in \mathbb{IR}$ for infinitely many n and in $\mathbb{Q}$ for only finitely many n, $f(x_n) - a \to 10 - a$ which is nonzero unless $a = 10$. Hence these type of sequences $f(x_n)$ converges to 10 and so 10 is a cluster point of f at any p.

If (x_n) is a sequence with infinitely many rational and irrational values, then

$$f(x_n) - a \quad \to \quad \begin{cases} -14 - a, & x_n \in \mathbb{Q} \\ 10 - a, & x_n \in \mathbb{IR} \end{cases}$$

Since these two values are distinct for all values of a, we see for none of these sequences does $f(x_n)$ converge. Thus, the only cluster points at any p are -14 and 10 and $\underline{\lim}_{x \to p} f(x) = -14$ and $\overline{\lim}_{x \to p} f(x) = 10$

We therefore suspect that limits of functions and continuity of functions are **pointwise** concepts, not **interval based** concepts. Of course, continuity for all points in an entire interval should have important consequences.

If the domain of f is **compact**, really powerful things happen!

6.1.4   Homework

Exercise 6.1.5

$$f(x) = \begin{cases} 2, & x \in \mathbb{Q} \\ -3, & x \in \mathbb{IR} \end{cases}$$

- *Find $S(p)$ for all p.*
- *Find $\underline{\lim}_{x \to p} f(x)$ and $\overline{\lim}_{x \to p} f(x)$ for all p.*
- *Where do you think this function is continuous?*

Exercise 6.1.6

$$f(x) = \begin{cases} 12, & x \in \mathbb{Q} \\ 5, & x \in \mathbb{IR} \end{cases}$$

- *Find $S(p)$ for all p.*
- *Find $\underline{\lim}_{x \to p} f(x)$ and $\overline{\lim}_{x \to p} f(x)$ for all p.*
- *Where do you think this function is continuous?*

Exercise 6.1.7

$$f(x) = \begin{cases} 22, & x \in \mathbb{Q} \\ -35, & x \in \mathbb{IR} \end{cases}$$

- *Find $S(p)$ for all p.*
- *Find $\underline{\lim}_{x \to p} f(x)$ and $\overline{\lim}_{x \to p} f(x)$ for all p.*
- *Where do you think this function is continuous?*

Exercise 6.1.8

$$f(x) = \begin{cases} 4, & x \in \mathbb{Q} \\ 5, & x \in \mathbb{IR} \end{cases}$$

- *Find $S(p)$ for all p.*
- *Find $\underline{\lim}_{x \to p} f(x)$ and $\overline{\lim}_{x \to p} f(x)$ for all p.*
- *Where do you think this function is continuous?*

Exercise 6.1.9

$$f(x) = \begin{cases} 22x, & x \in \mathbb{Q} \\ -35x, & x \in \mathbb{IR} \end{cases}$$

- *Find $S(p)$ for all p.*
- *Find $\underline{\lim}_{x \to p} f(x)$ and $\overline{\lim}_{x \to p} f(x)$ for all p.*
- *Where do you think this function is continuous?*

Exercise 6.1.10

$$f(x) = \begin{cases} 4x, & x \in \mathbb{Q} \\ 5x, & x \in \mathbb{IR} \end{cases}$$

- *Find $S(p)$ for all p.*

- *Find $\underline{\lim}_{x \to p} f(x)$ and $\overline{\lim}_{x \to p} f(x)$ for all p.*

- *Where do you think this function is continuous?*

6.1.5   Examples of Cluster Points of Functions

Let's find some cluster points!

Example 6.1.5 *Find the cluster points of*

$$f(x) \;\; = \;\; \sin(1/x), \quad x \neq 0$$

Solution *Note the domain of this function does not include 0! However, f is locally defined at 0, just not defined at 0 itself. Let $-1 \leq a \leq 1$. If $\theta_n = \sin^{-1}(a) + 2\pi n$ for $n \geq 1$, then $x_n = 1/\theta_n = 1/(\sin^{-1}(a) + 2\pi n)$ satisfies $x_n \to 0$ and $f(x_n) \to a$. Thus, here $S(0) = [-1, 1]$ and $\underline{\lim}_{x \to 0} f = -1$ and $\overline{\lim}_{x \to 0} f = +1$. Thus we would not expect the $\lim_{x \to 0} f$ to exist.*

Let $x \neq 0$. If $x_n \to x$, what happens to $f(x_n) = \sin(1/x_n)$? We don't really know how to do this yet, but no matter what x_n is you could look at $\sin(1/x_n) - \sin(1/x)$. In a bit, we will show the Mean Value Theorem (MVT) which says if f is differentiable on (a, b) and continuous on $[a, b]$, then $f(b) - f(a) = f'(c)\,(b - a)$ where c is some point in (a, b) and f' denotes the derivative of f at the point c.

So here we get $|\sin(y) - \sin(y_0)| = |\cos(c)|\,|y - y_0|$ for some c between y and y_0. Thus, using $y = 1/x_n$ and $y_0 = x$, we have $|\sin(1/x_n) - \sin(1/x)| = |\cos(c)|\,|1/x_n - 1/x|$ for some c between $1/x_n$ and $1/x$. Then, since $|\cos(c)| \leq 1$ always, we have given $|\sin(1/x_n) - \sin(1/x)| \leq |1/x_n - 1/x| = |x_n - x|/(|x_n\,x|)$. Since $x \neq 0$ and $x_n \to x$, from a previous result, we have there is an N so that if $n > N$, then $|x_n| > |x|/2$. Thus, $|\sin(1/x_n) - \sin(1/x)| \leq 2/(|x|^2)|x_n - x|$. Finally, since $x_n \to x$, this shows that $\sin(1/x_n) \to \sin(1/x)$.

This tells us the only cluster point at $x \neq 0$ is $\sin(1/x)$. So $S(x) = \{\sin(1/x)\}$, $\underline{\lim}_{x \to x} f = \overline{\lim}_{x \to x} f = \sin(1/x)$. This also suggests we should think of $\sin(1/x)$ as continuous at any point $x \neq 0$ even though it is not continuous at 0.

Example 6.1.6 *Find the cluster points of $f(x) = \tan(1/x)$ for $x \neq 0$.*

Solution *This time let $-\infty < a < \infty$. Then if $\theta_n = \tan^{-1}(a) + 2\pi n$ and $x_n = 1/\theta_n$, we have $\tan(1/x_n) = 1/(\tan^{-1}(a) + 2\pi n)$. We see $x_n \to 0$ and $f(x_n) \to a$. Thus $S(0) = (-\infty, \infty)$ here and $\underline{\lim}_{x \to 0} \tan(1/x) = -\infty$ and $\overline{\lim}_{x \to 0} \tan(1/x) = \infty$.*

Homework

Exercise 6.1.11 *Find the cluster points of*

$$f(x) \;\; = \;\; \sin(3/x), \quad x \neq 0$$

Exercise 6.1.12 *Find the cluster points of*

$$f(x) \;\; = \;\; \cos(2/x) + \sin(3/x), \quad x \neq 0$$

Exercise 6.1.13 *Find the cluster points of*

$$f(x) \;\; = \;\; \sec(5/x), \quad x \neq 0$$

Exercise 6.1.14 *Find the cluster points of*

$$f(x) = \sin(1/x) + x^2, \quad x \neq 0$$

Exercise 6.1.15 *Find the cluster points of*

$$f(x) = \sin(x + 1/x), \quad x \neq 0$$

6.1.6 C^∞ Bump Functions

We can construct really interesting examples of functions which are infinitely differentiable but which are nonzero on any chosen finite interval $[a, b]$ we want with all their derivatives zero at both a and b. We start by looking at the cluster points of

$$f(x) = \begin{cases} 0, & x \leq 0 \\ e^{-1/x}, & x > 0 \end{cases}$$

Let (x_n) be any sequence with $x_n \leq 0$ which converges to 0. Then $f(x_n) \to 0$ also. So 0 is in $S(0)$. Let (x_n) with $x_n > 0$ be any sequence with $x_n \to 0$. Then $f(x_n) = e^{-1/x_n}$. It is easy to see $-1/x_n$ with $x_n > 0$ always has an unbounded limit which we usually call $-\infty$. If you recall from calculus, $e^y \to 0$ as $y \to -\infty$. Thus, $f(x_n) \to 0$ here for these sequences.

If (x_n) was a sequence with infinitely many terms both positive and negative which still converged to 0, then the subsequence of positive terms, (x_n^1), converges to 0 and the subsequence of negative terms, (x_n^2), also converges to 0. Further, $f(x_n^1) \to 0$ and $f(x_n^2) \to 0$ as well. With a little work we can see $f(x_n) \to 0$ too.

We see the only cluster point at 0 is 0. Thus, $\underline{\lim}_{x \to 0} f = \overline{\lim}_{x \to 0} f = 0$. This function is going to be continuous at 0 and in fact, we can show it has infinitely many derivatives at each point x and if $f^{(n)}(x)$ denotes the nth derivative of f at x, we can show using a POMI argument that $f^{(n)}(0) = 0$ for all n.

This is a great example.

- f is so flat at 0 its Taylor Series expansion at 0 does not match f at all!

- Even though 0 is a global minimum of f occurring at 0 and other points, the second derivative test will fail there.

- And $f^{(n)}(x)$ is continuous for all n at 0!

Now shift this function over as follows:

$$f_a(x) = \begin{cases} 0, & x \leq a \\ e^{-1/(x-a)}, & x > a \end{cases}$$

A similar analysis shows f_a has a zero n^{th} order derivative at a for all n and has a continuous n^{th} order derivative at a for all n.

Next define a similar function from the other side:

$$g_b(x) = \begin{cases} 0, & x \geq b \\ e^{1/(x-b)}, & x < b \end{cases}$$

A similar analysis shows g_b has a zero n^{th} order derivative at b for all n and has a continuous n^{th} order derivative at b for all n.

Finally multiply these two types of functions together to create a very smooth bump.
We can then show for any $a < b$,

$$h_{a\,b}(x) = f_a(x)\, g_b(x) \;=\; \begin{cases} 0, & x \le a \\ e^{\frac{1}{x-b}}\, e^{-\frac{1}{x-a}}, & a < x < b \\ 0, & x \ge b \end{cases}$$

$$\;=\; \begin{cases} 0, & x \le a \\ e^{\frac{b-a}{(x-a)(x-b)}}, & a < x < b \\ 0, & x \ge b \end{cases}$$

A similar analysis shows h_{ab} has a zero n^{th} order derivative at a and b for all n which is continuous a and b for all n.

Now multiply h_{ab} by $e^{\frac{2}{a+b}}$ to get $H_{ab}(x) = e^{\frac{2}{a+b}}\, f_{ab}(x)$. $H_{ab}(x)$ is an infinitely differentiable function whose value is 0 off of the interval (a, b) and whose maximum value is $+1$. It is called a C^∞ **bump** function. The closure of the set of x values where $H_{ab}(x) > 0$ is then $[a, b]$ which is a compact set. This closure is called the **support** of $H_{ab}(x)$ and so $H_{ab}(x)$ is a C^∞ function with compact support. Very important!

Homework

Exercise 6.1.16 *Explain the analysis behind $h_{1,2}$ and sketch $h_{1,2}$ carefully. Where does the maximum of this function occur and what is its value?*

Exercise 6.1.17 *Explain the analysis behind $h_{-2,11}$ and sketch $h_{-2,11}$ carefully. Where does the maximum of this function occur and what is its value?*

Exercise 6.1.18 *Explain the analysis behind $f_{-0.5}$ and sketch $f_{-0.5}$ carefully.*

Exercise 6.1.19 *Explain the analysis behind $g_{-0.5}$ and sketch $g_{-0.5}$ carefully.*

Exercise 6.1.20 *Explain the analysis behind $f_{-0.05}$ and sketch $f_{-0.05}$ carefully.*

Exercise 6.1.21 *Explain the analysis behind $g_{0.05}$ and sketch $g_{0.05}$ carefully.*

6.2   Limits of Functions

It is now time to back up and decide how to define the limit of a function in terms of how f behaves at a point.

Definition 6.2.1 The Limit of a Function at a point

Let f be defined locally at p. It does not necessarily have to be defined at p itself. Thus, there is an $r > 0$ so that f is defined on $(p - r, p) \cup (p, p + r)$ which we call a punctured *or deleted circle about p and denote by $\hat{B}_r(p)$. We say $f(x)$ has the limit a as $x \to p$ if $\forall \epsilon > 0 \, \exists \, \delta > 0$ so that $0 < |x - p| < \delta \Rightarrow |f(x) - a| < \epsilon$.*

- This definition of function limit is quite similar to the definition of a sequence limit. The notion of being forced to look at x values close to p by finding a $\delta > 0$ is similar to the idea of finding a way to force the sequence value x_n to be close to the sequence limit. This is done by looking at $n > N$. So sequence limits are phrased in an $\epsilon - N$ way and function limits are phrased in an $\epsilon - \delta$ way, Again, note the similarities

$$\forall \epsilon > 0 \, \exists \delta > 0 \, \ni \, 0 < |x - p| < \delta \, \Rightarrow \, |f(x) - a| < \epsilon, \text{ function limit}$$
$$\forall \epsilon > 0 \, \exists N \, \ni \, n > N \, \Rightarrow \, |x_n - a| < \epsilon, \text{ sequence limit}$$

- Our notations for these limits are also similar. We say $\lim_{n \to \infty} x_n = a$ for a sequence limit and often just say $\lim_n x_n = a$ and we say $\lim_{x \to p} f(x) = a$ for the function limit.

- If $\lim_{x \to p} f(x) \neq a$, this means $\exists \epsilon_0 \, \ni \, \forall \delta > 0 \, \exists x_\delta \neq p \in (p - \delta, p + \delta)$ with $|f(x_\delta) - a| > \epsilon_0$ where x_δ must be in the circle $\hat{B}_r(p)$ where f is defined, of course.

We need to connect the ideas about limit with the ideas about cluster points of a function.

Theorem 6.2.1 The Function Limit Exists if and only if the Set of Cluster Points There has One Value Matching the Value of the Limit

Let A be a finite number.

$$\lim_{x \to p} f(x) = A \iff \left(\underline{\lim}_{x \to p} f(x) = \overline{\lim}_{x \to p} f(x) = A \right)$$

Proof 6.2.1

$(\Rightarrow)$:
We assume $\lim_{x \to p} f(x) = A$. Let a be any cluster point of f at p. Then there is a sequence (x_n) contained in the domain of f so that $x_n \to p$ and $f(x_n) \to a$. Let $\epsilon > 0$ be given. Then, for all n

$$|a - A| \;=\; |a - f(x_n) + f(x_n) - A| \le |a - f(x_n)| + |f(x_n) - A|.$$

Since $\lim_{n \to \infty} f(x_n) = a$, $\exists N_1 \ni n > N_1 \Rightarrow |a - f(x_n)| < \epsilon/2$.
Since $\lim_{x \to p} f(x) = A$, $\exists \delta > 0 \ni 0 < |x - p| < \delta \Rightarrow |f(x) - A| < \epsilon/2$. We also have $x_n \to p$, so $\exists N_2 \ni n > N_2 \Rightarrow |x_n - p| < \delta$.

So if we pick an $n > \max\{N_1, N_2\}$, we have $|x_n - p| < \delta$ which implies $|f(x_n) - A| < \epsilon/2$. Also, if $n > \max\{N_1, N_2\}$, $|a - f(x_n)| < \epsilon/2$.

Combining these two results, we have $|a - A| < \epsilon$. But this holds for all $\epsilon > 0$. This tells us $a = A$. Hence $S(p) = \{A\}$ and so $\underline{\lim}_{x \to p} f(x) = \overline{\lim}_{x \to p} f(x) = A$.
$(\Leftarrow)$:
Now we assume $\underline{\lim}_{x \to p} f(x) = \overline{\lim}_{x \to p} f(x) = A$ and show $\lim_{x \to p} f(x) = A$. This tells us $S(p) = \{A\}$. This then implies f is locally bounded near p as if not, we could find a sequence (x_n) converging to p with $|f(x_n)| \to \infty$. But such a sequence existing would say $\inf S(p) = -\infty$ or $\sup S(p) = \infty$ which is not possible as $S(p)$ is just the finite number A. So $|f(x)| \le B$ locally for some $B > 0$.

We will finish the proof by contradiction. Assume $\lim_{x \to p} f(x) \neq A$. Since f is locally defined, there is an $r > 0$ so that f is defined on $\hat{B}_r(p)$. Then there is some $\epsilon_0 > 0$ so that

$$\exists x_1 \neq p \in (p - r, p + r) \quad \ni \quad |f(x_1) - A| > \epsilon_0/2$$

$$\exists x_2 \neq p \in (p - r/2, p + r/2) \quad \ni \quad |f(x_2) - A| > \epsilon_0/2$$
$$\exists x_3 \neq p \in (p - r/3, p + r/3) \quad \ni \quad |f(x_3) - A| > \epsilon_0/2$$
$$\vdots$$
$$\exists x_n \neq p \in (p - r/n, p + r/n) \quad \ni \quad |f(x_n) - A| > \epsilon_0/2$$
$$\vdots$$

You can see $x_n \to p$. Now the sequence $(f(x_n))$ is a bounded set here, so by the Bolzano - Weierstrass Theorem, there is a subsequence $(f(x_{n_k}))$ that converges to some number a. This tells us a is a cluster point of f at p and hence $a = A$. But we also know by our construction, that $|f(x_{n_k}) - A| > \epsilon_0/2$ for all indices n_k and we also know $f(x_{n_k}) \to A$. This is not possible, so our assumption that $\lim_{x \to p} f(x) \neq A$ is wrong. We conclude $\lim_{x \to p} f(x) = A$, as desired. ∎

Homework

Exercise 6.2.1 *Let f be defined by*

$$f(x) = \begin{cases} 2x, & 0 \leq x < 1 \\ 5x, & 1 \geq x \leq 2 \end{cases}$$

- *Find $S(p)$ for all p.*

- *Find $\underline{\lim}_{x \to p} f(x)$ and $\overline{\lim}_{x \to p} f(x)$ for all p.*

- *Where do you think this function has a limit?*

- *Where do you think this function is continuous?*

Exercise 6.2.2 *Let f be defined by*

$$f(x) = \begin{cases} 2x + 1, & 0 \leq x < 1 \\ 5x - 3, & 1 \geq x \leq 2 \end{cases}$$

- *Find $S(p)$ for all p.*

- *Find $\underline{\lim}_{x \to p} f(x)$ and $\overline{\lim}_{x \to p} f(x)$ for all p.*

- *Where do you think this function has a limit?*

- *Where do you think this function is continuous?*

Exercise 6.2.3 *Let f be defined by*

$$f(x) = \begin{cases} 2x^2, & 0 \leq x < 1 \\ 5x^3, & 1 \geq x \leq 2 \end{cases}$$

- *Find $S(p)$ for all p.*

- *Find $\underline{\lim}_{x \to p} f(x)$ and $\overline{\lim}_{x \to p} f(x)$ for all p.*

- *Where do you think this function has a limit?*

- *Where do you think this function is continuous?*

Exercise 6.2.4 *Let f be defined by*

$$f(x) = \begin{cases} 2x - 4, & -2 \leq x \leq 5 \\ 5x + 5, & 5 < x \leq 12 \end{cases}$$

- *Find $S(p)$ for all p.*

- *Find $\underline{\lim}_{x \to p} f(x)$ and $\overline{\lim}_{x \to p} f(x)$ for all p.*

- *Where do you think this function has a limit?*

- *Where do you think this function is continuous?*

Exercise 6.2.5 *Let f be defined by*

$$f(x) = \begin{cases} 2x^2 - 4, & -2 \leq x \leq 5 \\ 5x^2 + 5, & 5 < x \leq 12 \end{cases}$$

- *Find $S(p)$ for all p.*

- *Find $\underline{\lim}_{x \to p} f(x)$ and $\overline{\lim}_{x \to p} f(x)$ for all p.*

- *Where do you think this function has a limit?*

- *Where do you think this function is continuous?*

Chapter 7

Continuity

7.1 Continuity

We are now ready to state formally what it means for a function to be continuous at a point.

Definition 7.1.1 The Continuity of a Function at a Point

> *Let f be defined locally at p and it now must be defined at p as well. Thus, there is an $r > 0$ so that f is defined on $(p - r, p + r)$. We say $f(x)$ is continuous at p if the limit a as $x \to p$ exists and equals a and $f(p) = a$ too. This is stated mathematically like this:*
> $\forall \epsilon > 0 \, \exists \, \delta > 0 \ni |x - p| < \delta \Rightarrow |f(x) - f(p)| < \epsilon.$
> *We often just say f is continuous at p means $\lim_{x \to p} f(x) = f(p)$.*

Comment 7.1.1 *So if $\underline{\lim}_{x \to p} f(x) \neq \overline{\lim}_{x \to p} f(x)$, we know $\lim_{x \to p}$ does not exist at p and that is enough to ensure f is not continuous at p.*

Comment 7.1.2 *If $\lim_{x \to p} f(x)$ exists and equals A, but f was not actually defined at p, this suggests we can make a new function $\hat{f}$ which is f off of p and is the new value A at p. An example is $f(x) = x \sin(1/x)$ which has limit 0 at 0 even though it is clearly not defined at 0. But*

$$\hat{f}(x) \quad = \quad \begin{cases} x \sin(1/x), & x \neq 0 \\ 0, & x = 0 \end{cases}$$

is a function which is continuous at 0. When this kind of behavior occurs, we say f has a **removeable discontinuity**.

- We have already seen examples of functions that are continuous nowhere because the limit inferior and superior never match and functions that are continuous at one and only one point because the limit inferior and superior only match at one point. It is also possible to have even stranger functions!

- Dirichlet's Function is defined like this: $f : [0, 1] \to \Re$ by

$$\hat{f}(x) \quad = \quad \begin{cases} 0, & x = 0 \\ \frac{1}{q}, & x = \frac{p}{q}, (p, q) = 1, p, q \in \mathbb{N} \\ 0, & x \in \mathbb{IR} \end{cases}$$

So $f(9/12) = 1/4$ as $9/12 = 3/4$ when common terms are removed. This is a very strange function and we can prove f is continuous at each $x \in \mathbb{IR} \cap [0, 1]$ and is not continuous at each

$x \in \mathbb{Q} \cap [0, 1]$. Hence, it is a function which fails to be continuous at a countable number of points, not just a finite number of points.

7.1.1   Homework

Exercise 7.1.1

$$f(x) \;=\; \begin{cases} 5x, & x \in \mathbb{Q} \cap [0, 1] \\ -7x, & x \in \mathbb{IR} \cap [0, 1] = \mathbb{Q}^C \cap [0, 1] \end{cases}$$

Explain why f is continuous only at 0 using our new theorems.

Exercise 7.1.2

$$f(x) \;=\; \begin{cases} 6, & x \in \mathbb{Q} \\ -11, & x \in \mathbb{IR} \end{cases}$$

Explain why f is continuous nowhere using our new theorems.

Exercise 7.1.3

$$f(x) \;=\; \begin{cases} 2, & x \le 4 \\ 5, & x > 4 \end{cases}$$

Find $\underline{\lim}_{x \to p} f(x)$ and $\overline{\lim}_{x \to p} f(x)$ and use that information to determine where f is continuous.

Exercise 7.1.4

$$f(x) \;=\; \begin{cases} 2x, & x \le 4 \\ 5 - 2x, & x > 4 \end{cases}$$

Find $\underline{\lim}_{x \to p} f(x)$ and $\overline{\lim}_{x \to p} f(x)$ and use that information to determine where f is continuous.

Exercise 7.1.5

$$f(x) \;=\; \begin{cases} 2x^3, & x \le 1 \\ 5 - 2x^2, & x > 1 \end{cases}$$

Find $\underline{\lim}_{x \to p} f(x)$ and $\overline{\lim}_{x \to p} f(x)$ and use that information to determine where f is continuous.

7.1.2   Dirichlet's Function: Lack of Continuity at Each Rational Number

Dirichlet's Function is defined like this: $f : [0, 1] \to \Re$ by

$$\hat{f}(x) \;=\; \begin{cases} 0, & x = 0 \\ \frac{1}{q}, & x = \frac{p}{q}, (p, q) = 1, p, q \in \mathbb{Q} \\ 0, & x \in \mathbb{IR} \end{cases}$$

We can prove f is continuous at each $x \in \mathbb{IR} \cap [0, 1]$ and is not continuous at each $x \in \mathbb{Q} \cap [01,]$.

Theorem 7.1.1 Dirichlet's Function is Continuous on the Irrationals Only

Dirichlet's function is defined on $[0, 1]$ and it is continuous at each irrational number in $[0, 1]$.

Proof 7.1.1

(Case $x_0 \in \mathbb{R}$):
We will show f is continuous at these x values. Choose $\epsilon > 0$ arbitrarily. Consider

$$|f(x) - f(x_0)| \quad = \quad \begin{cases} |0 - 0|, & x \in \mathbb{R} \cap [0,1] \\ |1/q - 0|, & x = p/q, \ (p,q) = 1 \end{cases}$$

Let

$$\begin{aligned} S_n &= \{x : 0 < x < 1, \ x = p/q, \ (p,q) = 1 \text{ with } 1/q \geq 1/n\} \\ &= \{0 < p/q < 1, \ (p,q) = 1, \ \text{with } n \geq q\} \end{aligned}$$

How many elements are in S_n? Consider the table below:

1/1	{1/2	1/3	...	...	...	1/n}
2/1	2/2	{2/3	2/4	...	...	2/n}
3/1	3/2	3/3	{3/4	...	...	3/n}
⋮	⋱					
j/1	j/2		j/j	{j/(j+1)...	...	j/n}
⋮						
(n-1)/n	(n-1)/2			(n-1)/(n-1)	{(n-1)/n}	
n/1	n/2			(n-1)/n	n/n	

The fractions in the braces in gray are the ones in $(0,1)$ although they are not necessarily in lowest terms. The first row has $n-1$ elements, the second $n-2$ and so one until we get to the $n-1$ row which just has one gray element. So the size of $S_n = |S_n| \leq \sum_{j=1}^{n-1} j = (n-1)n/2$.

The important thing is that S_n contains only a finite number of fractions in lowest terms. Now choose $N > 1/\epsilon$. Then $|S_N| \leq N(N-1)/2$. Label the elements of S_N as follows:

$$S_N \quad = \quad \{r_1, \ldots, r_p\}$$

where $p \leq N(N-1)/2$. Each fraction $r_i = p_i/q_i$ in lowest terms with $q_i \leq N$. One of these fractions is closest to x_0. Call this one $r_{min} = p_{min}/q_{min}$ and choose $\delta = (1/2)|r_{min} - x_0|$.

Note if $r = p/q, (p,q) = 1$ is any rational number in $(x_0 - \delta, x_0 + \delta)$, $r < r_{min}$ and so can't be in S_N!! (See the picture in the handwritten notes). Since $r \notin S_N$, the denominator of $r = p/q$ must satisfy $N < q$; i.e. $1/q < 1/N < \epsilon$. Combining, we see we have found a $\delta > 0$ so that

$$0 < |x - x_0| < \delta \quad \Longrightarrow \quad |f(x) - f(x_0)| = \begin{cases} 0 < \epsilon, & x \in \mathbb{R} \\ |1/q - 0| < 1/N < \epsilon, & x \in \mathbb{Q} \end{cases}$$

This shows f is continuous at any irrational number x_0 in $[0,1]$.
(Case $x_0 \in \mathbb{Q} \cap [0,1]$):
Then $x_0 = p_0/q_0, (p_0, q_0) = 1$. Consider

$$\begin{aligned} |f(x) - f(x_0)| &= \begin{cases} |0 - 1/q_0|, & x \in \mathbb{R} \cap [0,1] \\ |1/q - 1/q_0|, & x = p/q, \ (p,q) = 1 \end{cases} \\ &= \begin{cases} |0 - 1/q_0|, & x \in \mathbb{R} \cap [0,1] \\ |\frac{q_0 - q}{q \, q_0}|, & x = p/q, \ (p,q) = 1 \end{cases} \end{aligned}$$

Let $\epsilon_0 = (1/2)(1/q_0)$. We only have to look at the top part. The top says $|f(x) - f(x_0)| = 1/q_0 >$

$(1/2)(1/q_0) = \epsilon_0$. *Thus* $f(x) = 1/q_0 > \epsilon_0$ *for all* $x \in \mathbb{R}$. *Hence, no matter how small* $\delta > 0$ *we choose, we can always find irrational numbers* x_δ *in* $\hat{B}_\delta(x_0)$ *with* $|f(x_\delta) - f(x_0)| = 1/q_0 > \epsilon_2$. *This shows* f *can not be continuous at any* x_0 *in* $\mathbb{Q}$. ∎

This is a very interesting function! Again, these ideas of continuity and limit are **pointwise** concepts!!

Homework

Exercise 7.1.6 *Graph Dirichlet's function for* $x = p/1, p/2$ *and* $p/3$ *for all possibilities.*

Exercise 7.1.7 *Graph Dirichlet's function for* $x = p/1, p/2, p/3$ *and* $p/4$ *for all possibilities.*

Exercise 7.1.8 *Graph Dirichlet's function for* $x = p/1, p/2, p/3, p/4$ *and* $p/5$ *for all possibilities.*

Exercise 7.1.9 *Graph Dirichlet's function for* $x = p/1, p/2, p/3, p/4, p/5$ *and* $p/6$ *for all possibilities.*

Exercise 7.1.10 *Graph Dirichlet's function for* $x = p/1, p/2, p/3, p/4, p/5, p/6$ *and* $p/7$ *for all possibilities.*

7.2   Limit Examples

Let's do some examples.

Example 7.2.1 *Examine the limit behavior of the function*

$$f(x) \;=\; \begin{cases} x + 1, & x > 1 \\ 3, & x = 1 \\ -3x, & x < 1 \end{cases}$$

Solution *We can study this function's behavior a number of ways.*
(Case: $\epsilon - \delta$ *approach):*
Let's look at the point $x = 1$ *first.*

Maybe the limiting value here is 2.
We check

$$|f(x) - 2| \;=\; \begin{cases} |x + 1 - 2|, & x > 1 \\ |3 - 2|, & x = 1 \\ |-3x - 2|, & x < 1 \end{cases} = \begin{cases} |x - 1|, & x > 1 \\ |1|, & x = 1 \\ |-3(x - 1) - 5|, & x < 1 \end{cases}$$

where we are writing all the terms we have in terms of $x - 1$ *factors.*

Now by the backwards triangle inequality, $|3(x - 1) - 5| \geq 5 - 3|x - 1|$. *So*

$$\begin{cases} |f(x) - 2| = |x - 1|, & x > 1 \\ |f(x) - 2| = |1|, & x = 1 \\ |f(x) - 2| \geq 5 - 3|x - 1|, & x < 1 \end{cases}$$

If we make the top term small by restricting our attention to x *in* $(1 - \delta, 1) \cup (1, 1 + \delta)$, *then we have*

$$\begin{cases} |f(x) - 2| < \delta, & x \in (1, 1 + \delta) \\ |f(x) - 2| \geq 5 - 3\delta, & x \in (1 - \delta, 1) \end{cases}$$

where we have dropped what happens at $x = 1$ itself as it is not important for the existence of the limit. So no matter what ϵ we pick, for an $0 < \delta < \epsilon$, we have

$$\begin{cases} |f(x) - 2| < \epsilon, & x \in (1, 1 + \delta) \\ |f(x) - 2| \geq 5 - 3\epsilon, & x \in (1 - \delta, 1) \end{cases}$$

and although the top piece is small *the bottom piece is not! So the $\lim_{x \to 1} f(x)$ can not be 2.*

Maybe the limiting value here is -3.
We check

$$|f(x) - (-3)| = \begin{cases} |x + 1 + 3|, & x > 1 \\ |3 + 3|, & x = 6 \\ |-3x + 3|, & x < 1 \end{cases} = \begin{cases} |(x - 1) + 5|, & x > 1 \\ |6|, & x = 1 \\ |-3(x - 1)|, & x < 1 \end{cases}$$

where again we are writing all the terms we have in terms of $x - 1$ factors. Now by the backwards triangle inequality, $|(x - 1) - 5| \geq 5 - |x - 1|$. So

$$\begin{cases} |f(x) - (-3)| \geq 5 - |x - 1|, & x > 1 \\ |f(x) - (-3)| = |1|, & x = 1 \\ |f(x) - (-3)| = 3|x - 1|, & x < 1 \end{cases}$$

If we make the bottom term small by restricting our attention to x in $(1 - \delta, 1) \cup (1, 1 + \delta)$, then we have

$$\begin{cases} |f(x) - (-3)| \geq 5 - \delta, & x \in (1, 1 + \delta) \\ |f(x) - (-3)| \leq 3\delta, & x \in (1 - \delta, 1) \end{cases}$$

where we have dropped what happens at $x = 1$ itself as it is not important for the existence of the limit. So no matter what ϵ we pick, for any $0 < \delta < \epsilon/3$, we have

$$\begin{cases} |f(x) + 3| \geq 5 - \epsilon/3, & x \in (1, 1 + \delta) \\ |f(x) + 3| < \epsilon, & x \in (1 - \delta, 1) \end{cases}$$

and although the bottom piece is small *the top piece is not! So the $\lim_{x \to 1} f(x)$ can not be -3.*

Maybe the limiting value here is $a \neq 2, -3$.
We check

$$|f(x) - a| = \begin{cases} |x + 1 - a|, & x > 1 \\ |3 - a|, & x = 1 \\ |-3x - a|, & x < 1 \end{cases}$$

where we are writing all the terms we have in terms of $x - 1$ factors. In terms of $x - 1$ factors, this becomes

$$|f(x) - a| = \begin{cases} |x - 1 + 2 - a|, & x > 1 \\ |3 - a|, & x = 1 \\ |-3(x - 1) - (3 + a)|, & x < 1 \end{cases}$$

Now by the backwards triangle inequality, $|(x - 1) + (2 - a)| = |(x - 1) - (a - 2)| \geq |2 - a| - |x - 1|$ and $|-3(x - 1) - (3 + a)| \geq |3 + a| - 3|x - 1|$. Now the distance from a to 2 is $|a - 2|$ which we call d_1. The distance from a to -3 is $|a - (-3)| = |a + 3|$ which we call d_2. Using these estimates, *we find*

$$\begin{cases} |f(x) - a| \geq d_1 - |x - 1|, & x > 1 \\ |f(x) - a| = |3 - a|, & x = 1 \\ |f(x) - a| \geq d_2 - 3|x - 1|, & x < 1 \end{cases}$$

If we make the top term small by restricting our attention to x in $(1 - \delta, 1) \cup (1, 1 + \delta)$, then we have

$$\begin{cases} |f(x) - a| \geq d_1 - \delta, & x \in (1, 1 + \delta) \\ |f(x) - a| \geq d_2 - 3\delta, & x \in (1 - \delta, 1) \end{cases}$$

where yet again we have dropped what happens at $x = 1$ itself as it is not important for the existence of the limit. If we pick any positive $\epsilon < (1/2) \min\{d_1, d_2/3\}$ and set $\delta = \epsilon$, we have

$$\begin{cases} |f(x) - a| \geq d_1 - d_1/2 = d_1/2 > \epsilon, & x \in (1, 1 + \delta) \\ |f(x) - a| \geq d_2 - 3d_2/6 = d_2/2 > \epsilon, & x \in (1 - \delta, 1) \end{cases}$$

and both the top piece and the bottom piece are never small! So the $\lim_{x \to 1} f(x)$ can not be $a \neq 2, -3$ either.

Thus, $\lim_{x \to 1} f(x)$ does not exist and f can not be continuous at 1. *The method just done is tedious!*

Let's try using the $\underline{\lim}_{x \to 1} f(x)$ and $\overline{\lim}_{x \to 1} f(x)$ approach instead.

Any sequence (x_n) with $x_n \neq 1 \to 1$ from the left of 1 uses the bottom part of the definition of f. It is easy to see $-3x_n \to -3$ here so -3 is a cluster point of f at 1.

Any sequence (x_n) with $x_n \neq 1 \to 1$ from the right of 1 uses the top part of the definition of f. It is easy to see $x_n + 1 \to 2$ here so 2 is a cluster point of f at 1.

Any sequence (x_n) with $x_n \neq 1 \to 1$ containing an infinite number of points both to the left of 1 and to the right of 1, has a subsequence (x_n^1) converging to -3 and a subsequence (x_n^2) converging to 2 as

$$f(x_k) = \begin{cases} x_k^2 + 1, & x_k^2 \in (x_n), x_k^2 > 1 \\ -3x_k^1, & x_k^1 \in (x_n), x_k^1 < 1 \end{cases}$$

The top converges to 2 and the bottom converges to -3 and hence this type of subsequence (x_n) can not converge. We conclude $S(1) = \{-3, 2\}$ and so $\underline{\lim}_{x_n \to 1} f(x) = -3$ and $\overline{\lim}_{x_n \to 1} f(x) = 2$. Since these are not equal, we know $\lim_{x \to 1} f(x)$ does not exist. **Thus, f is not continuous at** 1.

It should be easy for you to see that the existence of the limit to the left of 1 and to the right of 1 is straightforward to establish. For example, at the point $x = 4$, $f(x) = x + 1$ and $f(4) = 5$. We have $|f(x) - 5| = |x + 1 - 5| = |x - 4|$. Given $\epsilon > 0$, if we choose $\delta = \epsilon$, we have $|x - 4| < \delta \Rightarrow |f(x) - 5| = |x - 4| < \epsilon$.
This shows f is continuous at $x = 4$. *Similar arguments work for all points $x \neq 1$. So this f is continuous at all x except 1.*

Homework

In the following exercises, use the careful analysis we have presented in this section.

Exercise 7.2.1 *Examine the limit behavior of the function*

$$f(x) \quad = \quad \begin{cases} 2x + 1, & x > 2 \\ 5, & x = 2 \\ -3x, & x < 2 \end{cases}$$

Exercise 7.2.2 *Examine the limit behavior of the function*

$$f(x) \quad = \quad \begin{cases} 5x + 2, & x > 2 \\ 6, & x = 2 \\ -2x + 1, & x < 2 \end{cases}$$

Exercise 7.2.3 *Examine the limit behavior of the function*

$$f(x) \quad = \quad \begin{cases} 3x, & x > -1 \\ 5, & x = 2 \\ -3x, & x < -1 \end{cases}$$

7.2.1 Right and Left Continuity at a Point

Another approach is to use right and left sided limits and continuity.

Definition 7.2.1 Right and Left Limits of a Function at a Point

Let f be locally defined near p. We say the **right hand limit** *of f as x approaches p exists and equals b if*

$$\forall \epsilon > 0 \exists \delta > 0 \ni p < x < p + \delta \Rightarrow |f(x) - b| < \epsilon$$

We denote the value b by the symbol $\lim_{x \to p^+} f(x) = b$. We say the **left hand limit** *of f as x approaches p exists and equals a if*

$$\forall \epsilon > 0 \exists \delta > 0 \ni p - \delta < x < p \Rightarrow |f(x) - a| < \epsilon$$

We denote the value a by the symbol $\lim_{x \to p^-} f(x) = a$.

We can then prove another way to characterize continuity at a point.

Theorem 7.2.1 The Limit of a Function Exists if and only if its Subsequential Limits Match the Limit Value

Let A be a real number. Then we have

$$\lim_{x \to p} f(x) = A \quad \Longleftrightarrow \quad \forall (x_n), x_n \neq p, x_n \to p, \lim_{n \to \infty} f(x_n) = A.$$

Proof 7.2.1

($\Rightarrow$): *We assume $\lim_{x \to p} f(x) = A$. Pick any $(x_n), x_n \neq p, x_n \to p$. Then for an arbitrary $\epsilon > 0$, we know there is a positive δ so that $x \in \hat{B}_\delta(p) \Rightarrow |f(x) - A| < \epsilon$. Since $x_n \to p$, there is a N so that $n > N \to |x_n - p| < \delta$. Combining, we see $n > N \Rightarrow |x_n - p| < \delta \Rightarrow |f(x_n) - A| < \epsilon$. Thus, $f(x_n) \to A$ too.*
($\Leftarrow$):
We assume $\forall (x_n), x_n \neq p, x_n \to p, \lim_{n \to \infty} f(x_n) = A$. Let's do this by contradiction. We

know f is locally defined in some $\hat{B}_r(p)$. Assume $\lim_{x \to p} f(x)$ does not equal A. Pick a sequence $\{1/N, 1/(N+1), \ldots\}$ so that $(p - 1/n, p + 1/n) \subset B_r(p)$ for all $n > N$.

Then we know there is a positive ϵ_0 so that for each $n > N$, there is an $x_n \neq p$ in $(p - 1/n, p + 1/n)$ with $|f(x_n) - A| > \epsilon_0$. This defines a sequence (x_n) which converges to p and each $x_n \neq p$. Thus, by assumption, we know $f(x_n) \to A$ which contradicts the construction we just did. So our assumption is wrong and $\lim_{x \to p} f(x) = A$. ∎

With this done, we can go to the next step.

Theorem 7.2.2 The Limit of a Function Exists with Value A if and only if Right and Left Hand Limits Match with Value A

Let A be a real number. Then

$$\lim_{x \to p} f(x) = A \quad \Longleftrightarrow \quad \left\{ \lim_{x \to p^+} f(x) = \lim_{x \to p^-} f(x) = A \right\}$$

Proof 7.2.2

($\Rightarrow$)
We assume $\lim_{x \to p} f(x) = A$. Let (x_n) be any sequence with $x_n \neq p$ that converges to p from below p. We then know by the previous theorem that $\lim_{n \to \infty} f(x_n) = A$ also. Since this is true for all such sequences, we apply the previous theorem again to the left hand limit of f at p to see $\lim_{x \to p^-} f(x) = A$. A similar argument shows $\lim_{x \to p^+} f(x) = A$.
($\Leftarrow$)
We assume $\lim_{x \to p^+} f(x) = \lim_{x \to p^-} f(x) = A$. We know f is locally defined in some $\hat{B}_r(p)$. Assume $\lim_{x \to p} f(x)$ does not equal A. Again pick a sequence $\{1/N, 1/(N+1), \ldots\}$ so that $(p - 1/n, p + 1/n) \subset B_r(p)$ for all $n > N$. Then we know there is a positive ϵ_0 so that for each $n > N$, there is an $x_n \neq p$ in $(p - 1/n, p + 1/n)$ with $|f(x_n) - A| > \epsilon_0$.

This defines a sequence (x_n) which converges to p and each $x_n \neq p$. This sequence contains infinitely many x_{n_k} to the left of p and/ or to the right of p. Hence, we can find a subsequence (x_{n_k}) with converges to p from either below or above. For this subsequence, we then know $f(x_{n_k}) \to A$ as both the right hand and left hand limits exist and equal A. But $|f(x_{n_k}) - A| > \epsilon_0$ for all k. This is not possible, so our assumption was wrong and $\lim_{x \to p} f(x) = A$. ∎

We can now finally define right and left hand continuity.

Definition 7.2.2 Right and Left hand Continuity at a Point

*Let f be locally defined at p. We say f is **continuous from the right** at p if $\lim_{x \to p^+} f(x) = f(p)$. We say f is **continuous from the left** at p if $\lim_{x \to p^-} f(x) = f(p)$.*

Example 7.2.2 *Let's look at our a new version of our old friend:*

$$f(x) \;=\; \begin{cases} x + 1, & x > 1 \\ 2, & x = 1 \\ -3x & x < 1 \end{cases}$$

Let's examine continuity at the point $x = 1$ again using these new tools.

Solution *It is easy to see* $\lim_{x \to 1^-} f(x) = -3 \neq f(1) = 2$ *and* $\lim_{x \to 1^+} f(x) = 2 = f(1)$. *So* f *is right continuous at* 1 *but* f *is not left continuous at* 1.

7.2.2 Homework

Do these exercises the easier way we presented here.

Exercise 7.2.4

$$f(x) \;=\; \begin{cases} 3x + 1, & x > 2 \\ 4, & x = 2 \\ 2x - 5 & x < 2 \end{cases}$$

Use the limit inferior and limit superior ideas to show f *is not continuous at* 2 *but it is continuous at any other* x.

Exercise 7.2.5

$$f(x) \;=\; \begin{cases} 3x + 1, & x > 2 \\ 4, & x = 2 \\ 2x - 5 & x < 2 \end{cases}$$

Use the limit inferior and limit superior ideas to show f *is not continuous at* 2 *but it is continuous at any other* x.

Exercise 7.2.6

$$f(x) \;=\; \begin{cases} 3x + 1, & x > 2 \\ 4, & x = 2 \\ 2x - 5 & x < 2 \end{cases}$$

Use the right and left limit ideas to show f *is not continuous at* 2 *but it is continuous at any other* x. *Determine if* f *is right or left continuous at* 2.

Exercise 7.2.7

$$f(x) \;=\; \begin{cases} 5x + 1, & x > 2 \\ 4, & x = 2 \\ 7x - 5 & x < 2 \end{cases}$$

Use the right and left limit ideas to show f *is not continuous at* 2 *but it is continuous at any other* x. *Determine if* f *is right or left continuous at* 2.

7.3 Limit and Continuity Proofs

Now let's do some proofs.

Example 7.3.1 *Prove* $\lim_{x \to 3} 4x^2 - 2x + 3 = 33$

Solution *Let* $\epsilon > 0$ *be arbitrary. Consider* $|f(x) - 33| = |4x^2 - 2x + 3 - 33|$. *Thus*

$$|f(x) - 33| \;=\; |4x^2 - 2x - 30| \;=\; |4x + 10|\,|x - 3|$$

Restrict our attention of the circle $\hat{B}_1(3)$. *Then* $2 < x < 4$ *implying* $18 < 4x + 10 < 26$. *We see*

$$|f(x) - 33| \;<\; 26\,|x - 3|$$

when $x \in \hat{B}_1(3)$. Hence if we choose $\delta < \min\{1, \epsilon/26\}$,

$$0 < |x - 3| < \delta \quad \Rightarrow \quad |f(x) - 33| \leq 26|x - 3| < 26\,(\epsilon/26) = \epsilon.$$

This shows $\lim_{x \to 3} 4x^2 - 2x + 3 = 33$.

Example 7.3.2 *Prove $f(x) = 3x^4$ is continuous at $x = 2$.*

Solution *Let $\epsilon > 0$ be arbitrary. Consider $|f(x) - f(2)| = |3x^4 - 48|$. Since 2 is a root of $3x^4 - 48$ we know $x - 2$ divides $3x^4 - 48$. This is an standard synthetic division.*

$$
\begin{array}{r}
3x^3 + 6x^2 + 12x + 24 \\
x - 2 \quad \overline{\smash{\big)}\, 3x^4 - 48} \\
\underline{3x^4 - 6x^3} \\
6x^3 - 48 \\
\underline{6x^3 - 12x^2} \\
12x^2 - 48 \\
\underline{12x^2 - 24x} \\
24x - 48 \\
\underline{24x - 48} \\
0
\end{array}
$$

So $|f(x) - f(2)| = |(x-2)||3x^3 + 6x^2 + 12x + 24|$. Restrict our attention of the circle $B_1(2)$. Then $1 < x < 3$ and $|3x^3 + 6x^2 + 12x + 24| \leq 3(27) + 6(9) + 12(3) + 24 = 195$ for $x \in B_1(2)$. Thus, so far we have $|f(x) - f(2)| \leq 195|x - 2|$. Hence if we choose $\delta < \epsilon/195$ we have

$$|x - 2| < \delta \quad \Rightarrow \quad |f(x) - f(2)| \leq 195|x - 2| < 195\,(\epsilon/195) = \epsilon.$$

This shows f is continuous at $x = 2$.

Homework

Exercise 7.3.1 *Prove $\lim_{x \to 3} 2x^2 + 3x + 7$ exists.*

Exercise 7.3.2 *Prove $2x^2 + 3x + 7$ is continuous at $x = 4$.*

Exercise 7.3.3 *Prove $\lim_{x \to 3} 4x^3 + 2x_10$ exists.*

Exercise 7.3.4 *Prove $4x^3 + 5x - 8$ is continuous at $x = 2$.*

Exercise 7.3.5 *Prove $\lim_{x \to 3} 4x^2 + 5x_9$ exists.*

Exercise 7.3.6 *Prove $x^2 - 7x + 13$ is continuous at $x = 1$.*

Theorem 7.3.1 Continuity at a Point Implies Local Bounds

Assume the $\lim_{x \to p} f(x)$ exists and equals a.

1. *There exists a $\delta > 0$ and a positive number B so that $|f(x)| \leq B$ if $0 < |x - p| < \delta$. In this case, we say f is **locally bounded**.*

2. *If $a \neq 0$, there exists a $\delta > 0$ so that $|f(x)| > |a|/2$ if $0 < |x - p| < \delta$. In this case, we say f is locally bounded away from 0.*

Proof 7.3.1

(1):
Since $\lim_{x \to p} f(x)$ *exists, we know* f *is locally defined near* p *in the circle* $\hat{B}_r(p)$ *for some* $r > 0$. *Let* $\epsilon = 1$. *Then there is a* $\delta > 0$ *with* $\delta < r$ *so that* $|f(x) - a| < 1$ *when* $0 < |x - p| < \delta$. *Thus, by the backwards triangle inequality,* $|f(x)| < 1 + |a|$ *when* $0 < |x - p| < \delta$. *This shows* $B = 1 + |a|$.
(2)
Let $\epsilon = |a|/2 > 0$ *by assumption. Then there is a* $\delta > 0$ *with* $\delta < r$ *so that* $|f(x) - a| < |a|/2$ *when* $0 < |x - p| < \delta$. *Thus, by the backwards triangle inequality,* $|a| - |f(x)| < |a|/2$ *when* $0 < |x - p| < \delta$. *Rearranging, we have* $|f(x)| > |a|/2$ *when* $0 < |x - p| < \delta$. ∎

Now let's use local boundedness.

Example 7.3.3 $\lim_{x \to 1} \frac{x+1}{x+2} = \frac{2}{3}$.

Solution *Let* $\epsilon > 0$ *be arbitrarily chosen. Consider*

$$|f(x) - f(1)| \quad = \quad \left| \frac{x+1}{x+2} - \frac{2}{3} \right| = \left| \frac{3(x+1) - 2(x+2)}{3(x+2)} \right|$$

So

$$|f(x) - f(1)| \quad = \quad \left| \frac{x-1}{3(x+2)} \right|$$

Let's restrict our attention to the circle $\hat{B}_1(1)$. *Thus* $0 < |x - 1| < 1$ *or* $x \in (0, 2)$ *with* $x \neq 1$. *Hence,* $x + 2 > 2$ *and so* $1/|x+2| = 1/(x+2) < 1/2$ *when* $0 < |x-1| < 1$. *Using this, we have*

$$|f(x) - f(1)| \quad = \quad \left| \frac{x-1}{3(x+2)} \right| < \left| \frac{x-1}{3(2)} \right| = (1/6)|x-1|$$

This is our **overestimate** *of* $|f(x) - f(1)|$. *Set this overestimate less than the arbitrary tolerance* $\epsilon > 0$ *and we have* $|x-1|/6 < \epsilon$ *which implies* $|x - 1| < 6\epsilon$. *Choose any* $\delta < \min\{1, 6\epsilon\}$ *and we have* $0 < |x - 1| < \delta$ *implies two things. First, since* $\delta < 1$, *our bound on* x *works and second,* $|x - 1| < 6\epsilon$ *which implies* $|x - 1|/6 < \epsilon$.
Thus,

$$|f(x) - f(1)| \quad < \quad (1/6)|x - 1| \quad < \quad \epsilon$$

which shows $\lim_{x \to 1} \frac{x+1}{x+2} = \frac{2}{3}$.

Example 7.3.4 $\lim_{x \to 2} \frac{x^2+1}{2x^3 + 3x + 4} = \frac{5}{26}$.

Solution *Let* $\epsilon > 0$ *be arbitrarily chosen. Consider*

$$|f(x) - f(2)| = \left| \frac{x^2+1}{2x^3+3x+4} - \frac{5}{26} \right| = \left| \frac{26(x^2+1) - 5(2x^3 + 3x + 4)}{26(2x^3+3x+4)} \right|$$

Thus,

$$|f(x) - f(2)| = \left| \frac{x^2+1}{2x^3+3x+4} - \frac{5}{26} \right| = \left| \frac{-10x^3 + 26x^2 - 15x + 6}{26(2x^3+3x+4)} \right|$$

Note $(-10x^3 + 26x^2 - 15x + 6)\big|_{x=2} = -80 + 104 - 30 + 6 = 0$ *so* $x - 2$ *is a factor. This has to be true if the limit as* $x \to 2$ *exists. So if it is not true, something has gone wrong in your arithmetic. Now do a synthetic division.*

$$\begin{array}{r} -10x^2 + 6x - 3 \\ x - 2 \overline{\smash{\big)}\ -10x^3 + 26x^2 - 15x + 6} \\ \underline{-10x^3 + 20x^2} \\ 6x^2 - 15x + 6 \\ \underline{6x^2 - 12x} \\ -3x + 6 \\ \underline{-3x + 6} \\ 0 \end{array}$$

So

$$|f(x) - f(2)| \quad = \quad \frac{|x - 2| \, |-10x^2 + 6x - 3|}{26 \, |2x^3 + 3x + 4|}$$

Now restrict our attention to $0 < |x - 2| < 1$ or $1 < x < 3$ with $x \neq 2$. In the numerator, we have $|-10x^2 + 6x - 3| \leq 10|x|^2 + 6|x| + 3 < 10(3^2) + 6(3) + 3 = 111$. In the denominator, for these x values, the absolute value is not needed and we have $|2x^3 + 3x + 4| = 2x^3 + 3x + 4 > 2 + 3 + 4 = 9$. This tells us $1/|2x^3 + 3x + 4| < 1/9$. We can now do our **overestimate**.

$$|f(x) - f(2)| \quad < \quad \frac{111 \, |x - 2|}{26(9)} = \frac{111}{234} \, |x - 2|$$

Now set the **overestimate** *$< \epsilon$. We want $(111/234)|x - 2| < \epsilon$ or $|x - 2| < (234/111)\epsilon$. Thus choose $\delta < \min(1, (234/111)\epsilon)$ and since $\delta < 1$, we have the bound on x works and since $\delta < (234/111)\epsilon$, we have*

$$|f(x) - f(2)| \quad < \quad \frac{111}{234} \, |x - 2| < \frac{111}{234} \frac{234}{111} \epsilon = \epsilon$$

Thus, $\lim_{x \to 2} \frac{x^2 + 1}{2x^3 + 3x + 4} = \frac{5}{26}$.

Homework

Exercise 7.3.7 *Prove $\lim_{x \to 1} \frac{x+2}{x+3}$ exists.*

Exercise 7.3.8 *Prove $\frac{x+2}{x+3}$ is continuous at $x = 1$.*

Exercise 7.3.9 *Prove $\lim_{x \to 2} \frac{5x+2}{7x+3}$ exists.*

Exercise 7.3.10 *Prove $\frac{5x+2}{7x+3}$ is continuous at $x = 2$.*

Exercise 7.3.11 *Prove $\lim_{x \to 2} \frac{5x^2+2}{7x+3}$ exists.*

Exercise 7.3.12 *Prove $\frac{5x^2+2}{7x+3}$ is continuous at $x = 2$.*

Exercise 7.3.13 *Prove $\lim_{x \to 3} \frac{5x^2+2}{7x^2+3x+4}$ exists.*

Exercise 7.3.14 *Prove $\frac{5x^2+2}{7x^2+3x+4}$ is continuous at $x = 3$.*

7.4   The Algebra of Limits

As you can see, the manipulations to show a limit exists can be intense! Let's see if we can find easier ways.

Theorem 7.4.1 The Algebra of Limits Theorem

Assume f and g are locally defined near p and $\lim_{x \to p} f(x) = a$ and $\lim_{x \to p} g(x) = b$.

   1. $\lim_{x \to p}(f(x) + g(x)) = a + b$.

   2. $\lim_{x \to p}(cf(x)) = ca$ *for any real number c.*

   3. $\lim_{x \to p}(f(x)g(x)) = ab$.

   4. *If $b \neq 0$, then* $\lim_{x \to p}(f(x)/g(x)) = a/b$

Proof 7.4.1

First, a note on how we handle the locally defined aspect of f and g. We know each is locally defined, so there is a common circle $\hat{B}_r(p)$ on which both are locally defined. In the arguments below, when we choose a $\delta > 0$ it is always with the understanding that we pick this $\delta < r$. It gets cumbersome to keep saying this, so just remember we are doing this in each of the proofs below where it is important. (1):
Let $\epsilon > 0$ be given. Then since $\lim_{x \to p} f(x) = a$, there is a $\delta_f > 0$ so that $0 < |x - p| < \delta_f$ implies $|f(x) - a| < \epsilon/2$, where the $\delta_f < r$.

Then since $\lim_{x \to p} g(x) = b$, there is a $\delta_g > 0$ so that $0 < |x - p| < \delta_g$ implies $|g(x) - b| < \epsilon/2$, where the $\delta_g < r$

Now let $\delta = \min\{\delta_f, \delta_g\}$. Then both conditions hold and we have $|(f(x) + g(x)) - (a + b)| \leq |f(x) - a| + |g(x) - b| < \epsilon/2 + \epsilon/2 = \epsilon$ when $0 < |x - p| < \delta$. Thus, $\lim_{x \to p}(f(x) + g(x)) = a + b$. (2):
We want to show $\lim_{x \to p}(cf(x)) = ca$ for all real numbers c. Choose $\epsilon > 0$ arbitrarily. Then there is a δ_f small enough so that $0 < |x - p| < \delta_f$ implies $|f(x) - a| < \epsilon/(1 + |c|)$ where we use $1 + |c|$ in case $c = 0$.

Then we have $0 < |x - p| < \delta_f$ implies $|cf(x) - ca| = |c|\,|f(x) - a| < (|c|)/(1 + |c|)\,\epsilon < \epsilon$. This shows $\lim_{x \to p}(cf(x)) = ca$.

   (3):
We want to show $\lim_{x \to p} f(x)g(x) = ab$. We will use the common **add and subtract** *trick here. Consider*

$$\begin{aligned} |f(x)g(x) - ab| &= |f(x)g(x) - ag(x) + ag(x) - ab| \\ &\leq |g(x)|\,|f(x) - a| + |a|\,|g(x) - b| \end{aligned}$$

Since $\lim_{x \to p} g(x) = b$, we know $g(x)$ is locally bounded. Hence, there is $\delta_g^1 > 0$ and a positive number D_g so that $|g(x)| \leq D_g$ if $x \in \hat{B}_{g\delta_g^1}(p)$. Thus, for arbitrary $\epsilon > 0$, if $0 < |x - p| < \delta_g^1$,

$$|f(x)g(x) - ab| \leq D_g\,|f(x) - a| + |a|\,|g(x) - b|$$

Since $\lim_{x \to p} f(x) = a$, $\exists \delta_f \ni x \in \hat{B}_{\delta_f}(p) \Rightarrow |f(x) - a| < \epsilon/(2D_g)$

Since $\lim_{x \to p} g(x) = b$, $\exists \delta_g^2 \ni x \in \hat{B}_{\delta_g^2}(p) \Rightarrow |g(x) - b| < \epsilon/(2(1 + |a|))$ where we use $1 + |a|$ in case $a = 0$.

So if $\delta = \min\{\delta_g^1, \delta_g^2, \delta_f\}$, $x \in \hat{B}_\delta(p)$ implies all three conditions hold and we have

$$|f(x)g(x) - ab| \quad < \quad D_g \frac{1}{2D_g}\epsilon + \frac{|a|}{2(1+|a|)}\epsilon = \epsilon$$

Thus, $\lim_{x \to p} f(x)g(x) = ab$.

(4):

We know assume $b \neq 0$ and show $\lim_{x \to p} f(x)/g(x) = a/b$. Since $\lim_{x \to p} g(x) = b \neq 0$, we know $g(x)$ is locally bounded below. Hence, there is $\delta_g^1 > 0$ so that so that $|g(x)| \geq |b|/2$ if $x \in \hat{B}_{g\,\delta_g^1}(p)$. Then for all $x \in \hat{B}_{g\,\delta_g^1}(p)$, $g(x)$ is not zero and the fractions below make sense:

$$\left|\frac{f(x)}{g(x)} - \frac{a}{b}\right| \quad = \quad \left|\frac{bf(x) - ag(x)}{b\,g(x)}\right| \leq \frac{2}{|b|^2}|bf(x) - ag(x)|$$

*Now do an **add and subtract** trick:*

$$\left|\frac{f(x)}{g(x)} - \frac{a}{b}\right| \quad \leq \quad \frac{2}{|b|^2}|bf(x) - ag(x)|$$

$$= \quad \frac{2}{|b|^2}|bf(x) - ab + ab - ag(x)|$$

$$\leq \quad \frac{2}{|b|^2}|b|\,|f(x) - a| + \frac{2}{|b|^2}|a|\,|b - g(x)|$$

*let $\epsilon > 0$ be arbitrary. We want the **overestimate** $\frac{2}{|b|^2}|b|\,|f(x) - a| < \epsilon/2$. This implies we want $|f(x) - a| < \frac{|b|}{4}\epsilon$. Since $f(x) \to a$, $\exists \delta_f \ni x \in \hat{B}_{\delta_f} \Rightarrow |f(x) - a| < \frac{|b|}{4}\epsilon$.*

*We want the **overestimate** $\frac{2|a|}{|b|^2}|g(x) - b| < \epsilon/2$ This implies we want $|g(x) - b| < \frac{|b|^2}{4(1+|a|)}\epsilon$ where again we use $1 + |a|$ in case $a = 0$. Since $g(x) \to b$, $\exists \delta_g^2 \ni x \in \hat{B}_{\delta_g^2} \Rightarrow |g(x) - b| < \frac{|b|^2}{4(1+|a|)}\epsilon$.*

So if $\delta = \min\{\delta_f, \delta_g^1, \delta_g^2\}$, all three conditions hold and we have

$$\left|\frac{f(x)}{g(x)} - \frac{a}{b}\right| \quad < \quad \frac{2}{|b|}\frac{|b|}{4}\epsilon + \frac{2|a|}{|b|^2}\frac{|b|^2}{4(1+|a|)}\epsilon = \epsilon.$$

Thus, $f(x)/g(x) \to a/b$. ∎

7.4.1   The Algebra of Continuity

The algebra of limits theorem can be applied right away to continuity.

Theorem 7.4.2 The Algebra of Continuity

Assume f and g are continuous at p

1. $f + g$ is also continuous at p.

2. cf is continuous at p for all real number c.

3. fg is continuous at p

4. If $g(p) \neq 0$, then f/g is continuous at p.

Proof 7.4.2
Just apply the algebra of limit theorem since we know $f(x) \to f(p)$ and $g(x) \to g(p)$ since f and g are continuous at p. Let $a = f(p)$ and $b = g(p)$. ∎

- When we say f is continuous on $[a, b]$ we mean f is continuous on (a, b), left continuous at b and right continuous at a.

- If we let $C([a, b]) = \{f : [a, b] \to \Re : f$ is continuous on $[a, b]\}$ $C([a, b])$ is a **vector space** over the real numbers as it is closed under the operation of $+$ and scalar multiplication. This is an example of a vector space which does not have a finite basis. In fact, there are many sequences of functions (f_n) which are linearly independent in $C([0, 1])$ and do not span $C([0, 1])$ such as the powers (t^n).

- Note our algebra of limits theorem for functions has almost the same proof as the algebra of sequences theorem.

7.4.2   Homework

Exercise 7.4.1 *If f and g are continuous at $x = 3$, prove $2f - 7g$ is continuous at $x = 3$ using an $\epsilon - \delta$ argument.*

Exercise 7.4.2 *If f and g are continuous at $x = 3$, prove $3f + 9g$ is continuous at $x = 3$ using an $\epsilon - \delta$ argument.*

Exercise 7.4.3 *If f and g are continuous at $x = 2$, prove $9fg$ is continuous at $x = 2$ using an $\epsilon - \delta$ argument.*

Exercise 7.4.4 *If f and g are continuous at $x = 2$ and $g(2)$ is not zero, prove $9f/g$ is continuous at $x = 2$ using an $\epsilon - \delta$ argument.*

Exercise 7.4.5 *Use POMI to prove x^n is continuous on $\Re$ for all positive integers n. You use the algebra of continuity theorem in the induction step.*

Exercise 7.4.6 *Use POMI to prove x^n is continuous on $\Re$ except 0 for all negative integers n. You use the algebra of continuity theorem in the induction step.*

Chapter 8

Consequences of Continuity on Intervals

There are many consequences when a function is continuous in intervals and other more extended domains.

8.1 Domains of Continuous Functions

In our discussions of continuity, we always assume that f is defined locally at the point p which means there is a radius r so that if $x \in B_r(p)$, $f(x)$ is defined. If at each point p in $dom(f)$, f is locally defined, this says each such p is an interior point of $dom(f)$ and so $dom(f)$ is an open set. Because of the requirement of f being locally defined at each p, we see $dom(f)$ can not have isolated boundary points.

Hence, any boundary points $dom(f)$ has must be accumulation points. This means if $p \in \partial(dom(f))$, there has to be a sequence (x_n) in $dom(f)$, each $x_n \neq p$, with $x_n \to p$. If for any sequence like this, we know $f(x_n) \to a$ for some value a, this implies $\lim_{x \to p} f(x)$ exists and equals a. Hence, we can define $f(p) = a$ and f therefore has a removeable discontinuity at such a p.

Now subsets of $\Re$ can be very complicated, so let's restrict our attention to **intervals**. An **interval** of $\Re$ is a subset I of the following form

$I = (-\infty, a)$	infinite open interval	$\partial I = \{-\infty, a\}$
$I = (-\infty, a]$	infinite closed interval	$\partial I = \{-\infty, a\}$
$I = (a, \infty)$	infinite open interval	$\partial I = \{a, \infty, \}$
$I = [a, \infty)$	infinite closed interval	$\partial I = \{a, \infty, \}$
$I = (-\infty, \infty)$	infinite open and closed interval	$\partial I = \{-\infty, \infty\}$
$I = (a, b),\ a < b$	finite open interval	$\partial I = \{a, b\}$
$I = [a, b),\ a < b$	finite half open interval	$\partial I = \{a, b\}$
$I = (a, b],\ a < b$	finite half open interval	$\partial I = \{a, b\}$
$I = [a, b],\ a < b$	finite closed interval	$\partial I = \{a, b\}$

Continuity at the points in ∂I is possible but not guaranteed. For example, $1/x$ is continuous on $(0, \infty)$ and the boundary point 0 can not be included. But $x \sin(1/x)$ on $(0, 1]$ can be extended to the boundary point 0 so that $x \sin(1/x)$ has a removeable discontinuity at 0. It is also possible to talk about boundary points at $\pm\infty$ but we won't do that here.

8.2   The Intermediate Value Theorem

Our first important theorem here is a bit difficult to prove.

Theorem 8.2.1 The Intermediate Value Theorem

> *Let $f : [a,b] \to \Re$ be continuous with $f(a) \neq f(b)$. Then if y is between $f(a)$ and $f(b)$, there is at least one c between a and b so that $f(c) = y$.*

Proof 8.2.1
For concreteness, let's assume $f(a) < f(b)$. We then have $f(a) < y < f(b)$ and we want to find a c so that $a < c < b$ and $f(c) = y$. Our argument here will be very abstract! So lot's of fun!

*Let $A = \{x \in [a,b] : f(x) \leq y\}$. Since $f(a) < y$ we know A is non empty. Further, since $x = b$ is the largest value allowed, we see A is bounded. So by the completeness axiom, $\sup(A)$ exists and is finite. **Let** $c = \sup(A)$. Since b is an upper bound of A, we see by the definition of supremum, that $c \leq b$. We now show $f(c) = y$.*
(Step 1: $f(c) \leq y$):
(Case (a)):
if $c \in A$, then by the definition of A, we must have $f(c) \leq y$.
(Case (b)):
Assume $c \notin A$. Then applying the Supremum Tolerance Lemma using the sequence of tolerances $\epsilon = 1/n$, we can construct a sequence (x_n) with $c-1/n < x_n \leq c$ and $x_n \in A$. Since we are in Case (b), none of these x_n can be c. Since f is continuous at c, we then must have $\lim_{n\to\infty} f(x_n) = f(c)$. But since $x_n \in A$, we also must have $f(x_n) \leq y$ for all n.

In general, if $a_n \to a$ and $b_n \to b$ with $a_n \leq b_n$, this implies $a \leq b$. To see this, note for any $\epsilon > 0$, there is an N_1 so that $n > N_1$ implies $a - \epsilon/2 < a_n < a + \epsilon/2$ and there is an N_2 so that $n > N_2$ implies $b - \epsilon/2 < b_n < b + \epsilon/2$. So for $n > \max\{N_1, N_2\}$, we have $a - \epsilon/2 < a_n \leq b_n < b + \epsilon/2$. But this tells us that $a - b < \epsilon$.
Now if $a > b$, we could choose $\epsilon = (a - b)/2$ and then we would have $a - b < (a - b)/2$ implying $(a - b)/2 < 0$ which contradicts our assumption. Hence, we must have $a \leq b$.

Applying this result here, $f(x_n) \leq y$ for all n thus implies $f(c) \leq y$. So in either Case (a) or Case (b). $f(c) \leq y$.
(Step 2: $f(c) = y$):
Assume $f(c) < y$. Since we know $y < f(b)$, this says $c < b$. Choose $\epsilon = (1/2)(y - f(c)) > 0$. Since f is continuous at c, for this ϵ, there is a $\delta > 0$ so that $B_\delta(c) \subset [a,b]$ and $|x - c| < \delta \Rightarrow |f(x) - f(c)| < \epsilon$. Rewriting, we have

$$c - \delta < x < c + \delta \quad \Rightarrow \quad f(c) - \epsilon < f(x) < f(c) + \epsilon$$

Since $B_\delta(c) \subset [a,b]$, we know $(c, c + \delta) \subset (c, b]$. Now pick a point z in $(c, c + \delta)$. Then, $f(z) < f(c) + \epsilon$.

But $\epsilon = (1/2)(y - f(c))$. Thus, $f(z) < (y + f(c))/2$. Now we assumed $f(c) < y$ so the average $(f(c) + y)/2 < y$ also. So we have $f(z) < y$. But this says $z \in A$ with $z > c$ which is the supremum of A.

This is not possible and so our assumption that $f(c) < y$ is wrong and we must have $f(c) = y$. ■

We can then specialize this result to intervals.

Theorem 8.2.2 The Intermediate Value Theorem for Intervals

Let $f : I \to \Re$ be continuous. Let a and b be any two points in I with $a \neq b$ and $f(a) \neq f(b)$. Then if y is between $f(a)$ and $f(b)$, there is at least one c between a and b so that $f(c) = y$.

Proof 8.2.2

Apply the Intermediate Value Theorem for $[a, b] \subset I$ and the result follows.   ■

We can do more. The consequences of continuity of the function on the entire interval are enormous!

Theorem 8.2.3 Continuous Images of Intervals are Intervals

Let I be an interval and assume f is continuous on I. Then $f(I)$ is also an interval.

Proof 8.2.3

Let u, v be in $f(I)$ with $u < v$. Then there are $s, t \in I$ so that $f(s) = u$ and $f(t) = v$. Choose any number y so that $u < y < v$.

(Case (a)):

If $s < t$, since f is continuous on $[s, t]$, by the Intermediate Value Theorem for finite closes intervals, there is a number c with $s < c < t$ with $f(c) = y$. Thus, $y \in f(I)$.

(Case (b))

If $t < s$, we apply the same sort of argument to again show $y \in f(I)$.

Combining these arguments, we see for any $u, v \in f(I)$ with $u < v$, all $u < y < v$ are also in $f(I)$. This shows $f(I)$ is an interval.   ■

Homework

Exercise 8.2.1 Let $f(x) = 2x^2$ on $[-1, 3]$. For $y = 3$ find the points where $f(c) = y$.

Exercise 8.2.2 Let $f(x) = -3x^2$ on $[-2, 3]$. For $y = 1$ find the points where $f(c) = y$.

Exercise 8.2.3 Let $f(x) = 2x^4 + 2x^2 + 5$ on $[-2, 4]$. For $y = 3$ find the points where $f(c) = y$.

Exercise 8.2.4 Let $f(x) = 2/(x^2 + 1)$ on $[-1, 3]$. For $y = 0.5$ find the points where $f(c) = y$.

Exercise 8.2.5 Let $f(x) = 5 \sin(3x)$ on $[-1, 3]$. For $y = 3$ find the points where $f(c) = y$.

Exercise 8.2.6 Let $f(x) = 3 \cos(4x)$ on $[-21, 30]$. For $y = 2$ find the points where $f(c) = y$.

8.3   Continuous Images of Compact Sets

What if the domain of a continuous function f is compact? What are the consequences?

Theorem 8.3.1 Continuous Images of Compact Sets are Compact

If $f : dom(f) \to \Re$ with $dom(f)$ a compact set, then $f(dom(f))$ is a compact set also.

Proof 8.3.1

For convenience, let $K = dom(f)$. Then $f(K) = \{f(x) : x \in K\}$. Let (y_n) be any sequence in

$f(K)$. Then there is a sequence $(x_n) \subset K$ with $y_n = f(x_n)$ for all n.

Since K is compact, there is a subsequence (x_n^1) of (x_n) so that $x_n^1 \to p$ for some $p \in K$. Since f is continuous at p, we then have $f(x_n^1) \to f(p)$. (Recall at a boundary point of K, we would interpret continuity at this point in terms of right or left continuity).

Hence (y_n) has a subsequence $(y_n^1) = (f(x_n^1))$ which converges to an element of $f(K)$. Thus, $f(K)$ is sequentially compact and topologically compact and closed and bounded. ∎

There are many consequences of this result!

- Let's recall what we know about minima and maxima of sets. First if a set Ω is nonempty and bounded, it has a finite infimum and supremum.

- It is a straightforward argument to show there are sequences (x_n^m) and (x_n^M) so that $x_n^m \to \inf(\Omega)$ and $x_n^M \to \sup(\Omega)$.

- If the $\inf(\Omega) \in \Omega$, then we say the **minimum** of the set Ω is achieved by at least the point $x_m = \inf(\Omega)$ and $x_n^m \to x_m$. This is also called the **absolute minimum** of Ω.

- If the $\sup(\Omega) \in \Omega$, then we say the **maximum** of the set Ω is achieved by at least the point $x_M = \sup(\Omega)$ and $x_n^M \to x_M$. This is also called the **absolute maximum** of Ω.

- So Ω compact implies Ω always has an absolute minimum and maximum value.

- If K is compact with f continuous, $f(K)$ is compact. Thus, $f(K)$ has an absolute minimum and maximum; i.e. there exist $\exists x_m, x_M \in K \ni f(x_m) \leq f(x)$ and $f(x_M) \geq f(x) \, \forall x \in K$. The points x_m and x_M need not be unique, of course.

Theorem 8.3.2 Continuous Functions on Compact Sets Have an Absolute Minimum and an Absolute Maximum

Let $f : [a, b] \to \Re$ be continuous. Then $f([a, b]) = [f(x_m), f(x_M)]$ where x_m, x_M are in $[a, b]$ and $f(x_m)$ and $f(x_M)$ are the absolute minimum and maximum of f on $[a, b]$, respectively.

Proof 8.3.2
We already know $f([a, b])$ is an interval and it is a compact set. and we have $f(x_m) \leq f(x) \leq f(x_M)$ for all $x \in [a, b]$. Thus $f([a, b]) = [f(x_m), f(x_M)]$. ∎

Homework

Exercise 8.3.1 *Let $f(x) = 2x^2$ on $[-1, 3]$. Find the range of f. Is it compact?*

Exercise 8.3.2 *Let $f(x) = -3x^2$ on $(-2, 3)$. Find the range of f. Is it compact?*

Exercise 8.3.3 *Let $f(x) = 2x^4 + 2x^2 + 5$ on $[-2, 4)$. Find the range of f. Is it compact?*

Exercise 8.3.4 *Let $f(x) = 2/(x^2 + 1)$ on $(-1, 3]$. Find the range of f. Is it compact?*

Exercise 8.3.5 *Let $f(x) = 5\sin(3x)$ on $(-1, 3)$. Find the range of f. Is it compact?*

Exercise 8.3.6 *Let $f(x) = 3\cos(4x)$ on $(-21, 30)$. Find the range of f. Is it compact?*

8.4 Continuity in Terms of Inverse Images of Open Sets

There are even more ways to look at continuity. We know that f is continuous at p if $\forall \epsilon > 0 \, \exists \delta > 0 \ni |x - p| < \delta \Rightarrow |f(x) - f(p)| < \epsilon$, where δ is small enough to fit inside the circle $B_r(p)$ on which f is locally defined. Now suppose V was an open set in the range of f. This means every point in V can be written as $y = f(x)$ for some point in the domain of f. Consider $f^{-1}(V)$, the inverse image of V under f. This is defined to be $f^{-1}(V) = \{x \in dom(f) : f(x) \in V\}$ Is $f^{-1}(V)$ open?

Let $y_0 = f(x_0) \in V$ for some $x_0 \in dom(f)$. Since V is open, y_0 is an interior point of V and so there is an $R > 0$ so that $B_R(y_0) \subset V$. Then since f is continuous at x_0, choose $\epsilon = R$ and we see there is a $\delta > 0$ so that if $|x - x_0| < \delta$, where we choose $\delta < r$, with $|f(x) - f(x_0)| < R$. This says $B_\delta(x_0) \subset f^{-1}(V)$ telling us x_0 is an interior point of $f^{-1}(V)$. Since x_0 is arbitrary, we see $f^{-1}(V)$ is open.

Thus, the inverse image $f^{-1}(V)$ of an open set V is open. We can use the idea of inverse images to rewrite continuity at a point p this way.

$$\forall \epsilon > 0 \, \exists \delta > 0 \ni x \in B_\delta(p) \quad \Rightarrow \quad f(x) \in B_\epsilon(f(p)) \text{ or}$$
$$\forall \epsilon > 0 \, \exists \delta > 0 \quad \ni \quad f(B_\delta(p)) \subset B_\epsilon(f(p)) \text{ or}$$
$$\forall \epsilon > 0 \, \exists \delta > 0 \quad \ni \quad B_\delta(p) \subset f^{-1}(B_\epsilon(f(p)))$$

We can state this as a nice Theorem.

Theorem 8.4.1 Inverse Images of Open Sets by Continuous Functions are Open

> *f is continuous on the set D if and only if $f^{-1}(V)$ is an open set whenever V is open.*

Proof 8.4.1
($\Rightarrow$):
We assume f is continuous on the set D. Let V in the range of f be an open set. By the arguments we have already shown you earlier, we know every point $x_0 \in f^{-1}V$ is an interior point and so $f^{-1}(V)$ is an open set.

($\Leftarrow$):
We assume $f^{-1}(V)$ is an open set when V is open. Since V is open, given y_0 in V, there is a radius R so that $B_R(y_0)$ is contained in V.

Choose any $\epsilon > 0$. If $\epsilon \geq R$, argue this way: The set $B_R(y_0)$ is open and so by assumption, $f^{-1}(B_R(y_0))$ is open. Thus there is an $x_0 \in f^{-1}(B_R(y_0))$ so that $y_0 = f(x_0)$. By assumption, x_0 must be an interior point of $f^{-1}(B_R(y_0))$. So there is a radius r so that $B_r(x_0)$ is contained in $f^{-1}(B_R(y_0))$. So $x \in B_r(x_0) \Rightarrow f(x) \in B_R(y_0)$. Said another way, $|x - x_0| < r \Rightarrow |f(x) - f(x_0)| < R \leq \epsilon$.

On the other hand, if $\epsilon < R$, we can argue almost the same way: instead of using the set $B_R(y_0)$, we just use the set $B_\epsilon(y_0)$ which is still open. The rest of the argument goes through as expected! ∎

Comment 8.4.1 *Note here we are* **explicitly** *talking about continuity on the full domain not just at one point! This suggests how to generalize the idea of continuity to very general situations. A* **topology** *is a collection of subsets T of a set of objects $\mathbb{X}$ which is closed under arbitrary unions, closed under finite intersections and which contains both the empty set and the whole set $\mathbb{X}$ also.*

Our open sets in $\Re$ as we have defined them are a collection like this. In this more general setting, we simply define the members of this set T to be what are called open sets even though they may not obey our usual definition of open. Now the set $\mathbb{X}$ with the topology T is called a **topological space** *and we denote it by the pair $(\mathbb{X}, T)$. If $(\mathbb{Y}, S)$ is another topological space, based on what we have seen, an obvious way to generalize continuity is to say $f : \Omega \subset \mathbb{X} \to \mathbb{Y}$ is continuous on Ω means $f^{-1}(V)$ is in T for all $V \in S$. Again, inverse images of open sets are open is a characterization of continuity. Note a set of objects can have many different topologies so continuity depends on the choice of topology really. For us we always use the topology of open sets in $\Re$ which is nice and comfortable! This is very abstract but it is generalizations like this that have let us solve some very hard problems!*

8.4.1 Homework

Exercise 8.4.1 *Let $f(x) = 14x + 25$. Prove $\lim_{x \to -1} f(x)$ exists using an $\epsilon - \delta$ argument. Now prove there is a ball about -1 which is mapped into a ball about $f(-1)$.*

Exercise 8.4.2 *Let $f(x) = (3x+2)/(4x^2+8)$. Prove $\lim_{x \to -2} f(x)$ exists using an $\epsilon - \delta$ argument. Now prove there is a ball about -2 which is mapped into a ball about $f(-2)$.*

Exercise 8.4.3 *Let $f(x) = 3x^2 - 5$. Prove $\lim_{x \to 2} f(x)$ exists using an $\epsilon - \delta$ argument. Now prove there is a ball about 2 which is mapped into a ball about $f(2)$.*

Exercise 8.4.4 *Let $f(x) = 4x + 25x^2$. Prove $\lim_{x \to 1} f(x)$ exists using an $\epsilon - \delta$ argument. Now prove there is a ball about 1 which is mapped into a ball about $f(1)$.*

Chapter 9

Lower Semicontinuous and Convex Functions

We are now going to expand our treatment of functions a bit and look at the class of **lower semicontinuous functions** and the class of **convex** functions. These functions gives us some new insights into how we can try to find extreme values of functions even when there is no compactness. The function $|x|$ clearly has an absolute minimum over $\Re$ of value 0 and its domain is not compact.

Note the function $f(x) = |x|$ does not have a derivative at 0 but the left hand derivative at 0 is -1 and the right hand derivative is 1. It turns out $|x|$ is a convex function and we can define an extension of the idea of derivative, the **subdifferential** ∂f which here would be $\partial f(0) = [-1, 1]$, a compact set! Individual values inside the subdifferential are called **subgradient** values. Note also as $|x| \to \infty$, $f(x) \to \infty$ too. Also note $0 \in \partial f(0)$ which is like the condition that extreme values may occur when the derivative is zero.

The function $f(x) = x^2$ also has an absolute minimum of value 0 where $f'(0) = 0$. It also satisfies $|x| \to \infty$ implies $f(x) \to \infty$, it is a convex (you probably know about concave up and concave down functions from your calculus class and concave up is what we mean right now by convex) function but its domain is not compact.

Let's study how much we can learn about the smallest and largest values of functions without traditional notions of continuity and compactness.

9.1   Lower Semicontinuous Functions

Let's start with a new result with continuous functions in the spirit of the two examples we just used.

Theorem 9.1.1 Extreme Values of a Continuous Function with Growth Conditions at Infinity

> *Let Ω be a nonempty unbounded closed set of real numbers and let $f : \Omega \to \Re$ be continuous. Assume $f(x) \to \infty$ when $|x| \to \infty$. Then $\inf(f(\Omega))$ is finite and there is a sequence $(x_n) \subset \Omega$ so that $x_n \to x_0 \in \Omega$ and $f(x_n) \to f(x_0) = \inf(f(\Omega))$.*

Proof 9.1.1
Let $\alpha = \inf(f(\Omega))$. First, there is a sequence $(y_n = f(x_n)) \subset \Omega$ that converges to α. If $(x_n) \subset \Omega$ satisfies $|x_n| \to \infty$, we would have $f(x_n) \to \infty$. Then, we would know $\inf(f(\Omega)) = \infty$ implying

$f(x) \geq \infty$ *for all x in Ω. But we assumed f is finite on Ω, so this is not possible. Hence, (x_n) must be bounded. Thus, by the Bolzano - Weierstrass Theorem, there is a subsequence (x_n^1) of (x_n) converging to x_0. This means x_0 is either an accumulation point in Ω or a boundary point of Ω. Since Ω is closed, Ω contains its boundary points. So we know $x_0 \in \Omega$. Since f is continuous at x_0, we must have $f(x_n^1) \to f(x_0)$ as $n^1 \to \infty$. Since $f(x_n) \to \alpha$ we also have $f(x_n^1) \to \alpha$. Thus, $f(x_0) = \alpha$ and we have shown f has an absolute minimum at the point x_0.* ∎

Note, the condition that $f(x) \to \infty$ when $|x| \to \infty$ allows us to bound the (x_n) sequence. This is how we get around the lack of compactness in the domain. We can relax the continuity assumption too. We can look at functions which are **lower semicontinuous**.

Definition 9.1.1 Lower Semicontinuous Functions

*Let $f : dom(f) \to \Re$ be finite. We say f is **lower semicontinuous** at p if $\underline{\lim} f(p) = f(p)$. By definition, this means for all sequences (x_n) with $x_n \to p$ and $f(x_n) \to a$, we have $\lim_{n \to \infty} f(x_n) \geq f(p)$.*

In Figure 9.1 we show two pairs of functions illustrating what the condition of lower semicontinuity means graphically.

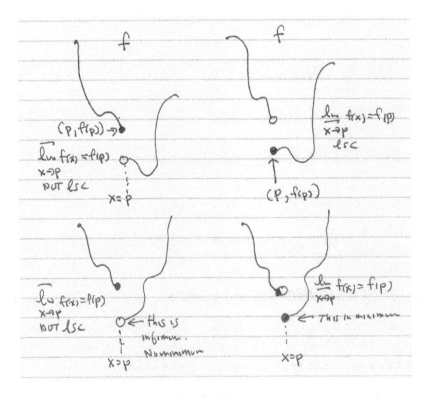

Figure 9.1: Examples of lower semicontinuity.

Let's relax our continuity condition into lower semicontinuity for the theorem we just proved.

Theorem 9.1.2 Extreme Values for Lower semicontinuous Functions with Growth Conditions at Infinity

> Let Ω be a nonempty unbounded closed set of real numbers and let $f : \Omega \to \Re$ be lower semicontinuous. Assume $f(x) \to \infty$ when $|x| \to \infty$. Then $\inf(f(\Omega))$ is finite and there is a sequence $(x_n) \subset \Omega$ so that $x_n \to x_0 \in \Omega$ and $f(x_n) \to f(x_0) = \inf(f(\Omega))$.

Proof 9.1.2

Again, let $\alpha = \inf(f(\Omega))$. There is a sequence $(y_n = f(x_n)) \subset \Omega$ that converges to α. The arguments we just used still show (x_n) must be bounded. Thus, by the Bolzano - Weierstrass Theorem, there is a subsequence (x_n^1) of (x_n) converging to x_0 and since Ω is closed, x_0 is in Ω. Since f is lower semicontinuous at x_0, we must have $\lim_{n^1 \to \infty} f(x_n^1) \geq f(x_0)$. Since $f(x_n) \to \alpha$ we also have $f(x_n^1) \to \alpha$. Thus, $\alpha \geq f(x_0)$. But $f(x_0) \geq \alpha$ by the definition of an infimum. Thus, $f(x_0) = \alpha = \inf(f(\Omega))$ and f has an absolute minimum at x_0. ∎

Homework

Exercise 9.1.1 *Let f be defined by*

$$\begin{cases} x^2, & -1 \leq x < 2 \\ -x^2 & 2 \leq x \leq 4 \end{cases}$$

- *Sketch f.*

- *Where is f lower semicontinuous?*

Exercise 9.1.2 *Let f be defined by*

$$\begin{cases} x^2, & -1 \leq x \leq 2 \\ -x^2 & 2 < x \leq 4 \end{cases}$$

- *Sketch f.*

- *Where is f lower semicontinuous?*

Exercise 9.1.3 *Let f be defined by*

$$\begin{cases} x^2, & -1 \leq x < 2 \\ 5, x = 2 \\ -x^2 & 2 < x \leq 4 \end{cases}$$

- *Sketch f.*

- *Where is f lower semicontinuous?*

Exercise 9.1.4 *Let f be defined by*

$$\begin{cases} x^2, & -1 \leq x < 2 \\ a, x = 2 \\ -x^2 & 2 < x \leq 4 \end{cases}$$

- *Sketch f.*

- *Is there a value of a that will make f lower semicontinuous at $x = 2$?*

Exercise 9.1.5 *Let f be defined by*

$$\begin{cases} 4x^2, & -1 \leq x < 3 \\ -2x & 3 \leq x \leq 5 \end{cases}$$

- *Sketch f.*

- *Where is f lower semicontinuous?*

Exercise 9.1.6 *Let f be defined by*

$$\begin{cases} 4x^2, & -1 \leq x \leq 3 \\ -2x & 3 < x \leq 5 \end{cases}$$

- *Sketch f.*

- *Where is f lower semicontinuous?*

Exercise 9.1.7 *Let f be defined by*

$$\begin{cases} 4x^2, & -1 \leq x < 3 \\ 5, x = 2 & \\ -2x & 3 < x \leq 5 \end{cases}$$

- *Sketch f.*

- *Where is f lower semicontinuous?*

Exercise 9.1.8 *Let f be defined by*

$$\begin{cases} 4x^2, & -1 \leq x < 3 \\ a, x = 2 & \\ -2x & 3 < x \leq 5 \end{cases}$$

- *Sketch f.*

- *Is there a value of a that will make f lower semicontinuous at $x = 2$?*

9.2   Convex Functions

We are now ready to define what we mean by convexity precisely. First, the **graph** of a function $f : dom(f) \to \Re$ is a subset of $dom(f) \times \Re$.

$$gr(f) \quad = \quad \{(x, a) : x \in dom(f) \text{ and } f(x) = a\}$$

and the **epigraph** of f is everything lying above the graph.

$$epi(g) \quad = \quad \{(x, b) : x \in dom(f) \text{ and } f(x) \leq b\}$$

We can now define what we mean by a convex function.

Definition 9.2.1 Convex Functions

---

*Let I be a nonempty interval of $\Re$. A function $f : I \to \Re$ is **convex** on I if*

$$f(tx + (1-t)y) \quad \leq \quad tf(x) + (1-t)f(y), \ \ \forall x, y \in I \text{ and } \forall t \in (0, 1)$$

*f is **strictly convex** if*

$$f(tx + (1-t)y) \quad < \quad tf(x) + (1-t)f(y), \ \ \forall x, y \in I \text{ and } \forall t \in (0, 1)$$

Let P_x be the point $(x, f(x))$ and P_y be the point $(y, f(y))$. The f is convex **means** the point $(u, f(u))$ on the graph of f, $gr(f)$, lies below the line segment $P_x P_y$ joining P_x and P_y. This is shown in Figure 9.2.

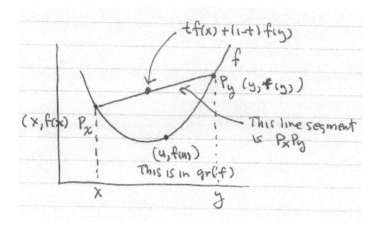

Figure 9.2: Convex functions.

Note the epigraph of f, $epi(f)$, is the **inside** of the curve in this sketch. This leads to a new definition of the **convexity** of f.

Definition 9.2.2 A Function is Convex if and only if its Epigraph is Convex

> *Let I be a nonempty interval of $\Re$. A function $f : I \to \Re$ is **convex** on I if and only if $epi(f)$ is a convex subset of $\Re^2$, where we recall a **convex subset** of $\Re^2$ is a set $\mathbb{C}$ so that given any two points P and Q in $\mathbb{C}$, the line segment PQ is contained in $\mathbb{C}$.*

Homework

Exercise 9.2.1 *Let f be defined by*

$$\begin{cases} x^2, & -1 \le x < 2 \\ -x^2 & 2 \le x \le 4 \end{cases}$$

- *Sketch f.*

- *Sketch the epigraph of f.*

- *Is f convex?*

Exercise 9.2.2 *Let f be defined by*

$$\begin{cases} x^2, & -1 \le x \le 2 \\ -x^2 & 2 < x \le 4 \end{cases}$$

- *Sketch f.*

- *Sketch the epigraph of f.*

- *Is f convex?*

Exercise 9.2.3 *Let f be defined by*

$$\begin{cases} x^2, & -1 \le x < 2 \\ 3x^2 - 8 & 2 < x \le 4 \end{cases}$$

- *Sketch f.*

- *Sketch the epigraph of f.*

- *Is f convex?*

Exercise 9.2.4 *Let f be defined by*

$$\begin{cases} 2x^2, & -1 \le x < 2 \\ a, x = 2 \\ 5x + 3 & 2 < x \le 4 \end{cases}$$

- *Sketch f.*

- *Sketch the epigraph of f.*

- *Is f convex?*

Exercise 9.2.5 *Let $f(x) = |x + 3|$.*

- *Sketch f.*

- *Sketch the epigraph of f.*

- *Is f convex?*

Exercise 9.2.6 *Let $f(x) = |x + 1| + 5|x - 2|$.*

- *Sketch f.*

- *Sketch the epigraph of f.*

- *Is f convex?*

Exercise 9.2.7 *Let $f(x) = |x + 1| + 5|x - 2| + 9|2x + 9|$.*

- *Sketch f.*

- *Sketch the epigraph of f.*

- *Is f convex?*

We can now prove an important lemma which is about how three points in the plane can appear.

Lemma 9.2.1 The Three Points Lemma

Let $P_x = (x, y)$, $P_{x'} = (x', y')$ and $P_u = (u, v)$ where $x < u < x'$ be three points in $\Re^2$. Then the following three properties are equivalent.

1. *P_u is below $P_x P_{x'}$.*

2. *The slope of $P_x P_u \le$ the slope of $P_x P_{x'}$.*

3. *The slope of $P_x P_{x'} \le$ the slope of $P_u P_{x'}$.*

Proof 9.2.1

The line segment $P_x P_{x'}$ in point slope form is $z = y + \frac{y'-y}{x'-x}(t-x)$ for $x \le t \le x'$. So at $t = u$, we
have $z_u = y + \frac{y'-y}{x'-x}(u-x)$.
((1) $\Longleftrightarrow$ (2)):
If Property (1) holds, we have $v \le y + \frac{y'-y}{x'-x}(u-x)$. This implies

$$\text{slope } P_x P_u = \frac{v-y}{u-x} \le \frac{y'-y}{x'-x} = \text{slope } P_x P_{x'}$$

which is Property (2). It is easy to see we can reverse this argument to show if Property (2) holds,
then so does Property (1). So we have shown Property (1) holds if and only if Property (2) holds.
((1) $\Longleftrightarrow$ (3)):
Next, if Property (1) holds, since P_u is below the line segment, the slope of the line segment $P_u P_{x'}$
is steeper than or equal to the slope of the line segment $P_x P_{x'}$. Look at the earlier picture which
clearly shows this although the role of y should be replaced by x' for our argument. Thus, the slope
of $P_x P_{x'} \le$ the slope of $P_u P_{x'}$ and so Property (3) holds.
The same picture shows us we can reverse the argument to show if the slope of $P_x P_{x'} \le$ the slope of
$P_u P_{x'}$ then P_u must be below the line segment $P_x P_{x'}$. So Property (3) implies Property (1).
((2) $\Longleftrightarrow$ (3)):
Hence, Property (2) implies Property (1) which implies Property (3). And it is easy to reverse the
argument to show Property (3) implies Property (2). So all of these statements are equivalent. ∎

Now if $x < u < x'$, then $u = tx + (1-t)x'$ for some $0 < t < 1$ and assume we have a convex
function f. By the convexity of f, we have $f(u) \le tf(x) + (1-t)f(x')$. Letting $P_x = (x, f(x))$
and $P_{x'} = (x', f(x'))$, convexity implies $P_u = (u, f(u))$ is below the line segment $P_x P_{x'}$ and by
Lemma 9.2.1,

$$\text{slope } P_x P_u \quad \le \quad \text{slope } P_x P_{x'} \le \text{slope } P_u P_{x'}$$

or

$$\frac{f(u) - f(x)}{u - x} \quad \le \quad \frac{f(x') - f(x)}{x' - x} \le \frac{f(x') - f(u)}{x' - u}$$

Since $u = tx + (1-t)x' = t(x - x') + x'$, we see $t = \frac{u-x'}{x-x'} = \frac{x'-u}{x'-x}$. Thus, convexity can be written

$$f(u) \quad \le \quad \left(\frac{x'-u}{x'-x}\right) f(x) + \left(\frac{u-x}{x'-x}\right) f(x')$$

For any $y \ne x_0$, let $S(y, x_0)$ denote the slope term $S(y, x_0) = \frac{f(y)-f(x_0)}{y-x_0}$. Recall, we showed

$$\frac{f(u) - f(x)}{u - x} \quad \le \quad \frac{f(x') - f(x)}{x' - x} \le \frac{f(x') - f(u)}{x' - u}$$

for any $x < u < x'$. Switching to $u = x_0$, we have

$$S(x, x_0) = \frac{f(x) - f(x_0)}{x - x_0} \quad \le \quad \frac{f(x')-f(x_0)}{x'-x_0} = S(x', x_0)$$

And this is true for all $x < x'$.
So f is convex on I implies the slope function $S(x, x_0)$ is increasing on the set $I \setminus \{x_0\}$ which is
all of I except the point x_0. This is called the **criterion of increasing slopes**. It is a straightforward
argument to see we can show the reverse: if the criterion of increasing slopes hold, then f is convex.

We have a Theorem!

Theorem 9.2.2 The Criterion of Increasing Slopes

Let I be a nonempty interval in $\Re$. Then $f : I \to \Re$ is **convex** $\Leftrightarrow$ *the slope function $S(x, x_0)$ is increasing on $I \setminus \{x_0\}$ for all $x_0 \in I$.*

Proof 9.2.2

We have just gone over this argument. ∎

Now let's look at what we can do with this information. We will define what might be considered a lower and upper form of a derivative next.

Theorem 9.2.3 Left Hand and Right Hand Secant Slopes Have Finite Supremum and Infimum

Let f be convex on the interval I and let x_0 be an interior point. Then $\sup_{x<x_0} \frac{f(x)-f(x_0)}{x-x_0}$ and $\inf_{x>x_0} \frac{f(x)-f(x_0)}{x-x_0}$ are both finite and

$$\sup_{x<x_0} \frac{f(x) - f(x_0)}{x - x_0} \leq \inf_{x>x_0} \frac{f(x) - f(x_0)}{x - x_0}.$$

Proof 9.2.3

By the criterion of increasing slopes, $\frac{f(x)-f(x_0)}{x-x_0}$ is increasing as x approaches x_0 from below. If $\sup_{x<x_0} \frac{f(x)-f(x_0)}{x-x_0} = \infty$, then there would be a sequence (x_n) in the interior of I, with $x_n \neq x_0 < x$ and $\frac{f(x_n)-f(x_0)}{x_n-x_0} > n$. Since x_0 is an interior point, choose a $y > x_0$ in I and apply the criterion of increasing slopes to see $n < \frac{f(x_n)-f(x_0)}{x_n-x_0} \leq \frac{f(y)-f(x_0)}{y-x_0}$.

But this tells us $f(y) \geq n(y - x_0) + f(x_0)$ implying $f(y)$ must be infinite in value. But f is finite on I so this is not possible. Thus, there is a constant $L > 0$ so that $\sup_{x<x_0} \frac{f(x)-f(x_0)}{x-x_0} = L$.

We also know that by the criterion of increasing slopes that $\frac{f(x)-f(x_0)}{x-x_0} \leq \frac{f(y)-f(x_0)}{y-x_0}$ for all $x < x_0 < y$ so we also have $\sup_{x<x_0} \frac{f(x)-f(x_0)}{x-x_0} \leq \frac{f(y)-f(x_0)}{y-x_0}$ for all $y > x_0$ too.

Thus, $\sup_{x<x_0} \frac{f(x)-f(x_0)}{x-x_0} \leq \inf_{y>x_0} \frac{f(y)-f(x_0)}{y-x_0}$

We can do an argument similar to the one above to also show $\inf_{y>x_0} \frac{f(y)-f(x_0)}{y-x_0}$ is finite and so there is a positive constant K so that $\inf_{y>x_0} \frac{f(y)-f(x_0)}{y-x_0} = K$. So we have shown

$$L = \sup_{x<x_0} \frac{f(x) - f(x_0)}{x - x_0} \leq \inf_{y>x_0} \frac{f(y) - f(x_0)}{y - x_0} = K.$$

∎

From Theorem 9.2.3, these finite limits allows us to replace the idea of a derivative at an interior point x_0 for a convex function f. We set the **lower derivative**, $D_- f(x_0)$ and **upper derivative** of f, $D_+ f(x_0)$ at an interior point to be the finite numbers

$$D_- f(x_0) = \sup_{x<x_0} \frac{f(x) - f(x_0)}{x - x_0}, \quad D_+ f(x_0) = \inf_{x>x_0} \frac{f(x) - f(x_0)}{x - x_0}$$

Next, if $[c, d] \subset int(I)$, choose $c < x < x' < d$. We can show $D_+ f(c) \leq D_- f(d)$ using a very long chain of inequalities. Be patient!

$$
\begin{aligned}
D_+ f(c) &= \inf_{\hat{x}>c} \frac{f(\hat{x}) - f(c)}{\hat{x} - c} \leq \frac{f(x) - f(c)}{x - c} = \frac{f(c) - f(x)}{c - x}, \text{ increasing slopes} \\
&\leq \sup_{w<x} \frac{f(w) - f(x)}{w - x} = D_- f(x) \leq D_+ f(x) \\
&= \inf_{\hat{x}>x} \frac{f(\hat{x}) - f(x)}{\hat{x} - x} \leq \frac{f(x') - f(x)}{x' - x} = \frac{f(x) - f(x')}{x - x'}.
\end{aligned}
$$

So continuing we have

$$
\begin{aligned}
&\leq \sup_{w<x'} \frac{f(w) - f(x')}{w - x'} = D_- f(x') \leq D_+ f(x') \\
&= \inf_{w>x'} \frac{f(w) - f(x')}{w - x'} \leq \frac{f(d) - f(x')}{d - x'} = \frac{f(x') - f(d)}{x' - d} \\
&\leq \sup_{w<d} \frac{f(w) - f(d)}{w - d} = D_- f(d).
\end{aligned}
$$

We conclude for $c < x < x' < d$, $D_+ f(c) \leq \frac{f(x') - f(x)}{x' - x} \leq D_- f(d)$.

If $\frac{f(x') - f(x)}{x' - x} > 0$, we have $\left| \frac{f(x') - f(x)}{x' - x} \right| \leq D_- f(d)$. If $\frac{f(x') - f(x)}{x' - x} < 0$, we have $\left| \frac{f(x') - f(x)}{x' - x} \right| \leq -D_+ f(c)$. Combining, we see

$$
\left| \frac{f(x') - f(x)}{x' - x} \right| \leq \max\{-D_+ f(c), D_- f(d)\}.
$$

If $x = c$ or $x' = d$, we can do a similar analysis to get the same result. So we have for $c \leq x < x' \leq d$

$$
\left| \frac{f(x') - f(x)}{x' - x} \right| \leq \max\{-D_+ f(c), D_- f(d)\}.
$$

We have proven the following result.

Theorem 9.2.4 The Derivative Ratio is Bounded for a Convex Function

If f is convex on the interval I, if $[c, d] \subset int(I)$, then there is a positive constant $L^{[c,d]}$ so that if $c \leq x < x' \leq d$, we have

$$
\left| \frac{f(x') - f(x)}{x' - x} \right| \leq L^{[c,d]}.
$$

Proof 9.2.4
We have just argued this. The constant is $L^{[c,d]} = \max\{-D_+ f(c), D_- f(d)\}$. ∎

Comment 9.2.1 *From the above, we have $|f(x') - f(x)| \leq L^{[c,d]} |x' - x|$ which immediately tells us f is continuous at each x in the interior of I. What happens at the boundary of I is not known yet. Here is the argument. Choose $\epsilon > 0$ arbitrarily. Let $\delta = \epsilon / L^{[c,d]}$. Then $|y - x| < \delta \Rightarrow |f(y) - f(x)| \leq L^{[c,d]} \epsilon / L^{[c,d]} = \epsilon$.*

Comment 9.2.2 *So this sort of a function can't be convex.*

$$f(x) = \begin{cases} x^2, & x \le 0 \\ x^2 - 2, & x > 0 \end{cases}$$

because f is not continuous at 0 and a convex function is continuous at each point in the interior of its domain. You should look at the epigraph here and make sure you understand why it is not a convex subset of $\Re^2$.

9.2.1   Homework

Exercise 9.2.8 *Let $f(x) = |x|$. Find $D_- f(0)$ and $D_+ f(0)$. This is just a matter of looking at slopes on the left and the right. You should get $D_- f(0) = -1$ and $D_+ f(0) = +1$.*

Exercise 9.2.9 *Let $f(x) = |x - 2|$. Find $D_- f(2)$ and $D_+ f(2)$.*

Exercise 9.2.10 *Let $f(x) = |x - 2| + |x_4|$. Find $D_- f$ and $D_+ f$ at all points x.*

Exercise 9.2.11 *Let $f(x) = |3x - 2| + |4 - 5x|$. Find $D_- f$ and $D_+ f$ at all points x.*

Exercise 9.2.12 *Let $f(x) = |16x + 3| + |3x - 2| + |4 - 5x|$. Find $D_- f$ and $D_+ f$ at all points x.*

9.3   More on Convex Functions

If f is convex on some domain Ω, given the definition of convexity, if f is defined on a and b in Ω, f is also defined at each $ta + (1 - t)b$ for $0 < t < 1$. So f is defined on $[a, b]$ as well. Hence, Ω must be an interval I. The choices of I are then:

- I is a finite interval and so $I = (a, b), [a, b), (a, b]$ or $[a, b]$ for some $a < b$.

- I is an infinite interval and so $I = (-\infty, a), (-\infty, a], (b, \infty), [b, \infty)$ or $(-\infty, \infty)$.

(Case One:) I is a finite interval with endpoints a and b.

Pick any interior point x_0 of I. The point $x_0 - t \to a$ from the right as $t \to (x_0 - a)^+$. Hence, the slope function

$$S(x_0 - t, x_0) = \frac{f(x_0 - t) - f(x_0)}{(x_0 - t) - x_0} = \frac{f(x_0 - t) - f(x_0)}{-t}$$

decreases as we increase t. Hence, the negative of $S(x_0 - t, x_0)$ increases. Then $q(t) = \frac{f(x_0 - t) - f(x_0)}{t}$ increases as $t \to (x_0 - a)^+$. Since this limit is of an increasing sequence, if the sequence is bounded above it will converge to the supremum of the sequence values or it will have limit ∞.

If the limit is finite, call it A. Note A can be any number, even zero. Then, from the definition of $q(t)$, we have $f(x_0 - t) = f(x_0) + tq(t)$. Thus, $\lim_{t \to (x_0 - a)^+} f(x_0 - t) = f(x_0) + (x_0 - a) \lim_{t \to (x_0 - a)^+} q(t)$. This gives $\lim_{t \to (x_0 - a)^+} f(x_0 - t) = f(x_0) + A(x_0 - a)$. This says the right hand limit $f(a^+)$ exists and equals $f(x_0) + A(x_0 - a)$. Note if $A = 0$, this says f is really a constant function.

If the limit is ∞, there must be a sequence of points (t_n) with $t_n \to (x_0 - a)^+$ and $\frac{f(x_0 - t_n) - f(x_0)}{t_n} > n$. This tells us $f(x_0 - t_n) > nt_n + f(x_0)$. Letting $n \to \infty$, we see $\lim_{n \to \infty} f(x_0 - t_n) = \infty$. So $f(a^+) = \infty$ here. We can do the same sort of argument at the endpoint b to get $f(b^-)$ exists and is either finite or ∞.

If the endpoint a is in I, we can say some more. We know $f(x_0 - t) = f(x_0) + tq(t)$ is increasing as t increases to $x_0 - a$. But if $a \in I$, we can use the number $t = x_0 - a$ in this expression to get $f(a) = f(x_0) + q(x_0 - a)(x_0 - a)$. We also know $q(t)$ increases, so we have for $0 < t < x_0 - a$ that $q(t) \leq q(x_0 - a) = \frac{f(a) - f(x_0)}{x_0 - a}$. Hence, we have $\lim_{t \to (x_0 - a)+} q(t) \leq \frac{f(a) - f(x_0)}{x_0 - a}$. This shows if $a \in I$, the limit $\lim_{t \to (x_0 - a)+} q(t)$ must be finite. Again, call this limit value A. Then, we have

$$A = \lim_{t \to (x_0 - a)+} \frac{f(x_0 - t) - f(x_0)}{x_0 - a} = \frac{f(a^+) - f(x_0)}{x_0 - a}.$$

Thus,

$$
\begin{aligned}
f(a^+) &= f(x_0) + A(x_0 - a) \\
&\leq f(x_0) + \frac{f(a) - f(x_0)}{x_0 - a}(x_0 - a) \\
&= f(x_0) + f(a) - f(x_0) = f(a).
\end{aligned}
$$

We conclude if $a \in I$, $f(a^+)$ is finite and $f(a^+) \leq f(a)$. A similar argument shows if $b \in I$, then $f(b^-)$ is finite and $f(b^-) \leq f(b)$.

(Case Two): The interval I is infinite. Let's assume $I = (a, \infty)$ or $I = [a, \infty)$. Our arguments in Case One have already shown us that $f(a^+)$ exists and if $I = [a, \infty)$, $f(a) \geq f(a^+)$.

Pick any interior point of I. Then the point $x_0 + t \to \infty$ as $t \to \infty$. Thus, setting

$$q(t) = S(x_0 + t, x_0) = \frac{f(x_0 + t) - f(x_0)}{t},$$

we see $q(t)$ increases with increasing t. This limit will be finite or ∞.

If the limit is finite, call it A. Remember A could be zero, positive or negative. Then, using the definition of $q(t)$, we have $f(x_0 + t) = f(x_0) + tq(t)$. Thus,

- $\lim_{t \to \infty} f(x_0 + t) = f(x_0)$ if $A = 0$

- $\lim_{t \to \infty} f(x_0 + t) = \infty$ if $A > 0$

- $\lim_{t \to \infty} f(x_0 + t) = -\infty$ if $A < 0$

If the limit is ∞, we also get $\lim_{t \to \infty} f(x_0 + t) = \infty$. Hence, we conclude $\lim_{x \to \infty} f(x) \in \{-\infty, 0, \infty\}$.

For the case $I = (-\infty, b)$ or $I = (-\infty, b]$, we already know $f(b^-)$ is finite and if $b \in I$, we have $f(b^-) \leq f(b)$. The argument at the endpoint $-\infty$ is essentially like the argument we just did. So we see $\lim_{x \to -\infty} f(x) \in \{-\infty, 0, \infty\}$. We can summarize these results with the pictures in Figure 9.3 which give example of convex functions satisfying these possibilities. The ∞ and $-\infty$ in the figure refer to the limiting values $\lim_{t \to \pm\infty} f(x)$ in the various cases.

9.4 Subdifferentials

Let's look at the behavior of $D_- f(x)$ and $D_+ f(x)$ when the domain of f is a finite interval $I = (a, b)$ with $f(a^+) = f(b^-) = \infty$. So this is the classic **bowl** shape.

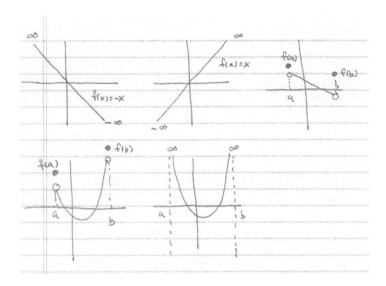

Figure 9.3: Convex functions at their endpoints.

Recall $D_- f(x) = \sup_{y<x} \frac{f(y)-f(x)}{y-x}$ and $D_+ f(x) = \inf_{y>x} \frac{f(y)-f(x)}{y-x}$. We can't define $D_\pm f(a)$ or $D_\pm f(b)$ because f is not defined at a or b. It is possible to analyze what happens at these boundary points but that will be for another class! But at all the interior points, we always have a finite $D_- f(x)$ and $D_+ f(x)$. It is time for another definition!!

Definition 9.4.1 The Subdifferential of a Convex Function

*Let f be convex on the interval I. We say s is a **subgradient** of $x \in I$ if $D_- f(x) \leq s \leq D_+ f(x)$. The **subdifferential** of f at $x \in I$ is denoted by $\partial f(x)$ and is defined to be $\partial f(x) = [D_- f(x), D_+ f(x)]$. Note this is a compact subset of real numbers.*

So at an interior point $x_0 \in I$ for the convex function f, if $s \in \partial f(x)$, $D_- f(x_0) = \sup_{x<x_0} \frac{f(x)-f(x_0)}{x-x_0}$ $\leq s \leq \inf_{x>x_0} \frac{f(x)-f(x)}{x-x_0} = D_+ f(x_0)$. Thus, $s \leq \frac{f(x)-f(x_0)}{x-x_0}$ for $x > x_0 \Rightarrow f(x) \geq f(x_0) + s(x - x_0)$.

But using the other inequality, we have $s \geq \frac{f(x)-f(x_0)}{x-x_0}$ for $x < x_0 \Rightarrow f(x) \geq f(x_0) + s(x - x_0)$ as the inequality flips because here $x - x_0 < 0$. The inequality is also true at $x = x_0$, so we have proven the following theorem.

Theorem 9.4.1 The Subgradient Growth Condition

Let f be convex on the interval I and let $x_0 \in I$ be an interior point. Then $f(x) \geq f(x_0) + s(x - x_0)$ for all $s \in \partial f(x_0)$.

Proof 9.4.1
This is what we just proved.   ∎

This leads to an important result.

Theorem 9.4.2 The Interior Minimum Condition for Convex Functions

Let f be convex on the interval I. Then there is an x_0 in the interior of I which minimizes $f \Leftrightarrow 0 \in \partial f(x_0)$.

Proof 9.4.2

($\Rightarrow$):
Let x_0 in the interior of I minimize f. Then $f(x) \geq f(x_0)$ for all $x \in I$. If $x > x_0$, we get $\frac{f(x)-f(x_0)}{x-x_0} \geq 0$ for $x > x_0$. Thus $\inf_{x>x_0} \frac{f(x)-f(x_0)}{x-x_0} = D_+f(x_0) \geq 0$. Also, $\frac{f(x)-f(x_0)}{x-x_0} \leq 0$ when $x < x_0$ implying $D_-f(x_0) = \sup_{x<x_0} \frac{f(x)-f(x_0)}{x-x_0} \leq 0$. So we have $D_-f(x_0) \leq 0 \leq D_+f(x_0)$ telling us $0 \in \partial f(x_0)$.
($\Leftarrow$):
If $0 \in \partial f(x_0)$, then from the previous theorem, $f(x) \geq f(x_0) + 0(x-x_0) = f(x_0)$ for all $x \in I$. Hence x_0 is an minimizer for f. ■

Comment 9.4.1 *Note this is more powerful than the First Derivative Test you know from Calculus. The FDT only gives us candidates for extreme values.*

Let's look at another minimization result.

Theorem 9.4.3 Minimization of Convex Functions with Growth Conditions

Let $I = (a, b)$ be a finite interval and let $f : I \to \Re$ be convex. Assume $f(a^+) = \infty$ and $f(b^-) = \infty$. Then $\inf(f(I))$ is finite and there is a sequence $(x_n) \subset \Omega$ with $x_n \to x_0 \in I$ and $f(x_n) \to f(x_0) = \inf(f(I))$.

Proof 9.4.3
Let $\alpha = \inf(f(\Omega))$. There is a sequence $(y_n = f(x_n)) \subset \Omega$ that converges to α. Since $(x_n) \subset (a, b)$, (x_n) is bounded and by the Bolzano - Weierstrass Theorem, there is then a subsequence (x_n^1) of (x_n) converging to x_0. If $x_n^1 \to a^+$, then by assumption $f(x_n^1) \to \infty$ implying $\alpha = \infty$. Thus f is not finite on (a, b) which is not true. So $x_n \not\to a^+$. A similar argument shows $x_n^1 \not\to b^-$. Thus x_0 is an interior point.

We then have for all $s \in \partial f(x_0)$ that $f(x_n^1) \geq f(x_0) + s(x_n^1 - x_0)$. Taking the limit as $n \to \infty$, we have $\alpha \geq f(x_0)$. We already know $\alpha \leq f(x_0)$ and so we have $\alpha = f(x_0)$. ■

Note if we let $I = (-\infty, \infty)$ in the Theorem 9.4.3 and assumed $f(x) \to \infty$ as $|x| \to \infty$ the argument would be similar and we would still find a minimum.

On the basis of our theorems, to minimize a convex function like this, we need the find where the x values where $0 \in \partial f(x)$. We can prove a theorem that tell us that $\partial(\sum_{i=1}^n r_i f_i) = \sum_{i=1}^n r_i \partial f_i$ for any positive values r_i. Then to solve a minimization problem, we find an x_0 so that 0 is in $\sum_{i=1}^n r_i \partial f_i(x_0)$.

Example 9.4.1 *Let $f(x) = |x - 1| + |x - 3|$. Find where f attains it minimum.*

Solution

$$f(x) = \begin{cases} 1 - x + 3 - x = 4 - 2x, & x < 1 \\ 2, & x = 1 \\ x - 1 + 3 - x = 2, & 1 < x < 3 \\ 2, & x = 3 \\ x - 1 + x - 3 = 2x - 4, & x > 3 \end{cases}$$

and by direct calculation

$$\partial f(x) = \begin{cases} -2, & x < 1 \\ [-2, 0], & x = 1 \\ 0, & 1 < x < 3 \\ [0, 2], & x = 3 \\ 2, & x > 3 \end{cases}$$

So $0 \in \partial f(x)$ when $x \in [1, 3]$ and for all these values of x, f has its minimum value of 2.

We can also say $\partial f(x) = \partial(|x - 1|)(x) + \partial(|x - 3|)(x)$. And so

$$\partial f(x) = \begin{cases} -1 + -1 = -2, & x < 1 \\ [-1, 1] - 1 = [-2, 0], & x = 1 \\ 1 + -1 = 0, & 1 < x < 3 \\ 1 + [-1, 1] = [0, 2], & x = 3 \\ 1 + 1 = 2, & x > 3 \end{cases}$$

Again $0 \in \partial f(x)$ when $x \in [1, 3]$ and for all these values of x, f has its minimum value of 2.

9.4.1   Homework

Exercise 9.4.1 *If f and g are convex on I, prove $f + g$ is also convex on I.*

Exercise 9.4.2 *If f and g are convex on I, prove $3f + 5g$ is also convex on I.*

Exercise 9.4.3 *Let $f(x) = (x - 1)^2 + (x - 3)^2$. Is f convex on $\Re$?*

Exercise 9.4.4 *Let $f(x) = |x - 2| + |x - 5|$. Is f convex on $\Re$?*

Exercise 9.4.5 *If $r > 0$ and f is convex, prove rf is also convex.*

Exercise 9.4.6 *Use POMI to show $f(x) = \sum_{i=1}^{n} r_i(x - x_i)^2$ is convex on $\Re$ where $\{r_1, \ldots, r_n\}$ are positive numbers and $\{x_1, \ldots, x_n\}$ are fixed points.*

Exercise 9.4.7 *Let $f(x) = |x - 2|$. Find ∂f at all points x and find where f attains its minimum.*

Exercise 9.4.8 *Let $f(x) = |x - 2| + |x_4|$. Find ∂f at all points x and find where f attains its minimum.*

Exercise 9.4.9 *Let $f(x) = |3x - 2| + |4 - 5x|$. Find ∂f at all points x and find where f attains its minimum.*

Exercise 9.4.10 *Let $f(x) = |16x + 3| + |3x - 2| + |4 - 5x|$. Find ∂f at all points x and find where f attains its minimum.*

Exercise 9.4.11 *Let $f(x) = |15x + 3| + |x - 2| + |4 - 5x| + 8|2x - 7|$. Find ∂f at all points x and find where f attains its minimum.*

Chapter 10

Basic Differentiability

10.1 An Introduction to Smoothness

Let's back up and approach the ideas of basic calculus by just looking at how far we can go with applying algebra alone. We are going to look carefully at a model of natural selection called **viability selection**. This is a model used in a typical Calculus for Biology course and is a nice example of applying limiting ideas. These discussions come from a great book on evolutionary biology **"Mathematical Models of Social Evolution: A Guide for the Perplexed"** by McElreath and Body (McElreath and Boyd (10) 2007) which we encourage you to pick up and study at some point. We are interested in understanding the long term effects of genes in a population. Obviously, it is very hard to even frame questions about this. One of the benefits of our use of mathematics is that it allows us to build a very simplified model which nevertheless helps us understand general principles. These are biological versions of the famous Einstein *gedanken* experiments: i.e. thought experiments which help develop intuition and clarity.

10.2 A Basic Evolutionary Model

- We start by assuming we have a population of $N(t)$ individuals at a given time t. It also seems reasonable to think of our time unit as **generations**. So we would say the population at generation t is given by $N(t)$.

- Each adult has Q chromosomes and the reproduction process does not mix genetic information from another adult and hence the zygote formed by what is evidently some form of asexual reproduction also has Q chromosomes.

- We also assume in each generation individuals go through their life cycle exactly the same: all individuals are born at the same time and all individuals reproduce at the same time. We will call this a **discrete** dynamic.

- Note a zygote does not have to live long enough to survive to an adult.

Since we want to develop a very simple model we assume there are only two genotypes, type **A** and type **B**. We also assume **A** is more likely to survive to an adult. We let $N_A(t)$ be the number of individuals of phenotype A and $N_B(t)$ be the number of individuals of phenotype B in a given generation. Then $N(t)$ is the number of individuals in the population and this number changes each generation as

$$N(t) \;=\; N_A(t) + N_B(t).$$

We also keep track of the fraction of individuals in the population that are genotype **A** or **B**. This fraction is also called the **frequency** of type **A** and **B** respectively.

$$P_A(t) \;=\; \frac{N_A(t)}{N_A(t) + N_B(t)}, \quad P_B(t) = \frac{N_B(t)}{N_A(t) + N_B(t)}$$

Individuals of each phenotype do not necessarily survive to adulthood so each survives to adulthood with a certain probability, V_A and V_B, respectively.

We will keep track of how the frequency P_A changes. Since $P_A(t)$ is the frequency for **A** at generation t. What is the frequency at the next generation $t + 1$?

- The number of zygotes from individuals of genotype **A** at generation t is assumed to be $z_A\, N_A(t)$ where z_A is the number of zygotes each individual of type **A** produces. Note that z_A plays the role of the **fertility** of individuals of type **A**. We assume individuals of type **B** also create z_B zygotes. Hence, their fertility is also z_B. So the number of zygotes of **B** type individuals is $z_B\, N_B(t)$.

- The frequency of **A** zygotes at generation t is then

$$P_{AZ}(t) \;=\; \frac{z_A\, N_A(t)}{z_A\, N_A(t) + z_B\, N_B(t)} = \frac{\frac{z_A}{z_B}\, N_A(t)}{\frac{z_A}{z_B}\, N_A(t) + N_B(t)} = \frac{\rho\, N_A(t)}{\rho\, N_A(t) + N_B(t)}$$

where $\rho = \frac{z_A}{z_B}$ is the fertility ratio and we add an additional subscript to indicate we are looking at zygote frequencies.

But not all zygotes survive to adulthood. If we multiply numbers of zygotes by their probability of survival, V_A or V_B, the number of **A** zygotes that survive to adulthood is $V_A\, N_A(t)$ and the number of **B** zygotes that survive to adulthood is $V_B\, N_B(t)$. We see the frequency of **A** zygotes that survive to adulthood to give the generation $t + 1$ must be

$$P_{AZS}(t+1) \;=\; \frac{z_A\, V_A N_A(t)}{z_A\, V_A N_A(t) + z_B\, V_b N_B(t)} = \frac{\rho\, V_A N_A(t)}{\rho\, V_A N_A(t) + V_B N_B(t)}$$

where we have added yet another subscript S to indicate survival. Note we add the generation label $t + 1$ to P_{AZS} because this number is for the next generation. Now for the final step. From the way we define stuff, notice that $z_A\, V_A N_A(t) - N_A(t+1)$ and $z_B\, V_B N_B(t) = N_B(t+1)$. Thus,

$$P_A(t+1) \;=\; \frac{N_A(t+1)}{N_A(t+1) + N_B(t+1)} = \frac{\rho\, V_A N_A(t)}{\rho\, V_A N_A(t) + V_B N_B(t)} = P_{AZS}(t+1)$$

Now replace the denominator by $N(t)$ and multiply both sides by this denominator to get

$$N(t)\, P_A(t) \;=\; N_A(t).$$

Then consider the frequency for **B**. Note

$$1 - P_A(t) \;=\; 1 - \frac{N_A(t)}{N_A(t) + N_B(t)} = 1 - \frac{N_A(t)}{N(t)}.$$

Getting a common denominator, we find

$$1 - P_A(t) \;=\; \frac{N(t) - N_A(t)}{N(t)}.$$

But $N(t) - N_A(t) = N_B(t)$. Using this in the last equation, we have found that

$$1 - P_A(t) = \frac{N_B(t)}{N(t)}$$

which leads to the identity we wanted: $N_B(t) = (1 - P_A(t)) N(t)$. This analysis works just fine at generation $t + 1$ too, but at that generation, we have

$$N_A(t+1) = z_A N(t) V_A P_A(t)$$
$$N_B(t+1) = z_B N(t) V_B (1 - P_A(t))$$

So we can say

$$P_A(t+1) = \frac{N_A(t+1)}{N_A(t+1) + N_B(t+1)}$$
$$= \frac{z_A P_A(t) N(t) V_A}{z_A P_A(t) N(t) V_A + (1 - P_A(t)) z_B N(t) V_B}.$$

Canceling the common $N(t)$, we get

$$P_A(t+1) = \frac{z_A V_A P_A(t)}{z_A P_A(t) V_A + (1 - P_A(t)) z_B V_B} = \frac{\rho V_A P_A(t)}{\rho P_A(t) V_A + (1 - P_A(t)) z_B V_B}$$

This tells us how the frequency of the A genotype changes each generation.

10.2.1 A Difference Equation

We can also derive a formula for the change in frequency at each generation by doing a subtraction. We consider

$$P_A(t+1) - P_A(t) = \frac{\rho V_A P_A(t)}{\rho P_A(t) V_A + (1 - P_A(t)) V_B} - P_A(t).$$

Get a common denominator next.

$$P_A(t+1) - P_A(t) = \frac{\rho V_A P_A(t)}{\rho P_A(t) V_A + (1 - P_A(t)) V_B}$$
$$- P_A(t) \frac{\rho P_A(t) V_A + (1 - P_A(t)) V_B}{\rho P_A(t) V_A + (1 - P_A(t)) V_B}.$$

Multiply everything out and put into one big fraction.

$$P_A(t+1) - P_A(t)$$
$$= \frac{\rho V_A P_A(t) - P_A(t) \left(\rho P_A(t) V_A + (1 - P_A(t)) V_B \right)}{\rho P_A(t) V_A + (1 - P_A(t)) V_B}$$
$$= \frac{\rho V_A P_A(t)(1 - P_A(t)) - P_A(t) (1 - P_A(t)) V_B}{\rho P_A(t) V_A + (1 - P_A(t)) V_B}.$$

Now factor to get

$$P_A(t+1) - P_A(t) = \frac{P_A(t)(1 - P_A(t))(\rho V_A - V_B)}{\rho P_A(t) V_A + (1 - P_A(t)) V_B}$$

This is what is called a **recursion** equation. For example, if the frequency of **A** in the population started at $P(0) = p_0 = 0.8$ and $\rho = 0.4$, the change in frequency of **A** in the next generation is given by

$$
\begin{aligned}
\boldsymbol{P_A}(1) - \boldsymbol{P_A}(0) &= \frac{\boldsymbol{P_A}(0)(1 - \boldsymbol{P_A}(0))(\rho \boldsymbol{V_A} - \boldsymbol{V_B})}{\rho \boldsymbol{P_A}(0)\,\boldsymbol{V_A} + (1 - \boldsymbol{P_A}(0))\,\boldsymbol{V_B}} \\
&= \frac{p_0(1 - p_0)(\rho \boldsymbol{V_A} - \boldsymbol{V_B})}{\rho p_0\,\boldsymbol{V_A} + (1 - p_0)\,\boldsymbol{V_B}} \\
&= \frac{0.8(0.2)(0.4\boldsymbol{V_A} - \boldsymbol{V_B})}{0.4(0.8)\,\boldsymbol{V_A} + (0.2)\,\boldsymbol{V_B}} = \frac{0.16(0.4\boldsymbol{V_A} - \boldsymbol{V_B})}{0.32\,\boldsymbol{V_A} + (0.2)\,\boldsymbol{V_B}}
\end{aligned}
$$

Substituting in the value p_0 we find we can solve for $\boldsymbol{P_A}(1)$ as follows:

$$
\boldsymbol{P_A}(1) = p_0 + \frac{p_0\,(1 - p_0)\left(\rho \boldsymbol{V_A} - \boldsymbol{V_B}\right)}{\rho p_0\,\boldsymbol{V_A} + (1 - p_0)\,\boldsymbol{V_B}}.
$$

and continuing in this vein, the frequency at generation 2 would be

$$
\boldsymbol{P_A}(2) = p_1 + \frac{p_1\,(1 - p_1)\left(\rho \boldsymbol{V_A} - \boldsymbol{V_B}\right)}{\rho p_1\,\boldsymbol{V_A} + (1 - p_1)\,\boldsymbol{V_B}}.
$$

where we denote $\boldsymbol{P_A}(1)$ more simply by p_1.

Example 10.2.1 *Find the frequency of type* **A** *individuals at generation 1 if initially the frequency of* **A** *individuals is $p_0 = .01$ with probabilities $\boldsymbol{V_A} = .8$ and $\boldsymbol{V_B} = .3$ and the fertility ratio is $\rho = 0.4$.*

Solution

$$
\begin{aligned}
\boldsymbol{P_A}(1) &= p_0 + \frac{p_0\,(1 - p_0)\,(\rho \boldsymbol{V_A} - \boldsymbol{V_B})}{\rho p_0\,\boldsymbol{V_A} + (1 - p_0)\,\boldsymbol{V_B}} \\
&= 0.01 + \frac{(0.01)\,(0.99)\,((0.4)0.8 - 0.3)}{(0.4)(0.01)\,(0.8) + 0.99\,(0.3)} = 0.01 + \frac{0.0002}{0.3} = 0.01 + 0.0007
\end{aligned}
$$

So the change in frequency of **A** *is 0.007 with the new frequency given by $\boldsymbol{P_A}(1) = 0.0107$.*

10.2.2   The Functional Form of the Frequency

We can also derive a functional form for $\boldsymbol{P_A}(t)$. We start with the number of **A** in the population initially, $\boldsymbol{N_A}(0)$. We know at the next generation, the number of **A** is

$$
\begin{aligned}
\boldsymbol{N_A}(1) = &\left(\text{Fertility of } \mathbf{A}\right) \times \left(\text{number of } \mathbf{A} \text{ at generation } 0\right) \\
&\times \left(\text{probability an } \mathbf{A} \text{ type lives to adulthood}\right) = z_A\,\boldsymbol{N_A}(0)\,\boldsymbol{V_A}.
\end{aligned}
$$

and

$$
\begin{aligned}
\boldsymbol{N_A}(2) = &\left(\text{Fertility of } \mathbf{A}\right) \times \left(\text{number of } \mathbf{A} \text{ at generation } 1\right) \\
&\times \left(\text{probability an } \mathbf{A} \text{ type lives to adulthood}\right)
\end{aligned}
$$

$$= z_A \, N_A(1) \, V_A = z_A \, (z_A \, N_A(0) \, V_A) \, V_A = N_A(0) \, (z_A \, V_A)^2.$$

We can do this over and over again. You should note this argument is a relaxed version of POMI! We find

$$N_A(3) = N_A(0) \, (z_A \, V_A)^3$$
$$N_A(4) = N_A(0) \, (z_A \, V_A)^4$$
$$\vdots$$

We can easily extrapolate from this to see that, in general,

$$N_A(t) = N_A(0) \, (z_A \, V_A)^t$$

A similar analysis shows us that

$$N_B(t) = N_B(0) \, (z_B \, V_B)^t.$$

From the definition of the frequency for **A**, we have

$$P_A(t) = \frac{N_A(t)}{N_A(t) + N_B(t)} = \frac{\frac{N_A(t)}{N_A(t)}}{\frac{N_A(t)+N_B(t)}{N_A(t)}} = \frac{1}{1 + \frac{N_B(t)}{N_A(t)}}.$$

Now plug in our formulae for $N_A(t)$ and $N_B(t)$. Note the fraction

$$\frac{N_B(t)}{N_A(t)} = \frac{N_B(0) \, (z_B \, V_B)^t}{N_A(0) \, (z_A \, V_A)^t} = \frac{N_B(0)}{N_A(0)} \left(\frac{z_B V_B}{z_A V_A} \right)^t.$$

Using this in our frequency formula, we have

$$P_A(t) = \frac{1}{1 + \frac{N_B(0)}{N_A(0)} \left(\frac{z_B V_B}{z_A V_A} \right)^t} = \frac{1}{1 + \frac{N_B(0)}{N_A(0)} \left(\frac{\theta}{\rho} \right)^t}$$

where $\theta = \frac{V_B}{V_A}$ is the B over A survival ratio. Note we have derived this formula using only algebra and a lot of thinking!

Example 10.2.2 *Find the frequency of type* **A** *individuals at generation* 1 *if initially the frequency of* **A** *individuals is* $p_0 = .01$ *and the probabilities are* $V_A = .8$ *and* $V_B = .3$ *and the fertility ratio is* $\rho = 0.4$. *We have* $\theta = 3/8 = 0.375$.

Solution *Since* $p_0 = .01$, *we can find the initial values* $N_A(0)$ *and* $N_B(0)$. $P_A(0) = 0.01 = \frac{N_A(0)}{N_A(0)+N_B(0)} = \frac{1}{1+\frac{N_B(0)}{N_A(0)}}$. *Thus,* $\frac{1}{1+\frac{N_B(0)}{N_A(0)}} = 0.01 \implies 1 + \frac{N_B(0)}{N_A(0)} = \frac{1}{0.01} = 100$ *and so* $\frac{N_B(0)}{N_A(0)} = 99$. *Using the probabilities, we find* $P_A(1) = \frac{1}{1+\frac{N_B(0)}{N_A(0)} \frac{\theta}{\rho}} = \frac{1}{1+99.0 \left(\frac{.375}{0.4} \right)} = \frac{1}{1+99.0 \, (0.9375)} \implies \frac{1}{93.8125} = 0.0107$. *This is the same solution we obtained in the previous example.*

Homework

Exercise 10.2.1 *Plot the frequency of* **A** *and* **B** *individuals on the same plot using MATLAB if initially the frequency of* **A** *individuals is* $p_0 = .01$ *and the probabilities are* $V_A = .8$ *and* $V_B = .3$ *and the fertility ratio is* $\rho = 0.4$.

Exercise 10.2.2 *Plot the frequency of* **A** *and* **B** **individuals on the same plot using MATLAB if initially the frequency of A individuals is** $p_0 = .06$ **and the probabilities are** $V_A = .4$ **and** $V_B = .43$ **and the fertility ratio is** $\rho = 0.35$**.**

Exercise 10.2.3 *Plot the frequency of* **A** *and* **B** **individuals on the same plot using MATLAB if initially the frequency of A individuals is** $p_0 = .05$ **and the probabilities are** $V_A = .55$ **and** $V_B = .45$ **and the fertility ratio is** $\rho = 0.6$**.**

Exercise 10.2.4 *Plot the frequency of* **A** *and* **B** **individuals on the same plot using MATLAB if initially the frequency of A individuals is** $p_0 = .08$ **and the probabilities are** $V_A = .2$ **and** $V_B = .7$ **and the fertility ratio is** $\rho = 0.3$**.**

10.2.3   Abstracting Generation Time

We are now ready to start discussing an important topic which is the **smoothness** of a function. Recall, the frequency of type **A** individuals at generation t was given by

$$P_A(t) \;=\; \frac{1}{1 + \frac{N_B(0)}{N_A(0)} \left(\frac{\theta}{\rho}\right)^t}.$$

and the change from generation t to $t + 1$ was governed by the formula

$$P_A(t+1) \;-\; P_A(t) \;=\; \frac{P_A(t)(1 - P_A(t))(\rho V_A - V_B)}{\rho P_A(t)\, V_A + (1 - P_A(t))\, V_B}$$

The length of a generation can vary widely. So far we are just letting t denote the generation number and we have not paid attention to the average time a generation lasts. Another way of looking at it is that we are implicitly thinking of our time unit as being a **generation**. The actual duration of the generation was not needed in our thinking. For convenience, let's switch variables now. You need to get comfortable with the idea that we can choose to name our variables of interest as we see fit. It is usually better to name them so that they mean something to you for the biology you are trying to study. So in viability selection, variables and parameters like P_A and N_B were chosen to remind us of what they stood for. However, that name choice is indeed arbitrary. Switch P_A to x and the ratio $N_B(0)/N_A(0)$ to simply a and the ratio θ/ρ to just b and we have

$$x(t) \;=\; \frac{1}{1 + a\, b^t}.$$

This looks a lot simpler even though it says the same thing. Our biological understanding of the viability selection problem tells us that b is not really arbitrary. We know $V_B/V_A < 1$ because V_B is smaller than V_A. We also know these values are between 0 and 1 as they are frequencies. The same can be said for a; the initial values of N_A and N_B are positive and so a is some positive number. So we should say a bit more

$$x(t) \;=\; \frac{1}{1 + a\, b^t}, \; where\, a > 0 \; and\, 0 < b < 1.$$

Currently, $x(t)$ is defined for a time unit that is generations, so it is easy to see that this formula defines a function which generates a value for any generation t whether t is measured in seconds, days, hours, weeks or years! It is really quite general. Our viability selection model gives us a function which models the frequency of a choice of action for any t regardless of which unit is used to measure t.

In fact, we could assume this formula works for any value of t. We stop thinking of t measured in terms of integer multiples of a time unit and extend our understanding of this formula by letting it be valid for any t. Make no mistake, this is an abstraction as our arguments only worked for generations! We often make this step into abstraction as it allows us to bring powerful tools to bear on understanding our models that we can't use if we are restricted to finite time measurements. In general,

- Build a model using units that are relevant for the biology. Use generations, seasonal cycles and so forth. Work hard to make the model realistic for your choice of units. This is a hard but important step.

- From your discrete time unit model, make the abstract jump to allow time to be any value at all.

- With the model extended to all time values, apply tools that allow you to manipulate these models to gain insight. These tools from Calculus include things like **limiting behavior**, **smoothness issues called continuity** and **rate of change smoothness issues called differentiability**.

Now **continuity** for our model roughly means the graph of our model with respect to time doesn't have jumps. Well, of course, our model for generational time has jumps! The $x(t)$ value simply jumps to the new value $x(t + 1)$ when we apply the formula. The continuity issue arises when we pass to letting t be any number at all. Another way of looking at it is that the generation time becomes smaller and smaller. For example, the generation time for a virus or a bacteria is very small compared to the generation time of a human! So letting our model handle shrinking generation time seems reasonable and as the generation time shrinks, it makes sense that the jumps we see get smaller. Continuity is roughly the idea that as the generation time shrinks to zero (yeah, odd concept!) the model we get has no jumps in it at all! Note how our earlier discussions involving limits and continuity using a traditional mathematical perspective sidestep all these complicated modeling issues!

10.3   Sneaking Up on Differentiability

To get a better feel for the **differentiability** idea, which you already know about from your Calculus courses, again let's resort to passing to an abstract model from a discrete time one. The frequency update law can be rewritten in terms of x also as

$$x(t + 1) \; - \; x(t) \;\; = \;\; \frac{x(t)\,(1 - x(t))\,(\rho c \; - \; d)}{\rho c\, x(t) \; + \; d\,(1 \; - \; x(t))}, \; c > d > 0$$

where we let $c = V_A$, $d = V_B$ and ρ is the usual fertility ratio. Now our time units are generations here, so t is really thought of as an integer. Let's assume the generation time is Δ and $t = n$ for some integer. Now rewrite what we have above as

$$\frac{x((n + 1)\Delta) \; - \; x(n\Delta)}{(n + 1)\Delta - n\Delta} \;\; = \;\; \frac{1}{\Delta}\left(\frac{x(t)\,(1 - x(t))\,(\rho c \; - \; d)}{\rho c\, x(t) \; + \; d\,(1 \; - \; x(t))} \right).$$

Now we are being a lot more clear about generation time. An obvious question is what happens as the generation time, now measures as Δ, gets smaller and smaller. If as we let Δ get really tiny, we find the fraction $\frac{x((n+1)\Delta)\, - \,x(n\Delta)}{\Delta}$ approaches some stable number, we can be pretty confident that this stable number somehow represents an abstraction of the rate of change of x over a generation. This shrinking of Δ and our look at the behavior of the fraction $\frac{x((n+1)\Delta)\, - \,x(n\Delta)}{\Delta}$ as Δ shrinks is the basic idea of what having a derivative at a point means. Of course, this model is very discrete and possesses jumps and this fraction makes no sense for time units smaller than the particular generation time we were using for our model. But we want our models to be independent of the generation time unit, so we wonder we what happens to this ratio as the generation time unit shrinks. Given our discussions so far, a reasonable thing to do is to ask what happens to this ratio as Δ gets smaller and smaller. Now this particular example is very hard, so let's make up a very simple example – strictly mathematical! – and see how it would go.

10.3.1   Change and More Change

Let's simplify our life and choose the function $f(x) = x^2$. Let's choose an h and consider the ratio

$$\frac{f(x+h) - f(x)}{(x+h) - (x)} = \frac{f(x+h) - f(x)}{h} = \frac{(x+h)^2 - x^2}{h}$$

Now we need to do some algebra. We have

$$(x+h)^2 - x^2 = x^2 + 2xh + h^2 - x^2 = 2xh + h^2.$$

Plugging this into our fraction, we find

$$\frac{f(x+h) - f(x)}{(x+h) - (x)} = \frac{f(x+h) - f(x)}{h} = \frac{(x+h)^2 - x^2}{h}$$

$$= \frac{2xh + h^2}{h} = 2x + h.$$

As h gets smaller and smaller, we see the *limiting value* is simply $2x$. We would say

$$\lim_{h \to 0} \left(\frac{f(x+h) - f(x)}{h} \right) = 2x.$$

This kind of limit works well for many functions f, including our function for the frequency of the gene of type **A** in the population if we assume our frequency function can be extended from being defined for just discrete generation times to all time (which we routinely do). It is such an important limiting process, it is given a special name: the **derivative** of f with respect to x. We often write this difference of function values of f divided by differences in x values as $\frac{\Delta f}{\Delta x}$, which of course **hides a lot**, and say $\lim_{\Delta x \to 0} \left(\frac{\Delta f}{\Delta x} \right) = 2x$.

The traditional symbol we use for this special limiting process is taken from the notation Δ. As h gets smaller and smaller, we use the symbol $\frac{df}{dx}$ to indicate we are taking this limit and write $\frac{df}{dx} = \lim_{\Delta x \to 0} \left(\frac{\Delta f}{\Delta x} \right) = 2x$. We thought it is important to see this limiting process from the point of view of an abstraction of a discrete model. But now it is time to look at it more mathematically using the tools we have been developing about limits.

Example 10.3.1 *Write down the limit formula for the derivative of $f(x) = x^3$ at $x = 2$ using both the h and Δx forms.*

Solution $\frac{df}{dx}(2) = \lim_{\Delta x \to 0} \left(\frac{(2+\Delta x)^3 - (2)^3}{\Delta x} \right) = \lim_{h \to 0} \left(\frac{(2+h)^3 - (2)^3}{h} \right)$

Example 10.3.2 *Write down the limit formula for the derivative of $f(x) = x^4$ at $x = 1$ using both the h and Δx forms.*

Solution $\frac{df}{dx}(1) = \lim_{\Delta x \to 0} \left(\frac{(1+\Delta x)^4 - (1)^4}{\Delta x} \right) = \lim_{h \to 0} \left(\frac{(1+h)^4 - (1)^4}{h} \right)$

Example 10.3.3 *Write down the limit formula for the derivative of $f(x) = 2x^3$ at $x = 4$ using both the h and Δx forms.*

Solution

$$
\begin{aligned}
\frac{df}{dx}(4) &= \lim_{\Delta x \to 0} \left(\frac{2(4 + \Delta x)^3 - 2(4)^3}{\Delta x} \right) \\
&= \lim_{h \to 0} \left(\frac{2(4 + h)^3 - 2(4)^3}{h} \right)
\end{aligned}
$$

10.3.2 Homework

Exercise 10.3.1 *Find the frequency of type A individuals at generation 1 given that initially the frequency of A individuals is $p_0 = .02$ and the probabilities of A and B are $V_A = .85$ and $V_B = .25$ and the fertility ratio is $\rho = 0.3$.*

Exercise 10.3.2 *Find the frequency of type A individuals at generation 1 and generation 2 given that initially the frequency of A individuals is $p_0 = .02$ and the probabilities of A and B are $V_A = .75$ and $V_B = .3$ and the fertility ratio is $\rho = 0.6$.*

Exercise 10.3.3 *Find the frequency of type A individuals at generation 10 given that initially the frequency of A individuals is $p_0 = .04$ and the probabilities of A and B are $V_A = .65$ and $V_B = .2$ and the fertility ratio is $\rho = 0.8$.*

Exercise 10.3.4 *Write down the limit formula for the derivative of $f(x) = 5x^2$ at $x = 1$ using both the h and Δx forms.*

Exercise 10.3.5 *Find the derivative of $f(x) = 2x^2 + 4x + 8$ using the definition at $x = 3$.*

Exercise 10.3.6 *Find the derivative of $f(x) = x^3 + 8$ using the definition at $x = 1$.*

10.4 Limits, Continuity and Right and Left Hand Slope Limits

Let's review the smoothness notion called **continuous**. This will allow us to transition easily into the new concept of differentiability of a function at a point.

Definition 10.4.1 Continuity of a Function at a Point: Reviewed

Let f be locally defined at p. We say f is said to be continuous at a point p in its domain if several conditions hold:

1. f is actually defined at p

2. The limit as x approaches p of f exists

3. The value of the limit above matches the value $f(p)$.

In more precise terms, $\forall \epsilon > 0 \, \exists \, \delta > 0, \, \ni \, |x - p| < \delta \Rightarrow |f(x) - f(p)| < \epsilon$, where $\delta > 0$ is smaller than the local radius of definition of f at p.

This is usually stated more succinctly as $f(p)$ exists and $\lim_{x \to p} f(x) = f(p)$, We can characterize the existence of this limit is several ways as we have different tools we can use to explore whether or not a function is continuous at a point. We also know

- $\lim_{x \to p} f(x)$ exist with value $A \iff \underline{\lim}_{x \to p} f(x) = \overline{\lim}_{x \to p} f(x) = A$.

- $\lim_{x \to p} f(x)$ exist with value $A \iff \lim_{x \to p^-} f(x) = \lim_{x \to p^+} f(x) = A$.

- $\lim_{x \to p} f(x)$ exist with value $f(p) \iff \underline{\lim}_{x \to p} f(x) = \overline{\lim}_{x \to p} f(x) = A$.

- $\lim_{x \to p} f(x)$ exist with value $f(p) \iff \lim_{x \to p^-} f(x) = \lim_{x \to p^+} f(x) = A$.

Example 10.4.1 *Examine the continuity of*

$$f(x) = \begin{cases} 1, & \text{if } 0 \leq x \leq 1 \\ 2 & \text{if } 1 < x \leq 2. \end{cases}$$

Solution *For this function, $\lim_{x \to 1} f(x)$ does not exist as $\underline{\lim}_{x \to 1} f(x) = 1 \neq \overline{\lim}_{x \to 1} f(x) = 2$. The same mismatch occurs if we look at the left and right sided limits. Hence, this function is not continuous at 1. We can also do an $\epsilon - \delta$ proof like so. Assume the limit is the value $A \neq 1, 2$ and let $d_1 = |A - 1|$, the distance from the proposed limiting value A to the right side possibility 1. Similarly, let $d_2 = |A - 2|$, the distance to the left side possibility. Pick $\epsilon_0 = (1/2) \min\{d_1, d_2\}$. Then*

$$|f(x) - A| = \begin{cases} |1 - A| = d_1 \geq \epsilon_0, & \text{if } 0 \leq x \leq 1 \\ |2 - A| = d_2 \geq \epsilon_0 & \text{if } 1 < x \leq 2. \end{cases}$$

no matter what $\delta > 0$ we choose. Hence the limit can not be $A \neq 1, 2$. If the limit was $A = 1$, then $d_1 = 0$ and $d_2 = 1$. We choose $\epsilon_0 = d_2/2$ and find

$$|f(x) - 1| = \begin{cases} |1 - 1| = 0, & \text{if } 0 \leq x \leq 1 \\ |2 - 1| = 1 \geq \epsilon_0 & \text{if } 1 < x \leq 2. \end{cases}$$

no matter what $\delta > 0$ we choose. Hence, the limit can not be $A = 1$. A similar argument works for the choice $A = 2$. So we know there is no value of A for which this limit will exist. Since the limit does not exist, f can not be continuous at 1.

Example 10.4.2 *Examine the continuity of*

$$f(x) = \begin{cases} 1, & \text{if } 0 \leq x < 1 \\ 10, & \text{if } x = 1 \\ 1 & \text{if } 1 < x \leq 2. \end{cases}$$

Solution *Note what happens now.*

- *The $\lim_{x \to 1} f(x)$ does exist and equals 1 because the right and left hand limits match and equal 1.*

- *The function value $f(1)$ is fine and $f(1) = 10$.*

- *So we know f is not continuous at 1 because the $\lim_{x \to 1} f(x)$ does exist but the value of that limit does not match the function value and that is enough to make the function fail to be continuous at 1.*

The lack of continuity is due to a flaw is the way we defined our function and it can easily be removed by simply redefining the function to be 1 at 1. So this kind of lack of continuity is called a **removeable discontinuity** *to reflect that. Also note even without convexity, we can still look at the terms $D_- f(1) = \sup_{x < 1} \frac{f(x) - f(1)}{x - 1}$ and $D_+ f(1) = \inf_{x > 1} \frac{f(x) - f(1)}{x - 1}$ in a sense. We no longer have the increasing nature of these slopes like we would with a convex function, so in this case, let's redefine as follows: $Df^-(1) = \lim_{x \to 1^-} \frac{f(x) - f(1)}{x - 1}$ and $Df^+(1) = \lim_{x \to 1^+} \frac{f(x) - f(1)}{x - 1}$. In the convex case these one sided limits become sup and inf because of the increasing/ decreasing nature of the slopes, but without convexity, we need to look at the one sided limits.*

- If $x < 1$, $\frac{f(x)-f(1)}{x-1} = \frac{1-10}{x-1} = -\frac{9}{x-1}$. So $Df^-(1) = \lim_{x\to 1^-} \frac{f(x)-f(1)}{x-1} = \lim_{x\to 1^-} -\frac{9}{x-1} = \infty$.

- If $x > 1$, $\frac{f(x)-f(1)}{x-1} = \frac{1-10}{x-1} = -\frac{9}{x-1}$. So $Df^+f(1) = \lim_{x\to 1^+} \frac{f(x)-f(1)}{x-1} = \lim_{x\to 1^+} -\frac{9}{x-1} = -\infty$.

Note we get the same results using $\underline{\lim}$ and $\overline{\lim}$ as well.

Example 10.4.3 *Examine the continuity of*

$$f(x) = \begin{cases} x^2, & \text{if } 0 \le x < 1 \\ 10, & \text{if } x = 1 \\ 2 + (x-1)^2 & \text{if } 1 < x \le 2. \end{cases}$$

Solution *Here $\lim_{x\to 1} f(x)$ does not exist because the right and left limits do not match. So not continuous at 1. Also $\underline{\lim}_{x\to 1} f(x) \ne \overline{\lim}_{x\to 1} f(x)$. Further,*

- If $x < 1$, $\frac{f(x)-f(1)}{x-1} = \frac{x^2-10}{x-1}$. *Let $x = 1 - e$ for small e. Then* $\frac{x^2-10}{x-1} = \frac{(1-e)^2-10}{-e} = \frac{e^2-2e-9}{-e} = -e + 2 + \frac{9}{e}$.

  Similarly, if $x > 1$, $\frac{f(x)-f(1)}{x-1} = \frac{2+(x-1)^2-10}{x-1}$, let $x = 1 + e$ and we have $\frac{(x-1)^2-8}{x-1} = \frac{e^2-8}{e} = e - \frac{8}{e}$.

- *So $Df^-(1) = \lim_{x\to 1^-} \frac{f(x)-f(1)}{x-1} = \lim_{e\to 0^+} -e + 2 + \frac{9}{e} = \infty$.*

- *So $Df^+f(1) = \lim_{x\to 1^+} \frac{f(x)-f(1)}{x-1} = \lim_{e\to 0^+} e - \frac{8}{e} = -\infty$.*

Example 10.4.4 *Examine the continuity of*

$$f(x) = \begin{cases} x^2, & \text{if } 0 \le x \le 1 \\ (x-1)^2 & \text{if } 1 < x \le 2. \end{cases}$$

Solution *Here $\lim_{x\to 1} f(x)$ does not exist because the right and left limits do not match. So not continuous at 1. Also $\underline{\lim}_{x\to 1} f(x) \ne \overline{\lim}_{x\to 1} f(x)$. Further,*

- If $x < 1$, $\frac{f(x)-f(1)}{x-1} = \frac{x^2-1}{x-1} = x + 1$. If $x > 1$, $\frac{f(x)-f(1)}{x-1} = \frac{(x-1)^2-1}{x-1}$, *let $x = 1 + e$ and we have* $\frac{(x-1)^2-1}{x-1} = \frac{e^2-1}{e} = e - \frac{1}{e}$.

- *So $Df^-(1) = \lim_{x\to 1^-} \frac{f(x)-f(1)}{x-1} = 2$.*

- *So $Df^+f(1) = \lim_{x\to 1^+} \frac{f(x)-f(1)}{x-1} = \lim_{e\to 0^+} e - \frac{1}{e} = -\infty$.*

Example 10.4.5 *Examine the continuity of*

$$f(x) = \begin{cases} x^2, & \text{if } 0 \le x \le 1 \\ x + (x-1)^2 & \text{if } 1 < x \le 2. \end{cases}$$

Solution *Here, the limit and the function value at 1 both match and so f is continuous at $x = 1$. Further,*

- If $x < 1$, $\frac{f(x)-f(1)}{x-1} = \frac{x^2-1}{x-1} = x + 1$. If $x > 1$, $\frac{f(x)-f(1)}{x-1} = \frac{x+(x-1)^2-1}{x-1}$, *let $x = 1 + e$ and we have* $\frac{x+(x-1)^2-1}{x-1} = \frac{1+e+e^2-1}{e} = 1 + e$.

- *So $Df^-(1) = \lim_{x \to 1^-} \frac{f(x) - f(1)}{x - 1} = 2$.*

- *So $Df^+f(1) = \lim_{x \to 1^+} \frac{f(x) - f(1)}{x - 1} = \lim_{e \to 0^+} 1 + e = 1$.*

Example 10.4.6 *For*

$$f(x) \;\; = \;\; \begin{cases} x^2 + 4x, & \text{if } 0 \le x \le 1 \\ -3x & \text{if } 1 < x \le 2. \end{cases}$$

determine if f is continuous at 1 and calculate $D^-f(1)$ and $D^+f(1)$.

Solution   • *The left hand limit is 5 and the right hand limit is -3. Since the limit does not exist at 1, the function is not continuous at $x = 1$.*

- $D^-f(1) = \lim_{x \to 1^-} \frac{x^2 + 4x - 5}{x - 1} = \lim_{x \to 1^-} \frac{(x + 5)(x - 1)}{x - 1} = \lim_{x \to 1^-}(x + 5) = 6.$

- $D^+f(1) = \lim_{x \to 1^+} \frac{-3x - 5}{x - 1}$. *Let $x = 1 + e$, then $D^+f(1) = \lim_{e \to 0^+} -3 - \frac{8}{e} = -\infty$.*

Example 10.4.7 *For the function*

$$f(x) \;\; = \;\; \begin{cases} x^2 + 3x + 2, & \text{if } 0 \le x < 2 \\ x^3 + 4 & \text{if } 2 \le x \le 5. \end{cases}$$

determine if f is continuous at 2 and calculate $D^-f(2)$ and $D^+f(2)$.

Solution   • *The left and right hand limits are 12 so the limit at 2 does exist and has the value 12. Since $f(2) = 12$, the function value at 2 matches the limit value at 2 and the function is continuous at 2.*

- $D^-f(2) = \lim_{x \to 2^-} \frac{x^2 + 3x + 2 - 12}{x - 2} = \lim_{x \to 2^-} \frac{(x + 5)(x - 2)}{x - 2} = \lim_{x \to 2^-}(x + 5) = 7.$

- $D^+f(2) = \lim_{x \to 2^+} \frac{x^3 + 4 - 12}{x - 2} = \lim_{x \to 2^+} \frac{(x^3 - 8)}{x - 2} = \lim_{x \to 2^+}(x^2 + 2x + 4) = 12.$

10.5   Differentiability

We are now ready to tackle differentiation carefully.

Definition 10.5.1 Differentiability of a Function at a Point

Let f be locally defined at p. f is said to be differentiable at a point p in its domain if the limit as x approaches p, $x \ne p$, of the quotients $\frac{f(x) - f(p)}{x - p}$ exists. When this limit exists with value A, the value of this limit is denoted by a number of possible symbols: $f'(p)$ or $\frac{df}{dx}(p)$.

This can also be phrased in terms of the right and left hand limits $f'(p^+)$ $=$ $\lim_{x \to p^+} \frac{f(x) - f(p)}{x - p}$ and $f'(p^-)$ $=$ $\lim_{x \to p^-} \frac{f(x) - f(p)}{x - p}$. If both exist and match at p, then $f'(p)$ exists and the value of the derivative is the common value.
Note $D^+f(p) = f'(p^+)$ and $D^-f(p) = f'(p^-)$.
Further, this is equivalent to saying there is a number A so that $\forall \epsilon > 0$, $\exists \delta > 0$, $\ni 0 < |x - p| < \delta \Rightarrow \left| \frac{f(x) - f(p)}{x - p} - A \right| < \epsilon$ where δ is smaller than the local radius of definition of f.

All of our usual machinery about limits can be brought to bear here. For example, all of the limit stuff could be rephrased in the $\epsilon - \delta$ framework but we seldom need to go that deep. However, the most useful way of all to view the derivative is to use an error term.

Let f have a derivative at p and denote its value by $f'(p)$. Define the error term $E(x,p)$ by

$$E(x,p) \quad = \quad f(x) - f(p) - f'(p)\,(x-p).$$

Note $E(x,p) = 0$ when $x = p$. Since $f'(p)$ exists, using the $\epsilon - \delta$ definition of differentiability, choose any $\epsilon > 0$. Then by definition, there is a δ so that

$$0 < |x-p| < \delta \quad \Longrightarrow \quad \left| \frac{f(x) - f(p)}{x - p} - f'(p) \right| < \epsilon.$$

We can rewrite this by getting a common denominator as

$$0 < |x-p| < \delta \quad \Longrightarrow \quad \left| \frac{f(x) - f(p) - f'(p)(x-p)}{x - p} \right| < \epsilon.$$

But the numerator here is the error, so we have

$$0 < |x-p| < \delta \quad \Longrightarrow \quad \left| \frac{E(x,p)}{x - p} \right| < \epsilon.$$

This tells us as $x \to p$, $|x-p| \to 0$ and $\frac{E(x,p)}{x-p} \to 0$ as well.
We can say more. Do as above for $\epsilon = 1$. Then there is a $\delta > 0$ so that

$$0 < |x-p| < \delta \quad \Longrightarrow \quad \left| \frac{E(x,p)}{x - p} \right| < 1.$$

This tells us

$$0 < |x-p| < \delta \quad \Longrightarrow \quad |E(x,p)| < |x-p|.$$

Thus as $x \to p$, $|x-p| \to 0$ and $E(x,p) \to 0$ as well. Combining, we see if f has a derivative at p, then $E(x,p) \to 0$ **and** $\frac{E(x,p)}{x-p} \to 0$ as $x \to p$.

We can argue the other way. Assume there is a number A so that $E(x,p) = f(x) - f(p) - A(x-p)$ and we know $E(x,p) \to 0$ and $\frac{E(x,p)}{x-p} \to 0$ as $x \to p$. Then, $\frac{f(x)-f(p)}{x-p} = \frac{E(x,p)+A(x-p)}{x-p} = \frac{E(x,p)}{x-p} + A$. But we are told that $\frac{E(x,p)}{x-p} \to 0$ as $x \to p$, so $\lim_{x \to p} \frac{f(x)-f(p)}{x-p} = A$ and f is differentiable at p with value A.

Theorem 10.5.1 Error Form for Differentiability of a Function at a Point

$f'(p)$ *exists if and only if*

- $E(x,p) \to 0$ *as* $x \to p$ *and*

- $E(x,p)/(x-p) \to 0$ *as* $x \to p$ *also.*

Proof 10.5.1
The arguments above have shown this. ■

Example 10.5.1 *Here is an example which should help. We will take our old friend* $f(x) = x^2$. *Let's look at the derivative of* f *at the point* x.

Solution *We have*

$$
\begin{aligned}
E(x + \Delta x, x) &= f(x + \Delta x) - f(x) - f'(x)\,\Delta x \\
&= (x + \Delta x)^2 - x^2 - 2x\,\Delta x \\
&= 2x\,\Delta x + (\Delta x)^2 - 2x\,\Delta x \\
&= (\Delta x)^2.
\end{aligned}
$$

See how $E(x + \Delta x, x) = (\Delta x)^2 \to 0$ *and* $E(x + \Delta x, x)/\Delta x = \Delta x \to 0$ *as* $\Delta x \to 0$? *This is the essential nature of the derivative.*

Since $f(x + \Delta x) = f(x) + f'(x)\Delta x + E(x + \Delta x, x)$, *replacing the original function value* $f(x + \Delta x)$ *by the value given by the straight line* $f(x) + f'(x)\,\Delta x$ *makes an error that is roughly proportional to* $(\Delta x)^2$. *Good to know.*

Also, since $f(x + \Delta x) = f(x) + f'(x)\Delta x + E(x + \Delta x, x)$ *as* $\Delta x \to 0$, $f(x + \Delta x) \to f(x)$ *telling us* f *is continuous at* x. *So* **differentiability at a point implies continuity at that point.**

Example 10.5.2 *We can show the derivative of* $f(x) = x^3$ *is* $3x^2$. *Using this, write down the definition of the derivative at* $x = 1$ *and also the error form at* $x = 1$. *State the two conditions on the error too.*

Solution *The definition of the derivative is*

$$
\frac{dy}{dx}(1) = \lim_{h \to 0} \frac{(1 + h)^3 - (1)^3}{h}.
$$

The error form is $E(x, 1) = f(x) - f(1) - f'(1)(x - 1) = x^3 - 1 - 3(x - 1)$. *Note by factoring,* $E(x, 1) = (x - 1)(x + x^2 - 2)$ *and so* $\frac{E(x,1)}{x-1} = x + x^2 - 2$ *and*

$$
x^3 = 1 + 3(x - 1) + E(x, 1)
$$

where you can see $E(x, 1)$ *and* $E(x, 1)/(x - 1)$ *both go to zero as* $x \to 1$.

10.5.1   Homework

Exercise 10.5.1 *For this function*

$$
f(x) = \begin{cases} 1, & \text{if } 0 \le x \le 1 \\ 5 & \text{if } 1 < x \le 2. \end{cases}
$$

- *Determine if* f *is continuous at* 1.

- *Calculate* $D^- f(1)$ *and* $D^+ f(1)$.

Exercise 10.5.2 *For this function*

$$
f(x) = \begin{cases} 4, & \text{if } 0 \le x \le 2 \\ 7 & \text{if } 2 < x \le 4. \end{cases}
$$

- *Determine if* f *is continuous at* 2.

- *Calculate* $D^- f(2)$ *and* $D^+ f(2)$.

Exercise 10.5.3 *For the function*

$$f(x) \;=\; \begin{cases} 2x + 4, & \text{if } 0 \le x \le 2 \\ x^2 + 4 & \text{if } 2 < x \le 4. \end{cases}$$

• *Determine if f is continuous at 2.*

• *Calculate $D^- f(2)$ and $D^+ f(2)$.*

Exercise 10.5.4 *For the function*

$$f(x) \;=\; \begin{cases} 2x^2, & \text{if } 0 \le x < 3 \\ x^3 & \text{if } 3 \le x \le 6. \end{cases}$$

• *Determine if f is continuous at 3.*

• *Calculate $D^- f(3)$ and $D^+ f(3)$.*

Exercise 10.5.5 *For the function*

$$f(x) \;=\; \begin{cases} 3x^2, & \text{if } 1 \le x < 2 \\ x^2 & \text{if } 2 \le x \le 4. \end{cases}$$

• *Determine if f is continuous at 1.5.*

• *Calculate $D^- f(1.5)$ and $D^+ f(1.5)$.*

• *Determine if f is continuous at 2.*

• *Calculate $D^- f(2)$ and $D^+ f(2)$.*

• *Determine if f is continuous at 3.*

• *Calculate $D^- f(3)$ and $D^+ f(3)$.*

Exercise 10.5.6 *Suppose a function has a jump at the point $x = -2$. Can this function have a derivative there?*

Exercise 10.5.7 *We can show the derivative of $f(x) = 2x^4 - 3x + 2$ is $8x^3 - 3$. Using this, write down the definition of the derivative at $x = 2$ and also the error form at $x = 2$. Show $E(x, 2)$ and $E(x, 2)/(x - 2)$ both go to 0 as x goes to 2.*

Exercise 10.5.8 *We can show the derivative of $f(x) = x^2 + 5$ is $2x$. Using this, write down the definition of the derivative at $x = 4$ and also the error form at $x = 4$. Show $E(x, 4)$ and $E(x, 4)/(x - 4)$ both go to 0 as x goes to 4.*

Exercise 10.5.9 *We can show the derivative of $f(x) = 3x^2 + 5$ is $6x$. Using this, write down the definition of the derivative at $x = 2$ and also the error form at $x = 2$. Show $E(x, 2)$ and $E(x, 2)/(x - 2)$ both go to 0 as x goes to 4.*

Chapter 11

The Properties of Derivatives

11.1 Simple Derivatives

We need some fast ways to calculate these derivatives. Let's start with constant functions. These never change and since derivatives are supposed to give rates of change, we would expect this to be zero. Here is the argument.

Let $f(x) = 5$ for all x. Then to find the derivative at any x, we calculate this limit

$$\frac{d}{dx}(5) \quad = \quad \lim_{h \to 0} \frac{f(x+h) - f(x)}{h} = \lim_{h \to 0} \frac{5 - 5}{h} = \lim_{h \to 0} 0 = 0.$$

We can also use an $\epsilon - \delta$ argument. Given any $\epsilon > 0$, note $0 < |y - x| < \delta \Rightarrow |\frac{f(y) - f(x)}{y - x} - 0| = |\frac{5 - 5}{y - x}| = 0 < \epsilon$. Here the value of δ does not matter at all. A little thought shows that the value 5 doesn't matter. So we have a general result which we dignify by calling it a theorem just because we can!

Theorem 11.1.1 The Derivative of a Constant

If c is any number, then the function $f(x) = c$ gives a constant function. The derivative of c with respect to t is then zero.

Proof 11.1.1
We just hammered this out! ∎

Let's do one more, the derivative of $f(x) = x$. We calculate

$$\frac{d}{dx}(x) \quad = \quad \lim_{h \to 0} \frac{f(x+h) - f(x)}{h} = \lim_{h \to 0} \frac{x + h - x}{h} = \lim_{h \to 0} \frac{h}{h} = \lim_{h \to 0} 1 = 1.$$

We can also use an $\epsilon - \delta$ argument. Given any $\epsilon > 0$, note $0 < |y - x| < \epsilon \Rightarrow |\frac{f(y) - f(x)}{y - x} - 1| = |\frac{y - x}{y - x} - 1| = 0 < \epsilon$. Here we choose $\delta = \epsilon$. We have a new theorem!

Theorem 11.1.2 The Derivative of x

The function $f(x) = x$ has derivative 1.

Proof 11.1.2
Just done. ∎

So now we know that $\frac{d}{dx}(x) = 1$ and it wasn't that hard. To find the derivatives of more powers of x, we are going to find an easy way. The easy way is to prove a general rule and then apply it for new functions. This general rule is called the **Product Rule**.

Homework

Exercise 11.1.1 *Prove using limits that the derivative of the function $f(x) = 5$ is zero.*

Exercise 11.1.2 *Prove using limits that the derivative of the function $f(x) = -1$ is zero.*

Exercise 11.1.3 *Prove using limits that the derivative of the function $f(x) = 5x$ is 5.*

Exercise 11.1.4 *Prove using limits that the derivative of the function $f(x) = -9x$ is -9.*

11.2   The Product Rule

The following theorem is one of our most useful tools.

Theorem 11.2.1 The Product Rule

> *If f and g are both differentiable at a point x, then the product fg is also differentiable at x and has the value $\left(f(x)\,g(x) \right)' = f'(x)\,g(x)\, +\, f(x)\,g'(x)$*

Proof 11.2.1

Consider this limit at x: $\frac{d}{dx}(fg)(x) = \lim_{h \to 0} \frac{f(x+h)\,g(x+h)\, -\, f(x)\,g(x)}{h}$. Let's add and subtract just the right term: $\frac{d}{dx}(fg)(x) = \lim_{h \to 0} \frac{f(x+h)g(x+h) - f(x)\,g(x+h) + f(x)\,g(x+h) - f(x)\,g(x)}{h}$. Now group the pieces like so:

$$\frac{d}{dx}(fg)(x) \;=\; \lim_{h \to 0} \left\{ \frac{(f(x+h)g(x+h) - f(x)g(x+h))}{h} \right.$$
$$\left. + \frac{(f(x)g(x+h) - f(x)g(x))}{h} \right\}$$

Now factor out common terms and rewrite as two separate limits:

$$\frac{d}{dx}(fg)(x) \;=\; \lim_{h \to 0} \left(\frac{f(x+h)\, -\, f(x)}{h} \right) g(x+h)$$
$$+\; \lim_{h \to 0} \left(\frac{g(x+h)\, -\, g(x)}{h} \right) f(x)$$

If both of these limits exist separately, then we know the sum of the limits equals the limit of the sum *and we are justified in writing these as two separate pieces.*

Look at the first limit: $\lim_{h \to 0} \left(\frac{f(x+h) - f(x)}{h} \right) g(x+h)$.

- *f is differentiable at x. Thus, $\lim_{h \to 0} \frac{f(x+h) - f(x)}{h} = f'(x)$.*

- *g is differentiable at x and so continuous at x giving $\lim_{h \to 0} g(x+h) = g(x)$.*

- *Since both limits exists, we know* the product of the limits is the limit of the product *giving $\lim_{h \to 0} \frac{f(x+h) - f(x)}{h} g(x+h) = f'(x)g(x)$.*

In the second limit, we consider $\lim_{h\to 0} \frac{g(x+h)-g(x)}{h} f(x)$.

- *We know g is differentiable at x. Thus, $\lim_{h\to 0} \frac{g(x+h)-g(x)}{h} = g'(x)$.*

- *$f(x)$ is constant here implying $\lim_{h\to 0} \frac{g(x+h)-g(x)}{h} f(x) = g'(x)f(x)$.*

Both limits exist, so using the sum of limits is the limit of the sum, we have $\frac{d}{dx}(f(x)\,g(x)) = f'(x)\,g(x) + f(x)\,g'(x)$.

*Let's redo this proof using the $\epsilon - \delta$ approach to show you how it is done. This is actually **much harder to do** than the limit approach. But you should see this. Remember, the more ways you can approach a problem, the better. Both the **limit** and the $\epsilon - \delta$ technique work, but the **limit** way is far more efficient.*

*Pick an arbitrary $\epsilon > 0$. Then $\frac{f(x+h)\,g(x+h)-f(x)\,g(x)}{h} =$
$\frac{f(x+h)g(x+h)-f(x)\,g(x+h)+f(x)\,g(x+h)-f(x)\,g(x)}{h}$. This simplifies to the sum of two terms:*

$\frac{f(x+h)-f(x)}{h} g(x+h) + f(x) \frac{g(x+h)-g(x)}{h}$. *We know since f and g are differentiable, they are locally bounded since they are continuous. So there are two positive numbers B_f and B_g so that $|f(y)| \leq B_f$ and $|g(y)| \leq B_g$ for all y in $B_r(x)$ for some $r > 0$.*

Now for convenience, let $\frac{\Delta(fg)}{h}$ denote our original expression. It is not very helpful to work with absolute values here (check this out and you'll see for yourself) so we are going to use paired $<$ and $<$ inequality chains.

- *Since g is continuous at x, given a tolerance $\xi_1 > 0$, there is a δ_1 so that $-\xi_1 < g(x+h) - g(x) < \xi_1$ if $|h| < \xi_1$.*

- *Since f and g are differentiable at x, for $\xi_2 > 0$ and $\xi_3 > 0$, there positive δ_2 and δ_3 so that $f'(x)-\xi_2 < \frac{f(x+h)-f(x)}{h} < f'(x)+\xi_2$ if $|h| < \delta_2$ and $g'(x)-\xi_3 < \frac{g(x+h)-g(x)}{h} < g'(x)+\xi_3$ if $|h| < \delta_3$*

So if $\delta < \min\{r, \delta_1, \delta_2, \delta_3\}$, all of our conditions hold. We have $(g(x) - \xi_1)(f'(x) - \xi_2) + f(x)(g'(x) - \xi_3) < \Delta(fg)/h < (g(x) + \xi_1)(f'(x) + \xi_2) + f(x)(g'(x) + \xi_3)$.

Simplifying, we find

$f(x)g'(x)+f'(x)g(x)-f'(x)\xi_1-g(x)\xi_2-f(x)\xi_3+\xi_1\xi_2 < \Delta(fg)/h < f(x)g'(x)+f'(x)g(x)+f'(x)\xi_1 + g(x)\xi_2 + f(x)\xi_3 + \xi_1\xi_2.$

Cancel the common term $\xi_1\xi_2$ and we find we can go back to using absolute values: $|\Delta(fg)/h - (f(x)g'(x) + f'(x)g(x))| < |f'(x)|\xi_1 + |g(x)|\xi_2 + |f(x)|\xi_3$.

It is now clear how to choose ξ_1, ξ_2 and ξ_3.

Set $\xi_1 = \epsilon/(3(|f'(x)| + 1))$, $\xi_2 = \epsilon/(3(B_g + 1))$ and $\xi_3 = \epsilon/(3(B_f + 1))$.

Then for the chosen δ, all three pieces are $< \epsilon/3$ and we have $|\Delta(fg)/h-(f(x)g'(x)+f'(x)g(x))| < \epsilon$ when $|h| < \delta$. Hence, $(fg)'(x) = f'(x)g(x) + f(x)g'(x)$. ∎

Homework

Exercise 11.2.1 *Use the limit approach to find the derivative of $(x^2 + 1)(5x + 3)$.*

Exercise 11.2.2 *Use the $\epsilon - \delta$ approach to find the derivative of $(x^2 + 1)(5x + 3)$.*

Exercise 11.2.3 *Use the limit approach to find the derivative of $(4x^2 + 1)(15x + 13)$.*

Exercise 11.2.4 *Use the $\epsilon - \delta$ approach to find the derivative of $(4x^2 + 1)(15x + 13)$.*

We are now ready to *blow your mind* as Jack Black would say.

- Let $f(x) = x^2$. Let's apply the product rule. The first function is x and the second one is x also.

$$\begin{aligned}
\frac{d}{dx}(x^2) &= \frac{d}{dx}(x)\,x + (x)\,\frac{d}{dx}(x) \\
&= (1)\,(x) + (x)\,(1) = 2\,x.
\end{aligned}$$

  This is just what we had before!

- Let $f(x) = x^3$. Let's apply the product rule here. The first function is x^2 and the second one is x.

$$\begin{aligned}
\frac{d}{dx}(x^3) &= \frac{d}{dx}(x^2)\,x + (x^2)\,\frac{d}{dx}(x) \\
&= (2\,x)\,(x) + (x^2)\,(1) = 3\,x^2.
\end{aligned}$$

- Let $f(x) = x^4$. Let's apply the product rule again. The first function is x^3 and the second one is x.

$$\begin{aligned}
\frac{d}{dx}(x^4) &= \frac{d}{dx}(x^3)\,x + (x^3)\,\frac{d}{dx}(x) \\
&= (3\,x^2)\,(x) + (x^3)\,(1) = 4\,x^3.
\end{aligned}$$

We could go on, but you probably see the pattern. If P is a positive integer, then $\frac{d}{dx}x^P = P\,x^{P-1}$. That's all there is too it. The actual proof is a POMI, of course.

Theorem 11.2.2 The Simple Power Rule: Positive Integers of x

> *If $f(x)$ is the function x^P for any positive integer P, then the derivative of f with respect to x satisfies $\left(x^P\right)' = P\,x^{P-1}$*

Proof 11.2.2
A careful proof would use POMI, but we did a relaxed version. ∎

Next, we don't want to overload you with a lot of tedious proofs of various properties, so let's just cut to the chase. From the way the derivative is defined as a limit, it is pretty clear the following properties hold:

- **the derivative of a sum of functions is the sum of the derivatives**: $(f + g)' = f' + g'$ holds at any x where both are differentiable.

- Second, **the derivative of a constant times a function is just constant times the derivative.** $(c\,f)' = c\,f'$ Armed with these tools, we can take a lot of derivatives!

Example 11.2.1 *Find $\left(t^2 + 4\right)'$.*

Solution

$$\left(t^2 + 4\right)' = (t^2)' + (4)' = 2t + 0 = 2t.$$

Example 11.2.2 *Find* $\left(3\,t^4\,+\,8\right)'$.

Solution

$$\left(3\,t^4\,+\,8\right)' \;=\; \left(3t^4\right)' + (8)' = 3(t^4)' + 0 = 3(4t^3) = 12\,t^3.$$

Example 11.2.3 *Find* $\left(3\,t^8\,+\,7\,t^5 - 2\,t^2 + 18\right)'$.

Solution

$$\left(3\,t^8\,+\,7\,t^5 - 2\,t^2 + 18\right)' \;=\; 24\,t^7 + 35\,t^4 - 4\,t.$$

Here are a few examples of product rule calculations.

Example 11.2.4 *Find* $\left((t^2\,+\,4)\,(5t^3\,+\,t^2\,-\,4)\right)'$.

Solution

$$\left((t^2 + 4)(5t^3 + t^2 - 4)\right)' \;=\; (2t)(5t^3 + t^2 - 4) + (t^2 + 4)(15t^2 + 2t)$$

and, of course, this expression above can be further simplified.

Example 11.2.5 *Find* $\left((2t^3 + t + 5)(-2t^6 + t^3 + 10)\right)'$.

Solution

$$\left((2t^3 + t + 5)(-2t^6 + t^3 + 10)\right)' \;=\; (6t^2 + 1)(-2t^6 + t^3 + 10)$$
$$+\; (2t^3 + t + 5)(-12t^5 + 3t^2)$$

Homework

Exercise 11.2.5 *If f and g are both differentiable at x, prove using an $\epsilon - \delta$ approach that $f + g$ is also differentiable at x.*

Exercise 11.2.6 *If f and g are both differentiable at x, prove using a limit approach that $f + g$ is also differentiable at x.*

Exercise 11.2.7 *If f and g are both differentiable at x, prove using an $\epsilon - \delta$ approach that $2f + 3g$ is also differentiable at x.*

Exercise 11.2.8 *If f and g are both differentiable at x, prove using a limit approach that $-5f + 6g$ is also differentiable at x.*

11.3   The Quotient Rule

Now let's do rational functions.

Theorem 11.3.1 The Derivative of $1/f$

If f is differentiable at a point x and $f(x) \neq 0$, then the quotient $1/f$ is also differentiable at x and has the value $-\frac{f'(x)}{f(x)^2}$.

Proof 11.3.1
Consider this limit at x: $\frac{d}{dx}(1/f)(x) = \lim_{h \to 0} \left(\frac{1}{f(x+h)} - \frac{1}{f(x)}\right)/h$. *Getting a common denomi-*

nator, we have $\frac{d}{dx}(1/f(x)) = \lim_{h \to 0} \left(\frac{f(x)-f(x+h)}{f(x)f(x+h)} \right) / h = \lim_{h \to 0} \left(\frac{f(x)-f(x+h)}{h} \right) \frac{1}{f(x)f(x+h)}$.

- *Since f is continuous at x and $f(x) \neq 0$, we know the fraction $1/f(x+h)$ is well defined locally at x since f is locally not zero at x.*

- *Since f is continuous, we know $\lim_{h \to 0} 1/f(x+h) = 1/f(x)$.*

We also know f is differentiable at x so $\lim_{h \to 0} \frac{f(x+h)-f(x)}{h} = f'(x)$. Combining, since each limit exists separately, the limit of the product is well defined and we have

$$\frac{d}{dx}(1/f(x)) = \lim_{h \to 0} \left(-\frac{f(x+h)-f(x)}{h} \right) \lim_{h \to 0} \frac{1}{f(x)f(x+h)} = -\frac{f'(x)}{f(x)^2}$$

∎

We can easily extend this result to more negative integer powers of x.

Theorem 11.3.2 The Simple Power Rule: Negative Integer Powers of x

> *If $f(x)$ is the function $1/x^P = x^{-P}$ for any positive integer P, then the derivative of f with respect to x satisfies $\left(x^P \right)' = -P\, x^{-P-1}$*

Proof 11.3.2
This is a standard POMI proof.   ∎

Example 11.3.1 *Find $(1/x)'$.*

Solution $(1/x)' = -(x)'/x^2 = -1/x^2 = -x^{-2}$.

Example 11.3.2 *Find $(1/x^3)'$.*

Solution $(1/x^3)' = -(x^3)'/(x^3)^2 = -3x^2/x^6 = -3x^{-4}$.

Theorem 11.3.3 The Quotient Rule

> *If f and g are differentiable at a point x and $g(x) \neq 0$, then the quotient f/g is also differentiable at x and has the value $\frac{f'(x)g(x)-f(x)g'(x)}{g^2(x)}$.*

Proof 11.3.3
$(f/g)' = (f \times 1/g)' = f' \times 1/g + f \times (-g'/g^2)$ by the product rule. Getting a common denominator, we have $(f/g)' = \frac{f'g-fg'}{g^2}$.   ∎

Example 11.3.3 *Prove $f(x) = 7x^2 - 6x + 8$ is differentiable at $x = 4$.*

Solution *We suspect the value of the derivative is $f'(4) = 14(4) - 6 = 50$. Let $\epsilon > 0$ be given. Consider*

$$\left| \frac{f(x)-f(4)}{x-4} - 50 \right| = \left| \frac{7x^2-6x+8-96}{x-4} - 50 \right| = \left| \frac{7x^2-6x-88-50(x-4)}{x-4} \right|$$

$$= \left| \frac{7x^2-56x+112}{x-4} \right| = \left| \frac{(7x-28)(x-4)}{x-4} \right| = 7|x-4|$$

Hence, if we choose $\delta < \epsilon/7$, we find

$$0 < |x - 4| < \delta \implies \left| \frac{f(x) - f(4)}{x - 4} - 50 \right| < \epsilon$$

which proves $f(x)$ is differentiable at $x = 4$.

Example 11.3.4 *Use an $\epsilon - \delta$ argument to prove if f and g are both differentiable at x, then $(6f - 9g)'(x) = 6f'(x) - 9g'(x)$.*

Solution *Let $\epsilon > 0$ be given. Consider*

$$\left| \left(\frac{(6f(x + h) - 9g(x + h)) - (6f(x) - 9g(x))}{h} \right) - (6f'(x) - 9g'(x)) \right|$$

$$= 6 \left| \left(\frac{f(x + h) - f(x)}{h} \right) - f'(x) \right| + 9 \left| \left(\frac{g(x + h) - g(x)}{h} \right) - g'(x) \right|$$

Since $f'(x)$ exists, there is a δ_1 so that

$$0 < |h| < \delta_1 \implies \left| \left(\frac{f(x + h) - f(x)}{h} \right) - f'(x) \right| < \frac{\epsilon}{2(6)}$$

and Since $g'(x)$ exists, there is a δ_2 so that

$$0 < |h| < \delta_2 \implies \left| \left(\frac{g(x + h) - g(x)}{h} \right) - g'(x) \right| < \frac{\epsilon}{2(9)}$$

So if $\delta < \min\{\delta_1, \delta_2\}$, both inequalities hold and we have

$$\left| \left(\frac{(6f(x + h) - 9g(x + h)) - (6f(x) - 9g(x))}{h} \right) - (6f'(x) - 9g'(x)) \right|$$

$$< 6 \frac{\epsilon}{2(6)} + 9 \frac{\epsilon}{2(9)} < \epsilon$$

Hence, $(6f - 9g)'(x) = 6f'(x) - 9g'(x)$.

11.3.1   Homework

Exercise 11.3.1 *Use an $\epsilon - \delta$ argument to prove $\frac{2x+1}{3x+2}$ is differentiable at $x = 2$.*

Exercise 11.3.2 *Use an $\epsilon - \delta$ argument to prove $\frac{2x^2+1}{3x^2+2}$ is differentiable at $x = 1$.*

Exercise 11.3.3 *Look at this function which is similar to one we have defined earlier.*

$$D(x) = \begin{cases} x^2, & x \in \mathbb{Q} \\ -x^2, & x \in \mathbb{R} \end{cases}$$

Prove that D is differentiable at only 0.

Exercise 11.3.4 *Look at this function which is similar to one we have defined earlier.*

$$D(x) = \begin{cases} 3x^2, & x \in \mathbb{Q} \\ -4x^2, & x \in \mathbb{R} \end{cases}$$

Prove that D is differentiable at only 0.

Exercise 11.3.5 *Use POMI to prove the* $(x^{-P})' = -Px^{-P-1}$ *for all positive integers* P.

Exercise 11.3.6 *Use an* $\epsilon - \delta$ *argument to prove* $f(x) = 3x^2 + 7x + 8$ *is differentiable at* $x = 2$.

Exercise 11.3.7 *Use an* $\epsilon - \delta$ *argument to prove* $f(x) = -2x^2 + 5x + 1$ *is differentiable at* $x = -1$.

11.4   Function Composition and Continuity

The composition of functions is actually a simple concept. You shove one function into another and calculate the result. Let's look at a simple example of this. We know how to find x^2 and u^2 for any x and u. So what about $(x^2 + 3)^2$? This just means take $u = x^2 + 3$ and square it. That is if $f(u) = u^2$ and $g(x) = x^2 + 3$, the *composition* of f and g is simply $f(g(x))$.

Let's talk about continuity first and then we'll go on to the idea of taking the derivative of a composition. If f and g are both continuous, then if you think about it, another way to phrase the continuity of f is that

$$\lim_{y \to x} f(y) \;\; = \;\; f(x) \implies f(x) = f(\lim_{y \to x} y).$$

So if f and g are both continuous, we can say

$$\lim_{y \to x} f(g(y)) \;\; = \;\; f(\lim_{y \to x} g(y)) = f(g(\lim_{y \to x} y)) = f(g(x)).$$

So the **composition of continuous functions is continuous**. This is a great result as smoothness has not been lost by pushing one smooth function into another smooth function!

Let's do this using the $\epsilon - \delta$ approach. We assume f is locally defined at the point $g(x)$ and g is locally defined at the point x. So there are radii r_f and r_g so that $f(w)$ is defined if $w \in B_{r_f}(g(x))$ and $g(y)$ is defined if $y \in B_{r_g}(x)$.

For any ϵ, there is a δ_1 so that $|f(u) - f(g(x))| < \epsilon$ if $|u - g(x)| < \delta_1$ because f is continuous at x. Of course, $\delta_1 < r_f$.

For the tolerance δ_1, there is a δ_2 so that $|g(y) - g(x)| < \delta_1$ if $|y - x| < \delta_2$ because f is continuous at x. Of course, $\delta_2 < r_g$. Thus, $|y - x| < \delta_2$ implies $|g(y) - g(x)| < \delta_1$ which implies $|f(g(y)) - f(g(x))| < \epsilon$.

This shows $f \circ g$ is continuous at x. You should be able to understand both types of arguments here!

11.5   Chain Rule

The next question is whether or not the composition of functions having derivatives gives a new function which has a derivative. And how could we calculate this derivative if the answer is yes? It turns out this is true but to see if requires a bit more work with limits.

So we want to know what $\frac{d}{dx}(f(g(x)))$ is. First, if g was always constant, the answer is easy. It is $\frac{d}{dx}(f(\mathbf{constant})) = 0$ which is a special case of the formula we are going to develop. So let's assume g is not constant locally. So there is a radius $r > 0$ so that $g(y) \neq g(x)$ for all $y \in B_r(x)$. Another way of saying this is $g(x + h) - g(x) \neq 0$ if $|h| < r$.

Then, we want to calculate

$$\frac{d}{dx}\left(f(g(x))\right) = \lim_{h \to 0} \frac{f(g(x+h)) - f(g(x))}{h}.$$

Rewrite by dividing and multiplying by $g(x+h) - g(x)$ which is ok to do as we assume g is not constant locally and so we don't divide by 0. We get

$$\left(f(g(x))\right)' = \lim_{h \to 0} \frac{f(g(x+h)) - f(g(x))}{g(x+h) - g(x)} \frac{g(x+h) - g(x)}{h}$$

Now $\lim_{h \to 0} \frac{g(x+h) - g(x)}{h} = g'(x)$ because we know g is differentiable at x. Now let's do some relabeling to make it easier to see what is going on.

Let $u = g(x)$ and $\Delta u = g(x+h) - g(x)$. Then, $g(x+h) = g(x) + \Delta u = u + \Delta u$.

Then, we have $\frac{f(g(x+h)) - f(g(x))}{g(x+h) - g(x)} = \frac{f(u+\Delta u) - f(u)}{\Delta u}$.

Since g is continuous because g has a derivative, we have $\lim_{h \to 0} \Delta u = \lim_{h \to 0} (g(x+h) - g(x)) = 0$.

Also, since f is differentiable at $u = g(x)$, we know $\lim \Delta u \to 0 \left(f(u + \Delta u) - f(u) \right)/\Delta u = f'(u) = f'(g(x))$.

Thus,

$$\lim_{h \to 0} \frac{f(g(x+h)) - f(g(x))}{g(x+h) - g(x)} = \lim_{h \to 0} \frac{f(u + \Delta u) - f(u)}{\Delta u}$$
$$= \lim_{\Delta u \to 0} \frac{f(u + \Delta u) - f(u)}{\Delta u} = f'(u) = f'(g(x)).$$

Since both limits above exist, we know

$$\lim_{h \to 0} \left(\frac{f(g(x+h)) - f(g(x))}{g(x+h) - g(x)} \frac{g(x+h) - g(x)}{h} \right) =$$
$$\left(\lim_{h \to 0} \frac{f(g(x+h)) - f(g(x))}{g(x+h) - g(x)} \right) \left(\lim_{h \to 0} \frac{g(x+h) - g(x)}{h} \right) = f'(g(x))\, g'(x).$$

This result is called the **Chain Rule**.

Theorem 11.5.1 Chain Rule for Differentiation: Limit Proof

If the composition of f and g is locally defined at a number x and if both $f'(g(x))$ and $g'(x)$ exist, then the derivative of the composition of f and g also exists and is given by $\frac{d}{dx}(f(g(x))) = f'(g(x))\, g'(x)$

Proof 11.5.1
We reasoned this out above. We usually think about this as follows $\frac{d}{dx}(f(\textbf{inside})) = f'(\textbf{inside})$ **inside**$'(x)$. ∎

Let's do a error term based proof. Make sure you pay attention to how each of our approaches get the job done but in a different way. Learning how to use different tools is an important part of your education.

Theorem 11.5.2 Chain Rule for Differentiation: Error Proof

> *If the composition of f and g is locally defined at a number x and if both $f'(g(x))$ and $g'(x)$ exist, then the derivative of the composition of f and g also exists and is given by $\frac{d}{dx}(f(g(x))) = f'(g(x)) \, g'(x)$*

Proof 11.5.2

Again, we do the case where $g(x)$ is not a constant locally. Like before, let $\Delta u = g(x + h) - g(x)$ with $u = g(x)$. Using error terms, since f has a derivative at $u = g(x)$ and g has a derivative at x, we can write $f(u+\Delta u) = f(u)+f'(u)\Delta u+E_f(u+\Delta u, u)$ where $\lim_{\Delta u \to 0}$ of both $E_f(u+\Delta u, u)$ and $E_f(u + \Delta u, u)/\Delta u$ are zero and $g(x + h) = g(x) + g'(x)\,h + E_g(x + h, x)$ where $\lim_{h \to 0}$ of both $E_g(x + h, x)$ and $E_g(x + h, x)/h$ are zero.

Then, since $u = g(x)$ and $\Delta u = g(x + h) - g(x)$, we have

$$
\begin{aligned}
\frac{d}{dx}\left(f(g(x))\right) &= \lim_{h \to 0} \frac{f(g(x + h)) - f(g(x))}{h} \\
&= \lim_{h \to 0} \frac{f(u + \Delta u) - f(u)}{h} \\
&= \lim_{h \to 0} \frac{f'(u)\,\Delta u + E_f(u + \Delta u, u)}{h} \\
&= \lim_{h \to 0} \left(\frac{f'(u)\,\Delta u + E_f(u + \Delta u, u)}{\Delta u} \frac{\Delta u}{h} \right)
\end{aligned}
$$

where we know $\Delta u \neq 0$ locally because g is not constant locally.

We see $\lim_{h \to 0} \frac{\Delta u}{h} = \lim_{h \to 0} \frac{g(x+h)-g(x)}{h} = g'(x)$ because g has a derivative at x.

Next, $\lim_{h \to 0} \frac{f'(u)\,\Delta u + E_f(u+\Delta u, u)}{\Delta u} = f'(u) + \lim_{h \to 0} \frac{E_f(u+\Delta u, u)}{\Delta u}$. But

$$
\lim_{h \to 0} \frac{E_f(u + \Delta u, u)}{\Delta u} = \lim_{\Delta u \to 0} \frac{E_f(u + \Delta u, u)}{\Delta u} = 0.
$$

Thus, $\lim_{h \to 0} \frac{f'(u)\,\Delta u + E_f(u+\Delta u, u)}{\Delta u} = f'(u)$. Since both limits exist, we have

$$
\begin{aligned}
\frac{d}{dx}\left(f(g(x))\right) &= \left(\lim_{h \to 0} \frac{f'(u)\,\Delta u + E_f(u + \Delta u, u)}{\Delta u} \right) \left(\lim_{h \to 0} \frac{\Delta u}{h} \right) \\
&= f'(g(x)) \, g'(x).
\end{aligned}
$$

∎

Finally, let's do an $\epsilon - \delta$ based proof.

Theorem 11.5.3 Chain Rule for Differentiation: $\epsilon - \delta$ Proof

> *If the composition of f and g is locally defined at a number x and if both $f'(g(x))$ and $g'(x)$ exist, then the derivative of the composition of f and g also exists and is given by $\frac{d}{dx}(f(g(x))) = f'(g(x))\, g'(x)$*

Proof 11.5.3

Again, we do the case where $g(x)$ is not a constant locally. Like before, let $\Delta u = g(x+h) - g(x)$ with $u = g(x)$. Since g is not constant locally at x, there is a radius r_g so that $g(x_h) - g(x) \neq 0$ when $|h| < r_g$. Again we write

$$\frac{d}{dx}(f(g(x))) = \lim_{h\to 0} \frac{f(g(x+h)) - f(g(x))}{g(x+h) - g(x)} \frac{g(x+h) - g(x)}{h}$$

Then, since $u = g(x)$ and $\Delta u = g(x+h) - g(x)$, we have

$$\frac{f(g(x+h)) - f(g(x))}{g(x+h) - g(x)} \frac{g(x+h) - g(x)}{h} = \frac{f(u+\Delta u) - f(u)}{\Delta u} \frac{\Delta u}{h}.$$

Since g is differentiable at x, for a given ξ_1, there is a δ_1 so that $|h| < \delta_1$ implies $g'(x) - \xi_1 < \frac{g(x+h)-g(x)}{h} = \frac{\Delta u}{h} < g'(x) + \xi_1$.

And since f is differentiable at $u = g(x)$, for a given ξ_2, there is a δ_2 so that $|\Delta u| < \delta_2$ implies $f'(u) - \xi_2 < \frac{f(u+\Delta u) - f(u)}{\Delta u} < f'(u) + \xi_2$.

Let $\frac{f(g(x+h)) - f(g(x))}{g(x+h) - g(x)} \frac{g(x+h)-g(x)}{h} = \frac{\Delta f}{\Delta u} \frac{\Delta u}{h}$ for convenience. Then, if $\delta < \min\{r_g, \delta_1, \delta_2\}$, all conditions hold and we have

$$(f'(u) - \xi_2)\,(g'(x) - \xi_1) < \frac{\Delta f}{\Delta u} \frac{\Delta u}{h} < (f'(u) + \xi_2)\,(g'(x) + \xi_1)$$

Multiplying out these terms and canceling the $\xi_1\xi_2$, we have

$$f'(u)g'(x) - \xi_1 f'(u) - \xi_2 g'(x) < \frac{\Delta f}{\Delta u} \frac{\Delta u}{h}$$
$$< f'(u)g'(x) + \xi_1 f'(u) + \xi_2 g'(x)$$

$$-\xi_1 f'(u) - \xi_2 g'(x) < \frac{\Delta f}{\Delta u} \frac{\Delta u}{h} - (\,f'(u)g'(x)\,) < \xi_1 f'(u) + \xi_2 g'(x)$$

Thus,

$$\left| \frac{\Delta f}{\Delta u} \frac{\Delta u}{h} - (\,f'(u)g'(x)\,) \right| < \xi_1 |f'(u)| + \xi_2 |g'(x)|$$

Choose $\xi_1 = \frac{\epsilon}{2(|f'(g(x))|+1)}$ and $\xi_2 = \frac{\epsilon}{2(|g'(x)|+1)}$.

Then, we have $|h| < \delta$ implies $\left| \frac{\Delta f}{\Delta u} \frac{\Delta u}{h} - (\,f'(u)g'(x)\,) \right| < \epsilon$.

This proves the result! ∎

Comment 11.5.1 *You should know how to attack this proof all three ways!*

Example 11.5.1 *Find the derivative of $(t^3 + 4)^3$.*

Solution *It is easy to do this if we think about it this way.*

$$\left((\text{thing})^{\text{power}}\right)' \;=\; \text{power} \times (\text{thing})^{\text{power - 1}} \times (\text{thing})'$$

Thus,

$$\left((t^3 + 4)^3\right)' \;=\; 3\,(t^3 + 4)^2\,(3t^2)$$

Example 11.5.2 *Find the derivative of* $1/(t^2 + 4)^3$.

Solution *This is also*

$$\left((\text{thing})^{\text{power}}\right)' \;=\; \text{power} \times (\text{thing})^{\text{power - 1}} \times (\text{thing})'$$

where **power** *is* -3 *and* **thing** *is* $t^2 + 4$. *So we get*

$$\left(1/(t^2 + 4)^3\right)' \;=\; -3\,(t^2 + 4)^{-4}\,(2t)$$

Example 11.5.3 *Find the derivative of* $(6t^4 + 9t^2 + 8)^6$.

Solution

$$\left((6t^4 + 9t^2 + 8)^6\right)' \;=\; 6\,(6t^4 + 9t^2 + 8)^5\,(24t^3 + 18t)$$

11.5.1 Homework

Exercise 11.5.1 *Use an* $\epsilon - \delta$ *argument to show* $\sqrt{x}$ *is continuous from the right at* $x = 0$.

Exercise 11.5.2 *Use an* $\epsilon - \delta$ *argument to show* $\sqrt{x}$ *is continuous at* $x = 1$.

Exercise 11.5.3 *Use an* $\epsilon - \delta$ *argument to show* $\sqrt{x}$ *is differentiable at* $x = 2$.

Exercise 11.5.4 *Use an* $\epsilon - \delta$ *argument to show* $\sqrt{x}$ *is differentiable at any* $x > 0$.

Recall in the error form for differentiation, $f(x + h) = f(x) + f'(x)h + E(x + h, x)$ where the linear function $T(x) = f(x) + f'(x)h$ is called the tangent line approximation to f at the base point x.

Exercise 11.5.5 *For* $f(x) = 2x^3 + 3x + 2$, *sketch* $f(x)$ *and* $T(x)$ *on the same graph carefully at the points* $x = -1$, $x = 0.5$ *and* $x = 1.3$. *Also draw a sample slope triangle* $\frac{f(x+h)-f(x)}{h}$ *for each base point. Finally, draw in the error function as a vertical line at the base points. Use multiple colors!! Also, write a MATLAB function to do these plots for you on the same plot.*

Exercise 11.5.6 *If* f *is continuous at* x, *why is it true that* $\lim_{n \to \infty} |f(x_n)| = |f(x)|$ *for any sequence* (x_n) *with* $x_n \to x$?

Exercise 11.5.7 *If* f *is continuous at* x, *why is it true that* $\lim_{n \to \infty} \sqrt{f(x_n)} = \sqrt{f(x)}$ *for any sequence* (x_n) *with* $x_n \to x$?

11.6 Sin, Cos and All That!

We will now go over some basic material about sin and cos functions. We want to take their derivatives and so forth. We will assume you know about sin and cos and their usual properties. But now we want to find the derivatives of the *sin* and *cos* functions and this leads us to limiting operations we can't handle using our tools so far. Hence, we will do this indirectly using geometric arguments. Look at Figure 11.1.

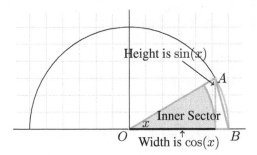

The circle here has radius 1 and the angle x determines three areas: the area of the inner sector, $\frac{1}{2} \cos^2(x) \, x$, the area of triangle $\triangle \mathbf{OAB}$, $\frac{1}{2} \sin(x)$ and the area of the outer sector, $\frac{1}{2} x$. We see the areas are related by

$$\frac{1}{2} \cos^2(x) \, x \; < \; \frac{1}{2} \sin(x) \; < \; \frac{1}{2} x.$$

Figure 11.1: The sin and cos geometry.

From it, we can figure out three important relationships.

- from high school times, you should know a number of cool things about circles. The one we need is the area of what is called a sector. Draw a circle of radius r in the plane. Measure an angle x counterclockwise from the horizontal axis. Then look at the pie shaped wedge formed in the circle that is bounded above by the radial line, below by the horizontal axis and to the side by a piece of the circle. It is easy to see this in the figure.

- Looking at the picture, note there is a first sector or radius $\cos(x)$ and a larger sector of radius 1. It turns out the area of a sector is $1/2 r^2 \theta$ where θ is the angle that forms the sector. From the picture, the area of the first sector is clearly less than the area of the second one. So we have

$$(1/2) \cos^2(x) \, x \quad < \quad (1/2) \, x$$

- Now if you look at the picture again, you'll see a triangle caught between these two sectors. This is the triangle you get with two sides having the radial length of 1. The third side is the straight line right below the arc of the circle cut out by the angle x. The area of this triangle is $(1/2) \sin(x)$ because the height of the triangle is $\sin(x)$. This area is smack dab in the middle of the two sector areas. So we have

$$(1/2) \cos^2(x) \, x \quad < \quad (1/2) \sin(x) < (1/2) \, x.$$

- These relationships work for all x and canceling all the $(1/2)$'s, we get

$$\cos^2(x) \, x \quad < \quad \sin(x) < x.$$

- Now as long as x is positive, we can divide to get

$$\cos^2(x) \;<\; \sin(x)/x \;<\; 1.$$

Homework

Exercise 11.6.1 *Using the ideas above, find* $\lim_{x \to 0} \frac{\sin(3x)}{5x}$.

Exercise 11.6.2 *Using the ideas above, find* $\lim_{x \to 0} \frac{\sin(2x)}{7x}$.

Exercise 11.6.3 *Follow the arguments we have used to find the derivative of* $\sin(7x)$.

Exercise 11.6.4 *Follow the arguments we have used to find the derivative of* $\sin(8x)$.

From our high school knowledge about cos, we know it is a very smooth function and has no jumps. So it is continuous everywhere and so $\lim_{x \to 0} \cos(x) = 1$ since $\cos(0) = 1$. If that is true, then the limit of the square is 1^2 or still 1. So $\lim_{x \to 0} \sin(x)/x$ is trapped between the limit of the $\cos^2$ term and the limit of the constant term 1. So we have to conclude

$$\lim_{x \to 0^+} \sin(x)/x \;=\; 1.$$

We can do the same thing for $x \to 0^-$, so we know $\lim_{x \to 0} \sin(x)/x = 1$. We can use this as follows. Note

$$\left(\sin(x)\right)'(0) \;=\; \lim_{h \to 0} \frac{\sin(h) - \sin(0)}{h} = \lim_{h \to 0} \frac{\sin(h)}{h}$$

because we know $\sin(0) = 0$. Now if $\lim_{x \to 0} \sin(x)/x = 1$, it doesn't matter if we switch letters! Hence, we also know $\lim_{h \to 0} \sin(h)/h = 1$. Using this, we see

$$\left(\sin(x)\right)'(0) \;=\; \lim_{h \to 0} \frac{\sin(h)}{h} = 1 = \cos(0)$$

as $\cos(0) = 1$! This result is the key. Consider the more general result

$$\left(\sin(x)\right)' \;=\; \lim_{h \to 0} \frac{\sin(x + h) - \sin(x)}{h}.$$

Now recall your sin identities. We know

$$\sin(u + v) \;=\; \sin(u)\,\cos(v) \,+\, \cos(u)\,\sin(v)$$

and so

$$\sin(x + h) \;=\; \sin(x)\,\cos(h) \,+\, \cos(x)\,\sin(h).$$

Using this we have

$$\begin{aligned}
\left(\sin(x)\right)' &= \lim_{h \to 0} \frac{\sin(x + h) - \sin(x)}{h} \\
&= \lim_{h \to 0} \frac{\sin(x)\,\cos(h) \,+\, \cos(x)\,\sin(h) - \sin(x)}{h} \\
&= \lim_{h \to 0} \frac{\sin(x)\,(-1 + \cos(h)) \,+\, \cos(x)\,\sin(h)}{h}.
\end{aligned}$$

Now regroup a bit to get

$$\left(\sin(x)\right)' = \lim_{h \to 0} \sin(x) \frac{(-1 + \cos(h))}{h} + \cos(x) \frac{\sin(h)}{h}.$$

We are about done. Rewrite $(1 - \cos(h))/h$ by multiplying top and bottom by $1 + \cos(h)$. This gives

$$\frac{\sin(x) (-1 + \cos(h))}{h} = \frac{\sin(x) (-1 + \cos(h))}{h} \frac{(1 + \cos(h))}{(1 + \cos(h))}$$

$$= -\sin(x) \frac{(1 - \cos^2(h))}{h} \frac{1}{(1 + \cos(h))}.$$

Now $1 - \cos^2(h) = \sin^2(h)$, so we have

$$\frac{\sin(x) (-1 + \cos(h))}{h} = -\sin(x) \frac{\sin^2(h)}{h} \frac{1}{(1 + \cos(h))}$$

$$= -\sin(x) \frac{\sin(h)}{h} \frac{\sin(h)}{1 + \cos(h)}.$$

We know $\sin(h)/h$ goes to 1 and $\sin(h)/(1 + \cos(h))$ goes to $0/2 = 0$ as h goes to zero. So the first term goes to $-\sin(x) \times 1 \times 0 = 0$. Since $\cos(0) = 1$ and cos is continuous, the first limit is $\sin(x) (0)$. We also know the second limit is $\cos(x) (1)$. So we conclude

$$\left(\sin(x)\right)' = \cos(x).$$

And all of this because of a little diagram drawn in Quadrant I for a circle of radius 1 plus some high school trigonometry!

What about cos's derivative? The easy way to remember that sin and cos are shifted versions of each other. We know $\cos(x) = \sin(x + \pi/2)$. So by the chain rule

$$\left(\cos(x)\right)' = \left(\sin(x + \pi/2)\right)' = \cos(x + \pi/2) (1).$$

Now remember another high school trigonometry thing.

$$\cos(u + v) = \cos(u) \cos(v) - \sin(u) \sin(v)$$

and so $\cos(x + \pi/2) = \cos(x) \cos(\pi/2) - \sin(x) \sin(\pi/2)$. We also know $\sin(\pi/2) = 1$ and $\cos(\pi/2) = 0$. So we find

$$\left(\cos(x)\right)' = -\sin(x).$$

Let's summarize:

$$(\sin(x))' = \cos(x)$$
$$(\cos(x))' = -\sin(x)$$

And, of course, we can use the chain rule too.

11.6.1   Examples

Example 11.6.1 *Simple chain rule! Find* $\left(\sin(4t)\right)'$.

Solution

$$\left(\sin(4t)\right)' = \cos(4t) \times 4 = 4\cos(4t).$$

Example 11.6.2 *Differentiate* $\sin^3(t)$

Solution *The derivative is*

$$\begin{aligned}\left(\sin^3(t)\right)' &= 3\sin^2(t)\left((\sin)'(t)\right)\\ &= 3\sin^2(t)\cos(t)\end{aligned}$$

Example 11.6.3 *Find* $\left(sin^3(x^2+4)\right)'$.

Solution

$$\left(sin^3(x^2+4)\right)' = 3\sin^2(x^2+4)\cos(x^2+4)\times(2x).$$

Example 11.6.4 *Chain and product rule*
Find $(\sin(2x)\cos(3x))'$.

Solution

$$\left(\sin(2x)\cos(3x)\right)' = 2\cos(2x)\cos(3x) - 3\sin(2x)\sin(3x).$$

Example 11.6.5 *Quotient rule!*
Find $(\tan(x))'$.

Solution *We usually don't simplify our answers, but we will this time as we are getting a new formula!*

$$\left(\tan(x)\right)' = \left(\frac{\sin(x)}{\cos(x)}\right)'$$
$$= \frac{\cos(x)\cos(x) - \sin(x)(-\sin(x))}{\cos^2(x)} = \frac{\cos^2(x) + \sin^2(x)}{\cos^2(x)}.$$

Homework

Exercise 11.6.5 *Find the derivative of* $\sec(8x)$.

Exercise 11.6.6 *Find the derivative of* $\cot(7x)$.

Exercise 11.6.7 *Find the derivative of* $\sin(x^2)$.

Exercise 11.6.8 *Find the derivative of* $\cos^2(8x)$.

11.6.2 A New Power Rule

Now that we have more functions to work with, let's use the chain rule in the special case of a power of a function. We have used this already, but now we state it as a theorem.

Theorem 11.6.1 Power Rule for Functions

If f is differentiable at the real number x, then for any integer p,

$$(f^p(x))' = p\left(f^{p-1}(x)\right) f'(x)$$

Example 11.6.6 *Differentiate* $(1 + \sin^3(2t))^4$

Solution $\frac{d}{dt}((1+\sin^3(2t))^4) = 4(1+\sin^3(2t))^3 3\sin^2(2t)\cos(2t)2$. *Next, recall* $\cos^2(x)+\sin^2(x) = 1$ *always and so*

$$\left(\tan(x)\right)' = \frac{1}{\cos^2(x)}.$$

Recall $1/\cos^2(x)$ *is called* $\sec^2(x)$. *So we have a new formula:*

$$(\tan(x))' = \sec^2(x)$$

Homework

Exercise 11.6.9 *Find the derivative of* $(\sec(8x) + \sin(3x))^4$.

Exercise 11.6.10 *Find the derivative of* $(\cot(7x) + 2\sin(x^2))^5$.

Exercise 11.6.11 *Find the derivative of* $(3\sin(x^2 + x + 1) + 2x^2 + 5)^7$.

Exercise 11.6.12 *Find the derivative of* $\sin(\cos^2(8x) + \sin^4(7x^3 + 2))$.

11.6.3 More Complicated Derivatives

Let $f(x) = x^{3/7}$. Then $(f(x))^7 = x^3$. Let $x_n \to x$ be any sequence converging to x. Then $\lim_{n\to\infty}(f(x_n))^7 = \lim_{n\to\infty} x_n^3 = x^3 = (f(x))^7$. Also, since the function u^7 is continuous, we have $(\lim_{n\to\infty} f(x_n))^7 = x^3 = f(x)^7$. Now take the root to see $(\lim_{n\to\infty} f(x_n)) = x^3 = f(x)$ which shows $f(x) = x^{3/7}$ is continuous. A little thought shows this argument works for $f(x) = x^{p/q}$ for any rational number p/q.

Now what about differentiability? We have for $x \neq 0$, $f(x + h) - f(x)$ is not zero for h sufficiently small and so we can write

$$3x^2 = \frac{d}{dx}(f(x))^7 = \lim_{h\to 0} \frac{(f(x+h))^7 - (f(x))^7}{f(x+h) - f(x)} \frac{f(x+h) - f(x)}{h}$$

Since we know $\lim_{h\to 0} \frac{(f(x+h))^7 - (f(x))^7}{f(x+h) - f(x)} = 7(f(x))^6$, the limit we seek is the ratio of two known limits and so

$$\lim_{h\to 0} \frac{f(x+h) - f(x)}{h} = \frac{3x^2}{7(f(x))^6} = \frac{3}{7}x^{-4/7}.$$

Note this argument fails for this next example. We let

$$f(x) \;=\; \begin{cases} 1, & x \in \mathbb{Q} \\ -1, & x \in \mathbb{IR} \end{cases}$$

Then $(f(x))^2 = 1$ which is differentiable. But we can't write

$$0 = \frac{d}{dx}(f(x))^2 \;=\; \lim_{h \to 0} \frac{(f(x+h))^2 - (f(x))^2}{f(x+h) - f(x)} \frac{f(x+h) - f(x)}{h}$$

because $f(x+h) - f(x)$ is never locally not zero.

We will often need to find the derivatives of more interesting things than simple polynomials. Finding rate of change is our mantra now! Let's look at a small example of how an excitable neuron transforms signals that come into it into output signals called **action potentials**. For now, think of an excitable neuron as a processing node which accepts inputs x and transforms them using a processing function we will call $\sigma(x)$ into an output y. As we know, neural circuits are built out of thousands of these processing nodes and we can draw them as a graph which shows how the nodes interact. Look at the simple example in Figure 11.2.

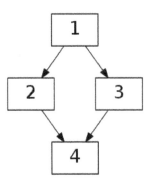

Figure 11.2: A simple neural circuit: $1 - 2 - 1$.

In the picture you'll note there are four edges connecting the neurons. We'll label them like this: $E_{1 \to 2}$, $E_{1 \to 3}$, $E_{2 \to 4}$ and $E_{3 \to 4}$. When an excitable neuron generates an action potential, the action potential is like a signal. The rest voltage of the cell is about -70 millivolts and if the action potential is generated, the voltage of the cell rises rapidly to about 80 millivolts or so and then falls back to rest. The shape of the action potential is like a scripted response to the conditions the neuron sees as the input signal. In many ways, the neuron response is like a digital **on** and **off** signal, so many people have modeled the response as a curve that rises smoothly from 0 (the **off**) to 1 (the **on**). Such a curve looks like the one shown in Figure 11.3.

The standard neural processing function has a high derivative value at 0 and small values on either side. You can see this behavior in Figure 11.4.

We can model this kind of function using many approaches: for example, $\sigma(x) = 0.5(1 + \tanh(x))$ works and we can also build σ using exponential functions which we will get to later. Now the action potential from a neuron is fed into the input side of other neurons and the strength of that interaction is modeled by the edge numbers $E_{i \to j}$ for our various i and j's. To find the input into a neuron, we take the edges going in and multiply them by the output of the node the edge is coming from. If we let Y_1, Y_2, Y_3 and Y_4 be the outputs of our four neurons, then if x is the input fed into neuron one, this is what happens in our small neural model

$$Y_1 \;=\; \sigma(x)$$

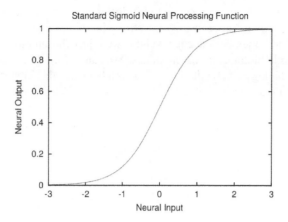

Figure 11.3: A simple neural processing function.

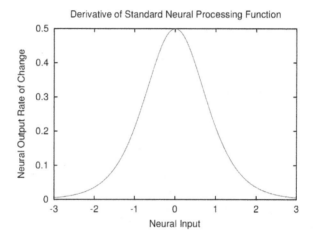

Figure 11.4: Neural processing derivatives.

$$
\begin{aligned}
Y_2 &= \sigma(E_{1\to 2}\, Y_1) \\
Y_3 &= \sigma(E_{1\to 3}\, Y_1) \\
Y_4 &= \sigma(E_{2\to 4}\, Y_2 \,+\, E_{3\to 4}\, Y_3).
\end{aligned}
$$

Note that Y_4 depends on the initial input x in a complicated way. Here is the *recursive* chain of calculations. First, plug in for Y_2 and Y_3 to get Y_4 in terms of Y_1.

$$
Y_4 \;=\; \sigma\!\left(E_{2\to 4}\, \sigma(E_{1\to 2}\, Y_1) \,+\, E_{3\to 4}\, \sigma(E_{1\to 3}\, Y_1) \right).
$$

Now plug in for Y_1 to see finally how Y_4 depends on x.

$$
Y_4(x) \;=\; \sigma\!\left(E_{2\to 4}\, \sigma\!\left(E_{1\to 2}\, \sigma(x) \right) \,+\, E_{3\to 4}\, \sigma\!\left(E_{1\to 3}\, \sigma(x) \right) \right).
$$

For the randomly chosen edge values $E_{1\to 2} = -0.11622$, $E_{1\to 3} = -1.42157$, $E_{2\to 4} = 1.17856$ and $E_{3\to 4} = 0.68387$, we can calculate the Y_4 output for this model for all x values from -3 to 3

and plot them. Now negative values correspond to **inhibition** and positive values are **excitation**. Our simple model generates outputs between 0.95 for strong inhibition and 0.65 for strong excitation. Probably not realistic! But remember the edge weights were just chosen randomly and we didn't try to pick them using realistic biologically based values. We can indeed do better. But you should see a bit of how interesting biology can be illuminated by mathematics that comes from this class! You can see the Y_4 output in Figure 11.5.

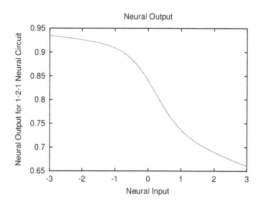

Figure 11.5: Y_4 output for our neural model.

Note that this is essentially a $\sigma(\sigma(\sigma))$ series of function compositions! So the idea of a composition of functions is not just some horrible complication mathematics courses throw at you. It is really used in biological systems. Also, while we know very well how to calculate the derivative of this monster, $Y_4'(x)$ using the chain rule, it requires serious effort and the answer we get is quite messy. Fortunately, over the years, we have found ways to get the information we need from models like this without finding the derivatives by hand! Also, just think, real neural subsystems have hundreds or thousands or more neurons interacting with a vast number of edge connections. Lots of sigmoid compositions going on!

Let's do a bit more here. We know we can approximate the derivative using a slope term. Here that is

$$Y_4(x) \quad = \quad Y_4(p) + Y_4'(p)(x - p) + E(x, p).$$

Since $E(h)/(h)$ is small near x too, we can say

$$\frac{Y_4(p + h) - Y_4(p)}{h} \quad \sim \quad Y_4'(p)$$

near x. We can use this idea to calculate the approximate value of the derivative of Y_4 and plot it. As you can see from Figure 11.6 the derivative is not that large, but it is always negative. Remember, derivatives are rates of change and looking at the graph of Y_4 we see it is always going down, so the derivative should always be negative.

We can show if f is twice differentiable at x, then $\lim_{h \to 0} \frac{f(x+2h)-2f(x+h)+f(x)}{h^2} = f''(x)$. This is called the forward difference approximation to the second derivative. Hence, $\frac{f(x+2h)-2f(x+h)+f(x)}{h^2}$ is an approximation to f''.

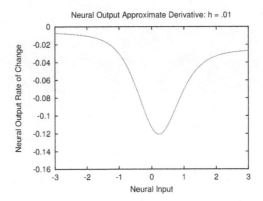

Figure 11.6: Y_4's approximate derivative using $h = .01$.

11.6.4   Homework

Exercise 11.6.13 *Let*

$$f(x) = \begin{cases} x, & x \in \mathbb{Q} \\ -x, & x \in \mathbb{IR} \end{cases}$$

Explain why $(f(x))^2$ is differentiable but $f(x)$ is differentiable nowhere.

Exercise 11.6.14 *Calculate Y_4' explicitly for our small neural circuit example.*

$$Y_4(x) = \sigma\left(E_{2 \to 4}\,\sigma\left(E_{1 \to 2}\,\sigma(x)\right) + E_{3 \to 4}\,\sigma\left(E_{1 \to 3}\,\sigma(x)\right)\right).$$

where $\sigma(x) = 0.5(1 + \tanh(x))$.

Exercise 11.6.15 *Use your favorite computational tool and generate plots of the true f' versus the forward difference approximation to f'' for $h = 0.5$, $h = 0.05$ and $h = 0.005$ for $f(x) = x^6 \sin^3(2x^2 + 5x + 8)$ for the interval $[-2, 2]$.*

Exercise 11.6.16 *Use your favorite computational tool and generate plots of the true f'' versus the forward difference approximation to f'' for $h = 0.5$, $h = 0.05$ and $h = 0.005$ for $f(x) = \cos(3x^5 + 4x^2 + 3)$ for the interval $[-2, 2]$.*

Chapter 12

Consequences of Derivatives

There are many things we can do with derivatives and now it is time to explore that.

12.1 Taylor Polynomials

We can approximate a function at a point using polynomials of various degrees. We can first find the constant that best approximates a function f at a point p. This is called the **zeroth order Taylor polynomial** and the equation we get is

$$f(x) \;=\; f(p) + E_0(x,p)$$

where $E_0(x,p)$ is the **error**. On the other hand, we could try to find the best straight line that does the job. We would find

$$f(x) \;=\; f(p) + f'(p)(x-p) + E_1(x,p)$$

where $E_1(x,p)$ is the **error** now. This straight line is the **first order Taylor polynomial** or the **tangent line**. This error is the same one we have discussed before. The quadratic case gives

$$f(x) \;=\; f(p) + f'(p)(x-p) + \frac{1}{2}\,f''(p)(x-p)^2 + E_2(x,p)$$

where $E_2(x,p)$ is the **error** in this case. This is called the **second order Taylor polynomial** or **quadratic approximation**.

Now let's dig into the theory behind this so that we can better understand the error terms. Let's consider a function which is defined locally at the point p. Let's also assume f' exists locally at p in this same interval. There are three tools we haven't mentioned yet but it is time to talk about them. They are **Critical Points**, **Rolle's Theorem** and the **Mean Value Theorem**. The first thing to think about is when a function might have a high or low place. This means that, at least locally, there is value for the function which is the highest one around or the lowest one around. Now if a function has a derivative at that high or low point – call this point p – we can say

$$f(x) \;=\; f(p) + f'(p)\,(x-p) \;+\; E(x,p).$$

Let's take the case where $f(p)$ is the highest value around. So we know $f(x)$ is smaller than $f(p)$ near p. We can then say $f(x) - f(p) < 0$ near p and this tells two things. If $x > p$, then the fraction $(f(x) - f(p))/(x - p) < 0$ as the numerator is negative and the denominator is positive. So we have $\frac{f(x)-f(p)}{x-p} = f'(p) + \frac{E(x-p)}{(x-p)} < 0$. Letting x go to p from the right hand side, we must have the right hand derivative $(f'(p))^+ = f'(p) \leq 0$. On the other hand, if $x < p$, we still have $f(x) < f(p)$ and the fraction $(f(x) - f(p))/(x - p) > 0$ now as both the numerator and denominator are negative. We then have for points on the left that $\frac{f(x) - f(p)}{x - p} = f'(p) + \frac{E(x - p)}{(x - p)} > 0$. Letting x go to p from the left hand side, we must have the left hand derivative $(f'(p))^- = f'(p) \geq 0$. But we started by assuming f had a derivative at p so the right and left hand derivatives match $f'(p)$. So $0 \leq f'(p) \leq 0$; i.e. $f'(p) = 0$.

The argument for the lowest point around is the same. We usually call the largest point locally, a **local maximum** and the smallest point locally, a **local minimum**. Our arguments have proven the following theorem.

Theorem 12.1.1 Critical Points

> *If f has a local maximum or local minimum at p where f has a derivative, then $f'(p) = 0$. Because of this, the places where f has a zero derivative are potentially places where f has maximum or minimum values. Points where the derivative does not exist and points on the boundary of the domain of the function are also called critical points. The minimum and maximum values are also called* **extreme** *values.*

We need two more theorems:

- **Rolle's Theorem** is pretty easy to understand. To see this proof, stand up and take a short piece of rope in your hands and follow the reasoning we present shortly.

- **The Mean Value Theorem** is a bit more technical but not too bad. It is also sort of common sense.

Theorem 12.1.2 Rolle's Theorem

> *Let $f : [a, b] \to \Re$ be a function defined on the interval $[a, b]$ which is continuous on the closed interval $[a, b]$ and is at least differentiable on the open interval (a, b). If $f(a) = f(b)$, then there is at least one point c, between a and b, so that $f'(c) = 0$.*

Proof 12.1.1

Take a piece of rope in your hand and make sure you hold the right and left side at the same height. When the rope is stretched tight, it represents the graph of a constant function: at all points c, $f'(c) = 0$. If we pull up on the rope in between our hands, the graph goes up to a maximum and at that point c, again we have $f'(c) = 0$. A similar argument holds if we pull down on the rope to get a minimum. ∎

The next theorem is very important to many types of approximation.

Theorem 12.1.3 The Mean Value Theorem

> *Let $f : [a, b] \to \Re$ be continuous on $[a, b]$ and is at least differentiable on (a, b). Then there is at least one point c, between a and b, so that $f'(c) = [f(b) - f(a)]/[b - a]$.*

Proof 12.1.2

Define $M = (f(b) - f(a))/(b - a)$ and the function $g(t) = f(t) - f(a) - M(t - a)$.
Note $g(a) = 0$ and $g(b) = f(b) - f(a) - [(f(b) - f(a))/(b - a)](b - a) = 0$. Apply **Rolle's Theorem** *to g. So there is $a < c < b$ so that $g'(c) = 0$. But $g'(t) = f'(t) - M$, so $g'(c) = 0$ tells us $f'(c) = M$. We conclude there is a point c, between a and b so that $f'(c) = (f(b) - f(a))/(b - a)$.* ∎

We can do this a bit more generally.

Theorem 12.1.4 The General Mean Value Theorem

> Let $f : [a, b] \to \Re$ be continuous on $[a, b]$ and is at least differentiable on (a, b). Then for each x, there is at least one point c, between a and x, so that $f'(c) = [f(x) - f(a)]/[x - a]$.

Proof 12.1.3
Define $M = (f(x) - f(a))/(x - a)$ and the function $g(t) = f(t) - f(a) - M(t - a)$. Note $g(a) = 0$ and $g(x) = f(x) - f(a) - [(f(x) - f(a))/(x - a)](x - a) = 0$. Apply **Rolle's Theorem** to g. So there is $a < c < x$ so that $g'(c) = 0$. But $g'(t) = f'(t) - M$, so $g'(c) = 0$ tells us $f'(c) = M$. We conclude there is a point c, between a and x so that $f'(c) = (f(x) - f(a))/(x - a)$. ∎

Homework

Exercise 12.1.1 Find the critical points of $f(x) = 1/(x + 1)$ on the domain $[-2, 4]$.

Exercise 12.1.2 Find the critical points of $f(x) = x^2 + 3x + 5$ on the domain $[-2, 4)$.

Exercise 12.1.3 For $f(x) = x^2 + 3x + 5$ on the domain $[-2, 4]$ find the points that satisfy Rolle's Theorem.

Exercise 12.1.4 For $f(x) = 3x^2 + 4x - 7$ on the domain $[-10, 10]$ find the points that satisfy the Mean Value Theorem.

We can apply this to our first polynomial approximation: the zeroth order Taylor Polynomial.

12.1.1 Zeroth Order Taylor Polynomials

Theorem 12.1.5 The Zeroth Order Taylor Polynomial

> Let $f : [a, b] \to \Re$ be continuous on $[a, b]$ and be at least differentiable on (a, b). Then for each x, there is at least one point c, between a and x, so that $f(x) = f(a) + f'(c)(x - a)$. The constant $f(a)$ is called the **zeroth order Taylor Polynomial** for f at a and we denote it by $p_0(x; a)$. The point a is called the **base point**. Note we are approximating $f(x)$ by the constant $f(a)$ and the error we make is $f'(c)(x - a)$.

Proof 12.1.4
This is just the Mean Value Theorem applied to f. ∎

Example 12.1.1 If $f(t) = t^3$, find the zeroth order polynomial approximation.

Solution We know on the interval $[1, 3]$ that $f(t) = f(1) + f'(c)(t - 1)$ where c is some point between 1 and 3. Thus, $t^3 = 1 + (3c^2)(t - 1)$ for some $1 < c < 3$. So here the **zeroth order Taylor Polynomial** is $p_0(t; 1) = 1$ and the **error** is $(3c^2)(t - 1)$.

Example 12.1.2 If $f(t) = e^{-1.2t}$, find the zeroth order polynomial approximation.

Solution We know on the interval $[0, 5]$ that $f(t) = f(0) + f'(c)(t - 0)$ where c is some point between 0 and 5. Thus, $e^{-1.2t} = 1 + (-1.2)e^{-1.2c}(t - 0)$ for some $0 < c < 5$ or $e^{-1.2t} = 1 - 1.2e^{-1.2c}t$ So here the **zeroth order Taylor Polynomial** is $p_0(t; 1) = 1$ and the **error** is $-1.2e^{-1.2c}t$.

Example 12.1.3 If $f(t) = e^{-.00231t}$, find the zeroth order polynomial approximation.

Solution *We know on the interval* $[0,8]$ *that* $f(t) = f(0) + f'(c)(t - 0)$ *where* c *is some point between 0 and 8. Thus,* $e^{-.00231t} = 1 + (-.00231)e^{-.00231c}(t - 0)$ *for some* $0 < c < 8$ *or* $e^{-.00231t} = 1 - .00231e^{-.00231c}t$. *So here the* **zeroth order Taylor Polynomial** *is* $p_0(t.1) = 1$ *and the* **error** *is* $-.00231e^{-.00231c}t$.

Homework

Exercise 12.1.5 *If* $f(t) = e^{rt}$ *for* $r > 0$ *find the zeroth order polynomial approximation and the estimate of the error on the interval* $[0, 4]$.

Exercise 12.1.6 *If* $f(t) = \sin(4t)$ *find the zeroth order polynomial approximation and the estimate of the error on the interval* $[-4, 4]$.

Exercise 12.1.7 *If* $f(t) = \cos(2t) + \sin(3t)$ *find the zeroth order polynomial approximation and the estimate of the error on the interval* $[-3, 4]$.

12.1.2   The Order One Taylor Polynomial

The Order One Taylor polynomial is the same thing as the tangent line approximation to a function f at a point.

Theorem 12.1.6 The Order One Taylor Polynomial

> *Let* $f : [a, b] \to \Re$ *be continuous on* $[a, b]$ *with* $'$ *continuous on* $[a, b]$ *and with* f'' *at least existing on* (a, b). *Then for each* x, *there is at least one point* c, *between* a *and* x, *so that* $f(x) = f(a) + f'(a)(x - a) + (1/2)f''(c)(x - a)^2$.

Proof 12.1.5
Define $M = (f(x) - f(a) - f'(a)(x - a))/[(x - a)^2]$ *and the function* $g(t) = f(t) - f(a) - f'(a)(t - a) - M(t - a)^2$. *So* $g(a) = 0$ *and*

$$
\begin{aligned}
g(x) &= f(x) - f(a) - f'(a)(x - a) \\
&\quad - \left(\frac{f(x) - f(a) - f'(a)(x - a)}{(x - a)^2} \right)(x - a)^2 \\
&= f(x) - f(a) - f'(a)(x - a) \\
&\quad - (f(x) - f(a) - f'(a)(x - a)) = 0.
\end{aligned}
$$

Apply **Rolle's Theorem** *to* g. *This tells us there is a point* d *with* $g'(d) = 0$ *and* d *between* a *and* x.

Next, we find $g'(t) = f'(t) - f'(a) - 2M(t - a)$. *Then* $g'(a) = 0$ *and we know* $g'(d) = 0$. *Apply* **Rolle's Theorem** *to* g' *on* $[a, d]$. *So there is a* $a < c < d$ *with* $g''(c) = 0$. *But* $g''(t) = f''(t) - 2M$, *so* $f''(c) = 2M$.

We conclude there is a point c, *between* a *and* b *so that* $(1/2)f''(c) = (f(x) - f(a) - f'(a)(x - a))/(x - a)^2$. *Solving, we see there is a point* c *between* a *and* x *so that* $f(x) = f(a) + f'(a)(x - a) + \frac{1}{2}f''(c)(x - a)^2$.   ∎

Comment 12.1.1 $f(a) + f'(a)(x - a)$ *is called the* **first order Taylor Polynomial** *for* f *at base point* a, $p_1(t; a)$. *It is also the traditional* **tangent line** *to* f *at* a. *Hence, we now know the* **error** $E(x, a) = (1/2)f''(c)(x - a)^2$.

Example 12.1.4 *For* $f(t) = e^{-1.2t}$ *on the interval* $[0,5]$ *find the tangent line approximation, the error and maximum the error can be on the interval using base point* $t = 0$.

Solution

$$
\begin{aligned}
f(t) &= f(0) + f'(0)(t - 0) + (1/2)f''(c)(t - 0)^2 \\
&= 1 + (-1.2)(t - 0) + (1/2)(-1.2)^2 e^{-1.2c}(t - 0)^2 \\
&= 1 - 1.2t + (1/2)(1.2)^2 e^{-1.2c}t^2.
\end{aligned}
$$

where c *is some point between* 0 *and* 5. *The* **first order Taylor Polynomial** *is* $p_1(t; 0) = 1 - 1.2t$ *which is also the tangent line to* $e^{-1.2t}$ *at* 0. *The* **error** *is* $(1/2)(-1.2)^2 e^{-1.2c}t^2$.
Now let **AE** *denote* **absolute value of the actual error** *and* **ME** *be* **maximum absolute error on a given interval**. *The largest the error can be on* $[0, 5]$ *is when* $f''(c)$ *is the biggest it can be on the interval. Here,*

$$
\begin{aligned}
\mathbf{AE} &= |(1/2)(1.2)^2 e^{-1.2c}t^2| \le (1/2)(1.2)^2 \times 1 \times (5)^2 \\
&= (1/2)1.44 \times 25 = \mathbf{ME} .
\end{aligned}
$$

Example 12.1.5 *Do this same problem on the interval* $[0, 10]$.

Solution *The approximations are the same but now* $0 < c < 10$ *and*

$$
\begin{aligned}
\mathbf{AE} &= |(1/2)(1.2)^2 e^{-1.2c}t^2| \le (1/2)(1.2)^2 \times 1 \times (10)^2 \\
&= (1/2)1.44 \times 100 = \mathbf{ME} .
\end{aligned}
$$

Example 12.1.6 *Do this same problem on the interval* $[0, 100]$.

Solution *The approximations are the same,* $0 < c < 100$ *and*

$$
\begin{aligned}
\mathbf{AE} &= |(1/2)(1.2)^2 e^{-1.2c} t^2| \le (1/2)(1.2)^2(1)(100)^2 \\
&= (1/2)1.44 \times 10^4 = \mathbf{ME}.
\end{aligned}
$$

Example 12.1.7 *Do this same problem on the interval* $[0, T]$.

Solution *The approximations are the same,* $0 < c < T$ *and*

$$
\begin{aligned}
\mathbf{AE} &= |(1/2)(1.2)^2 e^{-1.2c}t^2| \le (1/2)(1.2)^2(1)(T)^2 \\
&= (1/2)1.44 \times T^2 = \mathbf{ME}.
\end{aligned}
$$

Example 12.1.8 *If* $f(t) = e^{-\beta t}$, *for* $\beta = 1.2 \times 10^{-5}$, *find the tangent line approximation, the error and the maximum error on* $[0, 5]$ *using base point* $t = 0$.

Solution *On the interval* $[0, 5]$, *then*

$$
\begin{aligned}
f(t) &= f(0) + f'(0)(t - 0) + \frac{1}{2}f''(c)(t - 0)^2 \\
&= 1 + (-\beta)(t - 0) + \frac{1}{2}(-\beta)^2 e^{-\beta c}(t - 0)^2 \\
&= 1 - \beta t + \frac{1}{2}\beta^2 e^{-\beta c}t^2.
\end{aligned}
$$

where c *is some point between* 0 *and* 5. *The* **first order Taylor Polynomial** *is* $p_1(t; 0) = 1 - \beta t$ *which is also the tangent line to* $e^{-\beta t}$ *at* 0. *The* **error** *is* $\frac{1}{2}\beta^2 e^{-\beta c}t^2$. *The largest the error can be on*

$[0, 5]$ *is when* $f''(c)$ *is the biggest it can be on the interval. Here,*

$$\mathbf{AE} = |(1/2)(1.2 \times 10^{-5})^2 e^{-1.2 \times 10^{-5}c} \, t^2|$$
$$\leq (1/2)(1.2 \times 10^{-5})^2 (1)(5)^2 = (1/2)1.44 \times 10^{-10}(25) = \mathbf{ME}$$

Example 12.1.9 *Do this same problem on the interval* $[0, 10]$

Solution *The approximation is the same, c is between* 0 *and* 10 *and*

$$\mathbf{AE} = |(1/2)(1.2 \times 10^{-5})^2 e^{-1.2 \times 10^{-5}c} t^2| \leq$$
$$(1/2)(1.2 \times 10^{-5})^2 (1)(10)^2 = (1/2)1.44 \times 10^{-10} \times 100 = \mathbf{ME}.$$

Example 12.1.10 *Do this same problem on the interval* $[0, 100]$

Solution *The approximation is the same, c is between* 0 *and* 100 *and*

$$\mathbf{AE} = |(1/2)(1.2 \times 10^{-5})^2 e^{-1.2 \times 10^{-5}c} t^2| \leq$$
$$(1/2)(1.2 \times 10^{-5})^2 (1)(100)^2 = (1/2)1.44 \times 10^{-10} \, 10^4 = \mathbf{ME}.$$

Example 12.1.11 *Do this same problem on the interval* $[0, T]$

Solution *The approximation is the same, c is between* 0 *and* T *and*

$$\mathbf{AE} = |(1/2)(1.2 \times 10^{-5})^2 e^{-1.2 \, 10^{-5}c} t^2| \leq$$
$$(1/2)(1.2 \times 10^{-5})^2 (1)(T)^2 = (1/2)1.44 \times 10^{-10} T^2 = \mathbf{ME}.$$

Example 12.1.12 *If* $f(t) = \cos(3t) + 2\sin(4t)$, *then on the interval* $[0, 5]$ *find the tangent line approximation, the error and the maximum error using base point* $t = 0$.

Solution *We have* $f(0) = 1$ *and* $f'(t) = -3\sin(3t) + 8\cos(4t)$ *so that* $f'(0) = 8$. *Further,* $f''(t) = -9\cos(3t) - 32\sin(4t)$ *so* $f''(c) = -9\cos(3c) - 32\sin(4c)$. *Thus, for some* $0 < c < 5$

$$\begin{aligned} f(t) &= f(0) + f'(0)(t - 0) + \frac{1}{2}f''(c)(t - 0)^2 \\ &= 1 + (8)(t - 0) + \frac{1}{2}(-9\cos(3c) - 32\sin(4c))(t - 0)^2, \end{aligned}$$

The **first order Taylor Polynomial** *is* $p_1(t; 0) = 1 + 8t$ *which is also the tangent line to* $\cos(3t) + 2\sin(4t)$ *at* 0. *The* **error** *is* $(1/2)(-9\cos(3c) - 32\sin(4c))(t - 0)^2$. *The error is largest on* $[0, 5]$ *when* $f''(c)$ *is the biggest it can be on* $[0, 5]$. *Here,*

$$\mathbf{AE} = |(1/2)(-9\cos(3c) - 32\sin(4c)) \, t^2| \leq$$
$$(1/2) \, |-9\cos(3c) - 32\sin(4c)| \, 5^2$$
$$\leq (1/2) \, |9 + 32| \, 5^2 = (1/2)(41)(25) = \mathbf{ME}$$

Example 12.1.13 *Do this same problem on the interval* $[0, 10]$.

Solution *The approximations are the same,* $0 < c < 10$ *and*

$$\mathbf{AE} = |(1/2)(-9\cos(3c) - 32\sin(4c)) \, t^2| \leq$$
$$\leq (1/2) \, |-9\cos(3c) - 32\sin(4c)| \, (10)^2$$
$$= (1/2)(41)(100) = \mathbf{ME}.$$

Example 12.1.14 *Do this same problem on the interval* $[0, 100]$.

Solution *The approximations are the same, $0 < c < 100$ and*

$$\mathbf{AE} = |(1/2)(-9\cos(3c) - 32\sin(4c))\, t^2| \leq$$
$$\leq (1/2)\,|-9\cos(3c) - 32\sin(4c)|\,(100)^2$$
$$= (1/2)(41)(10^4) = \mathbf{ME}.$$

Example 12.1.15 *Do this same problem on the interval $[0, T]$.*

Solution *The approximations are the same, $0 < c < T$ and*

$$\mathbf{AE} = |(1/2)(-9\cos(3c) - 32\sin(4c))\, t^2| \leq$$
$$\leq (1/2)\,|-9\cos(3c) - 32\sin(4c)|\, T^2 = (1/2)(41)T^2 = \mathbf{ME}.$$

Homework

Exercise 12.1.8 *If $f(x) = 4x^3 - 2x^2 + 10x - 8$ using base point $x = 0$.*

- *Find the tangent line approximation, the error and the maximum error on $[-3, 5]$.*
- *Find the tangent line approximation, the error and the maximum error on $[-13, 15]$.*

Exercise 12.1.9 *If $f(x) = 4x^3 - 2x^2 + 10x - 8$ using base point $x = 1$.*

- *Find the tangent line approximation, the error and the maximum error on $[-3, 5]$.*
- *Find the tangent line approximation, the error and the maximum error on $[-13, 15]$.*

Exercise 12.1.10 *If $f(x) = xe^{-2x}$ using base point $x = 0$.*

- *Find the tangent line approximation, the error and the maximum error on $[0, 5]$.*
- *Find the tangent line approximation, the error and the maximum error on $[0, 15]$.*
- *Find the tangent line approximation, the error and the maximum error on $[-5, 5]$.*
- *Find the tangent line approximation, the error and the maximum error on $[-7, 15]$.*

Exercise 12.1.11 *If $f(x) = 2\cos(x)e^{-0.1x}$ using base point $x = 0$.*

- *Find the tangent line approximation, the error and the maximum error on $[0, 5]$.*
- *Find the tangent line approximation, the error and the maximum error on $[0, 15]$.*
- *Find the tangent line approximation, the error and the maximum error on $[-2, 5]$.*
- *Find the tangent line approximation, the error and the maximum error on $[-5, 15]$.*

12.1.3   Second Order Approximations

The Tangent line approximation is also the **best** linear function we can choose to approximate f near a. We can show the best linear function fit with $y = \alpha + \beta(x - a)$ is when $\alpha + \beta(x - a) = f(a) + f'(a)(x - a)$. We could also ask what quadratic function Q fits f best near a. The general quadratic is $y = \alpha + \beta(x - a) + \gamma(x - a)^2$ and we can show the best fit is when $\alpha + \beta(x - a) + \gamma(x - a)^2 = f(a) + f'(a)\,(x - a) + f''(a)\,\frac{(x - a)^2}{2}$. This is the **quadratic approximation** to f at a and the error is $E_2(x, p) = f(x) - f(a) - f'(a)\,(x - a) - f''(a)\,\frac{(x - a)^2}{2}$. If f is three times differentiable, we can argue like we did in the tangent line approximation (using the Mean Value Theorem and Rolle's theorem on an appropriately defined function g) to show there is a new point c_x^2 between p and c_x^1 with $E_2(x, p) = f'''(c_x^2)\,\frac{(x - p)^3}{6}$.

Theorem 12.1.7 The Second Order Taylor Polynomial

> *Let $f : [a, b] \to \Re$ be continuous on $[a, b]$ with f' and f'' continuous on $[a, .b]$ and with and $f^{(3)}$ is at least existing on (a, b). Then for each x, there is at least one point c, between a and x, so that*
>
> $$f(x) = f(a) + f'(a)(x - a) + (1/2)f''(a)(x - a)^2$$
> $$+ (1/6)f'''(c)(x - a)^3.$$
>
> *The function $f(a) + f'(a)(x - a) + (1/2)f''(a)(x - a)^2$ is the **second order Taylor polynomial** to f at a. It is also called the **second order approximation** to f at a. The **error** term is now $(1/6)f'''(c)(x - a)^3$.*

Proof 12.1.6

Define $M = (f(x) - f(a) - f'(a)(x - a) - f''(a)(x - a)^2/2)/[(x - a)^3]$ and $g(t) = f(t) - f(a) - f'(a)(t - a) - f''(a)(t - a)^2/2 - M(t - a)^3$. So $g(a) = 0$ and

$$
\begin{aligned}
g(x) &= f(x) - f(a) - f'(a)(x - a) - f''(a)(x - a)^2/2 - M(x - a)^3 \\
&\quad - \left(\frac{f(x) - f(a) - f'(a)(x - a) - f''(a)(x - a)^2/2}{(x - a)^3} \right)(x - a)^3 \\
&= f(x) - f(a) - f'(a)(x - a) - f''(a)(x - a)^2/2 \\
&\quad - (f(x) - f(a) - f'(a)(x - a) - f''(a)(x - a)^2/2) = 0.
\end{aligned}
$$

*Apply **Rolle's Theorem** to g. This tells us there is a point d with $g'(d) = 0$ and d between a and x.*

*Next, we find $g'(t) = f'(t) - f'(a) - f''(a)(t - a) - 3M(t - a)^2$. Then $g'(a) = 0$ and we already know $g'(d) = 0$. Apply **Rolle's Theorem** to g' on $[a, d]$. So there is $a < c < d$ with $g''(c) = 0$. Now $g''(t) = f''(t) - f''(a) - 6M(t - a)$ and so $g''(a) = 0$.*

*Apply **Rolle's Theorem** to g'' on $[a, c]$. So there is $a < \beta < c$ with $g'''(\beta) = 0$. But $g'''(\beta) = 0 = f'''(\beta) - 6M$ We conclude there is a point β, between a and b so that $(1/6)f'''(\beta) = [f(x) - f(a) - f'(a)(x - a) - f''(a)/(x - a)^2/2]/(x - a)^3$.*
Solving, we see there is a point β between a and x so that $f(x) = f(a) + f'(a)(x - a) + \frac{1}{2}f''(a)(x - a)^2 + \frac{1}{6}f'''(\beta)(x - a)^3$. ∎

Comment 12.1.2 *The function $f(a) + f'(a)(x - a) + (1/2)f''(a)(x - a)^2$ is called the **second order Taylor Polynomial** for f at base point a, $p_2(t; a)$. It is the **quadratic** approximation to f at a. There is a point c, with $a < c < b$ so that the error made is $E_2(x, a) = (1/6)f'''(c)(x - a)^2$.*

Example 12.1.16 *If $f(t) = e^{-\beta t}$, for $\beta = 1.2 \times 10^{-5}$, find the second order approximation, the error and the maximum error on $[0, 5]$ using the base point $t = 0$.*

Solution *On the interval $[0, 5]$, then there is some $0 < c < 5$ so that*

$$
\begin{aligned}
f(t) &= f(0) + f'(0)(t - 0) + (1/2)f''(0)(t - 0)^2 \\
&\quad + (1/6)f'''(c)(t - 0)^3 \\
&= 1 + (-\beta)(t - 0) + \frac{1}{2}(-\beta)^2(t - 0)^2 + \frac{1}{6}(-\beta)^3 e^{-\beta c}(t - 0)^3 \\
&= 1 - \beta t + (1/2)\beta^2 t^2 - (1/6)\beta^3 e^{-\beta c} t^3.
\end{aligned}
$$

The **second order Taylor Polynomial** *is $p_2(t; 0) = 1 - \beta t + (1/2)\beta^2 t^2$ and the error is $E_2(x, 0) = -\frac{1}{6}\beta^3 e^{-\beta c} t^3$. The error is largest on $[0, 5]$ when $f'''(c)$ is the biggest it can be on the interval. Here,*

$$\mathbf{AE} = |-(1/6)(1.2 \times 10^{-5})^3 e^{-1.2 \times 10^{-5} c} t^3|$$
$$\leq |-(1/6)(1.2 \times 10^{-5})^3 e^{-1.2 \times 10^{-5} c} t^3|$$
$$\leq (1/6)(1.2 \times 10^{-5})^3 (1) (5)^3$$
$$= (1/6) \, 1.728 \times 10^{-15} \, (125) = \mathbf{ME}$$

Example 12.1.17 *Do this same problem on the interval $[0, T]$.*

Solution *The approximations are the same, $0 < c < T$ and*

$$\mathbf{AE} = |-(1/6)(1.2 \times 10^{-5})^3 e^{-1.2 \times 10^{-5} c} t^3|.$$

Hence,

$$\mathbf{AE} \leq |-(1/6)(1.2 \times 10^{-5})^3 e^{-1.2 \times 10^{-5} c} t^3|$$
$$\leq (1/6)(1.2 \times 10^{-5})^3 T^3$$
$$= (1/6) \, 1.728 \times 10^{-15} \, (T^3) = \mathbf{ME}.$$

Example 12.1.18 *If $f(t) = 7\cos(3t) + 3\sin(5t)$, then on the interval $[0, 15]$ find the second order approximation, the error and the maximum error using the base point $t = 0$.*

Solution *We have $f(0) = 7$ and $f'(t) = -21\sin(3t) + 15\cos(5t)$ so that $f'(0) = 15$. Further, $f''(t) = -63\cos(3t) - 75\sin(5t)$ so $f''(0) = -63$ and finally $f'''(t) = 189\sin(3t) - 375\cos(5t)$ and $f'''(c) = 189\sin(3c) - 375\cos(5c)$. Thus, for some $0 < c < 15$*

$$\begin{aligned} f(t) &= f(0) + f'(0)(t - 0) + (1/2)f''(0)(t - 0)^2 \\ &\quad + (1/6)f'''(c)(t - 0)^3 \\ &= 7 + (15)(t - 0) + (1/2)(-63)(t - 0)^2 \\ &\quad + (1/6)(189\sin(3c) - 375\cos(5c))(t - 0)^3 \\ &= 7 + 15t - (1/2)63t^2 \\ &\quad + (1/6)(189\sin(3c) - 375\cos(5c))(t - 0)^3 \end{aligned}$$

The **second order Taylor Polynomial** *is $p_2(t; 0) = 7 + 15t - (63/2)t^2$ which is also the quadratic approximation to $7\cos(3t) + 3\sin(5t)$ at 0. The* **error** *is $(1/6)(189\sin(3c) - 375\cos(5c))(t - 0)^3$. The largest the error can be on the interval $[0, 15]$ is then*

$$\mathbf{AE} = |(1/6)(189\sin(3c) - 375\cos(5c))(t - 0)^3|$$
$$\leq (1/6)|(189\sin(3c) - 375\cos(5c))t^3|$$
$$\leq (1/6)(189 + 375)|t^3| \leq (1/6)(564)(15)^3 = \mathbf{ME}.$$

Example 12.1.19 *Do this same problem on the interval $[0, T]$.*

Solution *The approximations are the same, $0 < c < T$ and*

$$|\mathbf{AE}| = |(1/6)(189\sin(3c) - 375\cos(5c))(t - 0)^3|$$
$$\leq (1/6)(189 + 375)|t^3| \leq (1/6)(564)T^3 = \mathbf{ME}.$$

12.1.4   Homework

For these problems,

- find the Taylor polynomial of order two, i.e. the second order or quadratic approximation.

- state the error in terms of the third derivative.

- state the maximum error on the given intervals.

- given $\epsilon > 0$, find the interval centered at the base point for which the maximum error is less than the tolerance ϵ.

Exercise 12.1.12 $f(t) = e^{-2.8 \times 10^{-3} t}$ *at base point* 0, *the intervals are* $[0, 10]$, $[0, 100]$ *and* $[0, T]$ *and* $\epsilon = 0.1$.

Exercise 12.1.13 $f(t) = 6 \cos(2t) - 5 \sin(4t)$ *at base point* 0, *the intervals* $[0, 30]$, $[0, 300]$ *and* $[0, T]$ *and* $\epsilon - 0.01$.

Exercise 12.1.14 *If* $f(x) = 4x^3 - 2x^2 + 10x - 8$ *using base point* $x = 0$, *the intervals* $[-3, 5]$ *and* $[-13, 15]$ *and* $\epsilon = 0.3$.

Exercise 12.1.15 *If* $f(x) = 4x^3 - 2x^2 + 10x - 8$ *using base point* $x = 1$, *the intervals* $[-3, 5]$ *and* $[-13, 15]$ *and* $\epsilon = 0.4$.

Exercise 12.1.16 *If* $f(x) = xe^{-2x}$ *using base point* $x = 0$, *the intervals* $[0, 5]$, $[0, 15]$, $[-5, 5]$ *and* $[-7, 15]$ *and* $\epsilon = 0.06$.

Exercise 12.1.17 *If* $f(x) = 2 \cos(x)e^{-0.1x}$ *using base point* $x = 0$, *the intervals* $[0, 5]$, $[0, 15]$, $[-2, 5]$ *and* $[-5, 15]$ *and* $\epsilon = 0.3$.

Now a theory problem.

Exercise 12.1.18 *If* f *is continuous on* $[a, b]$ *and* f' *is continuous on* $[a, b]$, *prove there is a positive constant* L *so that* $|f(x) - f(y)| \le L|x - y|$ *for all* $x, y \in [a, b]$.

12.1.5   Taylor Polynomials in General

We can find higher order Taylor Polynomial approximations by applying Rolle's theorem repeatedly.

Theorem 12.1.8 The Order N Taylor Polynomial

Let $f^{(j)} : [a, b] \to \Re$ *be continuous on* $[a, b]$ *for* $0 \le j \le N$ *be continuous on* $[a, b]$, *where the notation* $f^{(j)}$ *denotes the* j^{th} *derivative of* f *and* $f^{(0)}$ *is identified with* f *itself, with* $f^{(N+1)}$ *at least existing on* (a, b). *Then for each* x, *there is at least one point* c_x, *between* a *and* x, *so that*

$$f(x) = f(a) + f^{(1)}(a)(x - a) + (1/2)f^{(2)}(a)(x - a)^2 + \ldots$$
$$+ (1/N!)f^{(N)}(a)(x - a)^N + (1/(N+1)!)f^{(N+1)}(c_x)(x - a)^{N+1}$$

The polynomial $p_N(x, a) = \sum_{i=0}^{N} \frac{1}{i!} f^{(i)}(a)(x - a)^i$ *is the* N^{th} *Taylor Polynomial based at* a *and the term* $\frac{1}{(N+1)!} f^{(N+1)}(c_x)(x - a)^{N+1}$ *is the* N^{th} *order error.*

Proof 12.1.7

Define $M = (f(x) - p_N(x, a))/[(x - a)^{N+1}]$ *and the function* $g(t) = f(t) - p_N(t, a) - M(t - a)^{N+1}$.

So $g(a) = 0$ and

$$
\begin{aligned}
g(x) &= f(x) - p_N(x, a) - \left\{ \frac{f(x) - p_N(x, a)}{(x-a)^{N+1}} \right\} (x-a)^{N+1} \\
&= (f(x) - p_N(x, a)) - (f(x) - p_N(x, a)) = 0.
\end{aligned}
$$

*Apply **Rolle's Theorem** to g. So there is $a < c_x^1 < d$ with $g'(c_x^1) = 0$. But since*

$$
\begin{aligned}
g'(t) &= f'(t) - p_N'(t, a) \\
&= f^{(1)}(t) - f^{(1)}(a) - f^{(2)}(t-a) - (1/2)f^{(3)}(a)(t-a)^2 \ldots \\
&\quad - (1/((N-1))!)f^{(N)}(a)(t-a)^N - (N+1)M(t-a)^N
\end{aligned}
$$

we see $g'(a) = 0$. Now apply Rolle's Theorem to g' to find a point c_x^2 with $a < c_x^2 < c_x^1$ with $g^{(2)}(c_x^2) = 0$.

You can see the pattern here. We always have the ith derivative is 0 at a and the previous step tells us this derivative is 0 at c_x^{i-1}.

Thus, Rolle's Theorem tells us there is a point $a < c_x^i < c_x^{i-1}$ with $g^{(i)}(c_x^i) = 0$. After doing this for $N+1$ steps, we have found a sequence of points $a < c_x^{N+1} < \ldots < c_x^2 < c_x^1$ and $g^{(N+1)}(c_x^{N+1}) = 0$. But

$$
g^{(N+1)}(t) = f^{(N+1)}(t) - ((N+1)!)M.
$$

Thus, $g^{(N+1)}(c_x^{N+1}) = 0$ tells us $M = \frac{1}{(N+1)!}f^{(N+1)}(c_x^{N+1})$ and after rearranging terms, we find

$$
\begin{aligned}
f(x) &= f(a) + f^{(1)}(a)(x-a) + (1/2)f^{(2)}(a)(x-a)^2 + \ldots \\
&\quad + (1/N!)f^{(N)}(a)(x-a)^N + (1/(N+1)!)f^{(N+1)}(c_x)(x-a)^{N+1}
\end{aligned}
$$

like we wanted. ■

Homework

For these problems,

- find the Taylor polynomial of order N as specified.

- state the error in terms of the t$N + 1$ derivative.

- state the maximum error on the given intervals.

- given $\epsilon > 0$, find the interval centered at the base point for which the maximum error is less than the tolerance ϵ.

Exercise 12.1.19 $f(x) = 4x^5 - 2x^2 + 10x - 8$ *using base point $x = 1$, $N = 3$, the interval in $[-2, 5]$ and $\epsilon = 0.2$.*

Exercise 12.1.20 $f(x) = \cos(4x)$ *using base point $x = 0$, $N = 5$, the interval in $[-2, 3]$ and $\epsilon = 0.3$.*

Exercise 12.1.21 $f(x) = \sin(5x)$ *using base point $x = 0$, $N = 7$, the interval in $[-1, 5]$ and $\epsilon = 0.1$.*

Exercise 12.1.22 $f(x) = e^{-0.03x}$ *using base point $x = 0$, $N = 4$, the interval in $[0, 5]$ and $\epsilon = 0.2$.*

12.2　Differences of Functions

Sometimes it is easier to approximate the differences between two close functions. Our first example is the difference between two closely related cosines.

12.2.1　Two Close Cosines

Let's look at how to approximate the difference $\cos((a+r)x) - \cos(ax)$ near $x = 0$. The first order expansion for $\cos(ax)$ at 0 is

$$
\begin{aligned}
\cos(ax) &= \cos(0) - a\sin(0)(x-0) - \frac{1}{2}a^2\cos(ac_x)(x-0)^2 \\
&= 1 - \frac{1}{2}\cos(ac_x)a^2x^2
\end{aligned}
$$

where c_x is between 0 and x. The second order expansion at 0 is

$$
\begin{aligned}
\cos(ax) &= \cos(0) - a\sin(0)(x-0) - a^2\frac{1}{2}\cos(0)(x-0)^2 \\
&+ a^3\frac{1}{6}\sin(ad_x)(x-0)^3 \\
&= 1 - \frac{1}{2}a^2x^2 + \frac{1}{6}\sin(ad_x)a^3x^3
\end{aligned}
$$

where d_x is between 0 and x. Thus, doing a similar expansion for $\cos((a+r)x)$ we see to first order

$$
\begin{aligned}
\cos((a+r)x) - \cos(ax) &= \left(1 - \frac{1}{2}\cos((a+r)c_x^1)(a+r)^2x^2\right) \\
&- \left(1 - \frac{1}{2}\cos(ac_x^2)a^2x^2\right) \\
&= \frac{1}{2}\left(\cos(ac_x^2)a^2 - \cos((a+r)c_x^1)(a+r)^2\right)x^2
\end{aligned}
$$

Thus, overestimating the error, we have

$$
\begin{aligned}
|\cos((a+r)x) - \cos(ax)| &\leq \frac{1}{2}\left(|\cos((a+r)c_x^1)|(a+r)^2 + |\cos(c_x^2)|a^2\right)x^2 \\
&\leq \frac{1}{2}\left((a+r)^2 + a^2\right)x^2 = (a^2 + ar + \frac{1}{2}r^2)x^2
\end{aligned}
$$

So $(\cos((a+r)x) - \cos(ax)) \approx 0$ with error $(a^2 + ar + \frac{1}{2}r^2)x^2$.

Then, doing a similar expansion for $\cos((a+r)x)$ we see to second order

$$
\begin{aligned}
\cos((a+r)x) - \cos(ax) &= \left(1 - \frac{1}{2}(a+r)^2x^2 + \frac{1}{6}\sin((a+r)d_x^1)(a+r)^3x^3\right) \\
&- \left(1 - \frac{1}{2}a^2x^2 + \frac{1}{6}\sin(d_x^2)a^3x^3\right) \\
&= \frac{1}{2}\left(a^2 - (a+r)^2\right)x^2 \\
&+ \frac{1}{6}\left(\sin((a+r)d_x^1)(a+r)^3 - \sin(ac_x^2)a^3\right)x^3
\end{aligned}
$$

So

$$(\cos((a+r)x) - \cos(ax)) - (1/2)(-2ar - r^2)x^2$$
$$= \frac{1}{6}\left(\sin((a+r)d_x^1)\,(a+r)^3x^3 - \sin(ad_x^2)\,a^3x^3\right)x^3$$

Thus, overestimating the error, we have

$$|\cos((a+r)x) - \cos(ax) - (1/2)(-2ar - r^2)x^2|$$
$$\leq \frac{1}{6}\left(|\sin((a+r)d_x^1)|\,(a+r)^3 + |\sin(ad_x^2)|\,a^3\right)|x|^3$$
$$\leq \frac{1}{6}\left((a+r)^3 + a^3\right)x^3$$

So $(\cos((a+r)x) - \cos(ax)) \approx -(1/2)(2ar + r^2)x^2$ with error $\frac{1}{6}((a+r)^3 + a^3)|x|^3$.

Example 12.2.1 *Approximate* $\cos(2.1x) - \cos(2x)$ *to both first and second order.*

Solution *We have* $a = 2$ *and* $r = .1$. *So to first order and second order*

$$\cos(2.1x) - \cos(2x) \approx 0, \text{ with error } (a^2 + ar + \frac{1}{2}r^2)x^2$$
$$= (4 + .2 + .005)x^2 = 4.205x^2.$$
$$\cos(2.1x) - \cos(2x) \approx -(2ar + r^2)x^2 = -(.4 + .01)x^2 = -.41x^2,$$
$$\text{with error} = \frac{1}{6}((a+r)^3 + a^3)x^3$$
$$= \frac{1}{6}(2.1^3 + 2^3)x^3 = 2.88x^3.$$

We can see these approximations graphically in Figure 12.1.

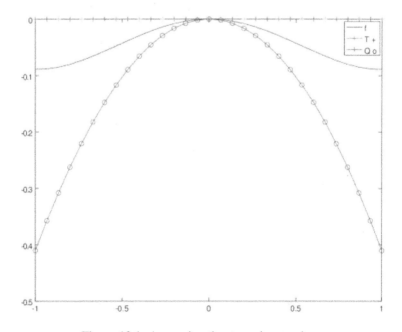

Figure 12.1: Approximating two close cosines.

The corresponding code is here.

Graphing The Approximation to two close cosines

Listing 12.1: **Code Fragment**

```
1 >> f = @(x) cos(2.1*x) - cos(2.0*x);
  >> T = @(x) 0;
  >> Q = @(x) -.41*x.^2;
  >> x = linspace(-1,1,31);
  >> plot(x,f(x),x,T(x),'+',x,Q(x),'o');
6 >> legend('f','T +','Q o');
```

We can see these errors graphically in Figure 12.2.

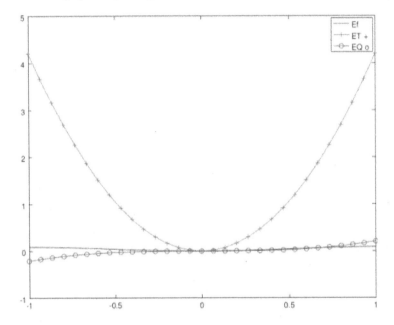

Figure 12.2: Graphing the error for two close cosines.

The Octave/ MATLAB code to do this is simple.

Graphing The error to two close cosines

Listing 12.2: **Code Fragment**

```
  >> x = linspace(-1,1,31);
  >> f = @(x) cos(2.1*x) - cos(2.0*x);
  >> Ef =@(x) abs(f(x));
4 >> ET = @(x) 4.205*x.^2;
  >> EQ = @(x) 1.88*abs(x).^3;
  >> plot(x,Ef(x),x,ET(x),'-+',x,EQ(x),'-o');
  >> legend('Ef','ET +','EQ o');
```

Homework

Exercise 12.2.1 *Using the techniques we have been discussing, find a nice approximation to* $\sin((a + r)t) - \sin(at)$ *and use it for the case* $a = 3.1$ *and* $r = 0.05$. *Do this for the base point* $t = 0$. *Do both the tangent and quadratic approximations.*

Exercise 12.2.2 *Approximate* $\cos(3.1x) - \cos(3.05x)$ *using both the tangent and quadratic approximations. Use* $a = 3$ *and* $r = 0.1$. *Do a MATLAB plot also.*

Exercise 12.2.3 *Approximate* $\cos(0.11x) - \cos(0.10x)$ *using both the tangent and quadratic approximations. Use* $a = 0.1$ *and* $r = 0.01$. *Do a MATLAB plot also.*

Exercise 12.2.4 *Approximate* $\sin(1.11x) - \sin(1.10x)$ *using both the tangent and quadratic approximations. Use* $a = 1.10$ *and* $r = 0.01$. *Do a MATLAB plot also.*

12.2.2 Two Close Exponentials

Here is a nice use of our approximation ideas which we can use when we study models in which exponential decay functions that differ slightly are found in the solutions. Let's look at the function $f(t) = e^{-rt} - e^{-(r+a)t}$ for positive r and a. To approximate this difference, we expand each exponential function into the second order approximation plus the error as usual. For $f(t) = e^{-rt} - e^{-(r+a)t}$ for positive r and a. we have

$$
e^{-rt} = 1 - rt + r^2\frac{t^2}{2} - r^3 e^{-rc_1}\frac{t^3}{6}
$$

$$
e^{-(r+a)t} = 1 - (r + a)t + (r + a)^2\frac{t^2}{2} - (r + a)^3 e^{-(r+a)c_2}\frac{t^3}{6}
$$

for some c_1 and c_2 between 0 and t. Subtracting and simplifying

$$
e^{-rt} - e^{-(r+a)t} = at - (a^2 + 2ar)\frac{t^2}{2} + \left(-r^3 e^{-rc_1} + (r + a)^3 e^{-(r+a)c_2}\right)\frac{t^3}{6} \qquad (12.1)
$$

We conclude

$$
e^{-rt} - e^{-(r+a)t} = at - (a^2 + 2ar)\frac{t^2}{2} + \mathcal{O}(t^3).
$$

We write the error is on the order of t^3 as $O(t^3)$ where O stands for order.

Example 12.2.2 *Approximate* $e^{-1.0t} - e^{-1.1t}$ *using Equation 12.1*

Solution *We know*

$$
e^{-rt} - e^{-(r+a)t} = at - (a^2 + 2ar)\frac{t^2}{2} + \left(-r^3 e^{-rc_1} + (r + a)^3 e^{-(r+a)c_2}\right)\frac{t^3}{6}
$$

and here $r = 1.0$ *and* $a = 0.1$. *So we have*

$$
e^{-1.0t} - e^{-1.1t} \approx 0.1t - (.01 + .2)\frac{t^2}{2} = 0.1t - 0.105t^2
$$

Example 12.2.3 *Approximate* $e^{-2.0t} - e^{-2.1t}$ *using Equation 12.1*

Solution *Again, we use our equation*

$$
e^{-rt} - e^{-(r+a)t} = at - (a^2 + 2ar)\frac{t^2}{2} + \left(-r^3 e^{-rc_1} + (r + a)^3 e^{-(r+a)c_2}\right)\frac{t^3}{6}
$$

and here $r = 2.0$ and $a = 0.1$. So we have

$$e^{-2.0t} - e^{-2.1t} \approx 0.1t - (.01 + 2(.1)(2))\frac{t^2}{2} = 0.1t - 0.21t^2$$

and the error is $O(t^3)$.

We can also approximate the function $g(t) = e^{-(r+a)t} - e^{-(r+b)t}$ for positive r, a and b.

$$e^{-(r+a)t} = 1 - (r+a)t + (r+a)^2 e^{-(r+a)c_1}\frac{t^2}{2}$$

$$e^{-(r+b)t} = 1 - (r+b)t + (r+b)^2 e^{-(r+a)c_2}\frac{t^2}{2}$$

for some c_1 and c_2 between 0 and t. Subtracting and simplifying, we find

$$e^{-(r+a)t} - e^{-(r+b)t} = (-a+b)t + \left((r+a)^2 e^{-(r+a)c_1} - (r+b)^2 e^{-(r+b)c_2}\right)\frac{t^2}{2} \quad (12.2)$$

Example 12.2.4 *Approximate $e^{-2.1t} - e^{-2.2t}$ using Equation 12.2.*

Solution *Now, we use our equation*

$$e^{-(r+a)t} - e^{-(r+b)t} = (-a+b)t$$
$$+ \left((r+a)^2 e^{-(r+a)c_1} - (r+b)^2 e^{-(r+b)c_2}\right)\frac{t^2}{2}$$

and here $r = 2.0$, $a = 0.1$ and $b = 0.2$. So we have

$$e^{-2.1t} - e^{-2.2t} \approx (0.2 - 0.1)t = 0.1t$$

and the error is $O(t^2)$.

Example 12.2.5 *Approximate $\left(e^{-2.1t} - e^{-2.2t}\right) - \left(e^{-1.1t} - e^{-1.2t}\right)$ using Equation 12.2.*

Solution *We have*

$$\left(e^{-2.1t} - e^{-2.2t}\right) - \left(e^{-1.1t} - e^{-1.2t}\right) \approx 0.1t - 0.1t = 0$$

plus $O(t^2)$ which is not very useful. Note if the numbers had been a little different everything would not have canceled out!

Solution *If approximate using Equation 12.1 we find*

$$\left(e^{-2.1t} - e^{-2.2t}\right) - \left(e^{-1.1t} - e^{-1.2t}\right) \approx$$
$$\left(0.1t - (0.01 + 2(0.1)(2.1))(t^2/2)\right) - \left(0.1t - (0.01 + 2(0.1)(1.1))(t^2/2)\right)$$
$$= -0.215t^2 + 0.230t^2 = 0.015t^2$$

plus $O(t^3)$ which is better.

Homework

Exercise 12.2.5 *Approximate $e^{-1.1t} - e^{-1.3t}$ using Equation 12.2 and estimate the maximum error on $[0, 10]$.*

Exercise 12.2.6 *Using the techniques we have been discussing, find a nice approximation to $\sigma(\alpha x) - \sigma(\beta x)$ and use it for the case $\alpha = 1.1$ and $\beta = 1.08$ where $\sigma(u) = 0.5(1 + \tanh(u))$. Do this for the base point $x = 0$. Do both the tangent and quadratic approximations.*

Exercise 12.2.7 *Approximate $e^{-3.1t} - e^{-3.0t}$ using Equation 12.2 and estimate the maximum error on $[0, 2]$.*

Exercise 12.2.8 *Approximate $e^{-0.0211t} - e^{-0.021t}$ using Equation 12.2 and Equation 12.1 and estimate the maximum errors on $[0, 1000]$.*

Exercise 12.2.9 *Approximate $(e^{-0.0211t} - e^{-0.021t}) - (e^{-0.0211t} - e^{-0.021t})$ using Equation 12.2 and Equation 12.1 and estimate the maximum errors on $[0, 10^4]$.*

12.2.3 Approximations in a Cancer Model

Here is another example taken from a colon cancer model. One of the variables in this model is $X_1(t)$ which is a fraction which is in $[0, 1]$; you can even interpret it as a probability that a cell is in the state called **1**. There are many parameters in this model: $u_1 = u_2 - 10^{-7}$, $N = 2000$ and $u_c = Ru_1$ where $R \approx 100$. So we can estimate the error over human lifetime $T = 3.65 \times 10^4$. For $r = u_c$, $a = u_1$ and $b = Nu_2$,

$$
\begin{aligned}
X_1(t) &= \frac{u_1}{Nu_2 - u_1}\left(e^{-(u_c + u_1)t} - e^{(u_c + Nu_2)t}\right) = \frac{u_1}{Nu_2 - u_1}\left((Nu_2 - u_1)t + \mathcal{O}(t^2)\right) \\
&= u_1 t + \mathcal{O}(t^2).
\end{aligned}
$$

The order t^2 term is

$$
\mathcal{O}(t^2) = \frac{u_1}{Nu_2 - u_1}\left((u_c + u_1)^2 e^{-(u_c + u_1)c_1} - (u_c + Nu_2)^2 e^{-(u_c + Nu_2)c_2}\right)\frac{t^2}{2}
$$

The parameters are $u_1 = u_2 = 10^{-7}$, $N = 2000$ and $u_c = Ru_1$ where $R \approx 100$. We can estimate the error over human lifetime, $T = 3.65 \times 10^4$ days. The maximum error in X_1 is then

$$
|E_1(T)| \leq \frac{u_1}{Nu_2 - u_1}\left((u_c + u_1)^2 + (u_c + Nu_2)^2\right)\frac{T^2}{2}
$$

Since $Nu_2 = 10^{-4}$ is dominant over $u_1 = 10^{-7}$ and $u_c = Ru_1 = 10^{-5}$ since $u_1 = u_2$, we have

$$
|E_1(T)| \approx \frac{u_1}{Nu_2}\left((1 + R)^2 u_1^2 + N^2 u_2^2\right)\frac{T^2}{2}
$$

$(1 + R)^2 u_1^2$ term is negligible in the second term so we have

$$
|E_1(T)| \approx \frac{u_1}{Nu_2}N^2 u_2^2\frac{T^2}{2} = Nu_1 u_2\frac{T^2}{2} = 2000 \times 10^{-14}\, 6.67\, 10^8 = 0.014.
$$

Another variable in the cancer model is X_2 which looks like this:

$$
X_2(t) = \frac{Nu_1 u_2}{Nu_2 - u_1}\left(\frac{1}{u_1}\left(e^{-u_c t} - e^{-(u_c + u_1)t}\right) - \frac{1}{Nu_2}\left(e^{-u_c t} - e^{-(u_c + Nu_2)t}\right)\right)
$$

$$= \frac{Nu_1 u_2}{Nu_2 - u_1} \left(\frac{1}{u_1} \left(u_1 t - (u_1^2 + 2u_1 u_c)\frac{t^2}{2} + \mathcal{O}(t^3) \right) \right.$$

$$\left. - \frac{1}{Nu_2} \left(Nu_2 t - ((Nu_2)^2 t + 2Nu_2 u_c)\frac{t^2}{2} + \mathcal{O}(t^3) \right) \right)$$

$$\approx Nu_1 u_2 \frac{t^2}{2} + \mathcal{O}(t^3).$$

The order t^3 term is very messy. Define $E(r, a, t) = \left(-r^3 e^{-rc_1} + (r+a)^3 e^{-(r+a)c_2} \right)\frac{t^3}{6}$. The maximum over human life time T is $E(r, a) = \left(2(r+a)^3 \right)\frac{T^3}{6}$. Then we have

$$E_2(T) = \left(\frac{Nu_2}{Nu_2 - u_1} E(u_c, u_1, t) - \frac{u_1}{Nu_2 - u_1} E(u_c, Nu_2, t). \right)$$

Thus,

$$E_2 \leq \frac{Nu_2}{Nu_2 - u_1} E(u_c, u_1) + \frac{u_1}{Nu_2 - u_1} E(u_c, Nu_2)$$

$$= \left(\frac{Nu_2}{Nu_2 - u_1} 2(u_1 + u_c)^3 + \frac{u_1}{Nu_2 - u_1} 2(u_c + Nu_2)^3 \right)\frac{T^3}{6}$$

The parameters now are $u_1 = u_2 - 10^{-7}$, $N = 4000$ and $u_c = Ru_1$ where $R \approx 100$. Since $u_1 = u_2 = 10^{-7}$, we see the term $Nu_2 - u_1 \approx Nu_2 = 4 \times 10^{-4}$. Also, $u_1 + u_c$ is dominated by u_c. Thus,

$$E_2 \approx \left(2u_c^3 + \frac{u_1}{Nu_2 - u_1}(Nu_2)^3 \right)\frac{T^3}{6}$$

$$\approx \left(2u_c^3 + \frac{u_1}{Nu_2}(Nu_2)^3 \right)\frac{T^3}{6}$$

$$\approx \left(2u_c^3 + u_1(Nu_2)^2 \right)\frac{T^3}{6}$$

But the term $u_1(Nu_2)^2$ is very small ($\approx 1.6e - 14$) and can also be neglected. So, we have

$$E_2 \approx 2u_c^3 \frac{T^3}{6}$$

We see

$$E_2 \approx 2 \times R^3 (10^{-7})^3 \, 8.1 \, 10^{12} = 16.2 \times 10^{-3} = .016.$$

We have shown $X_1(t) \approx u_1 t$ with error 0.014 over human life time with $N = 2000$ and $X_2(t) \approx Nu_1 u_2 \frac{t^2}{2}$ with error 0.016 over human life time with $N = 4000$.

12.2.4 Homework

Exercise 12.2.10 *Estimate the error in the X_1 and X_2 approximation if $u_1 = 10^{-6}$, $u_2 = 1.3 \times 10^{-6}$, $N = 25$ and $R = 2$ with $T = 2 \times 10^4$.*

Exercise 12.2.11 *Estimate the error in the* X_1 *and* X_2 *approximation if* $u_1 = 9 \times 10^{-6}$, $u_2 = 1.3 \times 10^{-7}$, $N = 11$ *and* $R = 3$ *with* $T = 5 \times 10^4$.

Chapter 13

Exponential and Logarithm Functions

We are now going to define the special number e and its consequences. We will find what is called the exponential function and its inverse, the natural logarithm function, and along the way use a lot of the tools we have been building. There is another way to do this using the theory of Riemann integration which we go over in Section 23.6 but that is another story.

13.1   The Number e: First Definition

We define this special number e as the limit of a particular sequence of rational numbers.

Theorem 13.1.1 The Number e

> For all positive integers n, let $a_n = (1 + 1/n)^n$ and $b_n = (1 + 1/n)^{n+1}$. Then
>
>   1. (a_n) is an increasing sequence bounded above and (b_n) is a decreasing sequence bounded below.
>
>   2. $\lim_{n \to \infty} a_n$ and $\lim_{n \to \infty} b_n$ both exist and have equal values.
>
> This common value is called e.

Proof 13.1.1
Recall the identity $(1 + x)^n > 1 + nx$ if $x < -1$ and $x \neq 0$ for all positive integers n. This is called Bernoulli's Identity. *We will use this in our calculations below.*

$$\frac{a_n}{b_{n-1}} = \frac{(1 + 1/n)^n}{1 + 1/(n-1)^n} = \left(\frac{(n+1)/n}{n/(n-1)} \right)^n = \left(\frac{(n+1)(n-1)}{n^2} \right)^n$$

After a bit of manipulation, we apply Bernoulli's Identity:

$$\frac{a_n}{b_{n-1}} = \left(\frac{(n^2 - 1)}{n^2} \right)^n = (1 - 1/n^2)^n > 1 + n(-1/n^2) = 1 - 1/n.$$

Next, look at another ratio:

$$\frac{b_{n-1}}{a_n} = \frac{(1 + 1/(n-1))^n}{(1 + 1/n)^n} = \left(\frac{(n)/(n-1)}{(n+1)/n} \right)^n = \left(\frac{n^2}{n^2 - 1} \right)^n$$

179

$$= \left(\frac{(n^2 - 1) + 1}{n^2 - 1} \right)^n = \left(1 + 1/(n^2 - 1) \right)^n > \left(1 + 1/n^2 \right)^n$$

Now apply Bernoulli's identity to see

$$\frac{b_{n-1}}{a_n} > 1 + n(1/n^2) = 1 + 1/n.$$

The inequality for a_n/b_{n-1} implies

$$\begin{aligned} a_n &> b_{n-1} \left(1 - 1/n \right) = \left(1 + 1/(n-1) \right)^n \left((n-1)/n \right) \\ &= \left(n/(n-1) \right)^n \left(n/(n-1) \right)^{-1} = \left(n/(n-1) \right)^{n-1} = a_{n-1}. \end{aligned}$$

This shows the sequence (a_n) is increasing. The inequality for b_{n-1}/a_n implies

$$\begin{aligned} b_{n-1} &> a_n \left(1 + 1/n \right) = \left(1 + 1/n \right)^n \left(1 + 1/n \right) \\ &= \left(1 + 1/n \right)^{n+1} = b_n \end{aligned}$$

This shows the sequence (b_n) is decreasing.

Next, we have the chain of inequalities

$$2 = a_1 < a_n < b_n < b_1 = 4.$$

This tells us the sequence (a_n) is bounded above and so $\sup(a_n)$ is finite. Since (a_n) is increasing, it is easy to see $a_n \to \sup(a_n)$. Also, the inequality chain tells us the sequence (b_n) is bounded below and so since b_n is decreasing, $b_n \to \inf(b_n)$.

Finally, we know

$$\begin{aligned} b_n &= \left(1 + 1/n \right)^n \left(1 + 1/n \right) = a_n \left(1 + 1/n \right) \\ &\Rightarrow \lim_{n \to \infty} b_n = \lim_{n \to \infty} \left(a_n \left(1 + 1/n \right) \right) \end{aligned}$$

Since each of the limits in the product above exists, the right hand side limit is $\lim_{n \to \infty} a_n \lim_{n \to \infty} (1 + 1/n)$. Thus, $\lim_{n \to \infty} b_n = \lim_{n \to \infty} a_n$. This common limit is the number e. ∎

Note

$$e^2 = \left(\lim_{n \to \infty} (1 + 1/n)^n \right)^2 = \lim_{n \to \infty} \left((1 + 1/n)^n \right)^2$$

because $(\ )^2$ is a continuous function. Thus,

$$e^2 = \left(\lim_{n \to \infty} (1 + 1/n)^n \right)^2 = \lim_{n \to \infty} \left((1 + 1/n)^n \right)^2 = \lim_{n \to \infty} (1 + 1/n)^{2n}$$

Let $m = 2n$ and make this change of variable to get

$$e^2 = \lim_{m \to \infty} (1 + 2/m)^m = \lim_{n \to \infty} (1 + 2/n)^n$$

as the choice of letter for the limit is not important. We could do this argument for any nonzero integer really. So we now know $e^p = \lim_{n \to \infty} (1 + p/n)^n$ for nonzero integers.

Since $(\ )^{1/5}$ is also continuous, we can use a similar argument to see

$$
\begin{aligned}
e^{1/5} &= \left(\lim_{n\to\infty}(1+1/n)^n\right)^{1/5} = \lim_{n\to\infty}((1+1/n)^n)^{1/5} \\
&= \lim_{n\to\infty}(1+1/n)^{n/5}.
\end{aligned}
$$

Now if $\lim_n$ exists here, we get the same value with any subsequential limit. So look at the subsequence of those integers where $m = n/5$.

$$
\begin{aligned}
e^{1/5} &= \lim_{5m\to\infty}(1+1/(5m))^{5m/5} = \lim_{5m\to\infty}(1+(1/5)/m)^m \\
&= \lim_{m\to\infty}(1+(1/5)/m)^m
\end{aligned}
$$

Now consider $e^{4/5}$. Since $(\ )^{4/5}$ is continuous, we can use a similar argument to see

$$
\begin{aligned}
e^{4/5} &= \left(e^4\right)^{1/5} = \left(\lim_{n\to\infty}(1+4/n)^n\right)^{1/5} \\
&= \lim_{n\to\infty}\left((1+4/n)^n\right)^{1/5} = \lim_{n\to\infty}(1+4/n)^{n/5}
\end{aligned}
$$

Look at the subsequence of those integers where $m = n/5$. Then we have

$$
e^{4/5} = \lim_{m\to\infty}(1+4/(5m))^m = \lim_{m\to\infty}(1+(4/5)/m)^m
$$

We could do this argument for any nonzero fraction really. So we now know $e^{p/q} = \lim_{n\to\infty}(1+(p/q)/n)^n$ for nonzero positive fractions.

What about negative fractions? Consider

$$
\begin{aligned}
e^{-4/5} &= \left(e^4\right)^{-1/5} = \left(\lim_{n\to\infty}(1+4/n)^n\right)^{-1/5} \\
&= \lim_{n\to\infty}\left((1+4/n)^n\right)^{-1/5} = \lim_{n\to\infty}(1+4/n)^{-n/5}
\end{aligned}
$$

Look at the subsequence of those integers where $m = -n/5$. Then we have

$$
e^{-4/5} = \lim_{m\to\infty}(1+4/(-5m))^m = \lim_{m\to\infty}(1-(4/5)/m)^m
$$

13.1.1 Homework

Exercise 13.1.1 *Prove $e^7 = \lim_{n\to\infty}(1+7/n)^n$*

Exercise 13.1.2 *Prove $e^{-5} = \lim_{n\to\infty}(1-5/n)^n$*

Exercise 13.1.3 *Prove $e^{2/3} = \lim_{n\to\infty}(1+(2/3)/n)^n$*

13.2 The Number e: Second Definition

We can also define the number e in other ways but the proofs are fairly technical.

Theorem 13.2.1 The Number e as a Series

(1) $e = \lim_{n \to \infty} \sum_{k=0}^{n} \frac{1}{k!}$
(2) e is an irrational number.
(3) $2.666.... < e < 2.7222....$
(4) If $S_n = \sum_{k=0}^{n} \frac{1}{k!}$, *the* $0 < e - S_n < \frac{1}{n!(n)}$.

Proof 13.2.1

$$\sum_{n=k+1}^{N} \frac{1}{k!} = \frac{1}{(k+1)!} + \ldots + \frac{1}{(k+(N-k))!}$$

$$= \frac{1}{k!} \left\{ \frac{1}{(k+1)} + \frac{1}{(k+1)(k+2)} + \ldots + \frac{1}{(k+1)\ldots(k+(N-k))} \right\}$$

Now since the terms in the summation S_N *are always positive, the* $\lim_{N \to \infty} S_N$ *is either finite or* ∞. *Thus, we have*

$$\sum_{n=k+1}^{N} \frac{1}{k!} = (S_N - S_k) = S - S_k$$

and so

$$S_N - S_k = \left\{ \frac{1}{(k+1)!} + \frac{1}{(k+2)!} + \ldots + \frac{1}{(k+(N-k))!} \right\}$$

$$= \frac{1}{k!} \left\{ \frac{1}{(k+1)} + \frac{1}{(k+1)(k+2)} + \ldots + \frac{1}{(k+1)\ldots(k+(N-k))} \right\}$$

Every term like $k+2$, $k+3$ *in the denominators of the fractions in the braces can be replaced by* $k+1$ *and the sum increases. So we have*

$$S_N - S_k < \frac{1}{k!} \left\{ \left(\frac{1}{(k+1)}\right)^1 + \ldots + \left(\frac{1}{(k+1)}\right)^{N-k} \right\}$$

Now recall if $x \neq 1$, $1 + x + x^2 + \ldots + x^q = \frac{1-x^{q+1}}{1-x}$. *Here* $x = 1/(k+1)$ *which is positive and less than 1. So the expansion works. Also, we now know* $\lim_{q \to \infty} x^{q+1} = 0$ *since* $0 < 1/(k+1) < 1$. *Thus, we have*

$$S_N - S_k < \frac{1}{k!} \left\{ \frac{1 - (1/(k+1))^{N+1}}{1 - 1/(k+1)} - 1 \right\}$$

Thus,

$$S_N - S_k < \frac{1}{k!} \left(\frac{1}{1 - 1/(k+1)} - 1 \right) = \frac{1}{k!} \left(\frac{k+1}{k} - 1 \right) = 1/(k!k)$$

Since $S_N - S_k > 0$, *we have* $0 < S_N - S_k < \frac{1}{k!k}$. *Thus,* S_N *is bounded as N goes to infinity and so* $\lim_{N \to \infty} S_N = S$ *for some positive number S. In particular, this says* $S < S_1 + 1 = 3$.

We know $a_n = (1 + 1/n)^n$ which we can expand using the Binomial Theorem.

$$
\begin{aligned}
a_n &= \sum_{k=0}^{n} \binom{n}{k} \frac{1}{n^k} = 1 + \sum_{k=1}^{n} \binom{n}{k} \frac{1}{n^k} = 1 + \sum_{k=1}^{n} \frac{n!}{k!(n-k)!} \frac{1}{n^k} \\
&= 1 + \sum_{k=1}^{n} \frac{n(n-1)(n-2)\ldots(n-(k-1))}{k!} \frac{1}{n^k} \\
&= 1 + \sum_{k=1}^{n} \left(\frac{n-1}{n} \right) \ldots \left(\frac{n-(k-1)}{n} \right) \frac{1}{k!}, \quad \textit{group the n terms} \\
&= 1 + \sum_{k=1}^{n} (1 - 1/n) \ldots (1 - (k-1)/n) \frac{1}{k!} < 1 + \sum_{k=1}^{n} \frac{1}{k!} = S_n
\end{aligned}
$$

Thus, $e = \lim_{n \to \infty} \leq \lim_{n \to \infty} S_n = S$.

To show the reverse inequality, pick an arbitrary $\epsilon > 0$. Since $S_n \to S$, there is a P so that if $n > P$, $S - \epsilon < S_n < S + \epsilon$. Pick any $p > P$ so that $S - \epsilon < S_p$. Then for any $n > p$, we have since (a_n) increases

$$
\begin{aligned}
e &> a_n > a_p = 1 + \sum_{k=1}^{p} (1 - 1/p) \ldots (1 - (k-1)/p) \frac{1}{k!} \\
&> 1 + \sum_{k=1}^{p} (1 - 1/n) \ldots (1 - (k-1)/n) \frac{1}{k!}
\end{aligned}
$$

Since this is true for all $n > p$, we can say

$$
e \geq \lim_{n \to \infty} a_n \geq \lim_{n \to \infty} \left\{ 1 + \sum_{k=1}^{p} (1 - 1/n) \ldots (1 - (k-1)/n) \frac{1}{k!} \right\}
$$

We conclude $e \geq 1 + \sum_{k=1}^{p} \frac{1}{k!} = S_p > S - \epsilon$. But if $e \geq S - \epsilon$ for all $\epsilon > 0$, this implies $e \geq S$. Combining these inequalities, we have $e = S$ and we have now shown (1) and (4) are true.

(e is irrational):
Suppose $e = m/n$ with m and n positive integers. Then, we would have

$$
0 < e - S_n < \frac{1}{n!n} \implies 0 < \frac{m}{n} - S_n < \frac{1}{n!n} \implies 0 < n!\frac{m}{n} - n!S_n < \frac{1}{n} \leq 1
$$

This says $0 < m(n-1)! - n!S_n < 1$. But it is easy to see the term $m(n-1)! - n!S_n$ is an integer. The inequality says this integer must be between 0 and 1 which is not possible. Hence, e must be irrational.

(Estimate (2)):
Note for $n = 3$, $0 < e - S_3 < \frac{1}{3!3} = 1/18$. But $S_3 = 8/3 = 2.666\ldots$ and $S_3 + 1.18 = 2.7222\ldots$ Thus $S_3 < e < S_3 + 1/18$ which is the result. ∎

13.3   The Exponential Function

Although we have shown

$$e \;=\; \lim_{n\to\infty}\left(1+\frac{1}{n}\right)^{n} = \lim_{n\to\infty}\sum_{k=0}^{n}\frac{1}{k!}$$

and we also know

$$e^{x} = \lim_{n\to\infty}\left(1+\frac{x}{n}\right)^{n},\,\forall x\in\mathbb{Q}$$

we do not know if e^{x} is continuous.

The question of continuity is a question of interchanging the order of limit operations. Let (x_p) be a sequence of nonzero rational numbers that converge to an irrational x. Then, we want to show

$$\lim_{p\to\infty}\;\lim_{n\to\infty}\left(1+\frac{x_p}{n}\right)^{n} \;=\; \lim_{n\to\infty}\;\lim_{p\to\infty}\left(1+\frac{x_p}{n}\right)^{n}$$

But this is the same as asking

$$\lim_{p\to\infty} e^{x_p} \;=\; \lim_{n\to\infty}\left(1+\frac{x}{n}\right)^{n}$$

and we don't know if the left hand side converges yet. If e^{x} was continuous as a function, we could say $\lim_{p\to\infty} e^{x_p} = e^{\lim_{p\to\infty} x_p} = e^{x}$, but we don't know this. Also, with more tools we can show

$$e^{x} \;=\; \lim_{n\to\infty}\sum_{k=0}^{n}\frac{x^{k}}{k!}$$

Actually, we have enough to show $\lim_{n\to\infty}\sum_{k=0}^{n}\frac{x^{k}}{k!}$ converges but we don't know enough yet to conclude the limit is the same as e^{x}.

If we denote $\lim_{n\to\infty}\sum_{k=0}^{n}\frac{x^{k}}{k!}$ by $\sum_{k=0}^{\infty}\frac{x^{k}}{k!}$, one way to see e^{x} is differentiable at x is to examine

$$\begin{aligned}
\frac{e^{y}-e^{x}}{y-x} &= \frac{\sum_{k=0}^{\infty}\frac{y^{k}}{k!}-\sum_{k=0}^{\infty}\frac{x^{k}}{k!}}{y-x} = \frac{1+\sum_{k=1}^{\infty}\frac{y^{k}}{k!}-1-\sum_{k=0}^{\infty}\frac{x^{k}}{k!}}{y-x}\\[2mm]
&= \frac{\sum_{k=1}^{\infty}\frac{y^{k}-x^{k}}{k!}}{y-x}
\end{aligned}$$

If we could interchange limit operations, we could write

$$\lim_{y\to x}\left(\frac{e^{y}-e^{x}}{y-x}\right) \;=\; \left(\sum_{k=1}^{\infty}\frac{1}{k!}\lim_{y\to x}\frac{y^{k}-x^{k}}{y-x}\right)$$

The inside limit is $(x^{k})' = kx^{k-1}$ and hence, at the end switching $k-1$ as k in the summation, we have

$$\lim_{y\to x}\left(\frac{e^{y}-e^{x}}{y-x}\right) \;=\; \sum_{k=1}^{\infty}\frac{kx^{k-1}}{k!} = \sum_{k=1}^{\infty}\frac{x^{k-1}}{(k-1)!} = \sum_{k=0}^{\infty}\frac{x^{k}}{k!} = e^{x}$$

So we have a ways to go to get the tools to show e^{x} is a continuous function and to show it is differentiable with $(e^x)' = e^x$. There are better ways to attack this and it is easiest if we do Riemann integration first, use that to define the natural logarithm of x, from that get yet another definition of e and so forth. So more stories to come! In Section 23.6, we find the number e is the unique number such that $\int_1^e \frac{1}{t} dt = 1$ and we can show $\ln(x) = \int_1^x \frac{1}{t} dt$ has as its inverse e^x with derivative itself. Then the Taylor polynomials for e^x converge to $\sum_{k=0}^{\infty} \frac{x^k}{k!}$ and so $e^1 = e = \sum_{k=0}^{\infty} \frac{1}{k!} = \lim_{n \to \infty} (1 + 1/n)^n$. And we have come full circle!

However, we can prove this results with the tools at our disposal now.

Theorem 13.3.1 The nth Term of the e^x Expansion Goes to Zero

$\lim_{k \to \infty} (x^k/k!) = 0$ for all x.

Proof 13.3.1
There is a k_0 with $|x|/k_0 < 1$. Thus,

$$\frac{|x|^{k_0+1}}{(k_0+1)!} = \frac{|x|}{k_0+1} \frac{|x|^{k_0}}{k_0!} \leq \frac{|x|}{k_0} \frac{|x|^{k_0}}{k_0!}$$

$$\frac{|x|^{k_0+2}}{(k_0+2)!} = \frac{|x|}{k_0+2} \frac{|x|}{k_0+1} \frac{|x|^{k_0}}{k_0!} < \left(\frac{|x|}{k_0}\right)^2 \frac{|x|^{k_0}}{k_0!}$$

$$\vdots$$

$$\frac{|x|^{k_0+j}}{(k_0+j)!} \leq \left(\frac{|x|}{k_0}\right)^j \frac{|x|^{k_0}}{k_0!}$$

Since $\frac{|x|}{k_0} < 1$, this shows $\lim_{k \to \infty} (x^k/k!) \to 0$ as $k \to \infty$. ∎

Although we do not have the best tools to analyze the function

$$S(x) = \lim_{n \to \infty} \left(1 + \frac{x}{n}\right)^n$$

let's see how far we can get. We start with this Theorem.

Theorem 13.3.2 The Series Expansion of e^x Converges

$\lim_{n \to \infty} \sum_{k=0}^{n} \frac{x^n}{k!}$ converges for all x.

Proof 13.3.2
Let $f(x) = \lim_{n \to \infty} \sum_{k=0}^{n} \frac{x^k}{k!}$ and $f_n(x) = \sum_{k=0}^{n} \frac{x^k}{k!}$. We need to show these limits exist and are finite. We will follow the argument we used in the earlier Theorem's proof with some changes.

$$\sum_{n=k+1}^{N} \frac{|x|^k}{k!} = \frac{|x|^k}{(k+1)!} + \frac{|x|^{k+1}}{(k+2)!} + \ldots + \frac{|x|^{N-k}}{(k+(N-k))!}$$

$$= \frac{|x|^k}{k!} \left\{ \frac{|x|}{(k+1)} + \ldots \frac{|x|^{N-k}}{(k+1)\ldots(k+(N-k))} \right\}$$

$$\leq \frac{|x|^k}{k!} \left\{ \frac{|x|}{(k+1)} + \ldots \frac{|x|^{N-k}}{(k+1)^{N-k}} \right\} \leq \frac{|x|^k}{k!} \sum_{k=0}^{N-k} r^k$$

where $r = \frac{|x|}{(k+1)}$. We can then use the same expansion of powers of r as before to find

$$\sum_{n=k+1}^{N} \frac{|x|^k}{k!} \leq \frac{|x|^k}{k!} \left\{ \frac{1 - \left(|x|/(k+1)\right)^{N-k+1}}{1 - |x|/(k+1)} \right\}$$

Since there is a K_1 so $|x|/(k+1) < 1/2$ for $k > K_1$, for all $k > K_1$, $1/(1 - |x|/(k+1)) < 2$ and so

$$\lim_{N \to \infty} \left\{ \sum_{n=k+1}^{N} \frac{|x|^k}{k!} \right\} \leq \frac{|x|^k}{k!} \lim_{N \to \infty} 2(1 - (1/2)^{N-k+1}) = 2\frac{|x|^k}{k!}$$

Now pick any $k > K_1$, then we see $f(x)$, the limit of an increasing sequence of terms, is bounded above which implies it converges to its supremum.

$$
\begin{aligned}
f(|x|) &= 1 + |x| + |x|/2! + \ldots + |x|^k/k! + \lim_{N \to \infty} \sum_{n=k+1}^{N} \frac{|x|^k}{k!} \\
&\leq 1 + |x| + |x|/2! + \ldots + |x|^k/k! + 2\frac{|x|^k}{k!}.
\end{aligned}
$$

So we know $f(|x|)$ has a limit for all x.

To see $f(x)$ converges also, pick an arbitrary $\epsilon > 0$ and note since $\frac{|x|^k}{k!} \to 0$ as $k \to \infty$, $\exists K \ni k > K$ implies $\frac{|x|^k}{k!} < \epsilon/2$.

So for $k > K$, we have

$$\lim_{N \to \infty} \sum_{n=k+1}^{N} \frac{|x|^k}{k!} = 2\frac{|x|^k}{k!} < \epsilon$$

Then, for $k > K$,

$$\left| f(x) - \sum_{n=0}^{k} \frac{x^k}{k!} \right| \leq \left| \lim_{N \to \infty} \sum_{n=k+1}^{N} \frac{x^k}{k!} \right| \leq \lim_{N \to \infty} \sum_{n=k+1}^{N} \frac{|x|^k}{k!} < \epsilon.$$

This says $\lim_{k \to \infty} \sum_{n=0}^{k} \frac{x^k}{k!} = f(x)$. ∎

Next, let's look at $S_n(x) = (1 + x/n)^n$ and $S(x) = \lim_{n \to \infty} S_n(x)$.

$$
\begin{aligned}
S_n(x) &= 1 + \sum_{k=1}^{n} \binom{n}{k} \frac{x^k}{n^k} = 1 + \sum_{k=1}^{n} \frac{n!}{k!(n-k)!} \frac{x^k}{n^k} \\
&= 1 + \sum_{k=1}^{n} \frac{n(n-1)(n-2)\ldots(n-(k-1))}{k!} \frac{x^k}{n^k} \\
&= 1 + \sum_{k=1}^{n} (1 - 1/n)\ldots(1 - (k-1)/n) \frac{x^k}{k!}
\end{aligned}
$$

Consider

$$
\left| \sum_{k=1}^{n} (1 - 1/n) \ldots (1 - (k-1)/n) \, \frac{x^k}{k!} - \sum_{k=1}^{n} \frac{x^k}{k!} \right| =
$$

$$
\left| \sum_{k=1}^{n} ((1 - 1/n) \ldots (1 - (k-1)/n) - 1) \, \frac{x^k}{k!} \right| < \left| \sum_{k=1}^{n} ((1 + 1/n) - 1) \, \frac{x^k}{k!} \right|
$$

Then we have

$$
\left| \sum_{k=1}^{n} (1 - 1/n) \ldots (1 - (k-1)/n) \, \frac{x^k}{k!} - \sum_{k=1}^{n} \frac{x^k}{k!} \right|
$$

$$
< \left| \sum_{k=1}^{n} \frac{x^k}{n \, k!} \right| \le (1/n) \sum_{k=1}^{n} \frac{|x|^k}{k!} = \frac{f(|x|)}{n}
$$

We know $f(|x|)$ is finite, so the ratio $\frac{f(|x|)}{n}$ goes to zero as $n \to \infty$.

Hence, for an $\epsilon > 0$, there is an N_1 so that $|f_n(x) - f(x)| < \epsilon/2$ if $n > N_1$ and there is an N_2 so that $|S_n(x) - f_n(x)| < \epsilon/2$ if $n > N2$. So if we pick an $n > \max(N_1, N_2)$, we have

$$
|S_n(x) - f(x)| \le |S_n(x) - f_n(x)| + |f_n(x) - f(x)| < \epsilon/2 + \epsilon/2 = \epsilon
$$

Hence, $S_n \to f$ and we have proven

Theorem 13.3.3 e^x is the Same as the Series Expansion

$$
S(x) = \lim_{n \to \infty} S_n(x) = 1 + \lim_{n \to \infty} \sum_{k=1}^{n} \frac{x^k}{k!} = f(x).
$$

Proof 13.3.3
We just did this argument. ∎

We usually abuse notation for these sorts of limits and we will start doing that now. The notation $\sum_{k=0}^{\infty} \frac{x^k}{k!}$ is equivalent to $\lim_{n \to \infty} \sum_{k=0}^{n} \frac{x^k}{k!}$. Also, since we are always letting n of N go to ∞, we will start using $\lim_n$ to denote $\lim_{n \to \infty}$ for convenience. Next, let's prove e^x is a continuous function.

Theorem 13.3.4 $S(x)$ is Continuous

$S(x)$ is continuous for all x

Proof 13.3.4
Fix x_0 and consider any $x \in B_1(x_0)$. Note $S_n'(x) = n(1 + x/n)^{n-1} (1/n) = (1 + x/n)^{n-1}$. We see $S_n'(x) = S_n(x) (1 + x/n)^{-1}$. For each n, apply the MVT to find c_x^n between x_0 and x so that $\frac{|S_n(x) - S_n(x_0)|}{|x - x_0|} = |S_n'(c_x^n)|$. Now $\lim_n (1 + c_x^n/n) = 1$, so for n sufficiently large, $1 < |1 + c_x^n/n|^{-1} < 2$. Hence, if n sufficiently large

$$
|(1 + c_x^n/n)^n| < |S_n'(c_x^n)| < 2 |(1 + c_x^n/n)^n|
$$

Also, if you look at $S_n(x)$ for $x_0 - 1 < x < x_0 + 1$,

$$1 + \frac{x_0 - 1}{n} \quad < \quad 1 + \frac{x}{n} < 1 + \frac{x_0 + 1}{n}$$

implying $S_n(x_0 - 1) < S_n(x) < S_n(x_0 + 1)$. Since these numbers could be negative, we see $|S_n(c_x^n)| \leq D(x_0) = \max\{|S_n(x_0 - 1)|, |S_n(x_0 - 1)|\}$ which we can choose to be positive. Thus, since $x_0 - 1 < c_x^n < x_0 + 1$, $|S_n'(c_x^n)| < 2D(x_0)$ for n sufficiently large. This tells us $\lim_n \frac{|S_n(x) - S_n(x_0)|}{|x - x_0|} \leq 2D(x_0)$. Letting $n \to \infty$, we find

$$\frac{|S(x) - S(x_0)|}{|x - x_0|} \quad \leq \quad 2D(x_0).$$

This shows $S(x)$ is continuous at x_0 as if $\epsilon > 0$ is chosen, if $|x - x_0| < \epsilon/D(x_0)$ we have $|S(x) - S(x_0)| < \epsilon$. ∎

Now pick any x that is irrational. Then let (x_p) be any sequence of rational numbers which converges to x. Then since $S(x) = \lim_{p \to \infty} S(x_p)$, we have $S(x) = \lim_{p \to \infty} (1 + x_p/n)^n = \lim_{p \to \infty} e^{x_p}$. Hence, for irrational x, we can define $e^x = \lim_{p \to \infty} e^{x_p}$ and we have extended the definition of e^x off of $\mathbb{Q}$ to $\mathbb{R}$.

We conclude

- $e^x = \lim_n (1 + x/n)^n = \sum_{k=0}^{\infty} \frac{x^k}{k!}$

- $e = \lim_n (1 + 1/n)^n = \sum_{k=0}^{\infty} \frac{1}{k!}$

- e^x is a continuous function of x.

Now let's look at differentiation of e^x.

Theorem 13.3.5 The Derivative of e^x is Itself

$(e^x)' = e^x$

Proof 13.3.5
Let (x_n) be any sequence converging to x_0. Then, we have

$$\frac{S_n(x_p) - S_n(x_0)}{x - x_0} \quad = \quad S_n(c_x^n) \left(1 + c_x^n/n\right)^{-1}.$$

where c_x^n is between x_0 and x_p implying $\lim_n c_n^x = x_0$.

Given $\epsilon > 0$ arbitrary, $\exists N \ni n > N \Rightarrow x_0 - \epsilon < c_x^n < x_0 + \epsilon$. We then have

$$1 + \frac{x_0 - \epsilon}{n} \quad < \quad 1 + \frac{c_x^n}{n} < 1 + \frac{x_0 + \epsilon}{n} \Rightarrow$$

$$\left(1 + \frac{x_0 - \epsilon}{n}\right)^n \quad < \quad \left(1 + \frac{c_x^n}{n}\right)^n < \left(1 + \frac{x_0 + \epsilon}{n}\right)^n$$

for n sufficiently large as then all terms will be positive. Taking the limit as $n \to \infty$, we have

$$e^{x_0 - \epsilon} \quad \leq \quad \lim_n S_n(c_x^n) \leq e^{x_0 + \epsilon}$$

Now, letting $\epsilon \to 0^+$, we find

$$e^{x_0} = \lim_{\epsilon \to 0^+} e^{x_0 - \epsilon} \;\leq\; \lim_n S_n(c_x^n) \leq \lim_{\epsilon \to 0^+} e^{x_0 + \epsilon} = e^{x_0}$$

as e^x is a continuous function. We conclude

$$\lim_n \left\{ \frac{S_n(x) - S_n(x_0)}{x - x_0} \right\} = \frac{S(x) - S(x_0)}{x - x_0} \;=\; S(x_0)\,(1) = e^{x_0}.$$

Since we can do this for any such sequence (x_n) we have shown e^x is differentiable with $(e^x)' = e^x$.
∎

13.3.1 The Properties of the Exponential Function

We can now prove the usual properties for the exponential function.

Theorem 13.3.6 The Properties of the Exponential function

> *1.* $e^{x+y} = e^x e^y$
>
> *2.* $e^{-x} = 1/e^x$
>
> *3.* $e^x e^{-y} = e^{x-y} = e^x/e^y$
>
> *4.* $(e^x)^r = e^{rx}$

Proof 13.3.6
(1):

$$
\begin{aligned}
e^x e^y &= \lim_n \left((1 + x/n)^n \right)\, \lim_n \left((1 + y/n)^n \right) \\
&= \lim_n \left((1 + x/n)\,(1 + y/n) \right)^n \\
&= \lim_n \left(1 + (x + y)/n + xy/n^2 \right)^n \\
&= \lim_n \left(1 + (x + y)/n + (xy/n)(1/n) \right)^n .
\end{aligned}
$$

Pick an $\epsilon > 0$. Then
$\exists N_1 \ni n > N_1 \Rightarrow |xy|/n < \epsilon$ and $\exists N_2 \ni n > N_2 \Rightarrow 1 - (x+y)/n > 0$. So if $n > \max\{N_1, N_2\}$, we have

$$1 + (x+y)/n - \epsilon/n \;<\; 1 + (x+y)/n + xy/n^2 < 1 + (x+y)/n + \epsilon/n$$

$$\left(1 + (x+y)/n - \epsilon/n \right)^n \;<\; \left(1 + (x+y)/n + xy/n^2 \right)^n$$

$$<\; \left(1 + (x+y)/n + \epsilon/n \right)^n .$$

because all terms are positive once n is large enough. Now let $n \to \infty$ to find

$$e^{x+y-\epsilon} \leq e^x e^y \leq e^{x+y+\epsilon}$$

Finally, let $\epsilon \to 0^+$ to find

$$e^{x+y} \le e^x e^y \le e^{x+y} \Rightarrow e^x e^y = e^{x+y}$$

which tells us (1) is true.

(2):

$$
\begin{aligned}
e^{-x} e^x &= \lim_n \left((1 - x/n)^n \right) \lim_n \left((1 + x/n)^n \right) \\
&= \lim_n \left((1 + x/n)(1 - x/n) \right)^n = \lim_n \left(1 - x^2/n^2 \right)^n \\
&= \lim_n \left(1 - (x^2/n)(1/n) \right)^n
\end{aligned}
$$

Now argue just like in (1). Given $\epsilon > 0$, $\exists N \ni n > N \Rightarrow x^2/n < \epsilon$. Thus,

$$
\begin{aligned}
1 - \epsilon/n \ &< \ 1 - x^2/n^2 < 1 + \epsilon/n \Rightarrow \\
\left(1 - \epsilon/n \right)^n \ &< \ \left(1 - x^2/n^2 \right)^n < \left(1 + \epsilon/n \right)^n \Rightarrow \\
\lim_n \left(1 - \epsilon/n \right)^n \ &\le \ \lim_n \left(1 - x^2/n^2 \right)^n \le \lim_n \left(1 + \epsilon/n \right)^n
\end{aligned}
$$

This says $e^{-\epsilon} \le e^{-x} e^x \le e^\epsilon$ and as $\epsilon \to 0$, we obtain (2).

(3): $e^x e^{-y} = e^{x+(-y)} = e^{x-y}$ and $e^x e^{-y} = e^x/e^y$ and so (3) is true.

(4): Let r be any real number and let $(r_n = p_n/q_n)$ be any sequence of rational numbers that converge to r. Then we know since e^x is a continuous function $e^{rx} = \lim_k e^{r_k x}$. Next, consider

$$
\lim_k (e^x)^{r_k} = \lim_k \left(\lim_n \left(1 + \frac{x}{n} \right)^n \right)^{r_k} = \lim_k \left(\lim_n \left(1 + \frac{x}{n} \right)^{r_k n} \right)
$$

Now let $m = r_k n$ which implies

$$
\lim_k (e^x)^{r_k} = \lim_k \left(\lim_m \left(1 + \frac{r_k x}{m} \right)^m \right) = \lim_k e^{r_k x}
$$

These two limits are the same and so $e^{rx} = (e^x)^r$. Hence, we can extend the **law** *$e^{rx} = (e^x)^r$ for $r \in \mathbb{Q}$ to the irrationals by defining $(e^x)^r \equiv \lim_n (e^x)^{p_n/q_n}$. The value we get does not depend on the sequence (r_n) we choose. So this is a well-defined value.* ∎

Also, note $e^x \ne 0$ ever. If it was, $\lim_n (1 + x/n)^n = 0$. Then for n sufficiently large, we have $(1 + x/n)^n < 1/2^n$. Then $1 + x/n < 1/2$. Letting $n \to \infty$, we have $1 \le 1/2$ which is impossible. Hence, $e^x \ne 0$ ever! This also means e^{-x} is always positive also. This means when you graph these functions you can never let e^x or e^{-x} hit the axis.

Also note that the first order Taylor polynomial with remainder about $x = 0$ gives $e^{-x} = 1 - x + e^{-c}(x^2/2)$ for some c between 0 and x. Thus, if $x > 0$, this means $e^{-x} > 1 - x$ or the tangent line at $x = 0$ is always below e^{-x}.

We have proven a large number of properties about the exponential function which we summarize below in a convenient form.

Theorem 13.3.7 The Exponential Function Properties

For all x, y and r:

1. $e^{x+y} = e^x e^y$

2. $e^{-x} e^x = 1 \Rightarrow e^{-x} = 1/e^x$.

3. $e^{x-y} = e^x e^{-y} = e^x/e^y$

4. $(e^x)^r = e^{rx}$.

5. e^x *is continuous and* $(e^x)' = e^x$.

6. *Since* $e^n \to \infty$ *as* $n \to \infty$, $\lim_{x \to \infty} e^x = \infty$.

7. *Since* $\lim_{n \to \infty} e^{-n} = 0$, $\lim_{x \to -\infty} e^x = 0$; 0 *is a horizontal asymptote.*

8. $e^x \neq 0$ *ever*

Proof 13.3.7
We have proven these results earlier. ∎

Thus, if $-x < 0$, $e^{-x} = \lim_n (1 - x/n)^n$ and for n sufficiently large, $1 - x/n > 0$ implying $e^{-x} \geq 0$. It is clear $e^x \geq 0$ if $x \geq 0$, so $e^x \geq 0$ for all x.

- If $f'(p) > 0$ this means f is increasing locally. The proof is simple. We already know if $\lim_{x \to p} f(x) = L > 0$, then there is a radius $r > 0$ so that $f(x) > L/2 > 0$ if $x \in B_r(p)$. Let $g(x) = \frac{f(x) - f(p)}{x - p}$, Then since $\lim_{x \to p} g(x) = f'(p) > 0$, there is an $r > 0$ so that $\frac{f(x) - f(p)}{x - p} > f'(p)/2 > 0$. So $f(x) - f(p) > 0$ if $p < x < p + r$ which tells us f is increasing locally.

- $(e^x)' = e^x > 0$ always and $e^x > 0$ always. So e^x is always increasing and so e^x is a 1-1 and onto function from its domain $(-\infty, \infty)$ to its range $(0, \infty)$.

13.4 Taylor Series

Let's look at Taylor polynomials for some familiar functions.
$\cos(x)$ has the following Taylor polynomials and error terms at the base point 0:

$$\cos(x) = 1 + (\cos(x))^{(1)}(c_0)x, c_0 \text{ between 0 and } x$$
$$\cos(x) = 1 + 0x + (\cos(x))^{(2)}(c_1)x^2/2, c_1 \text{ between 0 and } x$$
$$\cos(x) = 1 + 0x - x^2/2 + (\cos(x))^{(3)}(c_2)x^3/3!, c_2 \text{ between 0 and } x$$
$$\cos(x) = 1 + 0x - x^2/2 + 0x^3/3! + (\cos(x))^{(4)}(c_3)x^4/4!,$$
$$c_3 \text{ between 0 and } x$$
$$\cos(x) = 1 + 0x - x^2/2 + 0x^3/3! + x^4/4! + (\cos(x))^{(5)}(c_4)x^5/5!,$$
$$c_3 \text{ between 0 and } x$$

$$\cdots$$

$$\cos(x) = \sum_{k=0}^{n} (-1)^k x^{2k}/(2k)! + (\cos(x))^{(2n+1)}(c_{2n+1})x^{2n+1}/(2n+1)!,$$
$$c_{2n+1} \text{ between 0 and } x$$

Since the $(2n)^{th}$ Taylor Polynomial of $\cos(x)$ at 0 is $p_{2n}(x, 0) = \sum_{k=0}^{n}(-1)^k\, x^{2k}/(2k)!$, this tells us that

$$\cos(x) - p_{2n}(x, 0) \;=\; (\cos(x))^{(2n+1)}(c_{2n+1})x^{2n+1}/(2n+1)!.$$

The biggest the derivatives of $\cos(x)$ can be here in absolute value is 1. Thus $|\cos(x) - p_{2n}(x, 0)| \leq (1)|x|^{2n+1}/(2n+1)!$ and we know

$$\lim_n |\cos(x) - p_{2n}(x, 0)| \;=\; \lim_n |x|^{2n+1}/(2n+1)! = 0$$

Thus, the Taylor polynomials of $\cos(x)$ converge to $\cos(x)$ at each x. We would say $\cos(x) = \sum_{k=0}^{\infty}(-1)^k\, x^{2k}/(2k)!$ and call this the Taylor Series of $\cos(x)$ at 0.

We can do this for functions with easy derivatives; so we can show at $x = 0$:

- $\sin(x) = \sum_{k=0}^{\infty}(-1)^k\, x^{2k+1}/(2k+1)!$.

- $\sinh(x) = \sum_{k=0}^{\infty} x^{2k+1}/(2k+1)!$.

- $\cosh(x) = \sum_{k=0}^{\infty} x^{2k}/(2k)!$.

- $e^x = f(x) = \sum_{k=0}^{\infty} x^k/k!$.

It is very hard to do this for functions whose higher order derivatives require product and quotient rules. So what we know that we can do in theory is not necessarily what we can do in practice.

13.4.1   Homework

Exercise 13.4.1 *Prove the nth order Taylor polynomial of e^x at $x = 0$ is $f_n(x) = \sum_{k=0}^{n} x^k/k!$ and show the error term goes to zero as $n \to \infty$.*

Exercise 13.4.2 *Find the nth order Taylor polynomial of $\sin(x)$ at $x = 0$ and show the error term goes to zero as $n \to \infty$.*

Exercise 13.4.3 *Find the nth order Taylor polynomial of $\cosh(x)$ at $x = 0$ and show the error term goes to zero as $n \to \infty$.*

13.5   Inverse Functions

Let's backup and talk about the idea of an inverse function. Say we have a function $y = f(x)$ like $y = x^3$. Take the cube root of each side to get $x = y^{1/3}$. Just for fun, let $g(x) = x^{1/3}$; i.e., we switched the role of x and y in the equation $x = y^{1/3}$. Now note some interesting things:

$$f(g(x)) \;=\; f(x^{1/3}) \;=\; \left(x^{1/3}\right)^3 = x$$

$$g(f(x)) \;=\; g(x^3) \;=\; \left(x^3\right)^{1/3} = x.$$

Now the function $I(x) = x$ is called the identity because it takes an x as an input and does nothing to it. The output value is still x. So we have $f(g) = I$ and $g(f) = I$. When this happens, the function g is called the **inverse** of f and is denoted by the special symbol f^{-1}. Of course, the same is true going the other way: f is the inverse of g and could be denoted by g^{-1}. Another more abstract way of saying this is

$$f^{-1}(x) = y \iff f(y) = x.$$

Now look at Figure 13.1. We draw the function x^3 and its inverse $x^{1/3}$ in the unit square. We also draw the identity there which is just the graph of $y = x$; i.e. a line with slope 1. Note, if you take the point $(1/2, 1/8)$ on the graph of x^3 and draw a line from it to the inverse point $(1/8, 1/2)$ you'll note that this line is perpendicular to the line of the identity. This will always be true with a graph of a function and its inverse. Also note that x^3 has a positive derivative always and so is always increasing. It seems reasonable that if we had a function whose derivative was positive all the time, we could do this same thing. We could take a point on that function's graph, say (c, d), reverse the coordinates to (d, c) and the line connecting those two pairs would be perpendicular to the identity line just like in our figure. So we have a geometric procedure to define the inverse of any function that is always increasing.

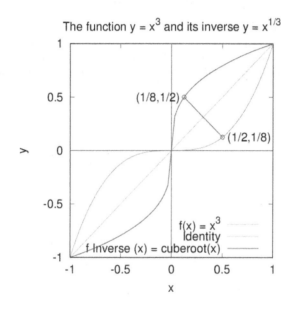

Figure 13.1: The function x^3 and its inverse.

So since it is always increasing, it is easy to see $f(x) = e^x$ has an inverse function $g(x)$ where $g : (0, \infty) \to (-\infty, \infty)$ with the properties $f(g(x)) = x$ for $x \in (0, \infty)$ and $g(f(x)) = x$ for $x \in (-\infty, \infty)$. We need to find out what it's properties are.

13.5.1 Homework

The next two problems are about inverses.

Exercise 13.5.1

The hyperbolic trigonometric functions are defined using $\sinh(x) = (e^x - e^{-x})/2$ *and* $\cosh(x) = (e^x + e^{-x})/2$ *and then defining the other four in the usual way. Find the derivatives of all 6 of these functions and graph all of them and their derivatives carefully.*

Exercise 13.5.2

Let $y = 0.5(1 + \tanh((x - O)/G))$ for a positive O and G. Find the inverse of this function and graph both this function and its inverse.

This problem is a return to theory!

Exercise 13.5.3 *We say f satisfies a Lipschitz condition of order p on the interval I if there is a positive constant L so that $|f(y) - f(x)| \leq L|y - x|^p$. We also say f is Lipschitz of order p on I.*
(1) Prove that if f is Lipschitz of order 1 on I, f is continuous on I.
(2) Prove that if f is Lipschitz of order 2 on I, f is a constant by proving its derivative is 0.

Exercise 13.5.4 *Prove e^x is Lipschitz on $[0, T]$ for any $T > 0$.*

Exercise 13.5.5 *Prove e^{-2x} is Lipschitz on $[0, \infty)$.*

13.5.2 Continuity of the Inverse Function

We can now investigate when the inverse will be continuous.

Theorem 13.5.1 The Continuity of the Inverse

Let I and J be intervals and let $f : I \to J$ be a continuous function which is $1-1$ and onto and is increasing or decreasing on I. Then the inverse function f^{-1} is also continuous.

Proof 13.5.1
For concreteness, assume f is increasing. The argument is similar if f is decreasing. Suppose there is a point $y_0 = f(x_0)$ where f^{-1} is not continuous. Then there is an $\epsilon_0 > 0$ so that for all n, there are points $y_n = f(x_n)$ with $|y_n - y_0| < 1/n$ and $|f^{-1}(y_n) - f^{-1}(y_0)| \geq \epsilon_0$. We also know the y_n are distinct as f is $1-1$.

Thus, for all n, $|x_n - x_0| \geq \epsilon_0$ and $\lim_n y_n = y_0$.

We know $x_n \leq x_0 - \epsilon_0$ or $x_n \geq x_0 + \epsilon_0$. Let Ω_1 be the set of indices n with $x_n \geq x_0 + \epsilon_0$ and Ω_2 be the set of indices n for the other case. At least one of these sets has an infinite number of indices in it. If it was Ω_2, this gives us a subsequence (x_n^1) with $x_n^2 \geq x_0 + \epsilon$ for all indices. Note also, since $x_n^2 \geq x_0 + \epsilon_0$, since f is increasing, $f(x_n^2) \geq f(x_0 + \epsilon_0)$.

However, we also know $y_n = f(x_n) \to y_0 = f(x_0)$ so the same is true for the subsequence: $\lim_n f(x_n^2) = f(x_0)$, However, we also have $\lim f(x_n^2) \geq f(x_0 + \epsilon_0)$ or $f(x_0) \geq f(x_0 + \epsilon_0)$ which is not possible as f is increasing.

If the index set Ω_1 was infinite, we would use a similar argument. We would have a subsequence $(x_n^1) \leq x_0 - \epsilon_0$. Then $f(x_n^2) \to y_0$ and we have $f(y_0) \leq f(x_0 - \epsilon_0)$ which is also not possible. ∎

Now let's look at differentiability of the inverse.

Theorem 13.5.2 Differentiability of the Inverse

Let I and J be intervals and let $f : I \to J$ be a differentiable function which is $1-1$ and onto and is increasing or decreasing on I. Then the inverse function f^{-1} is also differentiable at all points where $f'(x) \neq 0$ with $d/dy(f^{-1}(y)) = 1/f'(f^{-1}(y))$.

Proof 13.5.2
For concreteness, assume f is increasing. We know f^{-1} exists and is continuous. Consider for any

$y_0 = f(x_0)$ *where* $f'(x_0) \neq 0$.

$$\lim_{y \to y_0} \frac{f^{-1}(y) - f^{-1}(y_0)}{y - y_0} = \lim_{x \to x_0} \frac{x - x_0}{f(x) - f(x_0)} = 1/f'(x_0).$$

Hence, $d/dy(f^{-1})(y_0) = 1/f'(f^{-1}(y_0))$ ■

From the theorem above, since e^x is increasing and $1 - 1$ and onto from $(-\infty, \infty)$ to $(0, \infty)$ and continuous, it has a continuous inverse function which is normally called the natural logarithm of x, $\ln(x)$.

We see

- $\ln(x)$ is a continuous function.

- If $y = e^x$, then $\ln(y) = x$ and by the chain rule

$$\textbf{Left hand side } d/dx \ln(y) = d/dy(\ln(y)) \, dy/dx$$
$$= d/dx(x) = 1 \textbf{ Right hand side}$$

But $dy/dx = e^x = y$, so we have

$$d/dy(\ln(y)) \, y = 1 \implies d/dy(\ln(y)) = 1/y$$

Hence, switching variables, $(\ln(x))' = 1/x$ for all x positive.

- $e^x : (-\infty, \infty) \to (0, \infty)$ is $1 - 1$ and onto.

- $\ln(x) : (0, \infty) \to (-\infty, \infty)$ is $1 - 1$ and onto.

- $e^{\ln(x)} = x$ for all $x \in (0, \infty)$

- $ln(e^x) = x$ for all $x \in (-\infty, \infty)$

Also, we have a number of other properties:

- $\ln(e^n) = n$ so $\lim_{x \to \infty} \ln(x) = \infty$.

- $\ln(e^{-n}) = -n$ so $\lim_{x \to 0+} \ln(x) = -\infty$.

- Consider $w = \ln(uv)$ for a positive u and v. We know we can write $u = e^x$ and $v = e^y$ for some x and y. Thus, $e^w = e^{\ln(uv)} = uv = e^x e^y = e^{x+y}$. But then, $\ln(e^w) = \ln(e^{x+y})$ tells us $w = x + y = \ln(u) + \ln(v)$. So for any $u, v > 0$, $\ln(uv) = \ln(u) + \ln(v)$.

- For any $u > 0$, $0 = \ln(1) = \ln(u \, (1/u)) = \ln(u) + \ln(1/u)$ by the previous results. So we have $\ln(1/u) = -\ln(u)$ for all positive u.

- Thus $\ln(u/v) = \ln(u \, (1/v)) = \ln(u) + \ln(1/v) = \ln(u) - \ln(v)$ for all positive u and v.

- Let $w = \ln(u^r)$ for $u > 0$ and any real number r. Then we can write $u = e^x$ for some x. So $e^w = e^{\ln(u^r)} = u^r = (e^x)^r = e^{rx}$ implying $w = rx = r \ln(u)$. Thus, $\ln(u^r) = r \ln(u)$ for $u > 0$ and any r.

We can summarize these results as follows:

Theorem 13.5.3 Properties of the Natural Logarithm

The natural logarithm of the real number x satisfies

- $\ln$ *is a continuous function of x for positive x,*

- $\lim_{x \to \infty} \ln(x) = \infty$,

- $\lim_{x \to 0^+} \ln(x) = -\infty$,

- $\ln(1) = 0$,

- $\ln(e) = 1$,

- $(\ln(x))' = \frac{1}{x}$,

- *If x and y are positive numbers then $\ln(xy) = \ln(x) + \ln(y)$.*

- *If x and y are positive numbers then $\ln\left(\frac{x}{y}\right) = \ln(x) - \ln(y)$.*

- *If x is a positive number and y is any real number then $\ln\left(x^y\right) = y\,\ln(x)$.*

Proof 13.5.3
We have proven all of these results already.   ∎

You have seen how to graph the logarithm and exponential functions in earlier courses, but here are some reminders shown in Figure 13.2, Figure 13.3(a), Figure 13.3(b), Figure 13.4(a) and Figure 13.4(b).

Homework

Exercise 13.5.6 *Graph $e^{-2x}\cos(4x + 0.2)$ on $[0, 10]$. Note this uses the upper and lower envelope functions $\pm e^{-2x}$. Find explicitly where the function hits the upper and lower envelope function. Draw this by hand rather than with MATLAB and label the period of this damped cosine function. Note the phase shift of 0.2 also.*

Exercise 13.5.7 *Graph $e^{0.0023x}\sin(3x+0.6)$ on $[0, 15]$. Note this uses the upper and lower envelope functions $\pm e^{0.0023x}$. Find explicitly where the function hits the upper and lower envelope function. Draw this by hand rather than with MATLAB and label the period of this damped cosine function. Note the phase shift of 0.6 also.*

Exercise 13.5.8 *Graph $\ln(.2x)\sin(4x+0.3)$ on $[1, 10]$. Note this uses the upper and lower envelope functions $\pm \ln(.2x)$. Find explicitly where the function hits the upper and lower envelope function. Draw this by hand rather than with MATLAB and label the period of this damped cosine function. Note the phase shift of 0.3 also.*

13.6   L'Hôpital's Rules

Now that we thoroughly understand more interesting functions, let's look at some tools for finding limits in new more complicated situations. These are the L'Hôpital type rules of which we will prove just a few. You can easily look up more and follow their proofs as the need arises.

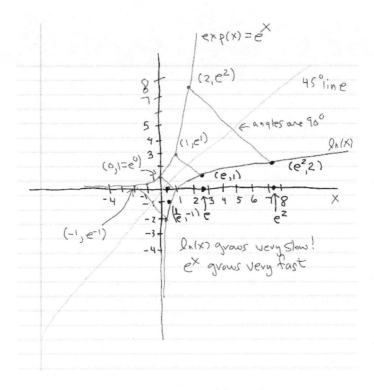

Figure 13.2: e^x and $\ln(x)$ together.

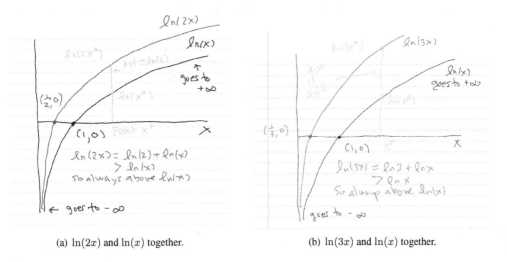

(a) $\ln(2x)$ and $\ln(x)$ together.

(b) $\ln(3x)$ and $\ln(x)$ together.

Figure 13.3: Logarithm function plots.

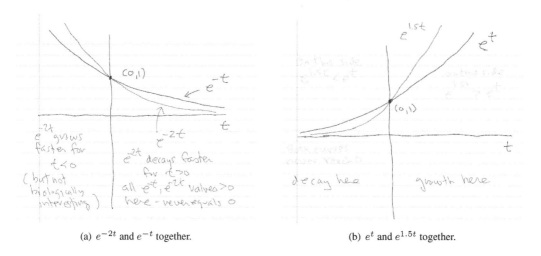

(a) e^{-2t} and e^{-t} together.   (b) e^t and $e^{1.5t}$ together.

Figure 13.4: Exponential function plots.

Assume f and g are defined locally at p and we want to evaluate $\lim_{x \to p} \frac{f(x)}{g(x)}$ when $f(p) = 0$ and $g(p) = 0$. In this case, our usual Algebra of Limits result fails since we have a zero divisor. However since we also have a zero in the denominator, we call this the $0/0$ undetermined case.

Assume f and g are differentiable at p. Then using the error form of differentiability, we have

$$\lim_{x \to p} \frac{f(p) + f'(p)(x - p) + E_f(x, p)}{g(p) + g'(p)(x - p) + E_g(x, p)}$$

But $f(p) = g(p) = 0$, so we can rearrange the numerator and denominator to get

$$\lim_{x \to p} \frac{f'(p) + E_f(x, p)/(x - p)}{g'(p) + E_g(x, p)/(x - p)}$$

Then as long as $g'(p) \neq 0$, we have the last terms in the numerator and denominator go to 0 as $x \to p$ giving $\lim_{x \to p} \frac{f(x)}{g(x)} = \frac{f'(p)}{g'(p)}$.

There are many such indeterminate forms and some are pretty hard to prove! Let's do one more which is on the **harder** side. First, we need a new Mean Value type theorem:

Theorem 13.6.1 Cauchy Mean Value Theorem

Let f and g be continuous on $[a, b]$ and differentiable on (a, b). Then, there is a c, $a < c < b$ so that

$$(f(b) - f(a)) g'(c) = (g(b) - g(a)) f'(c)$$

Proof 13.6.1
This one is easy. Define $h(x) = (f(b) - f(a))g(x) - (g(b) - g(a))f(x)$ and apply Rolle's Theorem to h. ∎

Our next L'Hôpital's result is this:

Theorem 13.6.2 ∞/∞ L'Hôpital's Result

Assume f and g are defined on an interval of the form (a, ∞) and they are differentiable on that interval. Further assume $\lim_{x \to \infty} f(x) = \infty$ and $\lim_{x \to \infty} g(x) = \infty$. Finally, assume $\lim_{x \to \infty} \frac{f'(x)}{g'(x)} = L$ with $g'(x) \neq 0$ for all x. Then, $\lim_{x \to \infty} \frac{f(x)}{g(x)} = \lim_{x \to \infty} \frac{f'(x)}{g'(x)} = L$.

Proof 13.6.2

Since $g'(x)$ is never zero, $g(x)$ is either increasing always or decreasing always. So terms like $g(y) - g(x)$ for $y \neq x$ are never 0. Since $\lim_{x \to \infty} \frac{f'(x)}{g'(x)} = L$, given $\epsilon > 0$ arbitrary, there is an $R > 0$ so that $\frac{f'(x)}{g'(x)} > L - \epsilon$ if $x > R$. Fix any number $S > R$. Apply the Cauchy Mean Value Theorem on the intervals $[S, S+k]$ to find points c_k so that $(f(S+k) - f(S))g'(c_k) = (g(S+k) - g(S))f'(c_k)$. Thus,

$$\frac{f(S+k) - f(S)}{g(S+k) - g(S)} = \frac{f'(c_k)}{g'(c_k)} \geq L - \epsilon$$

Let $S + k = S_k$ for convenience of notation. Now rearrange:

$$\frac{(f(S_k)/g(S_k)) - (f(S)/g(S_k))}{1 - (g(S)/g(S_k))} \geq L - \epsilon$$

Simplifying, we have

$$f(S_k)/g(S_k) \geq f(S)/g(S_k) + \left(1 - g(S)/g(S_k)\right)(L - \epsilon)$$

Now let $k \to \infty$ but keep S fixed. Since $g(S_k) \to \infty$, we have $\lim_{k \to \infty} f(S_k)/g(S_k) \geq L - \epsilon$.

We can do a similar argument from the other side to show $\lim_{k \to \infty} f(S_k)/g(S_k) \leq L + \epsilon$. Of course, this says $L - \epsilon \leq \lim_{x \to \infty} f(x)/g(x) \leq L + \epsilon$ which implies $\lim_{x \to \infty} f(x)/g(x) = L$. ∎

Example 13.6.1 *Find $\lim_{x \to \infty} (x^2 + 3x + 2)/e^x$.*

Solution

$$\lim_{x \to \infty} (x^2 + 3x + 2)/e^x = \infty/\infty, \text{ apply } L'H \Rightarrow$$
$$\lim_{x \to \infty} (x^2 + 3x + 2)/e^x = \lim_{x \to \infty} (2x + 3)/e^x = \infty/\infty, \text{ apply } L'H \Rightarrow$$
$$\lim_{x \to \infty} (2x + 3)/e^x = \lim_{x \to \infty} (2)/e^x = 0.$$

So the limit is 0.

13.6.0.1 Homework

Exercise 13.6.1 *Prove $\lim_{x \to \infty} p(x)/e^x = 0$ for any polynomial $p(x)$.*

Exercise 13.6.2 *Let p and q be polynomials of degree $n > 1$. Prove $\lim_{x \to \infty} p(x)/q(x) = r$ where r is the ratio of the coefficients of the highest order term in p and q.*

Exercise 13.6.3 *Prove $\lim_{x \to \infty} \ln(x)/p(x) = 0$ for any polynomial $p(x)$.*

Chapter 14

Extremal Theory for One Variable

The next thing we want to look at is to find ways to locate the minima and maxima of a given function. We know from theory that any continuous function on a compact set has a global minimum and a global maximum, but we need ways to find those values. This is where derivative information becomes useful. We have already seen we do not need differentiability really as convexity is enough to local minima, but if we do have a derivative it gives us extra tools.

14.1 Extremal Values

We know that a likely place for an extreme value is where the tangent line is flat as from simple drawings, we can see that where the tangent line is flat we often have a local minimum or local maximum of our function. We also know a function can have a minimum or maximum at a point where the functions has a corner or a cusp – in general where the function's derivative fails to exist. Finally, if the function was defined on a closed interval $[a, b]$, the function could have extreme behavior at an endpoint. These types of points are called *critical points* of the function.

Definition 14.1.1 Critical points of a function

> *The critical points of a function f are*
>
> - *Points p where $f'(p) = 0$*
>
> - *Points p where $f'(p)$ does not exist*
>
> - *Points p that are boundary points of the domain of f.*

We can be more precise. If p is a point where the tangent line to f is flat, then we know $f'(p) = 0$. The first order Taylor expansion is

$$f(x) \;=\; f(p) + f'(p)(x - p) + \frac{1}{2} f''(c)(x - p)^2,$$

for some c with c between p and x. Now since $f'(p) = 0$, this reduces to

$$f(x) \;=\; f(p) + \frac{1}{2} f''(c)(x - p)^2,$$

Now let's step back and talk about continuity and positive and negative values. Let's assume $f(p) > 0$. We have done this argument before, but it is not a bad thing to go over it again!

Let $\epsilon = f(p)/2$. Then from the definition of continuity, there is a radius $r > 0$ so that

$$p - r < x < p + r \quad \Rightarrow \quad -f(p)/2 < f(x) - f(p) < f(p)/2 \rightarrow$$
$$p - r < x < p + r \quad \Rightarrow \quad f(x) > f(p)/2 > 0.$$

We say if f is positive at a point p and continuous there, then it must be positive locally as well.

Let's assume $f(p) < 0$. Let $\epsilon = -f(p)/2 > 0$. Then from the definition of continuity, there is a radius $r > 0$ so that

$$p - r < x < p + r \quad \Rightarrow \quad f(p)/2 < f(x) - f(p) < -f(p)/2 \rightarrow$$
$$p - r < x < p + r \quad \Rightarrow \quad f(x) < f(p)/2 < 0.$$

We say if f is negative at a point p and continuous there, then it must be negative locally as well.

If f' is continuous at p we can say the same thing. If $f'(p) > 0$, let $\epsilon = f'(p)/2 > 0$. From the definition of continuity, there is a radius $r > 0$ so that

$$p - r < x < p + r \quad \Rightarrow \quad f'(x) > f'(p)/2 > 0.$$

So f' positive at a point p and continuous there, implies it must be positive locally as well.

Let's assume $f'(p) < 0$. Let $\epsilon = -f'(p)/2 < 0$. Then from the definition of continuity, there is a radius $r > 0$ so that

$$p - r < x < p + r \quad \Rightarrow \quad f(x) < f'(p)/2 < 0.$$

So f' negative at a point p and continuous there, implies it must be negative locally as well.

We can then do the same sort of thing for f''. This reasoning leads to several nice theorems:

Theorem 14.1.1 If f is Continuous and Not Zero at a Point, It is Locally Not Zero

If f is continuous at p and $f(p)$ is nonzero, there is a positive r so that $f(x)$ is nonzero with the same sign on $(p - r, p + r)$.

Proof 14.1.1
We have already sketched out the reasoning behind this result. ∎

Theorem 14.1.2 If f' is Continuous and Not Zero at a Point, It is Locally Not Zero

If f' is continuous at p and $f'(p)$ is nonzero, then there is a radius r where $f'(x)$ is nonzero with the same sign on $(p - r, p + r)$.

Proof 14.1.2
We have already sketched out the reasoning behind this result as well. ∎

Back to our problem! We have $f'(p) = 0$ and we have written down the first order Taylor expansion at p for f. So as long as f'' is continuous at p, we can say

- $f''(p) > 0$ implies $f''(c) > 0$ within some circle centered at p. This tells us $f(x) = f(p) +$ **a positive number** on this circle. Hence, $f(x) > f(p)$ locally which tells us p is a point where f has a local minimum.

- $f''(p) < 0$ implies $f''(c) < 0$ within some circle centered at p. This tells us $f(x) = f(p) -$ **a positive number** on this circle. Hence, $f(x) < f(p)$ locally which tells us p is a point where f has a local maximum.

This gives our second order test for maximum and minimum values.

Theorem 14.1.3 Second Order Test for Extremals

If f'' is continuous at p, $f'(p) = 0$, then $f''(p) > 0$ tells us f has a local minimum at p and $f''(p) < 0$ tells us f has a local maximum at p.

If $f''(p) = 0$, we don't know anything.

This fact comes from the examples $f(x) = x^4$ which $f''(0) = 0$ even though $f(0)$ is a minimum and $f(x) = -x^4$ which has a maximum at $x = 0$ even though $f''(0) = 0$.

Proof 14.1.3
We have already sketched out the reasoning behind this result too. ■

Example 14.1.1 *Show $f(x) = x^2 + 2x + 1$ has a minimum at $x = -1$.*

Solution *We have $f'(x) = 2x + 2$ which is zero when $2x + 2 = 0$ or $x = -1$. We also have $f''(x) = 2 > 0$ and so we have a minimum.*

Example 14.1.2 *Show $f(x) = 2x^2 + 5x + 1$ has a minimum at $x = -5/4$.*

Solution *We have $f'(x) = 4x + 5$ which is zero when $4x + 5 = 0$ or $x = -5/4$. We also have $f''(x) = 4 > 0$ and so we have a minimum.*

Example 14.1.3 *Show $f(x) = -2x^2 + 5x + 1$ has a maximum at $x = 5/4$.*

Solution *We have $f'(x) = -4x + 5$ which is zero when $-4x + 5 = 0$ or $x = 5/4$. We also have $f''(x) = -4 < 0$ and so we have a maximum.*

Example 14.1.4 *Show $f(x) = -2x^3 + 5x^2 + 1$ has a maximum at $x = 10/6$.*

Solution *We have $f'(x) = -6x^2 + 10x = x(-6x + 10)$ which is zero when $x(-6x + 10) = 0$ or when $x = 0$ or $x = 10/6$. We also have $f''(x) = -12x + 10$. Since $f''(0) = 10$ we see $x = 0$ is a minimum. Since $f''(10/6) = -12(10/6) + 10 < 0$, we have $x = 10/6$ is a maximum.*

Homework

Exercise 14.1.1 *Find the extremal values of $f(x) = 4x^3 + 7x^2 + 1$.*

Exercise 14.1.2 *Find the extremal values of $f(x) = 4x^3 + 7x^2 + 6x + 21$.*

Exercise 14.1.3 *Find the extremal values of $f(x) = 4x^3 + 7x^2 + 2x + 21$.*

Exercise 14.1.4 *Use MATLAB to find the extreme values of $f(x) = 4x^6 + 7x^3 + 2x^2 + 5x + 2$ by finding the critical points numerically. Make sure you document your code.*

14.2 The First Derivative Test

We have mentioned that the points where the tangent lines to a function are of interest. Let's make this more precise. Note

- We can have a maximum or minimum for the function f at p even if $f'(p)$ does not exist or has value 0. We still assume f is continuous at p though.

- Assume we can find a positive radius r so that $f'(x) < 0$ on $(p - r, p)$ and $f'(x) > 0$ on $(p, p + r)$. Applying the Mean Value Theorem on the interval $[x, p]$ with $x < p$, we have $\frac{f(x)-f(p)}{x-p} = f'(c)$ for some $x < c < p$. But here $f'(c) < 0$, so we have $\frac{f(x)-f(p)}{x-p} < 0$. Since $x < p$, this tells us $f(x) > f(p)$.

  Now apply the Mean Value Theorem on the interval $[p, x]$ with $p < x$. We have $\frac{f(x)-f(p)}{x-p} = f'(c)$ for some $p < c < x$. But here $f'(c) > 0$, so we have $\frac{f(x)-f(p)}{x-p} > 0$. Since $x > p$, this tells us $f(x) > f(p)$ again. Combining, we see $f(p)$ is a local minimum.

- If we can find a positive radius r so that $f'(x) > 0$ on $(p - r, p)$ and $f'(x) < 0$ on $(p, p + r)$, a similar analysis shows $f(p)$ is a local maximum.

This leads to the First Derivative Test:

Theorem 14.2.1 First Derivative Test

> *Assume there is an $r > 0$ so that*
>
>   *1. f' is $+$ on $(p - r, p)$ and f' is $-$ on $(p, p + r)$. Then f has a maximum at p.*
>
>   *2. f' is $-$ on $(p - r, p)$ and f' is $+$ on $(p, p + r)$. Then f has a minimum at p.*

Proof 14.2.1
We just finished arguing this.   ■

Things do not have to be so nice. Consider this example.

Example 14.2.1 *Examine $f(x) = x^2 + x^2 \sin^2(1/x)$ for extreme behavior.*

Solution *We see, letting $y = 1/x$, that*

$$\lim_{x \to 0} f(x) = \lim_{x \to 0} x^2 + \lim_{y \to \pm\infty} \sin^2(y)/y^2 = 0$$

Thus, f has a removeable discontinuity at 0 and the renamed function is continuous at 0.

Now consider the derivative of f for $x \neq 0$. We have

$$
\begin{aligned}
f'(x) &= 2x + 2x \sin^2(1/x) + x^2 (2) \sin(1/x) \cos(1/x) (-1/x^2) \\
&= 2x + 2x \sin^2(1/x) - 2 \sin(1/x) \cos(1/x) \\
&= 2x + 2\frac{\sin^2(1/x)}{1/x} - \sin(2/x)
\end{aligned}
$$

Now as $x \to 0$, we can find $r > 0$ so that

$$-1/2 < 2x + 2\frac{\sin^2(1/x)}{1/x} < 1/2$$

since these first two terms go to zero. However, in the interval $(-r, r)$ there are infinitely many points (x_n^1) where $-\sin(2/x_n^1) = 1$ and infinitely many points (x_n^2) where $-\sin(2/x_n^2) = -1$. Thus,

$f'(x_n^1) \in (-1/2 + 1, 1/2 + 1) = (1/2, 3/2)$ *implying* $f(x_n^1) > 0$.

Further, we find $f'(x_n^2) \in (-1/2 - 1, 1/2 - 1) = (-3/2, -1/2)$ *implying* $f(x_n^2) < 0$. *The sequences* (x_n^1) *and* (x_n^2) *both converge to 0 as* $x_n^1 = 2/(-\pi/2 + 2n\pi)$ *and* $x_n^2 = 2/(\pi/2 + 2n\pi)$. *Hence, it is* **not** *possible to find any* $r > 0$ *so that* $f' < 0$ *on* $(-r, 0)$ *and* $f' > 0$ *on* $(0, r)$.

But $f(0) = 0$ *is the minimum value of* f *globally at 0 because* $x^2 + x^2 \sin^2(1/x) > 0$ *for all* $x \neq 0$. **So** f **is a function the First Derivative Test fails on!**. *Note using the new definition of* f *at 0, we have*

$$\lim_{x \to 0} \frac{f(x) - f(0)}{x - 0} = \lim_{x \to 0} \frac{x^2 + x^2 \sin^2(1/x)}{x}$$
$$= \lim_{x \to 0} \left(x + \frac{\sin^2(1/x)}{1/x} \right) = 0$$

Thus, $f'(0)$ *exists and equals 0. But* $f'(x) = 2x + 2\frac{\sin^2(1/x)}{1/x} - \sin(2/x)$ *and* $\underline{\lim}_{x \to 0} f'(x) = -1$ *and* $\overline{\lim}_{x \to 0} f'(x) = 1$ *which tells us* f' *is not continuous at 0.*

So f' *always exists but is not continuous at the point 0.*

In general, the behavior of the functions $f(x) = x^p \sin(1/x^q)$, $f(x) = x^p \sin^2(1/x^q)$ **is always interesting for positive integers** p **and** q.

Homework

Exercise 14.2.1 *Use the FDT to find the extreme values of* $f(x) = |2x - 3|$.

Exercise 14.2.2 *Use the FDT to find the extreme values of* $f(x) = |2x - 3| + 5|x - 4|$.

Exercise 14.2.3 *Use the FDT to find the extreme values of* $f(x) = \cos(|2x + 3|)$.

Exercise 14.2.4 *Use the FDT to find the extreme values of* $f(x) = \sin(|5x|)$.

14.3 Cooling Models

Newton formulated a law of cooling by observing how the temperature of a hot object cooled. As you might expect, this is called *Newton's Law of Cooling*. If we let $T(t)$ represent the temperature of the liquid in some container and A denote the ambient temperature of the air around the container, then Newton observed that

$$T'(t) \quad \propto \quad (T(t) - A).$$

We will assume that the temperature outside the container, A, is smaller than the initial temperature of the hot liquid inside. So we expect the temperature of the liquid to go down with time. Let the constant of proportionality be k. Next, we solve

$$T'(t) = k(T(t) - A), \ T(0) = T_0$$

where T_0 is the initial temperature of the liquid. For example, we might want to solve

$$T'(t) = k(T(t) - 70), \ T(0) = 210$$

where all of our temperatures are measured in degrees Fahrenheit. We use traditional differential equation methods to solve this. Divide both sides by $T - 70$ to get $\frac{dT}{T-70} = k\,dt$. Then integrate both sides to get $\ln |T(t) - 70| = kt + C$. Then, exponentiate: $|T(t) - 70| = Be^{kt}$. Now, since we start at 210 degrees and the ambient temperature is 70 degrees, we know the temperature of our liquid will go down with time. Thus, we have

$$T(t) - 70 = Be^{kt}.$$

Solving, we find

$$T(t) = 70 + Be^{kt}.$$

Since the initial condition is $T(0) = 210$, we find

$$T(0) = 210 = 70 + Be^0 = 70 + B,$$

and so $B = 140$. Putting all of this together, we see the solution to the model is

$$T(t) = 70 + 140e^{kt}.$$

Now since the temperature of our liquid is going down, it is apparent that the proportionality constant k must be negative. In fact, our common sense tells us that as time increases, the temperature of the liquid approaches the ambient temperature 70 asymptotically from above.

Example 14.3.1 *Solve $T'(t) = k(T(t) - 70)$ with $T(0) = 210$ and then use the conditions $T(10) = 140$ to find k. Here time is measured in minutes.*

Solution *First, we solve as usual to find*

$$T(t) = 70 + 140e^{kt}.$$

Next, we know $T(10) = 140$, so we must have $T(10) = 140 = 70 + 140e^{10k}$. Thus

$$70 = 140e^{10k} \Rightarrow \frac{1}{2} = e^{10k} \Rightarrow -\ln(2) = 10k.$$

and so $k = -\ln(2)/10 = -.0693$.

14.3.1 Homework

Exercise 14.3.1 *Solve $T'(t) = k(T(t) - 90)$ with $T(0) = 205$ and then use the conditions $T(20) = 100$ to find k. Here time is measured in minutes.*

Exercise 14.3.2 *We now know the general cooling model is*

$$T'(t) = k(T(t) - A), T(0) = T_0.$$

which has solution

$$T(t) = A + (T_0 - A)e^{kt}.$$

The one parameter we don't know is the constant of proportionality, k. However, let's assume we have collected some data in the form $\{t_i, T_i\}$ for $1 \le i \le N$ for some finite integer N. Rewrite the *solution as*

$$\frac{T(t) - A}{T_0 - A} = e^{kt}.$$

Now take the logarithm of both sides to get

$$\ln\left(\frac{T(t) - A}{T_0 - A}\right) = kt.$$

This tells us how to estimate the value of k. Let the variable $U(t)$ be defined to be

$$U(t) = \ln\left(\frac{T(t) - A}{T_0 - A}\right)$$

Then, we have $U(t) = kt$. Thus, the variable U is linear in t; i.e. if we graph U versus t we should see a straight line with slope k. For the data below, find k by looking at the graph and estimating the slope value of the best straight line through this transformed data. Note we could also do this using linear regression! Assume the room temperature for our experiment is 76 degrees.

Listing 14.1: **Sample Cooling Data**

```
    0.0    205
    0.5    201
 3  2.0    201
    5.0    190
    8.0    178
    10.0   175
    13.0   167
 8  15.0   161
    25.0   141
    30.0   135
    40.0   127
    50.0   117
13  65.0   108
    80.0   100
    95.0   95
    110.0  89
    130.0  83
18  165.0  80
```

Exercise 14.3.3 *Analyze fully the function $f(x) = x^4 \sin(1/x^2)$ up to the second derivative.*

Exercise 14.3.4 *Let $f(x) = 2|x - 200/2| + 10|x - 180/10| + 20|x - 160/20|$. Use the methods of convex analysis to find where f attains its minimum. Then let $g(x) = 4(x - 200/2)^2 + 100(x - 180/10)^2 + 400(x - 160/20)^2$ and find the minimum of g using the SDT. Compare and comment on your results.*

Chapter 15

Differentiation in $\Re^2$ and $\Re^3$

The discussion of limits, continuity and differentiation on the real line $\Re$ is made easier because the way we measure distance between objects, here numbers, is quite easy. We just calculate $|x - y|$ for any two numbers x and y. The definition of a derivative requires we compute ratios such as $\frac{f(x) - f(y)}{x - y}$ for a given function f and although we don't think much about it, the presence of the $\frac{1}{x - y}$ implies we can find the inverse of the object $x - y$. In more general spaces of objects, this is not possible. The simplest case where this occurs is when we translate our ideas on smoothness to functions defined on subsets of $\Re^2$ and $\Re^3$. The distance between objects is now the distance between vectors written as $||x - y||$ where the $|| \cdot ||$ is the usual Euclidean norm of the difference between the vectors x and y. And of course, we know we can not invert the vector $x - y$!

So to extend our ideas, we have to revisit and redo some of our past ideas. Let's start with a quick review.

15.1 $\Re^2$ and $\Re^3$

Let's look at two dimensional column vectors. Let

$$V = \begin{bmatrix} V_1 \\ V_2 \end{bmatrix}$$

We graph this vector using its components as coordinates in the standard $x\,y$ plane by drawing a line from the origin $(0,0)$ to (V_1, V_2). This line has a length $\sqrt{(a)^2 + (c)^2}$ and we denote the length of V by $||V||$. This is called the **Euclidean norm** or **Euclidean metric** of V. The Ball of positive radius r centered at (x_0, y_0) is then $B_r(x_0, y_0) = \{(x, y) : \; || (x, y) - (x_0, y_0) || < r\}$ which is a full circle and its interior as a subset of $\Re^2$ in contrast to the simple open subset of the x - axis we had before for the ball $B_r(x)$ around a point $x \in \Re$. We show a typical vector in Figure 15.1.

Let V and W be column vectors of size 2×1:

$$V = \begin{bmatrix} V_1 \\ V_2 \end{bmatrix}, \; W = \begin{bmatrix} W_1 \\ W_2 \end{bmatrix}.$$

We define their **inner product** to be the real number

$$< V, W > = V_1 W_1 + V_2 W_2.$$

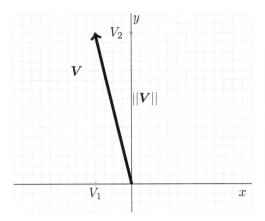

Figure 15.1: A typical two dimensional vector.

A vector V can be identified with an ordered pair (V_1, V_2). The components (V_1, V_2) are graphed in the usual Cartesian manner as an ordered pair in plane. The magnitude of V is $\| V \|$.

This is also denoted by $V \cdot W$. Next, look at three dimensional column vectors. Let

$$V \;=\; \begin{bmatrix} V_1 \\ V_2 \\ V_3 \end{bmatrix}$$

We graph this vector using its components as coordinates in the standard $x\,y\,z$ coordinate system by drawing a line from the origin $(0,0,0)$ to (V_1, V_2, V_3). This line has a length $\sqrt{(V_1)^2 + (V_2)^2 + (V_3)^2}$ and we denote the length of V by $\|V\|$. This is also called the **Euclidean norm** or **Euclidean metric** of V. We can usually tell which norm we need as context tells us if we are in a two or three dimensional situation. The Ball of positive radius r centered at (x_0, y_0, z_0) is $B_r(x_0, y_0, z_0) = \{(x,y,z) : \; \| (x,y,z) - (x_0, y_0, z_0) \| < r\}$ which is a sphere now in contrast to the circle of $\Re^2$ and the open interval in $\Re$. Let V and W be column vectors of size 3×1:

$$V \;=\; \begin{bmatrix} V_1 \\ V_2 \\ V_3 \end{bmatrix}, \; W = \begin{bmatrix} W_1 \\ W_2 \\ W_3 \end{bmatrix}.$$

We define their **inner product** to be the real number

$$< V, W > \;=\; V_1 W_1 + V_2 W_2 + V_3 W_3$$

This is also denoted by $V \cdot W$.

15.2   Functions of Two Variables

Now let's start looking at functions that map each ordered pair (x, y) into a number. Let's begin with an example. Consider the function $f(x,y) = x^2 + y^2$ defined for all x and y. Hence, for each x and y we pick, we calculate a number we can denote by z whose value is $f(x,y) = x^2 + y^2$. Using the same ideas we just used for the $x-y$ plane, we see the set of all such triples $(x, y, z) = (x, y, x^2 + y^2)$ defines a **surface** in $\Re^3$ which is the collection of all ordered triples (x, y, z). We can plot this surface in MATLAB with fairly simple code. Let's go through how to do these plots in a lot of detail so we can see how to apply this kind of code in other situations. To draw a portion of a surface, we pick a rectangle of x and y values. To make it simple, we will choose a point (x_0, y_0) as the center of our rectangle and then for a chosen Δx and Δy and integers n_x and n_y, we set up the rectangle

$$[x_0 - n_x \Delta x, \ldots, x_0, \ldots, x_0 + n_x \Delta x]$$
$$\times \quad [y_0 - n_y \Delta y, \ldots, y_0, \ldots, y_0 + n_y \Delta y]$$

The constant x and y lines determined by this grid result in a matrix of intersections with entries (x_i, y_j) for appropriate indices i and j. We will approximate the surface by plotting the triples $(x_i, y_j, z_{ij} = f(x_i, y_j))$ and then drawing a top for each rectangle. Right now though, let's just draw this base grid. In MATLAB, first set up the function we want to look at. We will choose a very simple one

Listing 15.1: **Defining the Function**

```
f = @(x,y) x.^2 + y.^2;
```

Now, we draw the grid by using the function `DrawGrid(f,delx,nx,dely,ny,x0,y0)`. This function has several arguments as you can see and we explain them in the listing below. So we are drawing a grid centered around $(0.5, 0.5)$ using a uniform 0.5 step in both directions. The grid is drawn at $z = 0$.

Listing 15.2: **Drawing the Base Grid**

```
% f is the surface function
% delx = 0.5, width of the delta x
% nx = 2, number of steps right and left from x0
% dely = 0.5, width of the delta y
5 % ny = 2, number of steps right and left from y0
% x0 = 0.5, y0 = 0.5
DrawGrid(f,0.5,2,0.5,2,0.5,0.5);
```

We show this grid in Figure 15.2. The code is available on the web site. Make sure you play with the

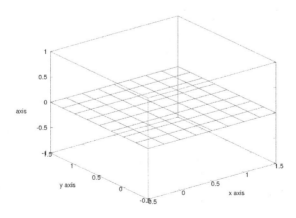

Figure 15.2: The base grid.

plot a bit. You can grab it and rotate it as you see fit to make sure you see all the detail. Right now, there is not much to see in the grid, but later when we plot the surface, the grid and other things, the ability to rotate in 3D is important to our understanding. So make sure you take the time to see how to do this! To draw the surface, we find the pairs (x_i, y_j) and the associated $f(x_i, y_j)$ values and then call the `DrawMesh(f,delx,nx,dely,ny,x0,y0)` command. The meaning in the arguments is the same as in `DrawGrid` so we won't repeat them here.

Listing 15.3: **Drawing the Surface Mesh**

```
DrawMesh( f ,0.5 ,2 ,0.5 ,2 ,0.5 ,0.5);
```

The resulting surface and grid are shown in Figure 15.3. Now we draw the **traces** corresponding to

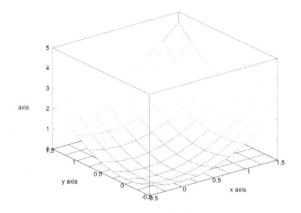

Figure 15.3: Drawing the surface mesh.

the values x_0 and y_0. The x_0 trace is the function $f(x_0, y)$ which is a *function* of the two variables y and z. The y_0 trace is the function $f(x, y_0)$ which is a *function* of the two variables x and z. We plot these curves using the function `DrawTraces`.

Listing 15.4: **Drawing the Traces**

```
DrawTraces( f ,0.5 ,2 ,0.5 ,2 ,0.5 ,0.5);
```

The resulting surface with grid and traces is shown in Figure 15.4. The traces are the thick parabolas on the surface. We are drawing the surface mesh on top of a rectangular grid in the $x - y$ plane. To help us see this visually, we can add the columns for the rectangular base having coordinates Lower Left (x_0, y_0), Lower Right $(x_0 + \Delta x, y_0)$, Upper Left $(x_0, y_0 + \Delta y)$ and Upper Right $(x_0 + \Delta x, y_0 + \Delta y)$. We draw and fill this base with the function `DrawBase`. We draw the vertical lines going from each of the four corners of the base to the surface with the code `DrawColumn` and we draw and fill the patch of surface this column creates in the full surface with the function `DrawPatch`.

First, we fill the base so we can see it better underneath the surface mesh. The code is pretty simple.

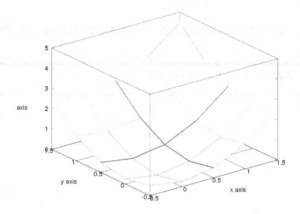

Figure 15.4: Drawing the traces.

Listing 15.5: **Filling the Base**

```
DrawBase(f,0.5,2,0.5,2,0.5,0.5);
```

You can see the filled in base in Figure 15.5. Then draw the corner columns using `DrawColumn`.

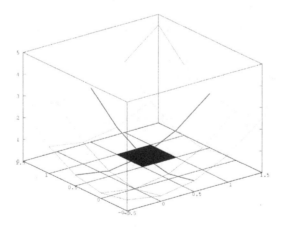

Figure 15.5: Filling in the base.

With this code, we draw four vertical lines from the base up to the surface. You'll note the figure is getting more crowded looking though. Make sure you grab the picture and rotate it around so you can see everything from different perspectives.

Listing 15.6: **Drawing the Columns for the Base**

```
DrawColumn(f,0.5,2,0.5,2,0.5,0.5);
```

We then draw the patch just like we drew the base.

Listing 15.7: Drawing the Patch of Surface Above the Base

```
DrawPatch(f,0.5,2,0.5,2,0.5,0.5);
```

You can see the columns with the patch in Figure 15.6. We have combined all of these functions into

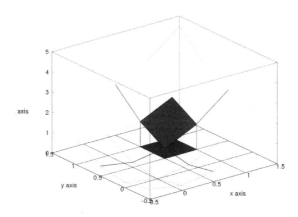

Figure 15.6: Adding the columns.

a utility function `DrawSimpleSurface` which manages these different graphing choices using boolean variables like `DoGrid` to turn a graph on or off. If the boolean variable `DoGrid` is set to one, the grid is drawn. The code is self-explanatory so we just lay it out here. We haven't shown all the code for the individual drawing functions, but we think you'll find it interesting to see how we manage the pieces in this one piece of code. So check this out. First, let's explain the arguments.

Listing 15.8: Explaining the DrawSimpleSurface Arguments

```
function DrawSimpleSurface(f,delx,nx,dely,ny,x0,y0,domesh,dotraces,
    dogrid,dopatch,docolumn,dobase)
% f is the function defining the surface
% delx is the size of the x step
% nx is the number of steps left and right from x0
5 % dely is the size of the y step
% ny is the number of steps left and right from y0
% (x0,y0) is the location of the column rectangle base
% domesh = 1 means do the mesh, dogrid = 1 means do the grid
% dopatch = 1 means add the patch above the column
10 % dobase = 1 means add the base of the column
% docolumn = 1 add the column, dotraces = 1 add the traces
%
% start hold
hold on
```

Now look at the code given below.

Listing 15.9: **The DrawSimpleSurfaces Code**

```
1  if  dotraces==1
   % set up x trace for x0, y trace for y0
      DrawTraces(f,delx,nx,dely,ny,x0,y0);
   end
   if  domesh==1 % plot the surface
6    DrawMesh(f,delx,nx,dely,ny,x0,y0);
   end
   if  dogrid==1 %plot x, y grid
      DrawGrid(f,delx,nx,dely,ny,x0,y0);
   end
11 if  dopatch==1
      % draw patch for top of column
      DrawPatch(f,delx,nx,dely,ny,x0,y0);
   end
   if  dobase==1
16    % draw patch for top of column
      DrawBase(f,delx,nx,dely,ny,x0,y0);
   end
   if  docolumn==1 %draw column
      DrawColumn(f,delx,nx,dely,ny,x0,y0);
21 end
   hold  off
   end
```

Hence, to draw everything for this surface, we would use the session:

Listing 15.10: **Drawing a Surface with DrawSimpleSurface**

```
  >> f = @(x,y) x.^2+y.^2;
2 >> DrawSimpleSurface(f,0.5,2,0.5,2,0.5,0.5,1,1,1,1,1,1);
```

This surface has circular cross sections for different positive values of z and it is called a *circular paraboloid*. If you used $f(x,y) = 4x^2 + 3y^2$, the cross sections for positive z would be ellipses and we would call the surface an *elliptical paraboloid*. Now this code is not perfect. However, as an exploratory tool it is not bad! Now it is time for you to play with it a bit in the exercises below.

15.2.1   Homework

Exercise 15.2.1 *Use the MATLAB tools on the website to explore the surface graph of* $f(x,y) = x^2 + 2y^2$.

Exercise 15.2.2 *Use the MATLAB tools on the website to explore the surface graph of* $f(x,y) = 3x^2 - 2y^2$.

Exercise 15.2.3 *Use the MATLAB tools on the website to explore the surface graph of* $f(x,y) = -x^2 - 2y^2$.

15.3   Continuity

Now let's look at continuity in this two dimensional setting. Let's recall the ideas of continuity for a function of one variable. Consider these three versions of a function f defined on $[0, 2]$.

$$f(x) \;=\; \begin{cases} x^2, & \text{if } 0 \le x < 1 \\ 10, & \text{if } x = 1 \\ 1 + (x-1)^2 & \text{if } 1 < x \le 2. \end{cases}$$

This function is not continuous at $x = 1$ because although the $\lim_{x \to 1} f(x)$ exists and equals 1 ($\lim_{x \to 1-} f(x) = 1$ and $\lim_{x \to 1+} f(x) = 1$), the value of $f(1)$ is 10 which does not match the limit. Hence, we know f here has a removeable discontinuity at $x = 1$. Note continuity failed because the limit existed but the value of the function did not match it.

The second version of f is given below.

$$f(x) \;=\; \begin{cases} x^2, & \text{if } 0 \le x \le 1 \\ (x-1)^2 & \text{if } 1 < x \le 2. \end{cases}$$

In this case, the $\lim_{x \to 1-} = 1$ and $f(1) = 1$, so f is continuous from the left. However, $\lim_{x \to 1+} = 0$ which does not match $f(1)$ and so f is not continuous from the right. Also, since the right and left hand limits do not match at $x = 1$, we know $\lim_{x \to 1}$ does not exist. Here, the function fails to be continuous because the limit does not exist.

The final example is below:

$$f(x) \;=\; \begin{cases} x^2, & \text{if } 0 \le x < 1 \\ x + (x-1)^2 & \text{if } 1 < x \le 2. \end{cases}$$

Here, the limit and the function value at 1 both match and so f is continuous at $x = 1$. To extend these ideas to two dimensions, the first thing we need to do is to look at the meaning of the limiting process. What does $\lim_{(x,y) \to (x_0,y_0)}$ mean? Clearly, in one dimension we can approach a point x_0 from x in two ways: from the left or from the right or jump around between left and right. Now, it is apparent that we can approach a given point (x_0, y_0) in an infinite number of ways. Draw a point on a piece of paper and convince yourself that there are many ways you can draw a curve from another point (x, y) so that the curve ends up at (x_0, y_0).

We still want to define continuity in the same way; i.e. f is continuous at the point (x_0, y_0) if $\lim_{(x,y) \to (x_0,y_0)} f(x, y) = f(x_0, y_0)$. If you look at the graphs of the surface $z = x^2 + y^2$ we have done previously, we clearly see that we have this kind of behavior. There are no jumps, tears or gaps in the surface we have drawn. Let's make this formal. We will lay out the definitions for a limit existing and for continuity at a point in both the two variable and three variable settings as a pair of matched definitions.

Let $z = f(x, y)$ be defined locally on $B_r(x_0, y_0)$ for some positive r. Here is the two dimensional extension of the idea of a limit of a function.

Definition 15.3.1 The Two Dimensional Limit

If $\lim_{(x,y) \to (x_0,y_0)} f(x, y)$ exists, this means there is a number L so that

$$\forall \epsilon > 0 \; \exists 0 < \delta < r \ni 0 < \| (x,y) - (x_0, y_0) \| < \delta \Rightarrow |f(x, y) - L| < \epsilon$$

We say $\lim_{(x,y) \to (x_0,y_0)} f(x, y) = L.$

We can now define continuity for a two dimensional function.

Definition 15.3.2 Continuity of a Two Dimensional Function

If $\lim_{(x,y)\to(x_0,y_0)} f(x,y)$ exists and matches $f(x_0,y_0)$, we say f is continuous at (x_0,y_0). That is

$$\forall \epsilon > 0 \; \exists 0 < \delta < r \; \ni || (x,y) - (x_0,y_0) || < \delta$$
$$\Rightarrow |f(x,y) - f(x_0,y_0)| < \epsilon$$

We say $\lim_{(x,y)\to(x_0,y_0)} f(x,y) = f(x_0,y_0)$.

Example 15.3.1 *Here is an example of a function which is not continuous at the point $(0,0)$. Let*

$$f(x,y) \;\; = \;\; \begin{cases} \dfrac{2x}{\sqrt{x^2+y^2}}, & \text{if } (x,y) \neq (0,0) \\ 0, & \text{if } (x,y) = (0,0). \end{cases}$$

Solution *If we show the limit as we approach $(0,0)$ does not exist, then we will know f is not continuous at $(0,0)$. If this limit exists, we should get the same value for the limit no matter what path we take to reach $(0,0)$.*

Let the first path be given by $x(t) = t$ and $y(t) = 2t$. We have two paths really; one for $t > 0$ and one for $t < 0$. We find for $t > 0$, $f(t,2t) = 2t/\sqrt{t^2 + 4t^2} = 2/\sqrt{5}$ and hence the limit along this path is $2/\sqrt{5}$.

We find for $t < 0$, $f(t,2t) = 2t/\sqrt{t^2 + 4t^2} = 2t/(|t| \sqrt{5}) = -2/\sqrt{5}$ and hence the limit along this path $-2/\sqrt{5}$. Since the limiting value differs on two paths, the limit can't exist. Hence, f is not continuous at $(0,0)$.

Example 15.3.2 *Prove $f(x,y) = 2x^2 + 3y^2$ is continuous at $(2,3)$.*

Solution *We find $f(2,3) = 8 + 27 = 35$. Pick $\epsilon > 0$. Consider*

$$
\begin{aligned}
|(2x^2 + 3y^2) - 35| &= |(2(x - 2 + 2)^2 + 3(y - 3 + 3)^2) - 35| \\
&= |2(x-2)^2 + 8(x-2) + 8 + 3(y-3)^2 + 18(y-3) + 27 - 35| \\
&= |2(x-2)^2 + 8(x-2) + 3(y-3)^2 + 18(y-3)| \\
&\leq 2|x-2|^2 + 8|x-2| + 3|y-3|^2 + 18|y-3|
\end{aligned}
$$

Next, start by choosing a positive $r < 1$. Then we note if (x,y) is in $B_r(2,3)$, we have $|x-2| \leq \sqrt{|x-2|^2 + |y-3|^2} < r$ implying $|x-2| < r$. In a similar way, we see $|y-3| < r$ also.

Thus, since $r < 1$ implies $r^2 < r$, we can say

$$|(2x^2 + 3y^2) - 35| < 2r^2 + 8r + 3r^2 + 18r = 5r^2 + 26r < 31r$$

Thus, if $\delta < \min\{1, \epsilon/31\}$, $|(2x^2 + 3y^2) - 35| < \epsilon$ and we have continuity at $(2,3)$.

Example 15.3.3 *Prove $f(x,y) = 2xy$ is continuous at $(1,4)$.*

Solution *We find $f(1,4) = 8$. Consider*

$$
\begin{aligned}
|2xy - 8| &= |2(x - 1 + 1)(y - 4 + 4) - 8| \\
&= |2(x-1)(y-4) + 2(y-4) + 8(x-1) + 8 - 8| \\
&= |2(x-1)(y-4) + 2(y-4) + 8(x-1)| \\
&\leq 2|x-1| \, |y-4| + 2|y-4| + 8|x-1|
\end{aligned}
$$

Next, start by choosing a positive $r < 1$. Then we note if (x, y) is in $B_r(1.4)$, we have $|x - 1| \le \sqrt{|x - 1|^2 + |y - 4|^2} < r$ implying $|x - 1| < r$. In a similar way, we see $|y - 4| < r$ also.

Thus, since $r < 1$ implies $r^2 < r$, we can say

$$|2xy - 8| < 2r^2 + 2r + 8r = 2r^2 + 10r < 12r$$

Thus, if $\delta < \min\{1, \epsilon/12\}$, $|2xy - 8| < \epsilon$ and we have continuity at $(1, 4)$.

15.3.1   Homework

Exercise 15.3.1 *Let f be a function whose second derivative is continuous. Prove*

1. *if $f''(p) > 0$ then $f'(x)$ increases locally at p, discuss how the graph of the function looks locally at p.*

2. *if $f''(p) < 0$ then $f'(x)$ decreases locally at p, discuss how the graph of the function looks locally at p.*

Exercise 15.3.2 *Let f be a function whose second derivative is continuous and it is positive at $x = -1$. Explain what the graph of f qualitatively could look like.*

Exercise 15.3.3 *Let $f(x, y) = 6x^2 + 9y^2$. Prove f is continuous at $(2, 3)$.*

Exercise 15.3.4 *Let $f(x, y) = 2xy + 3y^2$. Prove f is continuous at $(1, 2)$.*

Exercise 15.3.5 *Let $f(x, y) = 5xy + 4x^2$. Prove f is continuous at $(-2, -1)$.*

15.4   Partial Derivatives

Let's go back to our simple surface example and look at the traces again. In Figure 15.7, we show the traces for the base point $x_0 = 0.5$ and $y_0 = 0.5$. We have also drawn vertical lines down from the traces to the $x - y$ plane to further emphasize the placement of the traces on the surface. The surface itself is not shown as it is somewhat distracting and makes the illustration too busy.

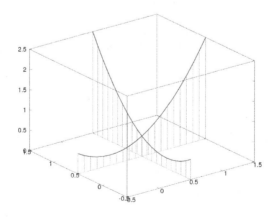

Figure 15.7: The full traces.

You can generate this type of graph yourself with the function **DrawFullTraces** as follows:

Listing 15.11: | **Drawing a Surface with Full Traces** |

```
DrawFullTraces(f,0.5,2,0.5,2,0.5,0.5);
```

Note, that each trace has a well-defined tangent line and derivative at the points x_0 and y_0. We have

$$\frac{d}{dx}\, f(x, y_0) \;=\; \frac{d}{dx}(x^2 + y_0^2) = 2x$$

as the value y_0 in this expression is a constant and hence its derivative with respect to x is zero. We denote this new derivative as $\frac{\partial f}{\partial x}$ which we read as *the partial derivative of f with respect to x*. It's value as the point (x_0, y_0) is $2x_0$ here. For any value of (x, y), we would have $\frac{\partial f}{\partial x} = 2x$. We also have

$$\frac{d}{dy}\, f(x_0, y) \;=\; \frac{d}{dy}(x_0^2 + y^2) = 2y$$

We then denote this new derivative as $\frac{\partial f}{\partial y}$ which we read as *the partial derivative of f with respect to y*. It's value as the point (x_0, y_0) is then $2y_0$ here. For any value of (x, y), we would have $\frac{\partial f}{\partial y} = 2y$. The tangent lines for these two traces are then

$$\begin{aligned}
T(x, y_0) &= f(x_0, y_0) + \frac{d}{dx}\, f(x, y_0)\Big|_{x_0} (x - x_0) \\
&= (x_0^2 + y_0^2) + 2x_0(x - x_0) \\
T(x_0, y) &= f(x_0, y_0) + \frac{d}{dy}\, f(x_0, y)\Big|_{y_0} (y - y_0) \\
&= (x_0^2 + y_0^2) + 2y_0(y - y_0).
\end{aligned}$$

We can also write these tangent line equations like this using our new notation for partial derivatives.

$$\begin{aligned}
T(x, y_0) &= f(x_0, y_0) + \frac{\partial f}{\partial x}(x_0, y_0)\,(x - x_0) \\
&= (x_0^2 + y_0^2) + 2x_0(x - x_0) \\
T(x_0, y) &= f(x_0, y_0) + \frac{\partial f}{\partial y}(x_0, y_0)\,(y - y_0) \\
&= (x_0^2 + y_0^2) + 2y_0(y - y_0).
\end{aligned}$$

We can draw these tangent lines in 3D. To draw $T(x, y_0)$, we fix the y value to be y_0 and then we draw the usual tangent line in the $x - z$ plane. This is a copy of the $x - z$ plane translated over to the value y_0; i.e. it is parallel to the $x - z$ plane we see at the value $y = 0$. We can do the same thing for the tangent line $T(x, y_0)$; we fix the x value to be x_0 and then draw the tangent line in the copy of the $y - z$ plane translated to the value x_0. We show this in Figure 15.8. Note the $T(x, y_0)$ and the $T(x_0, y)$ lines are determined by vectors as shown below.

$$\boldsymbol{A} \;=\; \begin{bmatrix} 1 \\ 0 \\ \frac{d}{dx}\, f(x, y_0)\big|_{x_0} \end{bmatrix} = \begin{bmatrix} 1 \\ 0 \\ 2x_0 \end{bmatrix} \text{ and } \boldsymbol{B} = \begin{bmatrix} 0 \\ 1 \\ \frac{d}{dy}\, f(x_0, y)\big|_{y_0} \end{bmatrix} = \begin{bmatrix} 0 \\ 1 \\ 2y_0 \end{bmatrix}$$

Note that if we connect the lines determined by the vectors $\boldsymbol{A}$ and $\boldsymbol{B}$, we determine a *flat* sheet which you can interpret as a piece of paper laid on top of these two lines. Of course, we can only envision a small finite subset of this sheet of paper as you can see in Figure 15.8. Imagine that the sheet extends infinitely in all directions! The sheet of paper we are plotting is called the **tangent plane** to our surface at the point (x_0, y_0). We will talk about this more formally later.

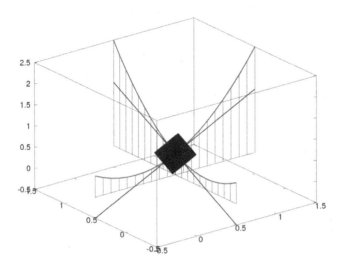

Figure 15.8: Drawing the tangent lines.

To draw this picture with the tangent lines, the traces and the tangent plane, we use the function `DrawTangentLines` with arguments `(f,fx,fy,delx,nx,dely,ny,r,x0,y0)`. There are three new arguments: `fx` which is $\partial f / \partial x$, `fy` which is $\partial f / \partial y$ and `r` which is the size of the tangent plane that is plotted. For Figure 15.9, we've removed the tangent plane because the plot was getting pretty busy.

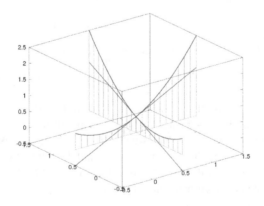

Figure 15.9: Drawing tangent lines: no tangent plane.

We did this by commenting out the line that plots the tangent plane. It is easy for you to go into the code and add it back in if you want to play around. The MATLAB command line is

Listing 15.12: | **Drawing the Tangent Lines** |

```
fx = @(x,y) 2*x;
fy = @(x,y) 2*y;
%
DrawTangentLines(f,fx,fy,0.5,2,0.5,2,.3,0.5,0.5);
```

If you want to see the tangent plane as well as the tangent lines, all you have to do is look at the following lines in **DrawTangentLines.m**.

Listing 15.13: | **Tangent Plane Code** |

```
1   % set up a new local mesh grid near (x0,y0)
    [U,V] = meshgrid(u,v)
    % set up the tangent plane at (x0,y0)
    W = f(x0,y0) + fx(x0,y0)*(U–x0) + fy(x0,y0)*(V–y0)
    % plot the tangent plane
6   surf(U,V,W,'EdgeColor','blue');
```

These lines set up the tangent plane and the tangent plane is turned off if there is a percent `%` in front of `surf(U,V,W,'EdgeColor','blue');`. We edited the file to take the `%` out so we can see the tangent plane. We then see the plane in Figure 15.10.

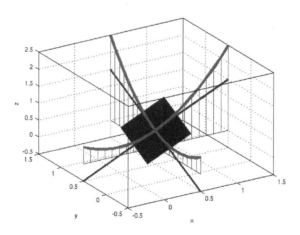

Figure 15.10: Adding the tangent plane.

The ideas we have been discussing can be made more general. When we take the derivative with respect to one variable while holding the other variable constant (as we do when we find the normal derivative along a trace), we say we are taking a **partial derivative of f**. Here there are two flavors: the partial derivative with respect to x and the partial derivative with respect to y. We can now state some formal definitions and introduce the notations and symbols we use for these things. We define the process of partial differentiation carefully below.

Definition 15.4.1 Partial Derivatives

Let $z = f(x, y)$ be a function of the two independent variables x and y defined on some domain. At each pair (x, y) where f is defined in a circle of some finite radius r, $B_r(x_0, y_0) = \{(x, y) \mid \sqrt{(x - x_0)^2 + (y - y_0)^2} < r\}$, it makes sense to try to find the limits

$$\frac{\partial f}{\partial x}(x_0, y_0) = \lim_{x \to x_0, y = y_0} \frac{f(x, y_0) - f(x_0, y_0)}{x - x_0}$$

$$\frac{\partial f}{\partial y}(x_0, y_0) = \lim_{x = x_0, y \to y_0} \frac{f(x_0, y) - f(x_0, y_0)}{y - y_0}$$

If these limits exists, they are called the partial derivatives of f with respect to x and y at (x_0, y_0), respectively. For these partial derivatives, we also use the symbols $f_x(x_0, y_0)$, $z_x(x_0, y_0)$ and $\frac{\partial z}{\partial x}(x_0, y_0)$ and $f_y(x_0, y_0)$, $z_y(x_0, y_0)$ and $\frac{\partial z}{\partial y}(x_0, y_0)$.

Comment 15.4.1 *We often use another notation for partial derivatives. The function f of two variables x and y can be thought of as having two arguments or slots into which we place values. So another useful notation is to let the symbol $D_1 f$ be f_x and $D_2 f$ be f_y. We will be using this notation later when we talk about the* **chain rule**.

Comment 15.4.2 *It is easy to take partial derivatives. Just imagine the one variable held constant and take the derivative of the resulting function just like you did in your earlier calculus courses.*

Let's do a few simple examples to remind you of how to do this.

Example 15.4.1 *Let $z = f(x, y) = x^2 + 4y^2$ be a function of two variables. Find $\frac{\partial z}{\partial x}$ and $\frac{\partial z}{\partial y}$.*

Solution *Thinking of y as a constant, we take the derivative in the usual way with respect to x. This gives*

$$\frac{\partial z}{\partial x} = 2x$$

as the derivative of $4y^2$ with respect to x is 0. So, we know $f_x = 2x$. In a similar way, we find $\frac{\partial z}{\partial y}$. We see

$$\frac{\partial z}{\partial y} = 8y$$

as the derivative of x^2 with respect to y is 0. So $f_y = 8y$.

Example 15.4.2 *Let $z = f(x, y) = 4x^2 y^3$. Find $\frac{\partial z}{\partial x}$ and $\frac{\partial z}{\partial y}$.*

Solution *Thinking of y as a constant, take the derivative in the usual way with respect to x: This gives*

$$\frac{\partial z}{\partial x} = 8xy^3$$

as the term $4y^3$ is considered a "constant" here. So $f_x = 8xy^3$. Similarly,

$$\frac{\partial z}{\partial y} = 12x^2 y^2$$

as the term $4x^2$ is considered a "constant" here. So $f_y = 12x^2 y^2$.

Homework

Exercise 15.4.1 *Let* $z = f(x, y) = 4x^4 y^5$. *Find* $\frac{\partial z}{\partial x}$ *and* $\frac{\partial z}{\partial y}$.

Exercise 15.4.2 *Let* $z = f(x, y) = 4x^2 y^3 + 2xy + x^2 + 3y^2 + 2x + 5y + 19$. *Find* $\frac{\partial z}{\partial x}$ *and* $\frac{\partial z}{\partial y}$.

Exercise 15.4.3 *Let* $z = f(x, y) = 4x^2 + 7y^4$. *Find* $\frac{\partial z}{\partial x}$ *and* $\frac{\partial z}{\partial y}$.

Exercise 15.4.4 *Let* $z = f(x, y) = 49x^2 - 3y^2$. *Find* $\frac{\partial z}{\partial x}$ *and* $\frac{\partial z}{\partial y}$.

15.5 Tangent Planes

It is very useful to have an analytical way to discuss tangent planes and their relationship to the real surface to which they are attached. Look at Figure 15.10 to remind yourself of what is happening. In this surface, it is clear the tangent plane deviates from the surface as you choose points on the surface that are further and further away from the attachment point of the tangent plane. We want to develop a way to make this statement precise.

Let's recall some facts about inner products. Any time the dot product of two vectors is 0, the vectors are perpendicular or 90° apart. We can use this fact to define a plane as follows.

Definition 15.5.1 Planes in Three Dimensions

A plane in 3D through the point (x_0, y_0, z_0) is defined as the set of all points (x, y, z) so that the angle between the vectors $\boldsymbol{D}$ and $\boldsymbol{N}$ is 90° where $\boldsymbol{D}$ is the vector we get by connecting the point (x_0, y_0, z_0) to the point (x, y, z). Hence, for

$$\boldsymbol{D} = \begin{bmatrix} x - x_0 \\ y - y_0 \\ z - z_0 \end{bmatrix} \text{ and } \boldsymbol{N} = \begin{bmatrix} N_1 \\ N_2 \\ N_3 \end{bmatrix}$$

the plane is the set of points (x, y, z) so that $< \boldsymbol{D}, \boldsymbol{N} >= 0$. The vector $\boldsymbol{N}$ is called the **normal vector** *to the plane.*

Recall the tangent plane to a surface $z = f(x, y)$ at the point (x_0, y_0) was the plane determined by the tangent lines $T(x, y_0)$ and $T(x_0, y)$. The $T(x, y_0)$ line was determined by the vector

$$\boldsymbol{A} = \begin{bmatrix} 1 \\ 0 \\ \frac{\partial f}{\partial x}(x_0, y_0) \end{bmatrix} = \begin{bmatrix} 1 \\ 0 \\ 2x_0 \end{bmatrix}$$

and the $T(x_0, y)$ line was determined by the vector

$$\boldsymbol{B} = \begin{bmatrix} 0 \\ 1 \\ \frac{\partial f}{\partial y}(x_0, y_0) \end{bmatrix} = \begin{bmatrix} 0 \\ 1 \\ 2y_0 \end{bmatrix}$$

We need to find a vector perpendicular to both $\boldsymbol{A}$ and $\boldsymbol{B}$. Let's try this one: $\boldsymbol{N} = [-f_x(x_0, y_0), -f_y(x_0, y_0), 1]^T$. The dot product of $\boldsymbol{A}$ with $\boldsymbol{N}$ is

$$< \boldsymbol{A}, \boldsymbol{N} >= 1 - f_x(x_0, y_0) + 0\left(-f_y(x_0, y_0)\right) + f_x(x_0, y_0)\, 1 = 0.$$

and the dot product of $\boldsymbol{B}$ with $\boldsymbol{N}$ is

$$< \boldsymbol{B}, \boldsymbol{N} >= 0\left(-f_x(x_0, y_0)\right) + 1\left(-f_y(x_0, y_0)\right) + f_y(x_0, y_0)\, 1 = 0.$$

So our N is perpendicular to both of these vectors and so we know the tangent plane to the surface $z = f(x, y)$ at the point (x_0, y_0) is then given by

$$-f_x(x_0, y_0)(x - x_0) - f_y(x_0, y_0)(y - y_0) + (z - f(x_0, y_0)) \quad = \quad 0.$$

This then gives the traditional equation of the tangent plane:

$$z \quad = \quad f(x_0, y_0) + f_x(x_0, y_0)(x - x_0) + f_y(x_0, y_0)(y - y_0).$$

We can use another compact definition at this point. We can define the **gradient** of the function f to be the vector ∇f which is defined as follows.

Definition 15.5.2 The Gradient

The gradient of the scalar function $z = f(x, y)$ is defined to be the vector ∇f where

$$\nabla f(x_0, y_0) \quad = \quad \begin{bmatrix} f_x(x_0, y_0) \\ f_y(x_0, y_0) \end{bmatrix}.$$

Note the gradient takes a scalar function argument and returns a vector answer. The word scalar *just means the function returns a number and not a vector.*

Using the gradient, the tangent plane equation can be rewritten as

$$\begin{aligned} z \quad &= \quad f(x_0, y_0) + \, < \nabla f, X - X_0 > \\ &= \quad f(x_0, y_0) + \nabla f^T(X - X_0) \end{aligned}$$

where $X - X_0 = \begin{bmatrix} x - x_0 & y - y_0 \end{bmatrix}^T$. **The obvious question to ask now is how much of a discrepancy is there between the value $f(x, y)$ and the value of the tangent plane?**

Example 15.5.1 *Find the gradient of $f(x, y) = x^2 + 4xy + 9y^2$ and the equation of the tangent plane to this surface at the point $(1, 2)$.*

Solution

$$\nabla f(x, y) \quad = \quad \begin{bmatrix} 2x + 4y \\ 4x + 18y \end{bmatrix}.$$

The equation of the tangent plane at $(1, 2)$ is then

$$\begin{aligned} z \quad &= \quad f(1, 2) + f_x(1, 2)(x - 1) + f_y(1, 2)(y - 2) \\ &= \quad 45 + 10(x - 1) + 40(y - 2) = -45 + 10x + 40y. \end{aligned}$$

Note this can also be written as $10x + 40y + z = 45$ which is also a standard form. However, in this form, the attachment point $(1, 2, 45)$ is hidden from view.

Example 15.5.2 *Find the gradient of $f(x, y) = 3x^2 + 2y^2$ and the equation of the tangent plane to this surface at the point $(2, 3)$.*

Solution

$$\nabla f(x, y) \quad = \quad \begin{bmatrix} 6x \\ 4y \end{bmatrix}.$$

The equation of the tangent plane at $(2, 3)$ *is then*

$$
\begin{aligned}
z &= f(2,3) + f_x(2,3)(x-2) + f_y(2,3)(y-3) \\
&= 30 + 12(x-2) + 12(y-3) \\
&= -30 + 12x + 12y.
\end{aligned}
$$

Note this can also be written as $12x + 12y + z = 35$ *which is also a standard form. However, in this form, the attachment point* $(2, 3, 30)$ *is hidden from view.*

15.5.1 Homework

Exercise 15.5.1 *If* $f(x, y)$ *has a local minimum or local maximum at* (x_0, y_0) *prove* $f_x(x_0, y_0) = f_y(x_0, y_0) = 0$.

Exercise 15.5.2 *If* $f_x(x, y_0)$ *is continuous locally and* $f_x(x_0, y_0)$ *is positive, prove* $f(x, y_0) > f(x_0, y_0)$ *locally on the right; i.e.* $f(x, y_0) > f(x_0, y_0)$ *if* $x \in (x_0, x_0 + r)$ *for some* $r > 0$.

Exercise 15.5.3 *If* $f_y(x_0, y)$ *is continuous locally and* $f_y(x_0, y_0)$ *is negative, prove* $f(x_0, y) < f(x_0, y_0)$ *locally on the right; i.e.* $f(x, y_0) < f(x_0, y_0)$ *if* $x \in (y_0, y_0 + r)$.

Exercise 15.5.4 *Find the tangent plane to* $f(x, y) = 3x^2 + 2xy + 5y^4$ *at* $(-1, 3)$.

Exercise 15.5.5 *Find the gradient of* $f(x, y) = -4x^2y^2 + 5xy + 3y + 2$ *at any* (x, y) *and at* $(2, 1)$. *Also find the tangent plane at* $(1, -2)$.

15.6 Derivatives in 2D!

We need to figure out what we mean by a derivative in two or more dimensions. Let's go back to one dimensional calculus for some motivation. If the function f is defined locally near x_0 that means that f is defined in a circle $B_r(x_0) = \{x : x_0 - r < x < x_0 + r\}$ for some positive value of r. In this case, we can attempt to find the usual limit as x approaches x_0 that defines the derivative of f and x_0: if this limit exists, it is called $f'(x_0)$ and

$$
f'(x_0) = \lim_{x \to x_0} \frac{f(x) - f(x_0)}{x - x_0}.
$$

This can be expressed in a different form. Recall that we can also use the $\epsilon - \delta$ notation to define a limit. In this case, it means that if we choose a positive ϵ, then there is a positive δ so that

$$
0 < |x - x_0| < \delta \implies \left| \frac{f(x) - f(x_0)}{x - x_0} - f'(x_0) \right| < \epsilon.
$$

Now define the error between the function value $f(x)$ and the tangent line value $f(x_0) + f'(x_0)(x - x_0)$ to be $E(x, x_0)$. The above statement can be rewritten as

$$
0 < |x - x_0| < \delta \implies \left| \frac{f(x) - f(x_0) - f'(x_0)(x - x_0)}{x - x_0} \right| < \epsilon.
$$

Then using the definition of error, $E(x, x_0)$, we see

$$
0 < |x - x_0| < \delta \implies \left| \frac{E(x, x_0)}{x - x_0} \right| < \epsilon.
$$

This is the same as saying

$$\lim_{x \to x_0} \frac{E(x, x_0)}{x - x_0} = 0.$$

Now rewrite the inequality again to have

$$0 < |x - x_0| < \delta \implies \left| E(x, x_0) \right| < \epsilon |x - x_0|.$$

Since we can do this for any positive ϵ, it works for the choice $\sqrt{\epsilon}$. Hence, there is a positive δ_1 so that

$$0 < |x - x_0| < \delta_1 \implies \left| E(x, x_0) \right| < \sqrt{\epsilon} |x - x_0| < \sqrt{\epsilon} \, \delta_1.$$

But this works as long as $0 < |x - x_0| < \delta_1$. So it also works if $0 < |x - x_0| < \delta_2 = \min(\delta_1, \sqrt{\epsilon}) \le \delta_1$! So

$$0 < |x - x_0| < \delta_2 \implies \left| E(x, x_0) \right| < \sqrt{\epsilon} |x - x_0|$$
$$< \sqrt{\epsilon} \, \delta_2 < \sqrt{\epsilon} \sqrt{\epsilon} < \epsilon.$$

So we can say $\lim_{x \to x_0} E(x, x_0) = 0$ as well. This leads to the following theorem which we have already seen in the one variable but we will restate it here for convenience.

Theorem 15.6.1 Error Form of Differentiability for One Variable: a Motivation

> *If f is defined locally at x_0, then f is differentiable at x_0 if the error function $E(x, x_0) = f(x) - f(x_0) - f'(x_0)(x - x_0)$ satisfies $\lim_{x \to x_0} E(x, x_0) = 0$ and $\lim_{x \to x_0} E(x, x_0)/(x - x_0) = 0$. Conversely, if there is a number L so that the error function $E(x, x_0) = f(x) - f(x_0) - L(x - x_0)$ satisfies the same behavior, then f is differentiable at x_0 with value $f'(x_0) = L$.*

Proof 15.6.1
If f is differentiable at x_0, we have already outlined the argument. The converse argument is quite similar. Since we know $\lim_{x \to x_0} E(x, x_0)/(x - x_0) = 0$, this tells us

$$\lim_{x \to x_0} \frac{f(x) - f(x_0) - L(x - x_0)}{x - x_0} = 0$$

or

$$\lim_{x \to x_0} \frac{f(x) - f(x_0)}{x - x_0} - L = 0.$$

But this states that f is differentiable at x_0 with value L. With this argument done, we have shown both sides of the statement are true. ∎

Note if f is differentiable at x_0, f must be continuous at x_0. This follows because $f(x) = f(x_0) + f'(x_0)(x - x_0) + E(x, x_0)$ and as $x \to x_0$, we have $f(x) \to f(x_0)$ which is the definition of f being continuous at x_0. Hence, we can say

Theorem 15.6.2 Differentiable Implies Continuous for One Variable: a Motivation

If f is differentiable at x_0 then f is continuous at x_0.

Proof 15.6.2
We have sketched the argument already. ∎

We apply this idea to the partial derivatives of $f(x, y)$. As long as $f(x, y)$ is defined locally at (x_0, y_0), we have $f_x(x_0, y_0)$ and $f_y(x_0, y_0)$ exist if and only if there are error functions $E_1(x, y, x_0, y_0)$ and $E_2(x, y, x_0, y_0)$ so that

$$f(x, y_0) = f(x_0, y_0) + f_x(x_0, y_0)(x - x_0) + E_1(x, x_0, y_0)$$
$$f(x_0, y) = f(x_0, y_0) + f_y(x_0, y_0)(y - y_0) + E_2(y, x_0, y_0)$$

with $E_1 \to 0$ and $E_1/(x - x_0) \to 0$ as $x \to x_0$ and $E_2 \to 0$ and $E_2/(y - x_0) \to 0$ as $y \to y_0$. Using the ideas we have presented here, we can come up with a way to define the differentiability of a function of two variables.

Definition 15.6.1 Error Form of Differentiability for Two Variables

If $f(x, y)$ is defined locally at (x_0, y_0), then f is differentiable at (x_0, y_0) if there are two numbers L_1 and L_2 so that the error function $E(x, y, x_0, y_0) = f(x, y) - f(x_0, y_0) - L_1(x - x_0) - L_2(y - y_0)$ satisfies $\lim_{(x,y) \to (x_0, y_0)} E(x, y, x_0, y_0) = 0$ and $\lim_{(x,y) \to (x_0, y_0)} E(x, y, x_0, y_0)/\|(x - x_0, y - y_0)\| = 0$. Recall, the term $\|(x - x_0, y - y_0)\| = \sqrt{(x - x_0)^2 + (y - y_0)^2}$.

Note if f is differentiable at (x_0, y_0), f must be continuous at (x_0, y_0). The argument is simple:

$$f(x, y) = f(x_0, y_0) + L_1(x_0, y_0)(x - x_0) + L_2(y - y_0) + E(x, y, x_0, y_0)$$

and as $(x, y) \to (x_0, y_0)$, we have $f(x, y) \to f(x_0, y_0)$ which is the definition of f being continuous at (x_0, y_0). Hence, we can say

Theorem 15.6.3 Differentiable Implies Continuous: Two Variables

If f is differentiable at (x_0, y_0) then f is continuous at (x_0, y_0).

Proof 15.6.3
We have sketched the argument already. ∎

From this definition, we can show if f is differentiable at the point (x_0, y_0), then $L_1 = f_x(x_0, y_0)$ and $L_2 = f_y(x_0, y_0)$. The argument goes like this: since f is differentiable at (x_0, y_0), we can say

$$\lim_{(x,y) \to (x_0, y_0)} \frac{f(x, y) - f(x_0, y_0) - L_1(x - x_0) - L_2(y - y_0)}{\sqrt{(x - x_0)^2 + (y - y_0)^2}} = 0.$$

We can rewrite this using $\Delta x = x - x_0$ and $\Delta y = y - y_0$ as

$$\lim_{(\Delta x, \Delta y) \to (0,0)} \frac{f(x_0 + \Delta x, y_0 + \Delta y) - f(x_0, y_0) - L_1 \Delta x - L_2 \Delta y}{\sqrt{(\Delta x)^2 + (\Delta y)^2}} = 0.$$

In particular, for $\Delta y = 0$, we find

$$\lim_{(\Delta x) \to 0} \frac{f(x_0 + \Delta x, y_0) - f(x_0, y_0) - L_1 \Delta x}{\sqrt{(\Delta x)^2}} = 0.$$

For $\Delta x > 0$, we find $\sqrt{(\Delta x)^2} = \Delta x$ and so

$$\lim_{\Delta x \to 0^+} \frac{f(x_0 + \Delta x, y_0) - f(x_0, y_0)}{\Delta x} \;=\; L_1.$$

Thus, the right hand partial derivative $f_x(x_0, y_0)^+$ exists and equals L_1. On the other hand, if $\Delta x < 0$, then $\sqrt{(\Delta x)^2} = -\Delta x$ and we find, with a little manipulation, that we still have

$$\lim_{(\Delta x) \to 0^-} \frac{f(x_0 + \Delta x, y_0) - f(x_0, y_0)}{\Delta x} \;=\; L_1.$$

So the left hand partial derivative $f_x(x_0, y_0)^-$ exists and equals L_1 also. Combining, we see $f_x(x_0, y_0) = L_1$. A similar argument shows that $f_y(x_0, y_0) = L_2$. Hence, we can say if f is differentiable at (x_0, y_0) then f_x and f_y exist at this point and we have

$$\begin{aligned} f(x, y) \;=\;& f(x_0, y_0) + f_x(x_0, y_0)(x - x_0) + f_y(x_0, y_0)(y - y_0) \\ &+ E_f(x, y, x_0, y_0) \end{aligned}$$

where $E_f(x, y, x_0, y_0) \to 0$ and $E_f(x, y, x_0, y_0)/\|(x - x_0, y - y_0)\| \to 0$ as $(x, y) \to (x_0, y_0)$. Note this argument is a pointwise argument. It only tells us that differentiability at a point implies the existence of the partial derivatives at that point.

Next, we look at 2D version of the chain rule.

15.7   Chain Rule

Now that we know a bit about two dimensional derivatives, let's go for gold and figure out the new version of the chain rule. The argument we make here is very similar in spirit to the one dimensional one. You should go back and check it out!

We assume there are two functions $u(x, y)$ and $v(x, y)$ defined locally about (x_0, y_0) and that there is a third function $f(u, v)$ which is defined locally around $(u_0 = u(x_0, y_0), v_0 = v(x_0, y_0))$. Now assume $f(u, v)$ is differentiable at (u_0, v_0) and $u(x, y)$ and $v(x, y)$ are differentiable at (x_0, y_0). Then we can say

$$\begin{aligned} u(x, y) &= u(x_0, y_0) + u_x(x_0, y_0)(x - x_0) + u_y(x_0, y_0)(y - y_0) + E_u(x, y, x_0, y_0) \\ v(x, y) &= v(x_0, y_0) + v_x(x_0, y_0)(x - x_0) + v_y(x_0, y_0)(y - y_0) + E_v(x, y, x_0, y_0) \\ f(u, v) &= f(u_0, v_0) + f_u(u_0, v_0)(u - u_0) + f_v(u_0, v_0)(v - v_0) + E_f(u, v, u_0, v_0) \end{aligned}$$

where all the error terms behave as usual as $(x, y) \to (x_0, y_0)$ and $(u, v) \to (u_0, v_0)$. Note that as $(x, y) \to (x_0, y_0)$, $u(x, y) \to u_0 = u(x_0, y_0)$ and $v(x, y) \to v_0 = v(x_0, y_0)$ as u and v are continuous at the (u_0, v_0) since they are differentiable there. Let's consider the partial of f with respect to x. Let $\Delta u = u(x_0 + \Delta x, y_0) - u(x_0, y_0)$ and $\Delta v = v(x_0 + \Delta x, y_0) - v(x_0, y_0)$. Thus, $u_0 + \Delta u = u(x_0 + \Delta x, y_0)$ and $v_0 + \Delta v = v(x_0 + \Delta x, y_0)$. Hence,

$$\begin{aligned} &\frac{f(u_0 + \Delta u, v_0 + \Delta v) - f(u_0, v_0)}{\Delta x} \\ &= \frac{f_u(u_0, v_0)(u - u_0) \;+\; f_v(u_0, v_0)(v - v_0) + E_f(u, v, u_0, v_0)}{\Delta x} \\ &= f_u(u_0, v_0) \frac{u - u_0}{\Delta x} \;+\; f_v(u_0, v_0) \frac{v - v_0}{\Delta x} \;+\; \frac{E_f(u, v, u_0, v_0)}{\Delta x} \end{aligned}$$

Continuing

$$\frac{f(u_0 + \Delta u, v_0 + \Delta v) - f(u_0, v_0)}{\Delta x}$$

$$= f_u(u_0, v_0) \frac{u_x(x_0, y_0)(x - x_0) + E_u(x, x_0, y_0)}{\Delta x}$$

$$+ f_v(u_0, v_0) \frac{v_x(x_0, y_0)(x - x_0) + E_v(x, x_0, y_0)}{\Delta x}$$

$$+ \frac{E_f(u, v, u_0, v_0)}{\Delta x}$$

$$= f_u(u_0, v_0) \, u_x(x_0, y_0) + f_v(u_0, v_0) \, v_x(x_0, y_0)$$

$$+ f_u(u_0, v_0) \frac{E_u(x, x_0, y_0)}{\Delta x} + f_v(u_0, v_0) \frac{E_v(x, x_0, y_0)}{\Delta x} + \frac{E_f(u, v, u_0, v_0)}{\Delta x}.$$

If f was locally constant, then

$$E_f(u_0, v_0, u, v) \quad = \quad f(u, v) - f(u_0, v_0) - f_u(u_0, v_0)(u - u_0) f_v((u_0, v_0) v - v_0) = 0$$

We know

$$\lim_{(a,b) \to (u_0, v_0)} \frac{E_f(u_0, v_0, u, v)}{\|(a, b) - (u_0, v_0)\|} \quad = \quad 0$$

and this is true no matter what sequence (a_n, b_n) we choose that converges to $(u(x_0, y_0), v(u_0, y_0))$. If f is not locally constant, a little thought shows locally about $(u(x_0, y_0), v(u_0, y_0))$ there are sequences $(u_n, v_n) = (u(x_n, y_n), v(x_n, y_n)) \to (u_0, v_0) = (u(x_0, y_0), v(u_0, y_0))$ with $(u_n, v_n) \neq (u_0, v_0)$ for all n. Also, by continuity, as $\Delta x \to 0$, $(u_n, v_n) \to (u_0, v_0)$ too. For (u_n, v_n) from this sequence, we then have

$$\lim_{\Delta x \to 0} \frac{E_f(u_0, v_0, u, v)}{\Delta x} \quad = \quad \left(\lim_{n \to \infty} \frac{E_f(u_0, v_0, u, v)}{\|(u_n, v_n) - (u_0, v_0)\|} \right) \left(\lim_{\Delta x \to 0} \frac{\|(u_n, v_n) - (u_0, v_0)\|}{\Delta x} \right)$$

We know the first term goes to zero by the properties of E_f. Expand the second term to get

$$\left(\lim_{\Delta x \to 0} \frac{\|(u_n, v_n) - (u_0, v_0)\|}{\Delta x} \right) =$$

$$\left(\lim_{\Delta x \to 0} \frac{\|u_x(x_0, y_0)\Delta x + E_u(x_0, y_0, x, y), v_x(x_0, y_0)\Delta x + E_v(x_0, y_0, x, y)\|}{\Delta x} \right) \leq$$

$$|u_x(x_0, y_0)| + |v_x(x_0, y_0)|$$

Hence, the product limit is zero. The other two error terms go to zero also as $(x, y) \to (x_0, y_0)$. Hence, we conclude

$$\frac{\partial f}{\partial x} \quad = \quad \frac{\partial f}{\partial u} \frac{\partial u}{\partial x} + \frac{\partial f}{\partial v} \frac{\partial v}{\partial x}.$$

A similar argument shows

$$\frac{\partial f}{\partial y} \quad = \quad \frac{\partial f}{\partial u} \frac{\partial u}{\partial y} + \frac{\partial f}{\partial v} \frac{\partial v}{\partial y}.$$

This result is known as the **Chain Rule**.

Theorem 15.7.1 The Chain Rule in Two Variables

Assume there are two functions $u(x, y)$ and $v(x, y)$ defined locally about (x_0, y_0) and that there is a third function $f(u, v)$ which is defined locally around $(u_0 = u(x_0, y_0), v_0 = v(x_0, y_0))$. Further assume $f(u, v)$ is differentiable at (u_0, v_0) and $u(x, y)$ and $v(x, y)$ are differentiable at (x_0, y_0). Then f_x and f_y exist at (x_0, y_0) and are given by

$$\frac{\partial f}{\partial x} = \frac{\partial f}{\partial u}\frac{\partial u}{\partial x} + \frac{\partial f}{\partial v}\frac{\partial v}{\partial x}$$
$$\frac{\partial f}{\partial y} = \frac{\partial f}{\partial u}\frac{\partial u}{\partial y} + \frac{\partial f}{\partial v}\frac{\partial v}{\partial y}.$$

Example 15.7.1 Let $f(x, y) = x^2 + 2x + 5y^4$. Then if $x = r\cos(\theta)$ and $y = r\sin(\theta)$, using the chain rule, we find

$$\frac{\partial f}{\partial r} = \frac{\partial f}{\partial x}\frac{\partial x}{\partial r} + \frac{\partial f}{\partial y}\frac{\partial y}{\partial r}$$
$$\frac{\partial f}{\partial \theta} = \frac{\partial f}{\partial x}\frac{\partial x}{\partial \theta} + \frac{\partial f}{\partial y}\frac{\partial y}{\partial \theta}$$

This becomes

$$\frac{\partial f}{\partial r} = \left(2x + 2\right)\cos(\theta) + \left(20y^3\right)\sin(\theta)$$
$$\frac{\partial f}{\partial \theta} = \left(2x + 2\right)\left(-r\sin(\theta)\right) + \left(20y^3\right)\left(r\cos(\theta)\right)$$

You can then substitute in for x and y to get the final answer in terms of r and θ (kind of ugly though!)

Example 15.7.2 Let $f(x, y) = 10x^2y^4$. Then if $u = x^2 + 2y^2$ and $v = 4x^2 - 5y^2$, using the chain rule, we find $f(u, v) = 10u^2v^4$ and so

$$\frac{\partial f}{\partial x} = \frac{\partial f}{\partial u}\frac{\partial u}{\partial x} + \frac{\partial f}{\partial v}\frac{\partial v}{\partial x}$$
$$\frac{\partial f}{\partial y} = \frac{\partial f}{\partial u}\frac{\partial u}{\partial y} + \frac{\partial f}{\partial v}\frac{\partial v}{\partial y}$$

This becomes

$$\frac{\partial f}{\partial x} = \left(20uv^4\right)2x + \left(40u^2v^3\right)8x$$
$$\frac{\partial f}{\partial \theta} = \left(20uv^4\right)4y + \left(40u^2v^3\right)(-10y)$$

You can then substitute in for u and v to get the final answer in terms of x and y (even more ugly though!)

15.7.1 Homework

Exercise 15.7.1 Let $f(x, y) = 2x^2 + 21x + 5y^3$. Then if $x = r\cos(\theta)$ and $y = r\sin(\theta)$. Find f_r and f_θ.

Exercise 15.7.2 Let $f(x, y) = 10x^2y^4$. Then if $u = x^2 + 2y^2$ and $v = 4x^2 - 5y^2$, find f_u and f_v.

Exercise 15.7.3 *Let* $f(x,y) = \frac{2xy}{\sqrt{x^2+y^2}}$. *Prove* $\lim_{(x,y)\to(0,0)} f(x,y) = 0$. *This means* $f(x,y)$ *has a removeable discontinuity at* $(0,0)$. *Hint:* $|x| \le \sqrt{x^2+y^2}$ *and* $|y| \le \sqrt{x^2+y^2}$. *Do an* $\epsilon - \delta$ *proof here.*

Exercise 15.7.4 *Let*

$$f(x,y) = \begin{cases} \frac{2xy}{\sqrt{x^2+y^2}}, & (x,y) \ne (0,0) \\ 0, & (x,y) = (0,0) \end{cases}$$

- *Find* $f_x(0,0)$ *and* $f_y(0,0)$. *(They are both* 0 *).*

- *For all* $(x,y) \ne (0,0)$ *find* f_x *and* f_y.

- *Look at the paths* (x, mx) *for* $m \ne 0$ *and show* $\lim_{(x,y)\to(0,0)} f_x(x,y)$ *and* $\lim_{(x,y)\to(0,0)}$ $f_y(x,y)$ *do not exist. Hint: use* $x \to 0^+$ *and* $x \to 0^-$. *You'll get limits that depend on both the sign of* x *and the value of* m.

- *Explain how the result above shows the partials of* f *are not continuous at* $(0,0)$.

Exercise 15.7.5 *Let*

$$f(x,y) = \begin{cases} \frac{2xy}{\sqrt{x^2+y^2}}, & (x,y) \ne (0,0) \\ 0, & (x,y) = (0,0) \end{cases}$$

We already know the partials of f *exist at all points but that* f_x *and* f_y *are not continuous at* $(0,0)$. *If* f *was differentiable at* $(0,0)$ *there would be numbers* L_1 *and* L_2 *so that the error term* $E(x,y,0,0) = \frac{2xy}{\sqrt{x^2+y^2}} - L_1 x - L_2 y$ *would satisfy* $\lim_{(x,y)\to(0,0)} E(x,y,0,0) = 0$ *and* $\lim_{(x,y)\to(0,0)} E(x,y,0,0)/$ $\|(x,y)\| = 0$.

- *Show* $\lim_{(x,y)\to(0,0)} E(x,y,0,0) = 0$ *does exist* $\forall L_1, L_2$.

- *Show* $\lim_{(x,y)\to(0,0)} E(x,y,0,0)/\|(x,y)\| = 0$ *does not exist* $\forall L_1, L_2$ *by looking at the paths* (x, mx) *like we did in HW 34.2.*

This example shows an f *where the partials exist at all points locally around a point* (x_0, y_0) *(here that point is* $(0,0)$*) but they fail to be continuous at* (x_0, y_0) *and* f *fails to be differentiable at* (x_0, y_0).

15.8 Tangent Plane Approximation Error

Now that we have the chain rule, we can quickly develop other results such as how much error we make when we approximate our surface $f(x,y)$ using a tangent plane at a point (x_0, y_0). The first thing we need is to know when a function of two variables is differentiable. Just because it's partials exist at a point is not enough to guarantee that! But we can prove that if the partials are continuous around that point, then the derivative does exist. And that means we can write the function in terms of its tangent plane plus an error. The arguments to do this are not terribly hard, but they are a bit involved. For now, let's go back to the old idea of a tangent plane to a surface. For the surface $z = f(x,y)$ if its partials are continuous functions (they usually are for our work!) then f is differentiable and hence we know that

$$f(x,y) = f(x_0,y_0) + f_x(x_0,y_0)(x - x_0) + f_y(x_0,y_0)(y - y_0) + E(x,y,x_0,y_0)$$

where $E(x,y,x_0,y_0)/\sqrt{(x - x_0)^2 + (y - y_0)^2} \to 0$ and $E(x,y,x_0,y_0)$ go to 0 as $(x,y) \to (x_0,y_0)$. We can characterize the error much better if we have access to what are called the second order partial derivatives of f. Roughly speaking, we take the partials of f_x and f_y to obtain the second order terms. Assuming f is defined locally as usual near (x_0, y_0), we can ask about the partial derivatives of the functions f_x and f_y with respect to x and y also.

We define the second order partials of f as follows.

Definition 15.8.1 Second Order Partials

If $f(x, y)$, f_x and f_y are defined locally at (x_0, y_0), we can attempt to find following limits:

$$\lim_{x \to x_0, y=y_0} \frac{f_x(x, y_0) - f_x(x_0, y_0)}{x - x_0} = \partial_x(f_x)$$

$$\lim_{x=x_0, y \to y_0} \frac{f_x(x_0, y) - f_x(x_0, y_0)}{y - y_0} = \partial_y(f_x)$$

$$\lim_{x \to x_0, y=y_0} \frac{f_y(x, y_0) - f_y(x_0, y_0)}{x - x_0} = \partial_x(f_y)$$

$$\lim_{x=x_0, y \to y_0} \frac{f_y(x_0, y) - f_y(x_0, y_0)}{y - y_0} = \partial_y(f_y)$$

Comment 15.8.1 *When these second order partials exist at (x_0, y_0), we use the following notations interchangeably: $f_{xx} = \partial_x(f_x)$, $f_{xy} = \partial_y(f_x)$, $f_{yx} = \partial_x(f_y)$ and $f_{yy} = \partial_y(f_y)$.*

The second order partials are often organized into a matrix called the **Hessian**.

Definition 15.8.2 The Hessian Matrix

If $f(x, y)$, f_x and f_y are defined locally at (x_0, y_0) and if the second order partials exist at (x_0, y_0), we define the Hessian, $\boldsymbol{H}(x_0, y_0)$ at (x_0, y_0) to be the matrix

$$\boldsymbol{H}(x_0, y_0) = \begin{bmatrix} f_{xx}(x_0, y_0) & f_{xy}(x_0, y_0) \\ f_{yx}(x_0, y_0) & f_{yy}(x_0, y_0) \end{bmatrix}$$

Comment 15.8.2 *It is also possible to prove that if the first and second order partials are continuous locally near (x_0, y_0) then the mixed order partials f_{xy} and f_{yx} must match at the point (x_0, y_0). We will work through that argument soon and many of our surfaces have this property. Hence, for these* **smooth** *surfaces, the Hessian is a symmetric matrix!*

Example 15.8.1 *Let $f(x, y) = 2x - 8xy$. Find the first and second order partials of f and its Hessian.*

Solution *The partials are*

$$\begin{aligned} f_x(x, y) &= 2 - 8y \\ f_y(x, y) &= -8x \\ f_{xx}(x, y) &= 0 \\ f_{xy}(x, y) &= -8 \\ f_{yx}(x, y) &= -8 \\ f_{yy}(x, y) &= 0. \end{aligned}$$

and so the Hessian is

$$\mathbf{H}(x,y) \;=\; \begin{bmatrix} f_{xx}(x,y) & f_{xy}(x,y) \\ f_{yx}(x,y) & f_{yy}(x,y) \end{bmatrix} = \begin{bmatrix} 0 & -8 \\ -8 & 0 \end{bmatrix}$$

Example 15.8.2 *Let* $f(x,y) = 2x^2 - 8y^3$. *Find the first and second order partials of f and its Hessian.*

Solution *The partials are*

$$\begin{aligned} f_x(x,y) &= 4x \\ f_y(x,y) &= -24y^2 \\ f_{xx}(x,y) &= 4 \\ f_{xy}(x,y) &= 0 \\ f_{yx}(x,y) &= 0 \\ f_{yy}(x,y) &= -48y. \end{aligned}$$

and so the Hessian is

$$\mathbf{H}(x,y) \;=\; \begin{bmatrix} f_{xx}(x,y) & f_{xy}(x,y) \\ f_{yx}(x,y) & f_{yy}(x,y) \end{bmatrix} = \begin{bmatrix} 4 & 0 \\ 0 & -48y \end{bmatrix}$$

Homework

Exercise 15.8.1 *Let* $f(x,y) = 2x^2 - 8x^2y$. *Find the first and second order partials of f and its Hessian.*

Exercise 15.8.2 *Let* $f(x,y) = 2x^4 + 8xy + 5y^4$. *Find the first and second order partials of f and its Hessian.*

Exercise 15.8.3 *Let* $f(x,y) = \sin(xy)$. *Find the first and second order partials of f and its Hessian.*

Exercise 15.8.4 *Let* $f(x,y) = \sqrt{x^2 + 2y^2}$. *Find the first and second order partials of f and its Hessian.*

Exercise 15.8.5 *Let* $f(x,y) = \sqrt{3x^2 + 4xy + 2y^4}$. *Find the first and second order partials of f and its Hessian.*

15.9   Hessian Approximations

We can now explain the most common approximation result for tangent planes. Let $h(t) = f(x_0 + t\Delta x, y_0 + t\Delta y)$. Then we know we can write $h(t) = h(0) + h'(0)t + h''(c)\frac{t^2}{2}$. Using the chain rule, we find

$$h'(t) \;=\; f_x(x_0 + t\Delta x, y_0 + t\Delta y)\Delta x + f_y(x_0 + t\Delta x, y_0 + t\Delta y)\Delta y$$

and

$$\begin{aligned} &h''(t) \\ &= \partial_x \Big(f_x(x_0 + t\Delta x, y_0 + t\Delta y)\Delta x + f_y(x_0 + t\Delta x, y_0 + t\Delta y)\Delta y \Big)\Delta x \\ &+ \partial_y \Big(f_x(x_0 + t\Delta x, y_0 + t\Delta y)\Delta x + f_y(x_0 + t\Delta x, y_0 + t\Delta y)\Delta y \Big)\Delta y \end{aligned}$$

$$= f_{xx}(x_0 + t\Delta x, y_0 + t\Delta y)(\Delta x)^2 + f_{yx}(x_0 + t\Delta x, y_0 + t\Delta y)(\Delta y)(\Delta x)$$
$$+ f_{xy}(x_0 + t\Delta x, y_0 + t\Delta y)(\Delta x)(\Delta y) + f_{yy}(x_0 + t\Delta x, y_0 + t\Delta y)(\Delta y)^2$$

We can rewrite this in matrix - vector form as

$$h''(t) = \begin{bmatrix} \Delta x & \Delta y \end{bmatrix} \begin{bmatrix} f_{xx}(x_0 + t\Delta x, y_0 + t\Delta y) & f_{yx}(x_0 + t\Delta x, y_0 + t\Delta y) \\ f_{xy}(x_0 + t\Delta x, y_0 + t\Delta y) & f_{yy}(x_0 + t\Delta x, y_0 + t\Delta y) \end{bmatrix} \begin{bmatrix} \Delta x \\ \Delta y \end{bmatrix}$$

Of course, using the definition of H, this can be rewritten as

$$h''(t) = \begin{bmatrix} \Delta x \\ \Delta y \end{bmatrix}^T H(x_0 + t\Delta x, y_0 + t\Delta y) \begin{bmatrix} \Delta x \\ \Delta y \end{bmatrix}$$

Thus, our tangent plane approximation can be written as

$$h(1) \quad = \quad h(0) + h'(0)(1 - 0) + h''(c)\frac{1}{2}$$

for some c between 0 and 1. Substituting for the h terms, we find

$$f(x_0 + \Delta x, y_0 + \Delta y) \quad = \quad f(x_0, y_0) + f_x(x_0, y_0)\Delta x + f_y(x_0, y_0)\Delta y$$
$$+ \frac{1}{2} \begin{bmatrix} \Delta x \\ \Delta y \end{bmatrix}^T H(x_0 + c\Delta x, y_0 + c\Delta y) \begin{bmatrix} \Delta x \\ \Delta y \end{bmatrix}$$

Clearly, we have shown how to express the error in terms of second order partials. There is a point c between 0 and 1 so that

$$E(x_0, y_0, \Delta x, \Delta y) \quad = \quad \frac{1}{2} \begin{bmatrix} \Delta x \\ \Delta y \end{bmatrix}^T H(x_0 + c\Delta x, y_0 + c\Delta y) \begin{bmatrix} \Delta x \\ \Delta y \end{bmatrix}$$

Note the error is a quadratic expression in terms of the Δx and Δy. We also now know for a function of two variables, $f(x, y)$, we can estimate the error made in approximating using the gradient at the given point (x_0, y_0) as follows: We have

$$f(x, y) = f(x_0, y_0) + < \nabla(f)(x_0, y_0), [x - x_0, y - y_0]^T >$$
$$+ (1/2)[x - x_0, y - y_0]H(x_0 + c(x - x_0), y_0 + c(y - y_0))[x - x_0, y - y_0]^T$$

Example 15.9.1 Let $f(x, y) = x^2y^4 + 2x + 3y + 10$. *Estimate the tangent plane error for various local circles about* $(0, 0)$.

Solution *We have*

$$\begin{aligned} f_x &= 2xy^4 + 2, \quad f_y = x^2 4y^3 + 3 \\ f_{xx} &= 2y^4, \quad f_{xy} = f_{yx} = 8xy^3 \\ f_{yy} &= 12x^2y^2 \end{aligned}$$

So at $(x_0, y_0) = (0, 0)$, *letting E denote the error, we have*

$$x^2y^4 + 2x + 3y + 10 = 10 + < [2, 3]^T, [x - 0, y - 0]^T > + E$$

Now

$$f_{xx} \quad = \quad 2y^4, \quad f_{xy} = f_{yx} = 8xy^3, \quad f_{yy} = 12x^2y^2$$

So we find these estimates:

$$\max_{(x,y)\in R_r} |f_{xx}| = 2r^4, \quad \max_{(x,y)\in R_r} |f_{xy}| = 8r^4, \quad \max_{(x,y)\in R_r} |f_{yy}| = 12r^4.$$

Now the error is

$$\frac{1}{2}\begin{bmatrix}\Delta x \\ \Delta y\end{bmatrix}^T H(x_0 + c\Delta x, y_0 + c\Delta y)\begin{bmatrix}\Delta x \\ \Delta y\end{bmatrix}$$
$$= (1/2)f_{xx}(x^*, y^*)(\Delta x)^2 + f_{xy}(x^*, y^*)(\Delta x)(\Delta y) + (1/2)f_{yy}(x^*, y^*)(\Delta y)^2.$$

where we are denoting the intermediate value $(x_0 + c(x - x_0), y_0 + c(y - y_0))$ by (x^*, y^*). For a circle $B_r(0,0)$, we know $\sqrt{(\Delta x)^2 + (\Delta y)^2} < r$ which tells us $|\Delta x| < r$ and $|\Delta y| < r$ too.

So the biggest the error can be in absolute value is $\frac{1}{2}(2r^4 \times r^2 + 2 \times 8r^4 \times r^2 + 12r^4 \times r^2) = 15r^6$ and the largest error for the approximation is

$$|(x^2y^4 + 2x + 3y + 10) - 10 - < [2,3]^T, [x,y]^T > | \leq 15r^6.$$

So for $r = .8$, the maximum error is overestimated by $15(.8)^6 = 3.93$ - probably bad! For $r = .4$, the maximum error is $15(.4)^6 = .06$ – better! To make the largest error $< 10^{-4}$, solve $15r^6 < 10^{-4}$. This gives $r^6 < 6.67x10^{-6}$. Thus, $r < .137$ will do the job.

Example 15.9.2 *We use the same function $f(x,y) = x^2y^4 + 2x + 3y + 10$. But now estimate the tangent plane error for various local circles about $(0.2, 0.8)$.*

Solution *As before, we have*

$$f_x = 2xy^4 + 2, \quad f_y = x^2 4y^3 + 3$$
$$f_{xx} = 2y^4, \quad f_{xy} = f_{yx} = 8xy^3$$
$$f_{yy} = 12x^2y^2$$

So at $(x_0, y_0) = (0.2, 0.8)$, letting E denote the error, we have

$$x^2y^4 + 2x + 3y + 10 = 12.82 + 2.16(x - 0.2) + 3.02(y - 0.8) + E$$

Now the in between point $(0.2 + c(x - 0.2), 0.8 + c(y - 0.8)) = (x^*, y^*)$ is on the line between (x, y) and $(0.2, 0.8)$, and at the point, we have

$$f_{xx}^* = 2(y^*)^4, \quad f_{xy}^* = f_{yx}^* = 8x^*(y^*)^3, \quad f_{yy}^* = 12(x^*)^2(y^*)^2$$

In the ball of radius r about $(0.2, 0.8)$, we have $|x^* - 0.2| < r$, $|y^* - 0.8| < r$ and so $|x^*| < 0.2 + r$ and $|y^*| < 0.8 + r$.

Further, we have $|x - 0.2| < r$ and $|y - 0.8| < r$. Now the error term is

$$\frac{1}{2}\begin{bmatrix}(x - 0.2) \\ (y - 0.8)\end{bmatrix}^T H(x^*, y^*)\begin{bmatrix}(x - 0.2) \\ (y - 0.8)\end{bmatrix}$$
$$= (1/2)f_{xx}^*(x - 0.2)^2 + f_{xy}^*(x - 0.2)(y - 0.8) + (1/2)f_{yy}^*(y - 0.8)^2.$$

On $B_r(0.2, 0.8)$, the absolute error is thus bounded by

$$|E(x, y, 0.2, 0.8)| \leq (1/2)|f_{xx}^*| |x - 0.2|^2 + |f_{xy}^*||x - 0.2| |y - 0.8|$$

$$+(1/2)|f_{yy}^*| \, |y - 0.8|^2$$

$$\leq \left((1/2)2|y^*|^4 + 8|x^*| \, |y^*|^3 + (1/2)12|x^*|^2 \, |y^*|^2 \right) r^2$$

$$\leq \left((1/2)2|0.8 + r|^4| + 8|0.2 + r| \, |0.8 + r|^3 \right.$$

$$\left. +(1/2)12|0.2 + r|^2 \, |0.8 + r|^2 \right) r^2$$

Now let's restrict attention to $r < 1$. Then we have

$$|E(x, y, 0.2, 0.8)| \leq \left((1/2)2|1.8|^4 + 8|1.2| \, |1.8|^3 + (1/2)12|1.2|^2 \, |1.8|^2 \right) r^2$$

$$= 94.49 r^2$$

The biggest the error can be in absolute value in $B_1(0.2, 0.8)$ is $94.49 \, r^2$. So for $r = .01$, the maximum error is overestimated 0.009456 which is not bad. For $r = .001$, the maximum error is 9.5×10^{-5} which is better!

15.9.1   Homework

Exercise 15.9.1 *Approximate $f(x, y) = x^2 + x^4 y^3$ near $(2, 1)$. Find the r where the error is less than 10^{-3}.*

Exercise 15.9.2 *Approximate $f(x, y) = x^2 + y^4 x^5 + 3x + 4y + 25$ near $(0, 0)$. Find the r where the error is less than 10^{-3}.*

Exercise 15.9.3 *Approximate $f(x, y) = 4x^4 y^4 + 3x + 40y + 5$ near $(1, 2)$. Find the r where the error is less than 10^{-6}.*

Exercise 15.9.4 *Approximate $f(x, y) = 3x^2 + y^2 x^5 + 3x + 4y^2 + 5$ near $(0, 0)$. Find the r where the error is less than 10^{-3}.*

Exercise 15.9.5 *Approximate $f(x, y) = 4x^2 y^2 + 5xy + 2x - 3y - 2$ near $(1, 1)$. Find the r where the error is less than 10^{-6}.*

15.10   Partials Existing Does Not Necessarily Imply Differentiable

Let's look at the problem you were asked to do in an earlier homework. We start with the function f defined by

$$f(x, y) = \begin{cases} \dfrac{2xy}{\sqrt{x^2 + y^2}}, & (x, y) \neq (0, 0) \\ 0, & (x, y) = (0, 0) \end{cases}$$

This function has a removeable discontinuity at $(0, 0)$ because

$$\left| \frac{2xy}{\sqrt{x^2 + y^2}} - 0 \right| = 2 \frac{|x| \, |y|}{\sqrt{x^2 + y^2}} \leq 2 \frac{\sqrt{x^2 + y^2} \sqrt{x^2 + y^2}}{\sqrt{x^2 + y^2}}$$

$$= 2\sqrt{x^2 + y^2}$$

because $|x| \leq \sqrt{x^2 + y^2}$ and $|y| \leq \sqrt{x^2 + y^2}$. Pick $\epsilon > 0$ arbitrarily. Then if $\delta < \epsilon/2$, we have

$$\left| \frac{2xy}{\sqrt{x^2 + y^2}} - 0 \right| < 2\frac{\epsilon}{2} = \epsilon$$

which proves that the limit exists and equals 0 as $(x, y) \to (0, 0)$. This tells us we can define $f(0, 0)$ to match this value. Thus, as said, this is a *removeable discontinuity*.

The first order partials both exist at $(0, 0)$ as is seen by an easy limit.

$$f_x(0, 0) = \lim_{x \to 0} \frac{f(x, 0) - f(0, 0)}{x} = \lim_{x \to 0} \frac{0 - 0}{x} = 0.$$
$$f_y(0, 0) = \lim_{y \to 0} \frac{f(0, y) - f(0, 0)}{y} = \lim_{y \to 0} \frac{0 - 0}{y} = 0.$$

We can also calculate the first order partials at any $(x, y) \neq (0, 0)$.

$$f_x(x, y) = \frac{2y(x^2 + y^2)^{1/2} - 2xy\,(1/2)(x^2 + y^2)^{-1/2}\,2x}{(x^2 + y^2)^1}$$
$$= \frac{2y(x^2 + y^2) - 2x^2 y}{(x^2 + y^2)^{3/2}} = \frac{2y^3}{(x^2 + y^2)^{3/2}}$$
$$f_y(x, y) = \frac{2x(x^2 + y^2)^{1/2} - 2xy\,(1/2)(x^2 + y^2)^{-1/2}\,2y}{(x^2 + y^2)^1}$$
$$= \frac{2x(x^2 + y^2) - 2xy^2}{(x^2 + y^2)^{3/2}} = \frac{2x^3}{(x^2 + y^2)^{3/2}}$$

We can show $\lim (x, y) \to (0, 0)\ f_x(x, y)$ and $\lim (x, y) \to (0, 0)\ f_y(x, y)$ do not exist as follows. We will find paths (x, mx) for $m \neq 0$ where we get different values for the limit depending on the value of m.

Because of the square roots here, these limits need to be handled separately for $x \to 0^+$ and $x \to 0^-$. We will do the 0^+ case and let you do the other one. We have on the path (x, mx), $|x| = x$ and so

$$\lim_{x \to 0^+} f_x(x, y) = \lim_{x \to 0^+} \frac{2m^3 x^3}{(1 + m^2)^{3/2}|x^2|^{3/2}} = \lim_{x \to 0^+} \frac{2m^3}{(1 + m^2)^{3/2}}$$

Then, for example, using $m = 2$ and $m = 3$, we get the limits $\frac{16}{5\sqrt{5}}$ and $\frac{54}{10\sqrt{10}}$ which are not the same. Hence, this limit does not exist.

A similar argument shows $\lim (x, y) \to (0, 0)\ f_y(x, y)$ does not exist either. So we have shown the first order partials exist locally around $(0, 0)$ but f_x and f_y are not continuous at $(0, 0)$.

Is f differentiable at $(0, 0)$? If so, there a numbers L_1 and L_2 so that two things happen:

$$\lim_{(x,y) \to (0,0)} \left(\frac{2xy}{\sqrt{x^2 + y^2}} - L_1 x - L_2 y \right) = 0$$
$$\lim_{(x,y) \to (0,0)} \frac{1}{\sqrt{x^2 + y^2}} \left(\frac{2xy}{\sqrt{x^2 + y^2}} - L_1 x - L_2 y \right) = 0$$

The first limit is 0 because

$$\lim_{(x,y)\to(0,0)} \left| \frac{2xy}{\sqrt{x^2+y^2}} - L_1 x - L_2 y - 0 \right| \leq$$

$$2\frac{|x||y|}{\sqrt{x^2+y^2}} + |L_1|\sqrt{x^2+y^2} + |L_2|\sqrt{x^2+y^2}$$

$$\leq 2\frac{\sqrt{x^2+y^2}\sqrt{x^2+y^2}}{\sqrt{x^2+y^2}} + |L_1|\sqrt{x^2+y^2} + |L_2|\sqrt{x^2+y^2}$$

$$\leq (2 + |L_1| + |L_2|)\sqrt{x^2+y^2}$$

which goes to 0 as $(x,y) \to 0$. So that shows the first requirement is met. However, the second requirement is not. Look at the paths (x, mx) and consider $\lim_{x\to 0+}$ like we did before. We find, since $|x| = x$ here

$$\lim_{x\to 0+} \frac{1}{x\sqrt{1+m^2}} \left(\frac{2mx^2}{x\sqrt{1+m^2}} - L_1 x - mL_2 x \right) =$$

$$\frac{1}{\sqrt{1+m^2}} \left(\frac{2mx}{x\sqrt{1+m^2}} - L_1 - L_2 m \right) = \frac{2m}{1+m^2} - \frac{L_1 + L_2 m}{\sqrt{1+m^2}}$$

Now look at path $(x, -mx)$ for $x > 0$. This is the limit from the other side. We have

$$\lim_{x\to 0+} \frac{1}{x\sqrt{1+m^2}} \left(\frac{-2mx^2}{x\sqrt{1+m^2}} - L_1 x + mL_2 x \right) =$$

$$\frac{1}{\sqrt{1+m^2}} \left(\frac{-2m}{\sqrt{1+m^2}} - L_1 + L_2 m \right) = \frac{-2m}{1+m^2} + \frac{-L_1 + L_2 m}{\sqrt{1+m^2}}$$

If these limits were equal we would have

$$\frac{2m}{1+m^2} + \frac{-L_1 - L_2 m}{\sqrt{1+m^2}} = \frac{-2m}{1+m^2} + \frac{-L_1 + L_2 m}{\sqrt{1+m^2}}$$

This implies $L_2 = 2/\sqrt{1+m^2}$ which tells us L_2 depends on m. This values should be independent of the choice of path and so this function cannot be differentiable at $(0,0)$.

We see that just because the first order partials exists locally in some $B_r(0,0)$ does not necessarily tell us f is differentiable at $(0,0)$. So **differentiable** at (x_0, y_0) implies the first order partials exist and (x_0, y_0) and $L_1 = f_x(x_0, y_0)$ and $L_2 = f_y(x_0, y_0)$. **But the converse is not true in general**.

We need to figure out the appropriate assumptions about the smoothness of f that will guarantee the existence of the derivative of f in this multivariable situations. We will do this for n dimensions as it is just as easy as doing the proof in $\Re^2$.

Theorem 15.10.1 Sufficient Conditions on the First Order Partials to Guarantee Differentiability

Let $f : D \subset \Re^2 \to \Re$ and assume $\nabla(f)$ exists and is continuous in a ball $B_r(\boldsymbol{p})$ of some radius $r > 0$ around the point $\boldsymbol{p} = (p_1, p_2)$ in $\Re^2$. Then f is differentiable at each point in $B_r(\boldsymbol{p})$.

Proof 15.10.1

By definition of $\frac{\partial f}{\partial x_i}$ at $\boldsymbol{p}$,

$$f(p_1 + h, p_2) - f(p_1, p_2) = \frac{\partial f}{\partial x_1}(\boldsymbol{p})h + E_1(p_1, p_2, h)$$

$$f(p_1, p_2 + h) - f(p_1, p_2) = \frac{\partial f}{\partial x_2}(\boldsymbol{p})h + E_2(p_1, p_2, h)$$

Now if $\boldsymbol{x}$ is in $B_r(\boldsymbol{p})$, all the first order partials of f are continuous. So

$$\boldsymbol{x} = (p_1, p_2) + (x_1 - p_1) + (x_2 - p_2)$$

Now let $r_i = x_i - p_i$. Note, $|r_i| < r$. Then we can write (this is the telegraphing sum trick!)

$$f(\boldsymbol{x}) - f(\boldsymbol{p}) = (f(p_1 + r_1, p_2 + r_2) - f(p_1 + r_1, p_2)) + (f(p_1 + r_1, p_2) - f(p_1, p_2))$$

Now apply the Mean Value Theorem in one variable to get

$$f(p_1 + r_1, p_2 + r_2) - f(p_1 + r_1, p_2) = \frac{\partial f}{\partial x_2}(p_1 + r_1, p_2 + s_2)\, r_2$$

$$f(p_1 + r_1, p_2) - f(p_1, p_2) = \frac{\partial f}{\partial x_1}(p_1 + s_1, p_2)\, r_1$$

where s_j is between p_j and $p_j + r_j$. Hence, $|s_j| < |r_j| < r$. So

$$f(\boldsymbol{x}) - f(\boldsymbol{p}) = \frac{\partial f}{\partial x_2}(p_1 + r_1, p_2 + s_2)\, r_2 + \frac{\partial f}{\partial x_1}(p_1 + s_1, p_2)\, r_1$$

Since $\frac{\partial f}{\partial x_j}$ is continuous on $B_r(\boldsymbol{p})$, given $\epsilon > 0$ arbitrary, there are $\delta_j > 0$ so that for $1 \leq j \leq 2$, $|\frac{\partial f}{\partial x_j}(\boldsymbol{x}) - \frac{\partial f}{\partial x_j}(\boldsymbol{p})| < \epsilon/2$. Thus, if $\|\boldsymbol{x} - \boldsymbol{p}\| < \hat{\delta} = \min\{\delta_1, \delta_2\}$, the points $(p_1 + r_1, p_2 + s_2)$ and $(p_1 + s_1, p_2)$ are in $B_r(\boldsymbol{p})$. Call these two pairs, $\boldsymbol{q}_1$ and $\boldsymbol{q}_2$ for convenience. So we have $|\frac{\partial f}{\partial x_j}(\boldsymbol{q}_j) - \frac{\partial f}{\partial x_j}(\boldsymbol{p})| < \epsilon/2$ Thus, we find

$$\frac{\partial f}{\partial x_j}(\boldsymbol{p}) - \epsilon/2 < \frac{\partial f}{\partial x_j}(\boldsymbol{q}_j) < \frac{\partial f}{\partial x_j}(\boldsymbol{p}) + \epsilon/2$$

Now, if r_1 and r_2 are positive, we find

$$\frac{\partial f}{\partial x_1}(\boldsymbol{p})r_1 + \frac{\partial f}{\partial x_2}(\boldsymbol{p})r_2 - \epsilon/2(r_1 + r_2) < \frac{\partial f}{\partial x_1}(\boldsymbol{q}_1)r_1 + \frac{\partial f}{\partial x_2}((\boldsymbol{q}_2)r_2$$

$$< \frac{\partial f}{\partial x_1}(\boldsymbol{p}) + \frac{\partial f}{\partial x_2}(\boldsymbol{p}) + \epsilon/2(r_1 + r_2)$$

If both are negative, we get

$$\frac{\partial f}{\partial x_1}(\boldsymbol{p})r_1 + \frac{\partial f}{\partial x_2}(\boldsymbol{p})r_2 + \epsilon/2(r_1 + r_2) < \frac{\partial f}{\partial x_1}(\boldsymbol{q}_1)r_1 + \frac{\partial f}{\partial x_2}(\boldsymbol{q}_2)r_2$$

$$< \frac{\partial f}{\partial x_1}(\boldsymbol{p}) + \frac{\partial f}{\partial x_2}(\boldsymbol{p}) - \epsilon/2(r_1 + r_2)$$

If $r_1 < 0$ and $r_2 > 0$, we find

$$\frac{\partial f}{\partial x_1}(\boldsymbol{p})r_1 + \frac{\partial f}{\partial x_2}(\boldsymbol{p})r_2 - \epsilon/2(-r_1 + r_2) < \frac{\partial f}{\partial x_1}(\boldsymbol{q}_1)r_1 + \frac{\partial f}{\partial x_2}(\boldsymbol{q}_2)r_2$$

$$< \frac{\partial f}{\partial x_1}(\boldsymbol{p}) + \frac{\partial f}{\partial x_2}(\boldsymbol{p}) + \epsilon/2(-r_1 + r_2)$$

and finally, if $r_1 > 0$ and $r_2 < 0$, we have

$$\frac{\partial f}{\partial x_1}(\boldsymbol{p})r_1 + \frac{\partial f}{\partial x_2}(\boldsymbol{p})r_2 - \epsilon/2(r_1 - r_2) \quad < \quad \frac{\partial f}{\partial x_1}(\boldsymbol{q}_1)r_1 + \frac{\partial f}{\partial x_2}(\boldsymbol{q}_2)r_2$$

$$< \frac{\partial f}{\partial x_1}(\boldsymbol{p}) + \frac{\partial f}{\partial x_2}(\boldsymbol{p}) + \epsilon/2(r_1 - r_2)$$

If we replace $-r_1$ by $sign(r_1)$ and so forth, all of these inequality chains can be combined into

$$\frac{\partial f}{\partial x_1}(\boldsymbol{p})sign(r_1)r_1 + \frac{\partial f}{\partial x_2}(\boldsymbol{p})sign(r)2)r_2 - \epsilon/2(sign(r_1)r_1 + sign(r_2)r_2) <$$
$$\frac{\partial f}{\partial x_1}(\boldsymbol{q}_1)sign(r_1)r_1 + \frac{\partial f}{\partial x_2}(\boldsymbol{q}_2)sign(r_2)r_2$$
$$< \frac{\partial f}{\partial x_1}(\boldsymbol{p}) + \frac{\partial f}{\partial x_2}(\boldsymbol{p}) + \epsilon/2(sign(r_1)r_1 + sign(r_2)r_2)$$

Recall,

$$f(\boldsymbol{x}) - f(\boldsymbol{p}) \quad = \quad \frac{\partial f}{\partial x_2}(\boldsymbol{q}_2)\, r_2 + \frac{\partial f}{\partial x_1}(\boldsymbol{q}_1)\, r_1$$

So

$$\left| (f(\boldsymbol{x}) - f(\boldsymbol{p})) - \left(\frac{\partial f}{\partial x_1}(\boldsymbol{p})sign(r_1)r_1 + \frac{\partial f}{\partial x_2}(\boldsymbol{p})sign(r)2)r_2 \right) \right| < \epsilon/2\,(|r_1| + |r_2|)$$

But $(\epsilon/2)\,(|r_1| + |r_2|) \le (\epsilon/2)\,2\|\boldsymbol{x} - \boldsymbol{p}\| = \epsilon\|\boldsymbol{x} - \boldsymbol{p}\|$. We therefore have the estimate

$$\left| (f(\boldsymbol{x}) - f(\boldsymbol{p})) - \frac{\partial f}{\partial x_1}(\boldsymbol{p})r_1 - \frac{\partial f}{\partial x_2}(\boldsymbol{p})r_2) \right| < \epsilon\|\boldsymbol{x} - \boldsymbol{p}\|$$

The term inside the absolute values is the error function $E(\boldsymbol{x}, \boldsymbol{p})$ and so we have $E(\boldsymbol{x}, \boldsymbol{p})/\|\boldsymbol{x} - \boldsymbol{p}\| < \epsilon$ if $\|\boldsymbol{x} - \boldsymbol{p}\| < \hat{\delta}$. This tells us both $E(\boldsymbol{x}, \boldsymbol{p})/\|\boldsymbol{x} - \boldsymbol{p}\|$ and $E(\boldsymbol{x}, \boldsymbol{p})$ go to 0 as $\boldsymbol{x} \to \boldsymbol{p}$. Hence, f is differentiable at $\boldsymbol{p}$. ∎

15.11   When Do Mixed Partials Match?

Next, we want to know under what conditions the mixed order partials match.

Theorem 15.11.1 Sufficient Conditions for the Mixed Order Partials to Match

Let $F : D \subset \Re^2 \to \Re$. Assume the first order partials $\frac{\partial F}{\partial X_1}$ and $\frac{\partial F}{\partial X_2}$ and the second order partials $\partial_{X_2}\left(\frac{\partial F}{\partial X_1} \right)$ and $\partial_{X_1}\left(\frac{\partial F}{\partial X_2} \right)$ all exist in $B_r(\boldsymbol{p})$ for some $r > 0$. Also assume $\partial_{X_2}\left(\frac{\partial F}{\partial X_1} \right)$ and $\partial_{X_1}\left(\frac{\partial F}{\partial X_2} \right)$ are continuous in $B_r(\boldsymbol{p})$. Then $\partial_{X_2}\left(\frac{\partial F}{\partial X_1}(\boldsymbol{p}) \right) = \partial_{X_1}\left(\frac{\partial F}{\partial X_2} \right)$ at $\boldsymbol{p}$.

Proof 15.11.1
Step 1*:*

Let $F_1 = \partial_{X_1} F$, $F_2 = \partial_{X_2} F$ and $F_{12} = \partial_{X_2}(\partial_{X_1} f)$ and $F_{21} = \partial_{X_1}(\partial_{X_2} F)$. Fix h small enough for that $(p_1 + h, p_2 + h)$ is in $B_r(\mathbf{p})$. Define

$$
\begin{aligned}
G(h) &= F(X_1 + h, X_2 + h) - F(X_1 + h, X_2) - F(X_1, X_2 + h) + F(X_1, X_2)\\
\phi(x) &= F(x, X_2 + h) - F(x, X_2), \quad \psi(y) = \phi(X_1 + y) - \phi(X_1)
\end{aligned}
$$

Now apply the Mean Value Theorem to ϕ. Note $\phi'(x) = \partial_{X_1} F(x, X_2 + h) - \partial_{X_1} F(x, X_2)$. Thus,

$$
\begin{aligned}
\phi(X_1 + h) - \phi(X_1) &= \phi'(X_1 + \theta_1 h)\, h, \quad \theta_1 \text{ between } X_1 \text{ and } X_1 + h\\
&= (F_{X_1}(X_1 + \theta_1 h, X_2 + h) - F_{X_1}(X_1 + \theta_1 h, X_2))h
\end{aligned}
$$

Now define for y between X_2 and $X_2 + h$, the function ξ by

$$
\begin{aligned}
\xi(z) &= F_{X_1}(X_1 + \theta_1 h, z)\\
\xi'(z) &= \partial_{X_2} F_{X_1}(X_1 + \theta_1 h, z)
\end{aligned}
$$

Apply the Mean Value Theorem to ξ: there is a θ_2 so that $X_2 + \theta_2 h$ is between X_2 and $X_2 + h$ with

$$
\xi(X_2 + h) - \xi(X_2) = \partial_{X_2} F_{X_1}(X_1 + \theta_1 h, X_2 + \theta_2 h)\, h
$$

Thus,

$$
\begin{aligned}
h\,(\xi(X_2 + h) - \xi(X_2)) &= h\,(F_{X_1}(X_1 + \theta_1 h, X_2 + h) - F_{X_1}(X_1 + \theta_1 h, X_2))\\
&= \partial_{X_2} F_{X_1}(X_1 + \theta_1 h, X_2 + \theta_2 h)\, h^2
\end{aligned}
$$

But,

$$
\begin{aligned}
\phi(X_1 + h) - \phi(X_1) &= (F(X_1 + h, X_2 + h) - F(X_1 + h, X_2))\\
&\quad -(F(X_1, X_2 + h) - F(X_1, X_2))\\
&= G(h)\\
&= (F_{X_1}(X_1 + \theta_1 h, X_2 + h) - F_{X_1}(X_1 + \theta_1 h, X_2))h\\
&= (\xi(X_2 + h) - \xi(X_2))h\\
&= \partial_{X_2} F_{X_1}(X_1 + \theta_1 h, X_2 + \theta_2 h)\, h^2
\end{aligned}
$$

So

$$
G(h) = \partial_{X_2} F_{X_1}(X_1 + \theta_1 h, X_2 + \theta_2 h)\, h^2
$$

Step 2*:*
Now start again with $L(w) = F(X_1 + h, w) - F(X_1, w)$, noting $G(h) = L(X_2 + h) - L(X_2)$. Apply the Mean Value Theorem to L like before to find

$$
\begin{aligned}
G(h) &= L(X_2 + h) - L(X_2) = L'(X_2 + \theta_3 h)h\\
&= (\partial_{X_2} F(X_1 + h, X_2 + \theta_3 h) - \partial_{X_2} F(X_1, X_2 + \theta_3 h))\, h
\end{aligned}
$$

for an intermediate value of θ_3 with $X_2 + \theta_3 h$ between X_2 and $X_2 + h$.

Now define for z between X_2 and $X_2 + h$, the function γ by

$$
\gamma(z) = F_{X_2}(z, X_2 + \theta_3 h)
$$

$$\gamma'(z) \;=\; \partial_{X_1} F_{X_2}(z, X_2 + \theta_3 h)$$

Now apply the Mean Value Theorem a fourth time: So there is a θ_4 with $X_1 + \theta_4 h$ between X_1 and $X_1 + h$ so that

$$
\begin{aligned}
\gamma(X_1 + h) - \gamma(X_1) &\;=\; \gamma'(X_1 + \theta_4 h)\, h \\
&\;=\; \partial_{X_1} F_{X_2}(X_1 + \theta_4 h, X_2 + \theta_3 h)\, h
\end{aligned}
$$

Using what we have found so far, we have

$$
\begin{aligned}
G(h) &\;=\; (\partial_{X_2} F(X_1 + h, X_2 + \theta_3 h) - \partial_{X_2} F(X_1, X_2 + \theta_3 h))\, h \\
&\;=\; (\gamma(X_1 + h) - \gamma(X_1))\, h = (\partial_{X_1} F_{X_2}(X_1 + \theta_4 h, X_2 + \theta_3 h))\, h^2
\end{aligned}
$$

So we have two representations for $G(h)$:

$$G(h) = \partial_{X_2} F_{X_1}(X_1 + \theta_1 h, X_2 + \theta_2 h)\, h^2 = (\partial_{X_1} F_{X_2}(X_1 + \theta_4 h, X_2 + \theta_3 h))\, h^2$$

Canceling the common h^2, we have

$$\partial_{X_2} F_{X_1}(X_1 + \theta_1 h, X_2 + \theta_2 h) \;=\; (\partial_{X_1} F_{X_2}(X_1 + \theta_4 h, X_2 + \theta_3 h))$$

Now let $h \to 0$ and since these second order partials are continuous at $\mathbf{p}$, we have $\partial_{X_2} F_{X_1}(X_1, X_2) = \partial_{X_1} F_{X_2}(X_1, X_2)$. This shows the result we wanted. ∎

Comment 15.11.1 *This theorem, while not overly difficult, is indeed messy. And you have to really pay attention to the notational details to get the arguments right.*

15.11.1   Homework

Let

$$
f(x,y) \;=\;
\begin{cases}
\dfrac{xy(x^2 - y^2)}{x^2 + y^2}, & (x,y) \neq (0,0) \\
0, & (x,y) = (0,0)
\end{cases}
$$

Exercise 15.11.1 *Prove $\lim_{(x,y) \to (0,0)} f(x,y) = 0$ thereby showing f has a removeable discontinuity at $(0,0)$ and so the definition for f above makes sense.*

Exercise 15.11.2 *Compute $f_x(0,0)$ and $f_y(0,0)$.*

Exercise 15.11.3 *Compute $f_x(x,y)$ and $f_y(x,y)$ for $(x,y) \neq (0,0)$. Simplify your answers to the best possible form.*

Exercise 15.11.4 *Compute $\partial_y(f_x)(0,0)$ and $\partial_x(f_y)(0,0)$. Are these equal?*

Exercise 15.11.5 *Compute $\partial_y(f_x)(x,y)$ and $\partial_x(f_y)(x,y)$ for $(x,y) \neq (0,0)$.*

Exercise 15.11.6 *Determine if $\partial_y(f_x)(x,y)$ and $\partial_x(f_y)(x,y)$ are continuous at $(0,0)$.*

Chapter 16

Multivariable Extremal Theory

We now explore how to find the extreme values of a function of two variables.

16.1 Extrema Ideas

To understand how to think about finding places where the minimum and maximum of a function to two variables might occur, all you have to do is realize it is a common sense thing. We already know that the tangent plane attached to the surface which represents our function of two variables is a way to approximate the function near the point of attachment. We have seen in our pictures what happens when the tangent plane is **flat**. This flatness occurs at the minimum and maximum of the function. It also occurs in other situations, but we will leave that more complicated event for other courses. The functions we want to deal with are quite nice and have great minima and maxima. However, we do want you to know there are more things in the world and we will touch on them only briefly. To see what to do, just recall the equation of the tangent plane error to our function of two variables $f(x, y)$.

$$f(x, y) = f(x_0, y_0) + \nabla(f)(x_0, y_0)[x - x_0, y - y_0]^T$$
$$+ (1/2)[x - x_0, y - y_0]H(x_0 + c(x - x_0), y_0 + c(y - y_0))[x - x_0, y - y_0]^T$$

where c is some number between 0 and 1 that is different for each x. We also know that the equation of the tangent plane to $f(x, y)$ at the point (x_0, y_0) is

$$z = f(x_0, y_0) + < \nabla f(x_0, y_0), X - X_0 >.$$

where $X - X_0 = [x - x_0.y - y_0]^T$. Now let's assume the tangent plane is flat at (x_0, y_0). Then the gradient $\nabla f(x_0, y_0)$ is the zero vector and we have $\frac{\partial f}{\partial x}(x_0, y_0) = 0$ and $\frac{\partial f}{\partial y}(x_0, y_0) = 0$. So the tangent plane error equation simplifies to

$$f(x, y) = f(x_0, y_0)$$
$$+ (1/2)[x - x_0, y - y_0]H(x_0 + c(x - x_0), y_0 + c(y - y_0))[x - x_0, y - y_0]^T$$

Now let's simplify this. The Hessian is just a 2×2 matrix whose components are the second order partials of f. Let

$$A(c) = \frac{\partial^2 f}{\partial x^2}(x_0 + c(x - x_0), y_0 + c(y - y_0))$$

$$B(c) = \frac{\partial^2 f}{\partial x \, \partial y}(x_0 + c(x - x_0), y_0 + c(y - y_0))$$

$$= \frac{\partial^2 f}{\partial y \, \partial x}(x_0 + c(x - x_0), y_0 + c(y - y_0))$$

$$D(c) = \frac{\partial^2 f}{\partial y^2}(x_0 + c(x - x_0), y_0 + c(y - y_0))$$

Then, we have

$$f(x,y) = f(x_0, y_0) + (1/2) \begin{bmatrix} x - x_0 & y - y_0 \end{bmatrix} \begin{bmatrix} A(c) & B(c) \\ B(c) & D(c) \end{bmatrix} \begin{bmatrix} x - x_0 \\ y - y_0 \end{bmatrix}$$

We can multiply this out (a nice simple pencil and paper exercise!) to find

$$f(x,y) = f(x_0, y_0)$$
$$+1/2 \left(A(c)(x - x_0)^2 + 2B(c)(x - x_0)(y - y_0) + D(c)(y - y_0)^2 \right)$$

Now it is time to remember an old technique from high school – completing the square. Remember if we had a quadratic like $u^2 + 3uv + 6v^2$, to complete the square we take half of the number in front of the mixed term uv and square it and add and subtract it times v^2 as follows.

$$u^2 + 3uv + 6v^2 = u^2 + 3uv + (3/2)^2 v^2 - (3/2)^2 v^2 + 6v^2.$$

Now group the first three terms together and combine the last two terms into one term.

$$u^2 + 3uv + 6v^2 = \left(u^2 + 3uv + (3/2)^2 v^2 \right) + \left(6 - (3/2)^2 \right) v^2.$$

The first three terms are a *perfect square*, $(u + (3/2)v)^2$. Simplifying, we find

$$u^2 + 3uv + 6v^2 = \left(u + (3/2)v \right)^2 + (135/4) \, v^2.$$

This is called *completing the square*. Now let's do this with the Hessian quadratic we have. First, factor out the $A(c)$. We will assume it is not zero so the divisions are fine to do. Also, for convenience, we will replace $x - x_0$ by Δx and $y - y_0$ by Δy. This gives

$$f(x,y) = f(x_0, y_0)$$
$$+\frac{A(c)}{2} \left((\Delta x)^2 + 2\frac{B(c)}{A(c)}\Delta x \, \Delta y + \frac{D(c)}{A(c)}(\Delta y)^2 \right).$$

One half of the $\Delta x \Delta y$ coefficient is $\frac{B(c)}{A(c)}$ so add and subtract $(B(c)/A(c))^2(\Delta y)^2$. We find

$$f(x,y) = f(x_0, y_0)$$
$$+\frac{A(c)}{2} \times \left((\Delta x)^2 + 2\frac{B(c)}{A(c)}\Delta x \Delta y + \left(\frac{B(c)}{A(c)} \right)^2 (\Delta y)^2 \right)$$
$$+\frac{A(c)}{2} \times \left(-\left(\frac{B(c)}{A(c)} \right)^2 (\Delta y)^2 + \frac{D(c)}{A(c)}(\Delta y)^2 \right).$$

Now group the first three terms together – the perfect square and combine the last two terms into one. We have

$$f(x,y) = f(x_0, y_0)$$

$$+\frac{A(c)}{2}\left(\left(\Delta x + \frac{B(c)}{A(c)}\Delta y\right)^2 + \left(\frac{A(c)\,D(c) - (B(c))^2}{(A(c))^2}\right)(\Delta y)^2\right).$$

Now we know if a function g is continuous at a point (x_0, y_0) and positive or negative, then it is positive or negative in a circle of radius r centered at (x_0, y_0). Here is the formal statement.

Theorem 16.1.1 Nonzero Values and Continuity

> *If $f(x_0, y_0)$ is a place where the function is positive or negative in value, then there is a radius r so that $f(x, y)$ is positive or negative in a circle of radius r around the center (x_0, y_0).*

Proof 16.1.1
We argue for the case $g(x_0, y_0) > 0$. Let $\epsilon = g(x_0, y_0)/2$. Since g is continuous at (x_0, y_0), there is a $\delta > 0$ so that $\sqrt{(x - x_0)^2 + (y - y_0)^2} < \delta$ implies $-g(x_0, y_0)/2 < g(x, y) - g(x_0, y_0) < g(x_0, y_0)/2$. Thus, $g(x, y) > g(x_0, y_0)/2$ if $(x, y) \in B_\delta(x_0, y_0)$. ■

Now getting back to our problem. We don't want to look at functions that are locally constant as they are not very interesting. So we are assuming $f(x, y)$, in addition to being differentiable, is not locally constant. We have at this point where the partials are zero, the following expansion

$$f(x, y) = f(x_0, y_0)$$
$$+\frac{A(c)}{2}\left(\left(\Delta x + \frac{B(c)}{A(c)}\Delta y\right)^2 + \left(\frac{A(c)\,D(c) - (B(c))^2}{(A(c))^2}\right)(\Delta y)^2\right).$$

The algebraic sign of the terms after the function value $f(x_0, y_0)$ are completely determined by the terms which are not squared. We have two simple cases:

- $A(c) > 0$ and $A(c)\,D(c) - (B(c))^2 > 0$ which implies the term after $f(x_0, y_0)$ is positive.

- $A(c) < 0$ and $A(c)\,D(c) - (B(c))^2 > 0$ which implies the term after $f(x_0, y_0)$ is negative.

Now let's assume all the second order partials are continuous at (x_0, y_0). We know $A(c) = \frac{\partial^2 f}{\partial x^2}(x_0 + c(x - x_0), y_0 + c(y - y_0))$ and from our theorem, if $\frac{\partial^2 f}{\partial x^2}(x_0, y_0) > 0$, then so is $A(c)$ in a circle around (x_0, y_0).
The other term $A(c)\,D(c) - (B(c))^2 > 0$ will also be positive is a circle around (x_0, y_0) as long as $\frac{\partial^2 f}{\partial x^2}(x_0, y_0)\frac{\partial^2 f}{\partial y^2}(x_0, y_0) - \frac{\partial^2 f}{\partial x \partial y}(x_0, y_0) > 0$. We can say similar things about the negative case.
Now to save typing let $\frac{\partial^2 f}{\partial x^2}(x_0, y_0) = f^0_{xx}$, $\frac{\partial^2 f}{\partial y^2}(x_0, y_0) = f^0_{yy}$ and $\frac{\partial^2 f}{\partial x \partial y}(x_0, y_0) = f^0_{xy}$. So we can restate our two cases as

- $f^0_{xx} > 0$ and $f^0_{xx} f^0_{yy} - (f^0_{xy})^2 > 0$ which implies the term after $f(x_0, y_0)$ is positive. This implies that $f(x, y) > f(x_0, y_0)$ in a circle of some radius r which says $f(x_0, y_0)$ is a minimum value of the function locally at that point.

- $f^0_{xx} < 0$ and $f^0_{xx} f^0_{yy} - (f^0_{xy})^2 > 0$ which implies the term after $f(x_0, y_0)$ is negative. This implies that $f(x, y) < f(x_0, y_0)$ in a circle of some radius r which says $f(x_0, y_0)$ is a maximum value of the function locally at that point.

So we have come up with a great condition to verify if a place where the partials are zero is a minimum or a maximum.

Theorem 16.1.2 Extreme Test

If the partials of f are zero at the point (x_0, y_0), we can determine if that point is a local minimum or local maximum of f using a second order test. We must assume the second order partials are continuous at the point (x_0, y_0).

- *If $f_{xx}^0 > 0$ and $f_{xx}^0 f_{yy}^0 - (f_{xy}^0)^2 > 0$ then $f(x_0, y_0)$ is a local minimum.*

- *If $f_{xx}^0 < 0$ and $f_{xx}^0 f_{yy}^0 - (f_{xy}^0)^2 > 0$ then $f(x_0, y_0)$ is a local maximum.*

We just don't know anything if the test $f_{xx}^0 f_{yy}^0 - (f_{xy}^0)^2 = 0$.

Recall the definition of the determinant of a 2×2 matrix:

$$\boldsymbol{A} = \begin{bmatrix} a & b \\ c & d \end{bmatrix} \implies det(\boldsymbol{A}) = ad - bc.$$

So, since we assume the mixed partials match

$$\boldsymbol{H}(x, y) = \begin{bmatrix} f_{xx}(x, y) & f_{xy}(x, y) \\ f_{yx}(x, y) & f_{yy}(x, y) \end{bmatrix} \implies$$

$$det(\boldsymbol{H}(x, y)) = f_{xx}(x, y) f_{yy}(x, y) - (f_{xy}(x, y))^2$$

and at a critical point (x_0, y_0) where $\nabla f(x_0, y_0) = \begin{bmatrix} 0 \\ 0 \end{bmatrix}$ we have

$$det(\boldsymbol{H}(x_0, y_0)) = f_{xx}(x_0, y_0) f_{yy}(x_0, y_0) - (f_{xy}(x_0, y_0))^2$$

Using our usual shorthand notations, we would write this as $det(\boldsymbol{H}^0) = f_{xx}^0 f_{yy}^0 - (f_{xy}^0)^2$. We can then rewrite out second order test for extremal values using this determinant idea:

Theorem 16.1.3 Two Dimensional Extreme Test for Minima and Maxima with Determinants

If the partials of f are zero at the point (x_0, y_0), we can determine if that point is a local minimum or local maximum of f using a second order test. We must assume the second order partials are continuous at the point (x_0, y_0).

- *If $f_{xx}^0 > 0$ and $det(\boldsymbol{H}^0) > 0$ then $f(x_0, y_0)$ is a local minimum.*

- *If $f_{xx}^0 < 0$ and $det(\boldsymbol{H}^0) > 0$ then $f(x_0, y_0)$ is a local maximum.*

We just don't know anything if the test $det(\boldsymbol{H}^0) = 0$.

Proof 16.1.2
We have shown this argument.   ∎

16.1.1   Saddle Points

Recall at a critical point (x_0, y_0), we found that

$$f(x, y) = f(x_0, y_0)$$
$$+ \frac{A(c)}{2} \left(\left(\Delta x + \frac{B(c)}{A(c)} \Delta y \right)^2 + \left(\frac{A(c) D(c) - (B(c))^2}{(A(c))^2} \right) (\Delta y)^2 \right).$$

And we have been assuming $A(c) \neq 0$ here. Now suppose we knew $A(c)\,D(c) - (B(c))^2 < 0$. Then, using the usual continuity argument, we know that there is a circle around the critical point (x_0, y_0) so that $A(c)\,D(c) - (B(c))^2 < 0$ when $c = 0$. This is the same as saying $\det(\boldsymbol{H}(x_0, y_0)) < 0$. But notice that on the line going through the critical point having $\Delta y = 0$, this gives

$$f(x, y) \;=\; f(x_0, y_0) + \frac{A(c)}{2}\left(\Delta x\right)^2.$$

and on the line through the critical point with $\Delta x + \frac{B(c)}{A(c)}\Delta y = 0$. we have

$$f(x, y) \;=\; f(x_0, y_0) + \frac{A(c)}{2}\left(\frac{A(c)\,D(c) - (B(c))^2}{(A(c))^2}\right)(\Delta y)^2$$

Now, if $A(c) > 0$, the first case gives $f(x, y) = f(x_0, y_0) + $ **a positive number** showing f has a minimum on that trace.

However, the second case gives $f(x, y) = f(x_0, y_0) - $ **a positive number** which shows f has a maximum on that trace.

The fact that f is minimized in one direction and maximized in another direction gives rise to the expression that we consider f to behave like a saddle at this critical point. The analysis is virtually the same if $A(c) < 0$, except the first trace has the maximum and the second trace has the minimum. Hence, the test for a saddle point is to see if $\det(\boldsymbol{H}(x_0, y_0)) < 0$.

If $A(c) = 0$, we have to argue differently. We are in the case where $\det(\boldsymbol{H}(x_0, y_0)) < 0$ which we know means we can assume $A(c)D(c) - (B(c))^2 < 0$ also. If $A(c) = 0$, we must $B(c) \neq 0$. We thus have

$$\begin{aligned}
f(x, y) &\;=\; f(x_0, y_0) + B(c)\Delta x\,\Delta y + (1/2)D(c)(\Delta y)^2 \\
&\;=\; f(x_0, y_0) + (1/2)\Big(2B(c)\Delta x + D(c)\Delta y\Big)\Delta y
\end{aligned}$$

If $D(c) = 0$, $f(x, y) = f(x_0, y_0) + B(c)\Delta x\,\Delta y$ and choosing the paths $\Delta x = \pm\Delta y$, we have $f(x, y) = f(x_0.y_0) \pm B(c)(\Delta y)^2$ which tell us we have a minimum on one path and a maximum on the other path; i.e. this is a saddle.

If $D(c) \neq 0$, since

$$f(x, y) \;=\; f(x_0, y_0) + (1/2)\Big(2B(c)\Delta x + D(c)\Delta y\Big)\Delta y$$

we can choose paths $2B(c)\Delta x = D(c)\Delta y$ and $2B(c)\Delta x = -3D(c)\Delta y$ to give $f(x, y) = f(x_0, y_0) + D(c)(\Delta y)^2$ or $f(x, y) = f(x_0, y_0) - D(c)(\Delta y)^2$ and again, on one path we have a minimum and on the other a maximum implying a saddle.

Now the second order test fails if $\det(\boldsymbol{H}(x_0, y_0)) = 0$ at the critical point as in that case, the surface can have a minimum, maximum or saddle.

- $f(x, y) = x^4 + y^4$ has a global minimum at $(0, 0)$ but at that point

$$\boldsymbol{H}(x, y) \;=\; \begin{bmatrix} 12x^2 & 0 \\ 0 & 12y^2 \end{bmatrix} \implies \det(\boldsymbol{H}(x_0, y_0)) = 144x^2y^2.$$

and hence, $\det(\boldsymbol{H}(x_0, y_0)) = 0$.

- $f(x, y) = -x^4 - y^4$ has a global maximum at $(0, 0)$ but at that point

$$\boldsymbol{H}(x, y) \;=\; \begin{bmatrix} -12x^2 & 0 \\ 0 & -12y^2 \end{bmatrix} \implies \det(\boldsymbol{H}(x_0, y_0)) = 144x^2 y^2.$$

and hence, $\det(\boldsymbol{H}(x_0, y_0)) = 0$ as well.

- Finally, $f(x, y) = x^4 - y^4$ has a saddle at $(0, 0)$ but at that point

$$\boldsymbol{H}(x, y) \;=\; \begin{bmatrix} 12x^2 & 0 \\ 0 & -12y^2 \end{bmatrix} \implies \det(\boldsymbol{H}(x_0, y_0)) = -144x^2 y^2.$$

and hence, $\det(\boldsymbol{H}(x_0, y_0)) = 0$ again.

Hence, since we have covered all the cases, we can state the full theorem.

Theorem 16.1.4 Two dimensional Extreme Test Including Saddle Points

If the partials of f are zero at the point (x_0, y_0), we can determine if that point is a local minimum or local maximum of f using a second order test. We must assume the second order partials are continuous at the point (x_0, y_0).

- *If $f_{xx}^0 > 0$ and $\det(\boldsymbol{H}^0) > 0$ then $f(x_0, y_0)$ is a local minimum.*
- *If $f_{xx}^0 < 0$ and $\det(\boldsymbol{H}^0) > 0$ then $f(x_0, y_0)$ is a local maximum.*
- *If $\det(\boldsymbol{H}^0) < 0$, then $f(x_0, y_0)$ is a local saddle.*

We just don't know anything if the test $\det(\boldsymbol{H}^0) = 0$.

Proof 16.1.3
We have shown this argument. ∎

Example 16.1.1 *Use our tests to show $f(x, y) = x^2 + 3y^2$ has a minimum at $(0, 0)$.*

Solution *The partials here are $f_x = 2x$ and $f_y = 6y$. These are zero at $x = 0$ and $y = 0$. The Hessian at this critical point is*

$$H(x, y) \;=\; \begin{bmatrix} 2 & 0 \\ 0 & 6 \end{bmatrix} = H(0, 0).$$

as H is constant here. Our second order test says the point $(0, 0)$ corresponds to a minimum because $f_{xx}(0, 0) = 2 > 0$ and $f_{xx}(0, 0)\, f_{yy}(0, 0) - (f_{xy}(0, 0))^2 = 12 > 0$.

Example 16.1.2 *Use our tests to show $f(x, y) = x^2 + 6xy + 3y^2$ has a saddle at $(0, 0)$.*

Solution *The partials here are $f_x = 2x + 6y$ and $f_y = 6x + 6y$. These are zero at when $2x + 6y = 0$ and $6x + 6y = 0$ which has solution $x = 0$ and $y = 0$. The Hessian at this critical point is*

$$H(x, y) \;=\; \begin{bmatrix} 2 & 6 \\ 6 & 6 \end{bmatrix} = H(0, 0).$$

as H is again constant here. Our second order test says the point $(0, 0)$ corresponds to a saddle because $f_{xx}(0, 0) = 2 > 0$ and $f_{xx}(0, 0)\, f_{yy}(0, 0) - (f_{xy}(0, 0))^2 = 12 - 36 < 0$.

Example 16.1.3 *Show our tests fail on $f(x, y) = 2x^4 + 4y^6$ even though we know there is a minimum value at $(0, 0)$.*

Solution *For $f(x, y) = 2x^4 + 4y^6$, you find that the critical point is $(0,0)$ and all the second order partials are 0 there. So all the tests fail. Of course, a little common sense tells you $(0,0)$ is indeed the place where this function has a minimum value. Just think about how it's surface looks. But the tests just fail. This is much like the curve $f(x) = x^4$ which has a minimum at $x = 0$ but all the tests fail on it also.*

Example 16.1.4 *Show our tests fail on $f(x, y) = 2x^2 + 4y^3$ and the surface does not have a minimum or maximum at the critical point $(0,0)$.*

Solution *For $f(x, y) = 2x^2 + 4y^3$, the critical point is again $(0,0)$ and $f_{xx}(0,0) = 4$, $f_{yy}(0,0) = 0$ and $f_{xy}(0,0) = f_{yx}(0,0) = 0$. So $f_{xx}(0,0)\, f_{yy}(0,0) - (f_{xy}(0,0))^2 = 0$ so the test fails. Note the $x = 0$ trace is $4y^3$ which is a cubic and so is negative below $y = 0$ and positive above $y = 0$. Not much like a minimum or maximum behavior on this trace! But the trace for $y = 0$ is $2x^2$ which is a nice parabola which does reach its minimum at $x = 0$. So the behavior of the surface around $(0,0)$ is not a maximum or a minimum. The surface acts a lot like a cubic. Do this in MATLAB.*

Listing 16.1: | **Plotting a Saddle** |

```
>> [X,Y] = meshgrid(-1:.2:1);
>> Z = 2*X.^2 + 4*Y.^3;
>> surf(Z);
```

This will give you the surface. In the plot that is shown go to the tool menu and click of the rotate 3D option and you can spin it around. Clearly like a cubic as we can see in Figure 16.1.

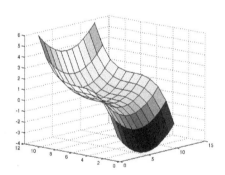

Figure 16.1: The saddle surface $2x^2 - 4y^3$.

16.1.2 Homework

Exercise 16.1.1 *This is a review of some ideas from statistics. Let $\{X_1, \ldots, X_n\}$ be some data for $N \gg 1$. The average of this set of data is $\overline{X} = (1/N) \sum_{i=1}^{N} x_i$. The average of the squares of the data is $\overline{X^2} = (1/N) \sum_{i=1}^{N} x_i^2$. Prove $0 \le \sum_{i=1}^{N} (x_i - \overline{X})^2 = N\overline{X^2} - N(\overline{X})^2$.*

Exercise 16.1.2 *Given data pairs $\{(X_1, Y_1), \ldots, (X_N, Y_N)\}$, the line of regression through this data is the line $y = mx + b$ which minimizes the error function $E(m, b) = \sum_{i=1}^{N} (Y_I - mX_i - b)^2$. Find the slope and intercept (m^*, b^*) which is a critical point for this minimization. The formulae you derive here for m^* and b^* give the optimal slope and intercept for the line of regression that best fits this data. However, the proof of this requires the next problem.*

Exercise 16.1.3 *Use the second order theory for the minimization of a function of two variables to show that the error is a global minimum at the critical point (m^*, b^*).*

16.2   Symmetric Problems

We can rewrite our tests for extremals in a form that is more easily extended to higher dimensions such as optimization in $\Re^3$ and so forth. This requires us to learn about positive and definite matrices. Let's start with a general 2×2 symmetric matrix A given by

$$A = \begin{bmatrix} a & b \\ b & d \end{bmatrix}$$

where a, b and d are arbitrary nonzero numbers. The eigenvalue equation here is $r^2 - (a+d)r + ad - b^2$ and $ad - b^2$ is the determinant of A. The roots are given by

$$
\begin{aligned}
r &= \frac{(a+d) \pm \sqrt{(a+d)^2 - 4(ad - b^2)}}{2} \\
&= \frac{(a+d) \pm \sqrt{a^2 + 2ad + d^2 - 4ad + 4b^2}}{2} \\
&= \frac{(a+d) \pm \sqrt{a^2 - 2ad + d^2 + 4b^2}}{2} \\
&= \frac{(a+d) \pm \sqrt{(a-d)^2 + 4b^2}}{2}
\end{aligned}
$$

It is easy to see the term in the square root is always positive implying two real roots.

Note we can find the eigenvectors with a standard calculation. For eigenvalue $\lambda_1 = \frac{(a+d) + \sqrt{(a-d)^2 + 4b^2}}{2}$, we must find the vectors V so that

$$
\begin{bmatrix} \frac{(a+d) + \sqrt{(a-d)^2 + 4b^2}}{2} - a & -b \\ -b & \frac{(a+d) + \sqrt{(a-d)^2 + 4b^2}}{2} - d \end{bmatrix} \begin{bmatrix} V_1 \\ V_2 \end{bmatrix} = \begin{bmatrix} 0 \\ 0 \end{bmatrix}
$$

We can use the top equation to find the needed relationship between V_1 and V_2. We have

$$\left(\frac{(a+d) + \sqrt{(a-d)^2 + 4b^2}}{2} - a \right) V_1 - bV_2 = 0.$$

Thus, we have for $V_2 = \frac{d - a + \sqrt{(a-d)^2 + 4b^2}}{2}$, $V_1 = b$. Thus, the first eigenvector is

$$E_1 = \begin{bmatrix} b \\ \frac{d - a + \sqrt{(a-d)^2 + 4b^2}}{2} \end{bmatrix}$$

The second eigenvector is a similar calculation. We must find the vector V so that

$$
\begin{bmatrix} \frac{(a+d) - \sqrt{(a-d)^2 + 4b^2}}{2} - a & -b \\ -b & \frac{(a+d) - \sqrt{(a-d)^2 + 4b^2}}{2} - d \end{bmatrix} \begin{bmatrix} V_1 \\ V_2 \end{bmatrix} = \begin{bmatrix} 0 \\ 0 \end{bmatrix}
$$

We find

$$\left(\frac{(a+d) - \sqrt{(a-d)^2 + 4b^2}}{2} - a \right) V_1 - b V_2 = 0.$$

Thus, we have for $V_2 = d + \frac{(a+d) - \sqrt{(a-d)^2 + 4b^2}}{2}$, $V_1 = b$. Thus, the second eigenvector is

$$E_2 = \begin{bmatrix} b \\ \frac{d - a - \sqrt{(a-d)^2 + 4b^2}}{2} \end{bmatrix}$$

Note that $< E_1, E_2 >$ is

$$< E_1, E_2 >$$
$$= b^2 + \left(\frac{d - a + \sqrt{(a-d)^2 + 4b^2}}{2} \right) \left(\frac{d - a - \sqrt{(a-d)^2 + 4b^2}}{2} \right)$$
$$= b^2 + \frac{(d-a)^2}{4} - \frac{(a-d)^2 + 4b^2}{4} = b^2 + \frac{(d-a)^2 - (a-d)^2}{4} - b^2 = 0.$$

Hence, these two eigenvectors are **orthogonal** to each other. Note, the two eigenvalues are

$$\lambda_1 = \frac{(a+d) + \sqrt{(a-d)^2 + 4b^2}}{2}, \quad \lambda_2 = \frac{(a+d) - \sqrt{(a-d)^2 + 4b^2}}{2}$$

The only way both eigenvalues can be zero is if both $a + d = 0$ and $(a-d)^2 + 4b^2 = 0$. That only happens if $a = b = d = 0$ which we explicitly ruled out at the beginning of our discussion because we said a, b and d were nonzero. However, both eigenvalues can be negative, both can be positive or they can be of mixed sign as our examples show.

Example 16.2.1 *Find the eigenvalues for*

$$A = \begin{bmatrix} 3 & 2 \\ 2 & 6 \end{bmatrix}$$

Solution *The eigenvalues are* $(9 \pm 5)/2$ *and both are positive.*

Example 16.2.2 *Find the eigenvalues for*

$$A = \begin{bmatrix} 3 & 5 \\ 5 & 6 \end{bmatrix}$$

Solution *The eigenvalues are* $(9 \pm \sqrt{9 + 100})/2$ *giving* $\lambda_1 = 9.72$ *and* $\lambda_2 = -5.22$ *and the eigenvalues have mixed sign.*

Homework

Exercise 16.2.1 *Find the eigenvalues and eigenvectors for*

$$A = \begin{bmatrix} 30 & -2 \\ -2 & 6 \end{bmatrix}$$

Exercise 16.2.2 *Find the eigenvalues and eigenvectors for*

$$A = \begin{bmatrix} -2 & 11 \\ 11 & 6 \end{bmatrix}$$

Exercise 16.2.3 *Find the eigenvalues and eigenvectors for*

$$A \; = \; \begin{bmatrix} 5 & 8 \\ 8 & 3 \end{bmatrix}$$

16.2.1   A Canonical Form for a Symmetric Matrix

Now let's look at 2×2 symmetric matrices more abstractly. Don't worry, there is a payoff here in understanding! Let A be a general 2×2 symmetric matrix. Then it has two distinct eigenvalues λ_1 and another one λ_2. Consider the matrix P given by

$$P \; = \; \begin{bmatrix} E_1 & E_2 \end{bmatrix}$$

whose transpose is then

$$P^T \; = \; \begin{bmatrix} E_1^T \\ E_2^T \end{bmatrix}$$

Thus,

$$P^T A P \; = \; \begin{bmatrix} E_1^T \\ E_2^T \end{bmatrix} A \begin{bmatrix} E_1 & E_2 \end{bmatrix} = \begin{bmatrix} E_1^T \\ E_2^T \end{bmatrix} \begin{bmatrix} AE_1 & AE_2 \end{bmatrix} = \begin{bmatrix} E_1^T \\ E_2^T \end{bmatrix} \begin{bmatrix} \lambda_1 E_1 & \lambda_2 E_2 \end{bmatrix}$$

After we do the final multiplications, we have

$$P^T A P \; = \; \begin{bmatrix} \lambda_1 < E_1, E_1 > & \lambda_2 < E_1, E_2 > \\ \lambda_1 < E_2, E_1 > & \lambda_2 < E_2, E_2 > \end{bmatrix}$$

We know the eigenvectors are orthogonal, so we must have

$$P^T A P \; = \; \begin{bmatrix} \lambda_1 < E_1, E_1 > & 0 \\ 0 & \lambda_2 < E_2, E_2 > \end{bmatrix}$$

Once last step and we are done! There is no reason, we can't choose as our eigenvectors, vectors of length one: here just replace E_1 by the new vector $E_1/\|E_1\|$ where $\|E_1\|$ is the usual Euclidean length of the vector. Similarly, replace E_2 by $E_2/\|E_2\|$. Assuming this is done, we have $< E_1, E_1 >= 1$ and $< E_2, E_2 >= 1$. We are left with the identity

$$P^T A P \; = \; \begin{bmatrix} \lambda_1 & 0 \\ 0 & \lambda_2 \end{bmatrix}$$

It is easy to see $P^T P = PP^T = I$ telling us $P^T = P^{-1}$. Thus, we can rewrite as

$$A \; = \; P \begin{bmatrix} \lambda_1 & 0 \\ 0 & \lambda_2 \end{bmatrix} P^T$$

This is an important thing. We have shown the 2×2 matrix A can be decomposed into the product $A = P \Lambda P^T$ where Λ is the diagonal matrix whose entries are the eigenvalues of A with the most positive one in the $(1, 1)$ position.

It is now clear how we solve an equation like $A X = b$. We rewrite as $P \Lambda P^T X = b$ which leads to the solution

$$X \; = \; P \Lambda^{-1} P^T b$$

and it is clear the reciprocal eigenvalue sizes determine how large the solution can get. The eigenvectors here are **independent vectors** in $\Re^2$ and since they **span** $\Re^2$, they form a **basis**. This is called an **orthonormal basis** because the vectors are perpendicular or orthogonal. Hence, any vector in $\Re^2$ can be written as a linear combination of this basis. That is, if V is such a vector, then we have

$$V = V_1 E_1 + V_2 E_2$$

and the components V_1 and V_2 are known as the components of V relative to the basis $\{E_1, E_2\}$. We often just refer to this basis as E. Hence, a vector V has many possible representations.

The one you are most used to is the one which uses the basis vectors

$$e_1 = \begin{bmatrix} 1 \\ 0 \end{bmatrix} \text{ and } e_2 = \begin{bmatrix} 0 \\ 1 \end{bmatrix}$$

which is called the **standard basis**. When we write $V = \begin{bmatrix} 3 \\ 5 \end{bmatrix}$ unless otherwise stated, we assume these are the components of V with respect to the standard basis. Now let's go back to V. Since the vectors E_1 and E_2 are orthogonal, we can take inner products on both sides of the representation of V with respect to the basis E to get

$$\begin{aligned} < V, E_1 > &= < V_1 E_1, E_1 > + < V_2 E_2, E_1 > \\ &= V_1 < E_1, E_1 > + V_2 < E_2, E_1 > \end{aligned}$$

as $< E_2, E_1 > = 0$ as the vectors are perpendicular and $< E_1, E_1 > = < E_2, E_2 > = 1$ as our eigenvectors have length one. So we have $V_1 = < V, E_1 >$ and similarly $V_2 = < V, E_2 >$. So we can decompose V as

$$V = < V, E_1 > E_1 + < V, E_2 > E_2$$

Another way to look at this is that the two eigenvectors can be used to find a representation of the data vector b and the solution vector X as follows:

$$\begin{aligned} b &= < b, E_1 > E_1 + < b, E_2 > E_2 = b_1 E_1 + b_2 E_2 \\ X &= < X, E_1 > E_1 + < X, E_2 > E_2 = X_1 E_1 + X_2 E_2 \end{aligned}$$

So $A X = b$ becomes

$$\begin{aligned} A \left(X_1 E_1 + X_2 E_2 \right) &= b_1 E_1 + b_2 E_2 \\ \lambda_1 X_1 E_1 + \lambda_2 X_2 E_2 &= b_1 E_1 + b_2 E_2. \end{aligned}$$

The only way this equation works is if the coefficients on the eigenvectors match. So we have $X_1 = \lambda_1^{-1} b_1$ and $X_2 = \lambda_2^{-1} b_2$. This shows very clearly how the solution depends on the size of the reciprocal eigenvalues. Thus, if our problem has a very small eigenvalue, we would expect our solution vector to be unstable. Also, if one of the eigenvalues is 0, we would have real problems! We can address this somewhat by finding a way to force all the eigenvalues to be positive.

Homework

Exercise 16.2.4 *Find the eigenvalues and eigenvectors for*

$$A = \begin{bmatrix} 4 & -2 \\ -2 & 6 \end{bmatrix}$$

and then find the representation $A = P\Lambda P^T$ as discussed in this section.

Exercise 16.2.5 *Find the eigenvalues and eigenvectors for*

$$A = \begin{bmatrix} -1 & 1 \\ 1 & 2 \end{bmatrix}$$

and then find the representation $A = P\Lambda P^T$ as discussed in this section.

Exercise 16.2.6 *Find the eigenvalues and eigenvectors for*

$$A = \begin{bmatrix} 4 & -3 \\ -3 & 5 \end{bmatrix}$$

and then find the representation $A = P\Lambda P^T$ as discussed in this section.

16.2.2   Signed Definite Matrices

A 2×2 matrix is said to be a **positive definite** matrix if $x^T A x > 0$ for all vectors x. If we multiply this out, we find the inequality below

$$ax_1^2 + 2bx_1x_2 + dx_2^2 > 0,$$

If we complete the square, we find

$$a\left(\left(x_1 + (b/a)x_2 \right)^2 + \left(((ad - b^2)/a^2)x_2^2 \right) \right) > 0$$

Now if the leading term $a > 0$, and if the determinant of $A = ad - b^2 > 0$, we have

$$\left(x_1 + (b/a)x_2 \right)^2 + \left(((ad - b^2)/a^2)x_2^2 \right)$$

is always positive. Note since the determinant is positive $ad > b^2$ which forces d to be positive as well. So in this case, a and d and $ad - b^2 > 0$. And the expression $x^T A x > 0$ in this case. Now recall what we found about the eigenvalues here. We had the eigenvalues were

$$r = \frac{(a + d) \pm \sqrt{(a - d)^2 + 4b^2}}{2}$$

Since $ad - b^2 > 0$, the term

$$(a - d)^2 + 4b^2 = a^2 - 2ad + d^2 + 4b^2 < a^2 - 2ad + 4ad + d^2 = (a + d)^2.$$

Thus, the square root is smaller than $a + d$ as a and d are positive. The first root is always positive.

The second root is too as $(a + d) - \sqrt{(a - d)^2 + 4b^2} > a + d - (a + d) = 0$. So both eigenvalues are positive if a and d are positive and $ad - b^2 > 0$. Note the argument can go the other way. If we assume the matrix is positive definite, the we are forced to have $a > 0$ and $ad - b^2 > 0$ which gives the same result.

We conclude our 2×2 symmetric matrix A is positive definite if and only if $a > 0$, $d > 0$ and the determinant of $A > 0$ too. **Note a positive definite matrix has positive eigenvalues**.

A similar argument holds if we have determinant of $A > 0$ but $a < 0$. The determinant condition will then force $d < 0$ too. We find that $x^T A x < 0$. In this case, we say the matrix is **negative definite**.

The eigenvalues are still

$$r = \frac{(a+d) \pm \sqrt{(a-d)^2 + 4b^2}}{2}.$$

But now, since $ad - b^2 > 0$, the term

$$(a-d)^2 + 4b^2 = a^2 - 2ad + d^2 + 4b^2 < a^2 - 2ad + 4ad + d^2 = |a+d|^2.$$

Since a and d are negative, $a+d < 0$ and so the second root is always negative. The first root's sign is determined by $(a+d) + \sqrt{(a-d)^2 + 4b^2} < (a+d) + |a+d| = 0$. So both eigenvalues are negative.

We have found the matrix A is negative definite if a and d are negative and the determinant of $A > 0$. **Note a negative definite matrix has negative eigenvalues**.

16.3 A Deeper Look at Extremals

We can now rephrase what we said about second order tests for extremals for functions of two variables: recall we had:

Theorem 16.3.1 Two Dimensional Extreme Test Review

> *If the partials of f are zero at the point (x_0, y_0), we can determine if that point is a local minimum or local maximum of f using a second order test. We must assume the second order partials are continuous at the point (x_0, y_0).*
>
> - *If $f_{xx}^0 > 0$ and $det(H^0) > 0$ then $f(x_0, y_0)$ is a local minimum.*
> - *If $f_{xx}^0 < 0$ and $det(H^0) > 0$ then $f(x_0, y_0)$ is a local maximum.*
> - *If $det(H^0) < 0$, then $f(x_0, y_0)$ is a local saddle.*
>
> *We just don't know anything if the test $det(H^0) = 0$.*

The Hessian at the critical point is $H^0 = \begin{bmatrix} A & B \\ B & D \end{bmatrix}$ and we see

- $f_{xx}^0 > 0$ and $det(H^0) > 0$ tells us H^0 is positive definite and both eigenvalues are positive.

- $f_{xx}^0 < 0$ and $det(H^0) > 0$ tells us H^0 is negative definite and both eigenvalues are negative.

We haven't proven it yet, but if the eigenvalues are nonzero and differ in sign, this gives us a saddle. Thus our theorem becomes

Theorem 16.3.2 Two Dimensional Extreme Test with Signed Hessians

> *If ∇f is zero at the point (x_0, y_0) and the second order partials are continuous at (x_0, y_0).*
>
> - *If H^0 is positive definite then $f(x_0, y_0)$ is a local minimum.*
>
> - *If H^0 is negative definite then $f(x_0, y_0)$ is a local maximum.*
>
> - *If the eigenvalues of $\det(H^0)$ are nonzero and of mixed sign, then $f(x_0, y_0)$ is a local saddle.*

16.3.1 Extending to Three Variables

The principle minors of a 2×2 matrix A as defined as follows:

$$
\begin{aligned}
(PM)_1 &= \det(A_{11}) = A_{11} \\
(PM)_2 &= \det\left(\begin{bmatrix} A_{11} & A_{12} \\ A_{12} & A_{22} \end{bmatrix} \right)
\end{aligned}
$$

The principle minors of a 3×3 matrix A as then defined as follows:

$$
\begin{aligned}
(PM)_1 &= \det(A_{11}) = A_{11} \\
(PM)_2 &= \det\left(\begin{bmatrix} A_{11} & A_{12} \\ A_{12} & A_{22} \end{bmatrix} \right) \\
(PM)_3 &= \det\left(\begin{bmatrix} A_{11} & A_{12} & A_{13} \\ A_{12} & A_{22} & A_{23} \\ A_{13} & A_{23} & A_{33} \end{bmatrix} \right)
\end{aligned}
$$

The principle minors are defined similarly for larger square matrices. We can state our extremal theorem for functions of many variables in terms of matrix minors:

Theorem 16.3.3 N Dimensional Extrema Test with Hessian Minors

> *We assume the second order partials are continuous at the point (x_0, y_0). Then if the Hessian $H(x_0, y_0)$ is positive definite, the critical point is a local minimum. If it is negative definite, it is a local maximum. Moreover, the Hessian $H(x_0, y_0)$ is positive definite if its principle minors are all positive (i.e. $H_{11}^0 > 0$ etc.) and $H(x_0, y_0)$ is negative definite if its principle minors alternate is sign with the first one negative (i.e. $H_{11}^0 < 0$.)*

Proof 16.3.1
The proof of this is done carefully in (Peterson (15) 2020).   ∎

16.3.2 Homework

The following 2×2 matrix is the Hessian H^0 at the critical point $(1, 1)$ of an extremal value problem.

- Determine if H^0 is positive or negative definite

- Determine if the critical point is a maximum or a minimum

- Find the two eigenvalues of H^0. Label the largest one as r_1 and the other as r_2

- Find the two associated eigenvectors as unit vectors

- Define

$$P = \begin{bmatrix} E_1 & E_2 \end{bmatrix}$$

- Compute $P^T H^0 P$

- Show $H^0 = P\Lambda P^T$ for an appropriate Λ.

Exercise 16.3.1

$$H^0 = \begin{bmatrix} 1 & -3 \\ -3 & 12 \end{bmatrix}$$

Exercise 16.3.2

$$H^0 = \begin{bmatrix} -2 & 7 \\ 7 & -40 \end{bmatrix}$$

Exercise 16.3.3

$$H^0 = \begin{bmatrix} -12 & 7 \\ 7 & 4 \end{bmatrix}$$

Exercise 16.3.4

$$H^0 = \begin{bmatrix} -1 & 17 \\ 17 & 3 \end{bmatrix}$$

Part III

Integration and Sequences of Functions

Chapter 17

Uniform Continuity

17.1 Uniform Continuity

You should already know what continuity means for a function. Let's review this:

Definition 17.1.1 Continuity at a Point

> *Let f be defined locally at p and it now must be defined at p as well. Thus, there is an $r > 0$ so that f is defined on $(p - r, p + r)$. We say $f(x)$ is continuous at p if the limit as $x \to p$ exists and equals a and $f(p) = a$ too. This is stated mathematically like this: $\forall \epsilon > 0 \exists \delta > 0 \ni |x - p| < \delta \Rightarrow |f(x) - f(p)| < \epsilon$. We often just say f is continuous at p means $\lim_{x \to p} f(x) = f(p)$.*

To be really careful, we should note the value of δ depends on ϵ. We also must make sure the value of δ is small enough to fit inside the ball of local definition for f. Also, all of this depends on the point p! So we could say

Definition 17.1.2 Careful Continuity at a Point

> *Let f be defined locally at p and it now must be defined at p as well. Thus, there is an $r > 0$ so that f is defined on $(p - r, p + r)$. We say $f(x)$ is continuous at p if the limit as $x \to p$ exists and equals a and $f(p) = a$ too. This is stated mathematically like this: $\forall \epsilon > 0 \exists 0 < \delta(p, \epsilon) < r \ni |x - p| < \delta(p, \epsilon) \Rightarrow |f(x) - f(p)| < \epsilon$.*

The next thing we want to explore is the idea that maybe the value of δ might not depend on the choice of p. This is, of course, a much stronger idea than continuity at a point.

Example 17.1.1 *Consider $f(x) = 1/x$ on the domain $(0, 1)$. Pick $\epsilon = 0.1$ (this choice of ϵ could be different, of course and you should work through the details of how the argument changes for a different ϵ choice). Can we find one choice of δ so that*

$$|x - y| < \delta \implies |f(x) - f(y)| < .1$$

where x and y are in $(0, 1)$?

Solution *This amounts to asking if we can make sure the change in function values is "small" for any "sufficiently small" change in domain values. In other words, there is a **sliding window** of size δ that we can move anywhere in $(0, 1)$ and be guaranteed the absolute change in function values is less than 0.1. Let's show this cannot work here. Pick any $\delta > 0$. We will show there are points less than δ apart in $(0, 1)$ where the absolute difference in function values is larger than $\epsilon = 0.1$. Let*

261

$x_n = 1/n$ and $y_n = 1/(n+1)$ for $n > 1$. These points are in $(0,1)$ and we have

$$|x_n - y_n| \quad = \quad |1/n - 1/(n+1)| = \frac{1}{n(n+1)} < 1/n^2$$

Since $1/n^2 \to 0$ there is N so the $1/n^2 < \delta$ if $n > N$. But then

$$|f(x_n) - f(y_n)| \quad = \quad |1/x_n - 1/y_n| = |n - (n+1)| = 1 > 0.1$$

We can do this argument for any choice of $\delta > 0$. Hence for this example, we cannot find a value of δ for which $|x - y| < \delta$ implies $|f(x) - f(y)| < 0.1$.

Functions for which we **can** do this sort of property are special and this kind of **smoothness** is given a special name: f is **uniformly continuous** because we can use one, **uniform** choice of δ for a given ϵ. Here is the definition: note this is defined on an interval not at a single point!

Definition 17.1.3 Uniform Continuity

Let f be defined on the interval I. We say $f(x)$ is uniformly continuous on I if $\forall \epsilon > 0 \exists \delta > 0 \ni |x - y| < \delta \Rightarrow |f(x) - f(y)| < \epsilon$.

Hence $f(x) = 1/x$ is **not** uniformly continuous on $(0,1)$.

Example 17.1.2 *Let's consider $f(x) = x^2 + 5$ on the interval $(2,3)$. Let's try to find one δ that will work for a given ϵ.*

Solution *Let $\epsilon > 0$ be arbitrary. Now consider*

$$|f(x) - f(y)| \quad = \quad |(x^2 + 5) - (y^2 + 5)| = |x^2 - y^2| = |x - y|\,|x + y|$$

for any x and y in $(2,3)$. We see $2 + 2 < x + y < 3 + 3 = 6$ on $(2,3)$ so $|x + y| \le 6$ on our interval. Thus we have

$$|f(x) - f(y)| \quad \le \quad 6\,|x - y|$$

*If we set this **overestimate** less than ϵ, we can find the δ we need. We want $6\,|x - y| < \epsilon$ which implies $|x - y| < \epsilon/6$. So any $\delta > 0$ with $\delta < \epsilon/6$ will work. Thus, $f(x) = x^2 + 5$ is **uniformly continuous** on $(2,3)$.*

Comment 17.1.1 *Note this argument will work for any finite interval I of the form (a,b), $[a,b)$, $(a,b]$ or $[a,b]$. Our estimate becomes $2a < x + y < 2b$. Since a and b could be negative, the estimate gets a bit harder. For example, if the interval is $(-2,-1)$ we have $-4 < x + y < 2$ implying $|x + y| < 4$. For the interval $(-3,1)$, we have $-6 < x + y < 2$ implying $|x + y| < 6$. So you have to be careful to take into account the signs of a and b!*

Our two examples are both on open intervals. The first function is **not** uniformly continuous on $(0,1)$ and the second one is uniformly continuous. Let's start abbreviating **uniformly continuous** by just **uc** also to make typing easier.

Example 17.1.3 *Look at $f(x) = x/(x^2 + 1)$ on $[3, \infty)$. This is our first example on an unbounded interval. If f uc on this interval?*

Solution *Let $\epsilon > 0$ be chosen. As usual, consider, for x and y in this interval*

$$|f(x) - f(y)| \quad = \quad |x/(x^2 + 1) - y/(y^2 + 1)| = \left| \frac{x(y^2 + 1) - y(x^2 + 1)}{(x^2 + 1)(y^2 + 1)} \right|$$

Rewrite this using an add and subtract trick:

$$|f(x) - f(y)| = \left|\frac{xy^2 - yx^2 + x - y}{(x^2+1)(y^2+1)}\right| = \left|\frac{xy^2 - x^3 + x^3 - yx^2 + x - y}{(x^2+1)(y^2+1)}\right|$$

$$= \left|\frac{x(y^2 - x^2) + x^2(x-y) + x - y}{(x^2+1)(y^2+1)}\right| = \left|\frac{x(y-x)(x+y) + x^2(x-y) + x - y}{(x^2+1)(y^2+1)}\right|$$

Then we have

$$|f(x) - f(y)| \leq \left|\frac{|x|\,|x-y|\,|x+y| + x^2|x-y| + |x-y|}{(x^2+1)(y^2+1)}\right| = |x-y|\left|\frac{|x|\,|x+y| + x^2 + 1}{(x^2+1)(y^2+1)}\right|$$

$$\leq |x-y|\left|\frac{|x|\,(|x|+|y|) + x^2 + 1}{(x^2+1)(y^2+1)}\right|$$

$$\leq |x-y|\left\{\left|\frac{|x|^2}{x^2+1}\frac{1}{y^2+1}\right| + \left|\frac{|x|}{x^2+1}\frac{|y|}{y^2+1}\right| + \left|\frac{x^2+1}{x^2+1}\frac{1}{y^2+1}\right|\right\}$$

But $x^2/(x^2+1) < 1$, $y^2/(y^2+1) < 1$, $|x|/(x^2+1) < 1$ and $|y|/(y^2+1) < 1$ for all x and y so we have

$$|f(x) - f(y)| \leq |x-y|\left\{\frac{1}{y^2+1} + (1)(1) + \frac{1}{y^2+1}\right\}$$

Finally $1/(y^2+1) < 1$ here so

$$|f(x) - f(y)| \leq |x-y|\,(1+1+1) = 3|x-y|.$$

We set this **overestimate** *less than ϵ to find $3|x-y| < \epsilon$. Thus, if we choose $0 < \delta < \epsilon/3$, we find $|f(x) - f(y)| < \epsilon$. This shows $f(x) = x/(x^2+1)$ is uc on $[3, \infty)$.*

Example 17.1.4 *Let $f(x) = 2x^2 + 6x + 3$ on $[3, 7]$. Show f is uc on this interval.*

Solution *Let $\epsilon > 0$ be chosen. As usual, consider, for x and y in this interval*

$$|f(x) - f(y)| = |(2x^2 + 6x + 3) - (2y^2 + 6y + 3)| = |2(x^2 - y^2) + 6(x-y)|$$

$$\leq 2|x-y|\,|x+y| + 6|x-y| = |x-y|\,(2|x+y| + 6)$$

Now on this interval $3 + 3 < x + y < 7 + 7$ implying $|x+y| < 14$. Thus,

$$|f(x) - f(y)| < (28 + 6)\,|x-y| = 34|x-y|$$

We set this **overestimate** *less than ϵ to find $34|x-y| < \epsilon$. Thus, if we choose $0 < \delta < \epsilon/34$, we find $|f(x) - f(y)| < \epsilon$. This shows $f(x) = 2x^2 + 6x + 3$ is uc on $[3, 7]$.*

17.1.1 Homework

Exercise 17.1.1

If f and g are uniformly continuous on $[2, 8]$, give and $\epsilon - \delta$ proof that $2f - 11g$ is uniformly continuous on $[2, 8]$.

Exercise 17.1.2

If f and g are uniformly continuous on $[-12, 18]$, give and $\epsilon - \delta$ proof that $7f + 2g$ is uniformly continuous on $[-12, 18]$.

Exercise 17.1.3

Prove $f(x) = 2/x$ is not uniformly continuous on $(0, 2)$.

Exercise 17.1.4

Prove $f(x) = \ln(x)$ is uniformly continuous on $[1, \infty)$.

Exercise 17.1.5

Prove $f(x) = \ln(x)$ is not uniformly continuous on $(0, 1)$.

Exercise 17.1.6

Prove $f(x) = 2x^2 + 6$ is uniformly continuous on $(-2, 1)$.

Exercise 17.1.7

Prove $f(x) = 3x/(x^2 + 2)$ is uniformly continuous on $(-5, \infty)$.

Exercise 17.1.8

Prove $f(x) = 3x^2 + 7x - 3$ is uniformly continuous on $[2, 7]$.

Exercise 17.1.9

Prove $f(x) = 3x^2$ is not uniformly continuous on $[2, \infty)$.

Exercise 17.1.10

Prove $f(x) = \sqrt{x}$ is uniformly continuous on $[1, \infty)$.

17.2   Uniform Continuity and Differentiation

Here is another way to prove uniform continuity.

Example 17.2.1 *We consider $f(x) = x^2 + 5$ on the interval $(2, 3)$. Prove f is uc on $(, 23)$*

Solution *Let $\epsilon > 0$ be arbitrary. Since f is differentiable, by the MVT,*

$$|f(x) - f(y)| \quad = \quad |f'(c)| \, |x - y| = |2c| \, |x - y|$$

*where c is between x and y. Now for any x and y in $(2, 3)$, $4 < 2c < 6 = 6$ and so $|f'(c)| \leq 6$ on our interval. Thus we have $|f(x) - f(y)| \leq 6 \, |x - y|$. We want $6 \, |x - y| < \epsilon$ which implies $|x - y| < \epsilon/6$. So any $\delta < \epsilon/6$ will work. Thus, $f(x) = x^2 + 5$ is **uniformly continuous** on $(2, 3)$.*

There is a more general condition of f which gives us uc. Recall

Definition 17.2.1 Lipschitz Conditions on a Function

> *We say $f(x)$ is Lipschitz on the interval I if there is an $L > 0$ so that*
>
> $$|f(x) - f(y)| \leq L \, |x - y| \text{ for all } x \text{ and } y \text{ in } I$$

Note if f has a bounded derivative on I, then using a MVT argument, we see $|f(x) - f(y)| \leq |f'(c)| \, |x - y| \leq L \, |x - y|$ where $L > 0$ is any upper bound for f' on I. So we have some theorems which are fairly easy to prove:

1. f differentiable with bounded derivative on I (I need not be bounded!) implies f is uc on I.

2. f Lipschitz on I implies f is uc on I.

Example 17.2.2 *We consider $f(x) = \ln(x)$ on the interval $[3, \infty)$. Is f uc on this interval?*

Solution *Let $\epsilon > 0$ be arbitrary. Since f is differentiable, by the MVT,*

$$|f(x) - f(y)| \quad = \quad |f'(c)| \, |x - y| = |1/c| \, |x - y|$$

where c is between x and y. Now for any x and y in $[3, \infty)$, $1/c \leq 1/3$ and so $|f'(c)| \leq 1/3$ on our interval. Thus we have $|f(x) - f(y)| \leq (1/3) \, |x - y|$. $\delta < 3\epsilon$ will work. Thus, $f(x) = \ln(x)$ is **uniformly continuous** *on $[3, \infty)$.*

Example 17.2.3 *However $f(x) = \ln(x)$ is not uc on $(0, \infty)$.*

Solution *Let $\epsilon = 1/2$. Let $x_n = 1/e^n$ and $y_n = 1/e^{n+1}$ for $n > 1$. Let $\delta > 0$ be arbitrary*

$$|x_n - y_n| = |1/e^n - 1/e^{n+1}| = |(e - 1)|/e^{n+1} < e/e^{n+1} = 1/e^n$$

Since $1/e^n \to 0$, there is an N so that $n > N$ implies $1/e^n < \delta$. Then

$$|\ln(x_n) - \ln(y_n)| = |-n + n + 1| = 1 > \epsilon = 1/2$$

even though $|x_n - y_n| < \delta$ when $n > N$. Since no δ works, we see $\ln(x)$ cannot be uc on $(0, \infty)$.

Just because a function is unbounded on I does not necessarily mean it fails to be uc on I.

Example 17.2.4 *Consider $f(x) = x$ on $[0, \infty)$. We can easily check f is uc on this interval.*

Solution *Then given $\epsilon > 0$, $|f(x) - f(y)| = |x - y| < \epsilon$ if we simply choose $\delta = \epsilon$. Hence this f is uc on the unbounded I even though it is unbounded. Note here f' is bounded.*

This goes back to our earlier theorem that f **differentiable with bounded derivative on I (I need not be bounded!) implies** f **is uc on I.** For $f(x) = x$, $f'(x) = 1$ and so the derivative is bounded. Even if f is continuous and bounded on I it doesn't have to be uc on I.

Example 17.2.5 *Consider $f(x) = \sin(1/x)$ on $(0, 1)$. Is f uc on this interval?*

Solution *Let $\epsilon = .1$ and let $\delta > 0$ be arbitrary. We know the biggest and smallest f can be is 1 and -1 respectively. We can use that to show the lack of uc. First, note $\sin(1/x) = 1$ when $1/x = \pi/2 + 2\pi n$ for $n > 1$ and $\sin(1/x) = -1$ when $1/x = 3\pi/2 + 2\pi n$ for $n > 1$. Let $x_n = 1/(\pi/2 + 2\pi n)$ and $y_n = 1/(3\pi/2 + 2\pi n)$. Then*

$$
\begin{aligned}
|x_n - y_n| \quad &= \quad |1/(\pi/2 + 2\pi n) - 1/(3\pi/2 + 2\pi n)| \\
&= \quad \left| \frac{(3\pi/2 + 2\pi n) - (\pi/2 + 2\pi n)}{(\pi/2 + 2\pi n)\,(3\pi/2 + 2\pi n)} \right| = \frac{\pi}{(\pi/2 + 2\pi n)\,(3\pi/2 + 2\pi n)} \\
&< \quad \pi/(4\pi^2 n^2) = 1/(4\pi n^2)
\end{aligned}
$$

Since $1/(4\pi n^2) \to 0$, there is N so that $n > N$ implies $1/(4\pi n^2) < \delta$. But $|\sin(x_n) - \sin(y_n)| = |1 - (-1)| = 2 > 0.1$. So f cannot be uc on I here.

We can also prove the usual algebraic rules.

Theorem 17.2.1 Addition of Uniformly Continuous Functions is Uniformly Continuous

> *If f and g are uc on I, so is $\alpha f + \beta g$.*

Proof 17.2.1
Case 1: $f + g$ is uc on I:

*Let $\epsilon > 0$ be chosen. Then since f is uc on I, there is a $\delta_1 > 0$ so that $|x - y| < \delta_1 \Longrightarrow$
$|f(x) - f(y)| < \epsilon/2$.*

    *Since g is uc on I, there is a $\delta_2 > 0$ so that $|x - y| < \delta_2 \Longrightarrow |g(x) - g(y)| < \epsilon/2$. Thus, if
$\delta = \min(\delta_1, \delta_2)$, we have $|x - y| < \delta$ implies*

$$
\begin{aligned}
|(\,f(x) + g(x)\,) - (\,f(y) - g(y)\,)| &\leq |f(x) - f(y)| + |g(x) - g(y)| \\
&< \epsilon/2 + \epsilon/2 = \epsilon
\end{aligned}
$$

which shows $f + g$ is uc on I.

*Case 2: αf is uc on I for all α. The choice $\alpha = 0$ is easy to see so we can assume $\alpha \neq 0$. Then since
f is uc on I, there is a $\delta > 0$ so that $|x - y| < \delta \Longrightarrow |f(x) - f(y)| < \epsilon/|\alpha|$.*

    Thus $|x - y| < \delta$ implies

$$
|\alpha f(x) - \alpha f(y)| = |\alpha||f(x) - f(y)| < |\alpha|\,\epsilon/|\alpha| = \epsilon
$$

*and we see αf is uc on I. Combining, we see $\alpha f + \beta g$ is uc on I because this is the sum of two uc
functions on I.* ∎

Comment 17.2.1 *Is it true that the product of two uc functions on I is also uc?*

No*: let $f(x) = x$ and $g(x) = x$. Both f and g are uc on $[0, \infty)$ but fg is the function x^2 and it is
not uc on $[0, \infty)$.*

17.2.1   Homework

Exercise 17.2.1 *Prove f differentiable with bounded derivative on I (I need not be bounded!) implies
f is uc on I.*

Exercise 17.2.2 *Prove f Lipschitz on I implies f is uc on I.*

Exercise 17.2.3 *If f and g are uniformly continuous on $[1, 4]$, give an $\epsilon - \delta$ proof that $2f - 5g$ is also
uniformly continuous on $[1, 4]$.*

Exercise 17.2.4 *If f and g are uniformly continuous on $[-1, 6]$, give an $\epsilon - \delta$ proof that $3f + 7g$ is
also uniformly continuous on $[-1, 6]$.*

Exercise 17.2.5 *If f and g are Lipschitz on $[1, 4]$, give an $\epsilon - \delta$ proof that $2f - 5g$ is also Lipschitz
on $[1, 4]$.*

Exercise 17.2.6 *If f and g are Lipschitz on $[-1, 6]$, give an $\epsilon - \delta$ proof that $3f + 7g$ is also Lipschitz
on $[-1, 6]$.*

Chapter 18

Cauchy Sequences of Real Numbers

18.1 Cauchy Sequences

Here is a new way to think about sequences of numbers.

Definition 18.1.1 Cauchy Sequence

> *We say the sequence (a_n) is a* **Cauchy Sequence** *is $\forall \epsilon > 0 \, \exists \, N \ni |a_n - a_m| < \epsilon$ when n and m exceed N.*

Note there is **no** mention of a **limit** here!

Theorem 18.1.1 If a Sequence Converges it is a Cauchy sequence

> *If (a_n) converges, it is a Cauchy sequence*

Proof 18.1.1
If (a_n) converges, there is a number a so that given $\epsilon > 0$, there is N so that $|a_n - a| < \epsilon/2$ when $n > N$. So consider $|a_n - a_m| = |a_n - a + a - a_m| \leq |a_n - a| + |a - a_m| < \epsilon/2 + \epsilon/2 = \epsilon$ when $n, m > N$. Thus a convergent sequence is a Cauchy sequence. ∎

Comment 18.1.1 *Consider the decimal expansion of*

$$\sqrt{2} = 1.41421 \cdots \Longrightarrow r_1 = 4/10, \ r_2 = 41/100, \ r_3 = 414/1000, \ldots$$

Then the sequence $(x_n = 1.r_1 \, r_2 \cdots r_n)$ is a Cauchy sequence as given $\epsilon > 0$, there is N so that $10^{-N} < \epsilon/2$. Then for $n > m > N$,

$$|x_n - x_m| = \frac{r_{m+1} \cdots r_n}{10^{n+m}} < \frac{2}{10^m} < \epsilon$$

Homework

Exercise 18.1.1 *Prove Cauchy sequences are bounded.*

Exercise 18.1.2 *Let $(2 + 3/n)$ be a sequence defined for $n \geq 1$. Prove it is a Cauchy sequence.*

Exercise 18.1.3 *Prove the decimal expansion of $\sqrt{3}$ is a Cauchy sequence.*

Exercise 18.1.4 *Prove the decimal expansion of $\sqrt{31}$ is a Cauchy sequence.*

Exercise 18.1.5 *If (x_n) and (y_n) are Cauchy sequences, give and $\epsilon - \delta$ argument that $(2x_n - 5y_n)$ is also a Cauchy sequence.*

Exercise 18.1.6 *If (x_n) and (y_n) are Cauchy sequences, give and $\epsilon - \delta$ argument that $(2x_n \, y_n)$ is also a Cauchy sequence.*

18.2   Completeness

The topic of completeness is quite abstract. It is really about a **set of objects** for which we have a way of **measuring** distance between objects and **what happens to Cauchy sequences** in this set. Let (X, d) be such a set of objects where d is the mechanism by which distance is measured. Formally, this means d is a function of two variables which assigns a nonnegative number to each pair of objects x and y from X. This is well defined and we see $d(x, y) = 0$ if and only if $x = y$.

- X is all the rational numbers. The distance between two rational numbers is just the usual $|x - y|$ where x and y are both rational.

- X is the set of all continuous functions on the interval $[0, 1]$. Given x and y in X, $x - y$ is continuous too. Since compositions of continuous functions are continuous, $|x - y|$ is another continuous function since $|\cdot|$ is continuous. Thus $|x - y|$ has a maximum value on $[0, 1]$. We define the distance between x and y by $d(x, y) = \max_{0 \leq t \leq 1} |x(t) - y(t)|$. This is well defined and we see $d(x, y) = 0$ if and only if $x = y$. X is usually called $C([0, 1])$ but we typically drop the enclosing parenthesis and just say $C([0, 1])$.

Given a set of objects, there is often more than one way to measure the distance between two objects. For the spaces mentioned above, here are some alternatives. We have not yet discussed Riemann integration in detail, but you have all had Calculus already, so these definitions should make perfect sense!

- For $C([0, 1])$, we can measure distance another way. Since $|x - y|$ is continuous, the Riemann Integral of $|x - y|$ on $[0, 1]$ is well defined. We can set the distance between x and y to be $d(x, y) = \int_0^1 |x(t) - y(t)| dt$. We can show the distance is zero if and only if $x = y$.

- For $C([0, 1])$, we can measure distance yet another way. Since $|x - y|$ is continuous, $|x - y|^2$ is continuous too and so the Riemann Integral of $|x - y|^2$ on $[0, 1]$ is well defined. We can set the distance between x and y to be $d(x, y) = \sqrt{\int_0^1 |x - y|^2 \, dt}$. We can show the distance is zero if and only if $x = y$.

There are many other such pairings between a set of objects X and a way to measure distance. Such a pairing needs to be described as the set plus its way of measuring distance; hence, we use the notation (X, d) to reflect this. We call d a **metric** on the space of objects X. We usually don't say *space of objects* and just say *space*. The pair is then called a **metric space**. Our four examples are then

- $(\mathbb{Q}, |\cdot|)$

- $(C([0, 1]), d_\infty)$ where $d_\infty(x, y) = \max_{0 \leq t \leq 1} |x(t) - y(t)|$. This is also denoted by $||x - y||_\infty$.

- $(C([0, 1]), d_1)$ where $d_1(x, y) = \int_0^1 |x(t) - y(t)| \, dt$. This is also denoted by $||x - y||_1$.

- $(C([0, 1]), d_2)$ where $d_2(x, y) = \sqrt{\int_0^1 |x(t) - y(t)|^2 \, dt}$. This is also denoted by $||x - y||_2$.

In $(\mathbb{Q}, |\cdot|)$, Cauchy sequences of rationals need not converge to a rational number. For example $(1 + 1/n)^n$ converges to the irrational number e. The set of rationals $\mathbb{Q}$ with the metric $|\cdot|$ is what is called **not complete** as some Cauchy sequences of rationals do not converge to a rational number. That is the limit of the Cauchy sequence of rationals is **not a member of the original set of objects**. These "*missing objects*" are called the **irrational numbers**, $\mathbb{I}$. Adding $\mathbb{I}$ to the set $\mathbb{Q}$ gives the real number line $\Re$. It turns out Cauchy sequences of objects from $\Re$ with the $|\cdot|$ metric do converge to objects in $\Re$. So $(\Re, |\cdot|)$ is called **complete**. We already know if a sequence in $\Re$ converges, it is a Cauchy sequence. Since $\Re$ is complete, we can also say all Cauchy sequence of real numbers converge. So we have a powerful theorem (which we can't prove yet at the level of abstraction we have in this course!).

Theorem 18.2.1 A Real Sequence is a Cauchy sequence if and only if it Converges

> (a_n) *is a Cauchy sequence in* $\Re$ *if and only if* (a_n) *converges.*

Proof 18.2.1
We know if a sequence converges, it is also a Cauchy sequence. The other direction can be proved. A full proof is given in (Peterson (14) 2020) and involves constructing what is called a field from equivalence classes of rational sequences in just the right way. ■

We can show $(C([0,1]), ||\cdot||_\infty)$ is also **complete** and we will do that in a bit. So a Cauchy sequence of continuous functions **converges** to another continuous function. The proof will use the idea that $\Re$ is complete.

We can show also $(C([0,1]), ||\cdot||_1)$ is **not complete**. We can easily construct Cauchy sequences in this metric which converge to a function which is not continuous. In fact, there are infinitely many such **limit** functions which are all zero distance apart in this metric. We will go through the details in (Peterson (14) 2020).

We can show $(C([0,1]), ||\cdot||_2)$ is also **not complete**. The arguments for this one are very similar to the ones we will use for the $||\cdot||_1$ way of measuring distance.

18.3 The Completeness of the Real Numbers

Let's look more carefully at what is called the completeness of $\Re$. Let $(\mathbb{Q}, |\cdot|) = X$. This is a nice **metric space** where $|x - y|$ measures the **distance** between the two rational numbers x and y. The arguments we use here are quite abstract but this space is a lot simpler than the ones with functions and so we can get across the basic ideas with as little mess as possible. We prove this carefully in (Peterson (14) 2020) by showing how we can extend the field $\mathbb{Q}$ to another field $\widetilde{\mathbb{Q}}$ which is totally ordered, satisfies the Completeness axiom (i.e. the least upper bound and greatest lower bound property) and in which Cauchy sequences of objects converge to an object in $\widetilde{\mathbb{Q}}$. This new field is then identified with $\Re$. There is a general process by which a metric space can be completed which we can illustrate by using the rational numbers as a guide. We actually don't use this process to construct the real numbers, but it will show you the steps we typically take.

Here are a few details of this construction. We already know Cauchy sequences of rational numbers need not converge to a rational number. A nice example is the sequence $x_n = (1 + 1/n)^n$ which we know converges to a number we call e. The proof of this is a bit difficult and you should go back and look at this material in Chapter 13 to refresh your understanding of this. Let's define a new metric space which we will call Y. Y is the set of all Cauchy sequences of rational numbers; i.e. the *objects* in our space are Cauchy sequences! Note each rational number p/q forms a nice constant sequence

$x_1 = p/q, x_2 = p/q, \ldots, x_n = p/q, \ldots$. We can denote this constant sequence by (p/q). So for example $(2/3)$ is the constant Cauchy sequence whose entries are all $2/3$. **Note** $(1 + 1/n)^n$ **is also a Cauchy sequence** as we know it converges (using all the arguments in Chapter 13) and so it must be a Cauchy Sequence.

We need a metric for Y. Define the distance between two Cauchy sequences in Y like this: $D((x_n), (y_n)) = \lim_{n \to \infty} |x_n - y_n|$. We can show this limit exists when we construct the field $\widetilde{\mathbb{Q}}$. The objects in Y divide naturally into **classes** called *equivalence classes*. You should go and look that up if you haven't seen that idea before. We just need you to know roughly what it means for our discussion here. Given any object from Y, (x_n), we let $[(x_n)]$ denote the collection of all other objects from Y, i.e. other Cauchy sequences of rational numbers, whose distance to (x_n) is zero. We call this set of equivalence classes $\tilde{Y}$ and we define the distance, $\tilde{D}$ between two equivalence classes as follows: $\tilde{D}([(x_n)], [(y_n)]) = \lim_{n \to \infty} |x_n - y_n|$. Of course, we would have to show the value of $\tilde{D}$ does not depend on the choice of representatives from the equivalence classes!

For example, the constant sequence $(3/5)$ is in Y and there are an infinite number of other sequences (a_n) so that $D((3/5), (a_n)) = 0$. Just let (b_n) be any sequence of rational numbers that converges to 0. Then $D((3/5), (3/5) + (b_n)) = 0$ and so $(3/5) + (b_n)$ is a member of $[(3/5)]$.

This is the big point now! The sequence $((1 + 1/n)^n)$ does not converge to a rational number and so it can not be in the equivalence class associated to any rational number $[p/q]$. Another way of saying this is that $D(((1 + 1/n)^n), (p/q)) \neq 0$ for all $p/q \in \mathbb{Q}$. The equivalence class $[((1 + 1/n)^n)]$ is thus **different** from the equivalence classes formed from constant sequences of rationals. The collection of all equivalence classes of objects from Y can thus be identified in a natural way with the numbers we see in $\Re$ using the field construction we mentioned earlier.

- Each constant rational sequence (p/q) is chosen as the representative of $[(p/q)]$.

- Each equivalence class that is different from the equivalence classes formed by constant rational sequences is identified with some representative from it. We call that α. Note it cannot come from a constant rational sequence so it can not be a rational number. We generally call this an **irrational number**. But remember, in this context, it is really a Cauchy sequence of rationals!

- It is hard, but in a more advanced class, we show Cauchy sequences in $(\tilde{Y}, \tilde{D})$ converge to an object in $(\tilde{Y}, \tilde{D})$. So we can prove $(\tilde{Y}, \tilde{D})$ is a complete metric space.

- We can do this construction process for any metric space (X, d) and build a new complete metric space $(\tilde{Y}, \tilde{D})$. We do this in a graduate course in analysis such as (Peterson (14) 2020).

So $\Re$ is the completion of the metric space $(\mathbb{Q}, |\cdot|)$ as outlined above.

We said $(C([0,1]), ||\cdot||_1)$ is not complete. In a first graduate course in analysis covered in (Peterson (14) 2020), we find the completion of $(C([0,1]), ||\cdot||_1)$ can be done following this construction process and generates the space $(\mathcal{L}_1, ||\cdot||_1)$ which is a space of equivalence classes of functions and to do this right we also have to extend our notion of Riemann Integration to something called Lebesgue Integration which is done in (Peterson (12) 2020).

We said $(C([0,1]), ||\cdot||_2)$ is not complete. We can also find the completion of $(C([0,1]), ||\cdot||_2)$ following this construction process. This generates the space $(\mathcal{L}_2, ||\cdot||_2)$ which is a space of equivalence classes of functions using Lebesgue Integration. This is real special and it turns out to be an inner product space which is complete. This is called a **Hilbert Space**.

The space $(C([0,1]), ||\cdot||_\infty)$ is complete as we will show in a bit. If we do the construction process as outlined earlier, we just get back the same space: (X, d) and (Y, D) will be the same here.

18.4   Uniform Continuity and Compact Domains

Let's look more carefully at continuous functions on compact domains. We can prove a nice theorem:

Theorem 18.4.1 Continuity on a Compact Domain implies Uniform Continuity

> *If f is continuous on the compact interval $I = [a,b]$, then f is uniformly continuous on I.*

Proof 18.4.1
We are going to prove this by **contradiction**. *If f is not uc on I, there is an ϵ_0 so that*

$$\forall \delta > 0, \ \exists x, y \in I \ni |x - y| < \delta \text{ and } |f(x) - f(y)| > \epsilon_0$$

In particular for the choice $\delta_n = 1/n$ for all $n \geq 1$, we have

$$\exists \, x_n, y_n \in I \ni |x_n - y_n| < 1/n \text{ and } |f(x_n) - f(y_n)| \geq \epsilon_0$$

Since (x_n) and (y_n) are contained in the compact set I, the Bolzano - Weierstrass Theorem tells us there are subsequences (x_n^1) and (y_n^1) and points x and y in I so that $x_n^1 \to x$ and $y_n^1 \to y$.
Claim 1: $x = y$
To see this, note for a tolerance ϵ', there are integers N_1 and N_2 so that

$$n > N_1 \implies |x_n^1 - x| < \epsilon'/6 \text{ when } n^1 > N_1$$
$$n > N_2 \implies |y_n^1 - y| < \epsilon'/6 \text{ when } n^1 > N_2$$

where n^1 indicates the subsequence index. Now pick any subsequence index greater than $\max(N_1, N_2)$. Call these subsequence elements $x_{\hat{n}}^1$ and $y_{\hat{n}}^1$. Also choose the subsequence index so that $1/\hat{n}^1 < \epsilon'/6$. So both conditions hold for this choice. Thus,

$$|x - y| \ = \ |x - x_{\hat{n}}^1 + x_{\hat{n}}^1 - y_{\hat{n}}^1 + y_{\hat{n}}^1 - y| \leq |x - x_{\hat{n}}^1| + |x_{\hat{n}}^1 - y_{\hat{n}}^1| + |y_{\hat{n}}^1 - y|$$

The first and last are less than $\epsilon'/6$, so we have

$$|x - y| \ \leq \ |x_{\hat{n}}^1 - y_{\hat{n}}^1| + \epsilon'/3$$

Now remember, we know $|x_{\hat{n}}^1 - y_{\hat{n}}^1| < 1/\hat{n}^1$. So we have

$$|x - y| \ \leq \ 1/\hat{n}^1 + \epsilon'/3 < \epsilon'/6 + \epsilon'/3 = 2\epsilon'/3 < \epsilon'$$

Since ϵ' is arbitrary, we see $x = y$ by a standard argument. Of course, this also means $f(x) = f(y)$ which says $|f(x) - f(y)| = 0$. Claim 2: $|f(x) - f(y)| \geq 2\epsilon_0/3$. Since $x_n^1 \to x$ and $y_n^1 \to y = x$ and f is continuous on I, we have

$$\exists M_1 \ni |f(x_n^1) - f(x)| < \epsilon_0/6 \ \ \forall n^1 > M_1 \quad \text{and} \quad \exists M_2 \ni |f(y_n^1) - f(y)| < \epsilon_0/6 \ \ \forall n^1 > M_2$$

where again the indices for these subsequences are denoted by n^1. Pick a fixed $n^1 > \max(M_1, M_2)$ and then both conditions hold. We can say

$$\begin{aligned}
\epsilon_0 \ &\leq \ |f(x_n^1) - f(y_n^1)| = |f(x_n^1) - f(x) + f(x) - f(y) + f(y) - f(y_n^1)| \\
&\leq \ |f(x_n^1) - f(x)| + |f(x) - f(y)| + |f(y) - f(y_n^1)| \leq |f(x) - f(y)| + \epsilon_0/3
\end{aligned}$$

This tells us $|f(x) - f(y)| \geq 2\epsilon_0/3$. But we also know $|f(x) - f(y)| = 0$. This contradiction tells us our assumption that f is not uc on I is wrong. Thus f is uc on I. ■

Comment 18.4.1 *This result is true for a continuous function on any compact set D of $\Re^n$ although we would have to use the Euclidean norm $\|\cdot\|$ to do the proof.*

So continuity and compactness are linked again. Recall continuous functions on compact sets must have an absolute minimum and absolute maximum too.

18.4.1   Homework

Provide a careful proof of these propositions.

Exercise 18.4.1 *Prove $\sqrt{x}$ is not Lipschitz on $[0, 1]$.*
Comment: the thing here is that you can't find an $L > 0$ that will work. You know if it works you have $|\sqrt{x} - \sqrt{y}| \leq L|x - y|$ holds for any x, y in $[0, 1]$. So let $y = 0$ and see what is happening there. Note it is easy to see why it fails but your job is to write your argument mathematically clear.

Exercise 18.4.2 *Prove $\sqrt{x}$ is continuous on $[0, 1]$ using an $\epsilon - \delta$ argument.*
Comments: there are two cases here: the case $p = 0$ and the others, $p \in (0, 1]$. for the first case, given ϵ, just pick $\delta = \epsilon^2$ (details left to you); for the other case, this is the multiply by $(\sqrt{x} + \sqrt{p})/(\sqrt{x} + \sqrt{p})$ trick.

Exercise 18.4.3 *Prove $\sqrt{x}$ is uniformly continuous on $[0, 1]$ the easy way.*

The exercise below shows a function can be uc on its domain even though it has an unbounded derivative there.

Exercise 18.4.4   • *prove $\sqrt{x}$ is uniformly continuous on $[0.5, \infty)$ using the MVT.*

• *prove $\sqrt{x}$ is uniformly continuous on $[0, 1]$ using our theorem.*

• *use these two facts to prove $\sqrt{x}$ is uniformly continuous on $(0, \infty)$ even though it has an unbounded derivative on $(0, \infty)$.*

Exercise 18.4.5 *Prove $\sqrt{x}$ is not Lipschitz on $[0, 1]$.*
Comment: the thing here is that you can't find an $L > 0$ that will work. You know if it works you have $|\sqrt{x} - \sqrt{y}| \leq L|x - y|$ holds for any x, y in $[0, 1]$. So let $y = 0$ and see what is happening there. Note it is easy to see why it fails but your job is to write your argument mathematically clear.

Exercise 18.4.6 *Prove $\sqrt{x}$ is continuous on $[0, 1]$ using an $\epsilon - \delta$ argument.*
Comments: there are two cases here: the case $p = 0$ and the others, $p \in (0, 1]$. for the first case, given ϵ, just pick $\delta = \epsilon^2$ (details left to you); for the other case, this is the multiply by $(\sqrt{x} + \sqrt{p})/(\sqrt{x} + \sqrt{p})$ trick.

Exercise 18.4.7 *Prove $\sqrt{x}$ is uniformly continuous on $[0, 1]$ the easy way.*

Exercise 18.4.8   • *prove $x^{1/3}$ is uniformly continuous on $[2, \infty)$ using the MVT.*

• *prove $x^{1/3}$ is uniformly continuous on $[0, 2]$ using our theorem.*

• *use these two facts to prove $x^{1/3}$ is uniformly continuous on $(0, \infty)$ even though it has an unbounded derivative on $(0, \infty)$.*

Exercise 18.4.9 *Prove $x^{1/3}$ is not Lipschitz on $[0, 1]$.*

Exercise 18.4.10 *Prove $x^{1/3}$ is continuous on $[0, 1]$ using an $\epsilon - \delta$ argument.*

Exercise 18.4.11 *Prove if f is differentiable on* $[1, 2]$ *with* f' *continuous on* $[a, b]$, *then* f' *is uniformly continuous on* $[1, 2]$.

Chapter 19

Series of Real Numbers

19.1 Series of Real Numbers

Given any sequence of real numbers, (a_n) for $n \geq 1$, we can construct from it a new sequence, called the sequence of **Partial Sums**, as follows:

$$
\begin{aligned}
S_1 &= a_1 \\
S_2 &= a_1 + a_2 \\
S_3 &= a_1 + a_2 + a_3 \\
&\vdots \\
S_n &= a_1 + a_2 + a_3 + \ldots + a_n
\end{aligned}
$$

This notation works really well when the sequence starts at $n = 1$. If the sequence was

$$
(a_n)_{n \geq -3} = \{a_{-3}, a_{-2}, a_{-1}, a_0, a_1, \ldots, a_n, \ldots\}
$$

we usually still start with S_1 but we let

$$
\begin{aligned}
S_1 &= a_{-3} \\
S_2 &= a_{-3} + a_{-2} \\
S_3 &= a_{-3} + a_{-2} + a_{-1}
\end{aligned}
$$

So when we talk about the first partial sum, S_1, in our minds we are thinking about a sequence whose indexing starts at $n = 1$ but, of course, we know it could apply to a sequence that starts at a different place. A similar problem occurs if the sequence is

$$
(a_n)_{n \geq 3} = \{a_3, a_4, \ldots, a_n, \ldots\}
$$

Then,

$$
\begin{aligned}
S_1 &= a_3 \\
S_2 &= a_3 + a_4 \\
S_3 &= a_3 + a_4 + a_5
\end{aligned}
$$

So the only time the indexing of the partial sums nicely matches the indexing of the original sequence is when the sequence starts at $n = 1$. This is usually not an issue: we just adjust mentally for the new start point. For our pedagogical purposes, we will just assume all the sequences in our discussions start at $n = 1$ for convenience!

For our more convenient sequences, we can use summation notation to write the partial sums. We have

$$S_n = \sum_{i=1}^{n} a_i$$

and note the choice of letter i here is immaterial. The use of i could have been changed to the use of the letter j and so forth. It is called a *dummy variable* of summation just like we have *dummy variables* of integration.

$$S_n = \sum_{i=1}^{n} a_i = \sum_{j=1}^{n} a_j = \sum_{k=1}^{n} a_k$$

and so forth. Now we are interested in whether or not the sequence of partial sums of a sequence converges. We need more notation.

Definition 19.1.1 The Sum of a Series or convergence of a Series

> *Let $(a_n)_{n \geq 1}$ be any sequence and let $(S_n)_{n \geq 1}$ be its associated sequence of partial sums.*
> *(a) If $\lim_{n \to \infty} S_n$ exists, we denote the value of this limit by S. Since this is the same as $\lim_{n \to \infty} \sum_{i=1}^{n} a_i = S$, we often use the symbol $\sum_{i=1}^{\infty} a_i$ to denote S. But remember it is a **symbol** for a limiting process. Again, note the choice of summation variable i is immaterial. We also say $\sum_{i=1}^{\infty} a_i$ is the **infinite series** associated with the sequence $(a_n)_{n \geq 1}$.*
> *(b) If the $\lim_{n \to \infty} S_n$ does not exist, we say the **series** $\sum_{i=1}^{\infty} a_i$ **diverges**.*

Comment 19.1.1 *Note the divergence of a series can be several things: we say the series $\sum_{i=1}^{\infty} (a_i = 1)$ **diverges** to ∞, the series $\sum_{i=1}^{\infty} (a_i = -i)$ **diverges** to $-\infty$ and the series $\sum_{i=1}^{\infty} (a_i = (-1)^i)$ **diverges by oscillation**.*

We can get a lot of information about a series by looking at the series we get by summing the absolute values of the terms of the base sequence used to construct the partial sums.

Definition 19.1.2 Absolute Convergence and Conditional Convergence

> *Let $(a_n)_{n \geq 1}$ be any sequence and let $\sum_{i=1}^{\infty} |a_i|$ be the series we construct from (a_n) by taking the absolute value of each term a_i in the sequence.*
> *(a) if $\sum_{i=1}^{\infty} |a_i|$ **converges** we say the series **converges absolutely**.*
> *(b) (a) if $\sum_{i=1}^{\infty} |a_i|$ **diverges** but $\sum_{i=1}^{\infty} a_i$ **converges**, we say the series **converges conditionally**.*

19.2   Basic Facts about Series

Theorem 19.2.1 The nth Term Test

> *Assume $\sum_{n=1}^{\infty} a_n$ converges, then $a_n \to 0$. Note this says if a_n does **not** converge to 0 or fails to converge at all, the original series must **diverge**.*

Proof 19.2.1

Since $\sum_{n=1}^{\infty} a_n$ converges, we know $S_n \to S$ for some S. Thus, given $\epsilon > 0$, there is an N so that

$$n > N \implies |S_n - S| < \epsilon/2$$

Now pick any $\hat{n} > N+1$. Then $\hat{n}-1$ and $\hat{n}$ are both greater than N. Thus, since $a_{\hat{n}}$ is the difference between two successive terms in (S_n), we have

$$|a_{\hat{n}}| \;=\; |S_{\hat{n}} - S + S - S_{\hat{n}-1}| \le |S_{\hat{n}} - S| + |S - S_{\hat{n}-1}| < \epsilon/2 + \epsilon/2 = \epsilon$$

Since the choice of $\hat{n} > N$ was arbitrary, we see we have shown $|a_n| < \epsilon$ for $n > N$ with the choice of $\epsilon > 0$ arbitrary. Hence, $\lim_{n \to \infty} a_n = 0$. ∎

Now if the sequence (S_n) converges to S, we also know the sequence is a Cauchy sequence. We also know since $\Re$ is **complete** that if (S_n) is a Cauchy sequence, it must converge. So we can say

$$(S_n) \text{ converges} \iff (S_n) \text{ is a Cauchy sequence}$$

Now (S_n) is a Cauchy sequence means given $\epsilon > 0$ there is an N so that

$$n, m > N \implies |S_n - S_m| < \epsilon$$

For the moment, assume $n > m$. Then

$$S_n - S_m \;=\; \left(\sum_{i=1}^{m} a_i + \sum_{i=m+1}^{n} a_i\right) - \sum_{i=1}^{m} a_i = \sum_{i=m+1}^{n} a_i$$

and for $m < n$, we would get

$$S_m - S_n \;=\; \left(\sum_{i=1}^{n} a_i + \sum_{i=n+1}^{m} a_i\right) - \sum_{i=1}^{n} a_i = \sum_{i=n+1}^{m} a_i$$

We can state this as a Theorem!

Theorem 19.2.2 The Cauchy Criterion for Series

$$\sum_{n=1}^{\infty} a_n \text{ converges} \iff \left(\forall \epsilon > 0 \; \exists \, N \ni \left|\sum_{i=m+1}^{n} a_i\right| < \epsilon \text{ if } n > m > N\right)$$

Now using this Cauchy criterion, we can say more about the consequences of absolute convergence.

Theorem 19.2.3 A Series that Converges Absolutely also Converges

If $\sum_{n=1}^{\infty} a_n$ converges absolutely then it also converges.

Proof 19.2.2

We know $\sum_{n=1}^{\infty} |a_n|$ converges. Let $\epsilon > 0$ be given. By the Cauchy Criterion for series, there is an N so that

$$\big|\, |a_{m+1}| + \ldots + |a_n| \,\big| \;<\; \epsilon \text{ if } n \ge m > N$$

The absolute values here are not necessary so we have

$$|a_{m+1}| + \ldots + |a_n| \; < \; \epsilon \; \textit{if } n > m > N$$

But the triangle inequality then tells us that

$$|a_{m+1} + \ldots + a_m| \leq |a_{m+1}| + \ldots + |a_n| \; < \; \epsilon \; \textit{if } n > m > N$$

Thus $\left| \sum_{i=m+1}^{n} a_i \right| < \epsilon$ *when* $n > m > N$. *Hence, the sequence of partial sums is a Cauchy sequence which tells us the* $\sum_{n=1}^{\infty} a_n$ *converges.* ∎

Comment 19.2.1 *Another note on notation. The difference between the limit of a series and a given partial sum is* $S - S_n$. *Using the series notation for S, this gives*

$$S - S_n \; = \; \sum_{i=1}^{\infty} a_i - \sum_{i=1}^{n} a_i$$

This is often just written as $\sum_{i=n+1}^{\infty} a_i$ *and is called the* $(n+1)^{st}$ *remainder term for the series. We have to be careful about manipulating this as it is really just a* **symbol**.

19.2.1   An ODE Example: Things to Come

We will show that a series of the form $\sum_{n=0}^{\infty} a_n t^n$ makes sense. Of course, this series **depends** on **t** which we haven't really discussed. Consider the ODE

$$(1+t)x''(t) + t^2 x'(t) + 2x(t) \; = \; 0, \quad x(0) = 1, \quad x'(0) = 3$$

We can show the solution of this problem can be written as $\sum_{n=0}^{\infty} a_n t^n$ for a unique choice of numbers a_n and we can also show this series will **converge** on an interval $(-R, R)$ for some $R > 0$. The number R is called the **radius of convergence** of the series $\sum_{n=0}^{\infty} a_n t^n$. Since we want $x(0) = 1$, we see

$$\left(a_0 + a_1 t + a_2 t^2 + \ldots \right)_{t=0} \; = \; 1 \implies a_0 = 1$$

We can also show

$$\left(\sum_{n=0}^{\infty} a_n t^n \right)' \; = \; \sum_{n=1}^{\infty} n\, a_n t^{n-1}, \quad \left(\sum_{n=0}^{\infty} a_n t^n \right)'' = \sum_{n=2}^{\infty} n\,(n-1)\, a_n t^{n-2}$$

and these series converge on $(-R, R)$ also. So we can rewrite the ODE as

$$(1+t)\left(\sum_{n=2}^{\infty} n\,(n-1)\, a_n t^{n-2} \right) + t^2 \left(\sum_{n=1}^{\infty} n\, a_n t^{n-1} \right) + 2\left(\sum_{n=0}^{\infty} a_n t^n \right) = 0$$

and the second initial condition becomes

$$\left(a_1 + 2a_2 t + \ldots \right)_{t=0} \; = \; 3 \implies a_1 = 3$$

We can show we can bring the outside powers of t inside the series. So the ODE becomes

$$\sum_{n=2}^{\infty} n(n-1) a_n t^{n-2} + \sum_{n=2}^{\infty} n(n-1) a_n t^{n-1} + \sum_{n=1}^{\infty} n a_n t^{n+1} + 2 \sum_{n=0}^{\infty} a_n t^n = 0$$

Now we want all powers of t inside the series to have the form t^n. So we make changes to the summation variables in the first three pieces above.

(Piece 1): Let $k = n - 2$ implying we get

$$\sum_{n=2}^{\infty} n(n-1) a_n t^{n-2} = \sum_{k=0}^{\infty} (k+2)(k+1) a_{k+2} t^k$$

(Piece 2): Let $k = n - 1$ implying we get

$$\sum_{n=2}^{\infty} n(n-1) a_n t^{n-1} = \sum_{k=1}^{\infty} (k+1)(k) a_{k+1} t^k$$

(Piece 3): Let $k = n + 1$ implying we get

$$\sum_{n=1}^{\infty} n a_n t^{n+1} = \sum_{k=2}^{\infty} (k-1) a_{k-1} t^k$$

The summation variables are all *dummy variables* so it doesn't matter what letter we use. Switch the first two pieces back to n. We get

$$\sum_{n=0}^{\infty} (n+2)(n+1) a_{n+2} t^n + \sum_{n=1}^{\infty} (n+1)(n) a_{n+1} t^n + \sum_{n=2}^{\infty} (n-1) a_{n-1} t^n + 2 \sum_{n=0}^{\infty} a_n t^n = 0$$

We can rewrite this as

$$(2a_2 + 2a_0)t^0 + (6a_3 + 2a_2 + 2a_1)t^1$$
$$+ \sum_{n=2}^{\infty} \left((n+2)(n+1) a_{n+2} + (n+1)(n) a_{n+1} + (n-1) a_{n-1} + 2a_n \right) t^n = 0$$

The coefficients of each power of t must be zero. So we get

$$2a_2 = -2a_0 = -2 \Longrightarrow a_2 = -1$$
$$6a_3 = -2a_2 - 2a_1 = 2 - 6 = -4 \Longrightarrow a_3 = -2/3$$

and

$$(n+2)(n+1) a_{n+2} + (n+1)(n) a_{n+1} + (n-1) a_{n-1} + 2a_n = 0, \quad \forall n \geq 2$$

The last equation is called a **recursion** equation for the coefficients. Solving for a_{n+2} we have

$$(n+2)(n+1) a_{n+2} = -(n+1)(n) a_{n+1} - 2a_n - (n-1) a_{n-1}$$
$$a_{n+2} = -\frac{(n+1)(n)}{(n+2)(n+1)} a_{n+1} - \frac{1}{(n+2)(n+1)} 2a_n - \frac{(n-1)}{(n+2)(n+1)} a_{n-1}$$
$$= -\frac{n}{n+2} a_{n+1} - \frac{2}{(n+2)(n+1)} a_n - \frac{n-1}{(n+2)(n+1)} a_{n-1}$$

So for $n = 2$, we have

$$a_4 \;=\; -\frac{2}{4}a_3 - \frac{2}{12}a_2 - \frac{1}{12}a_1 = -\frac{1}{2}\left(-\frac{2}{3}\right) - \frac{2}{12}\left(-1\right) - \frac{1}{12}\left(3\right) = \frac{1}{3} + \frac{1}{6} - \frac{1}{4} = \frac{1}{4}$$

and so on. What is remarkable is that we can prove a theorem that tells us this series, whose coefficients are defined recursively, converges! But you should know this is a pretty deep result!

19.2.2   Homework

Exercise 19.2.1 *If $\sum_{n=1}^{\infty} a_n$ and $\sum_{n=1}^{\infty} b_n$ both converge, give an $\epsilon - N$ proof that $\sum_{n=1}^{\infty}(2a_n + 5b_n)$ also converges.*

Exercise 19.2.2 *Show $\sum_{n=1}^{\infty}(-1)^n$ does not converge using the nth term test.*

Exercise 19.2.3 *Show $\sum_{n=1}^{\infty}(-1)^n$ has the partial sums $\{-1, 0, -1, 0, \ldots\}$. Find the $\underline{\lim}\ S_n$ and the $\overline{\lim}\ S_n$ and use that to show the series does not converge.*

Exercise 19.2.4 *Consider the ODE*

$$(3 + 2t)x''(t) + tx'(t) + 4x(t) \;=\; 0, \quad x(0) = -1, \quad x'(0) = 1$$

Find the recursion equation for this ODE.

Exercise 19.2.5 *Consider the ODE*

$$(3 + 5t)x''(t) + t^2 x'(t) + 4tx(t) \;=\; 0, \quad x(0) = 1, \quad x'(0) = -1$$

Find the recursion equation for this ODE.

19.3   Series with Non-Negative Terms

Recall there is a fundamental **axiom** about the behavior of the real numbers which is very important.

Axiom 2 The Completeness Axiom

> *Let S be a set of real numbers which is nonempty and bounded above. Then the supremum of S exists and is finite.*
>
> *Let S be a set of real numbers which is nonempty and bounded below. Then the infimum of S exists and is finite.*

There are powerful consequences of this axiom.

Lemma 19.3.1 The Infimum Tolerance Lemma

> *Let S be a nonempty set of real numbers that is bounded below. Let $\epsilon > 0$ be arbitrarily chosen. Then*
>
> $$\exists y \in S \ni \inf(S) \leq y < \inf(S) + \epsilon$$

Proof 19.3.1
We have done this earlier, but let's refresh your mind. We do this by contradiction. Assume this is not true for some $\epsilon > 0$. Then for all y in S, we must have $y \geq \inf(S) + \epsilon$. But this says $\inf(S) + \epsilon$ must be a lower bound of S. So by the definition of infimum, we must have $\inf(S) \geq \inf(S) + \epsilon$ for

a positive epsilon which is impossible. Thus our assumption is wrong and we must be able to find at least one y in S that satisfies $\inf(S) \le y < \inf(S) + \epsilon$. ∎

There is the corresponding supremum tolerance lemma which we use frequently.

Lemma 19.3.2 The Supremum Tolerance Lemma

> *Let S be a nonempty set of real numbers that is bounded above. Let $\epsilon > 0$ be arbitrarily chosen. Then*
> $$\exists\, y \in S \ni \sup(S) - \epsilon < y \le \sup(S)$$

Proof 19.3.2

Again, we refresh your mind. We do this by contradiction. Assume this is not true for some $\epsilon > 0$. Then for all y in S, we must have $y \le \sup(S) - \epsilon$. But this says $\sup(S) - \epsilon$ must be an upper bound of S. So by the definition of supremum, we must have $\sup(S) \le \sup(S) - \epsilon$ for a positive epsilon which is impossible. Thus our assumption is wrong and we must be able to find at least one y in S that satisfies $\sup(S) - \epsilon < y \le \sup(S)$. ∎

With these tools, we can attack some problems with series.

Theorem 19.3.3 Bounded Increasing or Decreasing Sequences Converge

> *Bounded increasing sequences (a_n) converge to their supremum.*
> *Bounded decreasing sequences (a_n) converge to their infimum.*

Proof 19.3.3

We do the case where (a_n) is defined for $n \ge 1$ and it is increasing and by assumption it is bounded above by some $B > 0$. By the completeness axiom, the set

$$U \;=\; \{a_1, a_2, a_3, \ldots, a_n, \ldots\}$$

is a nonempty set bounded above by B and so we know $\sup(U)$ is a finite number we will denote by A. Let $\epsilon > 0$ be given. Then by the Supremum Tolerance Lemma (henceforth abbreviated by STL for convenience), there is an element of U, a_N for which $A - \epsilon < a_N \le A$. But we also know this sequence is increasing; i.e. $a_N \le a_n$ for all $n > N$. Thus, we can say $A - \epsilon < a_N \le a_n < A$. This says $A - \epsilon < a_n < A + \epsilon$ for all $n > N$. This tells us $|a_n - A| < \epsilon$ for all $n > N$ and so $a_n \to A$.

The argument that shows a bounded decreasing sequence converges to its infimum is quite similar and it is left to you as a homework! ∎

Now let's look at a series of non-negative terms. It is easy to see the sequence of partial sums that corresponds to this sequence is an increasing sequence. We can prove this theorem:

Theorem 19.3.4 Non-negative Series Converge If and Only If They are Bounded

> *If $\sum_{n=1}^{\infty} a_n$ is a series with non-negative terms, i.e. $a_n \ge 0$ for all $n \ge 1$, then the series converges if and only if the sequence of partial sums is bounded.*

Proof 19.3.4

($\Longrightarrow$): Assume $\sum_{n=1}^{\infty} a_n$ converges. Let S denote the value of the limit. Then by definition $\lim_{n\to\infty} S_n = S$. Thus (S_n) is a convergent sequence and we proved in the first course that this implies (S_n) is bounded.

($\Longleftarrow$): If (S_n) is bounded, then there is a $B > 0$ so that $0 \leq S_n < B$ for all n. Thus, the set

$$U \;=\; \{S_1, S_2, S_3, \ldots, S_n, \ldots\}$$

is a nonempty bounded set and so $\sup(U)$ exists and is finite. As we have mentioned, since each $a_n \geq 0$, we see

$$S_1 = a_1 \leq S_2 = a_1 + a_2 \leq S_3 = S_2 + a_3$$

So the sequence of partial sums is an increasing and bounded above sequence. Hence, by a previous theorem, we must have $S_n \to \sup(U)$.   ∎

19.4   Comparison Tests

A powerful set of tools to check the convergence of series on non-negative terms are the comparison tests.

Theorem 19.4.1 Comparison Test for Non-Negative Series

> Assume $\sum_{n=1}^{\infty} a_n$ and $\sum_{n=1}^{\infty} b_n$ are two series with non-negative terms. If there is an N so that $a_n \leq b_n$ for all $n > N$ then
> (1) If $\sum_{n=1}^{\infty} b_n$ converges, so does $\sum_{n=1}^{\infty} a_n$.
> (2) If $\sum_{n=1}^{\infty} a_n$ diverges, so does $\sum_{n=1}^{\infty} b_n$.

Proof 19.4.1
Since $a_n \leq b_n$ for all $n > N$, we have

$$a_{N+1} + a_{N+2} + \ldots + a_{N+p} \;\leq\; b_{N+1} + b_{N+2} + \ldots + b_{N+p}$$

for any $p \geq 1$. Let the partial sums for the series $\sum_{n=1}^{\infty} a_n$ be denoted by S_n and the partials sums for the other series be called T_n. Then we can rewrite the inequality above as

$$S_{N+p} - S_N \;\leq\; T_{N+p} - T_N$$

for any $p \geq 1$.

Case (1): We assume $\sum_{n=1}^{\infty} b_n$ converges and hence the sequence of partial sums (T_n) is bounded above by some number $B > 0$.; i.e. $T_n \leq B$ for all n. But from the inequality above this says

$$S_{N+p} \;\leq\; S_N + T_{N+p} - T_n \leq S_N + 2B$$

for all $p \geq 1$. Thus, the set

$$U \;=\; \{S_{N+1}, S_{N+2}, \ldots, \}$$

is bounded above by $S_N + 2B$. Further, we also know $S_1 \leq S_2 \leq S_3 \leq \ldots \leq S_n$ as the terms a_n are non-negative. We see the set of all S_n is bounded above by $S_N + 2B$. So we see the series $\sum_{n=1}^{\infty} a_n$ must converge since its sequence of partial sums is bounded.

Case (2): Now assume the series $\sum_{n=1}^{\infty} a_n$ diverges. Since the sequence of partials (S_n) is increasing, if this sequence was bounded, the series would have to converge. Since we assume it diverges, we must have the sequence (S_n) is **NOT** bounded and in fact, $S_n \to \infty$. So, in this case, we could say $S = \infty$ and the series $\sum_{n=1}^{\infty} a_n = \infty$ instead of saying the series diverges. The arguments we *used in Case (1) still hold and using the same notation, we have*

$$S_{N+p} - S_N \leq T_{N+p} - T_N$$

or $S_{N+p} - S_N + T_N \leq T_{N+p}$ for all $p \geq 1$. Now take the limit as $p \to \infty$ on both sides. On the right side you get $\lim_{p \to \infty}(S_{N+p} - S_N + T_N) = \infty$ which tells us $\lim_{p \to \infty} T_{N+p} = \infty$ also. Thus $\lim_{n \to \infty} T_n = \infty$ too. Thus the sequence of partial sums of $\sum_{n=1}^{\infty} b_n$ is not bounded implying $\sum_{n=1}^{\infty} b_n$ diverges too. ∎

We can also use a limiting form of the comparison test which does not require us to manually show inequalities between terms.

Theorem 19.4.2 Limit form of the comparison test for non-negative series

Assume $\sum_{n=1}^{\infty} a_n$ and $\sum_{n=1}^{\infty} b_n$ are two series with positive terms. If $\lim_{n \to \infty} a_n/b_n$ exists and if $i0 < \lim_{n \to \infty} a_n/b_n < \infty$, then both series converge or both series diverge.

Proof 19.4.2
Since the $\lim_{n \to \infty} a_n/b_n$ exists, let's call it A. We assume $0 < A < \infty$. For the choice $\epsilon = A/2$, there is N so that $|a_n/b_n - A| < A/2$ if $n > N$. Rewriting, we have

$$-A/2 < a_n/b_n - A < A/2 \iff A/2 < a_n/b_n < 3A/2.$$

for $n > N$. So $a_n < (3A/2)b_n$ for all $n > N$. Now apply the comparison test. We see $\sum_{n=1}^{\infty} a_n$ converges if $\sum_{n=1}^{\infty} b_n$ converges and $\sum_{n=1}^{\infty} b_n$ diverges if $\sum_{n=1}^{\infty} a_n$ diverges. ∎

19.5 p Series

Let's look at a special type of series called the p series. Some examples will give you the gist of this idea.

- The series $\sum_{n=1}^{\infty}(1/n)$ is a p-series with $p = 1$ and is called the **harmonic series**.

- The series $\sum_{n=1}^{\infty}(1/n^2)$ is a p-series with $p = 2$.

- The series $\sum_{n=1}^{\infty}(1/\sqrt{n})$ is a p-series with $p = 1/2$.

Theorem 19.5.1 p Series convergence and divergence

A p-series converges if $p > 1$ and diverges if $p \leq 1$.

Proof 19.5.1
I: We show $\sum_{n=1}^{\infty} 1/n^p$ diverges for $p \leq 1$. First, we observe that $n^p \leq n$ as $p \leq 1$. The best thing to do is to show the sequence of partial sums in not bounded. To do this look at the terms S_{2^n}.

$$S_{2^3} = 1 + 1/2^p + \left(1/3^p + 1/4^p\right) + \left(1/5^p + \ldots + 1/8^p\right)$$

and

$$S_{2^4} = 1 + 1/2^p + \left(1/3^p + 1/4^p\right) + \left(1/5^p + \ldots + 1/8^p\right) + \left(1/9^p + \ldots + 1/(16)^p\right)$$

The last term above has the form $1/(2^3 + 1)^p + \ldots 1/(2^4)^p$ so we have the general formula:

$$
\begin{aligned}
S_{2^n} \;=\; & 1 + 1/2^p + \left(1/3^p + 1/4^p\right) + \left(1/5^p + \ldots + 1/8^p\right) + \\
& + \ldots + \left(1/(2^{n-1} + 1)^p + \ldots + 1/(2^n)^p\right)
\end{aligned}
$$

We can write this in terms of grouping G_i of terms:

$$
S_{2^n} \;=\; 1 + G_1 + G_2 + G_3 + \ldots + G_n
$$

where G_1, G_2 out to G_n are the groupings above. The pieces G_i can be counted. Note $|G_1| = 1 = 2^0$, $|G_2| = 2^1$, $G_3 = 4 = 2^2$ terms and so on. The last grouping gives $|G_n| = 2^{n-1}$ terms. Now

$$
3^p \le 3 \iff 1/3^p \ge 1/3, \quad 4^p \le 4 \iff 1/4^p \ge 1/4
$$

and so forth. Using these estimates, we can rewrite S_{2^n} as follows:

$$
\begin{aligned}
S_{2^n} \;\ge\; & 1 + 1/2 + \left(1/3 + 1/4\right) + \left(1/5 + \ldots + 1/8\right) \\
& + \ldots + \left(1/(2^{n-1} + 1) + \ldots + 1/(2^n)\right)
\end{aligned}
$$

Next underestimate some more:

$$
\begin{aligned}
1/3 + 1/4 \;\ge\; & 1/4 + 1/4 = 2\,(1/2) = 1/2 \\
1/5 + 1/6 + 1/7 + 1/8 \;\ge\; & 1/8 + 1/8 + 1/8 + 1/8 = 4/2^3 = 1/2 \\
& \vdots \\
1/(2^{n-1} + 1) + \ldots + 1/2^n \;\ge\; & 1/2 + \ldots + 1/2 = 2^{n-1}\,(1/2^n) = 1/2
\end{aligned}
$$

Thus adding up the group underestimates, we have $S_{2^n} \ge 1 + 1/2 + (1/2) + (1/2) + \ldots + (1/2)$ where there are n groupings we are adding. So we have $S_{2^n} \ge 1 + n/2$. This tells us immediately $S_{2^n} \to \infty$ and the sequence of partial sums is unbounded and these series must diverge.

II: We now show $\sum_{n=1}^{\infty} 1/n^p$ converges when $p > 1$. To do this, we show the sequence of partial sums is bounded. Pick any positive integer n. Choose another integer m so that $2^m - 1 > n$. Since the partial sums are increasing, we must have $S_n < S_{2^m - 1}$. Grouping the terms in these sums carefully, we have

$$
\begin{aligned}
S_{2^m - 1} \;=\; & 1 + \left(1/2^p + 1/3^p\right) + \ldots + \left(1/8^p + \ldots + 1/(15)^p\right) \\
& + \ldots + \left(1/(2^{m-1})^p + \ldots + 1/(2^m - 1)^p\right)
\end{aligned}
$$

Again, we can write this as groups:

$$
S_{2^m - 1} \;=\; 1 + G_1 + G_2 + \ldots + G_{m-1}
$$

where the cardinalities of the groups now are $|G_1| = 2^1$, $|G_2| = 2^2$ out to $|G_{m-1}| = 2^{m-1}$. Now let's do overestimates for the groups.

$$
1/2^p + 1/3^p \;<\; 1/2^p + 1/2^p = 2\,(1/2^p) = 1/2^{p-1}
$$

$$\begin{aligned}
1/4^p + \ldots + 1/7^p \quad &< \quad 1/4^p + \ldots + 1/4^p = 4\,(1/4^p) \\
&= \quad 1/4^{p-1} = 1/(2^{p-1})^2 \\
1/8^p + \ldots + 1/(15)^p \quad &< \quad 1/8^p + \ldots + 1/8^p = 8\,(1/8^p) \\
&= \quad 1/8^{p-1} = 1/(2^{p-1})^3
\end{aligned}$$

$$\vdots$$

$$\begin{aligned}
1/(2^{m-1})^p + \ldots + 1/(2^m - 1)^p \quad &\leq \quad 1/(2^{m-1})^p + \ldots + 1/(2^{m-1})^p \\
&= \quad 2^{m-1}\left(1/(2^{m-1})^p\right) = (2/2^p)^{m-1} \\
&= \quad 1/(2^{p-1})^{m-1}
\end{aligned}$$

where we have used the cardinality of each of the groups. Thus we have

$$S_{2^m - 1} \quad < \quad 1 + 1/(2^{p-1}) + 1/(2^{p-1})^2 + \ldots + 1/(2^{p-1})^{m-1}$$

Let $r = 1/2^{p-1}$. Then we have shown $S_{2^m - 1} < 1 + r + r^2 + \ldots + r^{m-1}$. Now we have seen in an earlier induction proof that $1 + r + r^2 + \ldots + r^{m-1} = (1 - r^m)/(1 - r) < 1/(1 - r)$ because $0 < r < 1$. So we have $S_{2^m - 1} < 1/(1 - (1/2)) = 2$. Since our choice of n was arbitrary, we have shown (S_n) is bounded and so the associated series must converge. ∎

19.6 Examples

Example 19.6.1 *Consider the series $\sum_{n=1}^{\infty} 1/(n^2 + 11)$. Discuss its convergence or divergence.*

Solution *This series is very similar to the p series for $p = 2$, $\sum_{n=1}^{\infty} 1/n^2$. We can use the **comparison test** here. We see $1/(n^2 + 11) < 1/n^2$ for all $n \geq 1$. Since $\sum_{n=1}^{\infty} 1/n^2$ converges, so does $\sum_{n=1}^{\infty} 1/(n^2 + 11)$.*

Example 19.6.2 *Consider the series $\sum_{n=4}^{\infty} 1/(n^2 - 11)$. Discuss its convergence or divergence.*

Solution *To use the **comparison test** here, we have to be more careful. We guess that $1/(n^2 - 11) < 2/n^2$ for sufficiently large n. Solving, we find $n^2 < 2n^2 - 22$ or $n^2 > 22$ implying $n > 5$. So for $n > 5$ we have the desired inequality and we can use the comparison test to see $\sum_{n=4}^{\infty} 1/(n^2 - 11)$ converges.*

Example 19.6.3 *Consider the series $\sum_{n=4}^{\infty} 1/(n^2 - 11)$. Discuss its convergence or divergence.*

Solution *This series converges by using the **limit comparison test** with $\sum_{n=4}^{\infty} 1/n^2$. We find $\lim_{n \to \infty}(1/(n^2 - 11) / (1/n^2)) = \lim_{n \to \infty} n^2/(n^2 - 11) = 1$. So both series converge or both series diverge. Since $\sum_{n=4}^{\infty} 1/n^2$ converges, so does $\sum_{n=4}^{\infty} 1/(n^2 - 11)$.*

Example 19.6.4 *Consider the series $\sum_{n=4}^{\infty} 1/(n - 3)$. Discuss its convergence or divergence.*

Solution *The series $\sum_{n=4}^{\infty} 1/(n - 3)$ should be compared to the series $\sum_{n=4}^{\infty} 1/n$ which diverges. Note the easiest thing to do is to use the **limit comparison test** because of the presence of the "$-$" in the denominator of the series $\sum_{n=4}^{\infty} 1/(n - 3)$. You can use the regular **comparison test** but you would have to find when $1/(n - 3) < 2/n$ to do so. Note the check by comparing with $2/n$ could have been done with another constant on top; i.e. you could check with $3/n$, $4/n$ and so forth.*

19.6.1 Homework

Exercise 19.6.1 *Bounded decreasing sequences (a_n) converge to their infimum.*

Exercise 19.6.2 *Use the comparison test to show $\sum_{n=1}^{\infty} 1/(n^2 + 14)$ converges.*

Exercise 19.6.3 *Use the limit comparison test to show $\sum_{n=3}^{\infty} 1/(n^2 - 8)$ converges.*

Exercise 19.6.4 *Use the comparison test to show $\sum_{n=1}^{\infty} 1/(n + 7)$ diverges.*

Exercise 19.6.5 *Use the limit comparison test to show $\sum_{n=10}^{\infty} 1/(n - 9)$ diverges.*

Exercise 19.6.6 *Determine convergence or divergence of $\sum_{n=1}^{\infty} 1/(4n^2 + 14n - 10)$*

Exercise 19.6.7 *Determine convergence or divergence of $\sum_{n=3}^{\infty} 1/(6n^2 + 7n + 2)$.*

Exercise 19.6.8 *Determine convergence or divergence of $\sum_{n=1}^{\infty} 1/\sqrt{n + 7}$.*

Exercise 19.6.9 *Determine convergence or divergence of $\sum_{n=10}^{\infty} 1/\sqrt{4n^3 + 10n - 2}$.*

Chapter 20

Series in General

20.1 Some Specific Tests for Convergence

We now discuss some interesting series of non-negative terms along with their corresponding series of absolute values and see what kind of tests are available to check for convergence. These ideas are very important and we will use them frequently!

20.1.1 Geometric Series

We have already seen the standard equality, proved by induction,

$$1 + r + r^2 + r^3 + \ldots + r^n \;=\; \frac{1 - r^{n+1}}{1 - r}$$

for all $r \neq 1$. Letting $n \to \infty$ leads us to what is called the geometric series.

Theorem 20.1.1 Geometric Series

$\sum_{n=0}^{\infty} r^n$ *converges if* $-1 < r < 1$ *to* $1/(1-r)$.
$\sum_{n=0}^{\infty} r^n$ *diverges if* $r \notin (-1, 1)$.

Proof 20.1.1
(1): We show $\sum_{n=0}^{\infty} r^n$ converges if $-1 < r < 1$ to $1/(1-r)$. We already know the partial sum $S_n = 1 + r + \ldots + r^n = (1 - r^{n+1})/(1 - r)$ form a previous induction argument. We also know that $r^{n+1} \to 0$ as $n \to \infty$ since $-1 < r < 1$. Thus, $\sum_{n=0}^{\infty} r^n$ converges to $1/(1-r)$.
(2): If $r = 1$, we have the series $1 + 1 + \cdots + 1 + \ldots$ which has partial sums $S_n = n$. The partial sums are unbounded and so the series does not converge.

If $r = -1$, we have seen already that $\sum_{n=0}^{\infty} (-1)^n$ diverges by oscillation.

If $r > 1$, we have $S_n = 1 + r + \ldots + r^n = (r^{n+1} - 1)/(r - 1)$ and we see $\lim_{n \to \infty} S_n = \infty$ since $r^{n+1} \to \infty$ and $r - 1 > 0$. So this series diverges.

If $r < -1$, the $\lim_{n \to \infty} r^n$ does not exist so by the nth term test, the series must diverge. ∎

Comment: note if $-1 < r < 1$
(1) $\sum_{n=1}^{\infty} r^n = 1/(1 - r) - 1$
(2) $\sum_{n=2}^{\infty} r^n = 1/(1 - r) - 1 - r$
(3) $\sum_{n=3}^{\infty} r^n = 1/(1 - r) - 1 - r - r^2$
So
(1) $\sum_{n=1}^{\infty} (1/3)^n = 1/(1 - (1/3)) - 1 = 3/2 - 1 = 1/2$

(2) $\sum_{n=2}^{\infty}(1/3)^n = 1/(1 - (1/3)) - 1 - (1/3) = 1/2 - 1/3 = 1/6$
(3) $\sum_{n=3}^{\infty} r^n = 1/(1 - (1/3)) - 1 - (1/3) - (1/9) = 1/6 - 1/9 = 1/(18)$

20.1.2 The Ratio Test

Theorem 20.1.2 The Ratio Test

> *Let $\sum_{n=1}^{\infty} a_n$ be a series with positive terms and assume $\lim_{n\to\infty} a_{n+1}/a_n = \rho$.*
> *(1) If $0 \leq \rho < 1$, the series converges.*
> *(2) If $\rho > 1$, the series diverges.*
> *(3) If $\rho = 1$, we don't know.*

Proof 20.1.2

*(1): We assume $0 \leq \rho < 1$. Then, we see $\rho < (1 + \rho)/2 < 1$. Let the value $(1 + \rho)/2$ be the "r" in the geometric series $\sum_{n=1}^{\infty} r^n$ which converges since $0 < r < 1$. We will use the **comparison test** to show our series converges. Let $\epsilon = (1 - \rho)/2$ which is a nice positive number. Then, $\epsilon + \rho = (1 + \rho)/2 = r < 1$. Since $\lim_{n\to\infty} a_{n+1}/a_n = \rho$, for this ϵ, there is an N so that*

$$n > N \implies \left| \frac{a_{n+1}}{a_n} - \rho \right| < \epsilon = \frac{1 - \rho}{2}$$

Thus, $a_{n+1}/a_n < \epsilon + \rho = (1 + \rho)/2 = r < 1$ for all $n > N$. We see

$$a_{N+2} < r\, a_{N+1}$$
$$a_{N+3} < r\, a_{N+2} < r\, (r\, a_{N+1}) = r^2 a_{N+1}$$
$$a_{N+4} < r\, a_{N+3} < r\, (r^2\, a_{N+1}) = r^3 a_{N+1}$$
$$\vdots$$
$$a_{N+\ell} < r^{\ell-1} a_{N+1} = \frac{a_{N+1}}{r^{N+1}} r^{\ell+N}$$

for $\ell > 1$.
Now let $k = N + \ell$ and $c = a_{N+1}/r^{N+1}$. We have shown therefore that $a_k < cr^k$ for $k > N + 1$. It is easy to see the series $c \sum_{k=N+2}^{\infty} r^n$ converges by the limit comparison test, so by the comparison test, we have shown $\sum_{k=N+2}^{\infty} a_k$ converges which also tells us $\sum_{k=1}^{\infty} a_k$ converges.
(2): If $\rho > 1$, choose $\epsilon = (\rho - 1)/2 > 0$ as $\rho > 1$. Then, there is an N so that

$$n > N \implies \left| \frac{a_{n+1}}{a_n} - \rho \right| < \epsilon = \frac{\rho - 1}{2}$$

This implies for $n > N$ that

$$\frac{1 + \rho}{2} = \rho - \epsilon < \frac{a_{n+1}}{a_n} < \rho + \epsilon$$

Thus, letting $r = (1 + \rho)/2 > 1$, we have

$$a_{N+2} > r\, a_{N+1}$$
$$a_{N+3} > r\, a_{N+2} > r\, (r\, a_{N+1}) = r^2 a_{N+1}$$
$$a_{N+4} > r\, a_{N+3} > r\, (r^2\, a_{N+1}) = r^3 a_{N+1}$$
$$\vdots$$
$$a_{N+\ell} > r^{\ell-1} a_{N+1} = \frac{a_{N+1}}{r^{N+1}} r^{\ell+N}$$

for $\ell > 1$. Let $C = a_{N+1}/r^{N+1}$ and $k = N + \ell$. Then, we have $a_k > C r^k$ for all $k > N + 1$. Now apply the comparison test. Since $\sum_{k=N+2}^{\infty} r^k$ diverges with $r > 1$ and we know $C \sum_{k=N+2}^{\infty} r^k$ diverges, by the comparison test $\sum_{k=N+2}^{\infty} a_k$ diverges.
(3): If $\rho = 1$, the series could converge or diverge.
(i) the series $\sum_{n=1}^{\infty} 1/n^2$ converges but $\rho = 1$.
(ii) the series $\sum_{n=1}^{\infty} 1/n$ diverges but $\rho = 1$.
So if $\rho = 1$, we just don't know. ∎

We can say more:

Theorem 20.1.3 The Ratio Test for Absolute Convergence

Let $\sum_{n=1}^{\infty} a_n$ be a series with nonzero terms. Assume $\lim_{n\to\infty} \left| \frac{a_{n+1}}{a_n} \right|$ converges to the value ρ.
(1) If $0 \leq \rho < 1$, the series converges.
(2) If $\rho > 1$, the series diverges.
(3) If $\rho = 1$, we don't know.

Proof 20.1.3

(1): If $0 \leq \rho < 1$, we can apply the previous theorem to see $\sum_{n=1}^{\infty} |a_n|$ converges. But if a series converges absolutely, it also converges.
(2): If $\rho > 1$, the arguments in the proof above show us there is an N so we can write $|a_k| \geq C r^k$ for $k > N + 1$ for $r = (1 + \rho)/2 > 1$. Now if $\sum_{k=N+2}^{\infty} a_k$ converged, this implies $a_k \to 0$. But if that were true, $|a_k| \to 0$ too. However, the inequality $|a_k| \geq C r^k$ with $r > 1$ says the sequence $(|a_k|)$ cannot converge. Thus, $\sum_{k=N+2}^{\infty} a_k$ must also diverge. Of course, this tells us $\sum_{n=1}^{\infty} a_n$ also diverges.
(3): The same examples as before show us why we cannot conclude anything about convergence or divergence of the series if $\rho = 1$. ∎

20.1.3   The Root Test

Theorem 20.1.4 The Root Test

Let $\sum_{n=1}^{\infty} a_n$ be a series with nonnegative terms. Assume $\lim_{n\to\infty}(a_n)^{1/n}$ exists and equals ρ.
(1) If $0 \leq \rho < 1$, the series converges.
(2) If $\rho > 1$, the series diverges.
(3) If $\rho = 1$, we don't know.

Proof 20.1.4

(1): We assume $0 \leq \rho < 1$. Choose $\epsilon = (1 - \rho)/2 > 0$. Then there is an N so that

$$n > N \implies |(a_n)^{1/n} - \rho| < \epsilon = (1 - \rho)/2$$

We can rewrite this: for $n > N$

$$-(1 - \rho)/2 \;<\; (a_n)^{1/n} - \rho < (1 - \rho)/2 \implies (a_n)^{1/n} < (1 + \rho)/2 < 1$$

Let $r = (1 + \rho)/2 < 1$.

$$(a_{N+1})^{1/(N+1)} < r \implies a_{N+1} < r^{N+1}$$
$$(a_{N+2})^{1/(N+2)} < r \implies a_{N+2} < r^{N+2}$$

$$\vdots$$

$$(a_{N+\ell})^{1/(N+\ell)} < r \implies a_{N+\ell} < r^{N+\ell}$$

for $\ell \geq 1$. Letting $k = N + \ell$, we have $a_k < r^k$ for $k > N$. Since $0 < r < 1$, the series $\sum_{k=N+1}^{\infty} r^k$ converges and by comparison, so does $\sum_{k=N+1}^{\infty} a_k$. This shows $\sum_{k=1}^{\infty} a_k$ converges.
(2): Now $\rho > 1$. This argument is very similar to the one we use for the proof of the ratio test. Let $\epsilon = (\rho - 1)/2 > 0$. Then there is an N so that

$$n > N \implies |(a_n)^{1/n} - \rho| < \epsilon = (\rho - 1)/2$$

We can rewrite this as

$$(1 - \rho)/2 \;\; < \;\; (a_n)^{1/n} - \rho < (\rho - 1)/2 \implies (a_n)^{1/n} > (1 + \rho)/2 > 1$$

for $n > N$. Let $r = (1 + \rho)/2$. Then, we have the inequality $a_n > r^n$ for all $n > N$. Since $\sum_{n=N+1}^{\infty} r^n$ diverges for $r > 1$, by the comparison test, we know $\sum_{n=N+1}^{\infty} a_n$ diverges. Hence the original series diverges too.
(3): If $\rho = 1$, we have $\lim_{n \to \infty} n^{1/n} = 1$. Look at the series $\sum_{n=1}^{\infty} 1/n$. We know this diverges and $\rho = 1$. Now look at the series $\sum_{n=1}^{\infty} 1/n^2$. This converges and $\lim_{n \to \infty} (n^2)^{1/n} = \lim_{n \to \infty} n^{2/n}$. But $\lim_{n \to \infty} n^{2/n} = \lim_{n \to \infty} (n^{1/n})^2$. Since the function $f(x) = x^2$ is continuous, we also know $\lim_{n \to \infty} (n^{1/n})^2 = (\lim_{n \to \infty} n^{1/n})^2 = 1$. So we just don't know whether the series converges or diverges if $\rho = 1$. ∎

We can restate this result for series whose terms are not necessarily non-negative.

Theorem 20.1.5 The Root Test for Absolute Convergence

> *Let $\sum_{n=1}^{\infty} a_n$ be a series. Assume $\lim_{n \to \infty} |a_n|^{1/n}$ exists and equals ρ.*
> *(1) If $0 \leq \rho < 1$, the series converges.*
> *(2) If $\rho > 1$, the series diverges.*
> *(3) If $\rho = 1$, we don't know.*

Proof 20.1.5
(1): If $0 \leq \rho < 1$, the previous theorem tells us $\sum_{n=1}^{\infty} |a_n|$ converges absolutely implying convergence.
(2): If $\rho > 1$, the earlier arguments imply there is an N so $|a_k| \geq r^k$ for $k > N$ for $r = (1+\rho)/2 > 1$. If $\sum_{k=N+1}^{\infty} a_k$ converged, then $a_k \to 0$ and $|a_k| \to 0$ too. But $|a_k| \geq r^k$ with $r > 1$ says the sequence $(|a_k|)$ cannot converge. Thus, $\sum_{k=N+1}^{\infty} a_k$ must also diverge and so the original series diverges too.
(3): The same examples as before show us why we can't conclude anything about convergence or divergence of the series if $\rho = 1$. ∎

Let's summarize what we know:

1. We can check convergence with comparison tests. If the terms are not non-negative, we can test for absolute convergence which will imply convergence.

2. The ratio and root test are also good tools. We can check for absolute convergence if the terms are not all non-negative to see if the series converge. But the ratio test requires that none of the terms are zero.

20.1.4   Examples

It is time to do some examples.

Example 20.1.1 *What is the sum of*

$$(3/4)^5 + (3/4)^6 + \ldots + (3/4)^n + \ldots?$$

Solution *This is a geometric series with $r = 3/4$ missing the first 5 terms. So the sum of the series is*

$$1/(1 - (3/4)) - 1 - (3/4) - (3/4)^2 - (3/4)^3 - (3/4)^4.$$

Note the five terms we subtract could be written

$$1 + (3/4) + (3/4)^2 + (3/4)^3 + (3/4)^4 = (1 - (3/4)^5)/(1 - (3/4))$$

but this is not necessarily an improvement!

Example 20.1.2 *What is the sum of $\sum_{k=1}^{\infty} (1/(3k) - 1/(4k))$?*

Solution *To do this we have to write the inner part as a single fraction. We find*

$$\sum_{k=1}^{\infty} (1/(3k) - 1/(4k)) = \sum_{k=1}^{\infty} k/(12k^2) = \sum_{k=1}^{\infty} 1/(12k)$$

which diverges as it is multiple of the harmonic series.

Example 20.1.3 *Does $\sum_{k=5}^{\infty} ((2k^3 + 3)/(k^5 - 4k^4))$ converge?*

Solution *This looks like the series $\sum 1/k^2$ so a good choice would be the limit comparison test. But that gives*

$$\lim_{k \to \infty} (2k^3 + 3)/(k^5 - 4k^4)/(1/k^2) = \lim_{k \to \infty} (2k^5 + 3k^2)/(k^5 - 4k^4) = 2$$

Thus, since the limit is positive and finite, our series converges. Note the ratio test will fail. We find

$$\lim_{k \to \infty} \frac{(2(k+1)^3 + 3)/((k+1)^5 - 4(k+1)^4)}{(2k^3 + 3)/(k^5 - 4k^4)} = \lim_{k \to \infty} \frac{(2(k+1)^3 + 3)(k^5 - 4k^4)}{((k+1)^5 - 4(k+1)^4)(2k^3 + 3)}$$

$$= \lim_{k \to \infty} \frac{2k^8 + \text{lower power terms of } k}{2k^8 + \text{lower power terms of } k} = 1$$

and so the ratio test fails. Of course, it would be insane to apply the root test here!

Example 20.1.4 *Does $\sum_{k=1}^{\infty} 1/((2k+1)2^k)$ converge?*

Solution *We'll try the ratio test here:*

$$\lim_{k \to \infty} \frac{1/((2(k+1) + 1)2^{k+1})}{1/((2k+1)2^k)} = \lim_{k \to \infty} \frac{(2k+1)2^k}{(2k+3)2^{k+1}} = \lim_{k \to \infty} \frac{2k+1}{(2k+3)2}.$$

Since the limit is $1/2$ and hence, this series converges by the ratio test.

Homework

Exercise 20.1.1 *Determine if $\sum_{n=1}^{\infty} n!/2^n$ converges.*

Exercise 20.1.2 *Determine if $\sum_{n=1}^{\infty} ((-1)^n(2n+5))/n!$ converges.*

Exercise 20.1.3 *Determine if $\sum_{n=1}^{\infty} 2/(n^3 + 5n - 2)$ converges.*

Exercise 20.1.4 *Determine if $\sum_{n=1}^{\infty} 2/(n^4 + 5n^{3/2} + 2n - 3)$ converges.*

Exercise 20.1.5 *Determine if $\sum_{n=1}^{\infty} 1/\ln(n)$ converges.*

20.1.5   A Taste of Power Series

A series of the form $\sum_{n=0}^{\infty} a_n x^n$ is called a **power series** centered at $x = 0$. Note no matter what the numbers a_n are, this series **always** converges at $x = 0$. Let's assume the terms a_n are not zero and that $\lim_{n\to\infty} |a_{n+1}/a_n| = \rho > 0$. We can check for the convergence and divergence of this series using the ratio test.

$$\lim_{n\to\infty} \left| \frac{a_{n+1} x^{n+1}}{a_n x^n} \right| \;=\; \lim_{n\to\infty} \left| \frac{a_{n+1}}{a_n} \right| |x| = \rho|x|$$

By the ratio test, the series converges when $\rho|x| < 1$ and diverges when $\rho|x| > 1$. So we have
(1) The series converges when $-1/\rho < x < 1/\rho$
(2) The series diverges when $x > 1/\rho$ or $x < -1/\rho$
(3) We don't know what happens at $x = \pm 1/\rho$. The number $1/\rho$ is called R, the **radius of convergence** of the power series. Hence, this power series defines a function $f(x)$ locally at $x = 0$ with radius of convergence $R > 0$. Any function $f(x)$ which has a power series that matches it locally at $x = 0$ is said to be **analytic** at $x = 0$.

The **Derived series** obtained from $f(x) = \sum_{n=0}^{\infty} a_n x^n$ is the series $\sum_{n=0}^{\infty} n\, a_n x^{n-1}$. We are still assuming $\lim_{n\to\infty} |a_{n+1}/a_n| = \rho > 0$, so we can check the convergence of this series using the ratio test too.

$$\lim_{n\to\infty} \left| \frac{(n+1)\, a_{n+1} x^n}{n\, a_n x^{n-1}} \right| \;=\; \lim_{n\to\infty} \left| \frac{a_{n+1}}{a_n} \right| \lim_{n\to\infty} \left(\frac{n+1}{n} \right) |x| = \rho|x|$$

Hence, this series converges when $|x| < R = 1/\rho$ also.

We see the derived series defines a function $g(x)$ on the circle $(-R, R)$. An obvious question is this: Is $f'(x) = g(x)$ on $(-R, R)$? That is does

$$f'(x) = \left(\sum_{n=0}^{\infty} a_n x^n \right)' \;=\; \sum_{n=0}^{\infty} \left(a_n x^n \right)' = \sum_{n=0}^{\infty} n\, a_n x^{n-1} = g(x)?$$

Note this is an **interchange of limit order** idea. Is it possible to interchange the limit called differentiation with the limit called series? When we can do this, this is called differentiating the series **term by term**.

We will find that we can differentiate and integrate a power series **term by term** as many times as we like and the radius of convergence remains the same! These are the kinds of series we construct using Taylor polynomials and the resulting series is called a **Taylor Series** expansion of $f(x)$ at $x = 0$. We have a long way to go to be able to prove this sort of theorem. But this is the kind of manipulation we must do to use power series to solve ODEs with non constant coefficients!

Another kind of series that is very useful is $a_0 + \sum_{n=1}^{\infty} (\, a_n \cos(n\pi x/L) + b_n \sin(n\pi x/L)\,)$ which is called a **Fourier series**. Note its **building blocks** are sin and cos functions rather than powers of x.

Homework

Exercise 20.1.6 *Determine for what values of x the series $\sum_{n=0}^{\infty} (-1)^n x^{2n} / ((2n)!)$ converges and for what values of x it diverges.*

Exercise 20.1.7 *Determine for what values of x the series $\sum_{n=0}^{\infty} (-1)^n x^n / ((3n)!)$ converges and for what values of x it diverges.*

Exercise 20.1.8 *Determine for what values of x the series $\sum_{n=0}^{\infty} (-1)^n x^n / ((n^2+1)(2n)!)$ converges and for what values of x it diverges.*

Exercise 20.1.9 *Determine for what values of x the series $\sum_{n=0}^{\infty} (-1)^n (2n^2+1) x^{2n} / ((5n+3)(2n)!)$ converges and for what values of x it diverges.*

20.2   The Cauchy - Schwartz Inequality and the Minkowski Inequality for Sequences

Now we will prove some fundamental relationships between the ℓ^p sequence spaces.

20.2.1   Conjugate Exponents

We say the positive numbers p and q are **conjugate exponents** if $p > 1$ and $1/p + 1/q = 1$. If $p = 1$, we define its conjugate exponent to be $q = \infty$. Conjugate exponents satisfy some fundamental identities. Clearly, if $p > 1$,

$$\frac{1}{p} + \frac{1}{q} \implies 1 = \frac{p+q}{pq}$$

From that it follows $pq = p + q$ and from that using factoring $(p - 1)(q - 1) = 1$. We will use these identities quite a bit.

Lemma 20.2.1 The $\alpha - \beta$ Lemma

> *Let α and β be positive real numbers and p and q be conjugate exponents. Then $\alpha \beta \leq \frac{\alpha^p}{p} + \frac{\beta^q}{q}$.*

Proof 20.2.1

To see this is a straightforward integration. We haven't discussed Riemann integration in depth yet, but all of you know how to integrate from your earlier courses. Let $u = t^{p-1}$. Then, $t = u^{1/(p-1)}$ and using the identity $(p - 1)(q - 1) = 1$, we have $t = u^{q-1}$. Now we are going to draw the curve $u = t^{p-1}$ or $t = u^{q-1}$ in the first quadrant and show you how this inequality makes sense. We will draw the curve $u = t^{p-1}$ as if it was concave up (i.e. like when $p = 3$) even though it could be concave down (i.e. like when $p = 3/2$). Whether the curve is concave up or down does not change how the argument goes. So make sure you can see that. A lot of times our pictures are just aids to helping us think through an argument. A placeholder, so to speak! In Figure 20.1(a), the area of the rectangle $\alpha \beta \leq$ the area of Region I + the area of Region II + the area vertically hatched. and in Figure 20.1(b), the area of the rectangle $\alpha \beta \leq$ the area of Region I + the area of Region II. In Figure 20.1(a),

(1): the area of Region I is the area under the curve $t = u^{q-1}$ from $u = 0$ to $u = \beta$. This is $\int_0^{\beta} u^{q-1} \, du$.

(2): the area of Region II + the area marked in vertical hatches is the area under the curve $u = t^{p-1}$

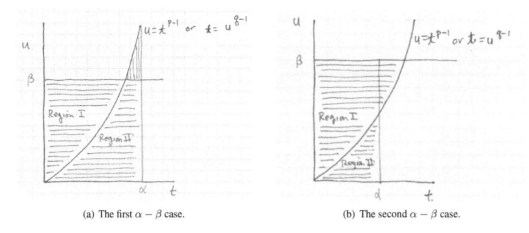

(a) The first $\alpha - \beta$ case. (b) The second $\alpha - \beta$ case.

Figure 20.1: The $\alpha - \beta$ diagrams.

from $t = 0$ to $t = \alpha$. This is $\int_0^\alpha t^{p-1} \, dt$.

In Figure 20.1(b),
(1): the area of Region I is still the area under the curve $t = u^{q-1}$ from $u = 0$ to $u = \beta$. This is $\int_0^\beta u^{q-1} \, du$.
(2): the area of Region II is the area under the curve $u = t^{p-1}$ from $t = 0$ to $t = \alpha$. This is $\int_0^\alpha t^{p-1} \, dt$.
So in both cases

$$\alpha \beta \; \leq \; \int_0^\beta u^{q-1} \, du + \int_0^\alpha t^{p-1} \, dt = \frac{\beta^q}{q} + \frac{\alpha^p}{p}.$$

∎

20.2.2 Hölder's Inequality

First, we define an important collection of sequences called the ℓ^p sequence spaces.

Definition 20.2.1 The ℓ^p Sequence Space

> *Let $p \geq 1$. The collection of all sequence, $(a_n)_{n=1}^\infty$ for which $\sum_{n=1}^\infty |a_n|^p$ converges is denoted by the symbol ℓ^p.*
> *(1) $\ell^1 = \{(a_n)_{n=1}^\infty : \sum_{n=1}^\infty |a_n| \text{ converges.}\}$*
> *(2) $\ell^2 = \{(a_n)_{n=1}^\infty : \sum_{n=1}^\infty |a_n|^2 \text{ converges.}\}$*
> *We also define $\ell^\infty = \{(a_n)_{n=1}^\infty : \sup_{n \geq 1} |a_n| < \infty\}$.*

There is a fundamental inequality connecting sequences in ℓ^p and ℓ^q when p and q are conjugate exponents called the **Hölder's Inequality**. It's proof is straightforward but has a few tricks.

Theorem 20.2.2 Hölder's Inequality

Let $p > 1$ and p and q be conjugate exponents. If $x \in \ell^p$ and $y \in \ell^q$, then

$$\sum_{n=1}^{\infty} |x_n\, y_n| \leq \left(\sum_{n=1}^{\infty} |x_n|^p\right)^{1/p} \left(\sum_{n=1}^{\infty} |y_n|^q\right)^{1/q}$$

where $x = (x_n)$ and $y = (y_n)$.

Proof 20.2.2

This inequality is clearly true if either of the two sequences x and y are the zero sequence. So we can assume both x and y have some nonzero terms in them. Then $x \in \ell^p$, we know

$$0 < u = \left(\sum_{n=1}^{\infty} |x_n|^p\right)^{1/p} < \infty, \quad 0 < v = \left(\sum_{n=1}^{\infty} |y_n|^q\right)^{1/q} < \infty$$

Now define new sequences, $\hat{x}$ and $\hat{y}$ by $\hat{x}_n = x_n/u$ and $\hat{y}_n = y_n/v$. Then, we have

$$\sum_{n=1}^{\infty} |\hat{x}_n|^p = \sum_{n=1}^{\infty} \frac{|x_n|^p}{u^p} = \frac{1}{u^p} \sum_{n=1}^{\infty} |x_n|^p = \frac{u^p}{u^p} = 1.$$

$$\sum_{n=1}^{\infty} |\hat{y}_n|^q = \sum_{n=1}^{\infty} \frac{|y_n|^q}{v^q} = \frac{1}{v^q} \sum_{n=1}^{\infty} |y_n|^q = \frac{v^q}{v^q} = 1.$$

Now apply the $\alpha - \beta$ Lemma to $\alpha = |\hat{x}_n|$ and $\beta = |\hat{y}_n|$ for any nonzero terms $\hat{x}_n$ and $\hat{y}_n$. Then $|\hat{x}_n\, \hat{y}_n| \leq |\hat{x}_n|^p/p + |\hat{y}_n|^q/q$.
This is also true, of course, if either $\hat{x}_n$ or $\hat{y}_n$ are zero although the $\alpha - \beta$ lemma does not apply!
Now sum over N terms to get

$$\sum_{n=1}^{N} |\hat{x}_n\, \hat{y}_n| \leq \frac{1}{p} \sum_{n=1}^{N} |\hat{x}_n|^p + \frac{1}{q} \sum_{n=1}^{N} |\hat{y}_n|^q$$

Since we know $x \in \ell^p$ and $y \in \ell^q$, we know

$$\sum_{n=1}^{N} |\hat{x}_n|^p \leq \sum_{n=1}^{\infty} |\hat{x}_n|^p = 1$$

$$\sum_{n=1}^{N} |\hat{y}_n|^q \leq \sum_{n=1}^{\infty} |\hat{y}_n|^q = 1$$

So we have

$$\sum_{n=1}^{N} |\hat{x}_n\, \hat{y}_n| \leq \frac{1}{p} + \frac{1}{q} = 1$$

This is true for all N so the partial sums $\sum_{n=1}^{N} |\hat{x}_n\, \hat{y}_n|$ are bounded above. Hence, the partial sums converge to this supremum which is denoted by $\sum_{n=1}^{\infty} |\hat{x}_n\, \hat{y}_n|$. We conclude $\sum_{n=1}^{\infty} |\hat{x}_n\, \hat{y}_n| \leq 1$. But $\hat{x}_n \hat{y}_n = 1/(u\, v)\, x_n y_n$ and so we have $\frac{1}{u\, v} \sum_{n=1}^{\infty} |x_n\, y_n| \leq 1$ which implies the result as

$$\sum_{n=1}^{\infty} |x_n\, y_n| \leq u\, v = \left(\sum_{n=1}^{\infty} |x_n|^p\right)^{1/p} \left(\sum_{n=1}^{\infty} |y_n|^q\right)^{1/q}$$

We can also do this inequality for the case $p = 1$ and $q = \infty$. It is a lot simpler to prove!

Theorem 20.2.3 Hölder's Theorem for $p = 1$ and $q = \infty$

If $x \in \ell^1$ and $y \in \ell^\infty$, then $\sum_{n=1}^{\infty} |x_n y_n| \leq \left(\sum_{n=1}^{\infty} |x_n| \right) \sup_{n \geq 1} |y_n|$.

Proof 20.2.3
We know since $y \in \ell^\infty$, $|y_n| \leq \sup_{k \geq 1} |y_k|$. Thus,

$$\sum_{n=1}^{N} |x_n y_n| \leq \left(\sum_{n=1}^{\infty} |x_n| \right) \sup_{k \geq 1} |y_k|$$

Thus the sequence of partial sums $\sum_{n=1}^{N} |x_n y_n|$ is bounded above by $\left(\sum_{n=1}^{\infty} |x_n| \right) \sup_{k \geq 1} |y_k|$. This gives us our result. ■

20.2.3   Minkowski's Inequality

We can use Hölder's Inequality to prove a fundamental inequality for sequences in a given ℓ^p.

Theorem 20.2.4 Minkowski's Inequality

Let $p \geq 1$ and let x and y be in ℓ^p, Then,

$$\left(\sum_{n=1}^{\infty} |x_n + y_n|^p \right)^{\frac{1}{p}} \leq \left(\sum_{n=1}^{\infty} |x_n|^p \right)^{\frac{1}{p}} + \left(\sum_{n=1}^{\infty} |y_n|^p \right)^{\frac{1}{p}}$$

and for x and y in ℓ^∞,

$$\sup_{n \geq 1} |x_n + y_n| \leq \sup_{n \geq 1} |x_n| + \sup_{n \geq 1} |y_n|$$

Note this shows if x and y are in ℓ^p so is their sum. Essentially this shows ℓ^p is a vector space.

Proof 20.2.4
(1): $p = \infty$
We know $|x_n + y_n| \leq |x_n| + |y_n|$ by the triangle inequality. So we have $|x_n + y_n| \leq \sup_{n \geq 1} |x_n| + \sup_{n \geq 1} |y_n|$. Thus, the right hand side is an upper bound for all the terms of the left side. We then can say $\sup_{n \geq 1} |x_n + y_n| \leq \sup_{n \geq 1} |x_n| + \sup_{n \geq 1} |y_n|$ which is the result for $p = \infty$.
(2): $p = 1$
Again, we know $|x_n + y_n| \leq |x_n| + |y_n|$ by the triangle inequality. Sum the first N terms on both sides to get

$$\sum_{n=1}^{N} |x_n + y_n| \leq \sum_{n=1}^{N} |x_n| + \sum_{n=1}^{N} |y_n| \leq \sum_{n=1}^{\infty} |x_n| + \sum_{n=1}^{\infty} |y_n|$$

The right hand side is an upper bound for the partial sums on the left. Hence, we have

$$\sum_{n=1}^{\infty} |x_n + y_n| \le \sum_{n=1}^{\infty} |x_n| + \sum_{n=1}^{\infty} |y_n|$$

(3) $1 < p < \infty$
We have

$$|x_n + y_n|^p = |x_n + y_n| \, |x_n + y_n|^{p-1} \le |x_n| \, |x_n + y_n|^{p-1} + |y_n| \, |x_n + y_n|^{p-1}$$

$$\sum_{n=1}^{N} |x_n + y_n|^p \le \sum_{n=1}^{N} |x_n| \, |x_n + y_n|^{p-1} + \sum_{n=1}^{N} |y_n| \, |x_n + y_n|^{p-1}$$

Let $a_n = |x_n|$, $b_n = |x_n + y_n|^{p-1}$, $c_n = |y_n|$ *and* $d_n = |x_n + y_n|^{p-1}$. *Hölder's Inequality applies just fine to finite sequences: i.e. sequences in* $\Re^N$. *So we have*

$$\sum_{n=1}^{N} a_n b_n \le \left(\sum_{n=1}^{N} a_n^p \right)^{\frac{1}{p}} \left(\sum_{n=1}^{N} b_n^q \right)^{\frac{1}{q}}$$

But $b_n^q = |x_n + y_n|^{q(p-1)} = |x_n + y_n|^p$ *using the conjugate exponents identities we established. So we have found*

$$\sum_{n=1}^{N} |x_n| \, |x_n + y_n|^{p-1} \le \left(\sum_{n=1}^{N} |x_n|^p \right)^{\frac{1}{p}} \left(\sum_{n=1}^{N} |x_n + y_n|^p \right)^{\frac{1}{q}}$$

We can apply the same reasoning to the terms c_n *and* d_n *to find*

$$\sum_{n=1}^{N} |y_n| \, |x_n + y_n|^{p-1} \le \left(\sum_{n=1}^{N} |y_n|^p \right)^{\frac{1}{p}} \left(\sum_{n=1}^{N} |x_n + y_n|^p \right)^{\frac{1}{q}}$$

We can use the inequalities we just figured out to get the next estimate

$$\sum_{n=1}^{N} |x_n + y_n|^p \le \left(\sum_{n=1}^{N} |x_n|^p \right)^{\frac{1}{p}} \left(\sum_{n=1}^{N} |x_n + y_n|^p \right)^{\frac{1}{q}}$$
$$+ \left(\sum_{n=1}^{N} |y_n|^p \right)^{\frac{1}{p}} \left(\sum_{n=1}^{N} |x_n + y_n|^p \right)^{\frac{1}{q}}$$

Now factor out the common term to get

$$\sum_{n=1}^{N} |x_n + y_n|^p \le \left(\left(\sum_{n=1}^{N} |x_n|^p \right)^{\frac{1}{p}} + \left(\sum_{n=1}^{N} |y_n|^p \right)^{\frac{1}{p}} \right) \left(\sum_{n=1}^{N} |x_n + y_n|^p \right)^{\frac{1}{q}}$$

Rewrite again as

$$\left(\sum_{n=1}^{N} |x_n + y_n|^p \right)^{1-\frac{1}{q}} \le \left(\sum_{n=1}^{N} |x_n|^p \right)^{\frac{1}{p}} + \left(\sum_{n=1}^{N} |y_n|^p \right)^{\frac{1}{p}}$$

But $1 - 1/q = 1/p$, *so we have* $\left(\sum_{n=1}^{N} |x_n + y_n|^p \right)^{\frac{1}{p}} \le \left(\sum_{n=1}^{N} |x_n|^p \right)^{\frac{1}{p}} + \left(\sum_{n=1}^{N} |y_n|^p \right)^{\frac{1}{p}}$.

Now apply the final estimate to find $\left(\sum_{n=1}^{N} |x_n + y_n|^p\right)^{\frac{1}{p}} \leq \left(\sum_{n=1}^{\infty} |x_n|^p\right)^{\frac{1}{p}} + \left(\sum_{n=1}^{\infty} |y_n|^p\right)^{\frac{1}{p}}.$
This says the right hand side is an upper bound for the partial sums on the left side. Hence, we know

$$\left(\sum_{n=1}^{\infty} |x_n + y_n|^p\right)^{\frac{1}{p}} \leq \left(\sum_{n=1}^{\infty} |x_n|^p\right)^{\frac{1}{p}} + \left(\sum_{n=1}^{\infty} |y_n|^p\right)^{\frac{1}{p}}$$

∎

Homework

Exercise 20.2.1 *If x and y are in ℓ^2, prove $2x + 3y$ is in ℓ^2.*

Exercise 20.2.2 *If x and y are in ℓ^∞, prove $-7x + 8y$ is in ℓ^∞.*

Exercise 20.2.3 *If x and y are in ℓ^4, prove $5x + 6y$ is in ℓ^4.*

Exercise 20.2.4 *If x and y are in ℓ^2, prove $x\,y$ is in ℓ^2.*

Exercise 20.2.5 *Prove ℓ^2 is a vector space over $\Re$.*

Exercise 20.2.6 *Define $\omega : \ell^2 \times \ell^2 \to \Re$ by $\omega(x, y) = \sum_{n=1}^{\infty} x_n y_n$ where x and y are arbitrary elements in ℓ^2 and we assume all sequences start counting at $n = 1$. Prove for all x, y and $c \in \Re$*

- $\omega(x, y) = \omega(y, x)$.

- $\omega(x + z, y) = \omega(x, y) + omega(z, y)$.

- $\omega(cx, y) = c\omega(x, y)$.

- $\omega(x, x) = 0 \iff x = 0$.

20.2.4   The ℓ^p Spaces and Their Metric

If $x \in \ell^p$, we can define a new function, called a **norm**, on ℓ^p like this:

$$||x||_p = \left(\sum_{n=1}^{\infty} |x_n|^p\right)^{\frac{1}{p}}, \quad 1 \leq p < \infty$$
$$||x||_\infty = \sup_{n \geq 1} |x_n|, \quad p = \infty$$

The Minkowski Inequality can then be rephrased as

$$||x + y||_p \leq ||x||_p + ||y||_p$$

and so we can use $|| \cdot ||_p$ as a way to measure both size of x and the distance between x and y. We let $d(x, y) = ||x - y||_p$ denote the distance between x and y. We note if $d(x, y) = 0$, then. in the case $1 \leq p < \infty$, we have $\sum_{n=1}^{\infty} |x_n - y_n|^p = 0$. But here you are adding up non-negative terms, so this must imply $|x_n - y_n|^p = 0$ for all n. This tells us $x_n = y_n$ for all n; i.e. $x = y$. On the other hand, when $p = \infty$, we would have $\sup_{n \geq 1} |x_n - y_n| = 0$. But this says, $|x_n - y_n| \leq 0$ for all n. Hence, $x_n = y_n$ for all n telling us $x = y$ again. Also, note the Minkowski Inequality tells us that ℓ^p is a vector space as if x and y are in ℓ^p, their sum $x + y$ defined by the sequence $(x_n + y_n)$ has $\sum_{n=1}^{\infty} |x_n + y_n|^p < \infty$ because of Minkowski's Inequality. A function like $|| \cdot ||_p$ on the set ℓ^p is called a **norm** and the set ℓ^p is called a **Normed Linear Space** or **Normed Vector Space**. A **vector** in ℓ^p is the sequence $x = (x_n)$ and its magnitude or size is $||x||_p$.

This is not a finite dimensional vector space and is another example of such things to add to your collection along with $(C([0,1]), d_1)$ and so forth.

20.2.5   Homework

Exercise 20.2.7 *Prove if x is in ℓ^1, it must be in ℓ^2 also.*

Exercise 20.2.8 *Prove if x is in ℓ^2, it must be in ℓ^4 also.*

Exercise 20.2.9 *Give an example of a sequence x which is in ℓ^2 but not in ℓ^1.*

Exercise 20.2.10 *Prove a geometric series x with common ratio r in $(-1, 1)$ is in ℓ^1 and find $||x||_1$.*

Exercise 20.2.11 *Prove a vector in $\Re^n$ can be identified with a sequence in ℓ^p for all p.*

Exercise 20.2.12 *If $x \in \Re^2$, the usual Euclidean norm is $|| \cdot ||_2$. Note $||x||_3$ and so on is also a norm on $\Re^2$. Compute $||x||_1, ||x||_2, ||x||_3, ..., ||x||_{10}$ for $x = [2, 5]'$. What do you think happens as $p \to \infty$?*

Exercise 20.2.13 *The Hölder's Inequality tells us that in $\Re^2$, $<x, y> /(||x||_p \, ||y||_q)$ is in $[-1, 1]$. So we can use this to define the angle between x and y. For $p = q = 2$ this is our usual angle, but for $p = 3$, $q = 3/2$ and so on the calculation changes. Calculate this angle for $p = 2$, $p = 3$, $p = 4$ and $p = 5$ where the q value is the exponent conjugate to p for the two vectors $x = [-1, 3]'$ and $y = [2, 4]'$.*

20.3   Convergence in a Metric Space

Definition 20.3.1 Convergence in a Metric Space

> Let (x_n) be a sequence in a metric space (X, d), then if there is an $x \in X$ so that given $\epsilon > 0$, there is an N so that $n > N$ implies $d(x_n, x) < \epsilon$ we say $x_n \to x$ in the metric d.

Comment 20.3.1 *In $\Re^2$, the standard metric is the Euclidean norm, and convergence of the vector sequence (x_n), where each $x_n = [x_{1,n}, x_{2,n}]'$, means*

$$\sqrt{(x_{1,n} - x_1)^2 + (x_{2,n} - x_2)^2} \quad \to \quad 0, \ as \ n \to \infty$$

Here the Euclidean norm is denoted by $|| \cdot ||_2$ and the metric is $d(x, y) = ||x - y||_2$.

Comment 20.3.2 *In $(C([a, b]), || \cdot ||_\infty)$, the norm is $||x||_\infty = \sup_{a \le t \le b} |x(t)|$ and the distance between two functions is $d(x, y) = ||x - y||_\infty$. The convergence of the sequence (x_n) to a function x in this metric means given $\epsilon > 0$, there is an N so that $||x_n - x||_\infty = \sup_{a \le t \le b} |x_n(t) - x(t)| < \epsilon$ when $n > N$.*

Here is a standard result about convergence of a sequence of continuous functions on a finite interval $[a, b]$. Using the $|| \cdot ||_\infty$ metric, the limit function retains the **property** of **continuous**. This is an important idea.

Theorem 20.3.1 Sequences of Continuous Functions Converge to a Continuous Function using the sup Metric

> If (x_n) converges to x in $C([a, b])$ with the sup norm, then x is continuous on $[a, b]$.

Proof 20.3.1
Pick t in $[a, b]$. Pick $\epsilon > 0$. Since (x_n) converges to x in $||\cdot||_\infty$ there is an N so that

$$\sup_{a \le t \le b} |x_n(t) - x(t)| \;\; < \;\; \epsilon/3$$

Now fix $\hat{n} > N$ and look at

$$
\begin{aligned}
|x(s) - x(t)| \;\;&=\;\; |x(s) - x_{\hat{n}}(s) + x_{\hat{n}}(s) - x_{\hat{n}}(t) + x_{\hat{n}}(t) - x(t)| \\
&\le\;\; |x(s) - x_{\hat{n}}(s)| + |x_{\hat{n}}(s) - x_{\hat{n}}(t)| + |x_{\hat{n}}(t) - x(t)|
\end{aligned}
$$

Now for this fixed $\hat{n}$, $x_{\hat{n}}$ is continuous, so there is a $\delta > 0$ so that $|s - t| < \delta$ with s and t in $[a, b]$ implies $|x_{\hat{n}}(s) - x_{\hat{n}}(t)| < \epsilon/3$. So, if $|s - t| < \delta$, we have

$$|x(s) - x(t)| \;\; \le \;\; |x(s) - x_{\hat{n}}(s)| + \epsilon/3 + |x_{\hat{n}}(t) - x(t)|.$$

But since $\hat{n} > N$, we also know the first and third piece are less than $\epsilon/3$ also. So we have $|x(s) - x(t)| < \epsilon/3 + \epsilon/3 + \epsilon/3 = \epsilon$ when $|s - t| < \delta$ and t and s are in $[a, b]$. This tells us x is continuous at t.

∎

20.4   The Completeness of Some Spaces

We can now show some spaces of objects are complete in the sense we have discussed. This is a hard idea to grasp so it is helpful to see more examples. Note in all these arguments we use the Completeness Axiom of the real numbers!

20.4.1   The Completeness of $(C([a, b]), ||\cdot||_\infty)$

Theorem 20.4.1 The Completeness of $C([a, b])$ with the sup norm

> $(C([a, b]), ||\cdot||_\infty)$ *is complete.*

Proof 20.4.1
Recall if $x \in C([a, b])$, then x is a continuous function on the interval $[a, b]$. To show this space is complete, we must show an arbitrary Cauchy sequence in it converges to a continuous function on $[a, b]$. So let $(x_n) \subset C([a, b])$ be a Cauchy sequence in the sup metric. Recall $||x||_\infty = \max_{a \le t \le b} |x(t)|$. Fix a point $\hat{s}$ in $[a, b]$. The sequence $(x_n(\hat{s}))$ is a sequence of real numbers. Does it converge? Since (x_n) is a Cauchy sequence in the sup-metric, given $\epsilon > 0$, there is an N_ϵ so that

$$m > n > N_\epsilon \;\; \implies \;\; \sup_{a \le t \le b} |x_n(t) - x_m(t)| < \epsilon/2$$

In particular, for $t = \hat{s}$, this says

$$m > n > N_\epsilon \;\; \implies \;\; |x_n(\hat{s}) - x_m(\hat{s})| < \epsilon/2$$

This tells us the sequence $(x_n(\hat{s}))$ is a Cauchy sequence of real numbers. Since $\Re$ is complete, there is a real number $a(\hat{s})$ so that $x_n(\hat{s}) \to a(\hat{s})$ as $n \to \infty$. The notation $a(\hat{s})$ is used because the value of this real number depends on the value of $\hat{s}$. This means we can use this number to define the limit *function $x : [a, b] \to \Re$ by*

$$x(\hat{s}) = \lim_{n \to \infty} x_n(\hat{s}) = a(\hat{s}).$$

So now we have a candidate for the limit function x. Does x_n converge to x at each t and is x continuous? Now look at the Cauchy sequence statement again:

$$m > n > N_\epsilon \implies \sup_{a \le t \le b} |x_n(t) - x_m(t)| < \epsilon/2$$

Dropping the sup *this says at each t in $[a, b]$, we have*

$$m > n > N_\epsilon \implies |x_n(t) - x_m(t)| < \epsilon/2$$

We know if f is continuous, then $\lim_{n \to \infty} f(x_n) = f(\lim_{n \to \infty} x_n)$. Now $|\cdot|$ is a nice continuous function. So we have, for any fixed $n > N_\epsilon$, letting $m \to \infty$

$$\lim_{m \to \infty} |x_n(t) - x_m(t)| = |\lim_{m \to \infty}(x_n(t) - x_m(t))| = |x_n(t) - \lim_{m \to \infty} x_m(t)|$$
$$= |x_n(t) - x(t)| \le \epsilon/2 < \epsilon$$

This shows the sequence $(x_n(t))$ converges to $x(t)$ for $a \le t \le b$. Does (x_n) converge to x in the sup *metric? The argument above says that*

$$|x_n(t) - x(t)| \le \epsilon/2, \ \forall a \le t \le b, \quad \text{if } n > N_\epsilon$$

and so if $n > N_\epsilon$

$$\sup_{a \le t \le b} |x_n(t) - x(t)| \le \epsilon/2 \implies ||x_n - x||_\infty \le \epsilon/2 < \epsilon$$

which is what we mean by $x_n \to x$ in the $|| \cdot ||_\infty$ metric. And from the previous theorem, this says x must be continuous. So $(C([a, b]), || \cdot ||_\infty)$ is a complete metric space. ∎

20.4.2 The Completeness of $(\ell^\infty, || \cdot ||_\infty)$

Theorem 20.4.2 The Completeness of ℓ^∞

$(\ell^\infty, || \cdot ||_\infty)$ *is complete.*

Proof 20.4.2
Here, the elements of ℓ^∞ are the sequences that are bounded. Given x and y, we know $||x - y||_\infty = \sup_n |a_n - b_n|$ where $x = (a_n)$ and $y = (b_n)$ are the sequences. Assume (x_n) is a Cauchy sequence. The element x_n is a sequence itself which we denote by $x_n = (x_{n,i})$ for $i \ge 1$ for convenience. Then for a given $\epsilon > 0$, there is an N so that

$$\sup_i |x_{n,i} - x_{m,i}| < \epsilon/2 \text{ when } m > n > N_\epsilon$$

So for each fixed index i, we have

$$|x_{n,i} - x_{m,i}| < \epsilon/2 \text{ when } m > n > N_\epsilon$$

This says for fixed i, the sequence $(x_{n,i})$ is a Cauchy sequence of real numbers and hence must converge to a real number we will call a_i. This defines a new sequence a. Is a a bounded sequence?

Does $x_n \to a$ in the $|| \cdot ||_\infty$ metric?

We use the continuity of the function $| \cdot |$ again to see for any $n > N_\epsilon$, we have

$$\lim_{m \to \infty} |x_{n,i} - x_{m,i}| \leq \epsilon/2 \Longrightarrow |x_{n,i} - \lim_{m \to \infty} x_{m,i}| \leq \epsilon/2$$

This argument works for all i and so $|x_{n,i} - a_i| \leq \epsilon/2$ when $n > N_\epsilon$ for all i which implies $\sup_i |x_{n,i} - a_i| \leq \epsilon/2 < \epsilon$ or $||x_n - a||_\infty < \epsilon$ when $n > N_\epsilon$. So $x_n \to a$ in $|| \cdot ||_\infty$ metric.

Finally, is a a bounded sequence? If we fix any $\hat{n} > N_\epsilon$, we have

$$\begin{aligned} |a_i| &= |a_i - x_{\hat{n},i} + x_{\hat{n},i}| \leq |a_i - x_{\hat{n},i}| + |x_{\hat{n},i}| \\ &\leq \sup_i |a_i - x_{\hat{n},i}| + ||x_{\hat{n}}||_\infty \\ &\leq \epsilon/2 + ||x_{\hat{n}}||_\infty \end{aligned}$$

This implies $\sup_i |a_i| \leq \epsilon/2 + ||x_{\hat{n}}||_\infty$ telling us (a_i) is a bounded sequence. Hence there is an $a \in \ell^\infty$ so that $x_n \to a$ in the $|| \cdot ||_\infty$ metric. Therefore this is a complete space. ■

20.4.3   The Completeness of $(\ell^p, || \cdot ||_p)$

Theorem 20.4.3 ℓ^p is Complete

ℓ^p *is complete in the $|| \cdot ||_p$ metric for $p \geq 1$ with p finite.*

Proof 20.4.3
Let (x_n) be a Cauchy sequence in ℓ^p with this metric. Recall if $x_n = (x_{n,i})$ is in ℓ^p, then $\sum_{i=1}^{\infty} |x_{n,i}|^p$ is finite and $||x_n||_p = \left(\sum_{i=1}^{\infty} |x_{n,i}|^p \right)^{1/p}$. Now let (x_n) be a Cauchy sequence in this metric. Then given $\epsilon > 0$, there is an N_ϵ so that $m > n > N_\epsilon$ implies

$$||x_n - x_m||_p = \left(\sum_{i=1}^{\infty} |x_{n,i} - x_{m,i}|^p \right)^{\frac{1}{p}} < \epsilon/2.$$

Thus, if $m > n > N_\epsilon$, $\sum_{i=1}^{\infty} |x_{n,i} - x_{m,i}|^p < (\epsilon/2)^p$. Since this is a sum on non-negative terms, each term must be less than $(\epsilon/2)^p$. So we must have $|x_{n,i} - x_{m,i}|^p < (\epsilon/2)^p$ or $|x_{n,i} - x_{m,i}| < (\epsilon/2)$ when $m > n > N_\epsilon$. This tells us immediately the sequence of real numbers $(x_{n,i})$ is a Cauchy sequence of real numbers and so must converge to a number we will call a_i. Now the Cauchy sequence inequality holds for any finite sum too. So we have

$$\sum_{i=1}^{k} |x_{n,i} - x_{m,i}|^p < (\epsilon/2)^p$$

when $m > n > N_\epsilon$. Now use the same trick as before. Since $| \cdot |^p$ is continuous, we can say

$$\begin{aligned} (\epsilon/2)^p &\geq \lim_{m \to \infty} \left(\sum_{i=1}^{k} |x_{n,i} - x_{m,i}|^p \right) = \left(\sum_{i=1}^{k} |x_{n,i} - \lim_{m \to \infty} x_{m,i}|^p \right) \\ &= \sum_{i=1}^{k} |x_{n,i} - a_i|^p \end{aligned}$$

Now let $k \to \infty$ and we have

$$\sum_{k=1}^{\infty} |x_{n,i} - a_i|^p = \lim_{k \to \infty} \left(\sum_{i=1}^{k} |x_{n,i} - a_i|^p \right) \leq (\epsilon/2)^p < \epsilon^p$$

This says immediately that $||x_n - a||_p < \epsilon$ when $n > N_\epsilon$ so $x_n \to a$ in $|| \cdot ||_p$ metric. But is $a \in \ell^p$?
Now we use Minkowski's inequality! Pick any $\hat{n} > N_\epsilon$. Then

$$||a|_p = \left(\sum_{i=1}^{\infty} |a_i|^p \right)^{1/p} = ||a - x_{\hat{n}} + x_{\hat{n}}||_p$$

$$\leq ||a - x_{\hat{n}}||_p + ||x_{\hat{n}}||_p$$

by Minkowski's Inequality. But from above we know $||a - x_{\hat{n}}||_p \leq \epsilon$ and so $||a||_p < \epsilon + ||x_{\hat{n}}||_p$ which is a finite number as $x_{\hat{n}}$ is in ℓ^p. We see then that $a \in \ell^p$, $x_n \to a$ in $|| \cdot ||_p$ and so ℓ^p is complete with respect to this metric. ■

20.4.4 Homework

Exercise 20.4.1 *Let (x_n) be the sequence in $C([0,2])$ defined by $x_n(t) = \sin(n^2 t)/n^2$ for all $n \geq 1$. Prove this is a Cauchy sequence in $(C([0,2]), || \cdot ||_\infty)$. What does this sequence converge to?*

Exercise 20.4.2 *Let (x_n) be the sequence in $C([-2,2])$ defined by $x_n(t) = \cos(n^2 t)/n$ for all $n \geq 1$. Prove this is a Cauchy sequence in $(C([-2,2]), || \cdot ||_\infty)$. What does this sequence converge to?*

Exercise 20.4.3 *Let (x_n) be the sequence in $C([-2,4])$ defined by $x_n(t) = 2 + 1/n + \cos(n^2 t)/n$ for all $n \geq 1$. Prove this is a Cauchy sequence in $(C([-2,4]), || \cdot ||_\infty)$. What does this sequence converge to?*

Exercise 20.4.4 *Let (x_n) be the sequence in $C([-3,5])$ defined by $x_n(t) = 2 + 1/n + e^{-nt}$ for all $n \geq 1$. Prove this is a Cauchy sequence in $(C([-3,5]), || \cdot ||_\infty)$. What does this sequence converge to?*

Chapter 21

Integration Theory

We will now develop the theory of the Riemann Integral for a bounded function f on the interval $[a, b]$. Another good source is the development in (Fulks (6) 1978).

21.1 Basics

To study the integration of a function f, there are two intellectually separate ideas: the primitive or antiderivative and the Riemann integral. We have the idea of a **Primitive** or **Antiderivative**: f is any function F which is differentiable and satisfies $F'(t) = f(t)$ at all points in the domain of f. Normally, the domain of f is a finite interval of the form $[a, b]$, although it could also be an infinite interval like all of $\Re$ or $[1, \infty)$ and so on. Note that an **antiderivative** does not require any understanding of the process of Riemann integration at all – only what differentiation is! The Riemann integral of a function is far more complicated to setup than the process of guessing a primitive or antiderivative. To define a Riemann integral properly, first, we start with a bounded function f on a finite interval $[a, b]$. This kind of function f need not be continuous!

21.1.1 Partitions of $[a, b]$

We select a finite number of points from the interval $[a, b]$, $\{t_0, t_1, \ldots, t_{n-1}, t_n\}$. We don't know how many points there are, so a different selection from the interval would possibly give us more or less points. But for convenience, we will just call the last point t_n and the first point t_0. These points are not arbitrary – t_0 is always a, t_n is always b and they are ordered like this:

$$t_0 = a < t_1 < t_2 < \ldots < t_{n-1} < t_n = b$$

The collection of points from the interval $[a, b]$ is called a Partition of $[a, b]$ and is denoted by some letter – here we will use the letter **P**. So if we say P is a partition of $[a, b]$, we know it will have $n + 1$ points in it, they will be labeled from t_0 to t_n and they will be ordered left to right with strict inequalities. But, we will not know what value the positive integer n actually is. The simplest Partition P is the two point partition $\{a, b\}$. Note these things also:

1. Each partition of $n + 1$ points determines n subintervals of $[a, b]$

2. The lengths of these subintervals always adds up to the length of $[a, b]$ itself, $b - a$.

3. These subintervals can be represented as $\{[t_0, t_1], [t_1, t_2], \ldots, [t_{n-1}, t_n]\}$ or more abstractly as $[t_i, t_{i+1}]$ where the index i ranges from 0 to $n - 1$.

4. The length of each subinterval is $t_{i+1} - t_i$ for the indices i in the range 0 to $n - 1$.

5. The largest subinterval length is called the **norm** of the partition and we denote it by the symbol $\| P \|$.

Now from each subinterval $[t_i, t_{i+1}]$ determined by the Partition P, select any point you want and call it s_i. This will give us the points s_0 from $[t_0, t_1]$, s_1 from $[t_1, t_2]$ and so on up to the last point, s_{n-1} from $[t_{n-1}, t_n]$. At each of these points, we can evaluate the function f to get the value $f(s_j)$. Call these points an **Evaluation Set** for the partition P. Let's denote such an evaluation set by the letter E. If the function f was nice enough to be positive always and continuous, then the product $f(s_i) \times (t_{i+1} - t_i)$ can be interpreted as the area of a rectangle.

21.1.2   Riemann Sums

If we add up all these rectangle areas we then get a sum which is useful enough to be given a special name: the **Riemann Sum** for the function f associated with the Partition P and our choice of evaluation set $E = \{s_0, \ldots, s_{n-1}\}$. This sum is represented by the symbol $S(f, P, E)$ where the arguments inside the parenthesis are there to remind us that this sum depends on our choice of the function f, the partition P and the evaluation set E. We have

$$S(f, P, E) \;\; = \;\; \sum_{i=0}^{n-1} f(s_i)\,(t_{i+1} - t_i)$$

Definition 21.1.1 The Riemann Sum

The Riemann sum for the bounded function f, the partition P and the evaluation set $E = \{s_0, \ldots, s_{n-1}\}$ from $P\{t_0,\, t_1,\,, \ldots, t_{n-1},\, t_n\}$ is defined by

$$S(f, P, E) \;\; = \;\; \sum_{i=0}^{n-1} f(s_i)\,(t_{i+1} - t_i)$$

It is pretty misleading to write the Riemann sum this way as it can make us think that the n is always the same when in fact it can change value each time we select a different P. So many of us write the definition this way instead $S(f, P, E) = \sum_{i \in P} f(s_i)\,(t_{i+1} - t_i)$ and we just remember that the choice of P will determine the size of n.

Let's look at an example of this. Here we see the graph of a typical function which is always positive on some finite interval $[a, b]$. Next, let's set the interval to be $[1, 6]$ and compute the Riemann Sum for a particular choice of Partition P and evaluation set E. The partition (gray) is $P = \{1.0, 1.5, 2.6, 3.8, 4.3, 5.6, 6.0\}$. Hence, we have subinterval lengths of $t_1 - t_0 = 0.5$, $t_2 - t_1 = 1.1, t_3 - t_2 = 1.2, t_4 - t_3 = 0.5, t_5 - t_4 = 1.3$ and $t_6 - t_5 = 0.4$, giving $\| P \| = 1.3$. For the evaluation set (red) $E = \{1.1, 1.8, 3.0, 4.1, 5.3, 5.8\}$ the Riemann sum is

$$
\begin{aligned}
S(f, P, E) \;\; &= \;\; \sum_{i=0}^{5} f(s_i)\,(t_{i+1} - t_i) \\
&= \;\; f(1.1)(0.5) + f(1.8)(1.1) + f(3.0)(1.2) \\
&+ \;\; f(4.1)(0.5) + f(5.3)(1.3) + f(5.8)(0.4)
\end{aligned}
$$

We can also interpret the Riemann sum as an approximation to the area under the curve.

- The partition (closed circles) is $P = \{1.0, 1.5, 2.6, 3.8, 4.3, 5.6, 6.0\}$.

- For the evaluation set (open circles) $E = \{1.1, 1.8, 3.0, 4.1, 5.3, 5.8\}$

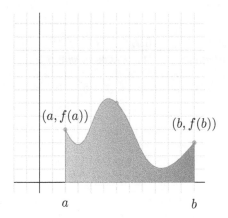

A generic curve f on the interval $[a, b]$ which is always positive. Note the area under this curve is the shaded region.

Figure 21.1: The area under the curve f.

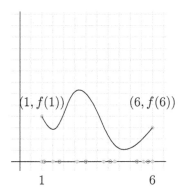

The partition points are closed circle and the evaluation points are open circles.

Figure 21.2: A simple Riemann sum.

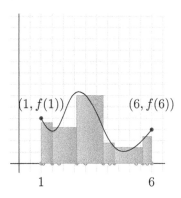

Figure 21.3: The Riemann sum as an approximate area.

Example 21.1.1 *Let* $f(t) = 3t^2$ *on the interval* $[-1, 2]$ *with* $P = \{-1, -0.3, 0.6, 1.2, 2.0\}$ *and* $E = \{-0.7, 0.2, 0.9, 1.6\}$. *Find the Riemann sum.*

Solution *The partition determines subinterval lengths of* $t_1 - t_0 = 0.7$, $t_2 - t_1 = 0.9$, $t_3 - t_2 = 0.6$, *and* $t_4 - t_3 = 0.8$, *giving* $\| P \| = 0.9$. *For the evaluation set* E *the Riemann sum is*

$$S(f, P, E) \quad = \quad f(-0.7) \times 0.7 + f(0.2) \times 0.9 + f(0.9) \times 0.6 + f(1.6) \times 0.8 = 8.739$$

Homework

Exercise 21.1.1 *Let* $f(t) = 2t^2$ *on the interval* $[-2, 2]$ *with* $P = \{-2.0, -1.0, -0.3, 0.6, 1.2, 2.0\}$ *and* $E = \{-1.4, -0.7, 0.2, 0.9, 1.6\}$. *Find the Riemann sum. Also find the norm of the partition.*

Exercise 21.1.2 *Let* $f(t) = t^2 + 3t + 4$ *on the interval* $[-1, 2]$ *with* $P = \{-1.0, -0.3, 0.6, 1.2, 2.0\}$ *and* $E = \{-0.7, 0.2, 0.9, 1.6\}$. *Find the Riemann sum. Also find the norm of the partition.*

Exercise 21.1.3 *Let* $f(t) = 4t^2 - 5t + 10$ *on the interval* $[1, 3]$ *with* $P = \{1.0, 1.3, 1.6, 2.2, 2.5, 3.0\}$ *and* $E = \{1.1, 1.4, 1.9, 2.3, 2.7\}$. *Find the Riemann sum. Also find the norm of the partition.*

Exercise 21.1.4 *Let* $f(t) = \sin(t)$ *on the interval* $[-2, 2]$ *with* $P = \{-2.0, -1.4, -1, -0.3, 0.6, 1.2, 2.0\}$ *and* $E = \{-1.7, -1.2, -0.7, 0.2, 0.9, 1.6\}$. *Find the Riemann sum. Also find the norm of the partition.*

Exercise 21.1.5 *Let* $f(t) = 1/(1 + t^2)$ *on the interval* $[-5, 5]$. *Let* $P = \{-5.0, -4.0, -3.0, -2.0, -1.0, -0.3, 0.6, 1.2, 2.0, 3.0, 4.0, 5, 0\}$ *and* $E = \{-4.3, -3.7, -2.8, -1.3, -0.7, 0.2, 0.9, 1.6, 2.4, 3.2, 4.3\}$. *Find the Riemann sum. Also find the norm of the partition.*

21.1.3   Riemann Sums in Octave

As you have seen, doing these by hand is tedious. Let's look at how we might do them using Octave. Here is a typical Octave session to do this. Let's calculate the Riemann sum for the function $f(x) = x^2$ on the interval $[1, 3]$ using the partition $P = \{1, 1.5, 2.1, 2.8, 3.0\}$ and evaluation set $E = \{1.2, 1.7, 2.5, 2.9\}$. First, set up our function. Octave allows us to define a function inside the Octave environment as follows

Listing 21.1: **Defining a function in Octave**

```
>> f = @(x) (x.^2);
```

This defines the function $f(x) = x^2$. If we had wanted to define $g(x) = 2x^2 + 3$, we would have used

Listing 21.2: **Defining another function in Octave**

```
>> g = @(x) (2*x.^2 + 3);
```

Now define the vector X and square it as follows: First define X by `X = [1;2;3];` Then to square X, we would write `X.^2` to square each *component* creating a new *vector* with each entry squared.

Listing 21.3: | **Square the Vector X** |

```
>> X.^2
ans =
     1
     4
     9
```

The way we set up the function `f = @(x)(x.^2);` in Octave makes use of this. The variable *X* may or may not be a *vector*. So we write `x.^2` so that if *x* is a vector, multiplication is done component wise and if not, it is just the squaring of a number. So for our function, to find *f* for all the values in *X*, we just type

Listing 21.4: | **Applying the Function f to X** |

```
>> f(X)
ans =
     1
     4
     9
```

Now let's setup the partition with the command

Listing 21.5: | **Setting Up the Partition P** |

```
>> P = [1;1.5;2.1;2.8;3.0]
P =
     1.0000
     1.5000
     2.1000
     2.8000
     3.0000
```

The command `diff` in Octave is applied to a vector to create the differences we have called the Δx_i's.

Listing 21.6: | **Finding the Subinterval Lengths of P** |

```
>> dx = diff(P)
dx =
     0.5000
     0.6000
     0.7000
     0.2000
```

Next, we set up the evaluation set *E*.

Listing 21.7: | **Set Up the Evaluation Set E** |

```
>> E = [1.2;1.7;2.5;2.9]
E =
     1.2000
     1.7000
     2.5000
     2.9000
```

Now we find $f(E)$, a new vector with the values $f(s_i)$'s. We use $f(E).*dx$ to create the new vector with components $f(s_i)\Delta x_i$.

Listing 21.8: | **Calculate the Areas of the Rectangles** |

```
>> g = f(E).*dx
g =
     0.7200
     1.7340
     4.3750
     1.6820
```

21.1.3.1   Homework

For the given function f, partition P and evaluation set E, do the following: use Octave to find $S(f, P, E)$ for the partition P and evaluation set E.

- Create a new **word** document for this homework.

- Do the document in single space.

- Do Octave fragments in bold font.

Then answer the problems like this:

- State the problem

- Insert into your doc the Octave commands you use to solve the problem. Do this in bold.

- Before each line of Octave add explanatory comments so we can check to see you know what you're doing.

Now here is the homework: For the given function f, partition P and evaluation set E, do the following.

1. Find $S(f, P, E)$ for the partition P and evaluation set E.

2. Find $||P||$.

3. Sketch a graph of this Riemann sum as an approximation to the area under the curve f. Do a nice graph with appropriate use of color.

Exercise 21.1.6 *Let* $f(t) = t^2 + 2$ *on the interval* $[1, 3]$ *with* $P = \{1, 1.5, 2.0, 2.5, 3.0\}$ *and* $E = \{1.2, 1.8, 2.3, 2.8\}$.

Exercise 21.1.7 *Let* $f(t) = t^2 + 3$ *on the interval* $[1, 3]$ *with* $P = \{1, 1.6, 2.3, 2.8, 3.0\}$ *and* $E = \{1.2, 1.9, 2.5, 2.85\}$.

Exercise 21.1.8 *Let* $f(t) = 3t^2 + 2t$ *on the interval* $[1, 2]$ *with* $P = \{1, 1.2, 1.5, 1.8, 2.0\}$ *and* $E = \{1.1, 1.3, 1.7, 1.9\}$

Exercise 21.1.9 *Let* $f(t) = 3t^2 + t$ *on the interval* $[1, 4]$ *with* $P = \{1, 1.2, 1.5, 2.8, 4.0\}$ *and* $E = \{1.1, 1.3, 2.3, 3.2\}$

Example 21.1.2 *Here is an example of this kind of problem:*
Your name, the Course Number, Section Number
today's date and HW Number,
Problem 1: Let $f(t) = \sin(5t)$ *on the interval* $[1, 3]$ *with* $P = \{1, 1.5, 2.0, 2.5, 3.0\}$ *and* $E = \{1.2, 1.8, 2.3, 2.8\}$. *Find* $S(f, P, E)$.

Listing 21.9: | **Sample Riemann Sum in Octave** |

```
% add explanation here
>> f = @(x) sin(5*x);
% add explanation here
>> P = [1; 1.5; 2.0; 2.5; 3.0];
% add explanation here
>> E = [1.2; 1.8; 2.3; 2.8];
% add explanation here
>> dx = diff(P);
% add explanation here
>> g = f(E).*dx;
% add explanation here
>> RS = sum(g)
RS = 0.1239
```

21.1.4   Graphing Riemann Sums

If we want to graph the Riemann sums we need to graph those rectangles we draw by hand. To graph a rectangle, we graph 4 lines. The command `plot([x1 x2], [y1 y2])` plots a line from the pair $(x1, y1)$ to $(x2, y2)$.

- So the command `plot([x(i)x(i+1)],[f(s(i))f(s(i))]);` plots the horizontal line which is the top of our rectangle.

- The command `plot([x(i)x(i)], [0 f(s(i))]);` plots a vertical line that starts on the x axis at x_i and ends at the function value $f(s_i)$.

- The command `plot([x(i+1)x(i+1)], [0 f(s(i))]);` plots a vertical line that starts on the x axis at x_{i+1} and ends at the function value $f(s_i)$.

- To plot rectangle, for the first pair of partition points, first we set the axis of our plot so we will be able to see it. We use the **axis** command in Octave – look it up using `help`! If the two x points are $x1$ and $x2$ and the y value is $f(s1)$ where $s1$ is the first evaluation point, we expand the x axis to $[x1 - 1, x2 + 1]$ and expand the y axis to $[0, f(s1)]$. This allows our rectangle to be seen. The command is `axis([x1-1 x2+1 0 f((s1))+1]);`.

Putting this all together, we plot the first rectangle like this:

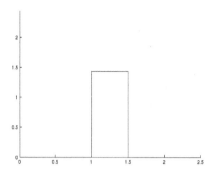

Figure 21.4: Simple rectangle.

Listing 21.10: **Graphing a Rectangle**

```
>> hold on
% set axis so we can see rectangle
>> axis([P(1)-1 P(2)+1 0 f(E(1))+1])
% plot top, LHS, RHS and bottom of rectangle
>> plot([P(1) P(2)],[f(E(1)) f(E(1))]);
>> plot([P(1) P(1)], [0 f(E(1))]);
>> plot([P(2) P(2)], [0 f(E(1))]);
>> plot([P(1) P(2)],[0 0]);
>> hold off
```

We have to force Octave to plot repeatedly without erasing the previous plot. We use **hold on** and **hold off** to do this. We start with **hold on** and then all plots are kept until the **hold off** is used. This generates the rectangle in Figure 21.4.

21.1.5 Uniform Partitions and Riemann Sums

To show the Riemann sum approximation as rectangles, we use a `for` loop in Octave. To put this all together, for an example of four rectangles, we use the loop:

Listing 21.11: **A Four Rectangle Loop**

```
hold on % set hold to on
for i = 1:4  % graph rectangles
  bottom = 0;
  top = f(E(i));
  plot([P(i) P(i+1)],[f(E(i)) f(E(i))]);
  plot([P(i) P(i)], [bottom top]);
  plot([E(i) E(i)], [bottom top],'r');
  plot([P(i+1) P(i+1)], [bottom top]);
  plot([P(i) P(i+1)],[0 0]);
end
hold off % set hold off
```

Of course, f could be negative, so we need to adjust our thinking as some of the rectangles might need to point down. We do that by setting the **bottom** and **top** of the rectangles using an **if** test. We use the following code:

Listing 21.12: | **Switching the Top and Bottom of a Rectangle** |

```
  bottom = 0;
  top = f(E(i));
  if f(E(i)) < 0
    top = 0;
    bottom = f(E(i));
  end
```

All together, we have the code block: the number of rectangles here is determined by the size of P which is found using the command `[sizeP,m] = size(P);` which returns the size of P in the variable `sizeP`. So the loop is up to `sizeP-1`.

Listing 21.13: | **Code Block for Drawing the Rectangles for Any P and E** |

```
hold on % set hold to on
[sizeP,m] = size(P);
for i = 1:sizeP-1  % graph all the rectangles
  bottom = 0;
  top = f(E(i));
  if f(E(i)) < 0
    top = 0;
    bottom = f(E(i));
  end
  plot([P(i) P(i+1)],[f(E(i)) f(E(i))]);
  plot([P(i) P(i)], [bottom top]);
  plot([E(i) E(i)], [bottom top],'r');
  plot([P(i+1) P(i+1)], [bottom top]);
  plot([P(i) P(i+1)],[0 0]);
end
hold off;
```

We also want to place the plot of f over these rectangles. This is done in the code at the bottom of the code block below.

Listing 21.14: | **Overlaying the Graph of f** |

```
hold on % set hold to on
[sizeP,m] = size(P);
for i = 1:sizeP-1  % graph all the rectangles
  bottom = 0;
  top = f(E(i));
  if f(E(i)) < 0
    top = 0;
    bottom = f(E(i));
  end
  plot([P(i) P(i+1)],[f(E(i)) f(E(i))]);
```

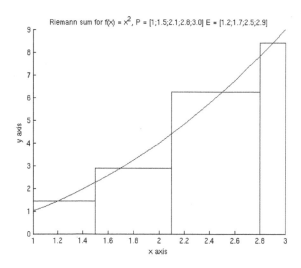

Figure 21.5: Riemann sum for $f(x) = x^2$ for Partition $\{1, 1.5, 2.1, 2.8, 3.0\}$.

```
    plot([P(i)  P(i)],  [bottom  top]);
    plot([E(i)  E(i)],  [bottom  top],'r');
    plot([P(i+1)  P(i+1)],  [bottom  top]);
    plot([P(i)  P(i+1)],[0  0]);
end
y = linspace(P(1),P(sizeP),  101);
plot(y,f(y));
xlabel('x axis');    ylabel('y axis');
title('Riemann  Sum  with  function  overlaid');
hold  off;
```

We generate Figure 21.5. We can place this code into a function called `RiemmanSum.m` which can be downloaded from the web site **The Primer On Analysis I: Part Two Home Page** . To use it is straightforward:

Listing 21.15: | **Using the RiemannSum.m Code** |

```
f = @(x)  sin(x);
P = [-3.3;  -2.1;-1.4;-.5;  0.2;1.4;2.6;2.9;3.6];
E = [-2.9;  -1.7;  -0.8;0.0;1.0;2.0;2.7;3.3];
RS = RiemannSum(f,P,E);
```

21.1.6   Refinements of Partitions

For each $j = 1, \ldots, n - 1$, we let $\Delta x_j = x_{j+1} - x_j$. The collection of all finite partitions of $[a, b]$ is denoted $\Pi[a, b]$. We now need to define a few terms. First, the refinement of a partition.

Definition 21.1.2 Partition Refinements

The partition $\pi_1 = \{y_0, \ldots, y_m\}$ is said to be a **refinement** of the partition $\pi_2 = \{x_0, \ldots, x_n\}$ if every partition point $x_j \in \pi_2$ is also in π_1. If this is the case, then we write $\pi_2 \preceq \pi_1$, and we say that π_1 is finer than π_2 or π_2 is coarser than π_1.

Next, we need the notion of the common refinement of two given partitions.

Definition 21.1.3 Common Refinement

Given $\pi_1, \pi_2 \in \Pi[a, b]$, there is a partition $\pi_3 \in \Pi[a, b]$ which is formed by taking the union of π_1 and π_2 and using common points only once. We call this partition the common refinement of π_1 and π_2 and denote it by $\pi_3 = \pi_1 \vee \pi_2$.

The relation $\preceq$ is a partial ordering of $\Pi[a, b]$. It is not a total ordering, since not all partitions are comparable. There is a coarsest partition, also called the trivial partition. It is given by $\pi_0 = \{a, b\}$. We may also consider uniform partitions of order k. Let $h = (b - a)/k$. Then $\pi = \{x_0 = a, x_0 + h, x_0 + 2h, \ldots, x_{k-1} = x_0 + (k-1)h, x_k = b\}$.

Proposition 21.1.1 Refinements and Common Refinements

If $\pi_1, \pi_2 \in \Pi[a, b]$, then $\pi_1 \preceq \pi_2$ if and only if $\pi_1 \vee \pi_2 = \pi_2$.

Proof 21.1.1

If $\pi_1 \preceq \pi_2$, then $\pi_1 = \{x_0, \ldots, x_p\} \subset \{y_0, \ldots, y_q\} = \pi_2$. Thus, $\pi_1 \cup \pi_2 = \pi_2$, and we have $\pi_1 \vee \pi_2 = \pi_2$. Conversely, suppose $\pi_1 \vee \pi_2 = \pi_2$. By definition, every point of π_1 is also a point of $\pi_1 \vee \pi_2 = \pi_2$. So, $\pi_1 \preceq \pi_2$. ∎

We have mentioned the norm of a partition before, but let's make it a formal definition now.

Definition 21.1.4 The Gauge or Norm of a Partition

For $\pi \in \Pi[a, b]$, we define the **gauge** or **norm** of π, denoted $\|\pi\|$, by $\|\pi\| = \max\{\Delta x_j : 1 \le j \le p\}$.

21.1.7 Homework

Exercise 21.1.10 *Prove that the relation $\preceq$ is a partial ordering of $\Pi[a, b]$.*

Exercise 21.1.11 *Fix $\pi_1 \in \Pi[a, b]$. The set $C(\pi_1) = \{\pi \in \Pi[a, b] : \pi_1 \preceq \pi\}$ is called the **core** determined by π_1. It is the set of all partitions of $[a, b]$ that contain (or are finer than) π_1.*

1. *Prove that if $\pi_1 \preceq \pi_2$, then $C(\pi_2) \subset C(\pi_1)$.*

2. *Prove that if $\|\pi_1\| < \epsilon$, then $\|\pi\| < \epsilon$ for all $\pi \in C(\pi_1)$.*

3. *Prove that if $\|\pi_1\| < \epsilon$ and $\pi_2 \in \Pi[a, b]$, then $\|\pi_1 \vee \pi_2\| < \epsilon$.*

Exercise 21.1.12 *Let $f(t) = 2t^2$ on the interval $[-2, 2]$ with $P = \{-2.0, -1.0, -0.3, 0.6, 1.2, 2.0\}$. Add the midpoints of each subinterval determined by P and write down the resulting refinement P'. Using the left hand endpoints of each subinterval determined by P' find the Riemann sum. Also find the norm of the refinement.*

Exercise 21.1.13 *Let $f(t) = t^2 + 3t + 4$ on the interval $[-1, 2]$ with $P = \{-1.0, -0.3, 0.6, 1.2, 2.0\}$. Add the midpoints of each subinterval determined by P and write down the resulting refinement P'. Using the right hand endpoints of each subinterval determined by P' find the Riemann sum. Also find the norm of the refinement.*

Exercise 21.1.14 *Let $f(t) = 4t^2 - 5t + 10$ on the interval $[1,3]$ with $P = \{1.0, 1.3, 1.6, 2.2, 2.5, 3.0\}$. Add the midpoints of each subinterval determined by P and write down the resulting refinement P'. Using the midpoints of each subinterval determined by P' find the Riemann sum. Also find the norm of the refinement.*

21.1.8   Uniform Partition Riemann Sums and Convergence

To see graphically how the Riemann sums converge to the Riemann integral, let's write a new function: Riemann sums using uniform partitions and midpoint evaluation sets. Here is the new function.

Listing 21.16:   **The Uniform Partition Riemann Sum Function**

```
function RS = RiemannUniformSum(f,a,b,n)
% set up a uniform partition with n+1 points
deltax = (b-a)/n;
P = [a:deltax:b]; % makes a row vector
for i=1:n
    start = a+(i-1)*deltax;
    stop = a+i*deltax;
    E(i) = 0.5*(start+stop);
end
% send in transpose of P and E so we use column vectors
% because original RiemannSum function uses columns
RS = RiemannSum(f,P',E');
end
```

We can then generate a sequence of Riemann sums for different values of n. Here is a typical session:

Listing 21.17:   **Typical Session**

```
f = @(x) sin(3*x);
RS = RiemannUniformSum(f,-1,4,10);
RS= RiemannUniformSum(f,-1,4,20);
RS = RiemannUniformSum(f,-1,4,30);
RS= RiemannUniformSum(f,-1,4,40);
```

These commands generate the plots of Figures 21.6 - Figure 21.9.

The actual value is $\int_{-1}^{4} \sin(3x)dx = -.611282$. The $n = 80$ case is quite close! The experiment we just did should help you understand better what we will mean by the **Riemann Integral**. What we have shown is

$$\lim_{n \to \infty} S(f, P_n, E_n) = -.611282...$$

for the particular sequence of uniform partitions P_n with the particular choice of the evaluation sets E_n being the midpoints of each of the subintervals determined by the partition. Note the $\|P_n\| = 5/n$ in each case. We will find that what we mean by the Riemann integral existing is that we get this value no matter what sequence of partitions we choose with associated evaluation sets as long as the norm of the partitions goes to 0.

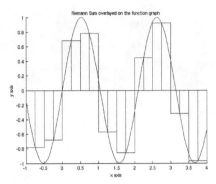

Figure 21.6: The Riemann sum with a uniform partition P_{10} of $[-1, 4]$ for $n = 10$. The function is $\sin(3x)$ and the Riemann sum is -0.6726.

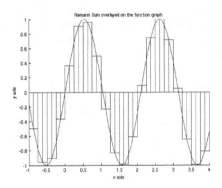

Figure 21.7: Riemann sum with a uniform partition P_{20} of $[-1, 4]$ for $n = 20$. The function is $\sin(3x)$ and the Riemann sum is -0.6258.

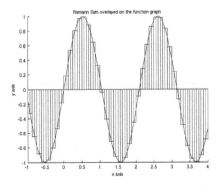

Figure 21.8: Riemann sum with a uniform partition P_{40} of $[-1, 4]$ for $n = 40$. The function is $\sin(3x)$ and the Riemann sum is -0.6149.

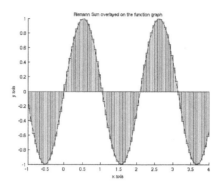

Figure 21.9: Riemann sum with a uniform partition P_{80} of $[-1, 4]$ for $n = 80$. The function is $\sin(3x)$ and the Riemann sum is -0.6122.

21.1.8.1   Homework

For the given function f, interval $[a, b]$ and choice of n, you'll calculate the corresponding uniform partition Riemann sum using the functions **RiemannSum** in file **RiemannSum.m** and **RiemannUniformSum** in file **RiemannUniformSum.m**. You can download these functions as files from the web site **www.ces.clemson.edu/~petersj/AdvancedCalcTwo.html.** Save them in your personal working directory.

- Create a new **word** document in single space with Octave fragments in bold font.

- Do the document in single space. Do Octave fragments in bold font.

- For each value of n, do a **save as** and save the figure with a filename like **HW#Problem#a[ ].png** where [ ] is where you put the number of the graph.

- Insert this picture into the doc resizing as needed to make it look good. Explain in the doc what the picture shows.

Your report would look something like this:
Your name
The date and HW number,
Problem 1:
Let $f(t) = \sin(5t)$ on the interval $[-1, 4]$. Find the Riemann Uniform sum approximations for $n = 10, 20, 40$ and 80.

Listing 21.18:   **Sample Problem Session**

```
% add explanation here
f = @(x) sin(5*x);
% add explanation here
RS  = RiemannUniformSum(f,-1,4,10)
% add explanation here
RS  = RiemannUniformSum(f,-1,4,20)
% add explanation here
RS  = RiemannUniformSum(f,-1,4,40)
% add explanation here
RS  = RiemannUniformSum(f,-1,4,80)
```

Exercise 21.1.15 *Let $f(t) = t^2 - 2t + 3$ on the interval $[-2, 3]$ with $n = 8, 16, 32$ and 48.*

Exercise 21.1.16 *Let $f(t) = \sin(2t)$ on the interval $[-1, 5]$ with $n = 10, 40, 60$ and 80.*

Exercise 21.1.17 *Let $f(t) = -t^2 + 8t + 5$ on the interval $[-2, 3]$ with $n = 4, 12, 30$ and 50.*

21.2   Defining The Riemann Integral

We can now define the Riemann Integral properly.

Definition 21.2.1 Riemann Integrability of a Bounded f

*We say $f \in B[a, b]$ is **Riemann Integrable** on $[a, b]$ if there exists a real number, I, such that for every $\epsilon > 0$ there is a partition, $\pi_0 \in \Pi[a, b]$ such that*

$$| S(f, \pi, \sigma) - I | < \epsilon$$

for any refinement, π, of π_0 and any evaluation set, $\sigma \subset \pi$. We denote this value, I, by

$$I \equiv RI(f; a, b)$$

We denote the set of Riemann integrable functions on $[a, b]$ by $RI[a, b]$. Also, it is readily seen that the number $RI(f; a, b)$ in the definition above, when it exists, is unique. So we can speak of the *Riemann Integral* of a function f. We also have the following conventions.

1. $RI(f; a, b) = -RI(f; b, a)$

2. $RI(f; a, a) = 0$

3. f is called the **integrand**.

Theorem 21.2.1 $RI[a, b]$ is a Vector Space

$RI[a, b]$ is a vector space over $\Re$ and the mapping $I_R : RI[a, b] \to \Re$ defined by

$$I_R(f) = RI(f; a, b)$$

is a linear mapping.

Proof 21.2.1
Let $f_1, f_2 \in RI[a, b]$, and let $\alpha, \beta \in \Re$. For any $\pi \in \Pi[a, b]$ and $\sigma \subset \pi$, we have

$$
\begin{aligned}
S(\alpha f_1 + \beta f_2, \pi, \sigma) &= \sum_{\pi} (\alpha f_1 + \beta f_2)(s_j) \Delta x_j \\
&= \alpha \sum_{\pi} f_1(s_j) \Delta x_j + \beta \sum_{\pi} f_2(s_j) \Delta x_j \\
&= \alpha S(f_1, \pi, \sigma) + \beta S(f_2, \pi, \sigma).
\end{aligned}
$$

Since f_1 is Riemann integrable, given $\epsilon > 0$, there is a real number $I_1 = RI(f_1, a, b)$ and a partition $\pi_1 \in \Pi[a, b]$ such that

$$| S(f_1, \pi, \sigma) - I_1 | < \frac{\epsilon}{2(| \alpha | + 1)} \tag{*}$$

for all refinements, π of π_1 and all $\sigma \subset \pi$.

Likewise, since f_2 is Riemann integrable, there is a real number $I_2 = RI(f_2; a, b)$ and a partition $\pi_2 \in \Pi[a, b]$ such that

$$| S(f_2, \pi, \sigma) - I_2 | \; < \; \frac{\epsilon}{2(| \beta | + 1)} \qquad (**)$$

for all refinements, π of π_2 and all $\sigma \subset \pi$.

Let $\pi_0 = \pi_1 \vee \pi_2$. Then π_0 is a refinement of both π_1 and π_2. So, for any refinement, π, of π_0, and any $\sigma \subset \pi$, we have Equation $*$ and Equation $**$ are valid. Hence,

$$| S(f_1, \pi, \sigma) - I_1 | \; < \; \frac{\epsilon}{2(| \alpha | + 1)}$$
$$| S(f_2, \pi, \sigma) - I_2 | \; < \; \frac{\epsilon}{2(| \beta | + 1)}.$$

Thus, for any refinement π of π_0 and any $\sigma \subset \pi$, it follows that

$$
\begin{aligned}
| S(\alpha f_1 + \beta f_2, \pi, \sigma) - (\alpha I_1 + \beta I_2) | \;&=\; | \alpha S(f_1, \pi, \sigma) + \beta S(f_2, \pi, \sigma) - \alpha I_1 - \beta I_2 | \\
&\leq\; | \alpha | \, | \, S(f_1, \pi, \sigma) - I_1 \, | + | \beta | \, | \, S(f_2, \pi, \sigma) - I_2 \, | \\
&<\; | \alpha | \frac{\epsilon}{2(| \alpha | + 1)} + | \beta | \frac{\epsilon}{2(| \beta | + 1)} \\
&<\; \epsilon.
\end{aligned}
$$

This shows that $\alpha f_1 + \beta f_2$ is Riemann integrable and that the value of the integral $RI(\alpha f_1 + \beta f_2; a, b)$ is given by $\alpha RI(f_1; a, b) + \beta RI(f_2; a, b)$. It then follows immediately that I_R is a linear mapping. ∎

Example 21.2.1 *If f and g are Riemann Integrable on $[a, b]$ prove $2f + 8g$ is Riemann Integrable on $[a, b]$.*

Solution *Let $\epsilon > 0$ be given. Since f is Riemann Integrable on $[a, b]$, there is a number I_f and a partition P_0 so that $|S(f, P, E) - I_f| < \frac{\epsilon}{4}$ when P refines P_0 for any evaluation set E of P.*

Since g is Riemann Integrable on $[a, b]$, there is a number I_g and a partition P_1 so that $|S(g, P, E) - I_f| < \frac{\epsilon}{16}$ when P refines P_1 for any evaluation set E of P.

So if P is a refinement of the common refinement $P_2 = P_0 \vee P_1$, both conditions hold and

$$|S(2f + 8g, P, E) - (2I_f + 8I_g)| \leq 2|S(f, P, E) - I_f| + 8|S(g, P, E) - I_g| < 2(\tfrac{\epsilon}{4}) + 8(\tfrac{\epsilon}{16}) = \epsilon$$

for all choices of evaluation set E in P. This shows the Riemann integral of $2f + 8g$ exists on $[a, b]$.

Homework:

Exercise 21.2.1 *If f is RI on $[1, 3]$ and g is RI on $[1, 3]$ prove $3f - 4g$ is RI on $[1, 3]$.*

Exercise 21.2.2 *If f is RI on $[-2, 5]$ and g is RI on $[-2, 5]$ prove $4f + 7g$ is RI on $[-2, 5]$.*

Exercise 21.2.3 *If f is RI on $[-12, 15]$ and g is RI on $[-12, 15]$ prove $8f - 2g$ is RI on $[-12, 15]$.*

Exercise 21.2.4 *Prove the constant function $f(x) = 7$ is RI for any finite interval $[a, b]$ with value $7(b - a)$.*

Exercise 21.2.5 *Prove the constant function $f(x) = 7x$ is RI for any finite interval $[a, b]$ with value $(7/2)(b^2 - a^2)$.*

21.2.1   Fundamental Integral Estimates

We need to find ways to estimate the value of these integrals.

Theorem 21.2.2 Fundamental Riemann Integral Estimates

Let $f \in RI[a, b]$. Let $m = \inf_{a \leq x \leq b} f(x)$ and let $M = \sup_{a \leq x \leq b} f(x)$. Then

$$m(b - a) \leq RI(f; a, b) \leq M(b - a).$$

Proof 21.2.2
If $\pi \in \Pi[a, b]$, then for all $\sigma \subset \pi$, we see that

$$\sum_{\pi} m\Delta x_j \leq \sum_{\pi} f(s_j)\Delta x_j \leq \sum_{\pi} M\Delta x_j.$$

But $\sum_{\pi} \Delta x_j = b - a$, so

$$m(b - a) \leq \sum_{\pi} f(s_j)\Delta x_j \leq M(b - a),$$

or

$$m(b - a) \leq S(f, \pi, \sigma) \leq M(b - a),$$

for any partition π and any $\sigma \subset \pi$.

Now, let $\epsilon > 0$ be given. Then there exist $\pi_0 \in \Pi[a, b]$ such that for any refinement π, of π_0 and any $\sigma \subset \pi$,

$$RI(f; a, b) - \epsilon < S(f, \pi, \sigma) < RI(f; a, b) + \epsilon.$$

Hence, for any such refinement, π, and any $\sigma \subset \pi$, we have

$$m(b - a) \leq S(f, \pi, \sigma) < RI(f; a, b) + \epsilon$$

and

$$M(b - a) \geq S(f, \pi, \sigma) > RI(f; a, b) - \epsilon.$$

Since $\epsilon > 0$ is arbitrary, it follows that

$$m(b - a) \leq RI(f; a, b) \leq M(b - a).$$

∎

Next we need an order preserving property.

Theorem 21.2.3 The Riemann Integral is Order Preserving

> The Riemann integral is order preserving. That is, if $f, f_1, f_2 \in RI[a, b]$, then
>
> (i)
> $$f \geq 0 \Rightarrow RI(f; a, b) \geq 0;$$
>
> (ii)
> $$f_1 \leq f_2 \Rightarrow RI(f_1; a, b) \leq RI(f_2; a, b).$$

Proof 21.2.3

If $f \geq 0$ on $[a, b]$, then $\inf_x f(x) = m \geq 0$. Hence, by Theorem 21.2.2

$$\int_a^b f(x) dx \geq m(b - a) \geq 0.$$

This proves the first assertion. To prove (ii), let $f = f_2 - f_1$. Then $f \geq 0$, and the second result follows from the first. ∎

Homework

Exercise 21.2.6 *If we assume $f(x) = x$ in integrable on $[1, 3]$, prove $2 \leq \int_1^3 x dx \leq 6$.*

Exercise 21.2.7 *If we assume $f(x) = \sin(x)$ in integrable on $[-\pi, \pi]$, prove $-2\pi \leq \int_{-\pi}^{\pi} \sin(x) dx \leq 2\pi$.*

Exercise 21.2.8 *If we assume $f(x) = 1/(1+x^2)$ in integrable on $[-2, 2]$, prove $(4/5) \leq \int_{-\pi}^{\pi} 1/(1+x^2) dx \leq 4$.*

Chapter 22

Existence of the Riemann Integral and Properties

Although we have a definition for what it means for a bounded function to be Riemann integrable, we still do not actually know that $RI([a,b])$ is nonempty! In this section, we will show how we prove that the set of Riemann integrable functions is quite rich and varied.

22.1 The Darboux Integral

To prove existence results for the Riemann integral, the Riemann sum approach is often awkward. Another way to approach it is to use Darboux sums.

22.1.1 Upper and Lower Darboux Sums

Definition 22.1.1 Darboux Upper and Lower Sums

Let $f \in B[a,b]$. Let $\pi \in \Pi[a,b]$ be given by $\pi = \{x_0 = a, x_1, \ldots, x_p = b\}$. Define

$$m_j = \inf_{x_{j-1} \leq x \leq x_j} f(x) \qquad 1 \leq j \leq p,$$

and

$$M_j = \sup_{x_{j-1} \leq x \leq x_j} f(x) \qquad 1 \leq j \leq p.$$

*We define the **Lower Darboux Sum** by*

$$L(f, \pi) = \sum_\pi m_j \Delta x_j$$

*and the **Upper Darboux Sum** by*

$$U(f, \pi) = \sum_\pi M_j \Delta x_j.$$

Comment 22.1.1 *A few comments*

    1. It is straightforward to see that

$$L(f, \pi) \quad \leq \quad S(f, \pi, \sigma) \leq U(f, \pi)$$

for all $\boldsymbol{\pi} \in \boldsymbol{\Pi}[a, b]$.

2. *We also have*

$$U(f, \boldsymbol{\pi}) - L(f, \boldsymbol{\pi}) = \sum_{\boldsymbol{\pi}}(M_j - m_j)\Delta x_j.$$

We can order the Darboux sums nicely.

Theorem 22.1.1 Lower and Upper Sum Orderings

If $\boldsymbol{\pi} \preceq \boldsymbol{\pi}'$, that is, if $\boldsymbol{\pi}'$ refines $\boldsymbol{\pi}$, then $L(f, \boldsymbol{\pi}) \leq L(f, \boldsymbol{\pi}')$ and $U(f, \boldsymbol{\pi}) \geq U(f, \boldsymbol{\pi}')$.

Proof 22.1.1

The general result is established by induction on the number of points added. It is actually quite an involved induction. Here are some of the details:

Step 1 *We prove the proposition for inserting points $\{z_1, \ldots, z_q\}$ into one subinterval of $\boldsymbol{\pi}$. The argument consists of*

    1. *The* Basis *Step where we prove the proposition for the insertion of a single point into one subinterval.*

    2. *The* Induction *Step where we assume the proposition holds for the insertion of q points into one subinterval and then we show the proposition still holds if an additional point is inserted.*

    3. *With the* Induction *Step verified, the Principle of Mathematical Induction then tells us that the proposition is true for any refinement of $\boldsymbol{\pi}$ which places points into one subinterval of $\boldsymbol{\pi}$.*

Basis:

Proof *Let $\boldsymbol{\pi} \in \boldsymbol{\Pi}[a, b]$ be given by $\{x_0 = a, x_1, \ldots, x_p = b\}$. Suppose we form the refinement, $\boldsymbol{\pi}'$, by adding a single point x' to $\boldsymbol{\pi}$. into the interior of the subinterval $[x_{k_0-1}, x_{k_0}]$. Let*

$$m' = \inf_{[x_{k_0-1}, x']} f(x)$$
$$m'' = \inf_{[x', x_{k_0}]} f(x).$$

Note that $m_{k_0} = \min\{m', m''\}$ and

$$
\begin{aligned}
m_{k_0}\Delta x_{k_0} &= m_{k_0}(x_{k_0} - x_{k_0-1}) \\
&= m_{k_0}(x_{k_0} - x') + m_{k_0}(x' - x_{k_0-1}) \\
&\leq m''(x_{k_0} - x') + m'(x' - x_{k_0-1}) \\
&\leq m''\Delta x'' + m'\Delta x',
\end{aligned}
$$

where $\Delta x'' = x_{k_0} - x'$ and $\Delta x' = x' - x_{k_0-1}$. It follows that

$$L(f, \boldsymbol{\pi}') = \sum_{j \neq k_0} m_j\Delta x_j + m'\Delta x' + m''\Delta x''$$

$$\geq \quad \sum_{j \neq k_0} m_j \Delta x_j + m_{k_0} \Delta x_{k_0}$$

$$\geq \quad L(f, \pi).$$

$\square$

Induction:

Proof *We assume that q points $\{z_1, \ldots, z_q\}$ have been inserted into the subinterval $[x_{k_0-1}, x_{k_0}]$. Let π' denote the resulting refinement of π. We assume that*

$$L(f, \pi) \quad \leq \quad L(f, \pi')$$

Let the additional point added to this subinterval be called x' and call π'' the resulting refinement of π'. We know that π' has broken $[x_{k_0-1}, x_{k_0}]$ into $q+1$ pieces. For convenience of notation, let's label these $q+1$ subintervals as $[y_{j-1}, y_j]$ where y_0 is x_{k_0-1} and y_{q+1} is x_{k_0} and the y_j values in between are the original z_i points for appropriate indices. The new point x' is thus added to one of these $q+1$ pieces; call it $[y_{j_0-1}, y_{j_0}]$ for some index j_0. This interval plays the role of the original subinterval in the proof of the Basis *Step. An argument similar to that in the proof of the* Basis *Step then shows us that*

$$L(f, \pi') \quad \leq \quad L(f, \pi'')$$

Combining with the first inequality from the Induction *hypothesis, we establish the result. Thus, the* Induction *Step is proved.* $\square$

Step 2 *Next, we allow the insertion of a finite number of points into a finite number of subintervals of π. The induction is now on the number of subintervals.*

1. *The* Basis *Step where we prove the proposition for the insertion of points into one subinterval.*

2. *The* Induction *Step where we assume the proposition holds for the insertion of points into q subintervals and then we show the proposition still holds if an additional subinterval has points inserted.*

3. *With the* Induction *Step verified, the Principle of Mathematical Induction then tells us that the proposition is true for any refinement of π which places points into any number of subintervals of π.*

Basis

Proof *Step 1 above gives us the* Basis *Step for this proposition.* $\square$

Induction

Proof *We assume the results holds for p subintervals and show it also holds when one more subinterval is added. Specifically, let π' be the refinement that results from adding points to p subintervals of π. Then the* Induction *hypothesis tells us that*

$$L(f, \pi) \quad \leq \quad L(f, \pi')$$

Let π'' denote the new refinement of π which results from adding more points into one more subinterval of π. Then π'' is also a refinement of π' where all the new points are added to one *subinterval of π'. Thus, Step 1 holds for the pair (π', π''). We see*

$$L(f, \pi') \;\leq\; L(f, \pi'')$$

and the desired result follows immediately. $\square$

A similar argument establishes the result for upper sums.   ■

We can use this refinement result to prove the next very important result which allows us to define lower and upper Darboux integrals.

Theorem 22.1.2 $L(f, \pi_1) \leq U(f, \pi_2)$

> *Let π_1 and π_2 be any two partitions in $\mathbf{\Pi}[a, b]$. Then $L(f, \pi_1) \leq U(f, \pi_2)$.*

Proof 22.1.6
Let $\pi = \pi_1 \vee \pi_2$ be the common refinement of π_1 and π_2. Then, by the previous result, we have

$$L(f, \pi_1) \leq L(f, \pi) \leq U(f, \pi) \leq U(f, \pi_2).$$

■

22.1.2   Upper and Lower Darboux Integrals and Darboux Integrability

Theorem 22.1.2 then allows us to define a new type of integrability for the bounded function f. We begin by looking at the infimum of the upper sums and the supremum of the lower sums for a given bounded function f.

Theorem 22.1.3 The Upper and Lower Darboux Integral are Finite

> *Let $f \in B[a, b]$. Let $\mathscr{L} = \{L(f, \pi) \,|\, \pi \in \mathbf{\Pi}[a, b]\}$ and $\mathscr{U} = \{U(f, \pi) \,|\, \pi \in \mathbf{\Pi}[a, b]\}$. Define $L(f) = \sup \mathscr{L}$, and $U(f) = \inf \mathscr{U}$. Then $L(f)$ and $U(f)$ are both finite. Moreover, $L(f) \leq U(f)$.*

Proof 22.1.7
By Theorem 22.1.2, the set $\mathscr{L}$ is bounded above by any upper sum for f. Hence, it has a finite supremum and so $\sup \mathscr{L}$ is finite. Also, by Theorem 22.1.2, the set $\mathscr{U}$ is bounded below by any lower sum for f. Hence, $\inf \mathscr{U}$ is finite. Finally, since $L(f) \leq U(f, \pi)$ and $U(f) \geq L(f, \pi)$ for all π, by definition of the infimum and supremum of a set of numbers, we must have $L(f) \leq U(f)$.   ■

Definition 22.1.2 Darboux Lower and Upper Integrals

> *Let f be in $B[a, b]$. The **Lower Darboux Integral** of f is defined to be the finite number $L(f) = \sup \mathscr{L}$, and the **Upper Darboux Integral** of f is the finite number $U(f) = \inf \mathscr{U}$.*

We can then define what is meant by a bounded function being *Darboux Integrable* on $[a, b]$.

Definition 22.1.3 Darboux Integrability

> *Let f be in $B[a, b]$. We say f is Darboux Integrable on $[a, b]$ if $L(f) = U(f)$. The common value is then called the Darboux Integral of f on $[a, b]$ and is denoted by the symbol $DI(f; a, b)$.*

Comment 22.1.2 *Not all bounded functions are Darboux Integrable. Consider the function* f : $[0,1] \to \Re$ *defined by*

$$f(t) \;=\; \begin{cases} 1 & t \in [0,1] \text{ and is rational} \\ -1 & t \in [0,1] \text{ and is irrational} \end{cases}$$

You should be able to see that for any partition of $[0,1]$*, the infimum of* f *on any subinterval is always* -1 *as any subinterval contains irrational numbers. Similarly, any subinterval contains rational numbers and so the supremum of* f *on a subinterval is* 1*. Thus* $U(f,\pi) = 1$ *and* $L(f,\pi) = -1$ *for any partition* π *of* $[0,1]$*. It follows that* $L(f) = -1$ *and* $U(f) = 1$*. Thus,* f *is bounded but not Darboux Integrable.*

Homework

Exercise 22.1.1 *Define* f *by*

$$f(t) \;=\; \begin{cases} 2 & t \in [0,1] \text{ and is rational} \\ -3 & t \in [0,1] \text{ and is irrational} \end{cases}$$

Compute the upper and lower Darboux Integrals and determine if f *is Darboux Integrable.*

Exercise 22.1.2 *Define* f *by*

$$f(t) \;=\; \begin{cases} -12 & t \in [0,1] \text{ and is rational} \\ -3 & t \in [0,1] \text{ and is irrational} \end{cases}$$

Compute the upper and lower Darboux Integrals and determine if f *is Darboux Integrable.*

Exercise 22.1.3 *Define* f *by*

$$f(t) \;=\; \begin{cases} -12 & t \in [0,1] \text{ and is rational} \\ 5 & t \in [0,1] \text{ and is irrational} \end{cases}$$

Compute the upper and lower Darboux Integrals and determine if f *is Darboux Integrable.*

22.2 The Darboux and Riemann Integrals are Equivalent

To show these two ways of defining integration, we start by defining a special criterion for integrability.

Definition 22.2.1 Riemann's Criterion for Integrability

Let $f \in B[a,b]$*. We say that* **Riemann's Criterion** *holds for* f *if for every positive* ϵ *there exists a* $\pi_0 \in \Pi[a,b]$ *such that* $U(f,\pi) - L(f,\pi) < \epsilon$ *for any refinement,* π*, of* π_0*. We use the phrase* the Riemann Criterion *also for this criterion.*

Theorem 22.2.1 The Riemann Integral Equivalence Theorem

> *Let $f \in B[a, b]$. Then the following are equivalent:*
>
>   *(i) $f \in RI([a, b])$.*
>
>   *(ii) f satisfies Riemann's Criterion.*
>
>   *(iii) f is Darboux Integrable, i.e, $L(f) = U(f)$, and $RI(f; a, b) = DI(f; a, b)$.*

Proof 22.2.1

$(i) \Rightarrow (ii)$

Proof *Assume $f \in RI([a, b])$ and let $\epsilon > 0$ be given. Let IR be the Riemann integral of f over $[a, b]$. Choose $\pi_0 \in \mathbf{\Pi}[a, b]$ such that $| S(f, \pi, \sigma) - IR | < \epsilon/3$ for any refinement, π, of π_0 and any $\sigma \subset \pi$. Let π be any such refinement, denoted by $\pi = \{x_0 = a, x_1, \ldots, x_p = b\}$, and let m_j, M_j be defined as usual. Using the Infimum and Supremum Tolerance Lemmas, we can conclude that, for each $j = 1, \ldots, p$, there exist $s_j, t_j \in [x_{j-1}, x_j]$ such that*

$$M_j - \frac{\epsilon}{6(b-a)} < f(s_j) \leq M_j$$

$$m_j \leq f(t_j) < m_j + \frac{\epsilon}{6(b-a)}.$$

It follows that

$$f(s_j) - f(t_j) > M_j - \frac{\epsilon}{6(b-a)} - m_j - \frac{\epsilon}{6(b-a)}.$$

Thus, we have

$$M_j - m_j - \frac{\epsilon}{3(b-a)} < f(s_j) - f(t_j).$$

Multiply this inequality by Δx_j to obtain

$$(M_j - m_j)\Delta x_j - \frac{\epsilon}{3(b-a)}\Delta x_j < \big(f(s_j) - f(t_j)\big)\Delta x_j.$$

Now, sum over π to obtain

$$\begin{aligned}
U(f, \pi) - L(f, \pi) &= \sum_\pi (M_j - m_j)\Delta x_j \\
&< \frac{\epsilon}{3(b-a)} \sum_\pi \Delta x_j + \sum_\pi \big(f(s_j) - f(t_j)\big)\Delta x_j.
\end{aligned}$$

This simplifies to

$$\sum_\pi (M_j - m_j)\Delta x_j - \frac{\epsilon}{3} < \sum_\pi \big(f(s_j) - f(t_j)\big)\Delta x_j. \qquad (*)$$

Now, we have

$$\begin{aligned}
\left| \sum_\pi \big(f(s_j) - f(t_j)\big)\Delta x_j \right| &= \left| \sum_\pi f(s_j)\Delta x_j - \sum_\pi f(t_j)\Delta x_j \right| \\
&= \left| \sum_\pi f(s_j)\Delta x_j - IR + IR - \sum_\pi f(t_j)\Delta x_j \right|
\end{aligned}$$

$$\leq \ \mid \sum_{\pi} f(s_j)\Delta x_j - IR \mid + \mid \sum_{\pi} f(t_j)\Delta x_j - IR \mid$$
$$= \ \mid S(f, \pi, \sigma_s) - IR \mid + \mid S(f, \pi, \sigma_t) - IR \mid$$

where $\sigma_s = \{s_1, \ldots, s_p\}$ and $\sigma_t = \{t_1, \ldots, t_p\}$ are evaluation sets of π. Now, by our choice of partition π, we know

$$\mid S(f, \pi, \sigma_s) - IR \mid \ < \ \frac{\epsilon}{3}$$
$$\mid S(f, \pi, \sigma_t) - IR \mid \ < \ \frac{\epsilon}{3}.$$

Thus, we can conclude that

$$\mid \sum_{\pi} (f(s_j) - f(t_j))\Delta x_j \mid < \frac{2\epsilon}{3}.$$

Applying this to the inequality in Equation ∗, we obtain

$$\sum_{\pi} (M_j - m_j)\Delta x_j < \epsilon.$$

Now, π was an arbitrary refinement of π_0, and $\epsilon > 0$ was also arbitrary. So this shows that f satisfies Riemann's Criterion.   □

$(ii) \Rightarrow (iii)$

Proof Assume that f satisfies Riemann's Criterion, and let $\epsilon > 0$ be given. Then there is a partition $\pi_0 \in \Pi[a, b]$ such that $U(f, \pi) - L(f, \pi) < \epsilon$ for any refinement, π, of π_0. Thus, by the definition of the upper and lower Darboux integrals, we have

$$U(f) \leq U(f, \pi) < L(f, \pi) + \epsilon \leq L(f) + \epsilon.$$

Since ϵ is arbitrary, this shows that $U(f) \leq L(f)$. The reverse inequality has already been established. Thus, we see that $U(f) = L(f)$.   □

$(iii) \Rightarrow (i)$

Proof Finally, assume f is Darboux integral which means $L(f) = U(f)$. Let ID denote the value of the Darboux integral. We will show that f is also Riemann integrable and that the value of the integral is ID. Let $\epsilon > 0$ be given. Now, recall that

$$ID \ = \ L(f) \ = \ \sup_{\pi} L(f, \pi)$$
$$= \ U(f) \ = \ \inf_{\pi} U(f, \pi)$$

Hence, by the Supremum Tolerance Lemma, there exists $\pi_1 \in \Pi[a, b]$ such that

$$ID - \epsilon = L(f) - \epsilon < L(f, \pi_1) \leq L(f) = ID$$

and by the Infimum Tolerance Lemma, there exists $\pi_2 \in \Pi[a, b]$ such that

$$ID = U(f) \leq U(f, \pi_2) < U(f) + \epsilon = ID + \epsilon.$$

Let $\pi_0 = \pi_1 \vee \pi_2$ be the common refinement of π_1 and π_2. Now, let π be any refinement of π_0, and let $\sigma \subset \pi$ be any evaluation set. Then we have

$$ID - \epsilon \;\; < \;\; L(f, \pi_1) \leq L(f, \pi_0) \leq L(f, \pi) \leq S(f, \pi, \sigma)$$
$$\leq \;\; U(f, \pi) \leq U(f, \pi_0) \leq U(f, \pi_2) < ID + \epsilon.$$

Thus, it follows that

$$ID - \epsilon \;\; < \;\; S(f, \pi, \sigma) \;\; < \;\; ID + \epsilon.$$

Since the refinement π of π_0 was arbitrary, as were the evaluation set σ and the tolerance ϵ, it follows that for any refinement, π, of π_0 and any $\epsilon > 0$, we have

$$\mid S(f, \pi, \sigma) - ID \mid < \epsilon.$$

This shows that f is Riemann Integrable and the value of the integral is ID. $\qquad \square$

$\blacksquare$

Comment 22.2.1 *By Theorem 22.2.1, we now know the Darboux and Riemann integral are equivalent. Hence, it is no longer necessary to use a different notation for these two approaches to what we call integration. From now on, we will use this notation*

$$RI(f; a, b) \equiv DI(f; a, b) \equiv \int_a^b f(t)\, dt$$

where the (t) in the new integration symbol refers to the name we wish to use for the independent variable and dt is a mnemonic to remind us that the $\parallel \pi \parallel$ is approaching zero as we choose progressively finer partitions of $[a, b]$. This is, of course, not very rigorous notation. A better notation would be

$$RI(f; a, b) \equiv DI(f; a, b) \equiv I(f; a, b)$$

where the symbol I denotes that we are interested in computing the integral of f using the equivalent approach of Riemann or Darboux. Indeed, the notation $I(f; a, b)$ does not require the uncomfortable lack of rigor that the symbol dt implies. However, for historical reasons, the symbol $\int_a^b f(t)\, dt$ will be used.

Also, the use of the $\int f(t)\, dt$ allows us to efficiently apply the integration techniques of substitution and so forth.

Homework

Exercise 22.2.1 *Define f by*

$$f(t) \;\; = \;\; \begin{cases} 2 & t \in [0, 1] \text{ and is rational} \\ -3 & t \in [0, 1] \text{ and is irrational} \end{cases}$$

Prove f does not satisfy the Riemann's Criterion.

Exercise 22.2.2 *Define f by*

$$f(t) \;\; = \;\; \begin{cases} -12 & t \in [0, 1] \text{ and is rational} \\ -3 & t \in [0, 1] \text{ and is irrational} \end{cases}$$

Prove f does not satisfy the Riemann's Criterion.

Exercise 22.2.3 *Define f by*

$$f(t) = \begin{cases} -12 & t \in [0,1] \text{ and is rational} \\ 5 & t \in [0,1] \text{ and is irrational} \end{cases}$$

Prove f does not satisfy the Riemann's Criterion.

22.3   Properties of the Riemann Integral

We can now prove a series of properties of the Riemann integral. Let's start with a lemma about infimum's and supremum's.

Lemma 22.3.1 Fundamental infimum and supremum equalities

If f is a bounded function on the finite interval $[a,b]$, then

1. $\sup_{a \le x \le b} f(x) = -\inf_{a \le x \le b}(-f(x))$ *and* $-\sup_{a \le x \le b}(-f(x)) = \inf_{a \le x \le b}(f(x))$

2.

$$\sup_{x,y \in [a,b]} (f(x) - f(y)) = \sup_{y \in [a,b]} \sup_{x \in [a,b]} (f(x) - f(y))$$

$$= \sup_{x \in [a,b]} \sup_{y \in [a,b]} (f(x) - f(y)) = M - m$$

   where $M = \sup_{a \le x \le b} f(x)$ and $m = \inf_{a \le x \le b} f(x)$.

3. $\sup_{x,y \in [a,b]} | f(x) - f(y) | = M - m$

Proof 22.3.1

First let $Q = \sup_{a \le x \le b}(-f)$ and $q = \inf_{a \le x \le b}(-f)$.
(1):
Let $(f(x_n))$ be a sequence which converges to M. Then since $-f(x_n) \ge q$ for all n, letting $n \to \infty$, we find $-M \ge q$.

Now let $(-f(z_n))$ be a sequence which converges to q. Then, we have $f(z_n) \le M$ for all n and letting $n \to \infty$, we have $-q \le M$ or $q \ge -M$.

Combining, we see $-q = M$ which is the first part of the statement; i.e. $\sup_{a \le x \le b} f(x) = -\inf_{a \le x \le b}(-f(x))$. Now just replace all the f's by $-f$'s in this to get $\sup_{a \le x \le b}(-f(x)) = -\inf_{a \le x \le b}(--f(x))$ or $-\sup_{a \le x \le b}(-f(x)) = \inf_{a \le x \le b}(f(x))$ which is the other identity.

(2):
We know

$$f(x) - f(y) \le \left(\sup_{a \le x \le b} f(x) \right) - f(y)$$

But $f(y) \geq \inf_{y \in [a,b]} f(y) = m$, so

$$f(x) - f(y) \quad \leq \quad M - f(y) \leq M - m$$

Thus,

$$\sup_{x,y \in [a,b]} (f(x) - f(y)) \quad \leq M - m$$

So one side of the inequality is clear. Now let $f(x_n)$ be a sequence converging to M and $f(y_n)$ be a sequence converging to m. Then, we have

$$f(x_n) - f(y_n) \quad \leq \quad \sup_{x,y \in [a,b]} (f(x) - f(y))$$

Letting $n \to \infty$, we see

$$M - m \quad \leq \quad \sup_{x,y \in [a,b]} (f(x) - f(y))$$

This is the other side of the inequality. We have thus shown that the equality is valid.
(3):
Note

$$| f(x) - f(y) | \quad = \quad \begin{cases} f(x) - f(y), & f(x) \geq f(y) \\ f(y) - f(x), & f(x) < f(y) \end{cases}$$

In either case, we have $| f(x) - f(y) | \leq M - m$ for all x, y using Part (2) implying that $\sup_{x,y} | f(x) - f(y) | \leq M - m$.

To see the reverse inequality holds, we first note that if $M = m$, we see the reverse inequality holds trivially as $\sup_{x,y} | f(x) - f(y) | \geq 0 = M - m$. Hence, we may assume without loss of generality that the gap $M - m$ is positive. Then, using the STL and ITL, given $0 < 1/j < 1/2(M - m)$, there exist, $s_j, t_j \in [a,b]$ such that $M - 1/(2j) < f(s_j)$ and $m + 1/(2j) > f(t_j)$, so that $f(s_j) - f(t_j) > M - m - 1/j$. By our choice of j, these terms are positive and so we also have $| f(s_j) - f(t_j) | > M - m - 1/j$.

It follows that

$$\sup_{x,y \in [x_{j-1}, x_j]} | f(x) - f(y) | \quad \geq \quad | f(s_j) - f(t_j) | > M - m - 1/j \,|.$$

Since we can make $1/j$ arbitrarily small, this implies that

$$\sup_{x,y \in [x_{j-1}, x_j]} | f(x) - f(y) | \quad \geq \quad M - m.$$

This establishes the reverse inequality and proves the claim.   ∎

We can use this fundamental set of inequalities to prove the following long list of useful results.

Theorem 22.3.2 Properties of the Riemann Integral

Let $f, g \in RI([a, b])$. Then

  (i) $|f| \in RI([a, b])$;

  (ii)
$$\left| \int_a^b f(x) dx \right| \leq \int_a^b |f| \, dx;$$

(iii) $f^+ = \max\{f, 0\} \in RI([a, b])$;

(iv) $f^- = \max\{-f, 0\} \in RI([a, b])$;

  (v)
$$\int_a^b f(x) dx = \int_a^b f^+(x) dx - \int_a^b f^-(x) dx$$
$$\int_a^b |f(x)| \, dx = \int_a^b f^+(x) dx + \int_a^b f^-(x) dx;$$

(vi) $f^2 \in RI([a, b])$;

(vii) $fg \in RI([a, b])$;

(viii) *If there exists m, M such that $0 < m \leq |f| \leq M$, then $1/f \in RI([a, b])$.*

Proof 22.3.2

(i):

Proof *Note, given a partition $\pi = \{x_0 = a, x_1, \ldots, x_p = b\}$, for each $j = 1, \ldots, p$, from Lemma 22.3.1, we know*

$$\sup_{x, y \in [x_{j-1}, x_j]} (f(x) - f(y)) = M_j - m_j$$

Now, let m_j' and M_j' be defined by

$$m_j' = \inf_{[x_{j-1}, x_j]} |f(x)|, \quad M_j' = \sup_{[x_{j-1}, x_j]} |f(x)|.$$

Then, applying Lemma 22.3.1 to $|f|$, we have

$$M_j' - m_j' = \sup_{x, y \in [x_{j-1}, x_j]} |f(x)| - |f(y)|.$$

For each $j = 1, \ldots, p$, we have

$$M_j - m_j = \sup_{x, y \in [x_{j-1}, x_j]} |f(x) - f(y)|.$$

So, since $|f(x)| - |f(y)| \leq |f(x) - f(y)|$ for all x, y, it follows that $M_j' - m_j' \leq M_j - m_j$, implying that $\sum_\pi (M_j' - m_j') \Delta x_j \leq \sum_\pi (M_j - m_j) \Delta x_j$. This means $U(|f|, \pi) - L(|f|, \pi) \leq U(f, \pi) - L(f, \pi)$ for the chosen π. Since f is integrable by hypothesis, by Theorem 22.2.1, we know Riemann's Criterion must also hold for f. Thus, given $\epsilon > 0$, there is a partition π_0 so that

$U(f, \boldsymbol{\pi}) - L(f, \boldsymbol{\pi}) < \epsilon$ *for any refinement $\boldsymbol{\pi}$ of $\boldsymbol{\pi_0}$. Therefore $|f|$ also satisfies Riemann's Criterion and so $|f|$ is Riemann integrable.* □

The other results now follow easily.
(ii):

Proof *We have $f \leq |f|$ and $f \geq -|f|$, so that*

$$\int_a^b f(x)dx \;\; \leq \;\; \int_a^b |f(x)|\, dx$$

$$\int_a^b f(x)dx \;\; \geq \;\; -\int_a^b |f(x)|\, dx,$$

from which it follows that

$$-\int_a^b |f(x)|\, dx \leq \int_a^b f(x)dx \leq \int_a^b |f(x)|\, dx$$

and so

$$\left| \int_a^b f \right| \leq \int_a^b |f|,$$

□

(iii) and (iv):

Proof *This follows from the facts that $f^+ = \frac{1}{2}(|f| + f)$ and $f^- = \frac{1}{2}(|f| - f)$ and the Riemann integral is a linear mapping.* □

(v):

Proof *This follows from the facts that $f = f^+ - f^-$ and $|f| = f^+ + f^-$ and the linearity of the integral.* □

(vi):

Proof *Note since f is bounded, there exists $K > 0$ such that $|f(x)| \leq K$ for all $x \in [a, b]$. Then, applying Lemma 22.3.1 to f^2, we have*

$$\sup_{x,y \in [x_{j-1}, x_j]} (f^2(x) - f^2(y)) \;\; = \;\; M_j(f^2) - m_j(f^2)$$

where $[x_{j-1}, x_j]$ is a subinterval of a given partition $\boldsymbol{\pi}$ and $M_j(f^2) = \sup_{x \in [x_{j-1}, x_j]} f^2(x)$ and $m_j(f^2) = \inf_{x \in [x_{j-1}, x_j]} f^2(x)$. Thus, for this partition, we have

$$U(f^2, \boldsymbol{\pi}) - L(f^2, \boldsymbol{\pi}) \;\; = \;\; \sum_{\boldsymbol{\pi}} (M_j(f^2) - m_j(f^2))\, \Delta x_j$$

But we also know

$$\sup_{x,y \in [x_{j-1}, x_j]} (f^2(x) - f^2(y)) \;\; = \;\; \sup_{x,y \in [x_{j-1}, x_j]} (f(x) + f(y))(f(x) - f(y))$$

$$\leq \;\; 2K \sup_{x,y \in [x_{j-1}, x_j]} (f(x) - f(y)) = 2K\,(M_j - m_j).$$

Thus,

$$U(f^2, \pi) - L(f^2, \pi) = \sum_{\pi} (M_j(f^2) - m_j(f^2)) \, \Delta x_j$$

$$\leq 2K \sum_{\pi} (M_j - m_j) \, \Delta x_j = 2K \, (U(f, \pi) - L(f\pi)).$$

Now since f is Riemann Integrable, it satisfies Riemann's Criterion and so given $\epsilon > 0$, there is a partition π_0 so that $U(f, \pi) - L(f\pi) < \epsilon/(2K)$ for any refinement π of π_0. Thus, f^2 satisfies Riemann's Criterion too and so it is integrable. □

(vii):

Proof *To prove that fg is integrable when f and g are integrable, simply note that*

$$fg = (1/2) \left((f + g)^2 - f^2 - g^2 \right).$$

Property (vi) and the linearity of the integral then imply fg is integrable. □

(viii):

Proof *Suppose $f \in RI([a, b])$ and there exist $M, m > 0$ such that $m \leq | f(x) | \leq M$ for all $x \in [a, b]$. Note that*

$$\frac{1}{f(x)} - \frac{1}{f(y)} = \frac{f(y) - f(x)}{f(x)f(y)}.$$

Let $\pi = \{x_0 = a, x_1, \ldots, x_p = b\}$ be a partition of $[a, b]$, and define

$$M_j' = \sup_{[x_{j-1}, x_j]} \frac{1}{f(x)}$$

$$m_j' = \inf_{[x_{j-1}, x_j]} \frac{1}{f(x)}.$$

Then we have

$$M_j' - m_j' = \sup_{x,y \in [x_{j-1}, x_j]} \frac{f(y) - f(x)}{f(x)f(y)}$$

$$\leq \sup_{x,y \in [x_{j-1}, x_j]} \frac{| f(y) - f(x) |}{| f(x) || f(y) |}$$

$$\leq \frac{1}{m^2} \sup_{x,y \in [x_{j-1}, x_j]} | f(y) - f(x) |$$

$$\leq \frac{M_j - m_j}{m^2}.$$

Since $f \in RI([a, b])$, given $\epsilon > 0$ there is a partition π_0 such that $U(f, \pi) - L(f, \pi) < m^2\epsilon$ for any refinement, pi, of π_0. Hence, the previous inequality implies that, for any such refinement, we have

$$U\left(\frac{1}{f}, \pi\right) - L\left(\frac{1}{f}, \pi\right) = \sum_{\pi} (M_j' - m_j')\Delta x_j$$

$$\leq \frac{1}{m^2} \sum_{\pi} (M_j - m_j)\Delta x_j$$

$$\leq \quad \frac{1}{m^2}\Big(U(f,\boldsymbol{\pi}) - L(f,\boldsymbol{\pi})\Big)$$

$$< \quad \frac{m^2\epsilon}{m^2} \; = \; \epsilon.$$

Thus $1/f$ satisfies Riemann's Criterion and hence it is integrable.   □

■

Homework

Exercise 22.3.1 *If f is integrable on $[a, b]$, prove any polynomial in f is also integrable using induction.*

Exercise 22.3.2 *If $1/f$ is integrable on $[a, b]$, prove any polynomial in $1/f$ is also integrable using induction.*

Exercise 22.3.3 *If f is integrable on $[a, b]$ and $1/g$ is integrable on $[a, b]$, prove any polynomial in f/g is also integrable using induction.*

22.4   What Functions are Riemann Integrable?

Now we need to show that the set $RI([a, b])$ is nonempty. We begin by showing that all continuous functions on $[a, b]$ will be Riemann Integrable.

Theorem 22.4.1 Continuous Implies Riemann Integrable

If $f \in C([a, b])$, then $f \in RI([a, b])$.

Proof 22.4.1
Since f is continuous on a compact set, it is uniformly continuous. Hence, given $\epsilon > 0$, there is a $\delta > 0$ such that $x, y \in [a, b]$, $\mid x - y \mid < \delta \Rightarrow \mid f(x) - f(y) \mid < \epsilon/(b-a)$. Let $\boldsymbol{\pi}_0$ be a partition such that $\parallel \boldsymbol{\pi}_0 \parallel < \delta$, and let $\boldsymbol{\pi} = \{x_0 = a, x_1, \ldots, x_p = b\}$ be any refinement of $\boldsymbol{\pi}_0$. Then $\boldsymbol{\pi}$ also satisfies $\parallel \boldsymbol{\pi} \parallel < \delta$. Since f is continuous on each subinterval $[x_{j-1}, x_j]$, f attains its supremum, M_j, and infimum, m_j, at points s_j and t_j, respectively. That is, $f(s_j) = M_j$ and $f(t_j) = m_j$ for each $j = 1, \ldots, p$. Thus, the uniform continuity of f on each subinterval implies that, for each j,

$$M_j - m_j = \mid f(s_j) - f(t_j) \mid < \frac{\epsilon}{b-a}.$$

Thus, we have

$$U(f, \boldsymbol{\pi}) - L(f, \boldsymbol{\pi}) = \sum_{\boldsymbol{\pi}}(M_j - m_j)\Delta x_j < \frac{\epsilon}{b-a}\sum_{\boldsymbol{\pi}}\Delta x_j = \epsilon.$$

Since $\boldsymbol{\pi}$ was an arbitrary refinement of $\boldsymbol{\pi}_0$, it follows that f satisfies Riemann's Criterion. Hence, $f \in RI([a, b])$.   ■

We can now actually integrate some simple functions.

Theorem 22.4.2 Constant Functions are Riemann Integrable

If $f : [a, b] \to \Re$ is a constant function, i.e. $f(t) = c$ for all t in $[a, b]$, then f is Riemann Integrable on $[a, b]$ and $\int_a^b f(t)dt = c(b-a)$.

Proof 22.4.2
Since f is a constant, for any partition π of $[a, b]$, all the individual m_j's and M_j's associated with π take on the value c. Hence, $U(f, \pi) - L(f, \pi) = 0$ always. It follows that f satisfies Riemann's Criterion and hence is Riemann Integrable. Finally, since f is integrable, by Theorem 21.2.2, we have

$$c(b - a) \le RI(f; a, b) \le c(b - a).$$

Thus, $\int_a^b f(t)dt = c(b - a)$. ∎

Another class of functions that are integrable is monotone functions.

Theorem 22.4.3 Monotone Implies Riemann Integrable

> *If f is monotone on $[a, b]$, then $f \in RI([a, b])$.*

Proof 22.4.3
As usual, for concreteness, we assume that f is monotone increasing. We also assume $f(b) > f(a)$, for if not, then f is constant and must be integrable by Theorem 22.4.2. Let $\epsilon > 0$ be given, and let π_0 be a partition of $[a, b]$ such that $\| \pi_0 \| < \epsilon/(f(b) - f(a))$. Let $\pi = \{x_0 = a, x_1, \ldots, x_p = b\}$ be any refinement of π_0. Then π also satisfies $\| \pi \| < \epsilon/(f(b) - f(a))$. Thus, for each $j = 1, \ldots, p$, we have

$$\Delta x_j < \frac{\epsilon}{f(b) - f(a)}.$$

Since f is increasing, we also know that $M_j = f(x_j)$ and $m_j = f(x_{j-1})$ for each j. Hence,

$$
\begin{aligned}
U(f, \pi) - L(f, \pi) &= \sum_\pi (M_j - m_j)\Delta x_j \\
&= \sum_\pi [f(x_j) - f(x_{j-1})]\Delta x_j \\
&< \frac{\epsilon}{f(b) - f(a)} \sum_\pi [f(x_j) - f(x_{j-1})].
\end{aligned}
$$

But this last sum is telescoping and sums to $f(b) - f(a)$. So, we have

$$U(f, \pi) - L(f, \pi) < \frac{\epsilon}{f(b) - f(a)}(f(b) - f(a)) = \epsilon.$$

Thus, f satisfies Riemann's criterion. ∎

Let's look at an interesting example.

Example 22.4.1 *Let f_n, for $n \ge 2$ be defined by*

$$
f_n(x) = \begin{cases} 1, & x = 1 \\ 1/(k+1), & 1/(k+1) \le x < 1/k, \ 1 \le k < n \\ 0, & 0 \le x < 1/n \end{cases}
$$

We know f_n is $RI([0, 1])$ because it is monotonic although we do not know what the value of the integral is.

Define f by $f(x) = \lim_{n \to \infty} f_n(x)$. Then given x, we can find an integer N so that $1/(N+1) \le x < 1/N$ telling us $f(x) = 1/(N+1)$. Moreover $f(x) = f_{N+1}(x)$. So if $x < y$, y is either in the interval $[1/(N+1), 1/N)$ or $y \in [1/N, 1]$ implying $f(x) \le f(y)$. Hence f is monotonic. At each

$1/N$, the right and left hand limits do not match and so f is not continuous at a countable number of points yet it is still Riemann Integrable because it is monotonic.

22.4.1   Homework

Exercise 22.4.1 *If you didn't know $f(x) = x$ was continuous, why would you know f is $RI([a, b])$ for any $[a, b]$?*

Exercise 22.4.2 *Use induction to prove $f(x) = x^n$ is $RI([a, b])$ for any $[a, b]$ without assuming continuity.*

Exercise 22.4.3 *Use induction to prove $f(x) = 1/x^n$ is $RI([a, b])$ on any $[a, b]$ that does not contain 0 without assuming continuity.*

Exercise 22.4.4 *For $f(x) = \sin(2x)$ on $[-2\pi, 2\pi]$, draw f^+ and f^-.*

Exercise 22.4.5 *Prove f is $RI([0, 1])$ where f is defined by*

$$f(x) \quad = \quad \begin{cases} x \, \sin(1/x), & x \in (0, 1] \\ 0, & x = 0 \end{cases}$$

Exercise 22.4.6 *Consider*

$$f_n(x) \quad = \quad \begin{cases} 1, & x = 1 \\ 1/(k+1), & 1/(k+1) \le x < 1/k, \ 1 \le k < n \\ 0, & 0 \le x < 1/n \end{cases}$$

- *Graph f_5 and f_8 and determine the cluster points $S(p)$.*

- *Calculate the upper and lower Darboux integrals of f_n and determine if f_n is integrable and if so what is its value.*

22.5   Further Properties of the Riemann Integral

We first want to establish the familiar summation property of the Riemann integral over an interval $[a, b] = [a, c] \cup [c, b]$. Most of the technical work for this result is done in the following Lemma.

Lemma 22.5.1 The Upper and Lower Darboux Integral is Additive on Intervals

> *Let $f \in B[a, b]$ and let $c \in (a, b)$. Let*
>
> $$\underline{\int_a^b} f(x) \, dx = L(f) \text{ and } \overline{\int_a^b} f(x) \, dx = U(f)$$
>
> *denote the lower and upper Darboux integrals of f on $[a, b]$, respectively. Then we have*
>
> $$\overline{\int_a^b} f(x)dx = \overline{\int_a^c} f(x)dx + \overline{\int_c^b} f(x)dx$$
>
> $$\underline{\int_a^b} f(x)dx = \underline{\int_a^c} f(x)dx + \underline{\int_c^b} f(x)dx.$$

Proof 22.5.1

We prove the result for the upper integrals as the lower integral case is similar. Let $\pi \in \Pi[a,b]$ be given by $\pi = \{x_0 = a, x_1, \ldots, x_p = b\}$. We first assume that c is a partition point of π. Thus, there is some index $1 \leq k_0 \leq p - 1$ such that $x_{k_0} = c$. For any interval $[\alpha, \beta]$, let $U_\alpha^\beta(f, \pi)$ denote the upper sum of f for the partition π over $[\alpha, \beta]$. Now, we can rewrite π as $\pi = \{x_0, x_1, \ldots, x_{k_0}\} \cup \{x_{k_0}, x_{k_0+1}, \ldots, x_p\}$. Let $\pi_1 = \{x_0, \ldots, x_{k_0}\}$ and $\pi_2 = \{x_{k_0}, \ldots, x_p\}$. Then $\pi_1 \in \Pi[a,c]$, $\pi_2 \in \Pi[c,b]$, and

$$
\begin{aligned}
U_a^b(f, \pi) &= U_a^c(f, \pi_1) + U_c^b(f, \pi_2) \\
&\geq \overline{\int_a^c} f(x)dx + \overline{\int_c^b} f(x)dx,
\end{aligned}
$$

by the definition of the upper sum. Now, if c is not in π, then we can refine π by adding c, obtaining the partition $\pi' = \{x_0, x_1, \ldots, x_{k_0}, c, x_{k_0+1}, \ldots, x_p\}$. Splitting up π' at c as we did before into π_1 and π_2, we see that $\pi' = \pi_1 \vee \pi_2$ where $\pi_1 = \{x_0, \ldots, x_{k_0}, c\}$ and $\pi_2 = \{c, x_{k_0+1}, \ldots, x_p\}$. Thus, by our properties of upper sums, we see that

$$
U_a^b(f, \pi) \geq U_a^b(f, \pi') = U_a^c(f, \pi_1) + U_c^b(f, \pi_2) \geq \overline{\int_a^c} f(x)dx + \overline{\int_c^b} f(x)dx.
$$

Combining both cases, we can conclude that for any partition $\pi \in \Pi[a,b]$, we have

$$
U_a^b(f, \pi) \geq \overline{\int_a^c} f(x)dx + \overline{\int_c^b} f(x)dx,
$$

which implies that

$$
\overline{\int_a^b} f(x)dx \geq \overline{\int_a^c} f(x)dx + \overline{\int_c^b} f(x)dx.
$$

Now we want to show the reverse inequality. Let $\epsilon > 0$ be given. By the definition of the upper integral, there exists $\pi_1 \in \Pi[a,c]$ and $\pi_2 \in \Pi[c,b]$ such that

$$
\begin{aligned}
U_a^c(f, \pi_1) &< \overline{\int_a^c} f(x)dx + \frac{\epsilon}{2} \\
U_c^b(f, \pi_2) &< \overline{\int_c^b} f(x)dx + \frac{\epsilon}{2}.
\end{aligned}
$$

Let $\pi = \pi_1 \cup \pi_2 \in \Pi[a,b]$. It follows that

$$
U_a^b(f, \pi) = U_a^c(f, \pi_1) + U_c^b(f, \pi_2) < \overline{\int_a^c} f(x)dx + \overline{\int_c^b} f(x)dx + \epsilon.
$$

But, by definition, we have

$$
\overline{\int_a^b} f(x)dx \leq U_a^b(f, \pi)
$$

for all π. Hence, we see that

$$
\overline{\int_a^b} f(x)dx < \overline{\int_a^c} f(x)dx + \overline{\int_c^b} f(x)dx + \epsilon.
$$

Since ϵ was arbitrary, this proves the reverse inequality we wanted. We can conclude, then, that

$$\overline{\int_a^b} f(x)dx = \overline{\int_a^c} f(x)dx + \overline{\int_c^b} f(x)dx.$$

∎

This immediately leads to the same result for Riemann integrals.

Theorem 22.5.2 The Riemann Integral Exists on Subintervals

If $f \in RI([a,b])$ and $c \in (a,b)$, then $f \in RI([a,c])$ and $f \in RI([c,b])$.

Proof 22.5.2

Let $\epsilon > 0$ be given. Then there is a partition $\pi_0 \in \Pi[a,b]$ such that $U_a^b(f,\pi) - L_a^b(f,\pi) < \epsilon$ for any refinement, π, of π_0. Let π_0 be given by $\pi_0 = \{x_0 = a, x_1, \ldots, x_p = b\}$. Define $\pi_0' = \pi_0 \cup \{c\}$, so there is some index k_0 such that $x_{k_0} \le c \le x_{k_0+1}$. Let $\pi_1 = \{x_0, \ldots, x_{k_0}, c\}$ and $\pi_2 = \{c, x_{k_0+1}, \ldots, x_p\}$. Then $\pi_1 \in \Pi[a,c]$ and $\pi_2 \in \Pi[c,b]$. Let π_1' be a refinement of π_1. Then $\pi_1' \cup \pi_2$ is a refinement of π_0, and it follows that

$$
\begin{aligned}
U_a^c(f,\pi_1') - L_a^c(f,\pi_1') &= \sum_{\pi_1'} (M_j - m_j)\Delta x_j \\
&\le \sum_{\pi_1' \cup \pi_2} (M_j - m_j)\Delta x_j \\
&\le U_a^b(f,\pi_1' \cup \pi_2) - L_a^b(f,\pi_1' \cup \pi_2).
\end{aligned}
$$

But, since $\pi_1' \cup \pi_2$ refines π_0, we have

$$U_a^b(f,\pi_1' \cup \pi_2) - L_a^b(f,\pi_1' \cup \pi_2) < \epsilon,$$

implying that

$$U_a^c(f,\pi_1') - L_a^c(f,\pi_1') < \epsilon$$

for all refinements, π_1', of π_1. Thus, f satisfies Riemann's Criterion on $[a,c]$, and $f \in RI([a,c])$. The proof on $[c,b]$ is done in exactly the same way. ∎

Now that we know if f is integrable on $[a,b]$ it is also integrable on $[a,c]$ and $[c,b]$, we are ready to prove the final result.

Theorem 22.5.3 The Riemann Integral is Additive on Subintervals

If $f \in RI([a,b])$ and $c \in (a,b)$, then

$$\int_a^b f(x)dx = \int_a^c f(x)dx + \int_c^b f(x)dx.$$

Proof 22.5.3

Since $f \in RI([a,b])$, we know that

$$\overline{\int_a^b} f(x)dx = \underline{\int_a^b} f(x)dx.$$

Further, we also know that $f \in RI([a,c])$ and $f \in RI([c,b])$ for any $c \in (a,b)$. Thus,

$$\overline{\int_a^c} f(x)dx = \underline{\int_a^c} f(x)dx$$

$$\overline{\int_c^b} f(x)dx = \underline{\int_c^b} f(x)dx.$$

So, applying Lemma 22.5.1, we conclude that, for any $c \in (a,b)$,

$$\int_a^b f(x)dx = \overline{\int_a^b} f(x)\,dx = \overline{\int_a^c} f(x)\,dx + \overline{\int_c^b} f(x)\,dx = \int_a^c f(x)\,dx + \int_c^b f(x)\,dx.$$

∎

Homework

Exercise 22.5.1 *We have shown in a previous exercise that*

$$f_n(x) = \begin{cases} 1, & x = 1 \\ 1/(k+1), & 1/(k+1) \le x < 1/k, \ 1 \le k < n \\ 0, & 0 \le x < 1/n \end{cases}$$

is integrable. We thus know f_n is integrable on subintervals between jumps of f_n. Calculate the Darboux integrals of f_5 on each appropriate subinterval which gives us another way to compute the integral of f_5.

Exercise 22.5.2 *We know*

$$f_n(x) = \begin{cases} 1, & x = 1 \\ 1/(k+1), & 1/(k+1) \le x < 1/k, \ 1 \le k < n \\ 0, & 0 \le x < 1/n \end{cases}$$

is integrable. We thus know f_n is integrable on subintervals between jumps of f_n. Calculate the Darboux integrals of f_{10} on each appropriate subinterval which gives us another way to compute the integral of f_{10}.

Exercise 22.5.3 *We know*

$$f_n(x) = \begin{cases} 1, & x = 1 \\ 1/(k+1), & 1/(k+1) \le x < 1/k, \ 1 \le k < n \\ 0, & 0 \le x < 1/n \end{cases}$$

is integrable. We thus know f_n is integrable on subintervals between jumps of f_n. Calculate the Darboux integrals of f_n in general on each appropriate subinterval which gives us another way to compute the integral of f_n. Your answer will now be a sum. What happens as $n \to \infty$? We know from a previous problem that as $n \to \infty$, the resulting function f we obtain is monotone ans so integrable. This gives us an example of whether $\int_0^1 f = \lim_{n\to\infty} \int_0^1 f_n$. We eventually will find conditions that make this true.

Chapter 23

The Fundamental Theorem of Calculus (FTOC)

The next result is the well-known **Fundamental of Theorem of Calculus**.

23.1   The Proof of the FTOC

Theorem 23.1.1 The Fundamental Theorem of Calculus

Let $f \in RI([a,b])$. Define $F : [a,b] \to \Re$ by

$$F(x) = \int_a^x f(t)dt.$$

Then

(i) $F \in C([a,b])$;

(ii) if f is continuous at $c \in [a,b]$, then F is differentiable at c and $F'(c) = f(c)$.

Proof 23.1.1
First, note that $f \in RI([a,b]) \Rightarrow f \in RI([a,x])$ for all $x \in [a,b]$, by our previous results. Hence, F is well-defined. We will prove the results in order.

(i) $F \in C([a,b])$: *Now, let $x, y \in [a,b]$ be such that $x < y$. Then*

$$\inf_{[x,y]} f(t)\,(y-x) \le \int_x^y f(t)dt \le \sup_{[x,y]} f(t)\,(y-x),$$

which implies that

$$\mid F(y) - F(x) \mid = \left| \int_x^y f(t)dt \right| \le \parallel f \parallel_\infty (y-x).$$

A similar argument shows that if $y, x \in [a,b]$ satisfy $y < x$, then

$$\mid F(y) - F(x) \mid = \left| \int_x^y f(t)dt \right| \le \parallel f \parallel_\infty (x-y).$$

Let $\epsilon > 0$ be given. Then if

$$| \ x - y \ |< \frac{\epsilon}{|| \ f \ ||_\infty +1},$$

we have

$$| \ F(y) - F(x) \ |\leq || \ f \ ||_\infty | \ y - x \ |< \frac{|| \ f \ ||_\infty}{|| \ f \ ||_\infty +1} \epsilon < \epsilon.$$

Thus, F is continuous at x and, consequently, on $[a, b]$.

(ii) (if) f is continuous at $c \in [a, b]$, then F is differentiable at c and $F'(c) = f(c)$: *Assume f is continuous at $c \in [a, b]$, and let $\epsilon > 0$ be given. Then there exists $\delta > 0$ such that $x \in (c - \delta, c + \delta) \cap [a, b]$ implies $| \ f(x) - f(c) \ |< \epsilon/2$. Pick $h \in \Re$ such that $0 <| \ h \ |< \delta$ and $c + h \in [a, b]$. Let's assume, for concreteness, that $h > 0$. Define*

$$m = \inf_{[c,c+h]} f(t) \qquad and \qquad M = \sup_{[c,c+h]} f(t).$$

If $c < x < c + h$, then we have $x \in (c - \delta, c + \delta) \cap [a, b]$ and $-\epsilon/2 < f(x) - f(c) < \epsilon/2$. That is,

$$f(c) - \frac{\epsilon}{2} < f(x) < f(c) + \frac{\epsilon}{2} \qquad \forall x \in [c, c + h].$$

Hence, $m \geq f(c) - \epsilon/2$ and $M \leq f(c) + \epsilon/2$. Now, we also know that

$$mh \leq \int_c^{c+h} f(t) dt \leq Mh.$$

Thus, we have

$$\frac{F(c + h) - F(c)}{h} = \frac{\int_a^{c+h} f(t) dt - \int_a^c f(t) dt}{h} = \frac{\int_c^{c+h} f(t) dt}{h}.$$

Combining inequalities, we find

$$f(c) - \frac{\epsilon}{2} \leq m \leq \frac{F(c + h) - F(c)}{h} \leq M \leq f(c) + \frac{\epsilon}{2}$$

yielding

$$\Rightarrow \left| \frac{F(c + h) - F(c)}{h} - f(c) \right| \leq \frac{\epsilon}{2} < \epsilon$$

if $x \in [c, c + h]$. Thus, since ϵ was arbitrary, this shows $F'^+(c) = f(c)$. The case where $h < 0$ is handled in exactly the same way which tells us $F'^-(c) = f(c)$. Combining, we have $F'(c) = f(c)$. Note that if $c = a$ or $c = b$, we need only consider the definition of the derivative from one side. ∎

Comment 23.1.1 *We call $F(x)$ the indefinite integral of f. F is always better behaved than f, since integration is a smoothing operation. We can see that f need not be continuous, but, as long as it is integrable, F is always continuous.*

Again, let's assume f is RI on $[a, b]$. Then if $G(x) = \int_x^b f(t) \ dt$, at each point x where f is continuous, F is differentiable. Since

$$F(x) + G(x) = \int_a^x f(t) \ dt + \int_x^b f(t) \ dt = \int_a^b f(t) \ dt.$$

we have G is also differentiable at x. It follows that $G'(x) = -F'(x) = -f(x)$.

Homework

Let's assume we know how to integrate polynomials even though we have not proven that yet.

Exercise 23.1.1 Let $f(x) = 2x^2 + 3x + 4$ on $[-1, 5]$. Compute $F(x) = \int_{-1}^{x} f(x)dx$ and verify $F' = f$ on $[-1, 5]$ and F is continuous.

Exercise 23.1.2 Let $f(x) = -2x^5 + 3x^3 + 14$ on $[1, 3]$. Compute $F(x) = \int_{1}^{x} f(x)dx$ and verify $F' = f$ on $[1, 3]$ and F is continuous.

Exercise 23.1.3 Let $f(x) = -2x^{10} + 33x^7 + 14x^2 - 10$ on $[0, 2]$. Compute $F(x) = \int_{0}^{x} f(x)dx$ and verify $F' = f$ on $[0, 2]$ and F is continuous.

Let's state this as a variant of the Fundamental Theorem of Calculus, the *Reversed Fundamental Theorem of Calculus* so to speak.

Theorem 23.1.2 The Fundamental Theorem of Calculus Reversed

Let f be Riemann Integrable on $[a, b]$. Then the function F defined on $[a, b]$ by $F(x) = \int_{x}^{b} f(t) \, dt$ satisfies

1. F is continuous on all of $[a, b]$

2. F is differentiable at each point x in $[a, b]$ where f is continuous and $F'(x) = -f(x)$.

Homework

Let's assume we know how to integrate polynomials even though we have not proven that yet.

Exercise 23.1.4 Let $f(x) = 2x^2 + 3x + 4$ on $[-1, 5]$. Compute $F(x) = \int_{x}^{5} f(x)dx$ and verify $F' = -f$ on $[-1, 5]$ and F is continuous.

Exercise 23.1.5 Let $f(x) = -2x^5 + 3x^3 + 14$ on $[1, 3]$. Compute $F(x) = \int_{x}^{3} f(x)dx$ and verify $F' = -f$ on $[1, 3]$ and F is continuous.

Exercise 23.1.6 Let $f(x) = -2x^{10} + 33x^7 + 14x^2 - 10$ on $[0, 2]$. Compute $F(x) = \int_{x}^{2} f(x)dx$ and verify $F' = -f$ on $[0, 2]$ and F is continuous.

23.2 The Cauchy FTOC

The next result is the standard means for calculating definite integrals in basic calculus. We start with a definition.

Definition 23.2.1 The Antiderivative of f

*Let $f : [a, b] \to \Re$ be a bounded function. Let $G : [a, b] \to \Re$ be such that G' exists on $[a, b]$ and $G'(x) = f(x)$ for all $x \in [a, b]$. Such a function is called an **antiderivative** or a **primitive** of f.*

Comment 23.2.1 *The idea of an antiderivative is intellectually distinct from the Riemann integral of a bounded function f. Consider the following function f defined on $[-1,1]$.*

$$f(x) \;=\; \begin{cases} x^2 \sin(1/x^2), & x \neq 0,\ x \in [-1,1] \\ 0, & x = 0 \end{cases}$$

It is easy to see that this function has a removable discontinuity at 0. Moreover, f is even differentiable on $[-1,1]$ with derivative

$$f'(x) \;=\; \begin{cases} 2x \sin(1/x^2) - (2/x) \cos(1/x^2), & x \neq 0,\ x \in [-1,1] \\ 0, & x = 0 \end{cases}$$

Note f' is not *bounded on $[-1,1]$ and hence it can not be Riemann Integrable. Now to connect this to the idea of antiderivatives, just relabel the functions. Let g be defined by*

$$g(x) \;=\; \begin{cases} 2x \sin(1/x^2) - (2/x) \cos(1/x^2), & x \neq 0,\ x \in [-1,1] \\ 0, & x = 0 \end{cases}$$

then define G by

$$G(x) \;=\; \begin{cases} x^2 \sin(1/x^2), & x \neq 0,\ x \in [-1,1] \\ 0, & x = 0 \end{cases}$$

We see that G is the antiderivative of g even though g itself does not have a Riemann integral. Again, the point is that the idea of the antiderivative of a function is intellectually distinct from that of being Riemann integrable.

Now if f was continuous on $[a,b]$, we know $F(x) = \int_a^x f(t)dt$ is an antiderivative of f since $F' = f$ on $[a,b]$. If G was any other antiderivatives of f, let $H = F - G$. Then we have

$$H'(x) \;=\; F'(x) - G'(x) = f(x) - f(x) = 0.$$

Thus, $H(x)$ is constant on $[a.b]$ and so $H(a) = H(b)$. So

$$-G(a) + F(a) \;=\; -G(b) + F(b) \implies F(b) = G(b) - G(a)$$

as $F(a) = 0$. We have

$$\int_a^b f(t)dt \;=\; G(t)\Big|_a^b = G(b) - G(a).$$

This is a version of the Cauchy FTOC. But we do not have to assume f is continuous on $[a,b]$. It is enough to assume f is RI.

Theorem 23.2.1 Cauchy's Fundamental Theorem

Let $f : [a,b] \to \Re$ be integrable. Let $G : [a,b] \to \Re$ be any antiderivative of f. Then

$$\int_a^b f(t)dt = G(t)\Big|_a^b = G(b) - G(a).$$

Proof 23.2.1
Since G' exists on $[a,b]$, G must be continuous on $[a,b]$. Let $\epsilon > 0$ be given. Since f is integrable,

there is a partition $\pi_0 \in \Pi[a, b]$ such that for any refinement, π, of π_0 and any $\sigma \subset \pi$, we have

$$\left| S(f, \pi, \sigma) - \int_a^b f(x)dx \right| < \epsilon.$$

Let π be any refinement of π_0 given by $\pi = \{x_0 = a, x_1, \ldots, x_p = b\}$. The Mean Value Theorem for differentiable functions then tells us that there is an $s_j \in (x_{j-1}, x_j)$ such that $G(x_j) - G(x_{j-1}) = G'(s_j)\Delta x_j$. Since $G' = f$, we have $G(x_j) - G(x_{j-1}) = f(s_j)\Delta x_j$ for each $j = 1, \ldots, p$. The set of points $\{s_1, \ldots, s_p\}$ is thus an evaluation set associated with π. Hence,

$$\sum_\pi [G(x_j) - G(x_{j-1})] = \sum_\pi G'(s_j)\Delta x_j = \sum_\pi f(s_j)\Delta x_j$$

The first sum on the left is a collapsing sum, hence we have

$$G(b) - G(a) = S(f, \pi, \{s_1, \ldots, s_p\}).$$

We conclude

$$\left| G(b) - G(a) - \int_a^b f(x)dx \right| < \epsilon.$$

Since ϵ was arbitrary, this implies the desired result. ∎

Comment 23.2.2 *Not all functions (in fact, most functions) will have closed form, or analytically obtainable, antiderivatives. So, the previous theorem will not work in such cases.*

Comment 23.2.3 This is huge! *This is what tells us how to integrate many functions. For example, if $f(t) = t^3$, we can guess the antiderivatives have the form $t^4/4 + C$ for an arbitrary constant C. Thus, since $f(t) = t^3$ is continuous, the result above applies. We can therefore calculate Riemann integrals like these:*

1.

$$\int_1^3 t^3 \, dt = \left. \frac{t^4}{4} \right|_1^3 = \frac{3^4}{4} - \frac{1^4}{4} = \frac{80}{4}$$

2.

$$\int_{-2}^4 t^3 \, dt = \left. \frac{t^4}{4} \right|_{-2}^4 = \frac{4^4}{4} - \frac{(-2)^4}{4} = \frac{256}{4} - \frac{16}{4} = \frac{240}{4}$$

Homework

Exercise 23.2.1 *Find the appropriate formula for $\int_a^b \cos(3x)dx$.*

Exercise 23.2.2 *Find the appropriate formula for $\int_a^b \sin(5x)dx$.*

Exercise 23.2.3 *Find the appropriate formula for $\int_a^b p(x)dx$ where p is any polynomial in x.*

Exercise 23.2.4 *Find the appropriate formula for $\int_a^b p'(x)/p(x)dx$ where p is any polynomial in x.*

Exercise 23.2.5 *Find the appropriate formula for $\int_a^b e^{p(x)} p'(x)dx$ where p is any polynomial in x.*

The Cauchy FTOC allows us to prove a critical result about when we can *recapture* a function from knowledge of its derivative.

Theorem 23.2.2 The Recapture Theorem

> *If f is differentiable on $[a, b]$, and if $f' \in RI([a, b])$, then*
> $$\int_a^x f'(t)dt = f(x) - f(a).$$

Proof 23.2.2
f is an antiderivative of f. Now apply Cauchy's Fundamental Theorem 23.2.1.   ■

23.3   Integral Mean Value Results

The next result is one of the many mean value theorems in the theory of integration. It is a more general form of the standard mean value theorem given in beginning calculus classes.

Theorem 23.3.1 The Mean Value Theorem for Riemann Integrals

> *Let $f \in C([a, b])$, and let $g \geq 0$ be integrable on $[a, b]$. Then there is a point, $c \in [a, b]$, such that*
> $$\int_a^b f(x)g(x)dx = f(c) \int_a^b g(x)dx.$$

Proof 23.3.1
Since f is continuous, it is also integrable. Hence, fg is integrable. Let m and M denote the lower and upper bounds of f on $[a, b]$, respectively. Then $mg(x) \leq f(x)g(x) \leq Mg(x)$ for all $x \in [a, b]$. Since the integral preserves order, we have

$$m \int_a^b g(x)dx \leq \int_a^b f(x)g(x)dx \leq M \int_a^b g(x)dx.$$

If the integral of g on $[a, b]$ is 0, then this shows that the integral of fg will also be 0. Hence, in this case, we can choose any $c \in [a, b]$ and the desired result will follow. If the integral of g is not 0, then it must be positive, since $g \geq 0$. Hence, in this case, we have

$$m \leq \frac{\int_a^b f(x)g(x)dx}{\int_a^b g(x)dx} \leq M.$$

Now, f is continuous implying that it attains the values M and m at some points. Hence, by the intermediate value theorem, there must be some $c \in [a, b]$ such that

$$f(c) = \frac{\int_a^b f(x)g(x)dx}{\int_a^b g(x)dx}.$$

This implies the desired result.   ■

The next result is another standard mean value theorem from basic calculus. It is a direct consequence of the previous theorem by simply letting $g(x) = 1$ for all $x \in [a, b]$. This result can be interpreted as stating that integration is an averaging process.

Theorem 23.3.2 Average Value for Riemann Integrals

> *If $f \in C([a,b])$, then there is a point $c \in [a,b]$ such that*
>
> $$\frac{1}{b-a} \int_a^b f(x)dx = f(c).$$

Homework

Exercise 23.3.1 *Find the average value of $f(x) = x^2 + 4$ on $[-2, 5]$.*

Exercise 23.3.2 *Find the average value of $f(x) = \sin(2x)$ on $[-\pi, \pi]$.*

Exercise 23.3.3 *Find the average value of $f(x) = x^3$ on $[-5, 6]$.*

23.4   Approximation of the Riemann Integral

Another way to evaluate Riemann integrals is to directly approximate them using an appropriate sequence of partitions. Theorem 23.4.1 is a fundamental tool that tells us when and why such approximations will work.

Theorem 23.4.1 Approximation of the Riemann Integral

> *If $f \in RI([a,b])$, then given any sequence of partitions $\{\pi_n\}$ with any associated sequence of evaluation sets $\{\sigma_n\}$ that satisfies $\| \pi_n \| \to 0$, we have*
>
> $$\lim_{n \to \infty} S(f, \pi_n, \sigma_n) = \int_a^b f(x)\, dx$$

Proof 23.4.1
Since f is integrable, given a positive ϵ, there is a partition π_0 so that

$$\left| S(f, \pi, \sigma) - \int_a^b f(x)dx \right| \; < \; \epsilon/2, \; \pi_0 \preceq \pi, \; \sigma \subseteq \pi. \qquad (*)$$

Let the partition π_0 be $\{x_0, x_1, \ldots, x_P\}$ and let ξ be defined to be the smallest Δx_j from π_0. Then since the norm of the partitions π_n goes to zero, there is a positive integer N so that

$$\| \pi_n \| \; < \; \min\left(\xi, \epsilon/(4P(\| f \|_\infty + 1))\right) \qquad (**)$$

Now pick any $n > N$ and label the points of π_n as $\{y_0, y_1, \ldots, y_Q\}$.

We see that the points in π_n are close enough together so that at most one point of π_0 lies in any subinterval $[y_{j-1}, y_j]$ from π_n. This follows from our choice of ξ. So the intervals of π_n split into two pieces: those containing a point of π_0 and those that do not have a point of π_0 inside.

Let $\mathscr{B}$ be the first collection of intervals and $\mathscr{A}$, the second. Note there are P points in π_0 and so there are P subintervals in $\mathscr{B}$. Now consider the common refinement $\pi_n \vee \pi_0$. The points in the common refinement match π_n except on the subintervals from $\mathscr{B}$. Let $[y_{j-1}, y_j]$ be such a subinterval *and let γ_j denote the point from π_0 which is in this subinterval. Let's define an evaluation set σ for this refinement $\pi_n \vee \pi_0$ as follows.*

1. *if we are in the subintervals labeled $\mathscr{A}$, we choose as our evaluation point, the evaluation point s_j that is already in this subinterval since $\sigma_n \subseteq \pi_n$. Here, the length of the subinterval will be denoted by $\delta_j(\mathscr{A})$ which equals $y_j - y_{j-1}$ for appropriate indices.*

2. *if we are in the the subintervals labeled $\mathscr{B}$, we have two intervals to consider as $[y_{j-1}, y_j] = [y_{j-1}, \gamma_j] \cup [\gamma_j, y_j]$. Choose the evaluation point γ_j for both $[y_{j-1}, \gamma]$ and $[\gamma, y_j]$. Here, the length of the subintervals will be denoted by $\delta_j(\mathscr{B})$. Note that $\delta_j(\mathscr{B}) = \gamma_j - y_{j-1}$ or $y_j - \gamma_j$.*

Then we have

$$
\begin{aligned}
S(f, \pi_n \vee \pi_0, \sigma) &= \sum_{\mathscr{A}} f(s_j)\delta_j(\mathscr{A}) + \sum_{\mathscr{B}} f(\gamma_j)\delta_j(\mathscr{B}) \\
&= \sum_{\mathscr{A}} f(s_j)(y_j - y_{j-1}) + \sum_{\mathscr{B}} \left(f(\gamma_j)(y_j - \gamma_j) + f(\gamma_j)(\gamma_j - y_{j-1}) \right) \\
&= \sum_{\mathscr{A}} f(s_j)(y_j - y_{j-1}) + \sum_{\mathscr{B}} \left(f(\gamma_j)(y_j - y_{j-1}) \right)
\end{aligned}
$$

Thus, since the Riemann sums over π_n and $\pi_n \vee \pi_0$ with these choices of evaluation sets match on $\mathscr{A}$, we have using Equation $**$ that

$$
\begin{aligned}
|S(f, \pi_n, \sigma_n) - S(f, \pi_n \vee \pi_0, \sigma)| &= \left| \sum_{\mathscr{B}} (f(s_j) - f(\gamma_j))(y_j - y_{j-1}) \right| \\
&\leq \sum_{\mathscr{B}} \left(|f(s_j)| + |f(\gamma_j)| \right) (y_j - y_{j-1})| \\
&\leq P\, 2\, \|f\|_\infty \|\pi_n\| \\
&< P\, 2\, \|f\|_\infty \frac{\epsilon}{4P(\|f\|_\infty + 1)} \\
&= \epsilon/2
\end{aligned}
$$

We conclude that for our special evaluation set σ for the refinement $\pi_n \vee \pi_0$ that

$$
\begin{aligned}
&\left| S(f, \pi_n, \sigma_n) - \int_a^b f(x)dx \right| \\
=&\left| S(f, \pi_n, \sigma_n) - S(f, \pi_n \vee \pi_0, \sigma) + S(f, \pi_n \vee \pi_0, \sigma) - \int_a^b f(x)dx \right| \\
\leq&\left| S(f, \pi_n, \sigma_n) - S(f, \pi_n \vee \pi_0, \sigma) \right| + \left| S(f, \pi_n \vee \pi_0, \sigma) - \int_a^b f(x)dx \right| \\
<&\, \epsilon/2 + \epsilon/2 = \epsilon
\end{aligned}
$$

using Equation $*$ as $\pi_n \vee \pi_0$ refines π_0. Since we can do this analysis for any $n > N$, we see we have shown the desired result. ∎

Example 23.4.1 *Let f be a continuously differentiable function on the finite interval $[a, b]$. Find a formula for the length of this curve using a Riemann Integral.*

Solution *The function f determines a curve in the plane whose coordinates are $(x, f(x))$ for $a \leq x \leq b$. Let π_n be the uniform partition of $[a, b]$ into subintervals of length $(b-a)/n$. The points in π_n have the form $x_i = a + i(b-a)/n$. By the MVT, we know there are points s_i in $[x_{i-1}, x_i]$ so that*

$$f(x_i) - f(x_{i-1}) = f'(s_i)(x_i - x_{i-1}) = f'(s_i)\Delta x_i.$$

We can approximate the length of the curve determined by f by adding up the hypotenuse lengths of the triangles formed by the points in π_n. We have the hypotenuse length H_n is

$$H_n \;=\; \sqrt{(x_i - x_{i-1})^2 + (f(x_i) - f(x_{i-1}))^2}$$

and the approximate arc length, L_n, is then

$$
\begin{aligned}
L_n \;&=\; \sum_{\pi_n} \sqrt{(x_i - x_{i-1})^2 + (f(x_i) - f(x_{i-1}))^2} \\
&=\; \sum_{\pi_n} \sqrt{1 + \frac{(f(x_i) - f(x_{i-1}))^2}{(x_i - x_{i-1})^2}} \,(x_i - x_{i-1}) \\
&=\; \sum_{\pi_n} \sqrt{1 + \left(\frac{\Delta f_i}{\Delta x_i}\right)^2} \,\Delta x_i \\
&=\; \sum_{\pi_n} \sqrt{1 + \left(f'(s_i)\right)^2} \,\Delta x_i
\end{aligned}
$$

But this says $L_n = S(h, \pi_n, \sigma_n)$ where σ_n is the Evaluation set determined by the points s_i where the function $h(x) = \sqrt{1 + (f'(x))^2}$.

We know h is continuous as f' is continuous, so h is RI on $[a, b]$. We know $\|\pi_n\| \to 0$ and so by the RI approximation theorem,

$$\int_a^b \sqrt{1 + (f'(x))^2}\, dx \;=\; \lim_n \sum_{\pi_n} \sqrt{1 + (f'(s_i))^2}\, \Delta x_i$$

So it is reasonable to define the length of the curve determined by f by this RI.

Example 23.4.2 *We are going to find a way to define the work done in moving a particle from a start point (x_s, y_s) to an end point (x_e, y_e) along a curve $\mathscr{C}$ under the influence of the force $\boldsymbol{F}$ given by*

$$\boldsymbol{F}(x, y) \;=\; \begin{bmatrix} M(x, y) \\ N(x, y) \end{bmatrix}$$

where M and N are continuous functions in the two variables x and y.

Solution *The curve $\mathscr{C}$ is parameterized by $x = f(t)$ and $y = g(t)$ on $a \le t \le b$ where f and g are continuously differentiable on that interval. Let π_n be a uniform partition of $[a, b]$ like in the previous example. Now apply the MVT to f and g on each subinterval to find u_i and v_i between t_{i-1} and t_i so that*

$$
\begin{aligned}
f(t_i) - f(t_{i-1}) &= f'(u_i)\,(t_i - t_{i-1}) \\
g(t_i) - g(t_{i-1}) &= g'(v_i)\,(t_i - t_{i-1})
\end{aligned}
$$

Now approximate the work done on in moving from (x_{i-1}, y_{i-1}) to (x_i, y_{i-1})

$$
\begin{aligned}
W_i^x &= M(f(u_i), g(u_i))\,(f(t_i) - f(t_{i-1})) \\
&= M(f(u_i), g(u_i))\,f'(u_i)\,(t_i - t_{i-1})
\end{aligned}
$$

the work done on in moving from (x_i, y_{i-1}) to (x_i, y_i)

$$W_i^y = N(f(v_i), g(v_i))\,(g(t_i) - g(t_{i-1}))$$
$$= N(f(v_i), g(v_i))\,g'(v_i)\,(t_i - t_{i-1})$$

Then the approximation to the work done along this path is the sum $\mathscr{W}_{\pi_n} = \sum_{\pi_n} W_i^x + \sum_{\pi_n} W_i^y$ or

$$\mathscr{W}_{\pi_n} \;=\; \sum_{\pi_n} M(f(u_i), g(u_i))\,f'(u_i)\,(t_i - t_{i-1})$$
$$+ \sum_{\pi_n} N(f(v_i), g(v_i))\,g'(v_i)\,(t_i - t_{i-1})$$

Now let $P(t) = M(f(t), g(t))f'(t)$ and $Q(t) = N(f(t), g(t))g'(t)$. Then both P and Q are continuous on $[a, b]$ and so their RI's are defined. We see we can rewrite the approximations to work above as

$$\mathscr{W}_{\pi_n} \;=\; S(P, \sigma_{u_n}, \pi_n) + S(Q, \sigma_{v_n}, \pi_n)$$

where σ_{u_n} and σ_{v_n} are the evaluation sets we found using the MVT on the subintervals determined by π_n. The approximation theorem then says

$$\mathscr{W}_{\pi_n} \;\to\; \int_a^b M(f(t), g(t))f'(t)dt + \int_a^b N(f(t), g(t))g'(t)dt$$

which is the usual line integral solution we see in calculus. So it seems reasonable to define the work done by moving a particle along this path using these RI's.

Note in this example, if you choose to approximate the force on each triangle differently than we did, you still get an nice approximation but it does not look like a Riemann sum. For example, if you used the force on the bottom of the triangle to be $M(x_{i-1}, y_{i-1}) = M(f(t_{i-1}), g(t_{i-1}))$ and the force on the vertical side of the triangle to be $N(x_i, y_{i-1}) = N(f(t_i), g(t_{i-1}))$.
The approximation would then be

$$\mathscr{W}_{\pi_n} \;=\; \sum_{\pi_n} M(f(t_{i-1}), g(t_{i-1}))\,f'(u_i)\,(t_i - t_{i-1})$$
$$+ \sum_{\pi_n} N(f(t_i), g(t_{i-1}))\,g'(v_i)\,(t_i - t_{i-1})$$

which is not in the right form for a Riemann sum as you can't write it with terms like $P(s_i)\,(t_i - t_{i-1}) + Q(s_i)\,(t_i - t_{i-1})$ for a suitable evaluation point s_i.

Example 23.4.3 We know $\int_0^1 1/(1 + x^2)dx = \arctan(1) = \pi/4$. Write this as a uniform sum limit.

Solution *Take uniformly spaced partitions of width $1/n$ of $[0, 1]$. Then points in the partition have the form k/n and we take as evaluation points the RH endpoint. The Riemann sum is then*

$$S(f(x) = 1/(1 + x^2), P_n, S_n) \;=\; \sum_{k=1}^{n} 1/(1 + k^2/n^2)(1/n)$$
$$= \sum_{k=1}^{n} 1/(1 + k^2/n^2)(1/n) = \pi/4.$$

So a question might be Prove $\lim_{n\to\infty} \sum_{k=1}^{n} n/(n^2 + k^2) = \pi$. Note we know $f(x) = 1/(1 + x^2)$

is RI on $[0, 1]$ *since it is continuous and our theorem says the limit of these Riemann sums converges to* $\int_0^1 1/(1 + x^2)dx$.

23.4.1   Homework

Exercise 23.4.1 *Let* $f(x) = x^2$ *on the interval* $[-1, 3]$. *Use Theorem 23.4.1 to prove that* $\int_{-1}^3 f(x)\,dx = 28/3$.

Hint 23.4.1 *We know* f *is Riemann integrable because it is continuous and so this theorem can be applied. Use the uniform approximations* $x_i = -1 + 4i/n$ *for* $i = 0$ *to* $i = n$ *to define partitions* π_n. *Then using left or right hand endpoints on each subinterval to define the evaluation set* σ_n, *you can prove directly that* $\int_{-1}^3 x^2 dx = \lim S(f, \pi_n, \sigma_n) = 28/3$. *Make sure you tell me all the reasoning involved.* □

Exercise 23.4.2 *If* f *is continuous, evaluate*

$$\lim_{x \to a} \frac{x}{x - a} \int_a^x f(t)dt$$

Exercise 23.4.3 *Prove if* f *is continuous on* $[a, b]$ *and* $\int_a^b f(x)g(x)dx = 0$ *for all choices of integrable* g, *then* f *is identically* 0.

Exercise 23.4.4 *Evaluate* $\lim_{n \to \infty} \sum_{k=1}^n \sqrt{k}/n^{3/2}$.

Exercise 23.4.5 *Evaluate* $\lim_{n \to \infty} \sum_{k=1}^n 1/(n + k)$.

Exercise 23.4.6 *Evaluate* $\lim_{n \to \infty} \sum_{k=1}^n k/(n^2 + k^2)$.

Exercise 23.4.7 *Evaluate* $\lim_{n \to infty} \sum_{k=1}^n 1/(n^2 + k^2)$.

Exercise 23.4.8 *Prove* $\lim_{n \to \infty} \sum_{k=0}^{n-1} 1/\sqrt{n^2 - k^2} = \pi/2$.

Hint 23.4.2 *The function you will find here is* **not** *Riemann integrable on* $[0, 1]$ *but it is on* $[L, 1]$ *for any* $0 < L < 1$. *So start with uniform partitions of the interval* $[L, 1]$ *and go from there.* □

23.5   Integration Methods

Using the Fundamental Theorem of Calculus, we can derive many useful tools. We begin with integration by parts.

Theorem 23.5.1 Integration by Parts

Assume $u : [a, b] \to \Re$ and $v : [a, b] \to \Re$ are differentiable on $[a, b]$ and u' and v' are integrable. Then

$$\int_a^x u(t)v'(t)\,dt = u(t)v(t)\Big|_a^x - \int_a^x v(t)u'(t)\,dt$$

Proof 23.5.1
Since u *and* v *are differentiable on* $[a, b]$, *they are also continuous and hence, integrable. Now apply the product rule for differentiation to obtain*

$$(u(t)v(t))' \;=\; u'(t)v(t) \,+\, u(t)v'(t)$$

By Theorem 22.3.2, we know products of integrable functions are integrable. Also, the integral is linear. Hence, the integral of both sides of the equation above is defined. We obtain

$$\int_a^x (u(t)v(t))' \, dt \;=\; \int_a^x u'(t)v(t) \, dt \,+\, \int_a^x u(t)v'(t) \, dt$$

Since $(uv)'$ is integrable, we can apply the Recapture Theorem to see

$$u(t)v(t)\Big|_a^x \;=\; \int_a^x u'(t)v(t) \, dt \,+\, \int_a^x u(t)v'(t) \, dt$$

This is the desired result.   ■

Homework

Exercise 23.5.1 *If f has sufficient derivatives, express the integral $\int_a^b f'' x^3 dx$ as an integral involving only f.*

Exercise 23.5.2 *If f has sufficient derivatives, express the integral $\int_a^b f''(2x^2 + 3x + 5)dx$ as an integral involving only f.*

Exercise 23.5.3 *Find $\int_a^b x \cos(2x) dx$.*

Exercise 23.5.4 *Find $\int_a^b x e^{-3x} dx$.*

Next, the technique of substitution.

Theorem 23.5.2 Substitution in Riemann Integration

Let f be continuous on $[c,d]$ and $u : [a,b] \to [c,d]$ be continuously differentiable on $[a,b]$ with $u(a) = c$ and $u(b) = d$. Then

$$\int_c^d f(u) \, du \;=\; \int_a^b f(u(t)) \, u'(t) \, dt$$

Proof 23.5.2
Let F be defined on $[c,d]$ by $F(u) = \int_c^u f(t)dt$. Since f is continuous, F is continuous and differentiable on $[c,d]$ by the Fundamental Theorem of Calculus. We know $F'(u) = f(u)$ and so

$$F'(u(t)) = f(u(t)), \; a \le t \le b$$

implying

$$F'(u(t)) \, u'(t) = f(u(t))u'(t) \,, \; a \le t \le b$$

By the Chain Rule for differentiation, we also know

$$(F \circ u)'(t) = F'(u(t))u'(t) \,, \; a \le t \le b.$$

and hence $(F \circ u)'(t) = f(u(t))u'(t)$ on $[a,b]$.

Now define g on $[a, b]$ by

$$\begin{aligned} g(t) &= (f \circ u)(t)\, u'(t) = f(u(t))\, u'(t) \\ &= (F \circ u)'(t). \end{aligned}$$

Since g is continuous, g is integrable on $[a, b]$. Now define G on $[a, b]$ by $G(t) = (F \circ u)(t)$. Then $G'(t) = f(u(t))u'(t) = g(t)$ on $[a, b]$ and G' is integrable. Now, apply the Cauchy Fundamental Theorem of Calculus to G to find

$$\int_a^b g(t)\, dt = G(b) - G(a)$$

or

$$\begin{aligned} \int_a^b f(u(t))\, u'(t)\, dt &= F(u(b)) - F(u(a)) \\ &= \int_c^{u(b)=d} f(t)dt - \int_c^{u(a)=c} f(t)dt \\ &= \int_c^d f(t)dt. \end{aligned}$$

■

Homework

Exercise 23.5.5 *Find $\int_1^2 f(p(x))p'(x)dx$ if f is integrable and p is any polynomial.*

Exercise 23.5.6 *Find $\int_1^2 1/(f(p(x)))p'(x)dx$ if $1/f$ is integrable and p is any polynomial.*

Another useful tool is Leibnitz's rule.

Theorem 23.5.3 Leibnitz's Rule

Let f be continuous on $[a, b]$, $u : [c, d] \to [a, b]$ be differentiable on $[c, d]$ and $v : [c, d] \to [a, b]$ be differentiable on $[c, d]$. Then

$$\left(\int_{u(x)}^{v(x)} f(t)\, dt \right)' = f(v(x))v'(x) - f(u(x))u'(x)$$

Proof 23.5.3

Let F be defined on $[a, b]$ by $F(y) = \int_a^y f(t)dt$. Since f is continuous, F is also continuous and moreover, F is differentiable with $F'(y) = f(y)$. Since v is differentiable on $[c, d]$, we can use the Chain Rule to find

$$\begin{aligned} (F \circ v)'(x) &= F'(v(x))\, v'(x) \\ &= f(v(x))\, v'(x) \end{aligned}$$

This says

$$\left(\int_a^{v(x)} f(t)\, dt \right)' = f(v(x))v'(x)$$

Next, define G on $[a, b]$ by $G(y) = \int_y^b f(t)dt$. Apply the Reversed Fundamental Theorem of Calculus to conclude

$$G'(y) = -\left(\int_a^y f(t)dt\right) = -f(y)$$

Again, apply the Chain Rule to see

$$
\begin{aligned}
(G \circ u)'(x) &= G'(u(x))\,u'(x) \\
&= -f(u(x))\,u'(x).
\end{aligned}
$$

We conclude

$$\left(\int_{u(x)}^b f(t)\,dt\right)' = -f(u(x))u'(x)$$

Now combine these results as follows:

$$\int_a^b f(t)dt = \int_a^{v(x)} f(t)dt + \int_{v(x)}^{u(x)} f(t)dt + \int_{u(x)}^b f(t)dt$$

or

$$
\begin{aligned}
(F \circ v)(x) + (G \circ u)(x) - \int_a^b f(t)dt &= -\int_{v(x)}^{u(x)} f(t)dt \\
&= \int_{u(x)}^{v(x)} f(t)dt
\end{aligned}
$$

Then, differentiate both sides to obtain

$$
\begin{aligned}
(F \circ v)'(x) + (G \circ u)'(x) &= f(v(x))v'(x) - f(u(x))u'(x) \\
&= \left(\int_{u(x)}^{v(x)} f(t)dt\right)'
\end{aligned}
$$

which is the desired result. ∎

Homework

Exercise 23.5.7 *Find $\left(\int_x^{x^2} \sin(t^4)dt\right)'$.*

Exercise 23.5.8 *Find $\left(\int_x^{x^2} \cos(t^4)dt\right)'$.*

Exercise 23.5.9 *Find $\left(\int_{2\sin(x)}^{5\cos(3x)} e^{-(t^4)}dt\right)'$.*

23.6 The Natural Logarithm Function

Now let's do something useful with all this complicated mathematics we have been learning. Look again at the continuous function $f(t) = 1/t$ for all $t \geq 1$. We'll do this in a step by step fashion explaining all the things we know and what they imply. A bit tedious but informative!! Let $F(x) = \int_1^x 1/t\, dt$ for any $0 < x$.

$F(x)$ is smooth for $x \geq 1$

We focus on the continuous function $f(t) = 1/t$ for all $1 \geq t$.

- Pick any $x \geq 1$.
- Pick any $L > x$.
- Look at $F(x) = \int_1^x 1/t \, dt$ on the interval $[1, L]$.
- The Fundamental Theorem of Calculus applies and so we know F is continuous on $[1, L]$ – in particular F is continuous at x.
- Since the integrand $1/t$ is continuous here, the Fundamental Theorem of Calculus also tells us $F'(x) = f(x) = 1/x$.
- We can do this argument for any $x \geq 1$. So we know our F is continuous at x and $F'(x) = 1/x$ for all $x \geq 1$.

$F(x)$ is smooth for $0 < x < 1$

We now look at the continuous function $f(t) = 1/t$ for all $0 < t \leq 1$.

- Pick any x with $0 < x \geq 1$.
- Pick any L so that $0 < L < x \leq 1$.
- Let $G(x) = \int_x^1 1/t \, dt$ on the interval $[L, 1]$.
- The Fundamental Theorem of Calculus applies and we know G is continuous on $[L, 1]$ – in particular G is continuous at x.
- Since the integrand $1/t$ is continuous here, the Reversed Fundamental Theorem of Calculus tells us $G'(x) = -f(x) = -1/x$.
- We can do this argument for any such x. So we know our G is continuous at x and $G'(x) = -1/x$ for all $0 < x \geq 1$.
- But for $0 < x < 1$, $G(x) = -F(x)$ and so $G'(x) = -F'(x)$. So we now know $-F'(x) = -(1/x)$ for $0 < x < 1$. Now just multiply through by -1 and we have the result we wanted: $F'(x) = 1/x$ on $0 < x < 1$.

So $F(x) = \int_1^x 1/t \, dt$ is continuous for $x > 0$ and $F'(x) = 1/x$ for $x > 0$. Very nice. This F is so special and useful it is given a special name: the **natural logarithm**. We denote this function by the symbol $\ln(x)$ and it is defined as follows:

$$\ln(x) \quad = \quad \int_1^x 1/t \, dt$$

As we have seen, this function is defined in two pieces:

$$\ln(x) \quad = \quad \begin{cases} F(x) = \int_1^x 1/t \, dt, & x \geq 1 \\ \int_1^x 1/t \, dt = -\int_x^1 1/t \, dt = -G(x), & 0 < x \leq 1. \end{cases}$$

We see $(\ln(x))' = 1/x$ for all positive x. Also, we now see that the number e is the unique number where $\ln(e) = 1$.

23.6.1 Logarithm Functions

We know that

$$\frac{d}{dx}\left(\ln(x)\right) \quad = \quad \frac{1}{x}$$

Let's start with a simple function, $u(x) = |x|$. Then $\ln(|x|)$ is nicely defined at all x not zero. Note we have

$$\ln(|x|) = \begin{cases} \ln(x) & \text{if } x > 0 \\ \ln(-x) & \text{if } x < 0 \end{cases}$$

and so since by the chain rule, if x is negative, using the chain rule with $u(x) = -x$, we have

$$\frac{d}{dx}(\ln(-x)) = \frac{1}{-x}\frac{d}{dx}(-x) = \frac{1}{-x}(-1) = \frac{1}{x}$$

Thus,

$$\frac{d}{dx}(\ln|x|) = \begin{cases} \frac{d}{dx}(\ln(x)) & \text{if } x > 0 \\ \frac{d}{dx}(\ln(-x)) & \text{if } x < 0 \end{cases} = \begin{cases} \frac{1}{x} & \text{if } x > 0 \\ \frac{1}{x} & \text{if } x < 0 \end{cases}$$
$$= \frac{1}{x} \text{ if } x \text{ is not } 0$$

We conclude that

$$\frac{d}{dx}(\ln(|x|)) = \frac{1}{x}, \text{ if } x \neq 0$$

It then follows for a more general $u(x)$, using the chain rule that

$$\frac{d}{dx}(\ln(|u(x)|)) = \frac{1}{u(x)}\frac{du}{dx}, \text{ if } x \neq 0 \tag{23.3}$$

23.6.2   Worked Out Examples: Integrals

We can also now find antiderivatives for a new class of functions. The chain rule for the derivatives of the logarithm function given in Equation 23.3 implies

$$\int \frac{1}{u(x)} u'(x)\, dx = \ln(|u(x)|) + C$$

Another way of saying this is that $\int du/u = \ln(|u|) + C$! Let's work some examples:

Example 23.6.1 *Find the integral $\int \frac{1}{x^2+x+1}(2x+1)dx$.*

Solution

$$\int \frac{1}{x^2+x+1}(2x+1)dx = \int \frac{1}{u}\, du, \text{ use substitution } u = x^2 + x + 1$$
$$= \ln(|u|) + C$$
$$= \ln(|x^2 + x + 1|) + C$$

Example 23.6.2 *Find the integral $\int \frac{1}{\tan(x)}\sec^2(x)\, dx$.*

Solution

$$\int \frac{1}{\tan(x)}\sec^2(x)\, dx = \int \frac{1}{u}\, du, \text{ use substitution } u = \tan(x); \, du = \sec^2(x)dx$$
$$= \ln(|u|) + C$$
$$= \ln(|\tan(x)|) + C$$

Example 23.6.3 *Find the integral* $\int \frac{2x}{4 + x^2} \, dx$.

Solution

$$
\begin{aligned}
\int \frac{2x}{4 + x^2} \, dx &= \int \frac{1}{u} \, du, \text{ use substitution } u = 4 + x^2; \, du = 2x dx \\
&= \ln(|\, u \,|) + C \\
&= \ln(|\, 4 + x^2 \,|) + C, \text{ but } 4 + x^2 \text{ is always positive} \\
&= \ln(4 + x^2) + C
\end{aligned}
$$

Example 23.6.4 *Find the integral* $\int \tan(w) dw$.

Solution

$$
\begin{aligned}
\int \tan(w) dw &= \int \frac{\sin(w)}{\cos(w)} \, dw \\
&= \int \frac{1}{u} (-du), \text{ use substitution } u = \cos(w); \, du = -\sin(w) dw \\
&= -\ln(|\, u \,|) + C \\
&= -\ln(|\, \cos(w) \,|) + C
\end{aligned}
$$

23.6.3 The Exponential Function

Let's backup and recall the idea of an inverse function. We discussed this earlier in Chapter 13 in Section 13.5. What about $\ln(x)$? It has derivative $1/x$ for all positive x, so it must be always increasing. So it has an inverse which we can call $\ln^{-1}(x)$ which is called the **exponential function** which is denoted by $\exp(x)$. In Theorem 13.5.1, we proved a continuous function which is increasing (our $\ln(x)$) has a continuous inverse. Then in Theorem 13.5.2, we proved if this function is also differentiable, so is its inverse and we showed $d/dy(\, f^{-1}(y)) = 1/f'(f^{-1}(y))$ where f is the function. So let's back up and refresh our memory here: the inverse is defined by the if and only relationship

$$(\ln)^{-1}(x) = y \quad \Leftrightarrow \quad \ln(y) = x$$

or, using the exp notation

$$\exp(x) = y \quad \Leftrightarrow \quad \ln(y) = x.$$

A little thought then tells us the range of $\ln(x)$ is all real numbers as for $x > 1$, $\ln(x)$ gets as large as we want and for $0 < x < 1$, as x gets closer to zero, the negative area $-\int_x^1 1/t \, dt$ approaches $-\infty$. By definition then

- $\ln(\exp(x)) = x$ for $-\infty < x < \infty$; i.e. for all x.
- $\exp(\ln(x)) = x$ for all $x > 0$.

Further since $\ln(\exp(x)) = x$ and we know the inverse is differentiable, we can take the derivative of both sides:

$$\left(\ln(\exp(x)) \right)' = \left(x \right)' = 1$$

Using the chain rule, for any function $u(x)$,

$$\left(\ln(u(x)) \right)' = \frac{1}{u(x)} u'(x).$$

So

$$\Big(\ln(\exp(x)) \Big)' \; = \; \frac{1}{\exp(x)} \Big(\exp(x) \Big)'.$$

Using this, we see $\frac{1}{\exp(x)} \Big(\exp(x) \Big)' = 1$ and so

$$\Big(\exp(x) \Big)' \; = \; \exp(x).$$

This is the only function whose derivative is itself!

23.6.4   Positive Powers of e

Now what is the real number c which satisfies $\ln(c) = 2$? This means

$$\ln(c) \; = \; \int_1^c \frac{1}{t} \, dt = 2.0?$$

We can see this graphically in Figure 23.1.

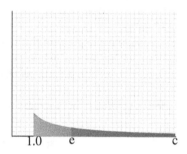

There is a point c on the t - axis where the area under the curve from 1 to c is exactly 2

Figure 23.1: There is a value c where the area under the curve from 1 to c is 2.

Since the area under the curve $\frac{1}{t}$ is increasing as t increases, we can immediately see that c must be larger than e. Thus, using properties of the Riemann Integral, we have

$$2 \; = \; \int_1^c \frac{1}{t} \, dt = \int_1^e \frac{1}{t} \, dt + \int_e^c \frac{1}{t} \, dt = 1.0 + \int_e^c \frac{1}{t} \, dt$$

Subtracting terms, we find $1 = \int_e^c \frac{1}{t} \, dt$. Now, make the change of variable $u = t/e$. Then

$$\int_e^c \frac{1}{t} \, dt \; = \; \int_1^{c/e} \frac{1}{u} \, du$$

Combining, we have $1 = \int_1^{c/e} \frac{1}{u} \, du$. Since the letter we use as the variable of integration is arbitrary, this can be rewritten as

$$1 \; = \; \int_1^{c/e} \frac{1}{t} \, dt$$

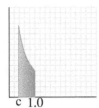

There is a point c on the t - axis where the area under the curve from c to 1 is exactly 1.

Figure 23.2: There is a value c where the area under the curve which ends at 1 is 1.

However, e is the unique number which marks the point on the t - axis where the area under this curve is 1. Thus, we can conclude that $c/e = e$ or $c = e^2$. A similar argument shows all of the following also to be true:

$$3 = \ln(e^3) = \int_1^{e^3} \frac{1}{t}\, dt, \quad 4 = \ln(e^4) = \int_1^{e^4} \frac{1}{t}\, dt, \quad 6 = \ln(e^6) = \int_1^{e^6} \frac{1}{t}\, dt.$$

We can thus interpret the number $\ln(e^p)$ for a positive integer power of p as an appropriate area under the graph of $\frac{1}{t}$ starting at 1 and moving to the right. That is e^p is the number that satisfies

$$p = \ln(e^p) = \int_1^{e^p} \frac{1}{t}\, dt.$$

23.6.5 Negative Integer Powers of e

We were able to interpret the number e^p for a positive integer power of p as follows:

$$p = \ln(e^p) = \int_1^{e^p} \frac{1}{t}\, dt$$

How can we interpret terms like $\frac{1}{e}$ or $\frac{1}{e^2}$? Let's start with e^{-1}. We can use a graphical analysis just like we did to define the number e. Look at Figure 23.2.

Since the area under the curve $\frac{1}{t}$ is increasing as t increases, we can immediately see that c must be less than 1. Further, we know that $1 = \int_c^1 \frac{1}{t}\, dt$ Now, make the change of variable $u = t/c$. Then

$$\int_c^1 \frac{1}{t}\, dt = \int_1^{1/c} \frac{1}{u}\, du$$

Combining, we have

$$1 = \int_1^{1/c} \frac{1}{u}\, du$$

Since the letter we use as the variable of integration is arbitrary, this can be rewritten as

$$1 = \int_1^{1/c} \frac{1}{t}\, dt$$

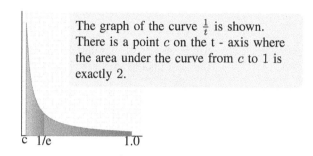

The graph of the curve $\frac{1}{t}$ is shown. There is a point c on the t - axis where the area under the curve from c to 1 is exactly 2.

Figure 23.3: There is a value c where the area under the curve which ends at 1 is 2.

However, e is the unique number which marks the point on the x - axis where the area under this curve is 1. Thus, we can conclude that $1/c = e$. This implies that $c = 1/e$. Hence, we have shown

$$-1 \;=\; \ln(e^{-1}).$$

What about the interpretation of the value e^{-2}? Consider Figure 23.3. Since the area under the curve $\frac{1}{t}$ is increasing as t increases, we can immediately see that c must be less than $1/e$ which is less than 1. Thus, using properties of the Riemann Integral, and the fact that $\int_{1/e}^{1} \frac{1}{t}\, dt = 1$, we see

$$2 \;=\; \int_{c}^{1} \frac{1}{t}\, dt = \int_{c}^{1/e} \frac{1}{t}\, dt + \int_{1/e}^{1} \frac{1}{t}\, dt = \int_{c}^{1/e} \frac{1}{t}\, dt + 1.0$$

Subtracting terms, we find $1 = \int_{c}^{\frac{1}{e}} \frac{1}{t}\, dt$. Now, make the change of variable $u = t/c$. Then $\int_{c}^{1/e} \frac{1}{t}\, dt = \int_{1}^{1/(ec)} \frac{1}{u}\, du$ Combining, we have $1 = \int_{1}^{1/(ec)} \frac{1}{u}\, du$. Since the letter we use as the variable of integration is arbitrary, this can be rewritten as

$$1 \;=\; \int_{1}^{1/(ec)} \frac{1}{t}\, dt$$

However, e is the unique number which marks the point on the t - axis where the area under this curve is 1. Thus, we can conclude that $\frac{1}{ec} = e$ or $c = e^{-2}$. Hence, we have shown $-2 = \ln(e^{-2})$. A similar argument shows all of the following also to be true:

$$-3 \;=\; \ln(e^{-3}) = \int_{1}^{e^{-3}} \frac{1}{t}\, dt, \quad -4 = \ln(e^{-4}) = \int_{1}^{e^{-4}} \frac{1}{t}\, dt, \quad -6 = \ln(e^{-6}) = \int_{1}^{e^{-6}} \frac{1}{t}\, dt.$$

We can thus interpret the number $\ln(e^{-p})$ for a positive integer power of p as an appropriate area under the graph of $\frac{1}{t}$ starting at e^{-p} and moving to the 1 That is e^{-p} is the number that satisfies

$$-p \;=\; \ln(e^{-p}) = \int_{1}^{e^{-p}} \frac{1}{t}\, dt.$$

23.6.6   Adding Natural Logarithms

What happens if we want to add $\ln(a) + \ln(b)$ for positive numbers a and b? There are several cases. We can

- We have a is less than one but b is bigger than one: for example, $a = .3$ and $b = 7$.

- We have a is less than one and b is less than one also: for example, $a = .3$ and $b = .7$.

- We have a is bigger than one and b is bigger than one: for example, $a = 4$ and $b = 8$.

We can work out how to do this in general but it is easier to work through some specific examples and extrapolate from them.

23.6.6.1   Adding Logarithms: Both Logarithms are Bigger than 1

Example 23.6.5 *Show* $\ln(3) + \ln(5) = \ln(3 \times 5) = \ln(15)$.

Solution   • *First, rewrite using the definition.*

$$\ln(3) + \ln(5) \;=\; \int_1^3 1/t \; dt \;+\; \int_1^5 1/t \; dt.$$

- *Now rewrite the second integral $\int_1^5 1/t \; dt$ so it starts at 3. This requires a change of variable.*

  - *Let $u = 3t$. Then when $t = 1$, $u = 3$ and when $t = 5$, $u = 3 \times 5 = 15$.*
  - *So $u = 3t$ and $du = 3dt$ or $1/3 \; du = dt$*
  - *Further, $u = 3t$ means $1/3 \; u = t$*
  - *Make the substitutions in the second integral.*

$$\int_1^5 1/t \; dt \;=\; \int_{t=1}^{t=5} \frac{1}{1/3 \; u} \frac{1}{3} \; du = \int_{t=1}^{t=5} \frac{3}{u} \frac{1}{3} \; du$$
$$=\; \int_{t=1}^{t=5} \frac{1}{u} \; du$$

- *Now switch the lower and upper limits of integration to u values. We didn't do this before although we could have. This gives*

$$\int_1^5 1/t \; dt \;=\; \int_{t=1}^{t=5} \frac{1}{u} \; du = \int_{u=3}^{u=15} \frac{1}{u} \; du.$$

- *Now, as we have said, the choice of letter for the name of variable in the Riemann integral does not matter. The integral above is the **area under the curve** $f(u) = 1/u$ **between** 3 **and** 15 which is exactly the same as **area under the curve** $f(y) = 1/y$ **between** 3 **and** 15 and indeed the same as **area under the curve** $f(t) = 1/t$ **between** 3 **and** 15. So we have*

$$\int_1^5 1/t \; dt \;=\; \int_3^{15} \frac{1}{u} \; du = \int_3^{15} \frac{1}{t} \; dt.$$

- *Now plug this new version of the second integral back into the original sum.*

$$\ln(3) + \ln(5) \;=\; \int_1^3 1/t \; dt \;+\; \int_1^5 1/t \; dt$$
$$=\; \int_1^3 1/t \; dt \;+\; \int_3^{15} 1/u \; du$$
$$=\; \int_1^3 1/t \; dt \;+\; \int_3^{15} 1/t \; dt$$

- But $\int_1^{15} 1/t\,dt = \ln(15)$ **so we have shown** $\ln(3) + \ln(5) = \ln(3 \times 5) = \ln(15)$.

Example 23.6.6 *Show* $\ln(4) + \ln(12) = \ln(4 \times 12) = \ln(48)$.

Solution • *First, rewrite using the definition.*

$$\ln(4) + \ln(12) = \int_1^4 1/t\,dt + \int_1^{12} 1/t\,dt.$$

- *Now rewrite the second integral $\int_1^{12} 1/t\,dt$ so it starts at 4. This requires a change of variable.*
  - *Let $u = 4t$. Then when $t = 1$, $u = 4$ and when $t = 12$, $u = 4 \times 12 = 48$.*
  - *So $u = 4t$ and $du = 4dt$ or $1/4\,du = dt$.*
  - *Further, $u = 4t$ means $1/4\,u = t$*
  - *Make the substitutions in the second integral.*

$$\int_1^{12} 1/t\,dt = \int_{t=1}^{t=12} \frac{1}{1/4\,u}\frac{1}{4}\,du = \int_{t=1}^{t=12} \frac{4}{u}\frac{1}{4}\,du$$
$$= \int_{t=1}^{t=12} \frac{1}{u}\,du$$

- *Now switch the lower and upper limits of integration to u values. We didn't do this before although we could have. This gives*

$$\int_1^{12} 1/t\,dt = \int_{t=1}^{t=12} \frac{1}{u}\,du = \int_{u=4}^{u=48} \frac{1}{u}\,du.$$

- *Now, as we have said, the choice of letter for the name of variable in the Riemann integral does not matter. The integral above is the **area under the curve** $f(u) = 1/u$ **between** 4 **and** 48 which is exactly the same as **area under the curve** $f(y) = 1/y$ **between** 4 **and** 48 *and indeed the same as **area under the curve** $f(t) = 1/t$ **between** 4 **and** 48*. So we have*

$$\int_1^{12} 1/t\,dt = \int_4^{48} \frac{1}{u}\,du = \int_4^{48} \frac{1}{t}\,dt.$$

- *Now plug this new version of the second integral back into the original sum.*

$$\ln(4) + \ln(12) = \int_1^4 1/t\,dt + \int_1^{12} 1/t\,dt$$
$$= \int_1^4 1/t\,dt + \int_4^{48} 1/u\,du$$
$$= \int_1^4 1/t\,dt + \int_4^{48} 1/t\,dt$$

- But $\int_1^{48} 1/t\,dt = \ln(48)$ **so we have shown** $\ln(4) + \ln(12) = \ln(4 \times 12) = \ln(48)$.

Homework

Follow the steps above to show

Exercise 23.6.1

$\ln(2) + \ln(7) = \ln(14)$

Exercise 23.6.2

$\ln(3) + \ln(9) = \ln(27)$

Exercise 23.6.3

$\ln(8) + \ln(10) = \ln(80)$

Exercise 23.6.4

$\ln(18) + \ln(3) = \ln(54)$

23.6.6.2   Adding Logarithms: One Logarithm Less than 1 and one bigger than 1

Example 23.6.7 *Show* $\ln(1/4) + \ln(5) = \ln((1/4) \times 5) = \ln((5/4))$.

Solution   • *First, rewrite using the definition.*

$$\ln(1/4) + \ln(5) \quad = \quad \int_1^{1/4} 1/t\, dt + \int_1^5 1/t\, dt.$$

• *Now rewrite the second integral* $\int_1^5 1/t dt$ *so it starts at* $1/4$. *This requires a change of variable.*

   – *Let* $u = 1/4\, t$. *Then when* $t = 1$, $u = 1/4$ *and when* $t = 5$, $u = 1/4 \times 5 = 5/4$.
   – *So* $u = 1/4t$ *and* $du = 1/4 dt$ *or* $4\, du = dt$
   – *Further,* $u = 1/4\, t$ *means* $4\, u = t$
   – *Make the substitutions in the second integral.*

$$\int_1^5 1/t\, dt \quad = \quad \int_{t=1}^{t=5} \frac{1}{4\,u}\, 4\, du = \int_{t=1}^{t=5} \frac{1}{u}\, du$$

• *Now switch the lower and upper limits of integration to* u *values. We didn't do this before although we could have. This gives*

$$\int_1^5 1/t\, dt \quad = \quad \int_{t=1}^{t=5} \frac{1}{u}\, du = \int_{u=1/4}^{u=5/4} \frac{1}{u}\, du.$$

• *Now, as we have said, the choice of letter for the name of variable in the Riemann integral does not matter. The integral above is the* **area under the curve** $f(u) = 1/u$ **between** $1/4$ **and** $5/4$ *which is exactly the same as* **area under the curve** $f(y) = 1/y$ **between** $1/4$ **and** $5/4$ *and indeed the same as* **area under the curve** $f(t) = 1/t$ **between** $1/4$ **and** $5/4$. *So we have*

$$\int_1^5 1/t\, dt \quad = \quad \int_{1/4}^{5/4} \frac{1}{u}\, du = \int_{1/4}^{5/4} \frac{1}{t}\, dt.$$

• *Now plug this new version of the second integral back into the original sum.*

$$\ln(1/4) + \ln(5) \quad = \quad \int_1^{1/4} 1/t\, dt + \int_1^5 1/t\, dt$$

$$= \quad \int_1^{1/4} 1/t\, dt + \int_{1/4}^{5/4} 1/u\, du$$

$$= -\int_{1/4}^{1} 1/t \, dt + \int_{1/4}^{5/4} 1/t \, dt$$

$$= -\int_{1/4}^{1} 1/t \, dt + \int_{1/4}^{1} 1/t \, dt + \int_{1}^{5/4} 1/t \, dt$$

$$= \int_{1}^{5/4} 1/t \, dt$$

- *But* $\int_{1}^{5/4} 1/t \, dt = \ln(5/4)$ **so we have shown** $\ln(1/4) + \ln(5) = \ln(1/4 \times 5) = \ln(5/4)$.

Homework

Follow the steps above to show

Exercise 23.6.5

$\ln(.5) + \ln(7) = \ln(3.5)$

Exercise 23.6.6

$\ln(.2) + \ln(3) = \ln(.6)$

Exercise 23.6.7

$\ln(.6) + \ln(9) = \ln(5.4)$

Exercise 23.6.8

$\ln(.3) + \ln(12) = \ln(3.6)$

23.6.6.3   Generalizing These Results

Consider $\ln(1/7 \times 7)$. Since $1/7 < 1$ and $7 > 1$, this is the case above.

- $\ln(1/7) + \ln(7) = \ln(1/7 \times 7) = \ln(1) = 0$.

- So $\ln(1/7) = -\ln(7)$.

- In general, if $a > 0$, then $\ln(1/a) = -\ln(a)$.

- So we can now handle the sum of two logarithms of numbers less than one. $\ln(1/5) + \ln(1/7) = -\ln(5) - \ln(7) = -(\ln(5) + \ln(7)) = -ln(35) = ln(1/35)$!

We can combine all these results into one general rule: for any $a > 0$ and $b > 0$,

$$\ln(a) + \ln(b) \quad = \quad \ln(a \times b).$$

Hence, now we can do subtractions:

- $\ln(7) - \ln(5) = \ln(7) + \ln(1/5) = \ln(7/5)$

- $\ln(1/9) - \ln(1/2) = -\ln(9) - \ln(1/2) = -\ln(9/2) = ln(2/9)$.

So in general, for any $a > 0$ and $b > 0$:

$$\ln(a) - \ln(b) \quad = \quad \ln(a/b).$$

We can summarize all these results as follows:

Proposition 23.6.1 Adding Natural Logarithms

If a and b are two positive numbers, then

$$\ln(a) + \ln(b) = \int_1^a \frac{1}{t}\,dt + \int_1^b \frac{1}{t}\,dt$$

$$= \int_1^{ab} \frac{1}{t}\,dt = \ln(ab).$$

What happens if we subtract areas? We handle this like before. For any $a > 0$, we know $0 = \ln(1) = \ln(a\,(1/a)) = \ln(a) + \ln((1/a))$. Hence, $\ln((1/a)) = -\ln(a)$. From this, for any a and b which are positive, we have $\ln(a) - \ln(b) = \ln(a) + \ln((1/b)) = \ln((a/b))$. This is our general subtraction law.

Proposition 23.6.2 Subtracting Natural Logarithms

If a and b are two positive numbers, then

$$\ln(a) - \ln(b) = \int_1^a \frac{1}{t}\,dt - \int_1^b \frac{1}{t}\,dt$$

$$= \int_1^{a/b} \frac{1}{t}\,dt = \ln(\frac{a}{b}).$$

23.6.6.4 Fractional Powers

We now have two basic rules: for any a and b which are positive, we can say $\ln(a) + \ln(b) = \ln(ab)$. and $\ln(a) - \ln(b) = \ln(a/b)$. We also know that $\ln(1/a) = -\ln(a)$. These simple facts give rise to many interesting consequences. What about fractions like $\ln(a^{p/q})$ for positive integers p and q? Let's do this for a specific fraction like $7/8$. Using Proposition 23.6.1, we see

$$\ln(a^2) = \ln(a \times a) = \ln(a) + \ln(a) = 2\ln(a), \quad \ln(a^3) = \ln(a^2 \times a) = \ln(a^2) + \ln(a) = 2\ln(a).$$

If we keep doing this sort of expansion, we find

$$\ln(a^7) = 7\ln(a).$$

Next, note that

$$\ln(a) = \ln\left(\left(a^{\frac{1}{8}}\right)^8\right) = 8\ln\left(a^{\frac{1}{8}}\right)$$

Dividing the last equation by 8, we find

$$\ln\left(a^{\frac{1}{8}}\right) = \frac{1}{8}\ln(a).$$

Combining these expressions, we have found that

$$\ln(a^{7/8}) = 7\ln(a^{1/8}) = \frac{7}{8}\ln(a).$$

Now the same thing can be done for the general number p/q. It is easy to see by a simple POMI that $\ln(a^p) = p\ln(a)$ when p is a positive integer. We also know by another simple POMI that for any positive integer q, $\ln(a^1) = \ln((a^{1/q})^q) = q\ln(a^{1/q})$. Thus, $\ln(a^{1/q}) = (1/q)\ln(a)$. Combining, we see

$$\ln(a^{p/q}) \;=\; p\ln(a^{1/q}) = (p/q)\ln(a)$$

The discussions above have led us to the proposition that if a is a positive number and $r = p/q$ is any positive rational number, then

$$\ln(a^r) \;=\; r\ln(a).$$

If r is any real number power, we know we can find a sequence of rational numbers which converges to r. We can use this to see $\ln(a^r) = r\ln(a)$. For example, consider $r = \sqrt{2}$. Let's consider how to figure out $\ln(4^{\sqrt{2}})$. We know $\sqrt{2} = 1.414214....$ So the sequence of fractions that converge to $\sqrt{2}$ is $\{p_1/q_1 = 14/10,\ p_2/q_2 = 141/100,\ p_3/q_3 = 1414/1000,\ p_4/q_4 = 14142/10000,\ p_5/q_5 = 141421/100000,\ p_6/q_6 = 1414214/1000000,\ldots\}$. So

$$\lim_{n\to\infty} \ln(4^{p_n/q_n}) \;=\; \lim_{n\to\infty} (p_n/q_n)\ln(4) \implies \sqrt{2}\,\ln(4).$$

Now define $4^x = \exp(\ln(4)\,x)$. Since $\exp(x)$ is continuous, so is $\exp(\ln(4)x)$. So 4^x is continuous and $\lim 4^{p_n/q_n} \longrightarrow 4^{\sqrt{2}}$. Thus, since $\ln$ is continuous at $4^{\sqrt{2}}$, we have by equating limits that

$$\lim_{n\to\infty} \ln(4^{p_n/q_n}) \;=\; \ln(4^{\sqrt{2}}) \implies \ln(4^{\sqrt{2}}) = \sqrt{2}\,\ln(4).$$

This suggests that $\ln(4^{\sqrt{2}}) = \sqrt{2}\,\ln(4)$. This can be shown to be true in general even if the power is negative. We can do this in more generality. Let (x_n) be any sequence of real numbers which converges to the number x. By the continuity of $\exp$ we then know

$$\lim_{n\to\infty} \exp(x_n \ln(a)) \;=\; \exp\left(\left(\lim_{n\to\infty} x_n\right)\ln(a)\right)$$

Thus, we can uniquely define the function a^x by $a^x = \lim_{n\to\infty} \exp(x_n \ln(a))$. In particular if $x_n \to \alpha$, we have $a^\alpha = \exp(\alpha \ln(a))$ is a uniquely defined continuous function which by the chain rule is differentiable. Also, this means the function $e^x = \exp(x\ln(e)) = \exp(x)$ is another way to write our inverse function $\exp(x)$. Let's state this as a Theorem.

Theorem 23.6.3 a^x for $a > 0$

> *For $a > 0$, the function defined by $a^x = exp(\ln(a)x)$ is continuous. In particular $e^x = \exp(\ln(e)x) = \exp(x)$. Note we have a^x is differentiable with $(a^x)' = \ln(a)a^x$.*

Proof 23.6.1
This is easy to see given our discussions above. ∎

Now let α be any real number and let (p_n/q_n) be a sequence of rational numbers which converges to α. We can always find such a sequence because not matter what circle $B_r(\alpha)$ we choose, we can find a rational number within a distance of r from α. We also know $\ln(x)$ is continuous on $(0, \infty)$. Then for a positive number a, we have

$$\lim_{n\to\infty} \ln(a^{p_n/q_n}) \;=\; \lim_{n\to\infty} (p_n/q_n)\ln(a) = \alpha\ln(a)$$

But the function a^x is continuous, so

$$\lim_{n \to \infty} \left(a^{p_n/q_n} \right) \;=\; a^\alpha \implies \ln(a^\alpha) = \alpha \ln(a)$$

This is called the power rule for the logarithm function: $\ln(a^r) = r \ln(a)$ for all positive a and any real number r.

Proposition 23.6.4 Powers

If a is a positive number and r is any real number, then $\ln(a^r) = r \ln(a)$.

Homework

Exercise 23.6.9 *Prove we now know that $x^{1/3}$ is a continuous function whose derivative is $(1/3)x^{-4/3}$.*

Exercise 23.6.10 *Prove we now know that $x^{4/5}$ is a continuous function whose derivative is $(4/5)x^{-1/5}$.*

Exercise 23.6.11 *Prove we now know that $x^{-7/3}$ is a continuous function whose derivative is $(-7/3)x^{-10/3}$.*

Exercise 23.6.12 *Prove we now know that $x^{1/6}$ is a continuous function whose derivative is $(1/6)x^{-5/6}$.*

23.6.7   Logarithm Properties

From our logarithm addition laws, we now know that for any positive integer p, we also see $\ln(e^p)$ is p. Thus, since e^p approaches ∞, we have

$$\lim_{x \to \infty} \ln(x) = \infty$$

Further, for any positive integer p, we also see $\ln(e^{-p})$ is $-p$. Thus, since e^{-p} approaches 0 in the limit as p goes to infinity, we infer

$$\lim_{x \to 0^+} \ln(x) = -\infty$$

We state all of these properties together for convenience.

Theorem 23.6.5 Properties of the Natural Logarithm

The natural logarithm of the real number x satisfies

- ln *is a continuous function of x for positive x,*

- $\lim_{x \to \infty} \ln(x) = \infty$,

- $\lim_{x \to 0+} \ln(x) = -\infty$,

- $\ln(1) = 0$,

- $\ln(e) = 1$,

- $\left(\ln(x)\right)' = \frac{1}{x}$,

- *If x and y are positive numbers then* $\ln(xy) = \ln(x) + \ln(y)$.

- *If x and y are positive numbers then* $\ln\left(\frac{x}{y}\right) = \ln(x) - \ln(y)$.

- *If x is a positive number and y is any real number then* $\ln\left(x^y\right) = y\,\ln(x)$.

Proof 23.6.2
We have already proven these properties.   ∎

23.6.8   The Exponential Function Properties

We already know that the inverse of the natural logarithm function, $e^x = \exp(x) = (\ln)^{-1}(x)$, is defined by the equation

$$e^x = \exp(x) = y \quad \Leftrightarrow \quad \ln(y) = x.$$

So if we wanted to graph the inverse function, exp, we can just take the table of values we compute for ln and switch the roles of the x and y variables. It is also traditional to call the inverse function $\exp(x)$ by the name e^x as we have discussed above. Note from our power rule for logarithms, for the number e and any power x, we have $\ln(e^x) = x\ln(e) = x$ as $\ln(e) = 1$ and $\exp(\ln(e^x)) = e^x$. Since the natural logarithm function has the properties listed in Proposition 23.6.5, we will also be able to derive interesting properties for e^x as well.

It is easy to see $\lim_{x \to \infty} \exp(x) = \infty$ and as the argument x becomes more negative, the value of $\exp(x)$ approaches 0 from above; that is, $\exp(x)$ has a horizontal asymptote of 0. Finally, consider the number $\exp(x + y)$. Now let $\exp(x) = a$, $\exp(y) = b$ and $\exp(x + y) = c$. Then, by the definition of the inverse function $\ln^{-1} = \exp$, we must have

$$\exp(x) = a \quad \Leftrightarrow \quad x = \ln(a), \quad \exp(y) = b \Leftrightarrow y = \ln(b)$$
$$\exp(x + y) = c \quad \Leftrightarrow \quad x + y = \ln(c)$$

Thus, using the properties of the natural logarithm,

$$x + y \;=\; \ln(a) + \ln(b) = \ln(a\,b) = \ln(c)$$

We conclude the c and ab must be the same. Thus,

$$\exp(x+y) \quad = \quad c = ab = \exp(x)\,\exp(y)$$

In a similar way, letting $\exp(x-y) = d$

$$x - y \quad = \quad \ln(a) - \ln(b) = \ln(\frac{a}{b}) = \ln(d)$$

and so d and $\frac{a}{b}$ must be the same. Thus,

$$\exp(x-y) \quad = \quad d = \frac{a}{b} = \frac{\exp(x)}{\exp(y)}$$

From this it immediately follows that

$$\exp(-x) \quad = \quad d = \frac{1}{\exp(x)}$$

Finally, we need to look at $(\exp(x))^y$ for any power y. Let $\exp(x) = a$. Then, as before,

$$\exp(x) \quad = \quad a \quad \Leftrightarrow \quad x = \ln(a)$$

Thus, $x = \ln(a)$ implies $xy = y\ln(a) = \ln(a^y)$. This immediately tells us that $\exp(xy) = a^y$. Combining, we have

$$(\exp(x))^y \quad = \quad a^y = \exp(xy)$$

We can summarize what we have just done in Proposition 23.6.6.

Proposition 23.6.6 Properties of the Exponential Function

The exponential function of the real number x, $\exp(x)$, satisfies

- exp *is a continuous function of x for all x,*
- $\lim_{x \to \infty} \exp(x) = \infty$ *and* $\lim_{x \to -\infty} \exp(x) = 0$,
- $\exp(0) = 1$,
- $(\exp(x))' = \exp(x)$,
- *If x and y are real numbers then* $\exp(x+y) = \exp(x)\,\exp(y)$ *and* $\exp(x-y) = \exp(x)\,\exp(-y) = \frac{\exp(x)}{\exp(y)}$.
- *If x and y are real numbers then* $(\exp(x))^y = \exp(xy)$.

Let's graph the natural logarithm function, $\ln(x)$ and its inverse $exp(x) = e^x$ on the same graph as we show in Figure 23.4 so we can put all this together. Note we have developed these two functions twice now but the results are the same.

Note, another way to derive the logarithm and exponential functions is to define the number e using a sequence limit: $e = \lim_{n \to \infty} (1 + 1/n)^n$. Then we derive the properties of e^x and show it has an inverse which we call $\ln(x)$. Let's assume our new way of doing this via integration gave the function $H(x) = \int_1^x 1/t\,dt$ with inverse $F(x)$ and f is the unique value where $H(f) = 1$. We know

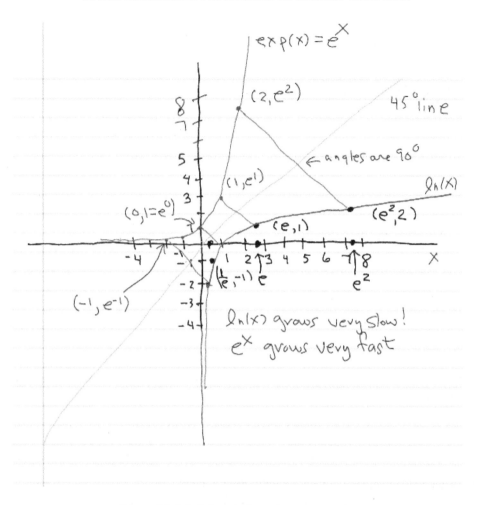

Figure 23.4: $\ln(x)$ and e^x on the same graph.

$F'(x) = F(x)$ always also. So $F^{(n)}(0) = F(0) = 1$ for all n and we also know from our previous work that $(e^x)' = e^x$ and $e^{(n)}(0) = e^0 = 1$ too. The functions $F(x)$ and e^x then have the same Taylor polynomials. We have

$$F(x) \;=\; \sum_{n=0}^{N} x^n/n! + F(c_{N+1})\, x^{N+1}/(N+1)!, \quad e^x = \sum_{n=0}^{N} x^n/n! + e^{d_{N+1}}\, x^{N+1}/(N+1)!$$

where c_{N+1} and d_{N+1} are points between 0 and x.

$$F(x) - e^x \;=\; (F(c_{N+1}) + e^{d_{N+1}})\, \frac{x^{N+1}}{(N+1)!}$$

Now both F and exp are continuous on the interval from 0 to x, so there are constants A_x and B_x which are bounds for them. So we have

$$|F(x) - e^x| \;\le\; (A_x + B_x)\, \frac{x^{N+1}}{(N+1)!} \to 0$$

as $N \to \infty$ since x is fixed. Thus, given $\epsilon > 0$, there is a Q so that $(A_x + B_x)\, \frac{|x|^{N+1}}{(N+1)!} < \epsilon$ if

$N + 1 > Q$. Hence, $|F(x) - e^x| < \epsilon$ with ϵ arbitrary. This tells us $F(x) = e^x$ for this fixed x. Since x was arbitrarily chosen, this means they are always the same. So $F(H(f)) = f$ implies $f = F(1) = e^1$ and so $f = e$.

Another easier way to see this (!!) is to note since $H'(x) = (\ln(x))' = 1/x$, $H(x)$ and $\ln(x)$ are both antiderivatives of $1/x$. Hence, they differ by a constant. But since $H(1) = \ln(1) = 0$, they must be the same. Thus, $H(f) = \ln(f)$ or $1 = \ln(f)$ telling us $f = e$ again. So $e = f$ even though we have used two different ways to find them!

Homework

Exercise 23.6.13 *Solve the initial value problem* $x' = 3x$, $x(0) = 2$ *using our discussions.*

Exercise 23.6.14 *Solve the initial value problem* $x' = -4x$, $x(0) = 8$ *using our discussions.*

Exercise 23.6.15 *If x is a function satisfying $|x(t)| \le Me^{-2t}$ for all $t \ge 0$, find the values of s where* $\lim_{n \to \infty} \int_0^n f(t)e^{-st}dt$ *exists.*

23.7   When Do Riemann Integrals Match?

Now we want to explore when two functions have the same Riemann Integral. You should already expect that changing a Riemann integrable function at a finite number of points should not affect the value of the integral as the "area under the curve does not change". We will now show how to make that conjecture precise.

23.7.1   When Do Two Functions Have the Same Integral?

We now want to find conditions under which the integrals of two functions, f and g, are equal.

Lemma 23.7.1 f **Zero on** (a, b) **Implies Zero Riemann Integral**

Let $f \in B([a, b])$, with $f(x) = 0$ on (a, b). Then f is integrable on $[a, b]$ and

$$\int_a^b f(x)dx = 0.$$

Proof 23.7.1
If f is identically 0, then the result follows easily. Now, assume $f(a) \ne 0$ and $f(x) = 0$ on $(a, b]$. Let $\epsilon > 0$ be given, and let $\delta > 0$ satisfy

$$\delta < \frac{\epsilon}{|f(a)|}.$$

Let $\pi_0 \in \Pi[a, b]$ be any partition such that $\| \pi_0 \| < \delta$. Let $\pi = \{x_0 = a, x_1, \ldots, x_p\}$ be any refinement of π_0. Then $U(f, \pi) = \max(f(a), 0)\Delta x_1$ and $L(f, \pi) = \min(f(a), 0)\Delta x_1$. Hence, we have

$$U(f, \pi) - L(f, \pi) = [\max(f(a), 0) - \min(f(a), 0)]\Delta x_1 = |f(a)| \Delta x_1.$$

But

$$|f(a)| \Delta x_1 < |f(a)| \delta < |f(a)| \frac{\epsilon}{|f(a)|} = \epsilon.$$

Hence, if π is any refinement of π_0, we have $U(f, \pi) - L(f, \pi) < \epsilon$. This shows that $f \in RI([a, b])$. Further, we have

$$U(f, \pi) = \max(f(a), 0)\Delta x_1 \Rightarrow U(f) = \inf_{\pi} U(f, \pi) = 0,$$

since we can make Δx_1 as small as we wish. Likewise, we also see that $L(f) = \sup_{\pi} L(f, \pi) = 0$, implying that

$$U(f) = L(f) = \int_a^b f(x)dx = 0.$$

The case where $f(b) \neq 0$ and $f(x) = 0$ on $[a, b]$ is handled in the same way. So, assume that $f(a), f(b) \neq 0$ and $f(x) = 0$ for $x \in (a, b)$. Let $\epsilon > 0$ be given, and choose $\delta > 0$ such that

$$\delta < \frac{\epsilon}{2 \max\{|f(a)|, |f(b)|\}}.$$

Let π_0 be a partition of $[a, b]$ such that $|\pi_0| < \delta$, and let π be any refinement of π_0. Then

$$U(f, \pi) = \max(f(a), 0)\Delta x_1 + \max(f(b), 0)\Delta x_p$$

$$L(f, \pi) = \min(f(a), 0)\Delta x_1 + \min(f(b), 0)\Delta x_p.$$

It follows that

$$
\begin{aligned}
U(f, \pi) - L(f, \pi) &= [\max(f(a), 0) - \min(f(a), 0)]\Delta x_1 + [\max(f(b), 0) - \min(f(a), 0)]\Delta x_p \\
&= |f(a)| \Delta x_1 + |f(b)| \Delta x_p \\
&< |f(a)| \delta + |f(b)| \delta \\
&< \epsilon.
\end{aligned}
$$

Since we can make Δx_1 and Δx_p as small as we wish, we see

$$\int_a^b f(x)dx = 0.$$

$\blacksquare$

Lemma 23.7.2 $f = g$ on (a, b) Implies Riemann Integrals Match

Let $f, g \in RI([a, b])$ with $f(x) = g(x)$ on (a, b). Then

$$\int_a^b f(x)dx = \int_a^b g(x)dx.$$

Proof 23.7.2
Let $h = f - g$, and apply the previous lemma. $\blacksquare$

Now extend this result to a finite number of mismatches.

Theorem 23.7.3 Two Riemann Integrable Functions Match at All But Finitely Many Points Implies Integrals Match

> Let $f, g \in RI([a,b])$, and assume that $f = g$ except at finitely many points $c_1, \ldots, c_k$.
> Then
> $$\int_a^b f(x)dx = \int_a^b g(x)dx.$$

Proof 23.7.3

We may re-index the points $\{c_1, \ldots, c_k\}$, if necessary, so that $c_1 < c_2 < \cdots < c_k$. Then apply Lemma 23.7.2 on the intervals (c_{j-1}, c_j) for all allowable j. This shows

$$\int_{c_{j-1}}^{c_j} f(t)dt = \int_{c_{j-1}}^{c_j} g(t)dt.$$

Then, since

$$\int_a^b f(t)dt = \sum_{j=1}^k \int_{c_{j-1}}^{c_j} f(t)dt$$

the results follows. ∎

Now let's look at functions which fail to be continuous at various points. The first result is this.

Theorem 23.7.4 f Bounded and Continuous at All But One Point Implies f is Riemann Integrable

> *If f is bounded on $[a,b]$ and continuous except at one point c in $[a,b]$, then f is Riemann integrable.*

Proof 23.7.4

For convenience, we will assume that c is an interior point, i.e. c is in (a,b). We will show that f satisfies the Riemann Criterion and so it is Riemann integrable. Let $\epsilon > 0$ be given. Since f is bounded on $[a,b]$, there is a real number M so that $f(x) < M$ for all x in $[a,b]$. We know f is continuous on $[a, c - \epsilon/(6M)]$ and f is continuous on $[c + \epsilon/(6M), b]$. Thus, f is integrable on both of these intervals and f satisfies the Riemann Criterion on both intervals. For this ϵ there is a partition π_0 of $[a, c - \epsilon/(6M)]$ so that

$$U(f, P) - L(f, P) \quad < \quad \epsilon/3, \ \ \textit{if } \pi_0 \preceq P$$

and there is a partition π_1 of $[c + \epsilon/(6M), b]$ so that

$$U(f, Q) - L(f, Q) \quad < \quad \epsilon/3, \ \ \textit{if } \pi_1 \preceq Q.$$

Let π_2 be the partition we get by combining π_0 with the points $\{c - \epsilon/(6M), c + \epsilon/(6M)\}$ and π_1. Then, we see

$$
\begin{aligned}
U(f, \pi_2) - L(f, \pi_2) &= U(f, \pi_0) - L(f, \pi_0) + \left(\sup_{x \in [c - \epsilon/(6M), c + \epsilon/(6M)]} f(x) \right) \epsilon/3 \\
&\quad + U(f, \pi_1) - L(f, \pi_1) \\
&< \epsilon/3 + M\epsilon/(3M) + \epsilon/3 = \epsilon
\end{aligned}
$$

Then if $\pi_2 \preceq \pi$ on $[a, b]$, we have

$$U(f, \pi) - L(f, \pi) \;\; < \;\; \epsilon$$

This shows f satisfies the Riemann criterion and hence is integrable if the discontinuity c is interior to $[a, b]$. The argument at $c = a$ and $c = b$ is similar but a bit simpler as it only needs to be done from one side. Hence, we conclude f is integrable on $[a, b]$ in all cases.. ∎

It is then easy to extend this result to a function f which is bounded and continuous on $[a, b]$ except at a finite number of points $\{x_1, x_2, \ldots, x_k\}$ for some positive integer k. We state this as Theorem 23.7.5.

Theorem 23.7.5 f Bounded and Continuous at All But Finitely Many Points Implies f is Riemann Integrable

> *If f is bounded on $[a, b]$ and continuous except at finitely many points $\{x_1, x_2, \ldots, x_k\}$ in $[a, b]$, then f is Riemann integrable.*

Proof 23.7.5
We may assume without loss of generality that the points of discontinuity are ordered as $a < x_1 < x_2 < \ldots < x_k < b$. Then f is continuous except at x_1 on $[a, x_1]$ and hence by Theorem 23.7.4 f is integrable on $[a, x_1]$. Now apply this argument on each of the subintervals $[x_{k-1}, x_k]$ in turn. ∎

Homework

Exercise 23.7.1 *Let f be defined by*

$$f(x) \;\; = \;\; \begin{cases} 1, & 0 \le x < 2 \\ 2, & x = 2 \end{cases}$$

and $g(x) = 1$ on $[0, 2]$.

- *Why is f integrable?*

- *Why is the integral of f and g the same and what is the common value?*

Exercise 23.7.2 *Let f be defined by*

$$f(x) \;\; = \;\; \begin{cases} 2x, & 0 \le x < 2 \\ 5, & x = 2 \end{cases}$$

and $g(x) = 2x$ on $[0, 2]$.

- *Why is f integrable?*

- *Why is the integral of f and g the same and what is the common value?*

Exercise 23.7.3 *Let f be defined by*

$$f(x) \;\; = \;\; \begin{cases} x, & 0 \le x < 2 \\ 2, & x = 2 \\ x^2, & 2 < x < 4 \\ 10, & x = 4 \end{cases}$$

- *Why is f integrable?*

- *What is value of the integral?*

Exercise 23.7.4 *Let f be defined by*

$$f(x) = \begin{cases} 9, & x = -2 \\ -2x^2, & -2 < x < 2 \\ 2, & x = 2 \\ x^2, & 2 < x < 4 \\ 10, & x = 4 \end{cases}$$

- *Why is f integrable?*

- *What is value of the integral?*

23.7.2   Calculating the FTOC Function

Now let's look at the Riemann integral of functions which have points of discontinuity. We are going to find the F's in the FTOC for various f's with discontinuities.

23.7.2.1   Removable Discontinuity

Consider the function f defined on $[-2, 5]$ by

$$f(t) = \begin{cases} 2t & -2 \leq t < 0 \\ 1 & t = 0 \\ (1/5)t^2 & 0 < t \leq 5 \end{cases}$$

Let's calculate $F(t) = \int_{-2}^{t} f(s)\, ds$. This will have to be done in several parts because of the way f is defined.

1. On the interval $[-2, 0]$, note that f is continuous except at one point, $t = 0$. Hence, f is Riemann integrable. Also, the function $2t$ is continuous on this interval and hence is also Riemann integrable. Then since f on $[-2, 0]$ and $2t$ match at all but one point on $[-2, 0]$, their Riemann integrals must match. Hence, if t is in $[-2, 0]$, we compute F as follows:

$$F(t) = \int_{-2}^{t} f(s)\, ds = \int_{-2}^{t} 2s\, ds = s^2 \Big|_{-2}^{t} = t^2 - (-2)^2 = t^2 - 4$$

2. On the interval $[0, 5]$, note that f is continuous except at one point, $t = 0$. Hence, f is Riemann integrable. Also, the function $(1/5)t^2$ is continuous on this interval and so is also Riemann integrable. Since f on $[0, 5]$ and $(1/5)t^2$ match at all but one point on $[0, 5]$, their Riemann integrals must match. Hence, if t is in $[0, 5]$, we compute F as follows:

$$F(t) = \int_{-2}^{t} f(s)\, ds = \int_{-2}^{0} f(s)\, ds + \int_{0}^{t} f(s)\, ds$$

$$= \int_{-2}^{0} 2s\, ds + \int_{0}^{t} (1/5)s^2\, ds$$

$$= s^2 \Big|_{-2}^{0} + (1/15)s^3 \Big|_{0}^{t} = -4 + t^3/15$$

Thus, we have found that

$$F(t) = \begin{cases} t^2 - 4 & -2 \le t \le 0 \\ t^3/15 - 4 & 0 \le t \le 5 \end{cases}$$

Since f is Riemann Integrable on $[-2, 5]$, we know from the Fundamental Theorem of Calculus, Theorem 23.1.1, that F must be continuous. Let's check. F is clearly continuous on either side of 0 and we note that $\lim_{t \to 0^-} F(t)$ which is $F(0^-)$ is -4 which is exactly the value of $F(0^+)$. Hence, F is indeed continuous at 0.

What about the differentiability of F? The Fundamental Theorem of Calculus guarantees that F has a derivative at each point where f is continuous and at those points $F'(t) = f(t)$. Hence, we know this is true at all t except perhaps at 0. Note at those t, we find

$$F'(t) = \begin{cases} 2t & -2 \le t < 0 \\ (1/5)t^2 & 0 < t \le 5 \end{cases}$$

which is exactly what we expect. Also, note $F'(0^-) = 0$ and $F'(0^+) = 0$ as well. Hence, since the right and left hand derivatives match, we see $F'(0)$ does exist and has the value 0. But this is not the same as $f(0) = 1$. Note, F is **not** the antiderivative of f on $[-2, 5]$ because of this mismatch.

23.7.2.2 Jump Discontinuity

Now consider the function f defined on $[-2, 5]$ by

$$f(t) = \begin{cases} 2t & -2 \le t < 0 \\ 1 & t = 0 \\ 2 + (1/5)t^2 & 0 < t \le 5 \end{cases}$$

Let's calculate $F(t) = \int_{-2}^{t} f(s)\, ds$. Again, this will have to be done in several parts because of the way f is defined.

1. On the interval $[-2, 0]$, note that f is continuous except at one point, $t = 0$. Hence, f is Riemann integrable. Also, the function $2t$ is continuous on this interval and therefore is also Riemann integrable. Then since f on $[-2, 0]$ and $2t$ match at all but one point on $[-2, 0]$, their Riemann integrals must match. Hence, if t is in $[-2, 0]$, we compute F as follows:

$$F(t) = \int_{-2}^{t} f(s)\, ds = \int_{-2}^{t} 2s\, ds = s^2 \Big|_{-2}^{t}$$
$$= t^2 - (-2)^2 = t^2 - 4$$

2. On the interval $[0, 5]$, note that f is continuous except at one point, $t = 0$. Hence, f is Riemann integrable. Also, the function $2 + (1/5)t^2$ is continuous on this interval and so is Riemann integrable. Then since f on $[0, 5]$ and $2 + (1/5)t^2$ match at all but one point on $[0, 5]$, their Riemann integrals must match. Hence, if t is in $[0, 5]$, we compute F as follows:

$$F(t) = \int_{-2}^{t} f(s)\, ds$$
$$= \int_{-2}^{0} f(s)\, ds + \int_{0}^{t} f(s)\, ds$$

$$= \int_{-2}^{0} 2s \, ds + \int_{0}^{t} (2 + (1/5)s^2) \, ds$$

$$= s^2 \Big|_{-2}^{0} + (2s + (1/15)s^3) \Big|_{0}^{t}$$

$$= -4 + 2t + t^3/15$$

Thus, we have found that

$$F(t) = \begin{cases} t^2 - 4 & -2 \le t \le 0 \\ -4 + 2t + t^3/15 & 0 \le t \le 5 \end{cases}$$

We can see F is continuous at $t = 0$ as expected from the FTOC. What about the differentiability of F? The Fundamental Theorem of Calculus guarantees that F has a derivative at each point where f is continuous and at those points $F'(t) = f(t)$. Hence, we know this is true at all t except 0. Note at those t, we find

$$F'(t) = \begin{cases} 2t & -2 \le t < 0 \\ 2 + (1/5)t^2 & 0 < t \le 5 \end{cases}$$

which is exactly what we expect. However, when we look at the one sided derivatives, we find $F'(0^-) = 0$ and $F'(0^+) = 2$. Since the right and left hand derivatives do not match, we see $F'(0)$ does not exist. Finally, note F is **not** the antiderivative of f on $[-2, 5]$ because of this mismatch.

23.7.2.3 Homework

Exercise 23.7.5 *Compute* $F(t) = \int_{-3}^{t} f(s) \, ds$ *for*

$$f(t) = \begin{cases} 3t & -3 \le t < 0 \\ 6 & t = 0 \\ (1/6)t^2 & 0 < t \le 6 \end{cases}$$

1. *Graph f and F carefully labeling all interesting points.*

2. *Verify that F is continuous and differentiable at all points but $F'(0)$ does not match $f(0)$ and so F is not the antiderivative of f on $[-3, 6]$.*

Exercise 23.7.6 *Compute* $F(t) = \int_{2}^{t} f(s) \, ds$ *for*

$$f(t) = \begin{cases} -2t & 2 \le t < 5 \\ 12 & t = 5 \\ 3t - 25 & 5 < t \le 10 \end{cases}$$

1. *Graph f and F carefully labeling all interesting points.*

2. *Verify that F is continuous and differentiable at all points but $F'(5)$ does not match $f(5)$ and so F is not the antiderivative of f on $[2, 10]$.*

Exercise 23.7.7 *Compute* $F(t) = \int_{-3}^{t} f(s) \, ds$ *for*

$$f(t) = \begin{cases} 3t & -3 \le t < 0 \\ 6 & t = 0 \\ (1/6)t^2 + 2 & 0 < t \le 6 \end{cases}$$

1. *Graph f and F carefully labeling all interesting points.*

2. *Verify that F is continuous and differentiable at all points except 0 and so F is not the antiderivative of f on $[-3, 6]$.*

Exercise 23.7.8 *Compute $F(t) = \int_2^t f(s)\, ds$ for*

$$
f(t) \;=\; \begin{cases} -2t & 2 \le t < 5 \\ 12 & t = 5 \\ 3t & 5 < t \le 10 \end{cases}
$$

1. *Graph f and F carefully labeling all interesting points.*

2. *Verify that F is continuous and differentiable at all points except 5 and so F is not the antiderivative of f on $[2, 10]$.*

Chapter 24

Convergence of Sequences of Functions

We have already discussed convergence of sequences in a metric space in Section 20.3 so it should be a familiar concept. In particular, we looked at convergence of sequences of continuous functions on the interval $[a, b]$ in the $|| \cdot ||_\infty$ metric and showed in Theorem 20.3.1 that such a sequence converges to a continuous function. We then showed $(C([a, b]), || \cdot ||_\infty)$ is a complete metric space in Theorem 20.4.1. This type of convergence, which is based on properties that uniformly continuous functions have on $[a, b]$ is also called **uniform convergence**. We next prove a fundamental approximation result for continuous functions called the **Weierstrass Approximation Theorem**.

24.1 The Weierstrass Approximation Theorem

This result is indispensable in modern analysis. Fundamentally, it states that a continuous real-valued function defined on a compact set can be uniformly approximated by a smooth function. This is used throughout analysis to prove results about various functions. We can often verify a property of a continuous function, f, by proving an analogous property of a smooth function that is uniformly close to f. We will only prove the result for a closed finite interval in $\Re$. The general result for a compact subset of a more general set called a *Topological Space* is a modification of this proof which is actually not that much more difficult, but that is another story. We follow the development of (Simmons (18) 1963) for this proof.

Theorem 24.1.1 Weierstrass Approximation Theorem

> *Let f be a continuous real-valued function defined on $[0, 1]$. For any $\epsilon > 0$, there is a polynomial, p, such that $|f(t) - p(t)| < \epsilon$ for all $t \in [0, 1]$, that is $|| p - f ||_\infty < \epsilon$*

Proof 24.1.1
We first derive some equalities. We will denote the interval $[0, 1]$ by I. By the binomial theorem, for any $x \in I$, we have

$$\sum_{k=0}^{n} \binom{n}{k} x^k (1-x)^{n-k} = (x + 1 - x)^n = 1. \tag{α}$$

Differentiating both sides of Equation α, we get

$$
\begin{aligned}
0 &= \sum_{k=0}^{n} \binom{n}{k}\left(kx^{k-1}(1-x)^{n-k} - x^k(n-k)(1-x)^{n-k-1}\right) \\
&= \sum_{k=0}^{n} \binom{n}{k}x^{k-1}(1-x)^{n-k-1}\left(k(1-x) - x(n-k)\right) \\
&= \sum_{k=0}^{n} \binom{n}{k}x^{k-1}(1-x)^{n-k-1}\left(k - nx\right)
\end{aligned}
$$

Now, multiply through by $x(1-x)$, to find

$$
0 = \sum_{k=0}^{n} \binom{n}{k}x^k(1-x)^{n-k}(k-nx).
$$

Differentiating again, we obtain

$$
0 = \sum_{k=0}^{n} \binom{n}{k}\frac{d}{dx}\left(x^k(1-x)^{n-k}(k-nx)\right).
$$

This leads to a series of simplifications. It is pretty messy and many texts do not show the details, but we think it is instructive.

$$
\begin{aligned}
0 &= \sum_{k=0}^{n} \binom{n}{k}\left[-nx^k(1-x)^{n-k} + (k-nx)\left((k-n)x^k(1-x)^{n-k-1} + kx^{k-1}(1-x)^{n-k}\right)\right] \\
&= \sum_{k=0}^{n} \binom{n}{k}\left[-nx^k(1-x)^{n-k} + (k-nx)(1-x)^{n-k-1}x^{k-1}\left((k-n)x + k(1-x)\right)\right] \\
&= \sum_{k=0}^{n} \binom{n}{k}\left(-nx^k(1-x)^{n-k} + (k-nx)^2(1-x)^{n-k-1}x^{k-1}\right) \\
&= -n\sum_{k=0}^{n} \binom{n}{k}x^k(1-x)^{n-k} + \sum_{k=0}^{n} \binom{n}{k}(k-nx)^2x^{k-1}(1-x)^{n-k-1}
\end{aligned}
$$

Thus, since the first sum is 1, we have $n = \sum_{k=0}^{n} \binom{n}{k}(k-nx)^2x^{k-1}(1-x)^{n-k-1}$ and multiplying through by $x(1-x)$, we have $nx(1-x) = \sum_{k=0}^{n} \binom{n}{k}(k-nx)^2x^k(1-x)^{n-k}$ or

$$
\frac{x(1-x)}{n} = \sum_{k=0}^{n} \binom{n}{k}\left(\frac{k-nx}{n}\right)^2 x^k(1-x)^{n-k}
$$

This last equality then leads to the

$$
\sum_{k=0}^{n} \binom{n}{k}\left(x - \frac{k}{n}\right)^2 x^k(1-x)^{n-k} = \frac{x(1-x)}{n} \tag{β}
$$

We now define the nth order Bernstein Polynomial *associated with f by*

$$
B_n(x) = \sum_{k=0}^{n} \binom{n}{k}x^k(1-x)^{n-k}f\left(\frac{k}{n}\right).
$$

Note that

$$f(x) - B_n(x) = \sum_{k=0}^{n} \binom{n}{k} x^k (1-x)^{n-k} \left[f(x) - f\left(\frac{k}{n}\right) \right].$$

Also note that $f(0) - B_n(0) = f(1) - B_n(1) = 0$, so f and B_n match at the endpoints. It follows that

$$|f(x) - B_n(x)| \leq \sum_{k=0}^{n} \binom{n}{k} x^k (1-x)^{n-k} \left| f(x) - f\left(\frac{k}{n}\right) \right|. \tag{γ}$$

Now, f is uniformly continuous on I since it is continuous. So, given $\epsilon > 0$, there is a $\delta > 0$ such that $|x - \frac{k}{n}| < \delta \Rightarrow |f(x) - f(\frac{k}{n})| < \frac{\epsilon}{2}$. Consider x to be fixed in $[0,1]$. The sum in Equation γ has only $n+1$ terms, so we can split this sum up as follows. Let $\{K_1, K_2\}$ be a partition of the index set $\{0,1,...,n\}$ such that $k \in K_1 \Rightarrow |x - \frac{k}{n}| < \delta$ and $k \in K_2 \Rightarrow |x - \frac{k}{n}| \geq \delta$. Then

$$|f(x) - B_n(x)| \leq \sum_{k \in K_1} \binom{n}{k} x^k (1-x)^{n-k} \left| f(x) - f\left(\frac{k}{n}\right) \right| + \sum_{k \in K_2} \binom{n}{k} x^k (1-x)^{n-k} \left| f(x) - f\left(\frac{k}{n}\right) \right|.$$

which implies

$$|f(x) - B_n(x)| \leq \frac{\epsilon}{2} \sum_{k \in K_1} \binom{n}{k} x^k (1-x)^{n-k} + \sum_{k \in K_2} \binom{n}{k} x^k (1-x)^{n-k} \left| f(x) - f\left(\frac{k}{n}\right) \right|$$

$$= \frac{\epsilon}{2} + \sum_{k \in K_2} \binom{n}{k} x^k (1-x)^{n-k} \left| f(x) - f\left(\frac{k}{n}\right) \right|.$$

Now, f is bounded on I, so there is a real number $M > 0$ such that $|f(x)| \leq M$ for all $x \in I$. Hence

$$\sum_{k \in K_2} \binom{n}{k} x^k (1-x)^{n-k} \left| f(x) - f\left(\frac{k}{n}\right) \right| \leq 2M \sum_{k \in K_2} \binom{n}{k} x^k (1-x)^{n-k}.$$

Since $k \in K_2 \Rightarrow |x - \frac{k}{n}| \geq \delta$, using Equation β, we have

$$\delta^2 \sum_{k \in K_2} \binom{n}{k} x^k (1-x)^{n-k} \leq \sum_{k \in K_2} \binom{n}{k} \left(x - \frac{k}{n} \right)^2 x^k (1-x)^{n-k} \leq \frac{x(1-x)}{n}.$$

This implies that

$$\sum_{k \in K_2} \binom{n}{k} x^k (1-x)^{n-k} \leq \frac{x(1-x)}{\delta^2 n}.$$

and so combining inequalities

$$2M \sum_{k \in K_2} \binom{n}{k} x^k (1-x)^{n-k} \leq \frac{2Mx(1-x)}{\delta^2 n}$$

We conclude then that

$$\sum_{k \in K_2} \binom{n}{k} x^k (1-x)^{n-k} \left| f(x) - f\left(\frac{k}{n}\right) \right| \leq \frac{2Mx(1-x)}{\delta^2 n}.$$

Now, the maximum value of $x(1-x)$ on I is $\frac{1}{4}$, so

$$\sum_{k \in K_2} \binom{n}{k} x^k (1-x)^{n-k} \left| f(x) - f\left(\frac{k}{n}\right) \right| \leq \frac{M}{2\delta^2 n}.$$

Finally, choose n so that $n > \frac{M}{\delta^2 \epsilon}$. Then $\frac{M}{n\delta^2} < \epsilon$ implies $\frac{M}{2n\delta^2} < \frac{\epsilon}{2}$. So, Equation γ becomes

$$| f(x) - B_n(x) | \leq \frac{\epsilon}{2} + \frac{\epsilon}{2} = \epsilon.$$

Note that the polynomial B_n does not depend on $x \in I$, since n only depends on M, δ, and ϵ, all of which, in turn, are independent of $x \in I$. So, B_n is the desired polynomial, as it is uniformly within ϵ of f. ∎

Comment 24.1.1 *A change of variable translates this result to any closed interval $[a, b]$.*

24.1.1 Bernstein Code Implementation

We can write code to implement the Bernstein polynomial approximations on the general interval $[a, b]$ as follows. We implement the function `Bernstein(f,a,b,n)` which calculates the polynomial $B_n(f)$ on the interval $[a, b]$. Note in this code, the $B_n(f)$ formula is altered to make it work on $[a, b]$ by taking $x \in [a, b]$ and shifting it to $y = (x - a)/(b - a)$. Also, the function evaluations $f(k/n)$ need to be adjusted so that we sample f on the interval $[a, b]$. So $f(k/n)$ is replaced by $f(a + k (b - a)/n)$. Since the value of n is not known, the function is defined recursively using a `for` loop. We use auxiliary functions in the loop to make it easier to build the successive p functions too.

Listing 24.1: The Bernstein Polynomial Function

```
   function p = Bernstein (f,a,b,n)
   % compute Bernstein polynomial approximation
   % of order n on the interval [a,b] to f
   % f is the function
 5 % n is the order of the Bernstein polynomial
   p = @(x) 0;
   for i = 1:n+1
     k = i-1;
     % convert the interval [a,b] to [0,1] here
10   y = @(x) (x-a)/(b-a);
     % convert the points we evaluate f at to be in [a,b]
     % instead of [0,1]
     z = a + k*(b-a)/n;
     q = @(x) nchoosek(n,k)*(y(x).^k).*((1-y(x)).^(n-k))*f(z);
15   p = @(x) (p(x) + q(x) );
   end
   end
```

This code is not very good as it uses the built in $\binom{n}{k}$ function `nchoosek` for the computations. On most laptops this will stall out at about $n \approx 200$ and we usually need a higher n value to do the job. An interesting exercise we encourage you to do is to rewrite this to make it better. A modest amount of work will lead you to code that can easily handle $n \approx 3000$. So give it a shot! We can then use this code to build a sample set of approximations. We use $f(x) = e^{-.3x} \cos(2x + 0.3)$

on the interval $[0, 10]$ as our example problem. We calculate $B_3(f)$, $B_{10}(f)$, $B_{25}(f)$, $B_{45}(f)$ and $B_{150}(f)$ and compare these polynomials to f in Figure 24.1. We deliberately tried to approximate an oscillating function with the Bernstein polynomials and we expected this to be difficult because we chose the interval $[0, 10]$ to work on.

The Bernstein polynomials require us to calculate the binomial coefficients. When n is large. such as in our example with $n = 150$, running this code generates errors about the possible loss of precision in the use of `nchoosek` which is the binomial coefficient function.

Listing 24.2: **Bernstein Polynomial Runtime Results**

```
   X = linspace(0,10,401);
   f = @(x) e.^(-.3*x).*cos(2*x + 0.3);
   B3 = Bernstein(f,0,10,3);
   plot(X,f(X),X,B3(X));
 5 B10 = Bernstein(f,0,10,10);
   plot(X,f(X),X,B3(X),X,B10(X));
   B25 = Bernstein(f,0,10,25);
   plot(X,f(X),X,B3(X),X,B10(X),X,B25(X));
   B45 = Bernstein(f,0,10,45);
10 plot(X,f(X),X,B3(X),X,B10(X),X,B25(X),X,B45(X));
   B150 = Bernstein(f,0,10,150);
   plot(X,f(X),X,B3(X),X,B10(X),X,B25(X),X,B45(X),X,B150(X));
   legend('f','B3','B10','B25','B45','B140');
   xlabel('x');   ylabel('y');
15 title('f(x) = e^{-3x}cos(2x + 0.3) on [0,10] and Bernstein Polynomials
        ');
```

Homework

Since we use the built in binomial coefficient function `nchoosek`, we have loss of precision warnings that show up starting at about $n = 200$ when you do the plots. You can fix this by reimplementing the binomial coefficient calculation using your own function. We encourage you to do this. You should be able to get it to work up to at least $N = 3000$!

Exercise 24.1.1 *For $f(x) = \sin^2(3x)$, graph the Bernstein polynomials $B_{150}(f)$, $B_{200}(f)$ and $B_{300}(f)$ along with f simultaneously on the interval $[-2, 4]$ on the same graph in MATLAB. This is a word doc report so write your code, document it and print out the graph as part of your report.*

Exercise 24.1.2 *For $f(x) = \sin(x) + \sqrt{x}$, graph the Bernstein polynomials $B_{200}(f)$, $B_{250}(f)$ and $B_{300}(f)$ along with f simultaneously on the interval $[0, 6]$ on the same graph in MATLAB. This is a word doc report so write your code, document it and print out the graph as part of your report.*

Exercise 24.1.3 *For $f(x) = e^{0.2x} \cos(06x+0.3)$, graph the Bernstein polynomials $B_{200}(f)$, $B_{250}(f)$ and $B_{400}(f)$ along with f simultaneously on the interval $[0, 6]$ on the same graph in MATLAB. This is a word doc report so write your code, document it and print out the graph as part of your report.*

24.1.2   A Theoretical Example

We can do interesting stuff with this approximation result.

Example 24.1.1 *If $\int_a^b f(s)s^n ds = 0$ for all n with f continuous, then $f = 0$ on $[a, b]$.*

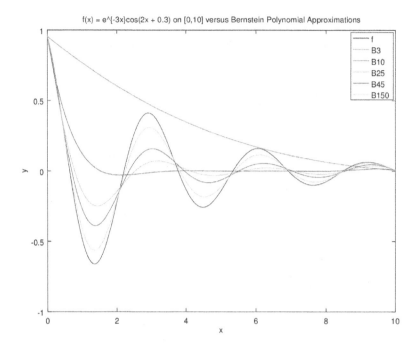

Figure 24.1: Bernstein polynomial approximations to $f(x) = e^{-.3x} \cos(2x + 0.3)$ on the interval $[0, 10]$.

Solution *Let $B_n(f)$ be the Bernstein polynomial of order n associated with f. From the assumption, we see $\int_a^b f(s)B_n(f)(s)ds = 0$ for all n also. Now consider*

$$\left| \int_a^b f^2(s)ds - \int_a^b f(s)B_n(f)(s)ds \right| = \left| \int_a^b f(s)(f(s) - B_n(f)(s))ds \right|$$

$$\leq \int_a^b \|f - B_n(f)\|_\infty |f(s)|ds$$

Then given $\epsilon > 0$, there is a N so that $n > N$ implies $\|f - B_n(f)\|\infty < \epsilon/((\|f\|_\infty + 1)(b - a))$. Then we have $n > N$ implies

$$\left| \int_a^b f^2(s)ds - \int_a^b f(s)B_n(f)(s)ds \right| < \frac{\epsilon}{(\|f\|_\infty + 1)(b - a)} \int_a^b |f(s)|ds$$

$$< \frac{\epsilon}{(\|f\|_\infty + 1)(b - a)} \|f\|_\infty (b - a) < \epsilon$$

This tells us $\int_a^b f(s)B_n(f)(s)ds \rightarrow \int_a^b f^2(s)ds$. Since $\int_a^b f(s)B_n(f)(s)ds = 0$ for all n, we see $\int_a^b f^2(s)ds = 0$. It then follows that $f = 0$ on $[a, b]$ using this argument. Assume f^2 is not zero at some point c in $[a, b]$. We can assume c is an interior point as the argument at an endpoint is similar. Since $f^2(c) > 0$ and f is continuous, there is r with $(r - c, r + c) \subset [a, b]$ and $f^2(x) > f^2(c)/2$ on that interval. Hence

$$0 = \int_a^b f^2(s)ds = \int_a^{c-r} f^2 ds + \int_{c-r}^{c+r} f^2(s)ds + \int_{c+r}^b f^2(s)ds > (2r)f^2(c)/2 > 0$$

This is a contradiction and so $f^2 = 0$ on $[a, b]$ implying $f = 0$ on $[a, b]$.

Homework

Exercise 24.1.4 *Let $f(x) = \sqrt{1 + \sin^2(x)}$.*

- *Approximate f to within 0.1 using a Bernoulli polynomial on $[0, 1]$. Use MATLAB to figure this out.*

- *Then approximate $\int_0^1 f(x)dx$ using this polynomial. You can do this exactly because you know how to integrate the polynomial.*

Exercise 24.1.5 *Let $f(x) = \sqrt{1 + x^2}$.*

- *Approximate f to within 0.1 using a Bernoulli polynomial on $[0, 1]$. Use MATLAB to figure this out.*

- *Then approximate $\int_0^1 f(x)dx$ using this polynomial. You can do this exactly because you know how to integrate the polynomial.*

Exercise 24.1.6 *Let $f(x) = e^{-0.05x^2}$.*

- *Approximate f to within 0.1 using a Bernoulli polynomial on $[-4, 4]$. Use MATLAB to figure this out.*

- *Then approximate $\int_{-4}^4 f(x)dx$ using this polynomial. You can do this exactly because you know how to integrate the polynomial.*

24.2 The Convergence of a Sequence of Functions

Now that we have explored some of the consequences and tools we can develop using uniform convergence, let's step back and distinguish the ideas of a sequence of functions converging at each point in their domain (this is called **pointwise convergence**) and the idea of uniform convergence we have already discussed. It is very helpful to look at a number of examples.

Example 24.2.1 *Consider the sequence of functions (x_n) on $[0, 1]$ by $x_n(t) = t^n$ for all integers $n \geq 0$. For each fixed t, does the sequence of real numbers $(x_n(t))$ converge?*

Solution *If $0 \leq t < 1$, $x_n(t) = t^n \to 0$. However, at $t = 1$, $x_n(1) = 1^n = 1$ for all n. Hence, we have*

$$\lim_{n \to \infty} x_n(t) = \begin{cases} 0, & 0 \leq t < 1 \\ 1, & t = 1 \end{cases}$$

Thus, this sequence of functions converges to the function

$$f(t) = \begin{cases} 0, & 0 \leq t < 1 \\ 1, & t = 1 \end{cases}$$

which is clearly not continuous at $t = 1$ even though each of the x_n is continuous on $[0, 1]$.

This leads to a definition.

Definition 24.2.1 Pointwise Convergence of a Sequence of Functions

> Let (x_n) be a sequence of real valued functions defined on a set S. We say $x : S \to \Re$
> is the **pointwise limit** of this sequence if at each $t \in S$, the sequence of numbers $(x_n(t))$
> converges to the number $x(t)$. In this case, if $\lim_{n\to\infty} x_n(t) = x(t)$ for all $t \in S$, we write
> $x_n \overset{pointwise}{\longrightarrow} x$. We also abbreviate this and just write $x_n \overset{ptws}{\to} x$.

Thus, so far we have seen $t^n \overset{ptws}{\to} x(t)$ on $[0,1]$ where

$$x(t) \;=\; \begin{cases} 0, & 0 \le t < 1 \\ 1, & t = 1 \end{cases}$$

is a discontinuous limit function. Note **pointwise** convergence of the sequence of functions (x_n)
means that at each t, given $\epsilon > 0$, there is $N_{\epsilon,t}$ so that

$$n \;>\; N_{\epsilon,t} \implies |x_n(t) - x(t)| < \epsilon.$$

Note that this value of integer $N_{\epsilon,t}$, in general, will depend on both the choice of ϵ and t which is
why it is written with the double subscript.

Example 24.2.2 *Consider the sequence of functions (x_n) on $[0,1]$ by $x_n(t) = t^n$ for all integers*
$n \ge 0$. For each fixed t, examine how the sequence converges at a given t.

Solution *Fix $t = 1/2$. We know $x_n(t) = t^n \to 0$ at this point. What is $N_{\epsilon,0.5}$? We want*

$$\left| \left(\frac{1}{2}\right)^n - 0 \right| \;<\; \epsilon \implies n > \log_2\left(\frac{1}{\epsilon}\right).$$

Thus, we would choose $N_{\epsilon,0.5} = \left\lceil \log_2(\frac{1}{\epsilon}) \right\rceil$, the smallest integer above or equal to $\log_2(\frac{1}{\epsilon})$.

On the other hand, at $t = 1/3$, we have $(1/3)^n \to 0$ also and we find $N_{\epsilon,1/3} = \left\lceil \log_3(\frac{1}{\epsilon}) \right\rceil$. Can we
find one N_ϵ that will work for all $t \in [0.1]$? If so, for the given ϵ, we want to find N_ϵ so that

$$n > N_\epsilon \implies |t^n - x(t)| < \epsilon, \quad 0 \le t \le 1.$$

We then have

$$|t^n - x(t)| \;=\; \begin{cases} |t^n| = t^n, & 0 \le t < 1 \\ |1^n - 1| = 0, & t = 1 \end{cases} \;=\; \begin{cases} t^n, & 0 \le t < 1 \\ 0, & t = 1 \end{cases}$$

If we could find such an N_ϵ, we would have $t^n < \epsilon$ if $n > N_\epsilon$ for $0 \le t < 1$. That would imply for a
fixed $n > N_\epsilon$ that $\sup_{0 \le t < 1}(t^n) \le \epsilon$. But this supremum is 1 which would force $1 \le \epsilon$ which would
fail if we chose $\epsilon < 1$. We conclude we can't find such a uniform *value of N_ϵ.*

Another way to look at this is that if we could find such a uniform choice of $N_\epsilon/2$, we could say

$$|x_n(t) - x(t)| \;<\; \epsilon/2, \quad 0 \le t \le 1$$

which would tell us $||x_n - x||_\infty \le \epsilon/2 < \epsilon$ on $[0,1]$. **But then Theorem 20.3.1 would tell us**
the limit function x would have to be continuous which we know is not true. Hence, we have
another way of seeing we can't choose one N_ϵ that will work for this example.

Let's look at $|| \cdot ||_\infty$ from another point of view.

Definition 24.2.2 The Uniform Convergence of a Sequence of Functions

Let (x_n) for $n \geq 1$ be a sequence of real valued functions defined on the set S. We say $x : S \to \Re$ is the **uniform limit** of this sequence if for all $\epsilon > 0$, there is a N_ϵ so that

$$n > N_\epsilon \implies \sup_{t \in S} |x_n(t) - x(t)| < \epsilon$$

Note this is the same as saying

$$n > N_\epsilon \implies ||x_n - x||_\infty < \epsilon$$

where $|| \cdot ||_\infty = \sup_{t \in S} |x_n(t) - x(t)|$. We also write $x_n \overset{uniformly}{\longrightarrow} x$ to denote this type of convergence. Often this is abbreviated to $x_n \overset{unif}{\to} x$ or even $x_n \overset{u}{\to} x$. In all of these statements the underlying set S is assumed and it is usually easy to know that from context.

We thus know, for our example, $x_n(t) = t^n \overset{ptws}{\to} x(t)$ but $x_n \overset{unif}{\not\to} x$.

Comment 24.2.1 *If the limit function is not continuous but all the functions x_n are, then the convergence can not be uniform. So this is a good test for the lack of uniform convergence.*

It is now time for another example.

Example 24.2.3 *Let $x_n(t) = e^{-nt}$ for $0 \leq t \leq 1$ and $n \geq 1$. Examine the convergence of this sequence.*

Solution *First, let's check for pointwise convergence to a limit function x. At $t = 0$, $x_n(0) = 1$ for all n, so*

$$\lim_{n \to \infty} x_n(t) = \begin{cases} 1, & t = 0 \\ 0, & 0 < t \leq 1 \end{cases}$$

We see the pointwise limit function x should be

$$x(t) = \begin{cases} 1, & t = 0 \\ 0, & 0 < t \leq 1 \end{cases}$$

which is clearly not continuous on $[0, 1]$. Hence $x_n \overset{unif}{\not\to} x$, but $e^{-nt} \overset{ptws}{\to} x(t)$. Note,

$$\sup_{0 \leq t \leq 1} |x_n(t) - x(t)| = \sup_{0 \leq t \leq 1} \begin{cases} |1 - 1|, & t = 0 \\ |e^{-nt} - 0|, & 0 < t \leq 1 \end{cases} = \sup_{0 \leq t \leq 1} \begin{cases} 0, & t = 0 \\ 1, & 0 < t \leq 1 \end{cases} = 1$$

This says $||x_n - x||_\infty = 1$ for all n! We could also repeat our arguments in the previous example to show we cannot find a single N_ϵ to use in the definition of a uniform limit.

A simpler way to do it is to note the limit function is not continuous even though the x_n are. Hence, the convergence can not be uniform.

24.2.1 Homework

Exercise 24.2.1 *Find the pointwise limit function for (x_n) where $x_n(t) = \sin(2t)/n$ on $[0, 3]$. Does this sequence converge uniformly to the limit function?*

Exercise 24.2.2 *Find the pointwise limit function for (x_n) where $x_n(t) = 1 + (2t)^n$ on $[0, 0.5]$. Does this sequence converge uniformly to the limit function?*

Exercise 24.2.3 *Let x_n be defined by*

$$x_n(t) \;=\; \begin{cases} 0, & 0 \le t \le 1/n \\ n, & 1/n < t < 2/n \\ 0, & 2/n \le t \le 1 \end{cases}$$

Graph x_n for $n = 4, 8$ and 10. Then find the pointwise limit of (x_n).
Hint: Fix any $t \in (0, 1]$. Then there is N so that $2/N < t$ implying $x_N(t) = 0$. Use this to figure out the pointwise limit for these t's. Then note you know what happens at $t = 0$ too.

Exercise 24.2.4 *Discuss the convergence of the sequence of functions (x_n), for suitable n, by*

$$x_n(t) \;=\; \begin{cases} n, & 0 \le t \le 1/n \\ 0, & 1/n < t \le 1 \end{cases}$$

Exercise 24.2.5 *If $x_n \overset{unif}{\longrightarrow} x$ and $y_n \overset{unif}{\longrightarrow} y$ uniformly on $[a, b]$, prove $2x_n - 5y_n \overset{unif}{\longrightarrow} 2x - 5y$ on $[a, b]$.*

Exercise 24.2.6 *If $x_n \overset{ptws}{\longrightarrow} x$ and $y_n \overset{ptws}{\longrightarrow} y$ uniformly on $[a, b]$, prove $3x_n - 15y_n \overset{ptws}{\longrightarrow} 3x - 15y$ on $[a, b]$.*

24.2.2 More Complicated Examples

Example 24.2.4 *Let (x_n) be the sequence of functions on $[0, 1]$ defined by $x_n(t) = \frac{2nt}{e^{nt^2}}$. Discuss pointwise and uniform convergence on $[0, 1]$.*

Solution *First, it is easy to see the pointwise limit function is $x(t) = 0$ on $[0, 1]$. Is the convergence uniform? Let's begin by graphing some of these functions in Octave. The code is straightforward.*

Listing 24.3: Sequence Plotting Code on $[0, 1]$

```
T = linspace(0,1,51);
f = @(t,n) 2*n*t./exp(n*t.^2);
plot(T,f(T,5),T,f(T,10),T,f(T,20),T,f(T,30));
xlabel('t'); ylabel('y');
title('x_n(t) = 2nt/e^{nt^2} for n = 5,10,20 and 30');
legend('x5','x10','x20','x30','location','north');
```

In Figure 24.2, you see a plot of x_5, x_{10}, x_{20} and x_{30} on the interval $[0, 1]$. You can clearly see the peaks of the functions are increasing with the maximums occurring closer and closer to 0. The derivative here, after some simplification, is $x'_n(t) = \frac{2n(1 - 2nt^2)}{e^{nt^2}}$ which is zero at $t = 0$ and $t = \pm 1/\sqrt{2n}$. The critical point at $t = 0$ is uninteresting and the maximum occurs at $t_n = 1/\sqrt{2n}$ and has value $\sqrt{2n/e}$.
Let's calculate $\|x_n - x\|_\infty$ here. We have

$$\|x_n - x\|_\infty \;=\; \sup_{0 \le t \le 1} |2nt/e^{nt^2} - 0| = \sqrt{2n/e}$$

For the convergence to be uniform, given $\epsilon > 0$, we would have to be able to find N_ϵ so that $n > N_\epsilon$ implies $\|x_n - x\|_\infty < \epsilon$. Here that means we want $\sqrt{2n/e} < \epsilon$ when $n > N_\epsilon$. But for $n > N_\epsilon$, $\sqrt{2n/e} \to \infty$. So this cannot be satisfied and the convergence is not uniform.

Now let's look at this same sequence on a new interval.

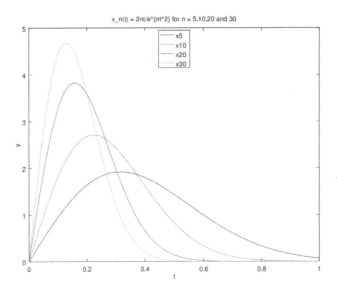

Figure 24.2: $x_n(t) = 2nt/e^{nt^2}$ for $n = 5, 10, 20$ and 30 on $[0, 1]$.

Example 24.2.5 *Examine the convergence of the sequence* $x_n(t) = \frac{2nt}{e^{nt^2}}$ *on* $[-2, 2]$.

Solution *It is easy to see the pointwise limit is still* $x(t) = 0$ *on* $[-2, 2]$*. We can graph some of the functions in this sequence on this new interval using the code below.*

Listing 24.4: **Sequence Plotting Code on** $\left[-2, 2\right]$

```
T = linspace(-2,2,101);
f = @(t,n) 2*n*t./exp(n*t.^2);
plot(T,f(T,5),T,f(T,10),T,f(T,20),T,f(T,30),T,f(T,50));
xlabel('t'); ylabel('y');
title('x_n(t) = 2nt/e^{nt^2} on [-2,2]');
```

In Figure 24.3, you see a plot of some of these functions on the interval $[-2, 2]$*. The critical point analysis is the same and you can clearly see the peaks of the functions are increasing with the minimums and maximums occurring closer and closer to 0. The minimum occur at* $-1/\sqrt{2n}$ *with value* $-\sqrt{2n/e}$*, while the maximums occur at* $1/\sqrt{2n}$ *with value* $\sqrt{2n/e}$*.*

- *Test convergence on* $[0.1, 0.5]$*.*
  Now take the small positive number 0.1 *and mentally imagine drawing a vertical line through Figure 24.3 at that point. The maximum's occur at* $1/\sqrt{2n}$*. If* $1/\sqrt{2n} < .1$ *or* $n > 50$*, the maximum values all occur before the value* $x = .1$*. We generate a plot for this as follows: (here we do not add the code for the axis labels etc.)*

Listing 24.5: **Sequence Plotting Code on** $[0.05, 0.5]$

```
T = linspace(0.05,0.5,101);
f = @(t,n) 2*n*t./exp(n*t.^2);
plot(T,f(T,500),T,f(T,700),T,f(T,900));
```

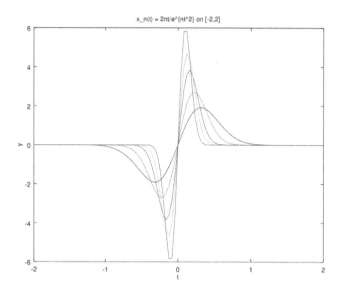

Figure 24.3: $x_n(t) = 2nt/e^{nt^2}$ on $[-2, 2]$.

On the interval $[0.1, 2.0]$, the functions have their maximum value at $x_n(0.1) = .2n/e^{.01n}$. Since $.2n/e^{.01n} \to 0$ as $n \to \infty$, we see given $\epsilon > 0$, there is N_ϵ so that $.2n/e^{.01n} < \epsilon$ when $n > N_\epsilon$. You can see this behavior clearly in Figure 24.4. There we graph x_{500}, x_{700} and x_{900} and you can easily see the value of these functions at 0.1 is decreasing. Hence, for $n > N_\epsilon$, $||x_n - 0||_\infty < \epsilon$ and so we can say $x_n \overset{unif}{\to} 0$ on $[.1, r]$ for any $r > 0.1$. A similar analysis works for any $a > 0$.

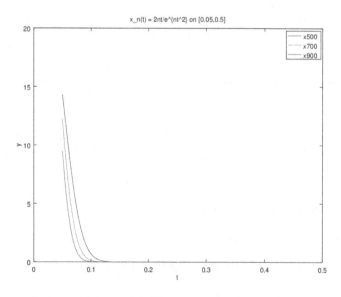

Figure 24.4: $x_n(t) = 2nt/e^{nt^2}$ on $[0.05, 0.5]$.

In fact, if we looked at the other side, we would show $x_n \overset{unif}{\to} 0$ on any interval of the form

$[-r, -a]$ with $a > 0$. Indeed, this convergence is uniform on any interval $[c, d]$ as long as $0 \notin [c, d]$. Now let's look at this same sequence on $[.1, 2]$ analytically. We know the minimum of $x_n(t)$ occur at $-1/\sqrt{2n}$ with value $-\sqrt{2n/e}$, while the maximum of $x_n(t)$ occurs at $1/\sqrt{2n}$ with value $\sqrt{2n/e}$. There is an N so that $n > N$ implies $< 1/\sqrt{2n} < .1$; i.e. the maximum value before $x = .1$. It occurs **before** the interval $[.1, 2]$. Since on the right of the maximum, $x_n(t)$ decreases, this tells us the maximum of $x_n(t)$ on $[.1, 2]$ is given by $x_n(.1)$. So

$$\sup_{t \in [.1.2]} |x_n(t) - 0|| = x_n(.1) = \frac{.2n}{e^{.01n}}$$

and we see this goes to zero with n. So convergence is uniform on $[.1, 2]$.

- For any interval $[a, b]$ that contains 0, the convergence will not be uniform because we can argue like we did for the interval $[0.1]$.

We can do some more examples like this to cement these ideas in place.

Example 24.2.6 Examine the pointwise and uniform convergence of (x_n) where $x_n(t) = 3nt/e^{4nt^2}$ on intervals of $\Re$.

Solution We know this type of convergence should be similar in spirit to the example we just did. Note the pointwise limit function is still 0. The extremal values of each x_n are at the critical points: we find

$$x'_n(t) = \frac{3n - 24n^2t^2}{e^{4nt^2}} = \frac{3n(1 - 8nt^2)}{e^{4nt^2}}$$

Hence, x_n also has a global minimum and global maximum occurring at the critical points $t = \pm 1/\sqrt{8n}$ of value $\pm 3n/\sqrt{8e}$. Following reasoning quite similar to what we did before, the convergence will be uniform on any interval $[c, d]$ which does not contain 0. For example, given $\epsilon > 0$, on the interval $[a, 6]$, the maximums will occur before a when $1/\sqrt{8n} < a$ or $n > 1/(8a^2)$. Let N_1 be chosen larger than $1/(8a^2)$. Then for $n > N_1$, the maximum on $[a, 6]$ always occurs at a with value $3na/e^{4na^2}$. Since $\lim_{n \to \infty} 3na/e^{4na^2} = 0$, there is a N_2 with $3na/e^{4na^2} < \epsilon$ if $n > \max\{N_1, N_2\}$. Then we have

$$\sup_{a \leq t \leq 6} |x_n(t) - 0| \leq 3na/e^{4na^2} < \epsilon$$

Thus, we have shown $||x_n - x||_\infty < \epsilon$ if $n > \max\{N_1, N_2\}$. This tells us the convergence is uniform on $[a, 6]$.

We can do a similar analysis on an interval like $[-8, -a]$ to see convergence is uniform on that interval. Finally if 0 is in the interval $[c, d]$, the extreme behavior of x_n now occurs at $\pm 1/\sqrt{8n}$ with absolute value $3n/\sqrt{8e}$. We see

$$\sup_{c \leq x \leq d} |x_n(t) - x(t)| = 3n/\sqrt{8e}$$

which goes to infinity as $n \to \infty$. So given $\epsilon > 0$, it is not possible for us to find any N so that $n > N$ implies $||x_n - x||_\infty < \epsilon$. We did not use Octave code to illustrate what is going on in this example, but you should easily be able to modify what we did before to get similar pictures.

Example 24.2.7 Examine the pointwise and uniform convergence of (x_n) where $x_n(t) = 2nt/e^{3nt^2}$ on intervals of $\Re$.

Solution *This is almost the same as what we just did for the last two examples and we will leave the details to you.*

Example 24.2.8 *Examine the convergence of (x_n) on $[0,1]$ where $x_n(t) = \sin(nt)/n$.*

Solution *First we can see the pointwise limit is $x(t) = 0$ on $[0,1]$ because $|x_n(t) - x(t)| = |\sin(nt)/n - 0| \le 1/n$. Hence, given $\epsilon > 0$, there is N so that $1/N < \epsilon$ which tells us right away that $|x_n(t) - x(t)| < \epsilon$ when $n > N$.*

This also proves the convergence is uniform as $n > N$ implies

$$\sup_{0 \le t \le 1} |\sin(nt)/n - 0| \le 1/n < \epsilon$$

Homework

Use the examples we have discussed here to answer these problems.

Exercise 24.2.7 *Examine the pointwise and uniform convergence of (x_n) where $x_n(t) = 3nt/e^{5nt^2}$ on intervals of $\Re$.*

Exercise 24.2.8 *Examine the pointwise and uniform convergence of (x_n) where $x_n(t) = 2nt/e^{8nt^2}$ on intervals of $\Re$.*

Exercise 24.2.9 *Examine the pointwise and uniform convergence of (x_n) where $x_n(t) = 9nt/e^{nt^2}$ on intervals of $\Re$.*

Exercise 24.2.10 *Examine the convergence of (x_n) where $x_n(t) = \cos(n^2 t)/n^2$ on $\Re$.*

Exercise 24.2.11 *Examine the convergence of (x_n) where $x_n(t) = 4t + \cos(n^2 t)/n^2$ on $\Re$.*

Exercise 24.2.12 *Examine the convergence of (x_n) where $x_n(t) = 1 + \sin(n^2 t + 3)/n^2$ on $\Re$.*

24.2.3   What Do Our Examples Show Us?

Let's collect what we know from these examples;

- On $[0,1]$,

$$t^n \xrightarrow{\text{ptws}} x(t) = \begin{cases} 0, & 0 \le t < 1 \\ 1, & t = 1 \end{cases}, \quad \sup_{0 \le t \le 1} |t^n - 0| = ||x_n - x||_\infty = 1$$

  which tell us $x_n \xrightarrow{\text{unif}} \mkern-18mu\diagup\ x$.

- On $[0,1]$,

$$e^{-nt} \xrightarrow{\text{ptws}} x(t) = \begin{cases} 1, & t = 0 \\ 0, & 0 < t \le 1 \end{cases}, \quad \sup_{0 \le t \le 1} |e^{-nt} - 0| = ||x_n - x||_\infty = 1$$

  which tell us $x_n \xrightarrow{\text{unif}} \mkern-18mu\diagup\ x$.

- On $[0,1]$,

$$\frac{2nt}{e^{nt^2}} \xrightarrow{\text{ptws}} x(t) = 0, \quad \sup_{0 \le t \le 1} \left| \frac{2nt}{e^{nt^2}} - 0 \right| = ||x_n - x||_\infty = \sqrt{2n/e}$$

  which tell us $x_n \xrightarrow{\text{unif}} \mkern-18mu\diagup\ x$.

- On $[0, 1]$,

$$\frac{\sin(nt)}{n} \xrightarrow{\text{ptws}} 0 = x(t). \quad \sup_{0 \le t \le 1} \left| \frac{\sin(nt)}{n} - 0 \right| = ||x_n - x||_\infty \le 1/n$$

which tell us $x_n \xrightarrow{\text{unif}} x$.

24.2.4 Sequence Convergence and Integration

Now let's integrate these sequences we collected above. We now have

a: On $[0, 1]$,

$$t^n \xrightarrow{\text{ptws}} x(t) = \begin{cases} 0, & 0 \le t < 1 \\ 1, & t = 1 \end{cases}, \quad \sup_{0 \le t \le 1} |t^n - 0| = ||x_n - x||_\infty = 1$$

which tell us $x_n \xrightarrow{\text{unif}} x$. Here each x_n is continuous but the limit function is not continuous. Also

$$\lim_{n \to \infty} \int_0^1 t^n \, dt = \lim_{n \to \infty} \frac{t^{n+1}}{n+1} \Big|_0^1 = \lim_{n \to \infty} \frac{1}{n+1} = 0$$

$$\int_0^1 \lim_{n \to \infty} x_n(t) \, dt = \int_0^1 0 \, dt = 0.$$

Thus, in this example, $\lim_{n \to \infty} \int_a^b x_n(t) dt = \int_a^b (\lim_{n \to \infty} x_n(t)) dt$. Just to make this easier to type, as we will be looking at this a lot, we will be sloppy and just say $\lim_n \int_a^b x_n = \int_a^b (\lim_n x_n)$. **This is our interchange theorem and here it does work with pointwise convergence even though the convergence is not uniform**.

b: On $[0, 1]$,

$$e^{-nt} \xrightarrow{\text{ptws}} x(t) = \begin{cases} 1, & t = 0 \\ 0, & 0 < t \le 1 \end{cases}, \quad \sup_{0 \le t \le 1} |e^{-nt} - 0| = ||x_n - x||_\infty = 1$$

which tell us $x_n \xrightarrow{\text{unif}} x$. Here each x_n is continuous but the limit function is not continuous. Also

$$\lim_{n \to \infty} \int_0^1 e^{-nt} \, dt = \lim_{n \to \infty} -\frac{e^{-nt}}{n} \Big|_0^1 = \lim_{n \to \infty} \frac{1 - e^{-n}}{n} = 0$$

$$\int_0^1 \lim_{n \to \infty} x_n(t) \, dt = \int_0^1 0 \, dt = 0.$$

Thus, in this example also, $\lim_n \int_a^b x_n = \int_a^b (\lim_n x_n)$. **This is our interchange theorem and here it does work with pointwise convergence even though the convergence is not uniform**.

c: On $[0, 1]$,

$$\frac{2nt}{e^{nt^2}} \xrightarrow{\text{ptws}} x(t) = 0, \quad \sup_{0 \le t \le 1} \left| \frac{2nt}{e^{nt^2}} - 0 \right| = ||x_n - x||_\infty = \sqrt{2n/e}$$

which tell us $x_n \overset{\text{unif}}{\not\longrightarrow} x$. Here each x_n is continuous and the limit function is continuous. Also

$$\lim_{n\to\infty} \int_0^1 \frac{2nt}{e^{nt^2}}\, dt \;=\; \lim_{n\to\infty} e^{-nt^2}\Big|_0^1 = \lim_{n\to\infty}(1 - e^{-n}) = 1$$

$$\int_0^1 \lim_{n\to\infty} x_n(t)\, dt \;=\; \int_0^1 0\, dt = 0.$$

In this example, $\lim_n \int_a^b x_n \neq \int_a^b (\lim_n x_n)$. **This relates to our interchange theorem and here it does not work with pointwise convergence. Note the convergence is still not uniform**.

d: On $[0,1]$,

$$\frac{\sin(nt)}{n} \overset{\text{ptws}}{\longrightarrow} 0 = x(t). \qquad \sup_{0\le t\le 1}\left|\frac{\sin(nt)}{n} - 0\right| = \|x_n - x\|_\infty \le 1/n$$

which tell us $x_n \overset{\text{unif}}{\longrightarrow} x$. Here each x_n is continuous and the limit function is continuous. Also

$$\lim_{n\to\infty} \int_0^1 \frac{\sin(nt)}{n}\, dt \;=\; \lim_{n\to\infty} \frac{-\cos(nt)}{n^2}\Big|_0^1 = \lim_{n\to\infty} \frac{1 - \cos(n)}{n^2} = 0$$

$$\int_0^1 \lim_{n\to\infty} x_n(t)\, dt \;=\; \int_0^1 0\, dt = 0.$$

In this example, $\lim_n \int_a^b x_n = \int_a^b (\lim_n x_n)$. **We note the interchange theorem works with uniform convergence**.

24.2.5   More Conclusions

From our carefully chosen examples, we can draw some conclusions:

1. $x_n \overset{\text{unif}}{\longrightarrow} x$ implies $x_n \overset{\text{ptws}}{\longrightarrow} x$.

2. $x_n \overset{\text{ptws}}{\longrightarrow} x$ and $x_n \overset{\text{unif}}{\not\longrightarrow} x$ implies x is continuous? **No**. Example (a) and Example (b) show us pointwise convergence sequences whose limit functions are not continuous. However. Example (c) says it can happen. It is just **not** guaranteed if the convergence is pointwise.

3. $x_n \overset{\text{ptws}}{\longrightarrow} x$ and $x_n \overset{\text{unif}}{\longrightarrow} x$ implies x is continuous? **Yes**. Example (d) suggests this is true and indeed we have proven this is true in Section 20.3 with Theorem 20.3.1.

4. If each x_n is Riemann integrable, does $x_n \overset{\text{ptws}}{\longrightarrow} x$ and $x_n \overset{\text{unif}}{\not\longrightarrow} x$ imply $\lim_n \int_a^b x_n \neq \int_a^b (\lim_n x_n)$? **No**. Example (c) shows us this does not have to be true. However, it can be true as we see in Example (a) and Example (b). But it is not guaranteed.

5. If each x_n is Riemann integrable, does $x_n \overset{\text{ptws}}{\longrightarrow} x$ and $x_n \overset{\text{unif}}{\longrightarrow} x$ imply $\lim_n \int_a^b x_n \neq \int_a^b (\lim_n x_n)$? **Probably**. See Example (d).

6. $\sup_{a\le t\le b}|x_n(t) - x(t)| \not\longrightarrow 0$ implies the convergence is not uniform? **Yes**. We have used this idea multiple times now. See Example (a), Example (b) and Example (c).

24.3 Basic Theorems

It is time to prove some Theorems about these things.

Theorem 24.3.1 The Weierstrass Sequence Test for Uniform Convergence

Let (x_n) be a sequence of functions defined on the set S. Assume there is a limit function x on S so that $x_n \xrightarrow{ptws} x$. $\sup_{t \in S} |x_n(t) - x(t)| \leq M_n$. Then $x_n \xrightarrow{unif} x$ if and only if there is a sequence (M_n) of nonnegative numbers so that $\lim_{n \to \infty} M_n = 0$.

Proof 24.3.1

$\Rightarrow$: *If $x_n \xrightarrow{unif} x$, given $\epsilon > 0$, there is N so that $\sup_{t \in S} |x_n(t) - x(t)| < \epsilon$ if $n > N$. Define $M_n = \sup_{t \in S} |x_n(t) - x(t)|$. Then $0 \leq M_n < \epsilon$ for $n > N$. This implies $M_n \to 0$.*

$\Leftarrow$: *If $M_n \to 0$, given $\epsilon > 0$, there is N so that $|M_n - 0| < \epsilon$ when $n > N$. Thus, $\sup_{t \in S} |x_n(t) - x(t)| \leq M_n < \epsilon$ if $n > N$. This says $||x_n - x||_\infty < \epsilon$ if $n > N$ telling us $x_n \xrightarrow{unif} x$.* ∎

We proved the next theorem already as Theorem 20.3.1 in Section 20.3. We restate it here slightly differently. Again, this says the property of continuity is preserved under uniform convergence. In Theorem 20.3.1 the context was the interval $[a, b]$. Now we let the domain simply be S.

Theorem 24.3.2 Sequences of Continuous Functions Converge to a Continuous Function under uniform convergence

If (x_n) is a sequence of continuous functions on the set S and (x_n) converges to x uniformly on S, then x is continuous on S.

Proof 24.3.2

Let $\epsilon > 0$ be given. Then since $x_n \xrightarrow{unif} x$ on S, there is N so that $|x_n(t) - x(t)| < \epsilon/3$ if $n > N$ for all $t \in S$. Now pick a particular $\hat{n} > N$. Fix $t_0 \in S$ and consider

$$
\begin{aligned}
|x(t) - x(t_0)| &= |x(t) - x_{\hat{n}}(t) + x_{\hat{n}}(t) - x_{\hat{n}}(t_0) + x_{\hat{n}}(t_0) - x(t_0)| \\
&\leq |x(t) - x_{\hat{n}}(t)| + |x_{\hat{n}}(t) - x_{\hat{n}}(t_0)| + |x_{\hat{n}}(t_0) - x(t_0)|
\end{aligned}
$$

Then since $\hat{n} > N$, we can say

$$|x(t) - x(t_0)| < \epsilon/3 + |x_{\hat{n}}(t) - x_{\hat{n}}(t_0)| + \epsilon/3$$

Finally, since $x_{\hat{n}}$ is continuous at t_0 in S, there is δ so that $|x_{\hat{n}}(t) - x_{\hat{n}}(t_0)| < \epsilon/3$ when $t \in S$ and $|t - t_0| < \delta$. This gives

$$|x(t) - x(t_0)| < 2\epsilon/3 + \epsilon/3 = \epsilon$$

when $t \in S$ and $|t - t_0| < \delta$. Since t_0 in S is arbitrary, we conclude x is continuous on S. ∎

Comment 24.3.1 *This proof is virtually identical to that of Theorem 20.3.1. The only difference is that the domain here is S and there it was $[a, b]$. The compactness of $[a, b]$ is not important here. It is the uniform convergence that is important. Also, as we have mentioned before, this is the result that lets us say that if $x_n \xrightarrow{ptws} x$ on S with x not continuous, then the convergence cannot be uniform.*

Now let's tackle limit interchange ideas involving integration. We start with the easiest one.

24.3.1   Limit Interchange Theorems for Integration

Theorem 24.3.3 **If the sequence (x_n) of continuous functions converges uniformly to x then the limit function is Riemann Integral and the limit of the integrals is the integral of the limit**

> If (x_n) is a sequence of continuous functions on $[a, b]$ and $x_n \xrightarrow{\text{unif}} x$, then x is also Riemann integrable and $\lim_{n \to \infty} \int_a^b x_n(t)dt = \int_a^b (\lim_{n \to \infty} x_n(t))dt = \int_a^b x(t)dt$.

Proof 24.3.3
Since each x_n is continuous on $[a, b]$ and the convergence is uniform, we know x is continuous too and so is Riemann Integrable. Further, given $\epsilon > 0$, there is N so that $\sup_{a \leq t \leq b} |x_n(t) - x(t)| < \epsilon/(b-a)$ if $n > N$. Thus, when $n > N$, we have

$$\left| \int_a^b x_n(t)dt - \int_a^b x(t)dt \right| \leq \int_a^b |x_n(t) - x(t)|dt < \int_a^b \epsilon/(b-a)dt = \epsilon/(b-a)\,(b-a) = \epsilon$$

This tells us $\int_a^b x_n(t)dt \to \int_a^b x(t)dt$. ∎

Example 24.3.1 *Discuss the convergence of (x_n) where $x_n(t) = n^2 t/(1 + n^3 t^2)$.*

Solution *When $t = 0$, we have $x_n(0) = 0$ always. but if $t \neq 0$, we have a different story. For $t > 0$, we see $x_n(t) \to 0$ from above and for $t < 0$, we have $x_n(t) \to 0$ from below. Thus, without even doing any work we know x_n has a global minimum for some negative t and a global maximum for some positive t. Taking the derivative, we have*

$$x_n'(t) \quad = \quad \frac{n^2(1 + n^3 t^2) - (n^2 t)(2n^3 t)}{(1 + n^3 t^2)^2} = \frac{n^2 + n^5 t^2 - 2n^5 t^2}{(1 + n^3 t^2)^2} = \frac{n^2(1 - n^3 t^2)}{(1 + n^3 t^2)^2}.$$

The critical points are $t = \pm 1/n^{3/2}$ and the maximum occurs at $t = 1/n^{3/2}$ with value $\sqrt{n}/2$. The minimum occurs at $t = -1/n^{3/2}$ with value $-\sqrt{n}/2$.

The behavior of this sequence is thus quite similar to that of the sequence $x_n(t) = 2nt/e^{nt^2}$. We have $x_n \xrightarrow{ptws} x$ where $x(t) = 0$ on any $[c, d]$ with the convergence uniform on any interval $[c, d]$ that does not contain 0. Note on an interval $[c, d]$ that does not contain 0, we have $\int_c^d x_n(t)dt \to \int_c^d x(t)dt$.

Example 24.3.2 *Discuss the convergence of (x_n) where $x_n(t) = t^{2n}/(1 + t^{2n})$.*

Solution *This sequence converges pointwise to the discontinuous function*

$$x(t) \quad = \quad \begin{cases} 0, & -1 < t < 1 \\ 1/2, & t = 1 \text{ or } t = -1 \\ 1, & t > 1 \text{ or } t < -1 \end{cases}$$

If you use Octave to do some plots, you will see x_n is very close to zero on the interval $(-1, 1)$ and rises asymptotically up to 1 on both the positive and negative axis. Its extreme behavior is very different from what we have seen before. A little thought shows that since x_n converges to 1 for both negative and positive t and is zero at $t = 0$, the fact that x_n is always nonnegative forces x_n to have a global minimum at $t = 0$. We see this in the critical point analysis. Taking the derivative, we find

$$x_n'(t) \quad = \quad \frac{2nt^{2n-1}(1 + t^{2n}) - t^{2n}(2nt^{2n-1})}{(1 + t^{2n})^2} = \frac{2nt^{2n-1}}{(1 + t^{2n})^2}$$

We see there is only one critical point which is $t = 0$ of value 0. Hence, these functions have a bottom value of 0 at the origin and then grow asymptotically and symmetrically towards 1 on either side. Of course, the value of x_n on $(-1, 1)$ is so small for large n it is difficult to plot!

a: *On any interval $[c, d]$ with $c > 1$, the limit function is continuous and $x(t) = 1$.*

b; *On the intervals $[c, d]$ with $d < -1$, the limit function is still continuous and $x(t) = 1$.*

c: *On the intervals $[c, d] \subset (-1, 1)$, the limit function is also continuous but now $x(t) = 0$.*

d: *On intervals $[c, d]$ that contain ± 1, the limit function is not continuous and so the convergence cannot be uniform.*

We don't know yet if the convergence is uniform in case (a), case (b) or case (c). Let's look at the interval $[2, 10]$ which is a typical interval from case(a). Consider

$$|x_n(t) - 1| \quad = \quad \left| \frac{t^{2n}}{1 + t^{2n}} - 1 \right| = \left| \frac{1}{1 + t^{2n}} \right| < \frac{1}{1 + 2^{2n}} < 2^{-2n}$$

We can clearly choose N so that $2^{-2n} < \epsilon$ when $n > N$, so we have shown convergence is uniform on $[2, 10]$. the argument for case (b) would be similar. For case (c), a typical interval would be $[-.9, .9]$. The pointwise limit is now $x(t) = 0$ and we have

$$|x_n(t) - 0| \quad = \quad \frac{t^{2n}}{1 + t^{2n}} \leq \frac{.9^{2n}}{1 + .9^{2n}} < .9^{2n}$$

and we can clearly choose N so that $.9^{2n} < \epsilon$ when $n > N$. Thus, we have shown convergence is uniform on $[-.9, .9]$.

So in case (a), case (b) and case (c), the integral interchange theorem works!

There is a more general limit interchange theorem for integration.

Theorem 24.3.4 If the sequence (x_n) of Riemann Integrable functions converges uniformly to x then the limit function is Riemann Integral and the limit of the integrals is the integral of the limit

---

Let (x_n) be a sequence of Riemann Integrable functions on $[a, b]$. Let $x_n \xrightarrow{unif} x$ on $[a, b]$. Then x is Riemann integrable on $[a, b]$ and

$$\lim_{n \to \infty} \int_a^b x_n(t)dt \quad = \quad \int_a^b (\lim_{n \to \infty} x_n(t))dt = \int_a^b x(t)dt.$$

Proof 24.3.4
If x is Riemann Integrable on $[a, b]$, the argument we presented in the proof of Theorem 24.3.3 works nicely. So to prove this theorem it is enough to show x is Riemann Integrable on $[a, b]$. Let $\epsilon > 0$ be given. Since the convergence is uniform, there is N so that

$$|x_n(t) - x(t)| \quad < \quad \frac{\epsilon}{5(b - a)}, \quad \forall n > N, \quad t \in [a, b] \tag{α}$$

Pick any $\hat{n} > N$. Since $x_{\hat{n}}$ is integrable, there is a partition π_0 so that

$$U(x_{\hat{n}}, \pi) - L(x_{\hat{n}}, \pi) < \frac{\epsilon}{5} \tag{β}$$

when π is a refinement of π_0.

Since $x_n \xrightarrow{unif} x$, we can easily show x is bounded. So we can define

$$M_j = \sup_{t_{j-1} \le t \le t_j} x(t), \quad m_j = \inf_{t_{j-1} \le t \le t_j} x(t)$$

$$\hat{M}_j = \sup_{t_{j-1} \le t \le t_j} x_{\hat{n}}(t), \quad \hat{m}_j = \inf_{t_{j-1} \le t \le t_j} x_{\hat{n}}(t)$$

where $\{t_0, t_1, \ldots, t_p\}$ are the points in the partition π. Using the infimum and supremum tolerance lemma, we can find points $s_j \in [t_{j-1}, t_j]$ and $u_j \in [t_{j-1}, t_j]$ so that

$$M_j - \frac{\epsilon}{5(b-a)} \quad < \quad x(s_j) \le M_j \tag{γ}$$

$$m_j \le x(u_j) < m_j + \frac{\epsilon}{5(b-a)} \tag{ξ}$$

Thus

$$U(x, \pi) - L(x, \pi) = \sum_{\pi} (M_j - m_j) \Delta t_j$$

$$= \sum_{\pi} \{ M_j - x(s_j) + x(s_j) - x_{\hat{n}}(s_j) + x_{\hat{n}}(s_j)$$

$$- x_{\hat{n}}(u_j) + x_{\hat{n}}(u_j) - x(u_j) + x(u_j) - m_j \} \Delta t_j$$

Now break this into individual pieces:

$$U(x, \pi) - L(x, \pi) = \sum_{\pi} (M_j - x(s_j)) \Delta t_j + \sum_{\pi} (x(s_j) - x_{\hat{n}}(s_j)) \Delta t_j$$

$$+ \sum_{\pi} (x_{\hat{n}}(s_j) - x_{\hat{n}}(u_j)) \Delta t_j + \sum_{\pi} (x_{\hat{n}}(u_j) - x(u_j)) \Delta t_j$$

$$+ \sum_{\pi} (x(u_j) - m_j) \Delta t_j$$

Using Equation γ and Equation ξ we can overestimate the first and fifth term to get

$$U(x, \pi) - L(x, \pi) < \sum_{\pi} \frac{\epsilon}{5(b-a)} \Delta t_j + \sum_{\pi} (x(s_j) - x_{\hat{n}}(s_j)) \Delta t_j$$

$$+ \sum_{\pi} (x_{\hat{n}}(s_j) - x_{\hat{n}}(u_j)) \Delta t_j + \sum_{\pi} (x_{\hat{n}}(u_j) - x(u_j)) \Delta t_j$$

$$+ \sum_{\pi} \frac{\epsilon}{5(b-a)} \Delta t_j$$

We know $\sum_{\pi} \Delta t_j = b - a$ and so simplifying a bit, we have

$$U(x, \pi) - L(x, \pi) < 2\frac{\epsilon}{5} + \sum_{\pi} (x(s_j) - x_{\hat{n}}(s_j)) \Delta t_j + \sum_{\pi} (x_{\hat{n}}(s_j) - x_{\hat{n}}(u_j)) \Delta t_j$$

$$+ \sum_{\pi} (x_{\hat{n}}(u_j) - x(u_j)) \Delta t_j$$

Now use Equation α in piece two and four above to get

$$U(x_{\hat{n}}, \pi) - L(x_{\hat{n}}, \pi) \quad < \quad 2\frac{\epsilon}{5} + \sum_{\pi} \frac{\epsilon}{5(b-a)} \Delta t_j + \sum_{\pi} (x_{\hat{n}}(s_j) - x_{\hat{n}}(u_j)) \Delta t_j$$

$$+ \sum_{\pi} \frac{\epsilon}{5(b-a)} \Delta t_j$$

$$< \quad 4\frac{\epsilon}{5} + \sum_{\pi} (x_{\hat{n}}(s_j) - x_{\hat{n}}(u_j)) \Delta t_j$$

since like before $\sum_{\pi} \frac{\epsilon}{5(b-a)} \Delta t_j = \frac{\epsilon}{5}$. Finally, $|x_{\hat{n}}(s_j) - x_{\hat{n}}(u_j)| \leq \hat{M}_j - \hat{m}_j$ and so

$$U(x, \pi) - L(x, \pi) \quad < \quad 4\frac{\epsilon}{5} + \sum_{\pi} (\hat{M}_j - \hat{m}_j) \Delta t_j = 4\frac{\epsilon}{5} + U(x_{\hat{n}}, \pi) - L(x_{\hat{n}}, \pi) < \epsilon$$

using Equation β. We conclude $U(x, \pi) - U(x, \pi) < \epsilon$ for all refinements of π_0 which shows x is Riemann Integrable on $[a, b]$.

It remains to show the limit interchange portion of the theorem. As mentioned at the start of this proof, this argument is the same as the one given in Theorem 24.3.3 and so it does not have to be repeated. ∎

24.3.2 Homework

Exercise 24.3.1 *Discuss the convergence of (x_n) where $x_n(t) = n^2 t/(1 + n^3 t^2)$ on the intervals $[-1, 1]$, $[2, 6]$ and $[-10, -2]$. Illustrate your arguments with nicely chosen Octave plots documenting your work carefully.*

Exercise 24.3.2 *Discuss the convergence of (x_n) where $x_n(t) = t^{2n}/(1 + t^{2n})$ on the intervals $[-0.5, 0.5]$, $[1.4, 12]$ and $[-20, -2]$. Illustrate your arguments with nicely chosen Octave plots documenting your work carefully.*

Exercise 24.3.3 *Discuss the convergence of (x_n) where $x_n(t) = 2n^2 t/(1 + 3n^3 t^2)$ on the intervals $[-1, 1]$, $[2, 6]$ and $[-10, -2]$. Illustrate your arguments with nicely chosen Octave plots documenting your work carefully.*

Exercise 24.3.4 *Discuss the convergence of (x_n) where $x_n(t) = t^{3n}/(1 + t^{3n})$ on the intervals $[-0.5, 0.5]$, $[1.4, 12]$ and $[-20, -2]$. Illustrate your arguments with nicely chosen Octave plots documenting your work carefully.*

Exercise 24.3.5 *Discuss the convergence of (x_n) where $x_n(t) = 4t^{3n}/(1 + 5t^{3n})$ on the intervals $[-0.5, 0.5]$, $[1.4, 12]$ and $[-20, -2]$. Illustrate your arguments with nicely chosen Octave plots documenting your work carefully.*

Exercise 24.3.6 *If $x_n \xrightarrow{unif} x$ on $[a, b]$ and each x_n is Riemann Integrable on $[a, b]$, prove x is bounded.*

24.4 The Interchange Theorem for Differentiation

We now consider another important theorem about the interchange of integration and limits of functions. We start with an illuminating example. Consider the sequence (x_n) on $[-1, 1]$ where $x_n(t) = t/e^{nt^2}$. It is easy to see $x_n \xrightarrow{ptws} x$ where $x(t) = 0$ on $[-1, 1]$. Taking the derivative, we see

$$x_n'(t) \quad = \quad \frac{1 - 2nt^2}{e^{nt^2}}$$

and the critical points of x_n are when $1 - 2nt^2 = 0$ or at $t = \pm 1/\sqrt{2n}$. We have studied this kind of sequence before but the other sequences had high peaks proportional to n which meant we could not get uniform convergence on any interval $[c, d]$ containing 0. In this sequence, the extreme values occur at $\pm 1/\sqrt{2ne}$. Hence, the *peaks* here decrease. We could illustrate this behavior with some carefully chosen Octave plots, but this time we will simply sketch the graph Figure 24.5 labeling important points. Let

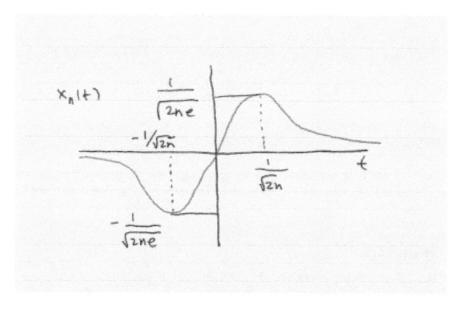

Figure 24.5: The sequence (x_n), $x_n(t) = t/e^{nt^2}$.

$$M_n \;=\; \sup_{-1 \le t \le 1} |x_n(t) - x(t)| \;=\; \sup_{-1 \le t \le 1} \left| \frac{t}{e^{nt^2}} \right| \;=\; \frac{1}{\sqrt{2ne}}$$

Since $M_n \to 0$, by the Weierstrass Theorem for Uniform Convergence, we have $x_n \overset{\text{unif}}{\longrightarrow} x$. At $t = 0$, $x_n'(0) = 1$ for all n and so $\lim_{n \to \infty} x_n'(t) \big|_{t=0} = 1$. But $x'(0) = 0$ as $x(t) = 0$ on $[-1, 1]$. Hence we conclude for this example that at $t = 0$ the differentiation interchange does not work:

$$\lim_{n \to \infty} x_n'(t) \;\neq\; \left(\lim_{n \to \infty} x_n(t) \right)',$$

that is the interchange of convergence and differentiation fails here **even though we have uniform convergence of the sequence**. Finally, since $x_n'(t) = \frac{1 - 2nt^2}{e^{nt^2}}$, we see the pointwise limit of the derivative sequence is

$$y(t) \;=\; \begin{cases} 1, & t = 0 \\ 0, & t \in [-1, 0) \cup (0, 1] \end{cases}$$

Since y is not continuous, we know $x_n' \overset{\text{unif}}{\not\longrightarrow} y$. The question is what conditions are required to guarantee the derivative interchange result? To discuss this carefully, we need additional tools.

Definition 24.4.1 The Uniform Cauchy Criterion

We say the sequence of functions (x_n) on the set S satisfies the **Uniform Cauchy Criterion** if

$$\forall \epsilon > 0, \exists N \ni |x_n(t) - x_m(t)| < \epsilon, \text{ for } n > m > N \text{ and } \forall t \in S$$

Note this can also be stated like this:

$$\forall \epsilon > 0, \exists N \ni \|x_n - x_m\|_\infty < \epsilon, \text{ for } n > m > N.$$

We will abbreviate this with **UCC**.

Using the UCC, we can prove another test for uniform convergence which is often easier to use.

Theorem 24.4.1 A Sequence of Functions Converges Uniformly if and only if It Satisfies the UCC

Let (x_n) be a sequence of functions on the set S. Then

$$\left(\exists x : S \to \Re \ni x_n \overset{unif}{\to} x\right) \iff \left((x_n) \text{ satisfies the UCC}\right).$$

Proof 24.4.1

$(\Rightarrow:)$

We assume there is an $x : S \to \Re$ so that $x_n \overset{unif}{\to} x$. Then, given $\epsilon > 0$, there is N so that $|x_n(t) - x(t)| < \epsilon/2$ for all $t \in S$. Thus, if $n > m > N$, we have

$$|x_n(t) - x_m(t)| = |x_n(t) - x(t) + x(t) - x_m(t)| \le |x_n(t) - x(t)| + |x(t) - x_m(t)|$$
$$< \epsilon/2 + \epsilon/2 = \epsilon$$

Thus, (x_n) satisfies the UCC.

$(\Leftarrow:)$

If (x_n) satisfies the UCC, then given $\epsilon > 0$, there is N so that $|x_n(t) - x_m(t)| < \epsilon$ for $n > m > N$. This says the sequence $(x_n(\hat{t}))$ is a Cauchy sequence for each $\hat{t} \in S$. Since $\Re$ is complete, there is a number $a_{\hat{t}}$ so that $x_n(\hat{t}) \to a_{\hat{t}}$. The number $a_{\hat{t}}$ defines a function $x : S \to \Re$ by $x(\hat{t}) = a_{\hat{t}}$.

Clearly, $x_n \overset{ptws}{\to} x$ on S. From the UCC, we can therefore say for the given ϵ,

$$\lim_{n \to \infty} |x_n(t) - x_m(t)| \le \epsilon/2, \text{ if } n > m > N, t \in S$$

But the absolute function is continuous and so

$$|\lim_{n \to \infty} x_n(t) - x_m(t)| \le \epsilon/2, \text{ if } n > m > N, t \in S$$

or $|x(t) - x_m(t)| < \epsilon$ when $m > N$. This shows $x_n \overset{unif}{\to} x$ as the choice of index m is not important. Note this argument is essentially the same as the one we used in the proof of Theorem 24.3.2. The difference here is that we do not know each x_n is continuous. We are simply proving the existence of a limit function which is possible as $\Re$ is complete. But the use of the completeness of $\Re$ is common in both proofs. ∎

We are ready to prove our next interchange theorem.

Theorem 24.4.2 The Derivative Interchange Theorem for Sequences of Functions

---

Let (x_n) be a sequence of functions defined on the interval $[a, b]$. Assume

1. *x_n is differentiable on $[a, b]$.*

2. *x'_n is Riemann Integrable on $[a, b]$.*

3. *There is at least one point $t_0 \in [a, b]$ such that the sequence $(x_n(t_0))$ converges.*

4. *$x'_n \xrightarrow{unif} y$ on $[a, b]$ and the limit function y is continuous.*

Then there is $x : [a, b] \to \Re$ which is differentiable on $[a, b]$ and $x_n \xrightarrow{unif}$ on $[a, b]$ and $x' = y$. Another way of saying this is

$$\left(\lim_{n \to \infty} x_n(t) \right)' = \lim_{n \to \infty} x'_n(t)$$

Proof 24.4.2

Since x'_n is integrable, by the recapture theorem,

$$x_n(t) = \int_{t_0}^{t} x'_n(s) \, ds + x_n(t_0).$$

Hence, for any n and m, we have

$$x_n(t) - x_m(t) = \int_{t_0}^{t} (x'_n(s) - x'_m(s)) \, ds + x_n(t_0) - x_m(t_0).$$

If $t > t_0$, then

$$|x_n(t) - x_m(t)| \leq \left| \int_{t_0}^{t} (x'_n(s) - x'_m(s)) \, ds \right| + |x_n(t_0) - x_m(t_0)|$$

$$\leq \int_{t_0}^{t} |x'_n(s) - x'_m(s)| + |x_n(t_0) - x_m(t_0)|$$

Since $x'_n \xrightarrow{unif} y$ on $[a, b]$, (x'_n) satisfies the UCC. For a given $\epsilon > 0$, we have there is N_1 so that

$$|x'_n(s) - x'_m(s)| < \frac{\epsilon}{4(b - a)}, \quad for \ \ n > m > N_1, \ s \in [a, b] \qquad (\alpha)$$

Thus, using Equation α,

$$|x_n(t) - x_m(t)| < \frac{\epsilon}{4(b - a)} \int_{t_0}^{t} ds + |x_n(t_0) - x_m(t_0)|$$

$$= \frac{\epsilon}{4} + |x_n(t_0) - x_m(t_0)|, \quad for \ \ n > m > N_1, \ t > t_0 \in [a, b]$$

A similar argument for $t < t_0$ gives the same result:

$$|x_n(t) - x_m(t)| < \frac{\epsilon}{4} + |x_n(t_0) - x_m(t_0)|, \quad for \ \ n > m > N_1, \ t < t_0 \in [a, b]$$

Hence, since the statement if clearly true at $t = t_0$, we can say

$$|x_n(t) - x_m(t)| \quad < \quad \frac{\epsilon}{4} + |x_n(t_0) - x_m(t_0)|, \quad \text{for} \quad n > m > N_1, \ t \in [a, b]$$

We also know $(x_n(t_0))$ converges and hence is a Cauchy sequence. Thus, there is N_2 so that $|x_n(t_0) - x_m(t_0)| < \epsilon/4$ for $n > m > N_2$. We conclude if $N > \max(N_1, N_2)$, both conditions apply and we can say

$$|x_n(t) - x_m(t)| \quad < \quad \frac{\epsilon}{4} + \frac{\epsilon}{4} = \frac{\epsilon}{2}, \quad \text{for} \quad n > m > N, \ t \in [a, b]$$

This shows (x_n) satisfies the UCC and there is $x : S \to \Re$ so that $x_n \xrightarrow{unif} x$. So far we know

$$x_n(t) \quad = \quad x_n(t_0) + \int_{t_0}^t x_n'(s) \, ds.$$

Since $x_n' \xrightarrow{unif} y$ and each x_n is Riemann Integrable on $[a, b]$, by the integral interchange theorem $\int_{t_0}^t x_n'(s) \, ds = \int_{t_0}^t y(s) \, ds$ on $[a, b]$. Also, $x_n \xrightarrow{unif} x$ implies $\lim_{n \to \infty} x_n(t) = x(t)$. We conclude

$$\lim_{n \to \infty} x_n(t) \quad = \quad \lim_{n \to \infty} x_n(t_0) + \lim_{n \to \infty} \int_{t_0}^t x_n'(s) \, ds.$$

and so

$$x(t) \quad = \quad x(t_0) + \int_{t_0}^t y(s) \, ds$$

Since we assume y is continuous on $[a, b]$, by the FTOC, $x'(t) = y(t)$ on $[a, b]$. ∎

Homework

Exercise 24.4.1 *For what intervals does the derivative interchange theorem hold for the functions (x_n) where $x_n(t) = 3nt/e^{5nt^2}$?*

Exercise 24.4.2 *For what intervals does the derivative interchange theorem hold for the functions (x_n) where $x_n(t) = 2nt/e^{8nt^2}$?*

Exercise 24.4.3 *For what intervals does the derivative interchange theorem hold for the functions (x_n) where $x_n(t) = 9nt/e^{nt^2}$?*

Exercise 24.4.4 *For what intervals does the derivative interchange theorem hold for the functions (x_n) where $x_n(t) = \cos(n^2 t)/n^2$?*

Exercise 24.4.5 *For what intervals does the derivative interchange theorem hold for the functions (x_n) where $x_n(t) = 4t + \cos(n^2 t)/n^2$?*

Exercise 24.4.6 *For what intervals does the derivative interchange theorem hold for the functions (x_n) where $x_n(t) = 1 + \sin(n^2 t + 3)/n^2$?*

Chapter 25

Series of Functions and Power Series

We are now ready to look at another type of sequence of functions: the partial sums we get when we add up the functions from a sequence.

25.1 The Sum of Powers of t Series

To help you understand this better, let's look at a simple example we know. Consider the sums

$$S_n(t) \quad = \quad 1 + t + t^2 + \ldots + t^n$$

This is the nth partial sum of the sequence of functions $(x_n)_{n=0}^{\infty}$ where $x_n(t) = t^n$ and $x_0(t) = 1$ for all t. We already have studied the convergence of this sequence on the interval $[0, 1]$ and we know it converges pointwise to a discontinuous x,

$$x(t) \quad = \quad \begin{cases} 0, & 0 \le t < 1 \\ 1, & t = 1 \end{cases}$$

and it does not converge uniformly as $||x_n - x||_\infty = 1$. However, we want to study the partial sums and ask similar convergence questions. Is there a function S which these partial sums converge to? Note, this is a series at each t and so it is very unlikely we can guess the pointwise limit. So our convergence analysis cannot rely on tools that require us to estimate $S_n(t) - S(t)|$ except in very rare instances. For this example, these partial sums come from the infinite series

$$S(t) \quad = \quad \sum_{n=0}^{\infty} t^n = \lim_{n \to \infty} S_n(t)$$

where we use $S(t)$ to denote the value this series might have. Of course, at this point the only point we know for sure the series converges is the point $t = 0$ because here all the partial sums are of value 1 and hence $S(0) = 1$.

We can apply the ratio test here.

$$\lim_{n \to \infty} \frac{|t|^{n+1}}{|t|^n} \quad = \quad \lim_{n \to \infty} |t| = |t|$$

This series converges absolutely if $t| < 1$, diverges if $|t| > 1$ and we have to check what happens at $t = \pm 1$.

- At $t = -1$, the partial sums oscillate between 1 and 0 and so the sequence (S_n) does not converge.

- At $t = 1$, the partial sums diverge to ∞ and so the sequence (S_n) does not converge.

Since we have absolute convergence for $|t| < 1$, we also know the series converges on the interval $(-1, 1)$ and diverges elsewhere.

We have shown $S_n \xrightarrow{\text{ptws}} S$ on $(-1, 1)$. Is the convergence uniform? We haven't yet figured out what $S(t)$ is, so how do we check for uniform convergence? The best thing to do is to see if the sequence (S_n) satisfies the UCC. We have for any $n > m$

$$
\begin{aligned}
|S_n(t) - S_m(t)| &= \left| \sum_{k=0}^{n} t^k - \sum_{k=0}^{m} t^k \right| = \left| \frac{1 - t^{n+1}}{1 - t} - \frac{1 - t^{m+1}}{1 - t} \right| \\
&= \left| \frac{t^{m+1} - t^{n+1}}{1 - t} \right| \leq \frac{|t|^{m+1}}{|1 - t|}
\end{aligned}
$$

From the backwards triangle inequality, we know $1 - |t| \leq |1 - t|$ and hence we can say

$$
|S_n(t) - S_m(t)| \leq \frac{|t|^{m+1}}{1 - |t|}
$$

Now restrict attention to the interval $[-r, r]$ for any $0 < r < 1$. This is a compact subset of $(-1, 1)$. On this interval, $|t| < r$ and so we have the estimate $|t| < r$ implying $1 - |t| > 1 - r$ giving

$$
|S_n(t) - S_m(t) \leq \frac{|r|^{m+1}}{1 - r}
$$

Since $|r| < 1$, given $\epsilon > 0$ there is N so that

$$
\frac{r^{m+1}}{1 - r} < \epsilon, \ \forall m > N.
$$

This implies for $m > N$,

$$
|S_n(t) - S_m(t)| \leq \frac{|t|^{m+1}}{1 - |t|} < \frac{r^{m+1}}{1 - r} < \epsilon
$$

We conclude $||S_m - S_m||_\infty < \epsilon$ on $[-r, r]$ and so there is a function U so that $S_n \xrightarrow{\text{unif}} U$ on $[-r, r]$. But limits are unique, so $U = S$ on $[-r, r]$. Since each S_n is continuous as it is a polynomial in t, we see the limit function $U = S$ is continuous on $[-r, r]$. In the context of the UCC criterion, we use $M_n = ||S_m - S_m||_\infty$. Note the sequence does not converge uniformly on the whole interval $(-1, 1)$ as

$$
\sup_{-1 \leq t \leq 1} |S_n(t) - S_m(t)| = \sup_{-1 \leq t \leq 1} |t^{m+1}(1 + t + t^2 + \ldots + t^{n-m-1})| = n - m
$$

Thus, $||S_n - S_m|| = n - m \to \infty$ as $n \to \infty$. Hence, convergence cannot be uniform.

Next, if you choose any arbitrary t_0 in $(-1, 1)$, you can find a r so that $t_0 \in [-r, r]$ and hence S is continuous at t_0 for arbitrary t_0. This shows S is actually continuous on the full interval $(-1, 1)$. Thus, we cannot argue directly for the continuity of the pointwise limit function; we must instead approach the issue indirectly by looking at convergence on compact subsets of the domain of S which can grow to include any arbitrary point in the domain of S.

Let's stop and see what we have done: we have a general strategy for proving the pointwise limit S is continuous on its domain.

- We start with a sequence of partial sums (S_n) and we find where this sequence converges. In general it converges to an interval (a, b) and it may or may not converge at $t = a$ and/ or $t = b$. We can use any test we want to find this out, but a common one is the ratio test.

- This shows we have a pointwise limit function S on (a, b).

- If the individual functions S_n are continuous on (a, b) and if we can show (S_n) satisfies the UCC on the interval $[a + r, b - r] \subset (a, b)$, we then know $S_n \xrightarrow{\text{unif}} U$ on $[a + r, b - r]$. By uniqueness of limits, then $U = S$ there and so is continuous there. and so S is continuous on $[a + r, b - r]$.

- To show S is continuous on (a, b) just note any t_0 in (a, b) is in some $[a + r, b - r]$ and so S is continuous at t_0. Since t_0 is arbitrary, S is actually continuous on (a, b).

We actually know more about S. Since for $t \in (-1, 1)$, this is a geometric series, we have

$$S_n(t) \quad = \quad 1 + t + \ldots + t^n = \frac{1 - t^{n+1}}{1 - t}$$

which converges to $U(t) = \frac{1}{1-t}$ on $(-1, 1)$. Clearly $U = S$ as limits must be unique. And we do know U is continuous on $(-1, 1)$. But the point of this procedure is it works even if we do not know such a function U.

We need another tool to determine convergence of a series we have not discussed yet.
Homework

Repeat the analysis of this section for the series below.

Exercise 25.1.1 *The series is $\sum_{n=0}^{\infty} (2t)^n$.*

Exercise 25.1.2 *The series is $\sum_{n=0}^{\infty} (3t)^n$.*

Exercise 25.1.3 *The series is $\sum_{n=1}^{\infty} (5t)^n$.*

Exercise 25.1.4 *The series is $\sum_{n=2}^{\infty} (4t)^n$.*

25.1.1 More General Series

We want to look at other types of series involving powers of t. Series of the form $\sum a_n t^n$ are called **power series** in t. So we want to be able to examine these. Our first example had the terms $a_n = 1$ always but we want to be able to understand more general choices of a_n's. For example, the familiar Taylor Series expansion for an infinitely differentiable function at a point p gives rise to a Series of the form $\sum_{n=0}^{\infty} a_n (t-p)^n$ for suitable a_n. At the point $p = 0$, this is a Series of the form $\sum_{n=0}^{\infty} a_n t^n$ and our sum of powers series is a simple example of that. We determine easily the open interval (a, b) where our series converge, but we always have to manually check what happens at $t = a$ and $t = b$. These checks require another tool to determine convergence of a series we have not discussed yet.

Theorem 25.1.1 The Alternating Series Test

Assume the sequence (a_n) is nonnegative, decreasing for sufficiently large n and $a_n \to 0$ as $n \to \infty$. Then the series $\sum_{n=1}^{\infty} (-1)^{n+1} a_n$ converges.

Proof 25.1.1

The even partial sums S_{2n} are

$$S_{2n} = (a_1 - a_2) + (a_3 - a_4) + \ldots + (a_{2n-1} - a_{2n}).$$

Since (a_n) is decreasing for n sufficiently large, there is N so that $a_{k-1} - a_k \geq 0$ if $k > N$. Hence for $2n > N$, we can say S_{2n} is always increasing. We can also write

$$S_{2n} = a_1 - (a_2 - a_3) - (a_4 - a_3) - \ldots - (a_{2n-2} - a_{2n-1}) - a_{2n} \leq a_1.$$

We have shown S_{2n} is an increasing sequence bounded above by a_1 for $2n > N$. This sequence converges to a finite supremum S. The odd partial sums for $2n > N$ are

$$S_{2n+1} = S_{2n} + a_{2n+1}.$$

We assume $\lim_{k \to \infty} a_k = 0$ and so

$$\lim_{n \to \infty} S_{2n+1} = \lim_{n \to \infty} S_{2n} + \lim_{n \to \infty} a_{2n+1} = S + 0 = S.$$

Given $\epsilon > 0$, we see there is M so that both $|S_{2m} - S| < \epsilon$ and $|S_{2m+1} - S| < \epsilon$ if $m > M$. Hence, for such m,

$$|S_n - S| = \begin{cases} |S_{2m} - S| < \epsilon, & n = 2m \\ |S_{2m+1} - S| < \epsilon, & n = 2m + 1 \end{cases}$$

We conclude $\sum_{n=1}^{\infty} (-1)^{n+1} a_n$ converges. ∎

Example 25.1.1 *Examine convergence of the series $\sum_{n=1}^{\infty} \frac{2}{5n \, 2^n} t^n$.*

Solution *From the ratio test, we find*

$$\lim_{n \to \infty} \frac{\frac{2}{5(n+1) \, 2^{n+1}} |t|^{n+1}}{\frac{2}{5n \, 2^n} |t|^n} = \lim_{n \to \infty} \frac{n}{n+1} \frac{1}{2} |t| = |t|/2.$$

This series thus converges when $|t|/2 < 1$ or $|t| < 2$.

- *at $t = 2$, we have $\sum_{n=1}^{\infty} \frac{2}{5n \, 2^n} 2^n = \sum_{n=1}^{\infty} \frac{2}{5n}$ which diverges by comparison to $\sum_{n=1}^{\infty} 1/n$.*

- *at $t = -2$, we have $\sum_{n=1}^{\infty} \frac{2}{5n \, 2^n} (-2)^n = \sum_{n=1}^{\infty} (-1)^n \frac{2}{5n}$ which converges by the alternating series test.*

Hence, this series converges to a function S pointwise on $[-2, 2)$. We have no idea what this function's representation is, but we do know it exists.

Now on the interval $[-2 + r, 2 - r]$, for $n > m$, consider

$$|S_n(t) - S_m(t)| = \left| \sum_{k=m+1}^{n} \frac{2}{5k \, 2^k} t^k \right| \leq \sum_{k=m+1}^{n} \frac{2}{5k \, 2^k} |t|^k$$

$$< \sum_{k=m+1}^{n} \frac{2}{5} |t/2|^k < \sum_{k=m+1}^{n} |t/2|^k$$

This next estimate is a standard geometric series one: Since $t \in [-2+r, 2-r]$, $|t|/2 \le 1 - r/2 = \rho < 1$. We see

$$\sum_{k=m+1}^{n} |t/2|^k = \sum_{k=m+1}^{n} \rho^k = \frac{1}{1-\rho} - \sum_{k=0}^{m} \rho^k = \frac{1}{1-\rho} - \frac{1 - \rho^{m+1}}{1-\rho} = \frac{\rho^{m+1}}{1-\rho}$$

Since $\rho < 1$, given $\epsilon > 0$, there is N so that $\frac{\rho^{m+1}}{1-\rho} < \epsilon/2$ if $m > N$. Thus we have shown on the interval $[-2+r, 2-r]$, for $n > m > N$, if $t \in [-2+r, 2-r]$,

$$|S_n(t) - S_m(t)| \quad < \quad \frac{\rho^{m+1}}{1-\rho} < \epsilon/2$$

So $\|S_n - S_m\|_\infty < \epsilon$ when $n > m > N$. We see the sequence (S_n) satisfies the UCC on the interval $[-2+r, 2-r]$. It then follows that there is a function U_r on $[-2+r, 2-r]$ so that $S_n \overset{unif}{\longrightarrow} U_r$ on $[-2+r, 2-r]$. We already know $S_n \overset{ptws}{\longrightarrow} S$ on $[-2+r, 2-r]$. Since limits are unique, we have $S = U_r$ on $[-2+r, 2-r]$. Also, since each S_n is continuous as it is a polynomial, the limit $U_r = S$ is continuous on $[-2+r, 2-r]$.

Then given any t_0 in $(-2, 2)$, t_0 is in some $[-2+r, 2-r]$ and so S must be continuous on $(-2, 2)$.

Note at the endpoint $t_0 = -2$, we have $\lim_{t \to -2} S_n(t) = S_n(-2)$ for each n which then implies

$$\lim_{n \to \infty} \lim_{t \to -2} S_n(t) = \lim_{n \to \infty} S_n(-2) = S(-2)$$

Hence, the question of whether or not $\lim_{t \to -2} S(t) = S(-2)$ is essentially a limit interchange issue: is

$$\lim_{n \to \infty} \lim_{t \to -2} S_n(t) = \lim_{t \to -2} \lim_{n \to \infty} S_n(t)?$$

and we don't know how to answer this! In this example, we also really do not know what the pointwise limit function is!

Example 25.1.2 *Examine convergence of the series $\sum_{n=1}^{\infty} \frac{1}{3^n} t^n$.*

Solution *This is the same as the series $\sum_{n=1}^{\infty} (t/3)^n$. Applying the ratio test, we find the series converges when $|t|/3 < 1$.*

- *at $t = 3$, we have $\sum_{n=1}^{\infty} (1)^n = \infty$. which diverges.*

- *at $t = -3$, we have $\sum_{n=1}^{\infty} (-1)^n$ which diverges by oscillation between 0 and 1.*

Hence, this series converges to a function S pointwise on $[-3, 3)$. We have shown $S_n \overset{ptws}{\longrightarrow} S$ on $(-3, 3)$.

Is the convergence uniform? The best thing to do is to see if the sequence (S_n) satisfies the UCC. We have for any $n > m$, letting $u = t/3$

$$|S_n(t) - S_m(t)| = \left| \sum_{k=0}^{n} u^k - \sum_{k=0}^{m} u^k \right| = \left| \sum_{k=m+1}^{n} u^k \right| \le \sum_{k=m+1}^{n} |u|^k$$

Now restrict attention to the interval $[-3 + r, 3 - r]$ *for suitable r. This is a compact subset of* $(-3, 3)$. *On this interval,* $|u| = |t/3| \le 1 - r/3 = \rho < 1$ *giving*

$$
\begin{aligned}
|S_n(t) - S_m(t)| &\le \sum_{k=m+1}^{n} |u|^k \le \sum_{k=m+1}^{n} \rho^k = \rho^{m+1} \sum_{k=0}^{n-m-1} \rho^k \\
&= \rho^{m+1} \frac{1 - \rho^{n-m}}{1 - \rho} < \frac{\rho^{m+1}}{1 - \rho}
\end{aligned}
$$

Since $|\rho| < 1$, *given* $\epsilon > 0$ *there is N so that*

$$
\frac{\rho^{m+1}}{1 - \rho} < \epsilon, \ \forall m > N.
$$

This implies for $m > N$,

$$
|S_n(t) - S_m(t)| \le \sum_{k=m+1}^{n} |u|^k < \frac{\rho^{m+1}}{1 - \rho} < \epsilon
$$

We conclude $\|S_n - S_m\|_\infty < \epsilon$ *on* $[-3 + r, 3 - r]$ *and so there is a function* U_r *so that* $S_n \xrightarrow{unif} U_r$ *on* $[-3 + r, 3 - r]$. *But limits are unique, so* $U_r = S$ *on* $[-3 + r, 3 - r]$. *Since each* S_n *is continuous as it is a polynomial in t, we see the limit function* $U_r = S$ *is continuous on* $[-3 + r, 3 - r]$. *Note the sequence does not converge uniformly on the whole interval* $(-3, 3)$ *as*

$$
\sup_{-3 \le t \le 3} |S_n(t) - S_m(t)| = \sup_{-3 \le t \le 3} |(t/3)^{m+1}(1 + (t/3) + (t/3)^2 + \ldots + (t/3)^{n-m-1})| = n - m
$$

Thus, $\|S_n - S_m\| = n - m \to \infty$ *as* $n \to \infty$. *Hence, convergence cannot be uniform.*

Then if you choose any arbitrary t_0 *in* $(-3, 3)$, *you can find a r so that* $t_0 \in [-3 + r, 3 - r]$ *and hence S is continuous at* t_0 *for arbitrary* t_0. *This shows S is actually continuous on the full interval* $(-3, 3)$.

25.1.2   Homework

You need to follow the full arguments we have done for the examples above for these problems. So lots of explaining!

Exercise 25.1.5 *Examine convergence of the series* $\sum_{n=1}^{\infty} \frac{1}{4^n} t^n$.

Exercise 25.1.6 *Examine convergence of the series* $\sum_{n=1}^{\infty} \frac{7}{3n \, 5^n} t^n$.

Exercise 25.1.7 *Examine convergence of the series* $\sum_{n=1}^{\infty} \frac{7}{3n \, 5^n} (t - 1)^n$.

Exercise 25.1.8 *Examine convergence of the series* $\sum_{n=1}^{\infty} \frac{2n+3}{(3n+2) \, 4^n} t^n$.

Exercise 25.1.9 *Examine convergence of the series* $\sum_{n=1}^{\infty} \frac{5n^2 + 4n + 10}{3n = 7 \, 5^n} t^n$.

25.2   Uniform Convergence Tests

We have already used the UCC in the context of series multiple times. It is time to state it in a form that is explicitly useful for series and to state and prove a variant of the Weierstrass Test for Uniform Convergence.

Definition 25.2.1 The Uniform Cauchy Criterion for Series

Let (u_n) be a sequence of functions defined on S and let (S_n) be the associated sequence of partial sums. For convenience, assume all indexing starts at $n = 1$. We say (S_n) satisfies **The Uniform Cauchy Criterion for Series** *if and only if*

$$\forall \epsilon > 0 \, \exists \, N \ni \left| \sum_{k=m}^{n} u_k(t) \right| < \epsilon, \; \text{for } n > m > N, \; \forall t \in S$$

This can be rephrased as (S_n) satisfies **The Uniform Cauchy Criterion for Series: UCC for series** *if and only if*

$$\forall \epsilon > 0 \, \exists \, N \ni \left\| \sum_{k=m}^{n} u_k(t) \right\|_{\infty} < \epsilon, \; \text{for } n > m > N$$

where it is understood the sup *norm is computed over S. Note this tells us the sequence $(u_n(x))$ is a Cauchy sequence for each x and hence there is a function $u(x)$ so that $u_n \overset{unif}{\longrightarrow} u$ on S.*

Using the UCC for series, we can prove another test for uniform convergence just like we did for a sequence of functions (x_n). The difference is we are specializing to the partial sum sequence.

Theorem 25.2.1 The Partial Sums of a Series of Functions converges Uniformly if and only if It Satisfies the UCC for Series

Let (x_n) be a sequence of functions on the set Ω with associated partials sums (S_n). Then

$$\left(\exists \, S : \Omega \to \Re \ni S_n \overset{unif}{\to} S \right) \iff \left((S_n) \text{ satisfies the UCC} \right).$$

Proof 25.2.1

$(\Rightarrow:)$

We assume there is an $S : \Omega \to \Re$ so that $S_n \overset{unif}{\to} S$. Then, given $\epsilon > 0$, there is N so that $|S_n(t) - S(t)| < \epsilon/2$ for all $t \in \Omega$. Thus, if $n > m > N$, we have

$$|S_n(t) - S_m(t)| = |S_n(t) - S(t) + S(t) - S_m(t)| \leq |S_n(t) - S(t)| + |S(t) - S_m(t)|$$
$$< \epsilon/2 + \epsilon/2 = \epsilon$$

Thus, (S_n) satisfies the UCC.

$(\Leftarrow:)$

If (S_n) satisfies the UCC, then given $\epsilon > 0$, there is N so that $|S_n(t) - S_m(t)| < \epsilon$ for $n > m > N$. This says the sequence $(S_n(\hat{t}))$ is a Cauchy sequence for each $\hat{t} \in \Omega$. Since $\Re$ is complete, there is a number $a_{\hat{t}}$ so that $S_n(\hat{t}) \to a_{\hat{t}}$. The number $a_{\hat{t}}$ defines a function $S : \Omega \to \Re$ by $S(\hat{t}) = a_{\hat{t}}$.

Clearly, $S_n \overset{ptws}{\to} S$ on Ω. From the UCC, we can therefore say for the given ϵ,

$$\lim_{n \to \infty} |S_n(t) - S_m(t)| \leq \epsilon/2, \; \text{if } n > m > N, \, t \in \Omega$$

But the absolute function is continuous and so

$$\left| \lim_{n \to \infty} S_n(t) - S_m(t) \right| \leq \epsilon/2, \; \text{if } n > m > N, \, t \in \Omega$$

or $|S(t) - S_m(t)| < \epsilon$ when $m > N$. This shows $S_n \overset{unif}{\to} S$ as the choice of index m is not important.
∎

This allows us to state a new test for uniform convergence specialized to series.

Theorem 25.2.2 Second Weierstrass Test for Uniform Convergence

> *Let (u_n) be a sequence of functions defined on the set Ω and let (S_n) be the associated sequence of partial sums. We assume all indexing starts at $n = 1$ for convenience. Let the sequence of constants (K_n) satisfy $\sup_{t \in \Omega} |u_n(t)| \leq K_n$. Then $\sum_{n=1}^{\infty} K_n$ converges implies there is a function $S : \Omega \to \Re$ so that $S_n \overset{unif}{\to} S$.*

Proof 25.2.2
If $\sum_{n=1}^{\infty} K_n$ converges, the sequence of partial sums $\sigma_n = \sum_{j=1}^{n} K_j$ converges and so it is a Cauchy sequence of real numbers. Thus, given $\epsilon > 0$, there is N so that

$$n > m > N \implies |\sigma_n - \sigma_m| = \sum_{j=m+1}^{n} K_j < \epsilon$$

Then if $n > m > N$, for all $t \in \Omega$

$$\left| \sum_{j=m+1}^{n} u_j(t) \right| \leq \sum_{j=m+1}^{n} |u_j(t)| \leq \sum_{j=m+1}^{n} K_j < \epsilon$$

Thus, (S_n) satisfies the UCC and so there is a function S so that $S_n \overset{unif}{\to} S$ on Ω where we usually denote this uniform limit by $S(t) = \sum_{n=1}^{\infty} u_n(t)$.   ∎

Comment 25.2.1 *Note this says if (S_n) does not converge uniformly on S, the associated series $\sum_{n=1}^{\infty} \sup_{t \in S} ||u_n||_{\infty}$ must diverge to infinity.*

Let's go back and do our original problems using these new tools.

Example 25.2.1 *Discuss the convergence of the series $\sum_{n=0}^{\infty} t^n$.*

Solution *We apply the ratio test here.*

$$\lim_{n \to \infty} \frac{|t|^{n+1}}{|t|^n} = \lim_{n \to \infty} |t| = |t|$$

This series converges absolutely if $t| < 1$, diverges if $|t| > 1$ and at $t = \pm 1$.

- *At $t = -1$, the partial sums oscillate between 1 and 0 and so the sequence (S_n) does not converge.*

- *At $t = 1$, the partial sums diverge to ∞ and so the sequence (S_n) does not converge.*

Thus, $\sum_{n=0}^{\infty} t^n$ converges pointwise to a function $S(t)$ on $(-1, 1)$. To determine uniform convergence, let $u_n(t) = t^n$ and for suitably small r, let $K_n = \sup_{t \in [-1+r, 1-r]} |t|^n = 1 - r = \rho < 1$.

Since $\sum_{n=1}^{\infty} K_n = \sum_{n=1}^{\infty} \rho^n$ is a geometric series with $0 \leq \rho < 1$, it converges. Hence by Theorem 25.2.2, the convergence of the series $\sum_{n=0}^{\infty} t^n$ is uniform on $[-1+r, 1-r]$ and so it converges to a function U_r on $[-1+r, 1-r]$. Since limits are unique, $S = U_r$ on $[-1+r, 1-r]$.

Further, each S_n is continuous as it is a polynomial in t, and so the limit function $U_r = S$ is continuous on $[-1+r, 1-r]$. Note the sequence does not converge uniformly on the whole interval $(-1, 1)$ as $K_n = \sup_{t \in (-1,1)} |t|^n = 1$ and so $\sum_{n=1}^{\infty} K_n$ diverges.

Finally if you choose any arbitrary t_0 in $(-1, 1)$, you can find a r so that $t_0 \in [-1+r, 1-r]$ and hence S is continuous at t_0 for arbitrary t_0. This shows S is actually continuous on the full interval $(-1, 1)$.

Example 25.2.2 *Examine convergence of the series $\sum_{n=1}^{\infty} \frac{2}{5n\,2^n} t^n$.*

Solution *From the ratio test, we find*

$$\lim_{n \to \infty} \frac{\frac{2}{5(n+1)\,2^{n+1}} |t|^{n+1}}{\frac{2}{5n\,2^n} |t|^n} = \lim_{n \to \infty} \frac{n}{n+1} \frac{1}{2} |t| = |t|/2.$$

This series thus converges when $|t|/2 < 1$ or $|t| < 2$.

- *at $t = 2$, we have $\sum_{n=1}^{\infty} \frac{2}{5n\,2^n} 2^n = \sum_{n=1}^{\infty} \frac{2}{5n}$ which diverges by comparison to $\sum_{n=1}^{\infty} 1/n$.*
- *at $t = -2$, we have $\sum_{n=1}^{\infty} \frac{2}{5n\,2^n} (-2)^n = \sum_{n=1}^{\infty} (-1)^n \frac{2}{5n}$ which converges by the alternating series test.*

Hence, this series converges to a function S pointwise on $[-2, 2)$. On the interval $[-2+r, 2-r]$, we have $|t/2| \leq 1 - r = \rho < 1$ and we can defined the sequence (K_n) by

$$\sup_{t \in [-2+r, 2-r]} \frac{2}{5n} |t/2|^n \leq \sup_{t \in [-2+r, 2-r]} \frac{2}{5} |t/2|^n \leq \sup_{t \in [-2+r, 2-r]} |t/2^n| = \rho^n = K_n.$$

Since $\sum_{n=1}^{\infty} K_n$ is a geometric series with $0 \leq \rho < 1$, it converges which implies $\sum_{n=1}^{\infty} \frac{2}{5n\,2^n} t^n$ converges uniformly on $[-2+r, 2-r]$ to a function U_r. Since limits are unique, we have $S = U_r$ on $[-2+r, 2-r]$. Also, since each S_n is continuous as it is a polynomial, the limit $U_r = S$ is continuous on $[-2+r, 2-r]$.

Then given any t_0 in $(-2, 2)$, t_0 is in some $[-2+r, 2-r]$ and so S must be continuous on $(-2, 2)$.

Note at the endpoint $t_0 = -2$, we have $\lim_{t \to -2} S_n(t) = S_n(-2)$ for each n which then implies

$$\lim_{n \to \infty} \lim_{t \to -2} S_n(t) = \lim_{n \to \infty} S_n(-2) = S(-2)$$

Hence, the question of whether or not $\lim_{t \to -2} S(t) = S(-2)$ is essentially a limit interchange issue: is

$$\lim_{n \to \infty} \lim_{t \to -2} S_n(t) = \lim_{t \to -2} \lim_{n \to \infty} S_n(t)?$$

and we don't know how to answer this!
Note in this example, we also really do not know what the pointwise limit function is!

Example 25.2.3 *Examine convergence of the series $\sum_{n=1}^{\infty} \frac{1}{3^n} t^n$.*

Solution *This is the same as the series $\sum_{n=1}^{\infty} (t/3)^n$. Applying the ratio test, we find the series converges when $|t|/3 < 1$.*

- *at $t = 3$, we have $\sum_{n=1}^{\infty} (1)^n = \infty$. which diverges.*
- *at $t = -3$, we have $\sum_{n=1}^{\infty} (-1)^n$ which diverges by oscillation between 0 and 1.*

Hence, this series converges to a function S pointwise on $(-3, 3)$. We have shown $S_n \xrightarrow{ptws} S$ on $(-3, 3)$.

Is the convergence uniform? Restrict attention to the interval $[-3 + r, 3 - r]$ for suitable r. On this interval, $|t/3| \leq 1 - r/3 = \rho < 1$, we define the sequence (K_n) as follows:

$$\sup_{t \in [-3+r, 3-r]} |t/3|^n \leq \rho^n = K_n.$$

Since $\sum_{n=1}^{\infty} K_n$ is a geometric series with $0 \leq \rho < 1$, it converges which implies $\sum_{n=1}^{\infty} (t/3)^n$ converges uniformly on $[-3 + r, 3 - r]$ to a function U_r. Since limits are unique, we have $S = U_r$ on $[-3 + r, 3 - r]$. Also, since each S_n is continuous as it is a polynomial, the limit $U_r = S$ is continuous on $[-3 + r, 3 - r]$. Then given any t_0 in $(-3, 3)$, t_0 is in some $[-3 + r, 3 - r]$ and so S must be continuous on $(-3, 3)$.

25.2.1 Homework

You need to follow the full arguments we have done for the examples above for these problems. Use our new tests here and do lots of explaining!

Exercise 25.2.1 *Examine convergence of the series $\sum_{n=1}^{\infty} \frac{1}{4^n} t^n$.*

Exercise 25.2.2 *Examine convergence of the series $\sum_{n=1}^{\infty} \frac{7}{3n\, 5^n} t^n$.*

Exercise 25.2.3 *Examine convergence of the series $\sum_{n=1}^{\infty} \frac{7}{3n\, 5^n} (t - 1)^n$.*

Exercise 25.2.4 *Examine convergence of the series $\sum_{n=1}^{\infty} \frac{2n+3}{(3n+2)\, 4^n} t^n$.*

Exercise 25.2.5 *Examine convergence of the series $\sum_{n=1}^{\infty} \frac{5n^2 + 4n + 10}{3n = 7\, 5^n} t^n$.*

25.3 Integrated Series

The series we obtain by integrating a series term by term is called the integrated series. Let's look at some examples.

Example 25.3.1 *Examine convergence of the integrated series from $\sum_{n=1}^{\infty} \frac{2n}{6^n} t^n$.*

Solution *The integrated series is $\sum_{n=1}^{\infty} \frac{2n}{(n+1)6^n} t^{n+1}$.*

$$\lim_{n \to \infty} \frac{\frac{2(n+1)}{(n+2)6^{n+1}} |t|^{n+2}}{\frac{2n}{(n+1)6^n} |t|^{n+1}} = \lim_{n \to \infty} \frac{(n+1)^2}{n(n+2)} \frac{1}{6} |t| = |t|/6.$$

Hence, this series converges on $(-6, 6)$. Also,

- *at $t = 6$, we have $\sum_{n=1}^{\infty} \frac{12n}{n+1} = \infty$. which diverges.*

- *at $t = -6$, we have $\sum_{n=1}^{\infty} \frac{12n}{n+1} (-1)^{n+1}$ which diverges by oscillation.*

We have shown $S_n \xrightarrow{ptws} S$ on $(-6, 6)$.
To determine if the convergence is uniform, restrict attention to the interval $[-6+r, 6-r]$ for suitable r. On this interval, $|t/6| \leq 1 - r/6 = \rho < 1$, we define the sequence (K_n) as follows:

$$\sup_{t \in [-6+r, 6-r]} \frac{12n}{n+1} (|t|/6)^{n+1} \leq \frac{12n}{n+1} \rho^{n+1} = K_n.$$

Using the ratio test, we see $\sum_{n=1}^{\infty} K_n$ converges as $0 \leq \rho < 1$. which implies $\sum_{n=1}^{\infty} \frac{2n}{(n+1)6^n} t^{n+1}$ converges uniformly on $[-6 + r, 6 - r]$ to a function U_r. Since limits are unique, we have $S = U_r$ on $[-6 + r, 6 - r]$. Also, since each partial sum of the integrated series is continuous as it is a polynomial, the limit $U_r = S$ is continuous on $[-6 + r, 6 - r]$. Then given any t_0 in $(-6, 6)$, t_0 is in some $[-6 + r, 6 - r]$ and so S must be continuous on $(-6, 6)$.

Example 25.3.2 *Examine convergence of the integrated series from $\sum_{n=1}^{\infty} \frac{5n^3+4}{(n^2+8)\,7^n} t^n$.*

Solution *The integrated series is $\sum_{n=1}^{\infty} \frac{5n^3+4}{(n+1)(n^2+8)\,7^n} t^{n+1}$. Using the ratio test, we find*

$$\lim_{n \to \infty} \frac{\frac{5(n+1)^3+4}{((n+2)((n+1)^2+8)\,7^{n+1}} |t|^{n+2}}{\frac{5n^3+4)}{(n+1)(n^2+8)\,7^n} |t|^{n+1}} =$$

$$\lim_{n \to \infty} \frac{5(n+1)^3 + 4}{(n+2)((n+1)^2 + 8)} \frac{(n+1)(n^2 + 8)}{5n^3 + 4} |t|/7 =$$

$$|t|/7.$$

Thus, this series converges on $(-7, 7)$. Also,
We have shown $S_n \xrightarrow{ptws} S$ on $(-7, 7)$.

- *at $t = 7$, we have $\sum_{n=1}^{\infty} \frac{7(5n^3+4)}{(n+1)(n^2+8)} = \infty$. which diverges.*

- *at $t = -7$, we have $\sum_{n=1}^{\infty} \frac{7(5n^3+4)}{(n+1)(n^2+8)} (-1)^{n+1}$ which diverges by oscillation.*

To determine if the convergence is uniform, restrict attention to the interval $[-7+r, 7-r]$ for suitable r. On this interval, $|t/7| \leq 1 - r/7 = \rho < 1$, we define the sequence (K_n) as follows:

$$\sup_{t \in [-7+r, 7-r]} \frac{7(5n^3 + 4)}{(n+1)(n^2 + 8)} (|t|/7)^{n+1} \leq \frac{7(5n^3 + 4)}{(n+1)(n^2 + 8)} \rho^{n+1}$$
$$= K_n.$$

Using the ratio test, we see $\sum_{n=1}^{\infty} K_n$ converges as $0 \leq \rho < 1$. which implies $\sum_{n=1}^{\infty} \frac{5n^3+4}{(n+1)(n^2+8)\,7^n} t^{n+1}$ converges uniformly on $[-7 + r, 7 - r]$ to a function U_r. Since limits are unique, we have $S = U_r$ on $[-7 + r, 7 - r]$. Also, since each partial sum of the integrated series is continuous as it is a polynomial, the limit $U_r = S$ is continuous on $[-7 + r, 7 - r]$. Then given any t_0 in $(-7, 7)$, t_0 is in some $[-7 + r, 7 - r]$ and so S must be continuous on $(-7, 7)$.

Comment 25.3.1 *In our examples, we have seen that the ratio test is an easy way to determine where our series converge and to determine continuity of the pointwise limit function we simply use the Weierstrass Uniform Convergence Theorem for Series.*

Comment 25.3.2 *Note that the integrated series has the same interval of convergence as the original series in our examples.*

Let's look at differentiating Series of functions and see what happens. This will require that we use the derivative interchange theorem.

25.3.1 Homework

Exercise 25.3.1 *Examine convergence of the series $\sum_{n=1}^{\infty} \frac{4}{7^n} t^n$ and its integrated series.*

Exercise 25.3.2 *Examine convergence of the series $\sum_{n=1}^{\infty} \frac{6}{8n^2\,(4^n)} t^n$ and its integrated series.*

Exercise 25.3.3 *Examine convergence of the series* $\sum_{n=1}^{\infty} \frac{1}{4^n} t^n$.

Exercise 25.3.4 *Examine convergence of the series* $\sum_{n=1}^{\infty} \frac{7}{3n\,5^n} t^n$.

Exercise 25.3.5 *Examine convergence of the series* $\sum_{n=1}^{\infty} \frac{7}{3n\,5^n} (t-1)^n$.

Exercise 25.3.6 *Examine convergence of the series* $\sum_{n=1}^{\infty} \frac{2n+3}{(3n+2)\,4^n} t^n$.

Exercise 25.3.7 *Examine convergence of the series* $\sum_{n=1}^{\infty} \frac{5n^2+4n+10}{3n=7\,5^n} t^n$.

25.4 Differentiating Series

Let's look at the problem of differentiating a Series of functions. We start with an old friend, the series $\sum_{n=0}^{\infty} t^n$ which we know converges pointwise to $S(t) = 1/(1-t)$ on $(-1,1)$. It also is known to converge uniformly on compact subsets of $[-1,1]$. We only showed the argument for sets like $[-1+r, 1-r]$ but it is easy enough to go through the same steps for $[c, d]$ in $(-1,1)$. We used the uniform convergence to prove the pointwise limit had to continuous on the compact subsets of $(-1,1)$ and easily extended the continuity to the whole interval $(-1,1)$. This type of argument did not need to know the pointwise limit was $1/(1-t)$ and so we worked through some other examples to illustrate how it works with series for which we cannot find the pointwise limit function. Our primary tools were the UCC which was a bit cumbersome to use, and the Second Weierstrass Uniform Convergence Theorem, which was much faster.

Now let's look at what is called the **derived series** of $\sum_{n=0}^{\infty} t^n$. This is the series we get by differentiating the original series term by term. Let's call this series $D(t) = \sum_{n=1}^{\infty} n\,t^{n-1}$. Note the $n = 0$ term is gone as the derivative of a constant is zero. This is the series constructed from the partial sums S_n'.
Using the ratio test, we find

$$\lim_{n\to\infty} \frac{(n+1)|t|^n}{n|t|^{n-1}} = \lim_{n\to\infty} \frac{(n+1)}{n} |t| = |t|$$

Thus the series converges when $|t| < 1$ or on the interval $(-1,1)$. At $t = 1$, the series is $\sum_{n=1}^{\infty} n$ which diverges and at $t = -1$, the series $\sum_{n=1}^{\infty} (-1)^{n-1} n$ diverges by oscillation. To find out about uniform convergence, note on the interval $[-1+r, 1-r]$ for sufficiently small r, we have $|t| \leq 1 - r = \rho < 1$. Define the sequence (K_n) by

$$\sup_{t \in [-1+r, 1-r]} n|t|^{n-1} \leq n\rho^{n-1} = K_n$$

Then $\sum_{n=1}^{\infty} n\rho^{n-1}$ converges by the ratio test. We conclude by the Second Weierstrass Uniform Convergence Theorem that the series $\sum_{n=1}^{\infty} n\,t^{n-1}$ converges uniformly on $[-1+r, 1-r]$ to a function V_r. Since each partial sum of $\sum_{n=1}^{\infty} n\,t^{n-1}$ is continuous, the uniform limit function V_r is continuous also and $D = V_r$ on $[-1+r, 1-r]$ since limits are unique. Finally, if t_0 in $(-1,1)$ is arbitrary, it is inside some $[-1+r, 1-r]$ and so $V_r = D$ is continuous there also.

Let's check the conditions of the derivative interchange theorem, Theorem 24.4.2 applied to the sequence of partial sums (S_n). Fix a r in $[0,1)$.

1. S_n is differentiable on $[-r,r]$: **True**.

2. S_n' is Riemann Integrable on $[-r,r]$: **True as each is a polynomial**.

3. There is at least one point $t_0 \in [-r,r]$ such that the sequence $(S_n(t_0))$ converges. **True as the series $\sum_{n=0}^{\infty} t^n$ converges at $t = 0$.**

4. $S'_n \xrightarrow{\text{unif}} y$ on $[-r, r]$ and the limit function y is continuous. **True as we have just shown** $S'_n \xrightarrow{\text{unif}} D$ **on** $[-r, r]$.

The conditions of Theorem 24.4.2 are satisfied and we can say there is a function W_r on $[-r, r]$ so that $S_n \xrightarrow{\text{unif}} W_r$ on $[-r, r]$ and $W'_r = D$. Since limits are unique, we then have $W_r = S$ with $S' = D$.

Then given any $t \in (-1, 1)$, $t \in [-r, r]$ for some r. Hence, $S' = D$ there also. We conclude $S' = D$ on $(-1, 1)$.

Since we also know $S(t) = 1/(1 - t)$, this proves that on $(-1, 1)$

$$\left(\sum_{n=0}^{\infty} t^n \right)' = \sum_{n=1}^{\infty} n t^{n-1} = \left(\frac{1}{1-t} \right)' = \frac{1}{(1-t)^2}$$

although in general we would not know a functional form for $S(t)$ and we could just say

$$S'(t) = \left(\sum_{n=0}^{\infty} t^n \right)' = \sum_{n=1}^{\infty} n t^{n-1} = D(t)$$

We need to do more examples!

Example 25.4.1 *Examine convergence of the derived series for* $\sum_{n=1}^{\infty} \frac{2}{5n \, (2^n)} t^n$.

Solution *The derived series here is* $\sum_{n=1}^{\infty} \frac{2}{5 \, (2^n)} t^{n-1}$. *Using the ratio test, we find*

$$\lim_{n \to \infty} \frac{\frac{1}{2^{n+1}} |t|^n}{\frac{1}{2^n} |t|^{n-1}} = \lim_{n \to \infty} \frac{1}{2} |t| = |t|/2.$$

This series thus converges when $|t|/2 < 1$ *or* $|t| < 2$.

- *at* $t = 2$, *we have* $\sum_{n=1}^{\infty} \frac{2}{5 \, 2^n} 2^{n-1} = \sum_{n=1}^{\infty} \frac{2}{10}$ *which diverges since it sums to infinity.*
- *at* $t = -2$, *we have* $\sum_{n=1}^{\infty} \frac{2}{5 \, 2^n} (-2)^{n-1} = \sum_{n=1}^{\infty} (-1)^{n-1} \frac{1}{5}$ *which diverges by oscillation.*

Hence, this series converges to a function D *pointwise on* $(-2, 2)$. *Recall the original series also converged at* $t = -2$ *but we do not have convergence of the derived series at this endpoint.*

To find out about uniform convergence, note on the interval $[-2 + r, 2 - r]$ *for sufficiently small* r, *we have* $|t|/2 \leq 1 - r = \rho < 1$. *Define the sequence* (K_n) *by*

$$\sup_{t \in [-2+r, 2-r]} \frac{2}{5 \, 2^n} |t|^{n-1} = \sup_{t \in [-2+r, 2-r]} \frac{1}{5} (|t|/2)^{n-1} \leq \rho^{n-1} = K_n$$

Then $\sum_{n=1}^{\infty} \rho^{n-1}$ *converges by the ratio test. We conclude by the Second Weierstrass Uniform Convergence Theorem that the derived series* $\sum_{n=1}^{\infty} \frac{2}{5 \, (2^n)} t^{n-1}$ *converges uniformly on* $[-2+r, 2-r]$ *to a function* V_r. *Since each partial sum of* $\sum_{n=1}^{\infty} \frac{2}{5 \, (2^n)} t^{n-1}$ *is continuous, the uniform limit function* V_r *is continuous also and* $D = V_r$ *on* $[-2 + r, 2 - r]$ *since limits are unique. Finally, if* t_0 *in* $(-2, 2)$ *is arbitrary, it is inside some* $[-2 + r, 2 - r]$ *and so* $V_r = D$ *is continuous there also. Note we have shown* $S'_n \xrightarrow{\text{unif}} D$ *on* $[-2 + r, 2 - r]$ *for all* r *with* $[-2 + r, 2 - r] \subset (-2, 2)$.

Let's check the conditions of the derivative interchange theorem, Theorem 24.4.2 applied to the sequence of partial sums (S_n). *Fix a* r *in* $[0, 2)$.

1. S_n *is differentiable on* $[-r, r]$: **True**.

2. S_n' *is Riemann Integrable on* $[-r, r]$: **True as each is a polynomial**.

3. *There is at least one point* $t_0 \in [-r, r]$ *such that the sequence* $(S_n(t_0))$ *converges.* **True as the series** $\sum_{n=1}^{\infty} \frac{2}{5n \, (2^n)} t^n$ **converges at** $t = 0$.

4. $S_n' \xrightarrow{unif} y$ *on* $[-r, r]$ *and the limit function* y *is continuous.* **True as we have just shown** $S_n' \xrightarrow{unif} D$ **on** $[-r, r]$.

The conditions of Theorem 24.4.2 are satisfied and we can say there is a function W_r *on* $[-r, r]$ *so that* $S_n \xrightarrow{unif} W_r$ *on* $[-r, r]$ *and* $W_r' = D$. *Since limits are unique, we then have* $W_r = S$ *with* $S' = D$.

Then given any $t \in (-2, 2)$, $t \in [-r, r]$ *for some* r. *Hence,* $S' = D$ *there also. We conclude* $S' = D$ *on* $(-2, 2)$. *Hence, on* (-2.2),

$$S'(t) = \left(\sum_{n=1}^{\infty} \frac{2}{5n \, (2^n)} t^n \right)' = \sum_{n=1}^{\infty} \frac{2}{5 \, (2^n)} t^{n-1} = D(t)$$

This is why we say we can **differentiate the series term by term**!

Example 25.4.2 *Examine convergence of the derived series for* $\sum_{n=1}^{\infty} \frac{1}{3^n} t^n$.

Solution *The derived series is* $\sum_{n=1}^{\infty} \frac{n}{3^n} t^{n-1}$. *Applying the ratio test, we find*

$$\lim_{n \to \infty} \frac{\frac{n+1}{3^{n+1}} |t|^n}{\frac{n}{3^n} |t|^{n-1}} = \lim_{n \to \infty} \frac{n+1}{n} \, |t|/3 = |t|/3.$$

This series thus converges when $|t|/3 < 1$ *or* $|t| < 3$.

- *at* $t = 3$, *we have* $\sum_{n=1}^{\infty} \frac{n}{3^n} 3^{n-1} = \sum_{n=1}^{\infty} \frac{1}{3n}$ *which diverges by comparison to* $\sum_{n=1}^{\infty} \frac{1}{n}$.
- *at* $t = -3$, *we have* $\sum_{n=1}^{\infty} \frac{n}{3^n} (-3)^{n-1} = \sum_{n=1}^{\infty} (-1)^{n-1} \frac{n}{3}$ *which diverges by oscillation.*

Hence, this series converges to a function D *pointwise on* $(-3, 3)$.

To find out about uniform convergence, note on the interval $[-3 + r, 3 - r]$ *for sufficiently small* r, *we have* $|t|/3 \leq 1 - r = \rho < 1$. *Define the sequence* (K_n) *by*

$$\sup_{t \in [-3+r, 3-r]} n|t/3|^{n-1} \leq n\rho^{n-1} = K_n$$

Then $\sum_{n=1}^{\infty} \rho^{n-1}$ *converges by the ratio test. We conclude by the Second Weierstrass Uniform Convergence Theorem that the derived series* $\sum_{n=1}^{\infty} \frac{2}{5 \, 2^n} t^{n-1}$ *converges uniformly on* $[-3 + r, 3 - r]$ *to a function* V_r. *Since each partial sum of* $\sum_{n=1}^{\infty} n(t/3)^{n-1}$ *is continuous, the uniform limit function* V_r *is continuous also and* $D = V_r$ *on* $[-3 + r, 3 - r]$ *since limits are unique. Finally, if* t_0 *in* $(-3, 3)$ *is arbitrary, it is inside some* $[-3 + r, 3 - r]$ *and so* $V_r = D$ *is continuous there also. Note we have shown* $S_n' \xrightarrow{unif} D$ *on* $[-3 + r, 3 - r]$ *for all* r *with* $[-3 + r, 3 - r] \subset (-3, 3)$.

Let's check the conditions of the derivative interchange theorem, Theorem 24.4.2 applied to the sequence of partial sums (S_n). *Fix a* r *in* $[0, 3)$.

1. S_n *is differentiable on* $[-r, r]$: **True**.

2. S_n' *is Riemann Integrable on* $[-r, r]$: **True as each is a polynomial.**

3. *There is at least one point* $t_0 \in [-r, r]$ *such that the sequence* $(S_n(t_0))$ *converges.* **True as the series** $\sum_{n=1}^{\infty} \frac{1}{3^n} t^n$ **converges at** $t = 0$.

4. $S_n' \xrightarrow{unif} y$ *on* $[-r, r]$ *and the limit function* y *is continuous.* **True as we have just shown** $S_n' \xrightarrow{unif} D$ **on** $[-r, r]$.

The conditions of Theorem 24.4.2 are satisfied and we can say there is a function W_r *on* $[0, r]$ *so that* $S_n \xrightarrow{unif} W_r$ *on* $[-r, r]$ *and* $W_r' = D$. *Since limits are unique, we then have* $W_r = S$ *with* $S' = D$.

Then given any $t \in (-3, 3)$, $t \in [-r, r]$ *for some* r. *Hence,* $S' = D$ *there also. We conclude* $S' = D$ *on* $(-3, 3)$. *Hence, on* $(-3, 3)$

$$S'(t) = \left(\sum_{n=0}^{\infty} (t/3)^n \right)' = \sum_{n=1}^{\infty} n(t/3)^{n-1} = D(t)$$

The series $\sum_{n=0}^{\infty} (t/3)^n$ *is a geometric series and converges to* $1/(1 - (t/3))$ *on* $(-3, 3)$, *Hence, we can also say*

$$\left(\frac{1}{1 - (t/3)} \right)' = \left(\sum_{n=0}^{\infty} (t/3)^n \right)' = \sum_{n=1}^{\infty} n(t/3)^{n-1} = \frac{1/3}{(1 - (t/3))^2}$$

25.4.1   Higher Derived Series

We can take higher derivatives of series. The series obtained by differentiating the derived series term by term is called the second derived series. The third derived series is the series derived from the second derived series and so forth. It is easiest to explain how this works in an example.

Example 25.4.3 *For the series* $\sum_{n=0}^{\infty} \frac{5}{8(n+1)^2} t^n$ *examine the first, second and third derived series for convergence.*

Solution *The first, second and third derived series are*

- *(First Derived Series:) This is* $\sum_{n=1}^{\infty} \frac{5n}{8(n+1)^2} t^{n-1}$.

- *(Second Derived Series:) This is* $\sum_{n=2}^{\infty} \frac{5(n-1)n}{8(n+1)^2} t^{n-2}$.

- *(Third Derived Series:) This is* $\sum_{n=3}^{\infty} \frac{5(n-2)(n-1)n}{8(n+1)^2} t^{n-3}$.

We find all of these series converge on $(-1, 1)$ *using the ratio test: we want all of these limits less than 1:*

- *(First Derived Series:)*

$$\lim_{n \to \infty} \frac{\frac{5(n+1)}{8(n+2)^2} t^n}{\frac{5n}{8(n+1)^2} t^{n-1}} = \lim_{n \to \infty} \frac{(n+1)^3}{n(n+2)^2} |t| = |t|$$

- *(Second Derived Series:)*

$$\lim_{n \to \infty} \frac{\frac{5(n)(n+1)}{8(n+2)^2} t^{n-1}}{\frac{5(n-1)n}{8(n+1)^2} t^{n-2}} = \lim_{n \to \infty} \frac{n(n+1)^3}{(n-1)n(n+2)^2} |t| = |t|$$

- *(Third Derived Series:)*

$$\lim_{n\to\infty} \frac{\frac{5(n-1)(n)(n+1)}{8(n+2)^2}\, t^{n-2}}{\frac{5(n-2)(n-1)n}{8(n+1)^2}\, t^{n-3}} = \lim_{n\to\infty} \frac{(n-1)n(n+1)^3}{(n-2)(n-1)(n+2)^2}|t| = |t|$$

It is clear from these limits that all three series converge pointwise to a limit function on $(-1,1)$. We could test if they converge at the endpoints but we will leave that to you. We check all three series for uniform convergence in the same way. On the interval $[-1+r, 1-r]$ for small enough r, we have $|t| \leq 1 - r = \rho < 1$. We then have these estimates

- *(First Derived Series:)*

$$\sum_{n=1}^{\infty} \frac{5n}{8(n+1)^2}|t|^{n-1} \leq \sum_{n=1}^{\infty} \frac{5(n+1)}{8(n+1)^2}\rho^{n-1} \leq \sum_{n=1}^{\infty} \frac{1}{(n+1)}\rho^{n-1}$$

So $K_n = \frac{1}{(n+1)}\rho^{n-1}$. Since $\sum_{n=1}^{\infty} K_n$ converges by the ratio test since $0 \leq \rho < 1$, we see the first derived series converges uniformly on $[-1+r, 1-r]$.

- *(Second Derived Series:)*

$$\sum_{n=2}^{\infty} \frac{5(n-1)n}{8(n+1)^2}|t|^{n-2} \leq \sum_{n=2}^{\infty} \frac{5(n+1)^2}{8(n+1)^2}\rho^{n-2} \leq \sum_{n=2}^{\infty} \rho^{n-2}$$

So $K_n = \rho^{n-2}$. Since $\sum_{n=1}^{\infty} K_n$ converges by the ratio test since $0 \leq \rho < 1$, we see the second derived series converges uniformly on $[-1+r, 1-r]$.

- *(Third Derived Series:)*

$$\sum_{n=3}^{\infty} \frac{5(n-2)(n-1)n}{8(n+1)^2}|t|^{n-3} \leq \sum_{n=3}^{\infty} \frac{5(n+1)^3}{8(n+1)^2}\rho^{n-3} \leq \sum_{n=3}^{\infty} (n+1)\rho^{n-3}$$

So $K_n = (n+1)\rho^{n-3}$. Since $\sum_{n=1}^{\infty} K_n$ converges by the ratio test since $0 \leq \rho < 1$, we see the third derived series converges uniformly on $[-1+r, 1-r]$.

By the Second Weierstrass Uniform Convergence Theorem, we see all these derived series converge uniformly. We have

- *(First Derived Series:) $S_n' \overset{unif}{\longrightarrow} D$ on $[-1+r, 1-r]$.*

- *(Second Derived Series:) $S_n'' \overset{unif}{\longrightarrow} E$ on $[-1+r, 1-r]$.*

- *(Third Derived Series:) $S_n''' \overset{unif}{\longrightarrow} F$ on $[-1+r, 1-r]$.*

Since the partial sums of these derived series are continuous and the convergence is uniform D, E and F are continuous on $[-1+r, 1-r]$. Further given any $t \in (-1,1)$, t is in some $[-1+r, 1-r]$ and so we know D, E and F are continuous on $(-1,1)$.

Now we check the derivative interchange for these derived series

- *(First Derived Series:)*

1. *Fix a r in $[0,1)$. S_n is differentiable on $[-r, r]$:* **True.**

2. S_n' is Riemann Integrable on $[-r, r]$: **True as each is a polynomial.**

3. There is at least one point $t_0 \in [-r, r]$ such that the sequence $(S_n(t_0))$ converges. **True as the original series converges at $t = 0$.**

4. $S_n' \overset{unif}{\longrightarrow} y$ on $[-r, r]$ and the limit function y is continuous. **True as we have just shown** $S_n' \overset{unif}{\longrightarrow} D$ **on** $[-r, r]$.

The conditions of derivative interchange theorem are satisfied and we can say there is a function W_r on $[-r, r]$ so that $S_n \overset{unif}{\longrightarrow} W_r$ on $[-r, r]$ and $W_r' = D$. Since limits are unique, we then have $W_r = S$ with $S' = D$.

Then given any $t \in (-1, 1)$, $t \in [-r, r]$ for some r. Hence, $S' = D$ there also. We conclude $S' = D$ on $(-1, 1)$. Hence, on $(-1, 1)$

$$S'(t) = \left(\sum_{n=1}^{\infty} \frac{5}{8(n+1)^2} t^{n-1} \right)' = \sum_{n=1}^{\infty} \frac{5n}{8(n+1)^2} t^{n-1} = D(t)$$

- *(Second Derived Series:)*

  1. *Fix a t in $[0, 1)$. S_n' is differentiable on $[-r, r]$:* **True.**

  2. *S_n'' is Riemann Integrable on $[-r, r]$:* **True as each is a polynomial.**

  3. *There is at least one point $t_0 \in [-r, r]$ such that the sequence $(S_n'(t_0))$ converges.* **True as the derived series converges at $t = 0$.**

  4. *$S_n'' \overset{unif}{\longrightarrow} y$ on $[-r, r]$ and the limit function y is continuous.* **True as we have just shown** $S_n'' \overset{unif}{\longrightarrow} E$ **on** $[-r, r]$.

The conditions of derivative interchange theorem are satisfied and we can say there is a function X_r on $[-r, r]$ so that $S_n' \overset{unif}{\longrightarrow} X_r$ on $[-r, r]$ and $X' = E$. Since limits are unique, we then have $X_r = D$ with $D' = E$. Using the results from the first derived series, we have $S'' = D' = E$.

Then given any $t \in (-1, 1)$, $t \in [-r, r]$ for some r. Hence, $S'' = D'$ there also. We conclude $S'' = D'$ on $(-1, 1)$. Hence, on $(-1, 1)$

$$D'(t) = \left(\sum_{n=1}^{\infty} \frac{5n}{8(n+1)^2} t^{n-1} \right)' = \sum_{n=2}^{\infty} \frac{5(n-1)n}{8(n+1)^2} t^{n-2} = E(t)$$

$$S''(t) = \left(\sum_{n=0}^{\infty} \frac{5}{8(n+1)^2} t^{n-1} \right)'' = \left(\sum_{n=1}^{\infty} \frac{5n}{8(n+1)^2} t^{n-1} \right)'$$

$$= \sum_{n=2}^{\infty} \frac{5(n-1)n}{8(n+1)^2} t^{n-2} = E(t)$$

- *(Third Derived Series:)*

  1. *Fix a t in $[0, 1)$. S_n'' is differentiable on $[-r, r]$:* **True.**

  2. *S_n''' is Riemann Integrable on $[-r, r]$:* **True as each is a polynomial.**

3. There is at least one point $t_0 \in [-r, r]$ such that the sequence $(S_n''(t_0))$ converges. **True as the second derived series converges at** $t = 0$.

4. $S_n''' \xrightarrow{\text{unif}} y$ on $[-r, r]$ and the limit function y is continuous. **True as we have just shown** $S_n''' \xrightarrow{\text{unif}} F$ **on** $[-r, r]$.

The conditions of derivative interchange theorem are satisfied and we can say there is a function Y_r on $[-r, r]$ so that $S_n'' \xrightarrow{\text{unif}} Y_r$ on $[0, t]$ and $Y_r' = F$. Since limits are unique, we then have $Y_r = E$ with $E' = F$. Using the results from the first and second derived series, we have $S''' = E' = F$.

Then given any $t \in (-1, 1)$, $t \in [-r, r]$ for some r. Hence, $S' = D$ there also. We conclude $S''' = E' = F$ on $(-1, 1)$. Hence, on $(-1, 1)$

$$E'(t) = \left(\sum_{n=2}^{\infty} \frac{5(n-1)n}{8(n+1)^2} t^{n-2} \right)' = \sum_{n=3}^{\infty} \frac{5(n-2)(n-1)n}{8(n+1)^2} t^{n-3} = F(t)$$

$$S'''(t) = \left(\sum_{n=0}^{\infty} \frac{5}{8(n+1)^2} t^{n-1} \right)''' = \left(\sum_{n=1}^{\infty} \frac{5n}{8(n+1)^2} t^{n-1} \right)''$$

$$= \left(\sum_{n=2}^{\infty} \frac{5(n-1)n}{8(n+1)^2} t^{n-2} \right)'$$

$$= \sum_{n=3}^{\infty} \frac{5(n-2)(n-1)n}{8(n+1)^2} t^{n-3} = F(t)$$

25.4.2   Homework

Exercise 25.4.1 *Examine the convergence of the derived Series of $\sum_{n=0}^{\infty} \frac{5}{n 4^n} t^n$.*

Exercise 25.4.2 *Examine the convergence of the derived series and the second derived Series of $\sum_{n=0}^{\infty} \frac{6}{n^2} t^n$.*

Exercise 25.4.3 *Examine the convergence of the derived series, second derived and third derived Series of $\sum_{n=0}^{\infty} \frac{6n}{(4n+2)^2} t^n$.*

Exercise 25.4.4 *Examine the convergence of the derived series, second derived and third derived Series of $\sum_{n=0}^{\infty} \frac{6n-3}{(4n^2+2n-5)^2} t^n$.*

Exercise 25.4.5 *Examine the convergence of the derived series, second derived and third derived Series of $\sum_{n=0}^{\infty} \frac{2n^2-5n+8}{(4n+2)3^n} t^n$.*

25.5   Power Series

With all these preliminaries behind us, we can now look at what are called general power series.

Definition 25.5.1 General Power Series

The **power series** *in t at base point p is* $\sum_{n=0}^{\infty} a_n (t - p)^n$ *where* (a_n) *is a sequence of real numbers.*

The largest interval $(p - R, p + R)$ *where this series converges is the* **interval of convergence** *of the power series and R is called the* **radius of convergence.**

The **derived series** *of this series is* $\sum_{n=1}^{\infty} n\, a_n (t - p)^{n-1}$ *and the* **integrated series** *of this series is* $\sum_{n=0}^{\infty} \frac{a_n}{n+1} (t - p)^{n+1}$.

From our work in the many examples we have done, we suspect the radius of convergence and interval of convergence for the derived and integrated series is the same as the original series. We also can see all power series converge at one point for sure: the base point p. We also suspect our tool of choice here is the ratio test and the Second Weierstrass Uniform Convergence Theorem. Here is the result.

Theorem 25.5.1 Power Series Convergence Results

Given the power series $\sum_{n=0}^{\infty} a_n (t - p)^n$, *the radius of convergence of the series is* $R = 1/\rho$ *where* $\rho = \lim_{n\to\infty} \frac{|a_{n+1}|}{|a_n|}$ *as long as this limit exists. If* $\rho = 0$, *then* $R = \infty$ *and the series converges everywhere. If* $\rho = \infty$, $R = 0$ *and the series only converges at the base point p.*

The radius of convergence of the derived and integrated series is also R.

The series, the derived series and the integrated series are continuous on the interval of convergence and these series converge uniformly on compact subsets of this interval. If $R = 0$, *then all three series converge at only the base point p and questions of continuity are moot.*

Further, on the interval of convergence, the derived series is the same as taking the derivative of the original series term by term and the integrated series is the same as integrating the original series term by term.

$$\left(\sum_{n=0}^{\infty} a_n(t - p)^n\right)' = \sum_{n=0}^{\infty} (a_n(t - p)^n)' = \sum_{n=0}^{\infty} n\, a_n(t - p)^{n-1}$$

$$\int_a^b \left(\sum_{n=0}^{\infty} a_n(t - p)^n\right) dt = \sum_{n=0}^{\infty} \int_a^b a_n(t - p)^n\, dt = \sum_{n=0}^{\infty} \int_a^b \frac{a_n}{n+1}(t - p)^{n+1} dt$$

Proof 25.5.1

Using the ratio test, we have

$$\lim_{n\to\infty} \frac{|a_{n+1}|\, |t - p|^{n+1}}{|a_n|\, |t - p|^n} = \lim_{n\to\infty} \frac{|a_{n+1}|}{|a_n|} |t - p| = \rho\, |t - p|.$$

where we assume this limit exists. The series thus converges when $|t - p| < 1/\rho$ *and letting* $R = 1/\rho$, *we see the interval of convergence if* $(p - R, p + R)$. *Let the limit function be S.*

The derived series has a different ratio test calculation: we get

$$\lim_{n\to\infty} \frac{(n+1)|a_{n+1}|\,|t-p|^{n+1}}{n|a_n|\,|t-p|^n} = \lim_{n\to\infty} \frac{n+1}{n}\,\frac{|a_{n+1}|}{|a_n|}\,|t-p| = \rho\,|t-p|.$$

The series thus converges when $|t-p| < 1/\rho$ and letting $R = 1/\rho$, we see the interval of convergence if $(p - R, p + r)$. Let the limit function be D.

The integrated series is handled similarly.

$$\lim_{n\to\infty} \frac{\frac{|a_{n+1}|}{n+2}\,|t-p|^{n+1}}{\frac{|a_n|}{n}\,|t-p|^n} = \lim_{n\to\infty} \frac{n}{n+1}\,\frac{|a_{n+1}|}{|a_n|}\,|t-p| = \rho\,|t-p|.$$

The series thus converges when $|t-p| < 1/\rho$ and letting $R = 1/\rho$, we see the interval of convergence if $(p - R, p + r)$. Let the limit function be J.

Thus, both the derived and integrated series have the same radius of convergence and interval of convergence as the original series.

The case where $\rho = \infty$ giving $R = 0$ means the series only converges at $t = p$. For this case, the question of uniform convergence is moot.

To examine uniform convergence of these series when $R > 0$, first consider the case where R is finite; i.e. $\rho > 0$ and finite. Look at the intervals $[p - R + r, p + R - r] \subset (p - R, p + R)$. We see $-R + r \leq t - p \leq R - r$ implies $|t - p| < R - r$. Then letting $\beta = 1 - r/R < 1$. we have

$$\sum_{n=0}^{\infty} |a_n|\,|t-p|^n \leq \sum_{n=0}^{\infty} |a_n|(R-r)^n = \sum_{n=0}^{\infty} |a_n|\beta^n R^n$$

Define $K_n = |a_n|\beta^n R^n$. Using the ratio test, we see

$$\lim_{n\to\infty} \frac{|a_{n+1}|\beta^{n+1} R^{n+1}}{|a_n|\beta^n R^n} = \lim_{n\to\infty} \frac{|a_{n+1}|}{|a_n|}\,\beta\,R = \rho\,\beta\,R = (1/R)\beta\,R = \beta < 1$$

and so $\sum_{n=0}^{\infty} K_n$ converges. This tells us the series converges uniformly on $[p - R + r, p + R - r]$ to a function U.

Since the partial sums of the series are polynomials, the limit function U is continuous by the Second Weierstrass Uniform Convergence Theorem for series. Further, by uniqueness of limits $S = U$ on $[p - R + r, p + R - r]$. Next, given any $t \in (p - R, p + R)$, t is in some $[p - R + r, p + R - r]$ and so S is continuous at t. Since t is arbitrary, S is continuous on $(p - R, p + R)$.

If $\rho = 0$ giving us $R = \infty$. Look at intervals of the form $[p - T, p + T]$ for any $T > 0$. Then $|t - p| < T$ and

$$\sum_{n=0}^{\infty} |a_n|\,|t-p|^n \leq \sum_{n=0}^{\infty} |a_n|T^n$$

Define $K_n = |a_n|T^n$. Using the ratio test, we see

$$\lim_{n\to\infty} \frac{|a_{n+1}|T^{n+1}}{|a_n|T^n} = \lim_{n\to\infty} \frac{|a_{n+1}|}{|a_n|}\,T = \rho\,T = 0$$

and so $\sum_{n=0}^{\infty} K_n$ converges. This tells us the series converges uniformly on $[p - T, p + T]$ to a function U.

Since the partial sums of the series are polynomials, the limit function U is continuous by the Second Weierstrass Uniform Convergence Theorem for series. Further, by uniqueness of limits $S = U$ on $[p - T, p + T]$. Next, given any $t \in \Re$, t is in some $[p - T, p + T]$ and so S is continuous at t. Since t is arbitrary, S is continuous on $\Re$.

The arguments to show the derived and integrated series converge to a continuous function on $(p - R, p + R)$ with the convergence uniform on compact subsets $[a, b] \subset (p - R, p + R)$ are quite similar and are omitted. Hence, we know

- *$S'_n \xrightarrow{\text{unif}} D$ on $[a, b]$ and $S'_n \xrightarrow{\text{ptws}} D$ on $(p - R, p + R)$ with D continuous on $(p - R, p + R)$.*

- *If we let J_n be the antiderivative of S_n with integration constant chosen to be zero, then $J_n \xrightarrow{\text{unif}} J$ on $[a, b]$ and $J_n \xrightarrow{\text{ptws}} J$ on $(p - R, p + R)$ with J continuous on $(p - R, p + R)$.*

To see that $S' = D$, we need to check the conditions of the Derivative Interchange Theorem:

1. *Fix a t in $[p - K, p + K]$ for $K \in [0, R)$. S_n is differentiable on $[p - K, p + K]$:* **True**.

2. *S'_n is Riemann Integrable on $[p - K, p + K]$:* **True as each is a polynomial**.

3. *There is at least one point $t_0 \in [p - K, p + K]$ such that the sequence $(S_n(t_0))$ converges.* **True as the original series converges at** $t = p$.

4. *$S'_n \xrightarrow{\text{unif}} y$ on $[p - K, p + K]$ and the limit function y is continuous.* **True as $S'_n \xrightarrow{\text{unif}} D$ on $[p - K, p + K]$ and D is continuous**.

The conditions of derivative interchange theorem are thus satisfied and we can say there is a function W_K on $[p - K, p + K]$ so that $S_n \xrightarrow{\text{unif}} W_K$ on $[p - K, p + K]$ and $W'_K = D$. Since limits are unique, we then have $W_K = S$ with $S' = D$.

Then given any $t \in (p - R, p + r)$, $t \in [p - K, p + K]$ for some K. Hence, $S' = D$ there also. We conclude $S' = D$ on $(p - R, p + R)$. Hence, on $(p - R, p + R)$, $S' = D$.

If we look at the integrated series, we have $J_n = \int S_n \xrightarrow{\text{unif}} J$ on $[a, b] \subset (p - R, p + R)$. Since convergence is uniform on $[a, b]$, we have $\lim_{n \to \infty} \int_a^b S_n(t) \, dt = \int_a^b (\lim_{n \to \infty} S_n(t)) \, dt$ or $\lim_{n \to \infty} \int_a^b S_n(t) \, dt = \int_a^b S(t) \, dt.$ ∎

Using these ideas we can define a new vector space of functions: the analytic functions.

Definition 25.5.2 Analytic Functions

> *A function f is called **analytic** at p if there is a power series $\sum_{n=0}^{\infty} a_n(t - p)^n$ with a positive radius of convergence R so that $f(t) = \sum_{n=0}^{\infty} a_n(t - p)^n$ on the interval of convergence $(p - R, p + R)$. We say f has a local power series expansion in this case.*

Comment 25.5.1 *There are lots of consequences here.*

- *The power series converges to a function S which is continuous on $(p - R, p + R)$ and so the function f must be continuous there as well. Hence, we are saying the function f has an alternate series form which matches it exactly on the interval of convergence.*

- We know the series expansion is differentiable and the derived series that results has the same radius of convergence. Hence, we know $f'(t) = \sum_{n=1}^{\infty} n\, a_n (t - p)^{n-1}$. This immediately says $f'(p) = a_1$.

We can find as many higher order derived series as we wish. Thus, we know f must be infinitely differentiable and that the kth derivative must be $f^{(k)} = \sum_{n=k}^{\infty} n(n-1) \ldots (n - k + 1)\, a_n (t - p)^{n-k}$. This tells us $f^{(k)}(p) = k(k-1) \ldots (k - k + 1)a_k = k!a_k$. Hence, we see $a_k = f^{(k)}(p)/k!$ which is the same coefficient we see in Taylor polynomials.

Comment 25.5.2 *It is easy to see sums, differences and scalar multiples of functions analytic at $t = p$ are still analytic. These combinations may have a different radius of convergence but they will still have a finite one and hence are analytic at $t = p$. The set of functions analytic at $t = p$ is thus a vector space of functions.*

Not all functions are analytic.

- $f(t) = \sqrt{t}$ is not analytic at $t = 0$ even though it is continuous on $[0, \infty)$ because it does not have derivatives at $t = 0$,

- The C^{∞} bump functions discussed in Section 6.1.6 are not analytic at their endpoints. Recall the function $h_{a\,b}(x)$ defined by for any $a < b$ is given by

$$
h_{a\,b}(x) = f_a(x)\, g_b(x) \;=\; \begin{cases} 0, & x \le a \\ e^{\frac{b-a}{(x-a)(x-b)}}, & a < x < b \\ 0, & x \ge b \end{cases}
$$

We can show the nth derivative $h_{a\,b}^{(n)}(x)$ is zero at $x = a$ and $x = b$ for all orders n. Hence, if $h_{a\,b}(x)$ was analytic at $x = a$, we would have

$$
h_{a\,b}(x) \;=\; \sum_{n=0}^{\infty} a_n (x - a)^p
$$

and by our earlier comments, we know $a_n = h_{a\,b}^{(n)}(a)/n! = 0$. Thus, the power series would sum to identically zero locally at $x = a$ and that contradicts the behavior of the bump function to the right of $x = a$. We can say similar things about what happens at $x = b$.

- Another example is $f(t) = |t|$ which is not differentiable at $x = 0$. It cannot have a local power series expansion at $x = 0$.

What about the Weierstrass Approximation Theorem? The functions $f(t) = |t|$ is continuous locally at $t = 0$ and so there is a sequence of Bernstein polynomials $p_n(t)$ so that $p_n \xrightarrow{\text{unif}} f$ on say $[-1, 1]$. Then we can write

$$
p_n(t) \;=\; \sum_{k=0}^{Q_n} a_{k,n}\, t^n = a_{0,n} + a_{1,n}t + a_{2,n}t^2 + \ldots + a_{Q_n,n}t^{Q_n}
$$

There is no guarantee in the Weierstrass Approximation Theorem that the degree of these polynomials increase monotonically. For example, we could have $Q_1 = 5$, $Q_2 = 3$, $Q_3 = 7$, $Q_4 = 11$ and so forth. To say it differently, suppose we stop at $n = 100$ and let $N = \max\{Q_1, \ldots, Q_{100}\}$. Then all the polynomials can be written as degree N polynomials by just letting the coefficients corresponding to powers of t not is the original polynomial be zero. For example, if $N = 200$

$$p_1(t) \quad = \quad a_{0,1} + a_{1,1}t + a_{2,1}t^2 + \ldots + a_{5,1}t^5$$

would be a polynomial of degree 200 by just setting the coefficients $a_{k,1} = 0$ if $k > 5$. We would then do a similar thing for p_2 and so on. We could have

$$p_2(t) \quad = \quad a_{0,2} + a_{1,2}t + a_{2,2}t^2 + \ldots + a_{7,2}t^7$$

which is also a polynomial of degree 200 by setting some coefficients to zero. If there was a power series expansion, we would have to have a sequence (b_n) so that

$$p_1(t) \quad = \quad a_{0,1} + a_{1,1}t + a_{2,1}t^2 + \ldots + a_{5,1}t^5 = b_0 + b_1t + \ldots + b_5t^5$$
$$p_2(t) \quad = \quad a_{0,2} + a_{1,2}t + a_{2,2}t^2 + \ldots + a_{7,2}t^7 = b_0 + b_1t + \ldots + b_5t^5 + b_6t^6 + b_7t^7$$

which implies equality in coefficients for the p_1 and p_2 which we do not know is true. Hence, the polynomials we find invoking the Weierstrass Approximation Theorem do not, in general, come from a power series expansion.

Homework

Exercise 25.5.1 *If f, g and h are analytic at $x = p$, prove $3f + 5g - 4h$ is analytic at $x = p$.*

Exercise 25.5.2 *Determine if $f(x) = \sin(x)$ is analytic on $\Re$ and if so find its expansion.*

Exercise 25.5.3 *Determine if $f(x) = \cos(x)$ is analytic on $\Re$ and if so find its expansion.*

Exercise 25.5.4 *Determine where $f(x) = \ln(x)$ is analytic and find its expansion.*

Exercise 25.5.5 *Determine where $f(x) = e^{-2x}$ is analytic and find its expansion.*

Comment 25.5.3 *It is easy to see the collection of all continuous functions locally defined on $\Re$ at the point p is a vector space. From our discussions above, we see the set of function analytic at p is a strict subset of this space.*

It should be clear now that the vector space of Riemann Integrable functions on the set $[a, b]$ is quite interesting. The vector space of all continuous functions on $[a, b]$ is a strict subset of it as there are many discontinuous functions which are Riemann Integrable.

The set of all polynomials is also a vector space on $[a, b]$ and from the Weierstrass Approximation Theorem the metric space $(C([a, b]), || \cdot ||_\infty)$ has within it a very special subset.

Definition 25.5.3 Dense Subsets of a Metric Space

We say E is a dense subset of the metric space (X, d) if given x in E, for all $\epsilon > 0$, there is an $y \in E$ so that $d(x, y) < \epsilon$. If the set E is countable, we say E is a countable dense subset. A metric space with a countable dense subset is called **separable***.*

Comment 25.5.4   • $\mathbb{Q}$ *is countably dense in $\Re$.*

   • *By the Weierstrass Approximation Theorem, the vector space of polynomials on $[a, b]$ is dense in $(C([a, b]), || \cdot ||_\infty)$. If we restrict our attention to polynomials with rational coefficients, we can show that it is a countably dense subset. Here is a sketch of the proof of this statement. Given $\epsilon > 0$, there is a polynomial p so that $||f - p||_\infty < \epsilon/2$ on $[a, b]$. We have $p(x) = \sum_{k=0}^N a_k x^k$ where N is the degree of the polynomial p. Now the function $\theta(x) = 1 + |x| + |x|^2 + \ldots + |x|^N$ is continuous on $[a, b]$, so there is a positive constant B so that $||\theta||_\infty < B$*

on $[a, b]$. Since $\mathbb{Q}$ is dense in $\mathfrak{R}$, there are rationals $\{r_0, \ldots, r_N\}$ so that $|r_i - a_i| < \epsilon/(2B)$. Then if q is the polynomial $q(x) = \sum_{k=0}^{N} r_k x^k$,

$$|q(x) - p(x)| \leq \sum_{k=0}^{N} |a_k - r_k||x|^k < \epsilon/(2B)\,\theta(x) < \epsilon/(2)$$

which implies $||q - p||_\infty \leq \epsilon/2$. Combining, we see $||f - q|| < \epsilon/2 + \epsilon/2 = \epsilon$.

Why is this set countable? We need a general theorem for this.

Theorem 25.5.2 The countable union of countable sets is countable

A finite union of countable sets is countable and an countably infinite union of countable sets is countable.

Proof 25.5.2

The first part is an induction. The base case for $k = 1$ is obvious, so let's assume any union of k countable sets is countable. Let A_1 to A_k be countable sets and then by assumption $B = \cup_{i=1}^{k} A_i$ is countable. Thus there is a $1-1$ and onto function ϕ between B and $\mathbb{N}$. Let $C = \cup_{i=1}^{k+1} A_i = B \cup A_{k+1}$. We know A_{k+1} is countable so there is a function θ that is $1-1$ and onto between A_{k+1} and $\mathbb{N}$. Write C as the disjoint union

$$C = C_1 \cup C_2 \cup C_3 = (B \cap A_{k+1}) \cup (B^C \cap A_{k+1}) \cup (B \cap A_{k+1}^C).$$

Each of these sets is countable so they can be labeled as $C_1 = \{r_i\}$, $C_2 = \{s_i\}$ and $C_3 = \{t_i\}$. Define the $1-1$ and onto function ψ by

$$\psi(x) = \begin{cases} 3n - 2, & x = r_n \in C_1 \\ 3n - 1, & x = s_n \in C_2 \\ 3n, & x = t_n \in C_3 \end{cases}$$

Hence, C is countable. This completes the POMI proof.

To show countably infinite unions of countable sets are countable, we can use a similar argument. We are given a collection $\{A_i\}$ of countable sets. We can show that we can rewrite the union $\cup_{i=1}^{\infty} A_i$ as a union of disjoint sets by defining the sequence of sets $\{B_i\}$ by $B_1 = A_1$ and $B_i = A_i \cap B_{i-1}^C$. Then $\cup_{i=1}^{\infty} A_i = \cup_{i=1}^{\infty} B_i$. Thus, each B_i is countable and can be enumerated as $B_i = \{a_{j,i}\}$. Define the $1-1$ and onto function ψ by

$$\psi(x) = \begin{cases} \{1/1, 2/1, 3/1, \ldots, j/1, \ldots\}, & x = a_{j,i} \in B_1 \\ \{1/2, 2/2, 3/2, \ldots, j/2, \ldots\}, & x = a_{j,2} \in B_2 \\ \vdots & \vdots \\ \{1/n, 2/n, 3/n, \ldots, j/n, \ldots\}, & x = a_{j,n} \in B_n \\ \vdots & \vdots \end{cases}$$

The range of ψ is the positive rationals with repeats and we already know this is a countable set. Thus, $\cup_{i=1}^{\infty} A_i$ is countable. ∎

With this theorem, we can show the set of polynomials with rational coefficients is countable. Call this set $\mathbb{P}$ and call the polynomials of degree less than or equal to n, with rational coefficients $\mathbb{P}_n$. Then $\mathbb{P} = \cup_{n=0}^{\infty} \mathbb{P}_n$. We can show each $\mathbb{P}_n$ is countable and so the union of all of them is countable also. Hence the polynomials with rational coefficients are dense in $(C([a, b]), || \cdot ||_\infty)$. This says

$(C([a,b]), ||\cdot||_\infty)$ is a **separable metric space** as it possesses a countable dense subset.

Since each A_i is countable, we know it has content zero and by arguments we have done earlier, we also know $\cup_{i=1}^\infty A_i$ has content zero. However, we do not know that a set of content zero must be countable so we cannot use this argument to prove the countability of $\cup_{i=1}^\infty A_i$. As a matter of fact, we can construct sets of content zero that are not countable! We do this in a course on new ways to define integration using what are called measures.

25.5.1 Homework

Exercise 25.5.6 *Show $\cup_{i=1}^\infty A_i = \cup_{i=1}^\infty B_i$ where $B_1 = A_1$ and $B_n = A_n \cap B_{n-1}^C$.*

Exercise 25.5.7 *Show $\mathbb{P}_1$ is countable.*

Exercise 25.5.8 *Show $\mathbb{P}_2$ is countable.*

Exercise 25.5.9 *Show $\mathbb{P}_3$ is countable.*

Exercise 25.5.10 *Show $\mathbb{P}_N$ is countable.*

Exercise 25.5.11 *Show $\cup_{N=0}^\infty \mathbb{P}_N$ is countable.*

Chapter 26

Riemann Integration: Discontinuities and Compositions

We now want to study how many discontinuities a Riemann Integrable function can have as well as a few more ideas that until now we have not had the tools to explore adequately: is the composition of two Riemann Integrable functions also Riemann integrable? We know compositions of continuous functions and differentiable functions preserve those properties, so this is a reasonable question. As for the answer, we will see...

26.1 Compositions of Riemann Integrable Functions

We already know that continuous functions and monotone functions are classes of functions which are Riemann Integrable on the interval $[a, b]$. A good reference for some of the material in this section is (Douglas (2) 1996) although it is mostly in problems and not in the text! Hence, since $f(x) = \sqrt{x}$ is continuous on $[0, M]$ for any positive M, we know f is Riemann integrable on this interval. What about the compositions of two Riemann integrable functions? Does this composition always result in a new integrable function?

Consider the composition $\sqrt{g}$ where g is just known to be nonnegative and Riemann integrable on $[a, b]$. If g were continuous, since compositions of continuous functions are also continuous, we would have immediately that $\sqrt{g}$ is Riemann Integrable. However, it is not so easy to handle the case where we only know g is Riemann integrable. Let's try this approach. Using Theorem 24.1.1, we know given a finite interval $[c, d]$, there is a sequence of polynomials $\{p_n(x)\}$ which converge uniformly to $\sqrt{x}$ on $[c, d]$. Of course, the polynomials in this sequence will change if we change the interval $[c, d]$, but you get the idea. To apply this here, note that since g is Riemann Integrable on $[a, b]$, g must be bounded. Since we assume g is nonnegative, we know that there is a positive number M so that $g(x)$ is in $[0, M]$ for all x in $[a, b]$. Thus, there is a sequence of polynomials $\{p_n\}$ which converge uniformly to $\sqrt{\cdot}$ on $[0, M]$.

Next, using Theorem 22.3.2, we know a polynomial in g is also Riemann integrable on $[a, b]$ ($f^2 = f \cdot f$ so it is integrable and so on). Hence, $p_n(g)$ is Riemann integrable on $[a, b]$. Then given $\epsilon > 0$, we know there is a positive N so that

$$| p_n(u) - \sqrt{u} | \quad < \quad \epsilon, \text{ if } n > N \text{ and } u \in [0, M].$$

Thus, in particular, since $g(x) \in [0, M]$, we have

$$| p_n(g(x)) - \sqrt{g(x)} | \quad < \quad \epsilon, \text{ if } n > N \text{ and } x \in [a, b].$$

We have therefore proved that $p_n \circ g$ converges uniformly to $\sqrt{g}$ on $[0, M]$. Then by Theorem 24.3.4, we see $\sqrt{g}$ is Riemann integrable on $[0, M]$.

If you think about it a bit, you should be able to see that this type of argument would work for any f which is continuous and g that is Riemann integrable. We state this as Theorem 26.1.1.

Theorem 26.1.1 f **Continuous and** g **Riemann Integrable Implies** $f \circ g$ **is Riemann Integrable**

> *If f is continuous on $g([a, b])$ where g is Riemann Integrable on $[a, b]$, then $f \circ g$ is Riemann Integrable on $[a, b]$.*

Proof 26.1.1

This proof is for you.   ∎

In general, the composition of Riemann Integrable functions is not Riemann integrable. Here is the standard counterexample. This great example comes from (Douglas (2) 1996). Define f on $[0, 1]$ by

$$f(y) \quad = \quad \begin{cases} 1 & \text{if } y = 0 \\ 0 & \text{if } 0 < y \le 1 \end{cases}$$

and g on $[0, 1]$ by

$$g(x) \quad = \quad \begin{cases} 1 & \text{if } x = 0 \\ 1/q & \text{if } x = p/q, (p, q) = 1, x \in (0, 1] \text{ and } x \text{ is rational} \\ 0 & \text{if } x \in (0, 1] \text{ and } x \text{ is irrational} \end{cases}$$

We see immediately that f is integrable on $[0, 1]$ by Theorem 23.7.4. We already know this function is continuous at each irrational point in $[0, 1]$ and discontinuous at all rational points in $[0, 1]$ from Section 7.1.2. We can show g is integrable with the following argument. First define g to be 0 at $x = 1$. If we show this modification of g is RI, g will be too. Let q be a prime number bigger than 2. Then form the uniform partition $\pi_q = \{0, 1/q, \ldots, (q-1)/q, 1\}$. On each subinterval of this partition, we see $m_j = 0$. Inside the subinterval $[(j-1)/q, j/q]$, the maximum value of g is $1/q$. Hence, we have $M_j = 1/q$. This gives

$$U(g, \pi_q) - L(g, \pi_q) \quad = \quad \sum_{\pi_q} (1/q)\Delta x_i = 1/q$$

Given $\epsilon > 0$, there is an N so that $1/N < \epsilon$. Thus if q_0 is a prime with $q_0 > N$, we have $U(g, \pi_{q_0}) - L(g, \pi_{q_0}) < \epsilon$.

So if π is any refinement of π_{q_0}, we have $U(g, \pi) - L(g, \pi) < \epsilon$ also, Thus g satisfies the Riemann Criterion and so g is RI on $[0, 1]$. It is also easy to see $\int_0^1 g(s)ds = 0$.

Now $f \circ g$ becomes

$$f(g(x)) \quad = \quad \begin{cases} f(1) & \text{if } x = 0 \\ f(1/q) & \text{if } x = p/q, (p, q) = 1, x \in (0, 1] \text{ and } x \text{ rational} \\ f(0) & \text{if } 0 < x < 1 \text{ and } x \text{ irrational} \end{cases}$$

$$= \begin{cases} 1 & \text{if } x = 0 \\ 0 & \text{if if } x \text{ rational} \in (0,1) \\ 1 & \text{if if } x \text{ irrational} \in (0,1) \end{cases}$$

The function $f \circ g$ above is not Riemann integrable as $U(f \circ g) = 1$ and $L(f \circ g) = 0$. Thus, we have found two Riemann integrable functions whose composition is not Riemann integrable!

Homework

Exercise 26.1.1 *Prove Theorem 26.1.1.*

Exercise 26.1.2 *Repeat the arguments for the counterexample using the same g and now use*

$$f(y) = \begin{cases} 2 & \text{if } y = 0 \\ 0 & \text{if } 0 < y \le 1 \end{cases}$$

26.2   Sets of Content Zero

We already know the length of the finite interval $[a, b]$ is $b - a$ and we exploit this to develop the Riemann integral when we compute lower, upper and Riemann sums for a given partition. We also know that the set of discontinuities of a monotone function is countable. We have seen that continuous functions with a finite number of discontinuities are integrable and in the last section, we saw a function which was discontinuous on a countably infinite set and still was integrable! Hence, we suspect that if a function is integrable, this should imply something about its discontinuity set. However, the concept of length doesn't seem to apply as there are no intervals in these discontinuity sets. With that in mind, let's introduce a new notion: the *content* of a set. We will follow the development of a set of content zero as it is done in (Sagan (17) 1974).

Definition 26.2.1 Sets of Content Zero

A subset S of $\Re$ is said to have content zero if and only if given any positive ϵ we can find a sequence of bounded open intervals $\{J_n^\epsilon = (a_n, b_n)\}$ either finite in number or infinite so that

$$S \subseteq \cup J_n,$$

with the total length

$$\sum (b_n - a_n) < \epsilon$$

If the sequence only has a finite number of intervals, the union and sum are written from 1 to N where N is the number of intervals and if there are infinitely many intervals, the sum and union are written from 1 to ∞.

Comment 26.2.1 *Here are some examples of sets of content zero.*

1. *A single point c in $\Re$ has content zero because $c \in (c - \epsilon/2, c + \epsilon/2)$ for all positive ϵ.*

2. *A finite number of points $S = \{c_1, \ldots, c_k\}$ in $\Re$ has content zero because $B_i = c_i \in (c_i - \epsilon/(2k), c_i + \epsilon/(2k))$ for all positive ϵ. Thus, $S \subseteq \cup_{i=1}^k B_i$ and the total length of these intervals is smaller than ϵ.*

3. *The rational numbers have content zero also. Let $\{c_i\}$ be any enumeration of the rationals. Let $B_i = (c_i - \epsilon/(2^i), c_i + \epsilon/(2^i))$ for any positive ϵ. The Q is contained in the union of these intervals and the length is smaller than $\epsilon \sum_{i=1}^\infty 1/2^i = \epsilon$.*

4. *Finite unions of sets of content zero also have content zero.*

5. *Subsets of sets of content zero also have content zero.*

We can also discuss the **content** of a general subset S. This is the definition:

Definition 26.2.2 The Content of a Set

Let S be a set of real numbers. Let $\mathbb{O} = \{O_n : n \in \mathbb{N}\}$ be a collection of open intervals $O_n = (a_n, b_n)$ so that $S \subset \cup_{n=1}^{\infty} O_n$. Let $m(\mathbb{O}) = \sum_{n=1}^{\infty}(b_n - a_n)$ be the generalized length of $\mathbb{O}$. We call such a collection an open cover of S. The content of S is

$$c(S) \quad = \quad \inf\{m(\mathbb{O}) : \mathbb{O} \text{ is an open cover of } S\}$$

It is clear that if S is a set of content zero, then $c(S) = 0$.

Comment 26.2.2 *If $S = (a, b)$ is a finite interval, the $c(S) = b - a$, the usual length of S. This is easy to see as $(a - \epsilon, b + \epsilon)$ is an open cover of (a, b) for any positive ϵ. Now these open covers form a sub collection of all possible open covers of $[a, b]$, so $c(S)$ must be smaller than or equal to the infimum over these covers which is $b - a$. We have $c(S) \leq b - a$.*

On the other hand, $m((a, b)) \leq m(\mathbb{O})$ for any cover of (a, b). Thus, $c(S) \geq b - a$ too. Combining $c(S) = b - a$.

The argument above also shows with little change that $c([a, b]) \leq b - a$. Next, note it is easy to see if $S_1 \subset S_2$, then $c(S_1) \leq s(S_2)$. Hence, we have $b - a = c((a, b)) \leq c([a, b]) \leq b - a$. We conclude $c([a, b]) = b - a$ also.

Comment 26.2.3 *It is also easy to see that $c(S_1 \cup S_2) \leq c(S_1) + c(S_2)$ for any two sets S_1 and S_2.*

The content of a set is an example of what is called an outer measure which is a fairly deep idea. In this deeper theory, not all subsets of an given set need to be nice enough to have a content like we expect. We need a definition to shows us which ones are the good ones.

We can use these ideas to prove a fundamental fact about intervals of the real line.

Theorem 26.2.1 The interval $[a, b]$ with $a < b$ is uncountable.

If $a < b$, $[a, b]$ is uncountable.

Proof 26.2.1

It is easy to see the interval $[a, b]$ cannot have content zero as the length of $[a, b]$ is $b - a$ and if we pick $\epsilon < b - a$, any collection of covering intervals will have to have a length that adds up to more than $b - a$. Since the tolerance cannot be arbitrarily small, $[a, b]$ is not content zero. Thus, if we assumed $[a, b]$ was countable, $[a, b]$ would be content zero which is not possible. We conclude $[a, b]$ must be uncountable. ∎

Further

Theorem 26.2.2 The set $\mathbb{I} \cap [a, b]$ has content $b - a$

$c(\mathbb{I} \cap [a, b]) = b - a$

Proof 26.2.2

Since $[a, b] = \mathbb{Q} \cap [a, b] \cup \mathbb{I} \cap [a, b]$, we know

$$b - a \quad = \quad c([a, b]) \leq c(\mathbb{Q} \cap [a, b]) + c(\mathbb{I} \cap [a, b]) = c(\mathbb{I} \cap [a, b])$$

But we also know since $\mathbb{I} \cap [a,b] \subset [a,b]$ *that* $c(\mathbb{I} \cap [a,b]) \leq b - a$. *Combining, we see* $c(\mathbb{I} \cap [a,b]) = b - a$. ∎

Homework

Exercise 26.2.1 *Prove that if* F_1, F_2 *and* F_3 *are subsets* $[a,b]$ *with content zero, then* $F_1 \cup F_2 \cup F_3$ *has content zero.*

Exercise 26.2.2 *Prove* $A = \{1,2,3,4,5\}$ *is a set of content zero.*

Exercise 26.2.3 *Prove that if* $F_n \subseteq [a,b]$ *has content zero for all* n, *then* $F = \cup F_n$ *also has content zero.*

Exercise 26.2.4 *Prove any countably infinite set has content zero.*

Hence, in our example of two integrable functions whose composition is not integrable, the function g was continuous on $[0,1]$ except on a set of content zero. We make this more formal with a definition.

Definition 26.2.3 Continuous Almost Everywhere

The function f *defined on the interval* $[a,b]$ *is said to be continuous almost everywhere if the set of discontinuities of* f *has content zero. We abbreviate the phrase* almost everywhere *by writing a.e.*

26.2.1   The Riemann - Lebesgue Lemma

We are now ready to prove an important theorem which is known as the **Riemann - Lebesgue Lemma**. This is also called **Lebesgue's Criterion for the Riemann Integrability of Bounded Functions**. We follow the proof given in (Sagan (17) 1974).

Theorem 26.2.3 Riemann - Lebesgue Lemma

> (i) $f \in B([a,b])$ *and continuous a.e. implies* $f \in RI([a,b])$.
>
> (ii) $f \in RI([a,b])$ *implies* f *is continuous a.e.*

Proof 26.2.3
The proof of this result is fairly complicated. So grab a cup of coffee, a pencil and prepare for a long battle!
(i)$f \in B([a,b])$ *and continuous a.e. implies* $f \in RI([a,b])$:
We will prove this by showing that for any positive ϵ, *we can find a partition* π_0 *so that the Riemann Criterion is satisfied. First, since* f *is bounded, there are numbers* m *and* M *so that* $m \leq f(x) \leq M$ *for all* x *in* $[a,b]$. *If* m *and* M *were the same, then* f *would be constant and it would therefore be continuous. If this is the case, we know* f *is integrable. So we can assume without loss of generality that* $M - m > 0$. *Let* D *denote the set of points in* $[a,b]$ *where* f *is not continuous. By assumption, the content of* D *is zero. Hence, given a positive* ϵ *there is a sequence of bounded open intervals* $J_n = (a_n, b_n)$ *(we will assume without loss of generality that there are infinitely many such intervals) so that*

$$D \subseteq \cup J_n, \quad \sum (b_n - a_n) < \epsilon/(2(M - m)).$$

Now if x is from $[a, b]$, x is either in D or in the complement of D, D^C. Of course, if $x \in D^C$, then f is continuous at x. The set

$$E = [a, b] \cap \left(\cup J_n \right)^C$$

is compact and so f must be uniformly continuous on E. Hence, for the ϵ chosen, there is a $\delta > 0$ so that

$$| f(y) - f(x) | < \epsilon/(8(b - a)), \tag{$*$}$$

if $x, y \in E$ and $|x - y| < \delta$. Now look at this collection

$$O = \{ J_n, B_{\delta/2}(x) \mid x \in E \}$$

The union of this collection is

$$\left(\cup J_n \right) \cup \left\{ [a, b] \cap \left(\cup J_n \right)^C \right\} \;\; = \;\; \left(\cup J_n \right) \cup [a, b]$$

Since D is in $[a, b]$, each of the subintervals J_n may contain points out of $[a, b]$ but we still see $[a, b] \subset \cup J_n \cup [a, b]$. Thus, O is an open cover of $[a, b]$ and hence must have a finite sub cover. Call this finite sub cover O' and label its members as follows:

$$O' = \{ J_{n_1}, \dots, J_{n_r}, B_{\delta/2}(x_1), \dots, B_{\delta/2}(x_s) \}$$

Then it is also true that we know that

$$[a, b] \subseteq O'' = \{ J_{n_1}, \dots, J_{n_r}, B_{\delta/2}(x_1) \cap E, \dots, B_{\delta/2}(x_s) \cap E \}$$

Note this is no longer a fsc as the sets $B_{\delta/2}(x_i) \cap E$ need not be open. Also, the union of the J_{n_i} sets is in E^C because of the definition of E. So the union of the $B_{\delta/2}(x_i)$ sets contains the E part; hence restricting to $B_{\delta/2}(x_i) \cap E$ does not lose any points of $[a, b]$. So this union still covers $[a, b]$ and

$$[a, b] \subseteq O'' = \{ J_{n_1}, \dots, J_{n_r}, B_{\delta/2}(x_1) \cap E, \dots, B_{\delta/2}(x_s) \cap E \}$$

All of the intervals in O'' have endpoints. Throw out any duplicates and arrange these endpoints in increasing order in $[a, b]$ and label them as $y_1, \dots, y_{p-1}$. Then, let

$$\pi_0 = \{ y_0 = a, y_1, y_2, \dots, y_{p-1}, y_p = b \}$$

be the partition formed by these points.

Suppose two successive points y_{j-1} and y_j are not in the union of the J_{n_k}'s. Then (y_{j-1}, y_j) is in $(\cup J_{n_k})^C$. Thus, (y_{j-1}, y_j) lies in some $\hat{B}_{\delta/2}(x_i)$ and $[y_{j-1}, y_j] \subset \hat{B}_{\delta}(x_i)$. On the other hand if the two successive points y_{j-1} and y_j are in the union of the J_{n_k}'s, then y_{j-1} and y_j are endpoints of possibly two J_{n_k}'s. Since there is no point from π_0 between them, this means they both belong to some J_{n_k}.

Now we separate the index set $\{1, 2, \dots, p\}$ into two disjoint sets. We define A_1 to be the set of all indices j so that (y_{j-1}, y_j) is contained in some J_{n_k}. Then we set A_2 to be the complement of A_1 in the entire index set, i.e. $A_2 = \{1, 2, \dots, p\} - A_1$. Note, by our earlier remarks, if j is in A_2,

$[y_{j-1}, y_j]$ is contained in some $B_\delta(x_i) \cap E$. Thus,

$$U(f, \boldsymbol{\pi_0}) - L(f, \boldsymbol{\pi_0}) = \sum_{j=1}^{n} \left(M_j - m_j \right) \Delta y_j$$

$$= \sum_{j \in A_1} \left(M_j - m_j \right) \Delta y_j + \sum_{j \in A_2} \left(M_j - m_j \right) \Delta y_j$$

Let's work with the first sum: we have

$$\sum_{j \in A_1} \left(M_j - m_j \right) \Delta y_j \leq \left(M - m \right) \sum_{j \in A_1} \Delta y_j$$

$$< (M - m) \, \epsilon / (2(M - m)) = \epsilon / 2$$

Now if j is in A_2, then $[y_{j-1}, y_j]$ is contained in some $B_\delta(x_i) \cap E$. So any two points u and v in $[y_{j-1}, y_j]$ satisfy $|u - x_i| < \delta$ and $|v - x_i| < \delta$. Since these points are this close, the uniform continuity condition, Equation ∗, holds. Therefore

$$|f(u) - f(v)| \leq |f(u) - f(x_i)| + |f(v) - f(x_i)| < \epsilon / (4(b - a)).$$

This holds for any u and v in $[y_{j-1}, y_j]$. In particular, we can use the Supremum and Infimum Tolerance Lemma to choose u_j and v_j so that

$$M_j - \epsilon / (8(b - a)) < f(u_j), \quad m_j + \epsilon / (8(b - a)) > f(v_j).$$

It then follows that

$$M_j - m_j < f(u_j) - f(v_j) + \epsilon / (4(b - a)).$$

Now, we can finally estimate the second summation term. We have

$$\sum_{j \in A_2} \left(M_j - m_j \right) \Delta y_j < \sum_{j \in A_2} \left(|f(u_j) - f(v_j)| + \epsilon / (4(b - a)) \right) \Delta y_j$$

$$< \sum_{j \in A_2} \left(|f(u_j) - f(v_j)| \right) \Delta y_j + \epsilon / (4(b - a)) \sum_{j \in A_2} \Delta y_j$$

$$< \epsilon / (4(b - a)) \sum_{j \in A_2} \Delta y_j + \epsilon / (4(b - a)) \sum_{j \in A_2} \Delta y_j$$

$$< \epsilon / 2$$

Combining our estimates, we have

$$U(f, \boldsymbol{\pi_0}) - L(f, \boldsymbol{\pi_0}) = \sum_{j \in A_1} \left(M_j - m_j \right) \Delta y_j + \sum_{j \in A_2} \left(M_j - m_j \right) \Delta y_j$$

$$< \epsilon / 2 + \epsilon / 2 = \epsilon.$$

Any partition $\boldsymbol{\pi}$ that refines $\boldsymbol{\pi_0}$ will also satisfy $U(f, \boldsymbol{\pi}) - L(f, \boldsymbol{\pi}) < \epsilon$. Hence, f satisfies the Riemann Criterion and so f is integrable.

(ii)$f \in RI([a, b])$ implies f is continuous a.e.:
We begin by noting that if f is discontinuous at a point x in $[a, b]$, if and only if there is a positive

integer m so that

$$\forall \delta > 0, \ \exists y \in (x - \delta, x + \delta) \cap [a, b] \ni \ |f(y) - f(x)| \geq 1/m.$$

This allows us to define some interesting sets. Define the set E_m by

$$E_m \ = \ \{x \in [a, b] \mid \forall \delta > 0 \ \exists y \in (x - \delta, x + \delta) \cap [a, b] \ni \ |f(y) - f(x)| \geq 1/m, \}$$

Then, the set of discontinuities of f, $\mathbf{D}$ can be expressed as $\mathbf{D} = \cup_{j=1}^{\infty} E_m$. Now let $\pi = \{x_0, x_1, \ldots, x_n\}$ be any partition of $[a, b]$. Then, given any positive integer m, the open subinterval (x_{k-1}, x_k) either intersects E_m or it does not. Define

$$A_1 \ = \ \left\{ k \in \{1, \ldots, n\} \mid (x_{k-1}, x_k) \cap E_m \neq \emptyset \right\},$$

$$A_2 \ = \ \left\{ k \in \{1, \ldots, n\} \mid (x_{k-1}, x_k) \cap E_m = \emptyset \right\}$$

By construction, we have $A_1 \cap A_2 = \emptyset$ and $A_1 \cup A_2 = \{1, \ldots, n\}$. We assume f is integrable on $[a, b]$. So, by the Riemann Criterion, given $\epsilon > 0$, and a positive integer m, there is a partition π_0 such that

$$U(f, \pi) - L(f, \pi) < \epsilon/(2m), \ \forall \pi_0 \preceq \pi. \tag{**}$$

It follows that if $\pi_0 = \{y_0, y_1, \ldots, y_n\}$, then its index set can be written as the union of disjoint sets A_1 and A_2 like above. We then have

$$U(f, \pi_0) - L(f, \pi_0) \ = \ \sum_{k=1}^{n} (M_k - m_k) \Delta y_k$$

$$= \ \sum_{k \in A_1} (M_k - m_k) \Delta y_k \ + \ \sum_{k \in A_2} (M_k - m_k) \Delta y_k$$

If k is in A_1, then by definition, there is a point u_k in E_m and a point v_k in (y_{k-1}, y_k) so that $|f(u_k) - f(v_k)| \geq 1/m$. Also, since u_k and v_k are both in (y_{k-1}, y_k),

$$M_k - m_k \geq |f(u_k) - f(v_k)|.$$

Thus,

$$\sum_{k \in A_1} (M_k - m_k) \Delta y_k \ \geq \ \sum_{k \in A_1} |f(u_k) - f(v_k)| \Delta y_k \ \geq \ (1/m) \sum_{k \in A_1} \Delta y_k.$$

Thus, using Equation **, we find

$$\epsilon/(2m) \ > \ U(f, \pi_0) - L(f, \pi_0) \geq (1/m) \sum_{k \in A_1} \Delta y_k.$$

which implies $\sum_{k \in A_1} \Delta y_k < \epsilon/2$.

The partition π_0 divides $[a, b]$ as follows:

$$[a, b] \ = \ \left(\cup_{k \in A_1} (y_{k-1}, y_k) \right) \ \cup \ \left(\cup_{k \in A_2} (y_{k-1}, y_k) \right) \ \cup \ \left(\{y_0, \ldots, y_n\} \right)$$

$$= \quad C_1 \cup C_2 \cup \pi_0$$

By the way we constructed the sets E_m, we know E_m does not intersect C_2. Hence, we can say

$$E_m \quad = \quad \left(C_1 \cap E_m \right) \cup \left(E_m \cap \pi_0 \right)$$

Therefore, we have $C_1 \cap E_m \subseteq \cup_{k \in A_1} (y_{k-1}, y_k)$ with $\sum_{k \in A_1} \Delta y_k < \epsilon/2$. Since ϵ is arbitrary, we see $C_1 \cap E_m$ has content zero. The other set $E_m \cap \pi_0$ consists of finitely many points and so it also has content zero by the comments at the end of Definition 26.2.1. This shows that E_m has content zero since it is the union of two sets of content zero. We finish by noting $D = \cup E_m$ also has content zero. The proof of this we leave as an exercise.

This completes the full proof. ∎

Homework

Exercise 26.2.5 Let

$$f(x) \quad = \quad \begin{cases} 1 & if\, x = 0 \\ 0 & if\, 0 < x \leq 1 \end{cases}$$

Explain why f is integrable using the Riemann - Lebesgue Lemma.

Exercise 26.2.6 Let

$$g(x) \quad = \quad \begin{cases} 1 & if\, x = 0 \\ 1/q & if\, x = p/q, (p,q) = 1, x \in (0,1]\ and\ x\ is\ rational \\ 0 & if\, x \in (0,1]\ and\ x\ is\ irrational \end{cases}$$

Explain why g is integrable using the Riemann - Lebesgue Lemma.

Exercise 26.2.7 Let

$$f(x) \quad = \quad \begin{cases} 1, & x \in \mathbb{Q} \cap [0,1] \\ -1, & x \in \mathbb{I} \cap [0,1] \end{cases}$$

Explain why f is not integrable using the Riemann - Lebesgue Lemma.

Exercise 26.2.8 Let

$$f(x) \quad = \quad \begin{cases} 2, & x \in \mathbb{Q} \cap [0,1] \\ -3, & x \in \mathbb{I} \cap [0,1] \end{cases}$$

Explain why f is not integrable using the Riemann - Lebesgue Lemma.

Exercise 26.2.9 Let

$$f(x) \quad = \quad \begin{cases} 2x, & x \in \mathbb{Q} \cap [0,1] \\ -3x, & x \in \mathbb{I} \cap [0,1] \end{cases}$$

Explain why f is not integrable using the Riemann - Lebesgue Lemma.

26.2.2   Equivalence Classes of Riemann Integrable Functions

If f is Riemann Integrable and 0 a.e. on $[a, b]$, we know f^+ and f^- are also Riemann integrable and 0 a.e. Let's start with f^+. Since it is Riemann integrable, given $\epsilon > 0$, there is a partition π_0 so that $U(f^+, \pi) - L(f^+, \pi) < \epsilon$ for any refinement π of π_0. Since $f^+ = 0$ a.e., all the lower sums are zero telling us the lower Darboux integral is zero. We then have $U(f^+, \pi) < \epsilon$ for all refinements π of π_0. Hence, the upper Darboux integral is zero too and since they match, $\int_a^b f^+(t)dt = 0$.

A similar argument shows $\int_a^b f^-(t)dt = 0$ too. Combining we have $0 = \int_a^b (f^+(t) - f^-(t))dt = \int_a^b f(t)dt$. This is an important theorem.

Theorem 26.2.4 If f is continuous and zero a.e., the Riemann integral of f is zero

> *If f is Riemann Integrable on $[a, b]$ and is 0 a.e., then $\int_a^b f(t)dt = 0$.*

Proof 26.2.4
We just proved this.   ∎

We can prove a sort of converse:

Theorem 26.2.5 If f is nonnegative and Riemann integrable with zero integrable, then f is zero a.e.

> *If $f \geq 0$ is Riemann Integrable on $[a, b]$ and $\int_a^b |f(s)|ds = 0$ then $f = 0$ a.e.*

Proof 26.2.5
Since $|f|$ is integrable, $|f|$ is continuous a.e. Suppose x is a point where $|f|$ is continuous with $|f(x)| > 0$. Then there is an $r > 0$ so that on $[x - r, x + r]$, $|f(s)| > \frac{|f(x)|}{2}$. Thus

$$
\begin{aligned}
0 &= \int_a^b |f(s)|ds = \int_a^{x-r} |f(s)|ds + \int_{x-r}^{x+r} |f(s)|ds + \int_{x=r}^b |f(s)|ds \\
&\geq \int_{x-r}^{x+r} |f(s)|ds > r|f(x)| > 0
\end{aligned}
$$

This is not possible, so at each point where $|f|$ is continuous, $|f(x)| = 0$. Hence $|f|$ us zero a.e. ∎

It then immediately follows that

Theorem 26.2.6 Riemann integrable f and g equal a.e. have the same integral

> *If f and g are both Riemann Integrable on $[a, b]$ and $f = g$ a.e., then $\int_a^b f(t)dt = \int_a^b g(t)dt$.*

Proof 26.2.6
Let $h = f - g$. Then h is Riemann Integrable and $h = 0$ a.e. Thus, $0 = \int_a^b h(t)dt = \int_a^b f(t)dt - \int_a^b g(t)dt$. This gives the result.   ∎

and

Theorem 26.2.7 Riemann integrable f and g and $d_1(f, g) = 0$ implies f is g a.e.

If f and g are both Riemann Integrable on $[a, b]$ and $\int_a^b |f(s) - g(s)| ds = 0$ then $f = g$ a.e.

Proof 26.2.7
Let $h = f - g$. Then h is Riemann Integrable and $\int_a^b |h| = 0$. Thus $|f - g| = 0$ a.e. This gives the result. ∎

Comment 26.2.4 *However, just because a Riemann integrable function f equals another function g a.e. does not mean g is also Riemann integrable. Here is an easy example: let $f(x) = 1$ on $[0, 1]$ and $g(x) = 1$ on $\mathbb{I} \cap [0, 1]$ and -1 on the complement, $\mathbb{Q} \cap [0, 1]$. Then g is not Riemann integrable even though it is equal to the Riemann integrable f a.e.*

Let $RI([a, b])$ denote the set of all Riemann Integrable functions on $[a, b]$. We can then define an equivalence relation $\sim$ on $(RI([a, b]), \| \cdot \|_1)$ by saying $f \sim g$ if $f = g$ a.e. This then implies $\int_a^b f(t)dt = \int_a^b g(t)dt$. This also gives

$$\int_a^b |f(t)|dt = \int_a^b |f(t) - g(t) + g(t)|dt \leq \int_a^b |f(t) - g(t)|dt + \int_a^b |g(t)|dt \leq \int_a^b |g(t)|dt$$

as $f = g$ a.e implies $\int_a^b |f(t) - g(t)|dt = 0$. Doing the same argument reversed shows $\int_a^b |g(t)|dt \leq \int_a^b |f(t)|dt$. Hence $\|f\|_1 = \|g\|_1$ when $f = g$ a.e. We conclude each function f in $RI([0, 1])$ determines an equivalence class. For example, the function

$$x(t) = \begin{cases} 0, & 0 \leq t \leq 1/2 \\ 1, & 1/2 < t \leq 1 \end{cases}$$

determines an equivalence class we call $[x]$ and all functions in it are equal to x a.e.

Consider the sequence of Riemann Integrable functions

$$x_n(t) = \begin{cases} 0, & 0 \leq t \leq 1/2 \\ n(t - 1/2), & 1/2 < t \leq 1/2 + 1/n \\ 1, & 1/2 + 1/n < t \leq 1 \end{cases}$$

We see the pointwise limit of this sequence is the x we defined above which is not continuous. These are all continuous functions on $[0, 1]$ and we can compute the limit of $\int_0^1 |x_n(t) - x(t)|dt$ easily.

$$\lim_{n \to \infty} \int_0^1 |x_n(t) - x(t)|dt = \lim_{n \to \infty} \int_0^{1/2} 0 \, dt + \lim_{n \to \infty} \int_{1/2}^{1/2+1/n} (1 - n(t - 1/2))dt$$

$$+ \lim_{n \to \infty} \int_{1/2+1/n}^1 (1 - 1)dt$$

$$= \lim_{n \to \infty} \int_{1/2}^{1/2+1/n} (1 - n(t - 1/2))dt$$

$$\leq \lim_{n \to \infty} \int_{1/2}^{1/2+1/n} 1 dt = \lim_{n \to \infty} 1/n = 0$$

Hence $\|x_n - x\|_1 \to 0$ as $n \to \infty$. Thus x_n converges to x in the $\| \cdot \|_1$ metric and (x_n) is a Cauchy sequence in $\| \cdot \|_1$.

This gives us an important example. The sequence (x_n) is a sequence in $C([0, 1])$ which is a Cauchy sequence in $\| \cdot \|_1$ which does not converge to a continuous function. Hence $(C([0, 1]), \| \cdot \|_1)$ cannot be complete! In fact, we have shown that (x_n) converges to any function in the equivalence class $[x]$.

Homework

Exercise 26.2.10 *Describe the equivalence class under $\sim$ for the function $f(x) = x^2$.*

Exercise 26.2.11 *Describe the equivalence class under $\sim$ for Dirichlet's function.*

Exercise 26.2.12 *Let*

$$f(x) \;=\; \begin{cases} 2, & x \in \mathbb{Q} \cap [0,1] \\ -3, & x \in \mathbb{I} \cap [0,1] \end{cases}$$

Is f in the equivalence under $\sim$ of $g(x) = -3$?

Exercise 26.2.13 *Let*

$$f(x) \;=\; \begin{cases} 2x, & x \in \mathbb{Q} \cap [0,1] \\ -3x, & x \in \mathbb{I} \cap [0,1] \end{cases}$$

Is f in the equivalence under $\sim$ of $g(x) = -3x$?

Chapter 27

Fourier Series

Now let's look at another type of series of functions: the Fourier Series. There are many deep mathematical issues here and we will just touch the surface with our discussions!

27.1 Vector Space Notions

Let's go back and look at Vector Spaces again. Recall what it means for two objects in a vector space to be called **independent**.

Definition 27.1.1 Two Linearly Independent Objects

> *Two Linearly Independent Objects:*
> *Let E and F be two objects in a vector space. We say E and F are* **linearly dependent** *if we can find nonzero constants α and β so that*
>
> $$\alpha E + \beta F = 0.$$
>
> *Otherwise, we say they are* **linearly independent.**

The constants come from a **field** which for us is $\Re$; however, another good choice is the complex numbers $\mathscr{C}$ or the quaternions $\mathscr{Q}$. We can then easily extend this idea to any finite collection of such objects as follows.

Definition 27.1.2 A Finite Number of Linearly Independent Objects

> *Finitely many Linearly Independent Objects:*
> *Let $\{E_i \; : \; 1 \leq i \leq N\}$ be N objects in a vector space. We say this set is* **linearly dependent** *if we can find nonzero constants α_1 to α_N, not all 0, so that*
>
> $$\alpha_1 E_1 + \ldots + \alpha_N E_N = 0.$$
>
> *Otherwise, we say the objects in this set are* **linearly independent.** *Note we have changed the way we define the constants a bit. When there are more than two objects involved, we can't say, in general, that* all *of the constants must be nonzero.*

Comment 27.1.1 *Hence to determine if a collection of objects $\{E_i \; : \; 1 \leq i \leq N\}$ in a vector space is linearly independent, we write down the equation*

$$\alpha_1 E_1 + \ldots + \alpha_N E_N = 0.$$

where 0 *is the zero object in the vector space and determine if the only solution to this equation is scalars* $\alpha_j = 0$. *We call this the* **linear independence equation**.

Now let's apply these ideas to functions f and g defined on some interval I. We would say f and g are linearly independent on the interval I if the equation

$$\alpha_1 f(t) + \alpha_2 g(t) = 0, \ \forall t \in I.$$

implies α_1 and α_2 must both be zero.

The linear independence condition for n functions $\{f_1, \ldots, f_b\}$ is that $\sum_{i=1}^{n} \alpha_i f_i(t) = 0$ for all $t \in I$ implies $\alpha_i = 0$ for all i.

Example 27.1.1 *The functions* $\sin(t)$ *and* $\cos(t)$ *are linearly independent on* $\Re$.

Solution *Write the* **linear independence equation**

$$\alpha_1 \cos(t) + \alpha_2 \sin(t) = 0, \ \forall t,$$

Take the derivative of both sides to get

$$-\alpha_1 \sin(t) + \alpha_2 \cos(t) = 0, \ \forall t,$$

This can be written as the system

$$\begin{bmatrix} \cos(t) & \sin(t) \\ -\sin(t) & \cos(t) \end{bmatrix} \begin{bmatrix} \alpha_1 \\ \alpha_2 \end{bmatrix} = \begin{bmatrix} 0 \\ 0 \end{bmatrix}$$

for all t. The determinant is $\cos^2(t) + \sin^2(t) = 1 \neq 0$. *This implies the unique solution is* $\alpha_1 = \alpha_2 = 0$ *for any t. Hence, these two functions are linearly independent on* I.

Example 27.1.2 *Show the three functions* $f(t) = t\cos(t)$, $g(t) = t\sin(t)$ *and* $h(t) = \sin(t)$ *are linearly independent on* $\Re$.

Solution *To see these three new functions are linearly independent, pick three points t from* $\Re$ *and solve the resulting linearly dependence equations. Since* $t = 0$ *does not give any information, let's try* $t = -\pi$, $t = \frac{\pi}{4}$ *and* $t = \frac{\pi}{2}$. *This gives the system*

$$\begin{bmatrix} \pi & 0 & 0 \\ \frac{\pi}{4}\frac{\sqrt{2}}{2} & \frac{\pi}{4}\frac{\sqrt{2}}{2} & \frac{\sqrt{2}}{2} \\ 0 & \frac{\pi}{2} & 1 \end{bmatrix} \begin{bmatrix} \alpha_1 \\ \alpha_2 \\ \alpha_3 \end{bmatrix} = \begin{bmatrix} 0 \\ 0 \\ 0 \end{bmatrix}$$

in the unknowns α_1, α_2 *and* α_3. *We see immediately* $\alpha_1 = 0$ *and the remaining two by two system has determinant* $-\frac{\sqrt{2}}{2}\frac{\pi}{4} \neq 0$. *Hence,* $\alpha_2 = \alpha_3 = 0$ *too. This shows* $t\sin(t)$, $t\cos(t)$ *and* $\sin(t)$ *are linearly independent on* $\Re$.

We can also talk about the linear Independence of an infinite set of vectors in a vector space. If $\mathscr{W} = \{u_i\}$ is an infinite set of vectors, we say it is **linearly independent** if any **finite** subset of it is linearly independent in the way we have defined above.

Vector spaces have two other important ideas associated with them. We have already talked about linearly independent objects. The first idea is that of the span of a set.

Definition 27.1.3 Span of a Set of Vectors

Given a finite set of vectors in a vector space $\mathcal{V}$, $\mathcal{W} = \{u_1, \ldots, u_N\}$ for some positive integer N, the span of $\mathcal{W}$ is the collection of all new vectors of the form $\sum_{i=1}^{N} c_i u_i$ for any choices of scalars $c_1, \ldots, c_N$. It is easy to see $\mathcal{W}$ is a vector space itself and since it is a subset of $\mathcal{V}$, we call it a vector subspace. The span of the set $\mathcal{W}$ is denoted by $Sp\mathbf{W}$. If the set of vectors $\mathcal{W}$ is not finite, the definition is similar but we say the span of $\mathcal{W}$ is the set of all vectors which can be written as $\sum_{i=1}^{N} c_i u_i$ for some finite set of vectors $u_1, \ldots u_N$ from $\mathcal{W}$.

Then there is the notion of a *basis* for a vector space.

Definition 27.1.4 Basis for a Vector Space

Basis of a Vector Space:
Given a set of vectors in a vector space $\mathcal{V}$, $\mathcal{W}$, we say $\mathcal{W}$ is a basis for $\mathcal{V}$ if the span of $\mathcal{W}$ is all of $\mathcal{V}$ and if the vectors in $\mathcal{W}$ are linearly independent. Hence, a basis is a linearly independent spanning set for $\mathcal{V}$. The number of vectors in $\mathcal{W}$ is called the dimension of $\mathcal{V}$.

If $\mathcal{W}$ is not finite is size, then we say $\mathcal{V}$ is an infinite dimensional vector space.

Example 27.1.3 *Let's look at the vector space $C([0,1])$, the set of all continuous functions on $[0,1]$. Let $\mathcal{W}$ be the set of all powers of t, $\{1, t, t^2, t^3, \ldots\}$. Show this is a linearly independent set.*

Solution *We can use the derivative technique to show this set is linearly independent even though it is infinite in size. Take any finite subset from $\mathcal{W}$. Label the resulting powers as $\{n_1, n_2, \ldots, n_p\}$. Write down the linear dependence equation*

$$c_1 t^{n_1} + c_2 t^{n_2} + \ldots + c_p t^{n_p} = 0.$$

Take n_p derivatives to find $c_p = 0$ and then backtrack *to find the other constants are zero also.*

Comment 27.1.2 *Hence $C([0,1])$ is an infinite dimensional vector space. It is also clear that $\mathcal{W}$ does not span $C([0,1])$ as if this was true, every continuous function on $[0,1]$ would be a polynomial of some finite degree. This is not true as $\sin(t)$, e^{-2t} and many others are not finite degree polynomials.*

If we have an object u in a Vector Space $\mathcal{V}$, we often want to find an *approximate* u using an element from a given subspace $\mathcal{W}$ of the vector space. To do this, we need to add another property to the vector space. This is the notion of an *inner product*. We already know what an inner product is in a simple vector space like $\Re^n$. Many vector spaces can have an inner product structure added easily.

27.1.1   Homework

Exercise 27.1.1

Prove e^t and e^{-t} are linearly independent on $\Re$.

Exercise 27.1.2

Prove 1, t and t^2 are linearly independent on $\Re$. Use the pick three points approach here.

Exercise 27.1.3 *Prove t and $\cos(t)$ are linearly independent on $\Re$.*

Exercise 27.1.4   • *What is the span of $\{t, t^2, t^3\}$.*

- *Find a basis for this span.*

27.1.2   Inner Products

In $C([a, b])$, since each object is continuous, each object is Riemann integrable. Hence, given two functions f and g from $C([a, b])$, the real number $\int_a^b f(s)g(s)ds$ is well defined. Note, this number would have been well defined if we used functions from $RI([a, b])$ also. It satisfies all the usual properties of the inner product in $\Re^n$.

In general for a vector space $\mathscr{V}$ with scalar multiplication $\odot$ and addition $\oplus$, an inner product is a mapping $\omega : \mathscr{V} \times \mathscr{V} \to F$, where F is the scalar field that satisfies certain properties. We will restrict our attention to the field $\Re$ here. An inner product over the $\Re$ field satisfies these properties:

Definition 27.1.5 Real Inner Products

Real Inner Product:
Let $\mathscr{V}$ be a vector space with the reals as the scalar field. Then a mapping ω which assigns a pair of objects to a real number is called an inner product on $\mathscr{V}$ if

    1. $\omega(\boldsymbol{u}, \boldsymbol{v}) = \omega(\boldsymbol{v}, \boldsymbol{u})$; that is, the order is not important for any two objects.

    2. $\omega(c \odot \boldsymbol{u}, \boldsymbol{v}) = c\omega(\boldsymbol{u}, \boldsymbol{v})$; that is, scalars in the first *slot can be pulled out.*

    3. $\omega(\boldsymbol{u} \oplus \boldsymbol{w}, \boldsymbol{v}) = \omega(\boldsymbol{u}, \boldsymbol{v}) + \omega(\boldsymbol{w}, \boldsymbol{v})$, for any three objects.

    4. $\omega(\boldsymbol{u}, \boldsymbol{u}) \geq 0$ and $\omega(\boldsymbol{u}, \boldsymbol{u}) = 0$ if and only if $u = 0$.

These properties imply that $\omega(\boldsymbol{u}, c \odot \boldsymbol{v}) = c\omega(\boldsymbol{u}, \boldsymbol{v})$ as well. A vector space $\mathscr{V}$ with an inner product is called an inner product space.

Comment 27.1.3 *The inner product is usually denoted with the symbol $<, >$ instead of $\omega(\ ,\ )$. We will use this notation from now on.*

Comment 27.1.4 *When we have an inner product, we can* measure *the* size *or* magnitude *of an object, as follows. We define the analogue of the Euclidean norm of an object $\boldsymbol{u}$ using the usual $||\ ||$ symbol as*

$$||\boldsymbol{u}|| \quad = \quad \sqrt{<\boldsymbol{u}, \boldsymbol{u}>}.$$

This is called the norm *induced by the inner product* of the object. *In $C([a, b])$, with the inner product $<f, g> = \int_a^b f(s)g(s)ds$, the norm of a function f is labeled $||f||_2 = \sqrt{\int_a^b f^2(s)ds}$. This is called the L_2 norm of f.*

If f is Riemann Integrable on $[a, b]$, then we can ask if the powers of f, f^p are also Riemann Integrable on $[a, b]$. We define new spaces of functions:

Definition 27.1.6 The $\mathscr{L}^p([a, b])$ Function Space

Let $p \geq 1$. The collection of functions f on $[a,b]$ is defined to be $\{f : [a,b] \to \Re \mid |f|^p \in R([a,b])\}$ and is denoted by $\mathscr{L}^p([a,b])$. Hence

- $\mathscr{L}^1([a,b]) = \{f : [a,b] \to \Re : |f| \in RI([a,b])\}$

- $\mathscr{L}^2([a,b]) = \{f : [a,b] \to \Re \mid |f|^2 \in RI([a,b])\}$
  and so forth with

- $\mathscr{L}^p([a,b]) = \{f : [a,b] \to \Re \mid |f|^p \in RI([a,b])\}$

We also define $\mathscr{L}^\infty([a,b]) = \{f : [a,b] \to \Re \mid ||f||_\infty < \infty\}$.
The p norm on $\mathscr{L}^p([a,b])$ is then defined by

$$||f||_p = \left(\int_a^b |f(s)|^p \, ds \right)^{\frac{1}{p}}$$

and in particular

$$||f||_2 = \sqrt{\int_a^b |f(s)|^2 \, ds}$$

Note not all functions are in $\mathscr{L}^1([a,b])$. If f is defined by

$$f(x) = \begin{cases} 2, & x \in \mathbb{Q} \cap [a,b] \\ 1, & x \in \mathbb{I} \cap [a,b] \end{cases}$$

then f is not Riemann integrable on $[a,b]$. We can prove new versions of the Hölder's and Minkowski's Inequalities for Riemann Integrable functions on $[a,b]$.

Theorem 27.1.1 Hölder's Inequality for $\mathscr{L}^p([a,b])$

Let $p > 1$ and p and q be conjugate exponents. If $x \in \mathscr{L}^p([a,b])$ and $y \in \mathscr{L}^q([a,b])$, then

$$\int_a^b |x(t)y(t)| \, dt \leq ||x||_p \, ||y||_q$$

For $p = 1$ with conjugate index $q = \infty$, if $x \in \mathscr{L}^1([a,b])$ and $y \in \mathscr{L}^\infty([a,b])$, then

$$\int_a^b |f(t)g(t)| \, dt \leq ||x||_1 \, ||y||_\infty$$

Proof 27.1.1
The proof is similar to the one we used for sequence spaces except that we use the norm $|| \cdot ||_p$ in the function space instead of the norm in the sequence space. ∎

and

Theorem 27.1.2 Minkowski's Inequality for $\mathscr{L}^p([a,b])$.

> *Let $p \geq 1$ and let x and y be in $\mathcal{L}^p([a,b])$, Then,*
>
> $$||x+y||_p \leq ||x||_p + ||y||_p$$

Proof 27.1.2

Again, the proof is similar for the one we used for sequence spaces except that we use the norm $||\cdot||_p$ in the function space instead of the norm in the sequence space. ∎

Comment 27.1.5 *When $p = 2$, Hölder's Inequality and Minkowski's Inequality immediately tell us that given $x, y \in \mathcal{L}^2([a,b])$,*

$$< x, y > \leq ||x||_2\, ||y||_2$$
$$||x+y||_2 \leq ||x||_2 + ||y||_2$$

The first inequality is the Cauchy - Schwartz Inequality in this setting which allows us to define the angle between functions which are in $\mathcal{L}^2([a,b])$. We have said $||\cdot||_2$ is the norm induced by the inner product in $\mathcal{L}^2([a,b])$. To be a norm, it must satisfy the usual triangle inequality for a metric; i.e. $d(x,y) \leq d(x,z) + d(z,y)$ for all x, y and z. This is true as the metric here is $d(x,y) = ||x - y||_2$ and

$$||x - y||_2 \quad = \quad ||(x - z) + (z - y)||_2 \leq ||x - z||_2 + ||z - y||_2$$

using the Minkowski Inequality.

Comment 27.1.6 *We define the angle θ between x and x in a vector space with an inner product via its cosine as usual.*

$$\cos(\theta) \quad = \quad \frac{< \boldsymbol{u}, \boldsymbol{v} >}{||\boldsymbol{u}||\, ||\boldsymbol{v}||}.$$

Hence, objects can be perpendicular or orthogonal even if we cannot interpret them as vectors in $\Re^2$. We see two objects are orthogonal if their inner product is 0.

Comment 27.1.7 *If $\mathcal{W}$ is a finite dimensional subspace, a basis for $\mathcal{W}$ is said to be an* **orthonormal basis** *if each object in the basis has L_2 norm 1 and all of the objects are mutually orthogonal. This means $< \boldsymbol{u}_i, \boldsymbol{u}_j >$ is 1 if $i = j$ and 0 otherwise. We typically let the Kronecker delta symbol δ_{ij} be defined by $\delta_{ij} = 1$ if $i = j$ and 0 otherwise so that we can say this more succinctly as $< \boldsymbol{u}_i, \boldsymbol{u}_j > = \delta_{ij}$.*

Now, let's return to the idea of finding the best object in an subspace $\mathcal{W}$ that approximates the given object . This is an easy theorem to prove.

Theorem 27.1.3 The Finite Dimensional Approximation Theorem

Let p be any object in the inner product space $\mathcal{V}$ with inner product $<,>$ and induced norm $\| \cdot \|$. Let $\mathcal{W}$ be a finite dimensional subspace with an orthonormal basis $\{w_1, \ldots w_N\}$ where N is the dimension of the subspace. Then there is an unique object p^* in $\mathcal{W}$ which satisfies

$$\|p - p^*\| \quad = \quad \min_{u \in \mathcal{W}} \|u - p\|$$

with

$$p^* \quad = \quad \sum_{i=1}^{N} <p, w_i> w_i.$$

Further, $p - p^*$ is orthogonal to the subspace $\mathcal{W}$.

Proof 27.1.3

Any object in the subspace has the representation $\sum_{i=1}^{N} a_i w_i$ for some scalars a_i. Consider the function of N variables

$$
\begin{aligned}
E(a_1, \ldots, a_N) \quad &= \quad \left\langle p - \sum_{i=1}^{N} a_i w_i, p - \sum_{j=1}^{N} a_j w_j \right\rangle \\
&= \quad <p, p> -2 \sum_{i=1}^{N} a_i <p, w_i> \\
&\quad + \sum_{i=1}^{N} \sum_{j=1}^{N} a_i a_j <w_i, w_j>.
\end{aligned}
$$

Simplifying using the orthonormality of the basis, we find

$$E(a_1, \ldots, a_N) \quad = \quad <p, p> -2 \sum_{i=1}^{N} a_i <p, w_i> + \sum_{i=1}^{N} a_i^2.$$

This is a quadratic expression and setting the gradient of E to zero, we find the critical points $a_j = <p, w_j>$. This is a global minimum for the function E. Hence, the optimal p^* has the form

$$p^* \quad = \quad \sum_{i=1}^{N} <p, w_i> w_i.$$

Finally, we see

$$
\begin{aligned}
<p - p^*, w_j> \quad &= \quad <p, w_j> - \sum_{k=1}^{N} <p, w_k> <w_k, w_j> \\
&= \quad <p, w_j> - <p, w_j> = 0,
\end{aligned}
$$

and hence, $p - p^*$ is orthogonal of $\mathcal{W}$.   ∎

Using inner products we can easily convert any basis into an orthonormal basis using a procedure called **Graham - Schmidt Orthogonalization** or **GSO**. If we had a finite number of linearly independent objects in an inner product space $\mathcal{V}$, then they form a basis for their span. It is easiest to show the Bf GSO process explicitly for three vectors. It is then easy to see how to generalize it to N objects. Let's assume we start with three linearly independent objects u, v and w. We will find the new basis as follows:

First Basis Object Set

$$g_1 \quad = \quad \frac{u}{||u||}.$$

Second Basis Object

- Subtract the part of v which lies along the object g_1.

$$h \quad = \quad v - <v, g_1> g_1.$$

- Find the length of h and set the second new basis object as follows:

$$g_2 \quad = \quad \frac{h}{||h||}.$$

Third Basis Object

- Subtract the part of w which lies along the object g_1 and the object g_2.

$$h \quad = \quad w - <v, g_1> g_1 - <v, g_2> g_2.$$

- Find the length of h and set the third new basis object as follows:

$$g_3 \quad = \quad \frac{h}{||h||}.$$

It is relatively easy to write code to do these computations.

27.1.3   Homework

Exercise 27.1.5

Let $f(t) = |t|$ on the interval $[-1, 1]$. Compute the inner products $< |t|, t^n >$ for all $n \geq 0$.

Exercise 27.1.6

*Let $f(t) = |t|$ on the interval $[-1, 1]$. Compute the best approximation of f to the subspace spanned by $u(t) = 1$ and $v(t) = 3t$. You will have to use **GSO** to find an o.n. basis for this subspace in order to find the best approximation.*

Exercise 27.1.7

Let $f(t) = 2t^2 + 4t + 2$ on the interval $[0, 10]$. Compute the best approximation of f to the subspace spanned by $\sin(\pi x/10)$ and $\sin(\pi x/10)$. Remember you need an o.n. basic in order to find the best approximation.

Exercise 27.1.8

Let $f(t) = 2t^2 + 4t + 2$ on the interval $[0, 10]$. Compute the best approximation of f to the subspace spanned by 1, $\cos(\pi x/10)$ and $\cos(\pi x/10)$. Remember you need an o.n. basic in order to find the best approximation.

Exercise 27.1.9

Show on the interval $[a, b]$, the functions $\sin(n\pi(x - a)/(b - a))$ are mutually orthogonal. **Hint:** *Note if we make the change of variable $y = (x - a)/(b - a)$, we have*

$$\int_a^b \sin\left(n\pi \frac{x - a}{b - a} \right) \sin\left(m\pi \frac{x - a}{b - a} \right) dx = (b - a) \int_0^1 \sin(n\pi y) \sin(m\pi y) dy$$

and we know about the family $\{\sin(n\pi y)\}$ on $[0, 1]$. Prove the length of these functions is $\sqrt{(b - a)/2}$.

Exercise 27.1.10

Let $f(t) = 2t^2 + 4t + 2$ on the interval $[-10, 10]$. Compute the best approximation of f to the subspace spanned by $\sin(\pi(x + 10)/20)$ and $\sin(\pi(x + 10)/20)$. Remember you need an o.n. basic in order to find the best approximation.

Exercise 27.1.11

Let $f(t) = 2t^2 + 4t + 2$ on the interval $[-10, 10]$. Compute the best approximation of f to the subspace spanned by 1, $\cos(\pi(x + 10)/20)$ and $\cos(\pi(x + 10)/20)$. Remember you need an o.n. basic in order to find the best approximation.

27.2 Fourier Series

A general trigonometric series $S(x)$ has the following form

$$S(x) = b_0 + \sum_{i=1}^{\infty} \left(a_i \sin\left(\frac{i\pi}{L} x \right) + b_i \cos\left(\frac{i\pi}{L} x \right) \right)$$

for any numbers a_n and b_n. Of course, there is no guarantee that this series will converge at any x! If we start with a function f which is continuous on the interval $[0, L]$, we can define the trigonometric series associated with f as follows

$$S(x) = \frac{1}{L} < f, 1 >$$
$$+ \sum_{i=1}^{\infty} \left(\frac{2}{L} \left\langle f(x), \sin\left(\frac{i\pi}{L} x \right) \right\rangle \sin\left(\frac{i\pi}{L} x \right) + \frac{2}{L} \left\langle f(x), \cos\left(\frac{i\pi}{L} x \right) \right\rangle \cos\left(\frac{i\pi}{L} x \right) \right).$$

The coefficients in the Fourier series for f are called the *Fourier coefficients* of f. Since these coefficients are based on inner products with the normalized sin and cos functions, we could call these the *normalized* Fourier coefficients. Let's be clear about this and a bit more specific. The nth *Fourier sin coefficient, $n \geq 1$, of f is as follows:*

$$a_n^L(f) = \frac{2}{L} \int_0^L f(x) \sin\left(\frac{i\pi}{L} x \right) dx$$

The nth *Fourier cos coefficient, $n \geq 0$, of f are defined similarly:*

$$b_0^L(f) = \frac{1}{L} \int_0^L f(x) dx$$

$$b_n^L(f) = \frac{2}{L} \int_0^L f(x) \cos\left(\frac{i\pi}{L} x \right) dx, \ n \geq 1.$$

Comment 27.2.1 *If we wanted to do this on the interval $[a, b]$ instead of $[0, L]$, we just make the change of variable $s \in [a, b] \to x \in [0, b - a]$ via $x = s - a$. Thus, $L = b - a$. The Fourier coefficients are thus*

$$a_n(f) = \frac{2}{b-a} \int_0^{b-a} f(x) \sin\left(\frac{i\pi}{b-a}(x-a)\right) dx$$

$$b_0(f) = \frac{1}{b-a} \int_0^{b-a} f(x) \, dx$$

$$b_n(f) = \frac{2}{b-a} \int_0^{b-a} f(x) \cos\left(\frac{i\pi}{L}(x-a)\right) dx, \quad n \geq 1.$$

27.3 Fourier Series Components

To prove that Fourier series converge pointwise for the kind of data functions f we want to use, requires a lot more work. So let's get started. Let's go back and look at the sin and cos functions we have been using. They have many special properties.

27.3.1 Orthogonal Functions

We will now look at some common sequences of functions on the domain $[0, L]$ that are very useful in solving models: the sequences of functions are $\left(\sin\left(\frac{i\pi}{L}x\right)\right)$ for integers $i \geq 1$ and $\left(\cos\left(\frac{i\pi}{L}x\right)\right)$ for integers $i \geq 0$. The second sequence is often written $\{1, \cos\left(\frac{\pi}{L}x\right), \cos\left(\frac{2\pi}{L}x\right), \ldots\}$. Here L is a positive number which often is the length of a cable or spatial boundary.

27.3.1.1 The Sine Sequence

Let's look carefully at this family the interval $[0, L]$. Using the definition of inner product on $C([0, L])$, by direct integration, we find for $i \neq j$

$$\left\langle \sin\left(\frac{i\pi}{L}x\right), \sin\left(\frac{j\pi}{L}x\right) \right\rangle = \int_0^L \sin\left(\frac{i\pi}{L}x\right) \sin\left(\frac{j\pi}{L}x\right) dx$$

Now use the substitution, $y = \frac{\pi x}{L}$ to rewrite this as

$$\left\langle \sin\left(\frac{i\pi}{L}x\right), \sin\left(\frac{j\pi}{L}x\right) \right\rangle = \frac{L}{\pi} \int_0^\pi \sin(iy) \sin(jy) dy$$

Now the trigonometric substitutions

$$\cos(u+v) = \cos(u)\cos(v) - \sin(u)\sin(v)$$
$$\cos(u-v) = \cos(u)\cos(v) + \sin(u)\sin(v)$$

imply

$$\sin(iy)\sin(jy) = \frac{1}{2}\left(\cos((i-j)y) - \cos((i+j)y)\right).$$

Using this identity in the integration, we see

$$\frac{L}{\pi} \int_0^\pi \sin(iy)\sin(jy)dy = \frac{L}{\pi} \int_0^\pi \left(\cos\left((i-j)y\right) - \cos\left((i+j)y\right) \right) dy$$

$$= \frac{L}{\pi} \left(\frac{\sin\left((i-j)y\right)}{i-j} \Big|_0^\pi + \frac{\sin\left((i+j)y\right)}{i+j} \Big|_0^\pi \right)$$

$$= 0.$$

Hence, the functions $\sin\left(\frac{i\pi}{L}x\right)$ and $\sin\left(\frac{j\pi}{L}x\right)$ are orthogonal on $[0, L]$ if $i \neq j$. On the other hand, if $i = j$, we have, using the same substitution $y = \frac{\pi x}{L}$, that

$$\left\langle \sin\left(\frac{i\pi}{L}x\right), \sin\left(\frac{i\pi}{L}x\right) \right\rangle \frac{L}{\pi} \int_0^\pi \sin(iy)\sin(iy)dy = \frac{L}{\pi} \int_0^\pi \left(\frac{1 - \cos(2iy)}{2} \right) dy$$

using the identify $\cos(2u) = 1 - 2\sin^2(u)$. It then follows that

$$\frac{L}{\pi} \int_0^\pi \sin(iy)\sin(iy)dy = \frac{L}{\pi} \left(\frac{y}{2} - \frac{\sin(2y)}{4} \right) \Big|_0^\pi$$

$$= \frac{L}{2}.$$

Hence, letting $u_n(x) = \sin\left(\frac{n\pi}{L}x\right)$, we have shown that

$$< u_i, u_j > = \begin{cases} \frac{L}{2}, & i = j \\ 0, & i \neq j. \end{cases}$$

Now define the new functions $\hat{u}_n$ by $\sqrt{\frac{2}{L}}u_n$. Then, $< \hat{u}_i, \hat{u}_j > = \delta_i^j$ and the sequence of functions $(\hat{u}_n)$ are all mutually orthogonal. It is clear $||\hat{u}_n||_2 = 1$ always. So the sequence of functions $(\hat{u}_n)$ are all mutually orthogonal and length one.

27.3.1.2   The Cosine Sequence

Let's look carefully at this family on the interval $[0, L]$. Using the definition of inner product on $C([0, L])$, by direct integration, we find for $i \neq j$ with both i and j at least 1, that

$$\left\langle \cos\left(\frac{i\pi}{L}x\right), \cos\left(\frac{j\pi}{L}x\right) \right\rangle = \int_0^L \cos\left(\frac{i\pi}{L}x\right) \cos\left(\frac{j\pi}{L}x\right) dx$$

Again use the substitution, $y = \frac{\pi x}{L}$ to rewrite this as

$$\left\langle \cos\left(\frac{i\pi}{L}x\right), \cos\left(\frac{j\pi}{L}x\right) \right\rangle = \frac{L}{\pi} \int_0^\pi \cos(iy)\cos(jy)dy$$

Again, the trigonometric substitutions

$$\cos(u+v) = \cos(u)\cos(v) - \sin(u)\sin(v)$$
$$\cos(u-v) = \cos(u)\cos(v) + \sin(u)\sin(v)$$

imply

$$\cos(iy)\cos(jy) \;=\; \frac{1}{2}\Big(\cos((i+j)y) + \cos((i-j)y)\Big).$$

Using this identity in the integration, we see

$$\frac{L}{\pi}\int_0^\pi \cos(iy)\cos(jy)\,dy \;=\; \frac{L}{\pi}\int_0^\pi \left(\cos\Big((i+j)y\Big) + \cos\Big((i-j)y\Big)\right)dy$$

$$=\; \frac{L}{\pi}\left(\frac{\sin\big((i+j)y\big)}{i+j}\Big|_0^\pi + \frac{\sin\big((i-j)y\big)}{i-j}\Big|_0^\pi\right)$$

$$=\; 0.$$

Hence, the functions $\cos\left(\frac{i\pi}{L}x\right)$ and $\cos\left(\frac{j\pi}{L}x\right)$ are orthogonal on $[0,L]$ if $i \neq j$, $i,j \geq 1$. Next, we consider the case of the inner product of the function 1 with a $\cos\left(\frac{j\pi}{L}x\right)$ for $j \geq 1$. This gives

$$\left\langle 1, \cos\left(\frac{j\pi}{L}x\right)\right\rangle \;=\; \int_0^L \cos\left(\frac{j\pi}{L}x\right)dx = \frac{L}{\pi}\int_0^\pi \cos(jy)\,dy = \frac{L}{\pi}\frac{\sin(jy)}{j}\Big|_0^\pi = 0.$$

Thus, the functions 1 and $\cos\left(\frac{j\pi}{L}x\right)$ for any $j \geq 1$ are also orthogonal. On the other hand, if $i = j$, we have for $i \geq 1$,

$$\left\langle \cos\left(\frac{i\pi}{L}x\right), \cos\left(\frac{i\pi}{L}x\right)\right\rangle \;=\; \frac{L}{\pi}\int_0^\pi \cos(iy)\cos(iy)\,dy = \frac{L}{\pi}\int_0^\pi \left(\frac{1 + \cos(2iy)}{2}\right)dy$$

using the identify $\cos(2u) = 2\cos^2(u) - 1$. It then follows that

$$\frac{L}{\pi}\int_0^\pi \cos(iy)\cos(iy)\,dy \;=\; \frac{L}{\pi}\left(\frac{y}{2} + \frac{\sin(2y)}{4}\right)\Big|_0^\pi = \frac{L}{2}.$$

We also easily find that $< 1, 1 > = L$. Hence, on $[0, L]$, letting

$$v_0(x) \;=\; 1$$

$$v_n(x) \;=\; \cos\left(\frac{n\pi}{L}x\right),\ n \geq 1,$$

we have shown that

$$< v_i, v_j > \;=\; \begin{cases} L, & i = j = 0, \\ L/2, & i = j,\ i \geq 1 \\ 0, & i \neq j. \end{cases}$$

Now define the new functions

$$\hat{v}_0(x) \;=\; \sqrt{\frac{1}{L}}$$

$$\hat{v}_n(x) \;=\; \sqrt{\frac{2}{L}}\,\cos\left(\frac{n\pi}{L}x\right),\; n \geq 1,$$

Then, $< \hat{v}_i, \hat{v}_j >= \delta_i^j$ and the sequence of functions $(\hat{v}_n)$ are all mutually orthogonal with length $\|v_n\| = 1$.

Note the projection of the data function f, $P_N(f)$, onto the subspace spanned by $\{\hat{u}_1, \ldots, \hat{u}_N\}$ is the minimal norm solution to the problem $\inf_{u \in U_n} \|f - u\|_2$ where U_n is the span of $\{\hat{u}_1, \ldots, \hat{u}_N\}$.

Further, projection of the data function f, $Q_N(f)$, onto the subspace spanned by $\{1, \hat{v}_1, \ldots, \hat{v}_N\}$ is the minimal norm solution to the problem $\inf_{v \in V_n} \|f - v\|_2$ where V_n is the span of $\{\hat{v}_0, \ldots, \hat{v}_N\}$. The Fourier Series of $f : [0, 2L]$ has partial sums that can be written as

$$S_N \;=\; P_N(f) + Q_N f = \sum_{n=1}^{N} < f, \hat{u}_n > \hat{u}_n + \; < f, \hat{v}_0 > \hat{v}_0 + \sum_{n=1}^{N} < f, \hat{v}_n > \hat{v}_n$$

and so the convergence of the Fourier Series is all about the conditions under which the projections of f to these subspaces converge pointwise to f.

27.3.1.3 Homework

Exercise 27.3.1

Show on the interval $[-1, 3]$, the functions $\sin(n\pi(x + 1)/4)$ are mutually orthogonal.

Exercise 27.3.2

Show on the interval $[2, 8]$, the functions $\cos(n\pi(x - 2)/6)$ are mutually orthogonal.

Exercise 27.3.3

Show on the interval $[a, b]$, the functions $\sin(n\pi(x - a)/(b - a))$ are mutually orthogonal.

Exercise 27.3.4

Show on the interval $[a, b]$, the functions $\cos(n\pi(x - a)/(b - a))$ are mutually orthogonal.

Exercise 27.3.5

Show on the interval $[0, 2L]$, the functions $\sin((n + 0.5)(\pi/L)x)$ are mutually orthogonal.

27.3.2 Fourier Coefficients Revisited

Let f be any continuous function on the interval $[0, L]$. Then we know f is Riemann integrable and so $\|f\| = \sqrt{\int_0^L f(t)^2 dt}$ is a finite number. From Section 27.3.1.1 and Section 27.3.1.2, recall the sequence of functions $\{\sin\left(\frac{\pi}{L}x\right), \sin\left(\frac{2\pi}{L}x\right), \ldots\}$ and $\{1, \cos\left(\frac{\pi}{L}x\right), \cos\left(\frac{2\pi}{L}x\right), \ldots\}$ are mutually orthogonal sequences on the interval $[0, L]$. Further, we know that we can divide each of the functions by their length to create the orthogonal sequences of length one, we called $(\hat{u}_n)$ for $n \geq 1$ and $(\hat{v}_n)$ for $n \geq 0$, respectively. Consider for any n

$$0 \;\leq\; < f - \sum_{i=1}^{n} < f, \hat{u}_i > \hat{u}_i, f - \sum_{j=1}^{n} < f, \hat{u}_j > \hat{u}_j >$$

$$= \; < f, f > -2\sum_{i=1}^{n}(< f, \hat{u}_i >)^2 + \sum_{i=1}^{n}\sum_{j=1}^{n} < f, \hat{u}_i >< f, \hat{u}_j >< \hat{u}_i, \hat{u}_j >$$

$$= \quad <f, f> -2\sum_{i=1}^{n}(<f, \hat{u}_i>)^2 + \sum_{i=1}^{n}(<f, \hat{u}_i>)^2$$

$$= \quad <f, f> -\sum_{i=1}^{n}(<f, \hat{u}_i>)^2,$$

where we use the fact that $<\hat{u}_i, \hat{u}_j> = \delta_i^j$. Hence, we conclude $\sum_{i=1}^{n}(<f, \hat{u}_i>)^2 \leq ||f||^2$ for all n. A similar argument shows that for all n, $\sum_{i=0}^{n}(<f, \hat{v}_i>)^2 \leq ||f||^2$. The numbers $<f, \hat{u}_i>$ and $<f, \hat{v}_i>$ are called the ith **Fourier sine** and i^{th} **Fourier cosine** coefficients of f, respectively. Note this is similar to the Fourier coefficients defined earlier but we are writing them in terms of the normalized sin and cos functions. In fact, in the context of the sums, they are the *same*; i.e.

$$<f, \hat{u}_i> \hat{u}_i \quad = \quad \left\langle f, \sqrt{\frac{2}{L}}u_i \right\rangle \sqrt{\frac{2}{L}}u_i = \frac{2}{L}<f, u_i> u_i$$

is the same term in both summations. We can do a similar expansion for the cos functions. So depending on where you read about these things, you could see them defined either way. Thus, we can rewrite the finite sums here in terms of the original sin and cos functions. We have

$$\sum_{i=1}^{n} <f, \hat{u}_i> \hat{u}_i \quad = \quad \sum_{i=1}^{n}\frac{2}{L}\left\langle f(x), \sin\left(\frac{i\pi}{L}x\right)\right\rangle \sin\left(\frac{i\pi}{L}x\right)$$

and

$$\sum_{i=0}^{n} <f, \hat{v}_i> \hat{v}_i \quad = \quad \frac{1}{L}<f(x), 1> 1 + \sum_{i=1}^{n}\frac{2}{L}\left\langle f(x), \cos\left(\frac{i\pi}{L}x\right)\right\rangle \cos\left(\frac{i\pi}{L}x\right).$$

We often write sums like this in terms of the original sine and cosine functions instead of the normalized ones.

The integrals which define the coefficients $\frac{2}{L}\left\langle f(x), \sin\left(\frac{i\pi}{L}x\right)\right\rangle$ are thus also called the ith **Fourier sine coefficients** of f. Further, the coefficients $\frac{1}{L}\left\langle f(x), 1\right\rangle$ and $\frac{2}{L}\left\langle f(x), \cos\left(\frac{i\pi}{L}x\right)\right\rangle$ are also the ith **Fourier cosine coefficients** of f. You can think of these as the *un normalized* Fourier coefficients for f, if you want as they are based on the *un normalized* sin and cos functions.

This ambiguity in the definition is easy to remember and it comes about because it is easy to write these sorts of finite expansions of f in terms of its Fourier coefficients in terms of the original sine and cosine functions. We are now ready to discuss the convergence of what are called Fourier series. We start with the sin and cos sequences discussed above. We can say more about them.

27.3.2.1   Fourier Coefficients Go to Zero

We will now show the nth Fourier sine and cosine coefficients must go to zero. From the previous section, we know that $\sum_{i=1}^{n}|<f, \hat{u}_i>|^2 \leq ||f||_2^2$ and $\sum_{i=0}^{n}|<f, \hat{v}_i>|^2 \leq ||f||_2^2$ which tells us that the sum of these series satisfies bounds $\sum_{i=1}^{\infty}|<f, \hat{u}_i>|^2 \leq ||f||_2^2$ and $\sum_{i=0}^{\infty}|<f, \hat{v}_i>|^2 \leq ||f||_2^2$. Hence, since these series of nonnegative terms are bounded above, they must converge. Therefore, their n^{th} terms must go to zero. Hence, we know $\lim_n <f, \hat{u}_n>^2 = 0$ and $\lim_n <f, \hat{v}_n>^2 = 0$ as well. This implies $\lim_n <f, \hat{u}_n> = 0$ and $\lim_n <f, \hat{v}_n> = 0$ also.

Another way to see this is to look at the difference between partial sums. Pick any positive tolerance ϵ. Let's focus on the sin series first. Since this series converges to say SS, there is a positive integer

N so that $n > N \implies |S_n - SS| < \frac{\epsilon}{2}$, where S_n is the usual partial sum. Thus, we can say for any $n > N + 1$, we have

$$
\begin{aligned}
|S_n - S_{n-1}| &= |S_n - SS + SS - S_{n-1}| \\
&\leq |S_n - SS| + |SS - S_{n-1}| \\
&< \frac{\epsilon}{2} + \frac{\epsilon}{2} \\
&= \epsilon.
\end{aligned}
$$

From the definition of partial sums, we know $S_n - S_{n-1} = (< f, \hat{u}_n >)^2$. Hence, we know $n > N \implies (< f, \hat{u}_n >)^2 < \epsilon$. But this implies $\lim_{n \to \infty} < f, \hat{u}_n >= 0$ which is the result we wanted to show. It is clear this is the same as

$$
\lim_{n \to \infty} \frac{2}{L} \int_0^L f(x) \sin\left(\frac{n\pi}{L} x\right) = 0.
$$

Now, a similar argument will show that $\lim_{n \to \infty} < f, \hat{v}_n >= 0$ or equivalently

$$
\lim_{n \to \infty} \frac{2}{L} \int_0^L f(x) \cos\left(\frac{n\pi}{L} x\right) = 0.
$$

27.3.2.2 Homework

Exercise 27.3.6 *Find the periodic extension for $f(t) = |t|$ on $[-1, 1]$. Draw several cycles.*

Exercise 27.3.7 *Find the periodic extension for $f(t) = 2t^2 - 3$ on $[0, 2]$. Draw several cycles.*

Exercise 27.3.8 *Find the periodic extension for $f(t) = t^2 - 1$ on $[-1, 1]$. Draw several cycles.*

Exercise 27.3.9 *Find the periodic extension for $f(t) = 2t^2 - 3t + 2$ on $[0, 4]$. Draw several cycles.*

Exercise 27.3.10 *Compute the first 5 Fourier Sine coefficients for $f(x) = |x|$ on $[-1, 1]$. Now recall a previous problem. The interval is $[a, b] = [-1, 1]$ and so the orthogonal functions to use are $\sin(n\pi(x - (-1)))/(1 - (-1)) = \sin(n\pi(x + 1)/2)$. You can show the length of these functions is $\sqrt{(b - a)/2} = 1$. Then $a_n = \frac{2}{b-a} \int_a^b f(x) \sin(n\pi(x + 1)/2)dx$.*

Exercise 27.3.11 *Compute the first 6 Fourier cos coefficients for $f(t) = |t|$ on $[-1, 1]$. The orthogonal functions here are $\cos(n\pi(x + 1)/2)$. Show the length of these functions is $\sqrt{(b - a)/2}$ also for $n \geq 1$ with length $\sqrt{b - a}$ for $n = 0$. Then $b_n = \frac{2}{b-a} \int_a^b f(x) \cos(n\pi(x + 1)/2)$ for $n \geq 1$ and $b_0 = \frac{1}{b-a} \int_a^b f(x)dx$ for $n = 0$.*

27.4 The Convergence of Fourier Series

The Fourier series associated to the continuous function f on $[0, L]$ is the series

$$
\begin{aligned}
S(x) &= \frac{1}{L} < f, 1 > \\
&+ \sum_{i=1}^{\infty} \left(\frac{2}{L} \left\langle f(x), \sin\left(\frac{i\pi}{L} x\right) \right\rangle \sin\left(\frac{i\pi}{L} x\right) + \frac{2}{L} \left\langle f(x), \cos\left(\frac{i\pi}{L} x\right) \right\rangle \cos\left(\frac{i\pi}{L} x\right) \right).
\end{aligned}
$$

We will sneak up on the analysis of this series by looking instead at the Fourier series of f on the interval $[0, 2L]$. The sin and cos functions we have discussed in Section 27.3.1.1 and Section 27.3.1.2

were labeled u_n and v_n respectively. On the interval $[0, L]$, their lengths were $\sqrt{\frac{2}{L}}$ for each u_n and v_n for $n \geq 1$ and $\sqrt{\frac{1}{L}}$ for v_0. We used these lengths to define the normalized functions $\hat{u}_n$ and $\hat{v}_n$ which formed mutually orthogonal sequences. On the interval $[0, 2L]$ the situation is virtually the same. The only difference is that on $[0, 2L]$, we still have

$$
\begin{aligned}
u_n(x) &= \sin\left(\frac{n\pi}{L}x\right), \; n \geq 1 \\
v_0(x) &= 1 \\
v_n(x) &= \cos\left(\frac{n\pi}{L}x\right), \; n \geq 1
\end{aligned}
$$

but now, although we still have orthogonality, the lengths change. We find

$$
< u_i, u_j > = \begin{cases} L, & i = j \\ 0, & i \neq j. \end{cases}
$$

$$
< v_i, v_j > = \begin{cases} 2L, & i = j = 0, \\ L, & i = j, \, i \geq 1 \\ 0, & i \neq j. \end{cases}
$$

The normalized functions are now

$$
\begin{aligned}
\hat{u}_n(x) &= \sqrt{\frac{1}{L}} \sin\left(\frac{n\pi}{L}x\right) \\
\hat{v}_0(x) &= \sqrt{\frac{1}{2L}} \\
\hat{v}_n(x) &= \sqrt{\frac{1}{L}} \cos\left(\frac{n\pi}{L}x\right), \; n \geq 1.
\end{aligned}
$$

The argument given in Section 27.3.2.1 still holds with just obvious changes. So these Fourier coefficients go to zero as $n \to \infty$ also. This series on $[0, 2L]$ does not necessarily converge to the value $f(x)$ at each point in $[0, 2L]$; in fact, it is known that there are continuous functions whose Fourier series does not converge at all. Still, the functions of greatest interest to us are typically functions that have continuous derivatives except at a finite number of points and for those sorts of functions, $S(x)$ and $f(x)$ usually match. We are going to prove this in the work below. Consider the difference between a typical partial sum $S_N(x)$ and our possible target $f(x)$.

$$
|S_N(x) - f(x)| = \left| \frac{1}{2L} < f, \mathbf{1} > \right.
$$

$$
\left. + \sum_{i=1}^{N} \left(\frac{1}{L}\left\langle f(x), \sin\left(\frac{i\pi}{L}x\right) \right\rangle \sin\left(\frac{i\pi}{L}x\right) + \frac{1}{L}\left\langle f(x), \cos\left(\frac{i\pi}{L}x\right) \right\rangle \cos\left(\frac{i\pi}{L}x\right) \right) - f(x) \right|
$$

$$
= \left| \frac{1}{2L} \int_0^{2L} f(t)\, dt + \sum_{i=1}^{N} \frac{1}{L} \int_0^{2L} f(t)\left(\sin\left(\frac{i\pi}{L}t\right)\sin\left(\frac{i\pi}{L}x\right) + \cos\left(\frac{i\pi}{L}t\right)\cos\left(\frac{i\pi}{L}x\right) \right) dt - f(x) \right|
$$

Now, $\sin(u)\sin(v) + \cos(u)\cos(v) = \cos(u - v)$ and hence we can rewrite the above as follows:

$$|S_N(x) - f(x)| = \left| \frac{1}{2L} \int_0^{2L} f(t)\, dt + \sum_{i=1}^{N} \frac{1}{L} \int_0^{2L} f(t) \left(\cos\left(\frac{i\pi}{L}(t-x) \right) \right) dt - f(x) \right|$$

$$= \left| \frac{1}{L} \int_0^{2L} f(t) \left(\frac{1}{2} + \sum_{i=1}^{N} \cos\left(\frac{i\pi}{L}(t-x) \right) \right) dt - f(x) \right|$$

27.4.1 Rewriting $S_N(x)$

Next, we use another identity. We know from trigonometry that

$$\cos(iy)\sin\left(\frac{y}{2} \right) = \sin\left(\left(i + \frac{1}{2} \right) y \right) - \sin\left(\left(i - \frac{1}{2} \right) y \right)$$

and so

$$\left(\frac{1}{2} + \sum_{i=1}^{N} \cos\left(\frac{i\pi}{L}(t-x) \right) \right) \sin\left(\frac{\pi}{2L}(t-x) \right)$$

$$= \frac{1}{2} \sin\left(\frac{\pi}{2L}(t-x) \right) + \sum_{i=1}^{N} \left(\sin\left(\left(i + \frac{1}{2} \right) \frac{\pi}{L}(t-x) \right) - \sin\left(\left(i - \frac{1}{2} \right) \frac{\pi}{L}(t-x) \right) \right)$$

$$= \frac{1}{2} \sin\left(\frac{\pi}{2L}(t-x) \right) + \frac{1}{2} \sin\left(\frac{3\pi}{2L}(t-x) \right) - \frac{1}{2} \sin\left(\frac{\pi}{2L}(t-x) \right)$$

$$+ \frac{1}{2} \sin\left(\frac{5\pi}{2L}(t-x) \right) - \frac{1}{2} \sin\left(\frac{3\pi}{2L}(t-x) \right)$$

$$+ \frac{1}{2} \sin\left(\frac{7\pi}{2L}(t-x) \right) - \frac{1}{2} \sin\left(\frac{5\pi}{2L}(t-x) \right)$$

$$\cdots$$

$$+ \frac{1}{2} \sin\left(\frac{(2N+1)\pi}{2L}(t-x) \right) - \frac{1}{2} \sin\left(\frac{(2N-1)\pi}{2L}(t-x) \right)$$

$$= \frac{1}{2} \sin\left(\left(N + \frac{1}{2} \right) \frac{\pi}{L}(t-x) \right)$$

We have found

$$\left(\frac{1}{2} + \sum_{i=1}^{N} \cos\left(\frac{i\pi}{L}(t-x) \right) \right) = \frac{\frac{1}{2} \sin\left(\left(N + \frac{1}{2} \right) \frac{\pi}{L}(t-x) \right)}{\sin\left(\frac{\pi}{2L}(t-x) \right)} \tag{27.1}$$

Note the argument $t - x$ is immaterial and for any y, the identity can also be written as

$$\left(\frac{1}{2} + \sum_{i=1}^{N} \cos\left(\frac{i\pi}{L} y \right) \right) = \frac{\frac{1}{2} \sin\left(\left(N + \frac{1}{2} \right) \frac{\pi}{L} y \right)}{\sin\left(\frac{\pi}{2L} y \right)} \tag{27.2}$$

Now, we use this to rewrite $S_N(x)$. We find

$$S_N(x) \;=\; \frac{1}{L}\int_0^{2L} f(t)\left(\frac{\frac{1}{2}\sin((N+\frac{1}{2})\frac{\pi}{L}(t-x))}{\sin(\frac{\pi}{2L}(t-x))}\right) dt$$

Making the change of variable $y = t - x$, we have

$$\frac{1}{L}\int_0^{2L} f(t)\left(\frac{\frac{1}{2}\sin((N+\frac{1}{2})\frac{\pi}{L}(t-x))}{\sin(\frac{\pi}{2L}(t-x))}\right) dt \;=\; \frac{1}{L}\int_{-x}^{2L-x} f(y+x)\left(\frac{\frac{1}{2}\sin((N+\frac{1}{2})\frac{\pi}{L}y)}{\sin(\frac{\pi}{2L}y)}\right) dy.$$

Now, switch the integration variable y back to t to obtain the form we want:

$$S_N(x) \;=\; \frac{1}{L}\int_{-x}^{2L-x} f(t+x)\left(\frac{\frac{1}{2}\sin((N+\frac{1}{2})\frac{\pi}{L}t)}{\sin(\frac{\pi}{2L}t)}\right) dt.$$

We can rewrite this even more. All of the individual sin terms are periodic over the interval $[0, 2L]$. We extend the function f to be periodic also by defining

$$\hat{f}(x + 2nL) \;=\; f(x),\; 0 < x < 2L.$$

and defining what happens at the endpoints 0, $\pm 2L$, $\pm 4L$ and so on using one sided limits. Since the original f is continuous on $[0, 2L]$, we know $f(0^+)$ and $f(2L^-)$ both exists and match $f(0)$ and $f(2L)$, respectively. Because these two values need not be the same, the periodic extension will always have a potential discontinuity at the point $2nL$ for all integers n. We will define the periodic extension at these points as

$$\hat{f}(2nL^-) \;=\; f(2L^-),\; \text{and } \hat{f}(2nL^+) = f(0^+)$$
$$\hat{f}(2nL) \;=\; \frac{1}{2}(f(0^+) + f(2L^-)).$$

For example, if f is the square wave

$$f(x) \;=\; \begin{cases} H, & 0 \le x \le L, \\ 0, & L < x \le 2L \end{cases}$$

which has a discontinuity at L, then the periodic extension will have discontinuities at each multiple $2nL$ as $\hat{f}(2L^-) = 0$, $f(0^+) = H$ and $f(2nL)$ is the average value.

For any periodic extension, the value of the integral $\int_{-x}^{2L-x}$ and $\int_0^{2L}$ are still the same. Hence, $S_N(x)$ can be written as

$$S_N(x) \;=\; \frac{1}{L}\int_0^{2L} \hat{f}(t+x)\left(\frac{\frac{1}{2}\sin((N+\frac{1}{2})\frac{\pi}{L}t)}{\sin(\frac{\pi}{2L}t)}\right) dt.$$

27.4.2 Rewriting $S_N - f$

Now we want to look at the difference between $S_N(x)$ and $f(x)$. Consider

$$
\begin{aligned}
\frac{1}{L}\int_0^{2L}\left(\frac{1}{2}+\sum_{i=1}^{N}\cos\left(\frac{i\pi}{L}(t)\right)\right)dt &= \frac{1}{L}\left(\frac{1}{2}t+\sum_{i=1}^{N}\frac{L}{i\pi}\sin\left(\frac{i\pi}{L}(t)\right)\right)\Bigg|_0^{2L} \\
&= \frac{1}{L}\left(L+\sum_{i=1}^{N}\frac{L}{i\pi}\left\{\sin\left(\frac{i\pi}{L}(2L)\right)-\sin\left(\frac{i\pi}{L}(0)\right)\right\}\right). \\
&= 1.
\end{aligned}
$$

Hence, we can say

$$
\begin{aligned}
f(x) &= f(x)\times 1 \\
&= \frac{1}{L}\int_0^{2L}\left(\frac{1}{2}+\sum_{i=1}^{N}\cos\left(\frac{i\pi}{L}(t)\right)\right)f(x)\,dt
\end{aligned}
$$

However, using Equation 27.2, we can rewrite again as

$$
\begin{aligned}
f(x) &= f(x)\times 1 \\
&= \frac{1}{L}\int_0^{2L}\left(\frac{\frac{1}{2}\sin((N+\frac{1}{2})\frac{\pi}{L}t)}{\sin(\frac{\pi}{2L}t)}\right)f(x)\,dt
\end{aligned}
$$

Using the periodic extension $\hat{f}$, the equation above is still valid. Hence, $S_N(x) - f(x) = S_N(x) - \hat{f}(x)$ and we find

$$
\begin{aligned}
&|S_N(x)-f(x)| \\
&= \left|\frac{1}{L}\int_0^{2L}\hat{f}(t+x)\left(\frac{\frac{1}{2}\sin((N+\frac{1}{2})\frac{\pi}{L}t)}{\sin(\frac{\pi}{2L}t)}\right)dt - \frac{1}{L}\int_0^{2L}\left(\frac{\frac{1}{2}\sin((N+\frac{1}{2})\frac{\pi}{L}t)}{\sin(\frac{\pi}{2L}t)}\right)\hat{f}(x)\,dt\right| \\
&= \left|\frac{1}{2L}\int_0^{2L}\left(\hat{f}(t+x)-\hat{f}(x)\right)\left(\frac{\sin((N+\frac{1}{2})\frac{\pi}{L}t)}{\sin(\frac{\pi}{2L}t)}\right)\right| \\
&= \left|\frac{1}{2L}\int_0^{2L}\frac{(\hat{f}(t+x)-\hat{f}(x))}{\sin\left(\frac{\pi}{2L}t\right)}\sin\left(\left(N+\frac{1}{2}\right)\frac{\pi}{L}t\right)\right|
\end{aligned}
$$

We can package this in a convenient form by defining the function h on $[0, 2L]$ by

$$
h(t) = \left(\hat{f}(t+x)-\hat{f}(x)\right)/\sin\left(\frac{\pi}{2L}t\right)
$$

Hence, the convergence of the Fourier series associated with f on $[0, 2L]$ is shown by establishing that

$$\lim_{N \to \infty} |S_N(x) - f(x)| = \lim_{N \to \infty} \left| \frac{1}{2L} \int_0^{2L} h(t) \sin\left(\left(N + \frac{1}{2}\right)\frac{\pi}{L}t\right) \right| = 0$$

We can simplify our arguments a bit more by noticing the function $t/\sin\left(\frac{\pi}{2L}t\right)$ has a removeable discontinuity at $t = 0$. This follows from a simple L'Hopital's rule argument. Hence, this function is continuous on $[0, 2L]$ and so there is a constant C so that $|t/\sin\left(\frac{\pi}{2L}t\right)| \leq C$. This implies $|1/\sin\left(\frac{\pi}{2L}t\right)| \leq C\,1/t$ on the interval and so we can establish the convergence we want by showing

$$\lim_{N \to \infty} \left| \frac{1}{2L} \int_0^{2L} \frac{\left(\hat{f}(t+x) - \hat{f}(x)\right)}{t} \sin\left(\left(N + \frac{1}{2}\right)\frac{\pi}{L}t\right) \right| = 0.$$

Hence, we need to look at the function $H(t) = \left(\hat{f}(t+x) - \hat{f}(x)\right)/t$ in our convergence discussions.

27.4.3 Convergence

It is straightforward to show that the sequence of functions $w_N(x) = \sin\left(\left(N + \frac{1}{2}\right)\frac{\pi}{L}t\right)$ is mutually orthogonal with length L on $[0, 2L]$ (see a previous problem). Hence, the functions $\hat{w}_N(x) = \sqrt{\frac{1}{L}}w_N(x)$ have length one and are orthogonal. Hence, we already know that if H is continuous on $[0, 2L]$, then the Fourier coefficients $< H, \hat{w}_N > \to 0$ using arguments just like we did in Section 27.3.2.1. For our purposes, we will only look at two kinds of functions f: the first is differentiable at x and the second has conditions on the derivatives to the right and left of x.

27.4.3.1 f is Differentiable

On the interval $[0, 2L]$, we can see that $H(t)$ is continuous at every point except possibly $t = 0$. At $t = 0$, continuity of H comes down to whether or not H has a removeable discontinuity at 0. First, note

$$\lim_{t \to 0} H(t) = \lim_{t \to 0} \frac{\hat{f}(t+x) - \hat{f}(x)}{t} = f'(x)$$

since we assume f is differentiable on $[0, 2L]$. So H has a removeable discontinuity at 0 as long as we define $H(0) = f'(x)$. Hence at each point where f has a derivative, we have shown that $S_N(x) \to f(x)$; i.e. the Fourier series of f on the interval $[0, 2L]$ converges to $f(x)$.

27.4.3.2 f' Has Nice Right and Left Behavior

Now we assume f is differentiable to the left of x and to the right of x. and $\lim_{y \to x^-} f'(y) = A$ for some finite value as well as $\lim_{y \to x^+} f'(y) = B$ for some finite value. This would include the case where f has a jump discontinuity at x as well as the case where f has a corner there.

Let's use this notation: $(f'(x))^+ = B$ and $(f'(x))^- = A$. We can handle this case by going back to our estimates and replacing $f(x)$ by the new function $\frac{1}{2}(f(x^+) + f(x^-))$ where $f(x^+)$ is the right hand value and $f(x^-)$ is the left hand value of the jump, respectively. Then, note

$$
\begin{aligned}
\frac{1}{L}\int_0^L \left(\frac{1}{2} + \sum_{i=1}^N \cos\left(\frac{i\pi}{L}(t)\right)\right) dt &= \frac{1}{L}\left(\frac{1}{2}t + \sum_{i=1}^N \frac{L}{i\pi}\sin\left(\frac{i\pi}{L}(t)\right)\right)\bigg|_0^L \\
&= \frac{1}{L}\left(\frac{L}{2} + \sum_{i=1}^N \frac{L}{i\pi}\left\{\sin\left(\frac{i\pi}{L}(L)\right) - \sin\left(\frac{i\pi}{L}(0)\right)\right\}\right). \\
&= \frac{1}{2}.
\end{aligned}
$$

and

$$
\begin{aligned}
\frac{1}{L}\int_L^{2L} \left(\frac{1}{2} + \sum_{i=1}^N \cos\left(\frac{i\pi}{L}(t)\right)\right) dt &= \frac{1}{L}\left(\frac{1}{2}t + \sum_{i=1}^N \frac{L}{i\pi}\sin\left(\frac{i\pi}{L}(t)\right)\right)\bigg|_L^{2L} \\
&= \frac{1}{L}\left(\frac{L}{2} + \sum_{i=1}^N \frac{L}{i\pi}\left\{\sin\left(\frac{i\pi}{L}(2L)\right) - \sin\left(\frac{i\pi}{L}(L)\right)\right\}\right). \\
&= \frac{1}{2}.
\end{aligned}
$$

Hence, we can say

$$
\begin{aligned}
f(x^+) &= 2f(x^+) \times \frac{1}{2} = \frac{2}{L}\int_0^L \left(\frac{1}{2} + \sum_{i=1}^N \cos\left(\frac{i\pi}{L}(t)\right)\right) f(x^+)\, dt \\
&= \frac{2}{L}\int_0^L \left(\frac{\frac{1}{2}\sin((N+\frac{1}{2})\frac{\pi}{L}t)}{\sin(\frac{\pi}{2L}t)}\right) f(x^+)\, dt
\end{aligned}
$$

and

$$
\begin{aligned}
f(x^-) &= 2f(x^-) \times \frac{1}{2} \\
&= \frac{2}{L}\int_L^{2L} \left(\frac{1}{2} + \sum_{i=1}^N \cos\left(\frac{i\pi}{L}(t)\right)\right) f(x^-)\, dt \\
&= \frac{2}{L}\int_L^{2L} \left(\frac{\frac{1}{2}\sin\left(\left(N+\frac{1}{2}\right)\frac{\pi}{L}t\right)}{\sin\left(\frac{\pi}{2L}t\right)}\right) f(x^-)\, dt
\end{aligned}
$$

Thus, $\frac{1}{2}(f(x^+) + f(x^-))$ can be rewritten as

$$
\frac{1}{2}(f(x^+) + f(x^-)) = \frac{1}{L}\int_0^L \left(\frac{\frac{1}{2}\sin((N+\frac{1}{2})\frac{\pi}{L}t)}{\sin(\frac{\pi}{2L}t)}\right) f(x^+)\, dt
$$

$$+ \frac{1}{L} \int_L^{2L} \left(\frac{\frac{1}{2}\sin((N+\frac{1}{2})\frac{\pi}{L}t)}{\sin(\frac{\pi}{2L}t)} \right) f(x^-)dt$$

We can also rewrite the $S_N(x)$ terms as two integrals, $\int_0^L$ and $\int_L^{2L}$ giving

$$S_N(x) = \frac{1}{L} \int_0^L \hat{f}(t+x) \left(\frac{\frac{1}{2}\sin((N+\frac{1}{2})\frac{\pi}{L}t)}{\sin(\frac{\pi}{2L}t)} \right) dt \; + \; \frac{1}{L} \int_L^{2L} \hat{f}(t+x) \left(\frac{\frac{1}{2}\sin((N+\frac{1}{2})\frac{\pi}{L}t)}{\sin(\frac{\pi}{2L}t)} \right) dt.$$

Combining, we obtain

$$\left| S_N(x) - \frac{1}{2}(f(x^+) + f(x^-)) \right| \leq \left| \frac{1}{L} \int_0^L \left(\hat{f}(t+x) - \hat{f}(x^+) \right) \left(\frac{\frac{1}{2}\sin((N+\frac{1}{2})\frac{\pi}{L}t)}{\sin(\frac{\pi}{2L}t)} \right) dt \right|$$
$$+ \left| \frac{1}{L} \int_L^{2L} \left(\hat{f}(t+x) - \hat{f}(x^-) \right) \left(\frac{\frac{1}{2}\sin((N+\frac{1}{2})\frac{\pi}{L}t)}{\sin(\frac{\pi}{2L}t)} \right) dt \right|$$

In the $[0, L]$ integrand, the appropriate function to look at is

$$h_0(t) = \frac{\hat{f}(t+x) - \hat{f}(x^+)}{\sin(\frac{\pi}{2L}t)}$$

which has a problem as $t \to 0^+$. If we can show h_0 has a Fourier series, then its Fourier coefficients go to zero as $N \to \infty$ and we have convergence of the first piece. However, using the same sort of approximation as before, instead of h_0, we can look at the function

$$H_0(t) = \frac{\hat{f}(t+x) - \hat{f}(x^+)}{t}$$

and if we can show this function has a Fourier Series we are done.

In the $[L, 2L]$ integrand, the appropriate function to look at is

$$h_1(t) = \frac{\hat{f}(t+x) - \hat{f}(x^-)}{\sin(\frac{\pi}{2L}t)}$$

which has a problem as $t \to (2L)^-$. If we can show h_1 has a Fourier series, then its Fourier coefficients go to zero as $N \to \infty$ and we have convergence of the second piece. Then, using the same sort of approximation as before, instead of h_1, we can look at the function

$$H_1(t) = \frac{\hat{f}(t+x) - \hat{f}(x^+)}{t - 2L}$$

and if we can show this function has a Fourier Series we are done.

Theorem 27.4.1 H_0 is continuous

H_0 *has a removeable discontinuity at $t = 0$ as*

$$\lim_{t \to 0^+} \frac{\hat{f}(t+x) - \hat{f}(x^+)}{t} = B$$

Proof 27.4.1

Since $\lim_{y \to x^+} f'(y) = B$ *given* $\epsilon > 0$, *there is a* $\delta_1 > 0$ *so that*

$$y \in (x_0, x_0 + \delta_1) \implies |f'(y) - B| < \epsilon/4$$

Since $f'(y_0)$ *exists, there is a* $\delta_2 > 0$ *so that*

$$y \in (y_0 - \delta_2, y_0 + \delta_2) \implies \left| \frac{f(y) - f(y_0)}{y - y_0} - f'(y_0) \right| < \epsilon/4$$

Thus, for any y_0 *and* y *in* $(x_0, x_0 + \delta)$, *where* $\delta < \min\{\delta_1/4, \delta_2/4\}$, *we have* $|y - x_0| \le \delta_1$ *and* $|y - y_0| < \delta_2$. *Thus,*

$$
\begin{aligned}
\left| \frac{f(y) - f(y_0)}{y - y_0} - B \right| &\le \left| \frac{f(y) - f(y_0)}{y - y_0} - f'(y_0) \right| + \left| f'(y_0) - B \right| \\
&\le \left| \frac{f(y) - f(y_0)}{y - y_0} - f'(y_0) \right| + \epsilon/4 \\
&\le \epsilon/4 + \epsilon/4 = \epsilon/2
\end{aligned}
$$

Then

$$\lim_{y \to x_0^+} \left| \frac{f(y) - f(y_0)}{y - y_0} - B \right| = \left| \frac{f(y) - f(x_0^+)}{y - x_0} - B \right| \le \epsilon/2 < \epsilon$$

Since $\epsilon > 0$ *is arbitrary, this proves the limit is* B. *The first integrand is continuous on* $[0, L]$ *if the integrand has a removeable discontinuity at* $t = 0$. *This is true as as*

$$\lim_{t \to x^+} \frac{\hat{f}(t + x) - \hat{f}(x^+)}{t} = B$$

This tells us H_0 *is continuous in these integrals as* H_0 *has a removeable discontinuity.* ∎

Since we can show the functions $\sin((N + \frac{1}{2})\frac{\pi}{L}t)$ are orthogonal and form an orthonormal sequence on $[0, L]$ just as they were on $[0, 2L]$, these integrals are the Fourier coefficients of a function continuous at x and so must go to zero. Hence, these terms go to 0 as $N \to \infty$.

In the second limit, we note the functions $\sin\left(\left(N + \frac{1}{2}\right)\frac{\pi}{L}t\right)$ are orthogonal and form an orthonormal sequence on $[L, 2L]$ also.

The integrand here is continuous on $[L, 2L]$ if the integrand has a removeable discontinuity at $t = 2L$. Using an argument similar to the earlier one, we can show

$$\lim_{t \to (2L)^-} \frac{\hat{f}(t + x) - \hat{f}(x^-)}{t - 2L} = A$$

using the periodicity of the extension $\hat{f}$.

This tells us H_1 is continuous in these integrals as H_1 has a removeable discontinuity at x^-. Since the functions $\sin((N + \frac{1}{2})\frac{\pi}{L}t)$ are orthogonal and form an orthonormal sequence on $[L, 2L]$, these integrals are the Fourier coefficients of a function continuous at x and so must go to zero.

We conclude that at any point x where f satisfies $\lim_{y \to x^+} f'(y)$ and $\lim_{y \to x^-} f'(y)$ are finite, then at x the series converges to the average of the jump values: $S_N(x) \to \frac{1}{2}(f(x^+) + f(x^-))$.

27.5   Fourier Sine Series

Let's look at the original Fourier Sine series and Fourier Cosine Series. These are

$$FS(x) \;=\; \sum_{i=1}^{\infty} \frac{2}{L}\left\langle f(x), \sin\left(\frac{i\pi}{L}x\right)\right\rangle \sin\left(\frac{i\pi}{L}x\right)$$

and

$$FC(x) \;=\; \frac{1}{L} <f,\mathbf{1}> + \sum_{i=1}^{\infty} \frac{2}{L}\left\langle f(x), \cos\left(\frac{i\pi}{L}x\right)\right\rangle \cos\left(\frac{i\pi}{L}x\right).$$

To understand the convergence of the Fourier Sine Series on $[0, L]$ we extend f from $[0, L]$ to $[0, 2L]$ as an **odd** function and then use the convergence analysis we have already completed on the interval $[0, 2L]$ to infer convergence of the Fourier Sine Series on $[0, L]$. Then to understand the convergence of the Fourier Cosine Series on $[0, L]$ we extend f from $[0, L]$ to $[0, 2L]$ as an **even** function and then use the convergence analysis we have already completed on the interval $[0, 2L]$ to infer convergence of the Fourier Cosine Series on $[0, L]$.

We start with the Fourier Sine Series. Let f be defined only on the interval $[0, L]$. Extend f to be an odd function f_o on $[0, 2L]$ as follows:

$$f_o(x) \;=\; \begin{cases} f(x), & 0 \le x \le L, \\ -f(2L - x), & L < x \le 2L. \end{cases}$$

Then extend f_o periodically as usual to $\hat{f}_o$. The Fourier coefficient for the sin terms are then

$$\begin{aligned}
a_n^{f_o,2L} \;&=\; \frac{1}{L}\int_0^{2L} f_o(t)\sin\left(\frac{i\pi}{L}t\right) dt \\
&=\; \frac{1}{L}\int_0^{L} f(t)\sin\left(\frac{i\pi}{L}t\right) dt + \frac{1}{L}\int_L^{2L} (-f(2L - t))\sin\left(\frac{i\pi}{L}t\right) dt.
\end{aligned}$$

Consider the second integration. Making the change of variable $y = 2L - t$, we find

$$\frac{1}{L}\int_L^{2L} (-f(2L - t))\sin\left(\frac{i\pi}{L}t\right) dt \;=\; \frac{1}{L}\int_L^{0} (-f(y))\sin\left(\frac{i\pi}{L}(2L - y)\right)(-dy).$$

But $\frac{i\pi}{L}(2L - y) = 2i\pi - \frac{i\pi}{L}y$ and since the sin function is 2π periodic, we have

$$\begin{aligned}
\frac{1}{L}\int_0^{L} (-f(y))\sin\left(\frac{i\pi}{L}(2L - y)\right) dy \;&=\; \frac{1}{L}\int_L^{0} (-f(y))\sin\left(\frac{i\pi}{L}(-y)\right)(-dy) \\
&=\; \frac{1}{L}\int_0^{L} f(y)\sin\left(\frac{i\pi}{L}(y)\right) dy.
\end{aligned}$$

This is the same as the first integral. Hence, we have shown

$$a_n^{f_o,2L} \;=\; \frac{1}{L}\int_0^{2L} f_o(t)\sin\left(\frac{i\pi}{L}t\right)dt = \frac{2}{L}\int_0^L f(t)\sin\left(\frac{i\pi}{L}t\right)dt = a_n^{f,L}$$

The terms corresponding to the cos parts will all then be zero. The argument is straightforward. For $i > 0$,

$$
\begin{aligned}
b_n^{f_o,2L} \;&=\; \frac{1}{L}\int_0^{2L} f_o(t)\cos\left(\frac{i\pi}{L}t\right)dt \\
&=\; \frac{1}{L}\int_0^L f(t)\cos\left(\frac{i\pi}{L}t\right)dt \;+\; \frac{1}{L}\int_L^{2L}(-f(2L-t))\cos\left(\frac{i\pi}{L}t\right)dt.
\end{aligned}
$$

Consider the second integration. Making the change of variable $y = 2L - t$, we find

$$\frac{1}{L}\int_L^{2L}(-f(2L-t))\cos\left(\frac{i\pi}{L}t\right)dt \;=\; \frac{1}{L}\int_L^0 (-f(y))\cos\left(\frac{i\pi}{L}(2L-y)\right)(-dy).$$

Again, $\frac{i\pi}{L}(2L-y) = 2i\pi - \frac{i\pi}{L}y$ and since cos term 2π periodic and cos is an even function, we have

$$
\begin{aligned}
\frac{1}{L}\int_L^{2L}(-f(2L-t))\cos\left(\frac{i\pi}{L}t\right)dt \;&=\; \frac{1}{L}\int_L^0 f(y)\cos\left(\frac{i\pi}{L}y\right)dy \\
&\quad \frac{1}{L}\int_0^L -f(y)\cos\left(\frac{i\pi}{L}y\right)dy
\end{aligned}
$$

which is the negative of the first integral. So all of the coefficients $b_n^{f_o,2L}$ are zero. The case $i = 0$ is $\int_0^{2L} f_o(t)dt = 0$ because f_0 is odd on this interval; so $b_0^{f_o,2L} = 0$. Thus, all the cos based terms in the Fourier series vanish. The Fourier series on the interval $[0, 2L]$ of the odd extension f_o becomes the standard Fourier sine series on the interval $[0, L]$ of the function f.

$$\sum_{i=1}^\infty <f, \hat{u}_i > \hat{u}_i \;=\; \sum_{i=1}^\infty \frac{2}{L}\left\langle f(x), \sin\left(\frac{i\pi}{L}x\right)\right\rangle \sin\left(\frac{i\pi}{L}x\right).$$

We know now from Section 27.4.3 this converges to $f(x)$ at each point x where f is differentiable and converges to the average $\frac{1}{2}(f(x^+) + f(x^-))$ at each point x where f satisfies $\lim_{y\to x\pm} f'(y)$ exists. Note because the sin functions are always 0 at the endpoints 0 and L, this series must converge to 0 at those points.

27.5.1 Homework

Exercise 27.5.1 *Show the sequence of functions* (w_N) *where* $w_N(x) = \sin\left(\left(N + \frac{1}{2}\right)\frac{\pi}{L}t\right)$ *is mutually orthogonal with length L on* $[0, 2L]$ *and so the functions* $\hat{w}_N(x) = \sqrt{\frac{1}{L}}w_N(x)$ *have length one and are orthogonal.*

Exercise 27.5.2 *Find the odd periodic extension* $\hat{f}_o$ *for* $f(t) = t^2 + 4t - 2$ *on* $[0, 6]$. *Draw several cycles.*

Exercise 27.5.3 *The function f is defined on* $[0, L] = [0, 4]$.

$$f(x) \;=\; \begin{cases} 3, & 0 \le x \le 2 \\ 0, & 2 < x \le 4 \end{cases}$$

- *Find its periodic extension to* $[0, 8]$ *and its odd periodic extension to* $[0, 8]$.

- *Find the first three terms of the Fourier sine series on* $[0, 4]$.

Exercise 27.5.4 *The function f is defined on* $[0, L] = [0, 5]$.

$$f(x) \;=\; \begin{cases} 0, & 0 \le x \le 2 \\ 5, & 2 < x \le 5 \end{cases}$$

- *Find its periodic extension to* $[0, 10]$ *and its odd periodic extension to* $[0, 10]$.

- *Find the first three terms of the Fourier sine series on* $[0, 5]$.

Exercise 27.5.5 *The function f is defined on* $[0, L] = [0, 5]$.

$$f(x) \;=\; \begin{cases} 0, & 0 \le x \le 3 \\ 5, & 3 < x \le 5 \end{cases}$$

- *Find its periodic extension to* $[0, 10]$ *and its odd periodic extension to* $[0, 10]$.

- *Find the first three terms of the Fourier sine series on* $[0, 5]$.

27.6   Fourier Cosine Series

Let f be defined only on the interval $[0, L]$. Extend f to be an even function f_e on $[0, 2L]$ as follows:

$$f_e(x) \;=\; \begin{cases} f(x), & 0 \le x \le L, \\ f(2L - x), & L < x \le 2L. \end{cases}$$

Then extend f_e periodically as usual to $\hat{f}_e$. The Fourier coefficient for the sin terms are now

$$
\begin{aligned}
a_n^{f_e,2L} &= \frac{1}{L} \int_0^{2L} f_e(t) \sin\left(\frac{i\pi}{L} t\right) dt \\
&= \frac{1}{L} \int_0^{L} f(t) \sin\left(\frac{i\pi}{L} t\right) dt + \frac{1}{L} \int_L^{2L} f(2L - t) \sin\left(\frac{i\pi}{L} t\right) dt.
\end{aligned}
$$

Consider the second integration. Making the change of variable $y = 2L - t$, we find

$$
\begin{aligned}
\frac{1}{L} \int_L^{2L} f(2L - t) \sin\left(\frac{i\pi}{L} t\right) dt &= \frac{1}{L} \int_L^{0} f(y) \sin\left(\frac{i\pi}{L}(2L - y)\right) (-dy) \\
&= \frac{1}{L} \int_0^{L} f(y) \sin\left(\frac{i\pi}{L}(-y)\right) dy.
\end{aligned}
$$

However, sin is an odd function and thus the second integral is the negative of the first and the coefficients $a_n^{f_e,2L}$ vanish.

Next, consider the first Fourier cos coefficient. This is

$$
\begin{aligned}
b_0^{f_e,2L} &= \frac{1}{2L}\int_0^{2L} f_e(t)dt = \frac{1}{2L}\int_0^L f(t)dt + \frac{1}{L}\int_L^{2L} f(2L-t)dt \\
&= \frac{1}{2L}\int_0^L f(t)dt + \frac{1}{L}\int_L^0 f(y)(-dy) = \frac{1}{L}\int_0^L f(t)dt = b_0^{f,L}
\end{aligned}
$$

Now let's look at the other cos based coefficients. We have

$$
\begin{aligned}
b_n^{f_e,2L} &= \frac{1}{L}\int_0^{2L} f_e(t)\cos\left(\frac{i\pi}{L}t\right)dt \\
&= \frac{1}{L}\int_0^L f(t)\cos\left(\frac{i\pi}{L}t\right)dt + \frac{1}{L}\int_L^{2L} f(2L-t)\cos\left(\frac{i\pi}{L}t\right)dt.
\end{aligned}
$$

Consider the second integration. Making the change of variable $y = 2L - t$, we find

$$
\begin{aligned}
\frac{1}{L}\int_L^{2L} f(2L-t)\cos\left(\frac{i\pi}{L}t\right)dt &= \frac{1}{L}\int_L^0 f(y)\cos\left(\frac{i\pi}{L}(2L-y)\right)(-dy) \\
&= \frac{1}{L}\int_0^L f(y)\cos\left(\frac{i\pi}{L}(2L-y)\right)dy
\end{aligned}
$$

Again, $\frac{i\pi}{L}(2L-y) = 2i\pi - \frac{i\pi}{L}y$ and since cos term 2π periodic and cos is an even function, we have

$$
\frac{1}{L}\int_L^{2L} (-f(2L-t))\cos\left(\frac{i\pi}{L}t\right)dt = \frac{1}{L}\int_0^L f(y)\cos\left(\frac{i\pi}{L}y\right)dy
$$

This is the same as the first integral. So

$$
b_0^{f_e,2L} = \frac{1}{2L}\int_0^{2L} f_e(t)dt = \frac{2}{L}\int_0^L f(y)\cos\left(\frac{i\pi}{L}y\right)dy = b_n^{f,L}
$$

Thus, the Fourier series on the interval $[0, 2L]$ of the even extension f_e becomes the standard Fourier cosine series on the interval $[0, L]$ of the function f

$$
\sum_{i=0}^{\infty} <f,\hat{v}_i> \hat{v}_i = \frac{1}{L}<f(x),1> + \sum_{i=1}^{\infty} \frac{2}{L}\left\langle f(x),\cos\left(\frac{i\pi}{L}x\right)\right\rangle \cos\left(\frac{i\pi}{L}x\right).
$$

We know from Section 27.4.3 this series converges to $f(x)$ at each point x where f is differentiable and converges to the average $\frac{1}{2}(f(x^+) + f(x^-))$ at each point x where f satisfies $\lim_{y\to x^\pm} f'(y)$ exists.

27.6.1   Homework

Exercise 27.6.1 *Show the sequence of functions $w_N(x) = \sin\left(\left(N + \frac{1}{2}\right)\frac{\pi}{L}t\right)$ are mutually orthogonal with length $L/2$ on $[L, 2L]$.*

Exercise 27.6.2 *Show the sequence of functions $w_N(x) = \sin\left(\left(N + \frac{1}{2}\right)\frac{\pi}{L}t\right)$ are mutually orthogonal with length $L/2$ on $[0, L]$ and $[0, 2L]$.*

Exercise 27.6.3 *Find the even periodic extension $\hat{f}_e$ for $f(t) = 3t^2 + 6t - 2$ on $[0, 4]$. Draw several cycles.*

Exercise 27.6.4 *The function f is defined on $[0, L] = [0, 3]$.*

$$f(x) \;=\; \begin{cases} 2, & 0 \le x \le 2 \\ 0, & 2 < x \le 3 \end{cases}$$

- *Find its periodic extension to $[0, 6]$ and its even periodic extension to $[0, 6]$.*

- *Find the first four terms of the Fourier cos series on $[0, 3]$.*

Exercise 27.6.5 *The function f is defined on $[0, L] = [0, 5]$.*

$$f(x) \;=\; \begin{cases} 2, & 0 \le x \le 3 \\ 0, & 3 < x \le 5 \end{cases}$$

- *Find its periodic extension to $[0, 10]$ and its even periodic extension to $[0, 10]$.*

- *Find the first four terms of the Fourier cos series on $[0, 5]$.*

Exercise 27.6.6 *The function f is defined on $[0, L] = [0, 5]$.*

$$f(x) \;=\; \begin{cases} 0, & 0 \le x \le 3 \\ 2 & 3 < x \le 5 \end{cases}$$

- *Find its periodic extension to $[0, 10]$ and its even periodic extension to $[0, 10]$.*

- *Find the first four terms of the Fourier cos series on $[0, 5]$.*

27.7   More Convergence Fourier Series

Let's look at some additional estimates. Assume we have a function f extended periodically on the interval $[0, 2L]$ to $\hat{f}$ as usual with Fourier series

$$S(x) \;=\; \frac{1}{L} <f, \mathbf{1}>$$
$$+ \sum_{i=1}^{\infty} \left(\frac{2}{L}\left\langle f(x), \sin\left(\frac{i\pi}{L}x\right) \right\rangle \sin\left(\frac{i\pi}{L}x\right) + \frac{2}{L}\left\langle f(x), \cos\left(\frac{i\pi}{L}x\right) \right\rangle \cos\left(\frac{i\pi}{L}x\right) \right).$$

and using $b_0^{2L} = \frac{1}{L} < f, 1 >$, $b_n^{2L} = \frac{2}{L}\left\langle f(x), \cos\left(\frac{i\pi}{L}x\right)\right\rangle$ and $a_n^{2L} = \frac{2}{L}\left\langle f(x), \sin\left(\frac{i\pi}{L}x\right)\right\rangle$ as usual, we can write

$$S(x) = b_0^{2L} + \sum_{i=1}^{\infty}\left(a_i^{2L}\sin\left(\frac{i\pi}{L}x\right) + b_i^{2L}\cos\left(\frac{i\pi}{L}x\right)\right).$$

Now if we assume f' exists in $[0, 2L]$, we know the Fourier series of f converges to f pointwise at each point and we have

$$f(x) = b_0^{2L} + \sum_{i=1}^{\infty}\left(a_i^{2L}\sin\left(\frac{i\pi}{L}x\right) + b_i^{2L}\cos\left(\frac{i\pi}{L}x\right)\right).$$

27.7.1 Bounded Coefficients

Let S_n be the nth partial sum of the series above. Then assuming $\int_0^{2L} f^2(x)dx$ is finite, we have

$$0 \leq\ < f - b_0^{2L} - \sum_{i=1}^{n}\left(a_i^{2L}\sin\left(\frac{i\pi}{L}x\right) + b_i^{2L}\cos\left(\frac{i\pi}{L}x\right)\right),$$

$$f - b_0^{2L} - \sum_{j=1}^{n}\left(a_j^{2L}\sin\left(\frac{i\pi}{L}x\right) + b_j^{2L}\cos\left(\frac{j\pi}{L}x\right)\right) >$$

As usual, we let $u_i(x) = \sin\left(\frac{i\pi}{L}x\right)$ and $v_i(x) = \cos\left(\frac{i\pi}{L}x\right)$ with $v_0(x) = 1$. Then, we can rewrite this as

$$0 \leq< f - b_0^{2L}v_0(x) - \sum_{i=1}^{n}(a_i^{2L}u_i(x) + b_i^{2L}v_i), f - b_0^{2L}v_0(x) - \sum_{j=1}^{n}(a_j^{2L}u_j(x) + b_j^{2L}v_j(x)) >$$

Thus, we find, after a lot of manipulation

$$0 \leq ||f - b_0^{2L}v_0(x) - \sum_{i=1}^{n}(a_i^{2L}u_i(x) + b_i^{2L}v_i(x))||^2$$

$$= < f, f > -2b_0^{2L} < f, v_0 > +(b_0^{2L})^2 < v_0, v_0 >$$

$$-2\sum_{i=1}^{n}a_i^{2L} < f, u_i > -2\sum_{i=0}^{n}b_i^{2L} < f, v_i > +\sum_{i=1}^{n}(a_i^{2L})^2 < u_i, u_i > +\sum_{i=1}^{n}(b_i^{2L})^2 < v_i, v_i >$$

because all the *cross terms* such as $< u_i, u_j >= 0$, $< v_i, v_j > -\delta_i^j$ and $< u_i, v_j >= 0$ on $[0, 2L]$. Hence, since $< f, v_i >= b_i^{2L}$ and $< f, u_i >= a_i^{2L}$ we have

$$0 \leq ||f - b_0^{2L}v_0(x) - \sum_{i=1}^{n}(a_i^{2L}u_i(x) + b_i^{2L}v_i)||^2$$

$$= < f, f > -(b_0^{2L})^2 - \sum_{i=1}^{n}(a_i^{2L})^2 - \sum_{i=1}^{n}(b_i^{2L})^2$$

We conclude that

$$(b_0^{2L})^2 + \sum_{i=1}^{n}((a_i^{2L})^2 + (b_i^{2L})^2) \;\; \leq \;\; \|f\|_2^2$$

This tells us that the series of positive terms, $(b_0^{2L})^2 + \sum_{i=1}^{\infty}((a_i^{2L})^2 + (b_i^{2L})^2)$ converges.

27.7.2   The Derivative Series

Let's assume f is differentiable on $[0, 2L]$ and we extend f' periodically as well. We can calculate the Fourier series of f' like usual. Now this series converges to f', if we assume f'' exists on $[0, 2L]$. Then we know the Fourier series of f' converges pointwise to $f'(x)$ at each point x. However, we can calculate the derivative Fourier series directly. Let the Fourier series of $f'(x)$ be $T(x)$. Then

$$
\begin{aligned}
T(x) \;\; &= \;\; \frac{1}{2L} < f', 1 > \\
&+ \sum_{i=1}^{\infty} \left(\frac{1}{L} \left\langle f'(x), \sin\left(\frac{i\pi}{L}x\right) \right\rangle \sin\left(\frac{i\pi}{L}x\right) + \frac{1}{L}\left\langle f'(x), \cos\left(\frac{i\pi}{L}x\right) \right\rangle \cos\left(\frac{i\pi}{L}x\right) \right).
\end{aligned}
$$

The first coefficient is

$$\frac{1}{2L} < f', \mathbf{1} > \;\; = \;\; \frac{1}{2L}(f(2L) - f(0))$$

Now let's consider the other Fourier coefficients carefully. We can rewrite each coefficient using integration by parts to find,

$$
\begin{aligned}
\frac{1}{L}\left\langle f'(x), \cos\left(\frac{i\pi}{L}x\right) \right\rangle \;\; &= \;\; \frac{1}{L}\int_0^{2L} f'(x), \cos\left(\frac{i\pi}{L}x\right) dx \\
&= \;\; \frac{1}{L}f(x)\cos\left(\frac{i\pi}{L}x\right)\Big|_0^{2L} + \frac{1}{L}\frac{i\pi}{L}\int_0^{2L} f(x)\sin\left(\frac{i\pi}{L}x\right) dx \\
\frac{1}{L}\left\langle f'(x), \sin\left(\frac{i\pi}{L}x\right) \right\rangle \;\; &= \;\; \frac{1}{L}\int_0^{2L} f'(x), \sin\left(\frac{i\pi}{L}x\right) dx \\
&= \;\; \frac{1}{L}f(x)\sin\left(\frac{i\pi}{L}x\right)\Big|_0^{2L} - \frac{1}{L}\frac{i\pi}{L}\int_0^{2L} f(x)\cos\left(\frac{i\pi}{L}x\right) dx
\end{aligned}
$$

A little thought shows we can rewrite this as

$$
\begin{aligned}
\frac{1}{L}\left\langle f'(x), \cos\left(\frac{i\pi}{L}x\right) \right\rangle \;\; &= \;\; \frac{1}{L}\left(f(2L)\cos\left(\frac{2Li\pi}{L}\right) - f(0)\cos\left(\frac{0i\pi}{L}\right) \right) + \frac{i\pi}{L}a_i^{2L} \\
&= \;\; \frac{1}{L}\left(f(2L) - f(0) \right) + \frac{i\pi}{L}a_i^{2L} \\
\frac{1}{L}\left\langle f'(x), \sin\left(\frac{i\pi}{L}x\right) \right\rangle \;\; &= \;\; \frac{1}{L}\left(f(2L)\sin\left(\frac{2Li\pi}{L}\right) - f(0)\sin\left(\frac{0\pi}{L}\right) \right) - \frac{i\pi}{L}b_i^{2L}
\end{aligned}
$$

Now if we assume $f(2L) = f(0)$, these reduce to

$$\hat{b}_0^{2L} = \frac{1}{2L} < f', \mathbf{1} > \;\; = \;\; \frac{1}{2L}(f(2L) - f(0)) = 0$$

$$\hat{b}_n^{2L} = \frac{1}{L}\left\langle f'(x), \cos\left(\frac{i\pi}{L}x\right)\right\rangle = \frac{1}{L}\left(f(2L) - f(0)\right) + \frac{i\pi}{L}a_i^{2L} = \frac{i\pi}{L}a_i^{2L}$$

$$\hat{a}_n^{2L} = \frac{1}{L}\left\langle f'(x), \sin\left(\frac{i\pi}{L}x\right)\right\rangle = -\frac{i\pi}{L}b_i^{2L}$$

Hence, if f is periodic, we find

$$T(x) = \sum_{i=1}^{\infty}\left(-\frac{\pi i}{L}b_i^{2L}\sin\left(\frac{i\pi}{L}x\right) + \frac{\pi i}{L}a_i^{2L}\cos\left(\frac{i\pi}{L}x\right)\right).$$

This is the same result we would have found if we differentiated the Fourier series for f term by term. So we conclude that the Fourier series of f' can be found by differentiating the Fourier series for f term by term and we know it converges to $f'(x)$ at points where f'' exists.

27.7.3 The Derivative Coefficient Bounds

We can apply the derivation we did above for f to the series expansion for f' we have just found. Assuming f' is integrable, we find

$$(\hat{b}_0^{2L})^2 + \sum_{i=1}^{n}((\hat{a}_i^{2L})^2 + (\hat{b}^{2L})_i^2) \leq \|f'\|_2^2$$

where here $\hat{b}_0^{2L} = 0$, $\hat{a}_i^{2L} = -\frac{\pi i}{L}b_i^{2L}$ and $\hat{b}_i = \frac{\pi i}{L}a_i^{2L}$. Hence, we have

$$\frac{\pi^2}{L^2}\sum_{i=1}^{n}i^2\left(a_i^2 + b_i^2\right) \leq \|f'\|_2^2$$

27.7.4 Uniform Convergence for f

We are almost at the point where we can see circumstances where the Fourier series expansion of f converges uniformly to f on the interval $[0, 2L]$. We assume f is continuous and periodic on $[0, 2L]$ and that f' exists. Further assume f' is integrable. Hence, we know the Fourier series of f exists and converges to $f(x)$. From our earlier arguments, if f has Fourier series

$$f(x) = b_0^{2L} + \sum_{i=1}^{\infty}\left(a_i^{2L}\sin\left(\frac{i\pi}{L}x\right) + b_i^{2L}\cos\left(\frac{i\pi}{L}x\right)\right).$$

then we know for all n

$$\frac{\pi^2}{L^2}\sum_{i=1}^{n}i^2\left((a_i^{2L})^2 + (b_i^{2L})^2\right) \leq \|f'\|_2^2$$

Let's look at the partial sums.

$$b_0^{2L} + \sum_{i=1}^{n}a_i^{2L}\sin\left(\frac{i\pi}{L}x\right) + \sum_{i=1}^{n}b_i^{2L}\cos\left(\frac{i\pi}{L}x\right)$$

$$= b_0^{2L} + \sum_{i=1}^{n}a_i^{2L}\sin\left(\frac{i\pi}{L}x\right) + \sum_{i=1}^{n}b_i^{2L}\cos\left(\frac{i\pi}{L}x\right)$$

$$= b_0^{2L} + \sum_{i=1}^{n} i\, a_i^{2L}\, \frac{1}{i} \sin\left(\frac{i\pi}{L} x\right) + \sum_{i=1}^{n} i\, b_i^{2L}\, \frac{1}{i} \cos\left(\frac{i\pi}{L} x\right).$$

Let T_n denote the nth partial sum here. Then, the difference of the nth and m^{th} partial sum for $m > n$ gives

$$T_m(x) - T_n(x) \;=\; \sum_{i=n+1}^{m} i\, a_i^{2L}\, \frac{1}{i} \sin\left(\frac{i\pi}{L} x\right) + \sum_{i=n+1}^{m} i\, b_i^{2L}\, \frac{1}{i} \cos\left(\frac{i\pi}{L} x\right).$$

Now apply our analogue of the Cauchy - Schwartz inequality for series.

$$
\begin{aligned}
T_m(x) - T_n(x) \;=\; & \sum_{i=n+1}^{m} \left(i\, |a_i^{2L}|\, \frac{1}{i} \right) \left| \sin\left(\frac{i\pi}{L} x\right) \right| + \sum_{i=n+1}^{m} \left(i\, |b_i^{2L}|\, \frac{1}{i} \right) \left| \cos\left(\frac{i\pi}{L} x\right) \right| \\[2mm]
\;\le\; & \sqrt{\sum_{i=n+1}^{m} i^2\, |a_i^{2L}|^2}\; \sqrt{\sum_{i=n+1}^{m} \frac{1}{i^2} \left| \sin\left(\frac{i\pi}{L} x\right) \right|^2} \\[2mm]
& + \sqrt{\sum_{i=n+1}^{m} i^2\, |b_i^{2L}|^2}\; \sqrt{\sum_{i=n+1}^{m} \frac{1}{i^2} \left| \cos\left(\frac{i\pi}{L} x\right) \right|^2} \\[2mm]
\;\le\; & \sqrt{\sum_{i=n+1}^{m} i^2\, |a_i^{2L}|^2}\; \sqrt{\sum_{i=n+1}^{m} \frac{1}{i^2}} + \sqrt{\sum_{i=n+1}^{m} i^2\, |b_i^{2L}|^2}\; \sqrt{\sum_{i=n+1}^{m} \frac{1}{i^2}}
\end{aligned}
$$

Now each of the front pieces satisfy

$$\sum_{i=1}^{\infty} i^2\, (a_i^{2L})^2 \;\le\; \frac{L^2}{\pi^2}\, \|f'\|_2^2, \qquad \sum_{i=1}^{\infty} i^2\, (b_i^{2L})^2 \le \frac{L^2}{\pi^2}\, \|f'\|_2^2$$

Hence,

$$\left| T_m(x) - T_n(x) \right| \;\le\; \left| \sum_{i=n+1}^{m} a_i \sin\left(\frac{i\pi}{L} x\right) + \sum_{i=n+1}^{m} b_i \cos\left(\frac{i\pi}{L} x\right) \right| \le \frac{2L \|f'\|_2}{\pi} \sqrt{\sum_{i=n+1}^{m} \frac{1}{i^2}}.$$

Since the series $\sum_{i=1}^{\infty} (1/i^2)$ converges, this says the sequence of partial sums (T_n) satisfies the UCC for series and hence by Theorem 25.2.1 we can conclude the Fourier series of f converges uniformly to a function T. But by uniqueness of limits, we must have $T = f$ and so we now know T_n converges uniformly to f under these circumstances! Let's summarize this result.

Theorem 27.7.1 Uniform Convergence of Fourier Series

> *Given f on $[0, 2L]$, assume f is continuous with $f(0) = f(2L)$ and f' exists. Further f' is integrable on $[0, 2L]$. Then the Fourier series of f converges uniformly to f on $[0, 2L]$.*

27.7.5 Extending to Fourier Sine and Fourier Cosine Series

The arguments above also apply to the Fourier Sine and Fourier Cosine series we have discussed. In those cases, we have a function f defined only on $[0, L]$. Assuming f' exists on $[0, L]$, consider the Fourier Sine series for f' on $[0, L]$. For the odd extensions of f, f_o and f_e, we know the Fourier sine and cosine coefficients of f_o, $A_{i,o}$, and f_e, $B_{i,o}$, satisfy

$$A_{i,o}^{2L} = \frac{1}{L} \int_0^{2L} f_o(s) \sin\left(\frac{i\pi}{L}s\right) ds = A_i^{2L} = \frac{2}{L} \int_0^L f(s) \sin\left(\frac{i\pi}{L}s\right) ds, \ i \geq 1$$

$$B_{0,0}^{2L} = \frac{1}{2L} \int_0^{2L} f_o(s) ds = B_0^{2L} = \frac{1}{L} \int_0^L f(s) ds, \ i = 0$$

$$B_{i,o}^{2L} = \frac{1}{L} \int_0^L f_o(s) \cos\left(\frac{i\pi}{L}s\right) = B_i^{2L} = \frac{2}{L} \int_0^L f(s) \cos\left(\frac{i\pi}{L}s\right) ds, \ i \geq 1$$

Note the Fourier Sine series for f_o' is

$$\sum_{i=0}^{\infty} \frac{2}{L} \left\langle f_o'(x), \sin\left(\frac{i\pi}{L}x\right) \right\rangle \sin\left(\frac{i\pi}{L}x\right)$$

We calculate the inner products by integrating by parts. If we assume f' is integrable, this leads to coefficient bounds. Let the Fourier coefficients for f_o' be $A_{i,o}'^{2L}$ and $B_{i,o}'^{2L}$. We have for $i > 0$

$$A_{i,o}'^{2L} = \frac{1}{L} \left\langle f_o', \sin\left(\frac{i\pi}{L}x\right) \right\rangle = \frac{1}{L} \int_0^{2L} f_o'(x) \sin(\frac{i\pi}{L}x) \, dx$$

$$= \frac{1}{L} \left(f_o(x) \cos(\frac{i\pi}{L}x) \Big|_0^{2L} - \int_0^{2L} f_o(x) \frac{i\pi}{L} \cos(\frac{i\pi}{L}x) \, dx \right)$$

$$= -\frac{1}{L} \int_0^{2L} f_o(x) \frac{i\pi}{L} \cos(\frac{i\pi}{L}x) \, dx = -\frac{i\pi}{L} \left\langle f_o(x), \cos\left(\frac{i\pi}{L}x\right) \right\rangle = -\frac{i\pi}{L} B_{i,o}^{2L}$$

We also have for $i > 0$

$$B_{i,o}'^{2L} = \frac{1}{L} \left\langle f_o', \cos\left(\frac{i\pi}{L}x\right) \right\rangle = \frac{1}{L} \int_0^{2L} f_o'(x) \cos(\frac{i\pi}{L}x) \, dx$$

$$= \frac{1}{L} \left(-f_o(x) \sin(\frac{i\pi}{L}x) \Big|_0^{2L} + \int_0^{2L} f_o(x) \frac{i\pi}{L} \sin(\frac{i\pi}{L}x) \, dx \right)$$

$$= \frac{1}{L} \int_0^{2L} f_o(x) \frac{i\pi}{L} \sin(\frac{i\pi}{L}x) \, dx = \frac{i\pi}{L} \left\langle f_o(x), \sin\left(\frac{i\pi}{L}x\right) \right\rangle = \frac{i\pi}{L} A_{i,o}^{2L}$$

Then letting $E_{n,o} = f_o' - \sum_{i=1}^n \frac{i\pi}{L} A_{i,o}^{2L} \cos(\frac{i\pi}{L}x)$

$$< E_{n,o}, E_{n,o} > = < f_o', f_o' > -2 \sum_{i=1}^n \frac{i\pi}{L} A_{i,o}^{2L} < f_o', \cos(\frac{i\pi}{L}x) >$$

$$+ \sum_{i=1}^n \sum_{j=1}^n \frac{i\pi}{L} \frac{j\pi}{L} A_{i,o}^{2L} A_{j,o}^{2L} < \cos(\frac{i\pi}{L}x), \cos(\frac{j\pi}{L}x) >$$

or

$$\left\langle f_o' - \left(\sum_{i=1}^n \frac{i\pi}{L} A_{i,o}^{2L} \cos\left(\frac{i\pi}{L}x\right) \right), f_o' - \left(\sum_{j=1}^n \frac{j\pi}{L} A_{j,o}^{2L} \cos\left(\frac{j\pi}{L}x\right) \right) \right\rangle =$$

$$< f_o', f_o' > -2 \sum_{i=1}^n \frac{i\pi}{L} A_{i,o}^{2L} \left\langle f_o', \cos\left(\frac{i\pi}{L}x\right) \right\rangle$$

$$+ \sum_{i=1}^{n} \frac{i^2\pi^2}{L^2} (A_{i,o}^{2L})^2 \left\langle \cos\left(\frac{i\pi}{L}x\right), \cos\left(\frac{i\pi}{L}x\right) \right\rangle$$

We know on $[0, 2L]$ that $< \cos(\frac{i\pi}{L}x), \cos(\frac{i\pi}{L}x) >= L$. Thus, we have

$$\left\langle f_o' - \sum_{i=1}^{n} \frac{i\pi}{L} A_{i,o}^{2L} \cos\left(\frac{i\pi}{L}x\right), f_o' - \sum_{j=1}^{n} \frac{j\pi}{L} A_{j,o}^{2L} \cos\left(\frac{j\pi}{L}x\right) \right\rangle$$

$$= < f_o', f_o' > -2 \sum_{i=1}^{n} \frac{i\pi}{L} A_{i,o}^{2L} \left\langle f_o', \cos\left(\frac{i\pi}{L}x\right) \right\rangle + \sum_{i=1}^{n} \frac{i^2\pi^2}{L} (A_{i,o}^{2L})^2$$

Since $\left\langle f_o', \cos\left(\frac{i\pi}{L}x\right) \right\rangle = i\pi A_{i,0}^{2L} = i\pi A_i^L$, we have

$$\left\langle f_o' - \sum_{i=1}^{n} \frac{i\pi}{L} A_{i,o}^{2L} \cos\left(\frac{i\pi}{L}x\right), f_o' - \sum_{j=1}^{n} \frac{j\pi}{L} A_{j,o}^{2L} \cos\left(\frac{j\pi}{L}x\right) \right\rangle$$

$$= < f_o', f_o' > -2 \sum_{i=1}^{n} \frac{i^2\pi^2}{L} (A_{i,o}^{2L})^2 + \sum_{i=1}^{n} \frac{i^2\pi^2}{L} (A_{i,o}^{2L})^2$$

$$= \; < f_o', f_o' > - \sum_{i=1}^{\infty} \frac{i^2\pi^2}{L} (A_{i,0}^{2L})^2 = < f_o', f_o' > - \sum_{i=1}^{\infty} \frac{i^2\pi^2}{L} (A_i^L)^2$$

But we know $\int_0^{2L} (f_o')^2(s)ds = 2\int_0^L (f')^2(s)ds$ and so for all n,

$$\sum_{i=1}^{n} \frac{i^2\pi^2}{2L} (A_i^L)^2 \; \leq \; \|f'\|_2^2.$$

Note this immediately tells us that $iA_i^L \to 0$ in addition to $A_i^L \to 0$. Hence, when f' is integrable we get more information about the rate at which these Fourier coefficients go to zero.

Note our calculations have also told us the Fourier Sine Series expansion of f' on $[0, L]$ is

$$\sum_{i=1}^{\infty} \frac{2}{L} \left\langle f', \sin\left(\frac{i\pi}{L}x\right) \right\rangle \sin\left(\frac{i\pi}{L}x\right) \; = \; -\sum_{i=1}^{\infty} \frac{i\pi}{L} B_i^L \sin\left(\frac{i\pi}{L}x\right)$$

For the even extension to f' to $[0, 2L]$, we have the Fourier Cosine series for f_e' is

$$\sum_{i=0}^{\infty} \frac{2}{L} \left\langle f_e'(x), \cos\left(\frac{i\pi}{L}x\right) \right\rangle \cos\left(\frac{i\pi}{L}x\right)$$

For the even extensions of f, f_o and f_e, we know the Fourier sine and cosine coefficients of f_o, $A_{i,o}$, and f_e, $B_{i,o}$, satisfy

$$A_{i,e}^{2L} = \frac{1}{L} \int_0^{2L} f_e(s) \sin\left(\frac{i\pi}{L}s\right) ds = A_i^L = \frac{2}{L} \int_0^L f(s) \cos\left(\frac{i\pi}{L}s\right) ds, \; i \geq 1$$

$$B_{0,e}^{2L} = \frac{1}{2L} \int_0^{2L} f_e(s) ds = B_0^L = \frac{1}{L} \int_0^L f(s) ds, \; i = 0$$

$$B_{i,e}^{2L} = \frac{1}{L} \int_0^L f_e(s) \cos\left(\frac{i\pi}{L}s\right) = B_i^L = \frac{2}{L} \int_0^L f(s) \cos\left(\frac{i\pi}{L}s\right) ds, \; i \geq 1$$

Let the Fourier coefficients for f_e' be $A_{i,e}'^{,2L}$ and $B_{i,e}'^{,2L}$. We have for $i > 0$

$$A_{i,e}'^{,2L} = \frac{1}{L} \left\langle f_e', \sin\left(\frac{i\pi}{L}x\right) \right\rangle = \frac{1}{L} \int_0^{2L} f_e'(x) \sin(\frac{i\pi}{L}x) \, dx$$

$$= \frac{1}{L} \left(f_e(x) \cos(\frac{i\pi}{L}x) \Big|_0^{2L} - \int_0^{2L} f_e(x)\frac{i\pi}{L} \cos(\frac{i\pi}{L}x) \, dx \right)$$

$$= -\frac{1}{L} \int_0^{2L} f_e(x)\frac{i\pi}{L} \cos(\frac{i\pi}{L}x) \, dx = -\frac{i\pi}{L} \left\langle f_e(x), \cos\left(\frac{i\pi}{L}x\right) \right\rangle = -\frac{i\pi}{L} B_i^{2L}$$

We also have for $i > 0$

$$B_{i,e}'^{,2L} = \frac{1}{L} \left\langle f_e', \cos\left(\frac{i\pi}{L}x\right) \right\rangle = \frac{1}{L} \int_0^{2L} f_e'(x) \cos(\frac{i\pi}{L}x) \, dx$$

$$= \frac{1}{L} \left(-f_e(x) \sin(\frac{i\pi}{L}x) \Big|_0^{2L} + \int_0^{2L} f_e(x)\frac{i\pi}{L} \sin(\frac{i\pi}{L}x) \, dx \right)$$

$$= \frac{1}{L} \int_0^{2L} f_e(x)\frac{i\pi}{L} \sin(\frac{i\pi}{L}x) \, dx = \frac{i\pi}{L} \left\langle f_e(x), \sin\left(\frac{i\pi}{L}x\right) \right\rangle = \frac{i\pi}{L} A_i^{2L}$$

Then

$$\left\langle f_e' - \sum_{i=1}^n B_{i,e}^{2L} \frac{i\pi}{L}(-) \sin\left(\frac{i\pi}{L}x\right), f_e' - \sum_{j=1}^n B_{j,e}^{2L} \frac{j\pi}{L}(-) \sin\left(\frac{j\pi}{L}x\right) \right\rangle =$$

$$\left\langle f_e' + \sum_{i=1}^n B_{i,e}^{2L} \frac{i\pi}{L} \sin\left(\frac{i\pi}{L}x\right), f_e' + \sum_{j=1}^n B_{j,e}^{2L} \frac{j\pi}{L} \sin\left(\frac{j\pi}{L}x\right) \right\rangle =$$

$$< f_e', f_e' > +2 \sum_{i=1}^n B_{i,e}^{2L} \frac{i\pi}{L} \left\langle f_e', \sin\left(\frac{i\pi}{L}x\right) \right\rangle$$

$$+ \sum_{i=1}^n (B_{i,e}^{2L})^2 \frac{i^2\pi^2}{L^2} \left\langle \sin\left(\frac{i\pi}{L}x\right), \sin\left(\frac{j\pi}{L}x\right) \right\rangle$$

$$= < f_e', f_e' > -2 \sum_{i=1}^n \frac{i^2\pi^2}{L} (B_{i,e}^{2L})^2 + \sum_{i=1}^n \frac{i^2\pi^2}{L} (B_{i,e}^{2L})^2$$

$$= \ < f_e', f_e' > - \sum_{i=1}^n \frac{i^2\pi^2}{L} (B_{i,e}^{2L})^2$$

as $< \sin(\frac{i\pi}{L}x), \sin(\frac{i\pi}{L}x) >= L$. Since $\int_0^{2L} (f_e')^2(s)ds = 2\int_0^L (f')^2(s)ds$, we conclude since $B_{i,e}^{2L} = B_i^L$ that for all n,

$$\sum_{i=1}^n \frac{i^2\pi^2}{2L} (B_i^L)^2 \ \leq \ \|f'\|_2^2.$$

Again, this immediately tells us that $iB_i^L \to 0$ in addition to $B_i^L \to 0$. Hence, when f' is integrable we get more information about the rate at which these Fourier coefficients go to zero.

We are now ready to show the Fourier Sine Series converges uniformly on $[0, L]$. Let T_n denote the nth partial sum of the Fourier Sine Series on $[0, L]$. Then, the difference of the nth and m^{th} partial sum for $m > n$ gives

$$T_m(x) - T_n(x) \;=\; \sum_{i=n+1}^{m} A_i^L \sin\left(\frac{i\pi}{L}x\right) = \sum_{i=n+1}^{m} \left(\frac{i\pi}{L}A_i^L\right)\left(\frac{L}{i\pi}\sin\left(\frac{i\pi}{L}x\right)\right)$$

Now apply our analogue of the Cauchy - Schwartz inequality for series.

$$\left|T_m(x) - T_n(x)\right| \leq \sum_{i=n+1}^{m}\left(\frac{i\pi}{L}|A_i^L|\right)\frac{L}{i\pi}\left|\sin\left(\frac{i\pi}{L}x\right)\right|$$

$$\leq \sqrt{\sum_{i=n+1}^{m}\frac{i^2\pi^2}{L^2}|A_i^L|^2}\sqrt{\sum_{i=n+1}^{m}\frac{L^2}{i^2\pi^2}\left|\sin\left(\frac{i\pi}{L}x\right)\right|^2}$$

$$\leq \sqrt{\sum_{i=n+1}^{m}\frac{i^2\pi^2}{L^2}|A_i^L|^2}\sqrt{\sum_{i=n+1}^{m}\frac{L^2}{i^2\pi^2}}$$

Since our Fourier Series derivative arguments gave us the estimate

$$\sum_{i=1}^{n}\frac{i^2\pi^2}{2L}A_i^2 \;\leq\; \|f'\|_2^2.$$

we have

$$\left|T_m(x) - T_n(x)\right| \;\leq\; \|f'\|\sqrt{\frac{L}{2}}\sqrt{\sum_{i=n+1}^{m}\frac{L^2}{i^2\pi^2}}.$$

Since the series $\sum_{i=1}^{\infty}\frac{L^2}{i^2\pi^2}$ converges, this says the sequence of partial this says (T_n) satisfies the UCC for series and so the sequence of partial sums converges uniformly to a function T which by uniqueness of limits must be f except possibly at the points 0 and L. We know if $f(0) \neq f(L)$, the Fourier Sine series will converge to a function with a jump at those points. Hence, we know the Fourier Sine Series of f converges uniformly to f on compact subsets of $(0, L)$ as long as f' exists.

The same sort of argument works for the Fourier Cosine Series on $[0, L]$ using the bounds we found from the Fourier Sine Series expansion for f'. So there is a function Y which this series converges to uniformly. Since limits are unique, we must have $Y = f$ except possibly at the points 0 and L. We know if $f(0) \neq f(L)$, the Fourier Cosine series will converge to a function with a jump at those points. Hence, we know the Fourier Cosine Series of f converges uniformly to f on compact subsets of $(0, L)$ as long as f' exists.

Homework

Exercise 27.7.1 *Compute the first five Fourier sine coefficients of f' on $[0, 5]$ for $f(x) = x^2 + 3$.*

Exercise 27.7.2 *Compute the first five Fourier cosine coefficients of f' on $[0, 5]$ for $f(x) = x^2 + 3$.*

Exercise 27.7.3 *Compute the first five Fourier sine coefficients of f' on $[0, 5]$ for $f(x) = -x^2 + 3x + 2$.*

Exercise 27.7.4 *Compute the first five Fourier cosine coefficients of f' on $[0, 5]$ for $f(x) = -x^2 + 3x + 2$.*

27.8   Code Implementation

We often need to approximate our Fourier series expansions for arbitrary functions f by finding approximations to the desired Fourier Coefficients. All of these approximations require that we find inner products of the form $< f, u_n >$ for some function u_n which is a sin or cos term. In general, if we are given the linearly independent vectors/ functions $\{f_1, \ldots, f_N\}$, we can apply Graham - Schmidt Orthogonalization (GSO) to find the orthogonal linearly independent vectors of length one $\{g_1, \ldots, g_N\}$. Of course, the needed inner products are computed numerically and therefore there is computational error. A simple test to see if our numerical inner product calculations are sufficiently accurate is to compute the $N \times N$ matrix $d = (< g_i, g_j >)$ which should be essentially the $N \times N$ identify matrix as the off diagonal entries should all be zero. However, if the inner product computations are inaccurate, off diagonal values need not be zero and the GSO method will fail. So we begin with a search for a better inner product.

27.8.1   Inner Products Revisited

A standard tool to handle the approximation of integrals such as these, is to use the Newton-Cotes formulae. We will not discuss that here and instead, simply remark that the errors we get when we use the Newton-Cotes formulae can be unacceptable. Hence, we will handle our inner product calculations using a Riemann sum approximation to the needed integral. This is not very sophisticated, but it works well and is easy to program. The function `innerproduct` to do this is shown below. This code approximates $\int_a^b f(t)g(t)dt$ using a Riemann sum with N terms formed from a uniform partition based on the subinterval length $\frac{b-a}{N}$. This performs well in practice for us.

Listing 27.1: | **Riemann Sum Inner Product** |

```
function c = innerproduct(f,g,a,b,N)
%
%
h = @(t) f(t)*g(t);

delx = (b-a)/N;
x = linspace(a,b,N+1);
c = 0;
for i=1:N
    c = c+ h(x(i))*delx;
end

end
```

We can check how accurate we are with these inner product calculations by using them to do a Graham - Schmidt orthogonalization on the functions $1, t, t^2, \ldots, t^N$ on the interval $[a, b]$. The code to do this is in the function `GrahamSchmidtTwo`. This function returns the new orthogonal of length one functions g_i and also prints out the matrix with terms $< g_i, g_j >$ which should be an identity. The input `NIP` is the number of terms to use in the Riemann sum approximations to the inner product and `N` is size of the number of functions we perform GSO on.

Listing 27.2: | **Using Riemann Sums for GSOs: GrahamSchmidtTwo** |

```
function g = GrahamSchmidtTwo(a,b,N,NIP)
```

```
%
% Perform Graham - Schmidt Orthogonalization
% on a set of functions t^0,..., t^N
5 %
%Setup function handles
f = SetUpFunctions(N);
g = cell(N+1,1);

10 nf = sqrt(innerproduct(f{1},f{1},a,b,NIP));
g{1} = @(x) f{1}(x)/nf;
d = zeros(N+1,N+1);
for k=2:N+1
   %compute next orthogonal piece
15   phi = @(x) 0;
   for j = 1:k-1
     c = innerproduct(f{k},g{j},a,b,NIP);
     phi = @(x) (phi(x)+c*g{j}(x));
   end
20   psi = @(x) (f{k}(x) - phi(x));
   nf = sqrt(innerproduct(psi,psi,a,b,NIP));
   g{k} = @(x) (psi(x)/nf);
   end

25 for i=1:N+1
   for j=1:N+1
     d(i,j) = innerproduct(g{i},g{j},a,b,NIP);
   end
   end
30 d
   end
```

Here is a typical run.

Listing 27.3: | **Sample GrahamSchmidtTwo Run** |

```
g = GrahamSchmidtTwo(0,2,2,50);
```

This computes the GSO of the functions $1, t, t^2$ on the interval $[0, 2]$ using Riemann sum approximations with 5000 points. The matrix $< g_i, g_j >$ which is calculated is

Listing 27.4: | **GSO Orthogonality Results: First 3 Powers of t** |

```
d =

      1.0000e+00   -1.3459e-13    1.7656e-13
4    -1.3459e-13    1.0000e+00    6.1171e-13
      1.7656e-13    6.1171e-13    1.0000e+00
```

This is the 3×3 identify that we expect. Next, we do GSO on the functions $1, t, \ldots, t^6$.

Listing 27.5: | **Running GSO on the First 5 Powers of** t |

```
g = GrahamSchmidtTwo(0,2,6,5000);
```

This also generates the identity matrix we expect for $< g_i, g_j >$.

Listing 27.6: | **GSO Orthogonality Results: First 5 Powers of t** |

```
d =

      1.0000e+00   -1.3459e-13    1.7656e-13   -2.0684e-13    1.8885e-13
     -1.3459e-13    1.0000e+00    6.1171e-13   -1.8011e-12    3.6344e-12
      1.7656e-13    6.1171e-13    1.0000e+00   -1.0743e-11    4.8201e-11
     -2.0684e-13   -1.8011e-12   -1.0743e-11    1.0000e+00    3.8416e-10
      1.8885e-13    3.6344e-12    4.8201e-11    3.8416e-10    1.0000e+00
```

To compute the first N terms of the Fourier Cosine series of a function f on the interval $[0, L]$, we first need a way to encode all the functions $\cos\left(\frac{n\pi}{L}x\right)$. We do this in the function `SetUpCosines`.

Listing 27.7: | **SetUpCosines** |

```
function f = SetUpCosines(L,N)
%
% Setup function handles
%
f = cell(N+1,1);
for i=1:N+1
   f{i} = @(x) cos( (i-1)*pi*x/L );
end
```

This generates handles to the functions 1, $\cos\left(\frac{\pi}{L}x\right)$, $\cos\left(\frac{2\pi}{L}x\right)$ and so forth ending with $\cos\left(\frac{N\pi}{L}x\right)$. A similar function encodes the corresponding sin functions we need for the first N terms of a Fourier Sine series. The function is called `SetUpSines` with code:

Listing 27.8: | **SetUpSines** |

```
function f = SetUpSines(L,N)
%
% Setup function handles
%
f = cell(N,1);
for i=1:N
   f{i} = @(x) sin( i*pi*x/L );
end
```

This generates handles to the functions $\sin\left(\frac{\pi}{L}x\right)$, $\sin\left(\frac{2\pi}{L}x\right)$ and so forth ending with $\sin\left(\frac{N\pi}{L}x\right)$. Let's check how accurate our inner product calculations are on the sin and cos terms. First, we modify the functions which set up the sin and cos functions to return functions of length one on the interval $[0, L]$. This is done in the functions `SetUpOrthogCos` and `SetUpOrthogSin`.

Listing 27.9: **SetUpOrthogCos**

```
function f = SetUpOrthogCos(L,N)
%
% Setup function handles
%
f = cell(N+1,1);
f{1} = @(x) sqrt(1/L);
for i=2:N+1
    f{i} = @(x) sqrt(2/L)*cos( (i-1)*pi*x/L );
end
```

and

Listing 27.10: **SetUpOrthogSin**

```
function f = SetUpOrthogSin(L,N)
%
% Setup function handles
%
f = cell(N,1);
for i=1:N
    f{i} = @(x) sqrt(2/L)*sin( (i)*pi*x/L );
end
```

Then, we can check the accuracy of the inner product calculations by computing the entries of the matrix $d = (< f_i, f_j >)$. We do this with the new function `CheckOrtho`. We input the function `f` and interval $[a, b]$ endpoints `a` and `b` and the number of terms to use in the Riemann sum approximation of the inner product, `NIP`.

Listing 27.11: **Checking Orthogonality of the GSO: CheckOrtho**

```
function CheckOrtho(f,a,b,NIP)
%
% Perform Graham - Schmidt Orthogonalization
% on a set of functions f
%
%Setup function handles
N = length(f);

for i=1:N
    for j=1:N
        d(i,j) = innerproduct(f{i},f{j},a,b,NIP);
    end
end
```

```
     d
  15 end
```

Let's try it with the cos functions. We use a small number of terms for the Riemann sums, `NIP` = 50 and compute $< g_i, g_j >$ for the first 7 functions.

Listing 27.12: | **Code to Check Cosine Orthogonality** |

```
f = SetUpOrthogCos(5,7);
CheckOrtho(f,0,5,50);
```

We do not get the identity matrix as we expect some of the off diagonal values are too large.

Listing 27.13: | **Checking Orthogonality for Cosine Functions: NIP = 50** |

```
    d =

 3  1.0e+00    2.8e−02  −5.5e−17    2.8e−02  −1.2e−16    2.8e−02  −4.5e−17    2.8e
          −02
    2.8e−02    1.0e+00    4.0e−02  −1.2e−16    4.0e−02    4.8e−17    4.0e−02    1.1e
          −16
   −5.5e−17    4.0e−02    1.0e+00    4.0e−02  −6.2e−17    4.0e−02  −2.0e−16    4.0e
          −02
    2.8e−02  −1.2e−16    4.0e−02    1.0e+00    4.0e−02  −5.5e−17    4.0e−02  −6.9e
          −18
   −1.1e−16    4.0e−02  −6.2e−17    4.0e−02    1.0e+00    4.0e−02  −9.7e−17    4.0e
          −02
 8  2.8e−02    4.8e−17    4.0e−02  −5.5e−17    4.0e−02    1.0e+00    4.0e−02    2.8e
          −17
   −4.5e−17    4.0e−02  −2.0e−16    4.0e−02  −9.7e−17    4.0e−02    1.0e+00    4.0e
          −02
    2.8e−02    1.1e−16    4.0e−02  −6.9e−18    4.0e−02    2.8e−17    4.0e−02    1.0e
          +00
```

Resetting `NIP` = 200, we find a better result.

Listing 27.14: | **Checking Orthogonality for Cosine Functions: NIP = 200** |

```
    d =

    1.0e+00    7.1e−03  −6.1e−17    7.1e−03  −7.1e−17    7.1e−03  −3.9e−17    7.1e−03
    7.1e−03    1.0e+00    1.0e−02  −5.5e−17    1.0e−02  −1.1e−16    1.0e−02  −3.8e−17
 5  6.1e−17    1.0e−02    1.0e+00    1.0e−02  −7.8e−17    1.0e−02  −1.6e−16    1.0e−02
    7.1e−03  −5.5e−17    1.0e−02    1.0e+00    1.0e−02  −2.7e−16    1.0e−02  −3.3e−17
    7.1e−17    1.0e−02  −7.8e−17    1.0e−02    1.0e+00    1.0e−02  −1.0e−16    1.0e−02
    7.1e−03  −1.2e−16    1.0e−02  −2.6e−16    1.0e−02    1.0e+00    1.0e−02    2.1e−17
    3.9e−17    1.0e−02  −1.6e−16    1.0e−02  −1.0e−16    1.0e−02    1.0e+00    1.0e−02
 10 7.1e−03  −3.8e−17    1.0e−02  −3.3e−17    1.0e−02    2.1e−17    1.0e−02    1.0e+00
```

We can do even better by increasing `NIP`, of course. We encourage you to do these experiments yourself. We did not show the results for the sin terms, but they will be similar.

Homework

Exercise 27.8.1 *Find the GSO of the functions $f_1(t) = t^2$, $f_2(t) = \cos(2t)$ and $f_3(t) = 4t^3$ on the interval $[-1, 2]$ using the implementations given in this section for various values of* `NIP`. *Check to see the matrix d is the diagonal.*

Exercise 27.8.2 *Find the GSO of the functions $f_1(t) = 1$, $f_2(t) = t$ and $f_3(t) = t^2$ on the interval $[-1, 1]$ using the implementations given in this section for various values of* `NIP`. *Check to see the matrix d is the diagonal.*

Exercise 27.8.3 *Find the GSO of the functions $f_1(t) = 1$, $f_2(t) = t$ and $f_3(t) = t^2$ on the interval $[-1, 4]$ using the implementations given in this section for various values of* `NIP`. *Check to see the matrix d is the diagonal.*

Exercise 27.8.4 *Find the GSO of the functions $f_1(t) = 1$, $f_2(t) = t$ and $f_3(t) = t^2$ and $f_4(t) = t^3$ on the interval $[-2, 6]$ using the implementations given in this section for various values of* `NIP`. *Check to see the matrix d is the diagonal.*

27.8.2   General GSO

To find the GSO of arbitrary functions, use the `GrahamScmidtThree` function.

Listing 27.15: **GSO for Arbitrary Functions**

```
function g = GrahamSchmidtThree(f,a,b,NIP)
% Perform Graham - Schmidt Orthogonalization
% on a set of functions f
%Setup function handles
5 N = length(f);
g = cell(N,1);

nf = sqrt(innerproduct(f{1},f{1},a,b,NIP));
g{1} = @(x) f{1}(x)/nf;
10 d = zeros(N,N);
for k=2:N
    %compute next orthogonal piece
    phi = @(x) 0;
    for j = 1:k-1
15      c = innerproduct(f{k},g{j},a,b,NIP);
        phi = @(x) (phi(x)+c*g{j}(x));
    end
    psi = @(x) (f{k}(x) - phi(x));
    nf = sqrt(innerproduct(psi,psi,a,b,NIP));
20  g{k} = @(x) (psi(x)/nf);
    end
% check orthogonality
for i=1:N
    for j=1:N
25    d(i,j) = innerproduct(g{i},g{j},a,b,NIP);
    end
end
```

```
    d
  end
```

To use this, we have to setup a cell structure to hold our list of functions. For example, if $f(t) = t^2$, $g(t) = \sin(4t + 3)$ and $h(t) = 1/(1 + t^2)$, we would do the GSO as follows:

Listing 27.16: **GSO for Three Functions**

```
1 f = @(t)  t.^2;
  g = @(t)  sin(4*t+3);
  h = @(t)  1./(1+t.^2);
  F = cell(3,1);
  F{1} = f;
6 F{2} = g;
  F{3} = h;
  G = GrahamSchmidtThree(F,-1,1,300);
  d =
        1.0000      0.0000     -0.0000
11      0.0000      1.0000      0.0000
       -0.0000      0.0000      1.0000
```

We can then see the graphs of the original functions in Figure 27.1 using the basic code:

Listing 27.17: **Graphing the Original Functions**

```
  T = linspace(-1,1,51);
  plot(T,F{1}(T),T,F{2}(T),T,F{3}(T));
3 xlabel('Time');
  ylabel('Original Functions');
  title('Original Functions on [-1,1]');
```

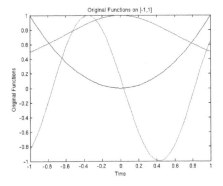

Figure 27.1: The original functions on $[-1, 1]$.

We can then plot the new orthonormal basis using the code:

Listing 27.18: New Functions from GSO

```
T = linspace(-1,1,51);
plot(T,G{1}(T),T,G{2}(T),T,G{3}(T));
xlabel('Time');
ylabel('Basis Functions');
title('Basis Functions on [-1,1]');
```

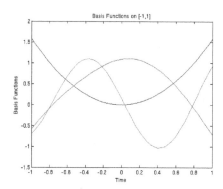

Figure 27.2: The new GSO functions on $[-1, 1]$.

27.8.3　Fourier Cosine and Sine Approximation

We can then calculate the first $N + 1$ terms of a Fourier series or the first N terms of a Fourier Sine series using the functions `FourierCosineApprox` and `FourierSineApprox`, respectively. Let's look at the Fourier Cosine approximation first. This function returns a handle to the function `p` which is the approximation

$$p(x) \;=\; \sum_{n=0}^{N} <f, g_i> \, g_i(x)$$

where $g_i(x) = \cos\left(\frac{i\pi}{L}x\right)$. It also returns the Fourier cosine coefficients $A_1 = \frac{1}{L} <f, g_0>$ through $A_{N+1} = \frac{2}{L} <f, g_N>$. We must choose how many points to use in our Riemann sum inner product estimates – this is the input variable `N`. The other inputs are the function handle `f` and the length `L`. We also plot our approximation and the original function together so we can see how we did.

Listing 27.19: FourierCosineApprox

```
function [A,p] = FourierCosineApprox(f,L,M,N)
%
% p is the Nth Fourier Approximation
% f is the original function
% M is the number of Riemann sum terms in the inner product
% N is the number of terms in the approximation
% L is the interval length
```

```
     % get the first N+1 Fourier cos approximations
10   g = SetUpCosines(L,N);

     % get Fourier Cosine Coefficients
     A = zeros(N+1,1);
     A(1) = innerproduct(f,g{1},0,L,M)/L;
15   for i=2:N+1
        A(i) = 2*innerproduct(f,g{i},0,L,M)/L;
     end

     % get Nth Fourier Cosine Approximation
20   p = @(x) 0;
     for i=1:N+1
        p = @(x) (p(x) + A(i)*g{i}(x));
     end

25   x = linspace(0,L,101);
     for i=1:101
        y(i) = f(x(i));
     end
     yp = p(x);
30
     figure
     s = [' Fourier Cosine Approximation with ',int2str(N+1),' term(s)'];
     plot(x,y,x,yp);
     xlabel('x axis');
35   ylabel('y axis');
     title(s);

     end
```

We will test our approximations on two standard functions: the sawtooth curve and the square wave. To define these functions, we will use an auxiliary function `splitfunc` which defines a new function z on the interval $[0, L]$ as follows

$$z(x) = \begin{cases} f(x) & 0 \leq x < \frac{L}{2}, \\ g(x) & \frac{L}{2} \leq x \leq L \end{cases}$$

The arguments to `splitfunc` are the functions `f` and `g`, the value of `x` for we wish to find the output `z` and the value of `L`.

Listing 27.20: **Splitfunc**

```
     function z = splitfunc(x,f,g,L)
     %
     if x < L/2
        z = f(x);
5    else
        z = g(x);
     end
```

It is easy then to define a sawtooth and a square wave with the following code. The square wave, Sq has value H on $[0, \frac{L}{2})$ and value 0 on $[\frac{L}{2}, L]$. In general, the sawtooth curve, Saw, is the straight line connecting the point $(0,0)$ to $(\frac{L}{2}, H)$ on the interval $[0, \frac{L}{2}]$ and the line connecting $(\frac{L}{2}, H)$ to $(L, 0)$ on the interval $(\frac{L}{2}, L)$. Thus,

$$Sq(x) \;=\; \left\{ \begin{array}{ll} H & 0 \le x < \frac{L}{2}, \\ 0 & \frac{L}{2} \le x \le L \end{array} \right.$$

and

$$Saw(x) \;=\; \left\{ \begin{array}{ll} \frac{2}{L}Hx & 0 \le x < \frac{L}{2}, \\ 2H - \frac{2}{L}Hx & \frac{L}{2} \le x \le L \end{array} \right.$$

We implement the sawtooth in the function `sawtooth`.

Listing 27.21: **Sawtooth**

```
   function  f  =  sawtooth(L,H)
   %
   %
   %  fleft  = H/(L/2)  x
 5 %         = 2Hx/L
   %  check  fleft(0)  =  0
   %         fleft(L/2)  =  2HL/(2L)  =  H
   %
   %  fright  = H +  (H −  0)/(L/2  −  L)(x  −  L/2)
10 %          = H −  (2H/L)  *x  +  (2H/L)*(L/2)
   %          = H −  2H/L  *  x  +  H
   %          = 2H −  2Hx/L
   %  check  fright(L/2)  =  2H −  2HL/(2L)  =  2H −H = H
   %         fright(L)     =  2H −  2HL/L  =  0
15 %
   fleft   = @(x)  2*x*H/L;
   fright  = @(x)  2*H −  2*H*x/L;
   f       = @(x)  splitfunc(x,fleft,fright,L);

20 end
```

As an example, we build a square wave and sawtooth of height 10 on the interval $[0, 10]$. It is easy to plot these functions as well, although we won't show the plots yet. We will wait until we can compare the functions to their Fourier series approximations. However, the plotting code is listed below for your convenience. Note the plot must be set up in a **for** loop as the inequality checks in the function **splitfunc** do not handle a vector argument such as the **x** from a **linspace** command correctly.

Listing 27.22: **Build a Sawtooth and Square Wave**

```
   f  = @(x)  10;
   g  = @(x)  0;
   Saw  =  sawtooth(10,10);
   Sq  = @(x)  splitfunc(x,f,g,10);
 5 x  =  linspace(0,10,101);
   for  i=1:101
     ySq(i)  =  Sq(x(i));
```

```
      ySaw(i) = Saw(x(i));
      end
10   plot(x,ySq);
      axis([−.1,10.1 −.1 10.1]);
      plot(x,YSaw);
```

27.8.3.1 Homework

Exercise 27.8.5 *Find the GSO of the functions* $f_1(t) = 1$, $f_2(t) = t + 2$ *and* $f_3(t) = t^2 + 3t + 4$ *on the interval* $[-2, 5]$ *using the implementations given in this section for various values of* **NIP**. *Check to see the matrix d is the diagonal.*

Exercise 27.8.6 *Write the functions needed to generate a periodic square wave on the intervals* $[0, L] \cup [L, 2L] \cup [2L, 3L] \cup \ldots \cup [(N-1)L, NL]$ *of height H. Generate the need graphs also.*

Exercise 27.8.7 *Write the functions needed to generate a periodic square wave on the intervals* $[0, L] \cup [L, 2L] \cup [2L, 3L] \cup \ldots \cup [(N-1)L, NL]$ *with the high and low value reversed; i.e., the square wave is 0 on the front part and H on the back part of each chunk of length L. Generate the need graphs also.*

Exercise 27.8.8 *Write the functions needed to generate a periodic sawtooth wave on the intervals* $[0, L] \cup [L, 2L] \cup [2L, 3L] \cup \ldots \cup [(N-1)L, NL]$. *Generate the need graphs also.*

Exercise 27.8.9 *Write the functions needed to generate a periodic sawtooth - square wave on multiple intervals of length* $[0, 2L]$ *where the first* $[0, L]$ *is the sawtooth and the second* $[L, 2L]$ *is the square. Generate the need graphs also.*

27.8.4 Testing the Approximations

We can then test the Fourier Cosine Approximation code. This code generates the plot you see in Figure 27.3.

Listing 27.23: **Fourier Cosine Approximation to a Sawtooth**

```
    f = @(x) 10;
    g = @(x) 0;
    Saw = sawtooth(10,10);
    Sq = @(x) splitfunc(x,f,g,10);
5   [Acos,pcos] = FourierCosineApprox(Saw,10,100,5);
```

We can then do the same thing and approximate the square wave with the command:

Listing 27.24: **Fourier Cosine Approximation to a Square Wave with 6 Terms**

```
    [Acos,pcos] = FourierCosineApprox(Sq,10,100,5);
```

Note these approximations are done using only 100 terms in the Riemann sum approximations to the inner product. This generates the relatively poor approximation shown in Figure 27.4.
We do better if we increase the number of terms to 11.

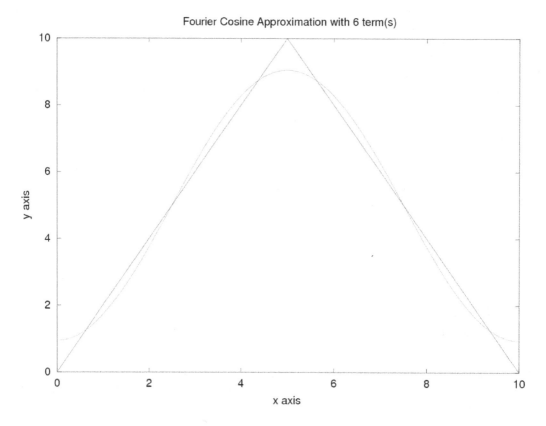

Figure 27.3: 6 term Fourier cosine series approximation to a sawtooth: NIP= 100.

Listing 27.25:

Fourier Cosine Approximation to the Square Wave with 11 Terms

```
[Acos,pcos] = FourierCosineApprox(Sq,10,100,10);
```

This generates the improvement we see in Figure 27.5. The code for `FourierSineApprox` is next. It is quite similar to the cosine approximation code and so we will say little about it. It returns the Fourier sine coefficients as **A** with $A_1 = \frac{2}{L} < f, g_1 >$ through $A_N = \frac{2}{L} < f, g_N >$ where $g_i(x) = \sin\left(\frac{i\pi}{L}x\right)$. It also returns the handle to the Fourier sine approximation

$$p(x) \;=\; \sum_{n=1}^{N} < f, g_i > g_i(x).$$

Listing 27.26:   **FourierSineApprox**

```
function [A,p] = FourierSineApprox(f,L,M,N)
%
```

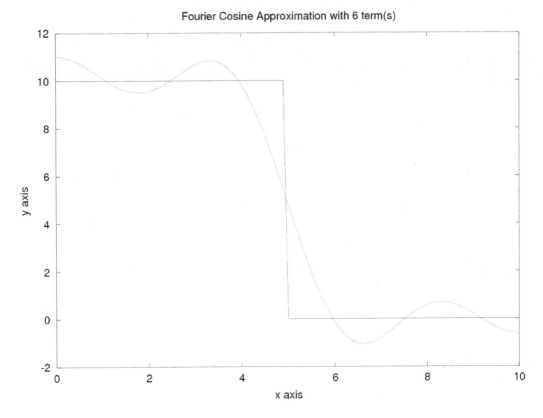

Figure 27.4: 6 term Fourier Cosine Series Approximation to a Square wave: NIP= 100.

```
% p is the Nth Fourier Approximation
% f is the original function
% M is the number of Riemann sum terms in the inner product
% N is the number of terms in the approximation
% L is the interval length

% get the first N Fourier sine approximations
g = SetUpSines(L,N);

% get Fourier Sine Coefficients
A = zeros(N,1);
for i=1:N
    A(i) = 2*innerproduct(f,g{i},0,L,M)/L;
end

% get Nth Fourier Sine Approximation
p = @(x) 0;
for i=1:N
    p = @(x) (p(x) + A(i)*g{i}(x));
end

x = linspace(0,L,101);
for i=1:101
    y(i) = f(x(i));
```

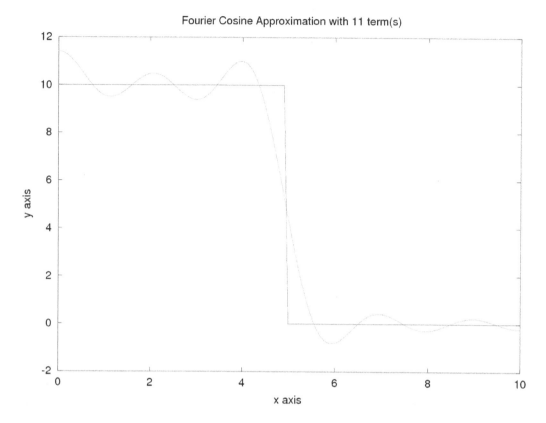

Figure 27.5: 11 term Fourier cosine series approximation to a square wave: NIP= 100.

```
    end
    yp = p(x);

30  figure
    s = [' Fourier Sine Approximation with ',int2str(N),' term(s)'];
    plot(x,y,x,yp);
    xlabel('x axis');
    ylabel('y axis');
35  title(s);
    end
```

Let's test the approximation code on the square wave. using 22 terms. The sin approximations are not going to like the starting value at 10 as you can see in Figure 27.6. The approximation is generated with the command:

Listing 27.27: **Fourier Sine Approximation to a Square Wave with 22 Terms**

```
[Asin,psin] = FourierSineApprox(Sq,10,100,22);
```

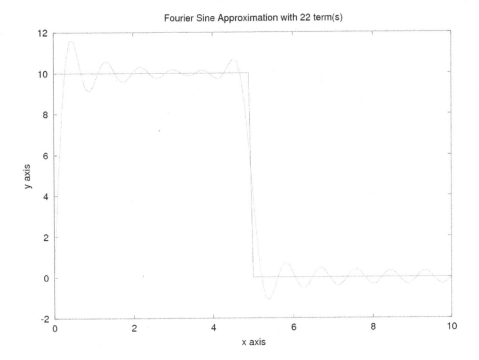

Figure 27.6: 22 term Fourier sine series approximation to a square wave: NIP= 100.

We could do similar experiments with the sawtooth function. Note, we are not using many terms in the inner product calculations. If we boost the number of terms, NIP, to 500, we would obtain potentially better approximations. We will leave that up to you.

Here is another generic type of function to work with: a pulse. This is a function constant of an interval of the form $[x_0 - r, x_0 + r]$ and 0 off of that interval. We implement it in `pulsefunc`.

Listing 27.28: **A Pulse Function**

```
function z = pulsefunc(x,x0,r,H)
%
if x > x0-r && x < x0+r
    z = H;
else
    z = 0;
end
```

It is easy to use. Here is a pulse centered at 3 of radius 0.5 on the interval $[0, 10]$.

Listing 27.29:

Approximating a Pulse with a Fourier Sine Series Approximation

```
P = @(x) pulsefunc(x,3,0.5,10);
[Asin,psin] = FourierSineApprox(P,10,100,22);
```

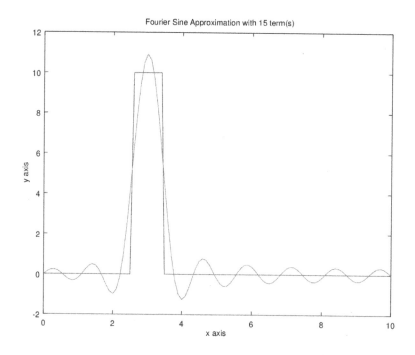

Figure 27.7: Approximating a pulse with a Fourier sine Series approximation.

The resulting approximation is shown in Figure 27.7.

We can combine the Fourier Sine and Fourier Cosine approximations by adding them and dividing by two. Here is the code with the plotting portion removed to make it shorter.

Listing 27.30: **Combined Fourier Sine and Cosine Approximation Code**

```
function [A,B] = FourierApprox (f,L,M,N)
%
g = SetUpCosines(L,N);
B = zeros(N+1,1);
B(1) = innerproduct(f,g{1},0,L,M)/L;
for i=2:N+1
   B(i) = 2*innerproduct(f,g{i},0,L,M)/L;
end
p = @(x) 0;
for i=1:N+1
   p = @(x) (p(x) + B(i)*g{i}(x));
end
%
h = SetUpSines(L,N);
A = zeros(N,1);
for i=1:N
   A(i) = 2*innerproduct(f,h{i},0,L,M)/L;
end
q = @(x) 0;
for i=1:N
   q = @(x) (q(x) + A(i)*h{i}(x));
```

```
    end
23  x = linspace(0,L,101);
    for i=1:101
      y(i) = f(x(i));
    end
    yp = 0.5*(p(x)+q(x));
28  end
```

We can use this code easily with a command like this:

Listing 27.31:

Pulse Approximation: Combined Fourier Sine, Cosine Approximations

```
   P = @(x) pulsefunc(x,3,0.5,10);
2  [A,B] = FourierApprox(P,10,100,22);
```

This generates Figure 27.8.

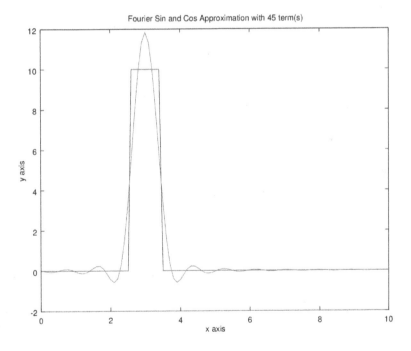

Figure 27.8: Pulse approximation: combined Fourier sine, cosine approximations.

27.8.4.1 Homework

Exercise 27.8.10 *Generate the Fourier Cosine Approximation with varying number of terms and different values of NIP for the periodic square wave. Draw figures as needed.*

Exercise 27.8.11 *Generate the Fourier Sine Approximation with varying number of terms and different values of NIP for the periodic square wave. Draw figures as needed.*

Exercise 27.8.12 *Generate the Fourier Cosine Approximation with varying number of terms and different values of* NIP *for the periodic sawtooth wave. Draw figures as needed.*

Exercise 27.8.13 *Generate the Fourier Sine Approximation with varying number of terms and different values of* NIP *for the periodic sawtooth wave. Draw figures as needed.*

Exercise 27.8.14 *Generate the Fourier Cosine Approximation with varying number of terms and different values of* NIP *for the periodic sawtooth-square wave. Draw figures as needed.*

Exercise 27.8.15 *Generate the Fourier Sine Approximation with varying number of terms and different values of* NIP *for the periodic sawtooth-square wave. Draw figures as needed.*

Exercise 27.8.16 *Generate the Fourier Sine Approximation with varying number of terms and different values of* NIP *for pulses applied at various locations. Draw figures as needed.*

Exercise 27.8.17 *Generate the Fourier Cosine Approximation with varying number of terms and different values of* NIP *for pulses applied at various locations. Draw figures as needed.*

Chapter 28

Applications

We will now look at some applications using Fourier series and the tools we use for proving various types of convergence. We begin by looking briefly at a problem in ordinary differential equations.

28.1 Solving Differential Equations

Some kinds of ordinary differential equation models are amenable to solution using Fourier series techniques. The following problem is a simple one but illustrates how we can use Fourier expansions in this context. Consider the model

$$\begin{aligned} u''(t) &= f(t) \\ u(0) &= 0, \quad u'(0) = u'(L) \end{aligned}$$

where $f \in C([0, L])$ with $f(0) = 0$. This is an example of a **Boundary Value Problem** or **BVP** and problems of this sort need not have solutions at all. Note the easiest thing to do is to integrate.

$$\begin{aligned} u'(t) &= A + \int_0^t f(s)ds \\ u(t) &= B + At + \int_0^t u'(s)\,ds = B + At + \int_0^t \int_0^s f(z)dz \end{aligned}$$

Now apply the boundary conditions.

$$u(0) = 0 \quad \longrightarrow \quad 0 = B + \int_0^0 \int_0^s f(z)dz = B$$

For the other condition, we have

$$A + \int_0^0 f(s)ds = A + \int_0^L f(s)ds \quad \longrightarrow \quad \int_0^L f(s)ds = 0$$

and we see we do not have a way to determine the value of A. Hence, we have an infinite family of solutions S,

$$S = \left\{ \phi(t) = At + \int_0^t \int_0^s f(z)dz \; : \; A \in \Re \right\}$$

as long as the external data f satisfies $\int_0^L f(s)ds = 0$. To find a way to resolve the constant A, we can use Fourier Series.

To motivate how this kind of solution comes about, look at the model below for $\theta > 0$.

$$u''(t) + \theta^2 u(t) = 0$$
$$u(0) = 0, \quad u'(0) = u'(L)$$

This can be rewritten as

$$
\begin{aligned}
u''(t) &= -\theta^2 u(t) \\
u(0) &= 0, \quad u'(0) = u'(L)
\end{aligned}
$$

Let Y denote the vector space of functions

$$Y = \{x : [0, L] \to \Re : x \in C^2([0, L]), \ x(0) = 0, \ x'(0) = x'(L)\}$$

where $C^2([0, L])$ is the vector space of functions that have two continuous derivatives. Then this equation is equivalent to the differential operator $\mathscr{D}(u) = u''$ whose domain in Y satisfying

$$\mathscr{D}(u) = -\theta^2 u$$

which is an example of an **eigenvalue** problem. Nontrivial solutions (i.e. solutions which are not identically zero on $[0, L]$) and the corresponding values of $-\theta^2$ are called **eigenvalues** of the differential operator $\mathscr{D}$ and the corresponding solutions u for a given eigenvalue $-\theta^2$ are called **eigenfunctions**. The eigenfunctions are not unique and there is an infinite family of eigenfunctions for each eigenvalue. Finding the eigenvalues amounts to finding the general solution of this model. The general solution is $u(t) = \alpha \cos(\theta t) + \beta \sin(\theta t)$ and applying the boundary conditions, we find

$$
\begin{aligned}
\alpha &= 0 \\
\beta\, \theta(1 - \cos(\theta L)) &= 0
\end{aligned}
$$

This has unique solution $B = 0$ giving us the trivial solution $u = 0$ on $[0, L]$ unless $1 - \cos(\theta L) = 0$. This occurs when $\theta_n = 2n\pi/L$ for integers n. Thus, the eigenvalues of $\mathscr{D}$ are $-\theta_n^2 = -4n^2\pi^2/L^2$ with eigenfunctions $u_n(t) = \beta \sin(2n\pi\, t/L)$. On $[0, L]$ we already know these eigenfunctions are mutually orthogonal with length L on the interval $[0, L]$ and indeed these are the functions we previously saw but with only even, $2n$, arguments. No $\sin((2n + 1)\pi t/L)$ functions! Hence the eigenfunctions $\{\sqrt{2/L}\,\sin(2n\pi t/L)\}$ form an orthonormal sequence in Y.

Since f is continuous on $[0, L]$ if we also assume f is differentiable everywhere or is differentiable except at a finite number of points where it has finite right and left hand derivative limits, we know it has a Fourier Sine Series

$$f(t) = \sum_{n=1}^{\infty} A_n \sin\left(\frac{2n\pi t}{L}\right)$$

where $A_n = \frac{2}{L}\int_0^L f(t) \sin(2n\pi t/L)\, dt$. We know any finite combination of the functions $\sin(\frac{2n\pi t}{L})$ satisfies the boundary conditions but we still don't know what to do about the data condition. If we set

$$\left(\sum_{i=1}^{n} B_i \sin(\frac{2i\pi t}{L})\right)'' = \sum_{i=1}^{n} -\frac{4i^2\pi^2}{L^2} B_i \sin(\frac{2i\pi t}{L}) = f(t)$$

then taking inner products for the interval $[0, L]$, we have

$$\frac{2}{L}\int_0^L \sum_{i=1}^n -\frac{4i^2\pi^2}{L^2}B_i\sin(\frac{2i\pi t}{L})\,\sin(\frac{j\pi t}{L})dt \;=\; \frac{2}{L}\int_0^L f(t)\sin(\frac{j\pi t}{L})dt = A_j$$

The orthonormality of the functions $\sin(\frac{j\pi t}{L})$ gives

$$\sum_{i=1}^n -\frac{4i^2\pi^2}{L^2}B_i\frac{2}{L}\int_0^L \sin(\frac{2i\pi t}{L})\,\sin(\frac{j\pi t}{L})dt \;=\; \sum_{i=1}^n -\frac{4i^2\pi^2}{L^2}B_i\delta_j^{2i}$$

Hence, if $j = 2k$, we have a match and $B_k = -A_{2k}\frac{L^2}{4k^2\pi^2}$. But if $j = 2k+1$ (i.e. it is odd), the left hand side is always 0 leading us to $0 = A_{2k+1}$. This suggests we need a function f with a certain type of Fourier Sine Series:

- $f(0) = 0$ and be periodic on $[0, L]$ which makes $f(L) = 0$.

- $\int_0^L f(t)dt = 0$

- The Fourier Sine Series for f on $[0, L]$ must have all its odd terms zero.

A good choice is this: for a given function g on $[0, L/2]$ with $g(0) = g(L/2) = 0$ define

$$f(t) \;=\; \begin{cases} g(t), & 0 \le t \le L/2 \\ -g(L-t), & L/2 < t \le L \end{cases}$$

We can check the condition that the odd coefficients are zero easily.

$$\frac{2}{L}\int_0^L f(t)\sin(\frac{n\pi t}{L})dt =$$

$$\frac{2}{L}\int_0^{L/2} g(t)\sin(\frac{n\pi t}{L})dt + \frac{2}{L}\int_{L/2}^L -g(L-t)\sin(\frac{n\pi t}{L})dt$$

Letting $s = L - t$ in the second integral, we have

$$\frac{2}{L}\int_0^L f(t)\sin(\frac{n\pi t}{L})dt =$$

$$\frac{2}{L}\int_0^{L/2} -g(t)\sin(\frac{n\pi t}{L})dt + \frac{2}{L}\int_{L/2}^0 -g(s)(-)\sin(\frac{n\pi s}{L})\cos(n\pi)(-ds)$$

$$\begin{cases} 2\frac{2}{L}\int_0^{L/2} g(t)\sin(\frac{n\pi t}{L})dt, & n \text{ is even } = A_{2n} \\ 0, & n \text{ is odd } = A_{2n+1} \end{cases}$$

For the f as described above, since f is continuous, periodic on $[0, L]$ if we also assume f is differentiable everywhere or is differentiable except at a finite number of points where it has finite right and left hand derivative limits, we know it has a Fourier Sine Series

$$f(t) \;=\; \sum_{n=1}^\infty A_n\sin\left(\frac{n\pi t}{L}\right) = \sum_{n=1}^\infty A_{2n}\sin\left(\frac{2n\pi t}{L}\right)$$

where $A_{2n} = \frac{2}{L}\int_0^L f(t)\sin(2n\pi t/L)\,dt$ and $A_{2n+1} = 0$.

To solve the boundary value problem, assume the solution u also has a Fourier Sine Series expansion which is differentiable term by term so that we have

$$u(t) = \sum_{n=1}^{\infty} B_n \sin\left(\frac{2n\pi t}{L}\right)$$

$$u'(t) = \sum_{n=1}^{\infty} B_n \frac{2n\pi}{L} \cos\left(\frac{2n\pi t}{L}\right)$$

$$u''(t) = \sum_{n=1}^{\infty} -B_n \left(\frac{2n\pi}{L}\right)^2 \sin\left(\frac{2n\pi t}{L}\right)$$

This solution u is built from functions which satisfy the BVP and so u does also. The last condition is that $u'' = f$ which gives

$$\sum_{n=1}^{\infty} -B_n \left(\frac{2n\pi}{L}\right)^2 \sin\left(\frac{2n\pi t}{L}\right) = \sum_{n=1}^{\infty} A_n \sin(2n\pi t/L)$$

As discussed, this suggests we choose $B_n = -A_{2n}\frac{L^2}{4n^2\pi^2}$. For this choice, we have

$$u(t) = \sum_{n=1}^{\infty} -A_{2n} \frac{L^2}{4k^2\pi^2} \sin(\frac{2n\pi t}{L})$$

Now let $\theta_n = \frac{2n\pi}{L}$ and $w_{2n}(t) = \sin(\frac{2n\pi t}{L})$. We know $\sum_{n=1}^{\infty} A_{2n}^2$ converges and so for $n > m$ we have

$$\left\langle \sum_{k=m+1}^{n} -\frac{A_{2k}}{\theta_k^2} w_{2k}, \sum_{j=m+1}^{n} -\frac{A_{2j}}{\theta_j^2} w_{2j} \right\rangle = \sum_{k=m+1}^{n} \frac{A_k^2}{\theta_k^4} \frac{L}{2}$$

The series

$$\sum_{k=1}^{n} \frac{A_{2k}^2}{\theta_k^4} \frac{L}{2} = \sum_{k=1}^{n} A_{2k}^2 \frac{L^4}{16k^4\pi^4} \frac{L}{2} < \sum_{k=1}^{n} A_{2k}^2 \frac{L^5}{2}$$

and so it converges by comparison. Hence, its partial sums are a Cauchy sequence and we therefore know $\sum_{n=1}^{\infty} -A_{2n}/\theta_n^2 \, w_{2n}$ satisfies the UCC and converges uniformly on $[0, L]$ to a function u on $[0, L]$. In addition,

$$\lim_{N\to\infty} \left\langle w_{2j}, \sum_{k=1}^{N} -\frac{A_{2k}}{\theta_k^2} w_{2k} \right\rangle = \left\langle w_{2j}, \lim_{N\to\infty} \sum_{k=1}^{N} -\frac{A_{2k}}{\theta_k^2} w_{2k} \right\rangle$$

$$= \left\langle w_{2j}, \sum_{k=1}^{\infty} -\frac{A_{2k}}{\theta_k^2} w_{2k} \right\rangle$$

Thus

$$\left\langle w_{2j}, \sum_{k=1}^{\infty} -\frac{A_{2k}}{\theta_k^2} w_{2k} \right\rangle = -\frac{A_{2j}}{\theta_j^2} \frac{L}{2}$$

We already knew $\sum_{n=1}^{\infty} A_{2n} w_{2n}$ converged uniformly on $[0, L]$ and so we can conclude our original guess that $B_n = -A_{2n}/\theta_n^2$ is justified as

$$\left\langle w_{2j}, \sum_{k=1}^{\infty} -\frac{B_k}{\theta_k^2} w_{2k} \right\rangle = \left\langle w_{2j}, \sum_{k=1}^{\infty} A_{2k} w_{2k} \right\rangle$$

implies $-\frac{B_j}{\theta_j^2} = A_{2j}$.

The last question is whether

$$u'(t) = \left(\sum_{n=1}^{\infty} -A_{2n}/\theta_n^2 \, w_{2n}(t) \right)' = \sum_{n=1}^{\infty} -A_{2n}/\theta_n^2 \, w_{2n}'(t)$$

$$= \sum_{n=1}^{\infty} -\left(A_{2n}/\theta_n^2 \right) \theta_n \cos(\theta_n t)$$

$$= \sum_{n=1}^{\infty} -A_{2n}/\theta_n \, \cos(\theta_n t)$$

The partial sums of the $u'(t)$ series are $T_n(t) = \sum_{k=1}^{n} -A_{2n}/\theta_n \, \cos(\theta_n t)$. We see for $n > m$,

$$\left| \sum_{k=m+1}^{n} A_{2k}/\theta_k \, \cos(\theta_k t) \right| \leq \sqrt{\sum_{k=m+1}^{n} \left(\frac{L}{2\pi k} \right)^2} \sqrt{\sum_{k=m+1}^{n} |A_{2k}|^2}$$

and since $\sum_{k=1}^{\infty} \frac{1}{k^2}$ and $\sum_{k=1}^{\infty} |A_k|^2$ converge, each has partial sums which form Cauchy sequences and hence the partial sums of $u_n'(t)$ satisfy the UCC on $[0, L]$.

Thus, there is a continuous function D so that $u_n' \xrightarrow{\text{unif}} D$ on $[0, L]$. Since the u'' series is just the Fourier Sine Series for f, we also know the partial sums of u'' converge uniformly to a function E on $[0, L]$. To apply the derivative interchange theorem for u', we let $u_n(t) = \sum_{k=1}^{n} A_{2n}/\theta_n^2 \, \sin(\theta_n t)$ be the nth partial sum of u.

1. u_n is differentiable on $[0, L]$: **True**.

2. u_n' is Riemann Integrable on $[0, L]$: **True as each is a polynomial of** cos **functions**.

3. There is at least one point $t_0 \in [0, L]$ such that the sequence $(u_n(t_0))$ converges. **True as the series converges on** $[0, L]$.

4. $u_n' \xrightarrow{\text{unif}} y$ on $[0, t]$ and the limit function y is continuous. **True as we have just shown** $u_n' \xrightarrow{\text{unif}} D$ **on** $[0, L]$.

The derivative interchange theorem conditions are satisfied and so there is a function W on $[0, L]$ with $u_n \xrightarrow{\text{unif}} W$ on $[0, t]$ and $W' = D$. Since limits are unique, we then have $W = u$ with $u' = D$. Thus, we have

$$u'(t) = \left(\sum_{n=1}^{\infty} -A_{2n}/\theta_n^2 \, w_{2n}(t) \right)' = \sum_{n=1}^{\infty} -A_{2n}/\theta_n^2 \, w_{2n}'(t) = D(t)$$

Also, it is true

$$u''(t) = \left(\sum_{n=1}^{\infty} -A_{2n}/\theta_n^2 \, w_{2n} \right)'' = \sum_{n=1}^{\infty} -A_{2n}/\theta_n^2 \, w_{2n}''(t)$$

$$= \sum_{n=1}^{\infty} \left(A_{2n}/\theta_n^2 \right) \theta_n^2 w_{2n}(t)$$

$$= \sum_{n=1}^{\infty} A_{2n} w_{2n}(t)$$

as these derivative interchanges are valid as the Derivative Interchange Theorem can be justified here as well. Hence, we have used the Fourier Sine Series expansions to solve this ordinary differential equation model with just modest assumptions about the external data f.

Example 28.1.1 *Find the first four terms of the solution u to*

$$u''(t) \;=\; f(t); \quad u(0) = 0, \quad u'(0) = u'(L)$$

for $L = 4$ and f is the function which is $g(t) = 10t(2 - t)$ on $[0, 2]$ and is $-g(t - 2)$ on $[2, 4]$.

Solution *Let's code the data function.*

Listing 28.1: | **Coding the Data Function** |

```
  f = @(x)  10*x.*(2-x);
  f2 = @(x)  -f(4-x);
  g = @(x)  splitfunc(x,f,f2,4);
  X = linspace(0,4,101);
5 for i = 1:101
  G(i) = g(X(i));
  end
  plot(X,G);
```

The graph of the data function is shown in Figure 28.1(a). Next, we find the first eight Fourier Sine coefficients on $[0, 4]$. Then we find the B coefficients for the solution to the ODE.

Listing 28.2: | **Finding the BVP Coefficients** |

```
   theta = @(n,L)  2*n*pi/L;
2  [A,p8] = FourierSineApprox(g,4,200,8);
   theta = @(n,L)  2*n*pi/L;
   B = [];
   for i = 1:4
     B(i) = -A(2*i)/(theta(i,4))^2;
7  end
   >> A
     -6.0474e-16
      1.0320e+01
      4.5325e-16
12   -1.9543e-16
     -6.3843e-16
      3.8224e-01
     -4.4631e-16
     -1.6621e-16
17 >> B
   B =
     -4.1827e+00    1.9801e-17   -1.7213e-02    4.2101e-18
```

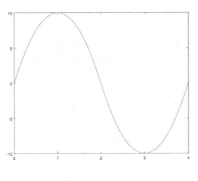

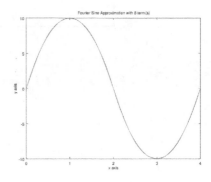

(a) The data function.
(b) The Fourier sine series approximation to the data.

Figure 28.1: The data function and its approximation.

The Fourier Sine approximation to the data function is shown in Figure 28.1(b)
We can then construct the approximation to the solution using these coefficients.

Listing 28.3: **Finding the Approximate Solution**

```
1  u4 = @(t) 0;
   for i = 1:4
     u4 = @(t) (u4(t) + B(i)*sin(2*i*pi*t/4));
   end
   plot(X,u4(X));
6  xlabel('Time');
   ylabel('Solution');
   title('u''''(t) = f(t), u(0)=0, u''(0)=u''(4)');
```

We show this approximate solution in Figure 28.2.

28.1.1 Homework

Exercise 28.1.1 *Find the first four terms of the solution u to*

$$u''(t) = f(t); \quad u(0) = 0, \quad u'(0) = u'(L)$$

for $L = 6$ and f is the function which is $g(t) = 5t(3 - t)$ on $[0, 3]$ and is the on $-g(6 - t)$ on $[3, 6]$. Plot the data function, the approximation to the data function and the approximate solution.

Exercise 28.1.2 *Find the first four terms of the solution u to*

$$u''(t) = f(t); \quad u(0) = 0, \quad u'(0) = u'(L)$$

for $L = 10$ and $f(t) = P(t)$ on $[0, 5]$ where P is the pulse P on $[0, 5]$ with height 100 applied at $t_0 = 1.5$ for a duration of 1.3 and f is the function $-P(10 - t)$ on the interval $[5, 10]$. Plot the data function, the approximation to the data function and the approximate solution. Note that f is not differentiable at the jumps but f has simple jumps at these points. Hence the Fourier Sine series for

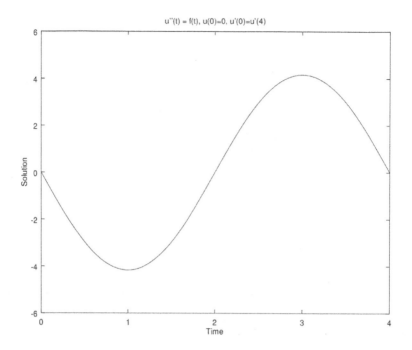

Figure 28.2: The approximate solution.

f converges to the average value of the jump at those points. Also, the original differential equation is not defined at those jumps either. We could eliminate these issues by using C^∞ bump functions instead of pulse functions.

28.2   Some ODE Techniques

We eventually want to look at a simple linear partial differential equation called the cable equation which arises in information transmission such as in a model of a neuron. there are many places you can get the requisite background for this kind of model, but here we will concentrate on the use of analysis in physical modeling. First, we need to talk about a technique in differential equations called **Variation of Parameters**.

28.2.1   Variation of Parameters

We can use the ideas of the linear independence of functions in the solution of nonhomogeneous differential equations. Here the vector space is $C^2[a, b]$ for some interval $[a, b]$ and we use the linear independence of the two solutions of the homogeneous linear second order model to build a solution to the nonhomogeneous model. Rather than doing this in general, we will focus on a specific model:

$$\beta^2 \, \frac{d^2 x}{dt^2} \; - \; x(t) = f(t)$$
$$x(0) = 1, \quad x(5) = 4.$$

where β is a nonzero number. The homogeneous solution x_h solves

$$\beta^2 \frac{d^2 x}{dt^2} - x(t) = 0$$

and has the form

$$x_h(t) = A e^{-\frac{t}{\beta}} + B e^{\frac{t}{\beta}}$$

We want to find a particular solution, called x_p, to the model. Hence, we want x_p to satisfy

$$\beta^2 x_p''(t) - x_p(t) = f(t)$$

Since we don't know the explicit function f we wish to use in the nonhomogeneous equation, a common technique to find the particular solution is the one called **Variation of Parameters** or **VoP**. In this technique, we take the homogeneous solution and replace the constants A and B by unknown functions $u_1(t)$ and $u_2(t)$. Then we see if we can derive conditions that the unknown functions u_1 and u_2 must satisfy in order to work.

So we start by assuming

$$
\begin{aligned}
x_p(t) &= u_1(t) e^{-\frac{t}{\beta}} + u_2(t) e^{\frac{t}{\beta}} \\
x_p'(t) &= u_1'(t) e^{-\frac{t}{\beta}} - u_1(t) \frac{1}{\beta} e^{-\frac{t}{\beta}} + u_2'(t) e^{\frac{t}{\beta}} + u_2(t) \frac{1}{\beta} e^{\frac{t}{\beta}} \\
&= \left(u_1'(t) e^{-\frac{t}{\beta}} + u_2'(t) e^{\frac{t}{\beta}} \right) + \left(-u_1(t) \frac{1}{\beta} e^{-\frac{t}{\beta}} + u_2(t) \frac{1}{\beta} e^{\frac{t}{\beta}} \right)
\end{aligned}
$$

We know there are two solutions to the model, $e^{-\frac{t}{\beta}}$ and $e^{\frac{t}{\beta}}$ which are *linearly independent* functions. Letting $x_1(t) = e^{-\frac{t}{\beta}}$ and $x_2(t) = e^{\frac{t}{\beta}}$, consider the system

$$\begin{bmatrix} x_1(t) & x_2(t) \\ x_1'(t) & x_2'(t) \end{bmatrix} \begin{bmatrix} \phi(t) \\ \psi(t) \end{bmatrix} = \begin{bmatrix} 0 \\ f(t) \end{bmatrix}$$

For each fixed t, the determinant of this system is not zero and so is a unique solution for the value of $\phi(t)$ and $\psi(t)$ for each t. Hence, the first row gives us a condition we must impose on our unknown functions $u_1'(t)$ and $u_2'(t)$. We must have

$$u_1'(t) e^{-\frac{t}{\beta}} + u_2'(t) e^{\frac{t}{\beta}} = 0$$

This simplifies the first derivative of x_p to be

$$x_p'(t) = -u_1(t) \frac{1}{\beta} e^{-\frac{t}{\beta}} + u_2(t) \frac{1}{\beta} e^{\frac{t}{\beta}}$$

Now take the second derivative to get

$$x_p''(t) = -u_1'(t) \frac{1}{\beta} e^{-\frac{t}{\beta}} + u_2'(t) \frac{1}{\beta} e^{\frac{t}{\beta}} + u_1(t) \frac{1}{\beta^2} e^{-\frac{t}{\beta}} + u_2(t) \frac{1}{\beta^2} e^{\frac{t}{\beta}}$$

Now plug these derivative expressions into the nonhomogeneous equation to find

$$f(t) = \beta^2 \left(u_2'(t) \frac{1}{\beta} e^{\frac{t}{\beta}} - u_1'(t) \frac{1}{\beta} e^{-\frac{t}{\beta}} \right) + \beta^2 \left(u_2(t) \frac{1}{\beta^2} e^{\frac{t}{\beta}} + u_1(t) \frac{1}{\beta^2} e^{-\frac{t}{\beta}} \right)$$

$$-\left(u_2(t)e^{\frac{t}{\beta}} + u_1(t)e^{-\frac{t}{\beta}}\right)$$

Now factor out the common $u_1(t)$ and $u_2(t)$ terms to find after a bit of simplifying that

$$f(t) \;=\; \beta^2\left(u_2'(t)\,\frac{1}{\beta}e^{\frac{t}{\beta}} - u_1'(t)\frac{1}{\beta}e^{-\frac{t}{\beta}}\right) + u_2(t)(e^{\frac{t}{\beta}} - e^{\frac{t}{\beta}}) + u_1(t)(e^{-\frac{t}{\beta}} - e^{-\frac{t}{\beta}})$$

We see the functions u_1 and u_2 must satisfy

$$\frac{f(t)}{\beta^2} \;=\; u_2'(t)\,\frac{1}{\beta}e^{\frac{t}{\beta}} - u_1'(t)\frac{1}{\beta}e^{-\frac{t}{\beta}}$$

This gives us a second condition on the unknown functions u_1 and u_2. Combining we have

$$u_1'(t)e^{-\frac{t}{\beta}} + u_2'(t)e^{\frac{t}{\beta}} \;=\; 0$$
$$-u_1'(t)\frac{1}{\beta}e^{-\frac{t}{\beta}} + u_2'(t)\frac{1}{\beta}e^{\frac{t}{\beta}} \;=\; \frac{f(t)}{\beta^2}$$

This can be rewritten in a matrix form:

$$\begin{bmatrix} e^{-\frac{t}{\beta}} & e^{\frac{t}{\beta}} \\ -\frac{1}{\beta}e^{-\frac{t}{\beta}} & \frac{1}{\beta}e^{\frac{t}{\beta}} \end{bmatrix} \begin{bmatrix} u_1'(t) \\ u_2'(t) \end{bmatrix} \;=\; \begin{bmatrix} 0 \\ \frac{f(t)}{\beta^2} \end{bmatrix}$$

We then use Cramer's Rule to solve for the unknown functions u_1' and u_2'. Let W denote the matrix

$$W \;=\; \begin{bmatrix} e^{-\frac{t}{\beta}} & e^{\frac{t}{\beta}} \\ -\frac{1}{\beta}e^{-\frac{t}{\beta}} & \frac{1}{\beta}e^{\frac{t}{\beta}} \end{bmatrix}$$

Then the determinant of W is $det(W) = \frac{2}{\beta}$ and by Cramer's Rule

$$u_1'(t) \;=\; \frac{\begin{bmatrix} 0 & e^{\frac{t}{\beta}} \\ \frac{f(t)}{\beta^2} & \frac{1}{\beta}e^{\frac{t}{\beta}} \end{bmatrix}}{det(W)} \;=\; -\frac{1}{2\beta}f(t)e^{\frac{t}{\beta}}$$

and

$$u_2'(t) \;=\; \frac{\begin{bmatrix} e^{-\frac{t}{\beta}} & 0 \\ -\frac{1}{\beta}e^{-\frac{t}{\beta}} & \frac{f(t)}{\beta^2} \end{bmatrix}}{det(W)} \;=\; \frac{1}{2\beta}f(t)e^{-\frac{t}{\beta}}$$

Thus, integrating, we have

$$u_1(t) \;=\; -\frac{1}{2\beta}\int_0^t f(u)\,e^{\frac{u}{\beta}}\,du$$
$$u_2(t) \;=\; \frac{1}{2\beta}\int_0^t f(u)\,e^{-\frac{u}{\beta}}\,du$$

where 0 is a convenient starting point for our integration. Hence, the particular solution to the non-homogeneous time independent equation is

$$x_p(t) \;=\; u_1(t)\,e^{-\frac{t}{\beta}} + u_2(t)\,e^{\frac{t}{\beta}}$$

$$= \left(-\frac{1}{2\beta} \int_0^t f(u)\, e^{\frac{u}{\beta}}\, du \right) e^{-\frac{t}{\beta}} + \left(\frac{1}{\beta} \int_0^t f(u)\, e^{-\frac{u}{\beta}}\, du \right) e^{\frac{t}{\beta}}.$$

The general solution is thus

$$
\begin{aligned}
x(t) &= x_h(t) + x_p(t) \\
&= A_1 e^{-\frac{t}{\beta}} + A_2 e^{\frac{t}{\beta}} - \frac{1}{2\beta}\, e^{-\frac{t}{\beta}} \int_0^t f(u)\, e^{\frac{u}{\beta}}\, du + \frac{1}{2\beta}\, e^{\frac{t}{\beta}} \int_0^t f(u)\, e^{-\frac{u}{\beta}}\, du
\end{aligned}
$$

for any real constants A_1 and A_2. Finally, note we can rewrite these equations as

$$x(t) = A_1 e^{-\frac{t}{\beta}} + A_2 e^{\frac{t}{\beta}} - \frac{1}{2\beta} \int_0^t f(u)\, e^{\frac{u-t}{\beta}}\, du + \frac{1}{2\beta} \int_0^t f(u)\, e^{-\frac{u-t}{\beta}}\, du$$

or

$$x(t) = A_1 e^{-\frac{t}{\beta}} + A_2 e^{\frac{t}{\beta}} - \frac{1}{\beta} \int_0^t f(u) \left(\frac{e^{\frac{u-t}{\beta}} - e^{-\frac{u-t}{\beta}}}{2} \right) du$$

In applied modeling work, the function $\frac{e^w - e^{-w}}{2}$ arises frequently enough to be given a name. It is called the *hyperbolic sine function* and is denoted by the symbol $\sinh(w)$. You should have seen these functions in earlier work, of course. So just a friendly reminder. Hence, we can rewrite once more to see

$$x(t) = A_1 e^{-\frac{t}{\beta}} + A_2 e^{\frac{t}{\beta}} - \frac{1}{\beta} \int_0^t f(u) \sinh\left(\frac{u-t}{\beta} \right) du$$

Finally, sinh is an odd function, so we can pull the minus side inside by reversing the argument into the sinh function. This gives

$$x(t) = A_1 e^{-\frac{t}{\beta}} + A_2 e^{\frac{t}{\beta}} + \frac{1}{\beta} \int_0^t f(u) \sinh\left(\frac{t-u}{\beta} \right) du$$

Since $\sinh(t/\beta)$ and $\cosh(t/\beta)$ are just linear combinations of $e^{t/\beta}$ and $e^{-t/\beta}$, we can also rewrite as

$$x(t) = A_1 \cosh(\frac{t}{\beta}) + A_2 \sinh(\frac{t}{\beta}) + \frac{1}{\beta} \int_0^t f(u) \sinh\left(\frac{t-u}{\beta} \right) du$$

Now let's use the initial conditions $x(0) = -1$ and $x(5) = 4$. Hence,

$$
\begin{aligned}
1 &= x(0) = A_1 \\
4 &= x(5) = \cosh(\frac{5}{\beta}) + A_2 \sinh(\frac{5}{\beta}) + \frac{1}{\beta} \int_0^5 f(u) \sinh\left(\frac{5-u}{\beta} \right) du
\end{aligned}
$$

which is solvable for A_1 and A_2. We find

$$
\begin{aligned}
A_1 &= 1 \\
A_2 &= \frac{1}{\sinh(\frac{5}{\beta})} \left(4 - \cosh(\frac{5}{\beta}) - \frac{1}{\beta} \int_0^5 f(u) \sinh\left(\frac{5-u}{\beta} \right) du \right)
\end{aligned}
$$

We can use this technique on lots of models and generate similar solutions. Here is another example.

Example 28.2.1 *Solve*

$$\begin{aligned} u''(t) + 9u(t) &= 2t \\ u(0) &= 1 \\ u'(0) &= -1 \end{aligned}$$

Solution *The characteristic equation is $r^2 + 9 = 0$ which has the complex roots $\pm 3i$. Hence, the two linearly independent solutions to the homogeneous equation are $u_1(t) = \cos(3t)$ and $u_2(t) = \sin(3t)$. We set the homogeneous solution to be*

$$u_h(t) = A\cos(3t) + B\sin(3t)$$

where A and B are arbitrary constants. The nonhomogeneous solution is of the form

$$u_p(t) = \phi(t)\cos(3t) + \psi(t)\sin(3t).$$

We know the functions ϕ and ψ must then satisfy

$$\begin{bmatrix} \cos(3t) & \sin(3t) \\ -3\sin(3t) & 3\cos(3t) \end{bmatrix} \begin{bmatrix} \phi'(t) \\ \psi'(t) \end{bmatrix} = \begin{bmatrix} 0 \\ 2t \end{bmatrix}$$

Applying Cramer's rule, we have

$$\phi'(t) = \frac{1}{3}\begin{bmatrix} 0 & \sin(3t) \\ 2t & 3\cos(3t) \end{bmatrix} = -\frac{2}{3}t\sin(3t)$$

and

$$\psi'(t) = \frac{1}{3}\begin{bmatrix} \cos(3t) & 0 \\ -3\sin(3t) & 2t \end{bmatrix} = \frac{2}{3}t\cos(3t)$$

Thus, integrating, we have

$$\phi(t) = -\frac{2}{3}\int_0^t u\sin(3u)\,du$$

$$\psi(t) = \frac{2}{3}\int_0^t u\cos(3u)\,du.$$

The general solution is therefore

$$u(t) = A\cos(3t) + B\sin(3t) - \left(\frac{2}{3}\int_0^t u\sin(3u)du\right)\cos(3t) + \left(\frac{2}{3}\int_0^t u\cos(3u)du\right)\sin(3t).$$

This can be simplified to

$$\begin{aligned} u(t) &= A\cos(3t) + B\sin(3t) + \frac{2}{3}\int_0^t u\{\sin(3t)\cos(3u) - \sin(3u)\cos(3t)\}\,du \\ &= A\cos(3t) + B\sin(3t) + \frac{2}{3}\int_0^t u\,\sin(3t - 3u)\,du. \end{aligned}$$

Applying Leibnitz's rule, we find

$$u'(t) = -3A\sin(3t) + 3B\cos(3t) + \frac{2}{3}t\,\sin(3t - 3t) + \frac{2}{3}\int_0^t 3u\cos(3t - 3u)\,du$$

$$= -3A\sin(3t) + 3B\cos(3t) + \frac{2}{3}\int_0^t 3u\cos(3t - 3u)\,du$$

Applying the boundary conditions and using Leibnitz's rule, we find

$$
\begin{aligned}
1 &= u(0) = A \\
-1 &= u'(0) = 3B
\end{aligned}
$$

Thus, $A = 1$ and $B = -\frac{1}{3}$ giving the solution

$$u(t) = \cos(3t) + \frac{1}{3}\sin(3t) + \frac{2}{3}\int_0^t u\,\sin(3t - 3u)\,du.$$

Example 28.2.2 *Solve*

$$
\begin{aligned}
u''(t) + 9u(t) &= 2t \\
u(0) &= 1 \\
u(4) &= -1
\end{aligned}
$$

Solution *The model is the same as the previous example except for the boundary conditions. We have*

$$
\begin{aligned}
1 &= u(0) = A \\
-1 &= u(4) = A\cos(12) + B\sin(12) + \frac{2}{3}\int_0^4 u\,\sin(12 - 3u)\,du.
\end{aligned}
$$

Thus, since $A = 1$, we have

$$B\sin(12) = -1 - \cos(12) - \frac{2}{3}\int_0^4 u\,\sin(12 - 3u)\,du.$$

and so

$$B = -\frac{1 + \cos(12) + \frac{2}{3}\int_0^4 u\,\sin(12 - 3u)\,du}{\sin(12)}.$$

We can then assemble the solution using these constants.

28.2.2 Homework

Exercise 28.2.1 *Solve*

$$
\begin{aligned}
u''(t) + 4u(t) &= 2t \\
u(0) &= 2 \\
u(4) &= -6
\end{aligned}
$$

expressing the particular solution in integral form.

Exercise 28.2.2 *Solve*

$$
\begin{aligned}
u''(t) - u'(t) - 6u(t) &= 2t \\
u(0) &= 10 \\
u(4) &= -5
\end{aligned}
$$

expressing the particular solution in integral form.

Exercise 28.2.3 *Solve*

$$
\begin{aligned}
u''(t) - 4u(t) &= 2t \\
u(0) &= 2 \\
u'(0) &= -1
\end{aligned}
$$

expressing the particular solution in integral form.

28.2.3   Boundary Value Problems

The next model we study is on the interval $[0, L]$ for convenience. Since the boundary conditions involve derivatives, we call these *derivative boundary conditions*.

$$
u'' + \omega^2 u = f, \ 0 \le x \le L,
$$
$$
u'(0) = 0, \quad u'(L) = 0
$$

We examined a similar model earlier with periodic boundary conditions in the derivative when we looked at using Fourier Series expansions for f to find a solution. We will now use **VoP** instead and a more specific form of the periodic boundary conditions. The general solution to the homogeneous equation is given by

$$
u_h(x) \ = \ A\cos(\omega x) + B\sin(\omega x).
$$

Applying Variation of Parameters, we look for a particular solution of the form

$$
u_p(x) \ = \ \phi(x)\cos(\omega x) + \psi(x)\sin(\omega x).
$$

We seek functions ϕ and ψ which satisfy

$$
\begin{bmatrix} \cos(\omega x) & \sin(\omega x) \\ -\omega\sin(\omega x) & \omega\cos(\omega x) \end{bmatrix} \begin{bmatrix} \phi'(x) \\ \psi'(x) \end{bmatrix} = \begin{bmatrix} 0 \\ f(x) \end{bmatrix}
$$

Let W denote the 2×2 matrix above. Note that $\det(W) = \omega$ which is not zero by assumption. Hence, by Cramer's rule, we have

$$
\phi'(x) \ = \ \frac{\begin{vmatrix} 0 & \sin(\omega x) \\ f(x) & \omega\cos(\omega x) \end{vmatrix}}{\omega} = -\frac{f(x)\sin(\omega x)}{\omega}
$$

$$
\psi'(x) \ = \ \frac{\begin{vmatrix} \cos(\omega x) & 0 \\ -\omega\sin(\omega x) & f \end{vmatrix}}{\omega} = \frac{f(x)\cos(\omega x)}{\omega}
$$

Hence,

$$
\phi(x) \ = \ -\frac{1}{\omega} \int_0^x f(s)\sin(\omega s)\, ds
$$

$$
\psi(x) \ = \ \frac{1}{\omega} \int_0^x f(s)\cos(\omega s)\, ds
$$

The general solution $u(x) = u_h(x) + u_p(x)$ and so

$$
\begin{aligned}
u(x) &= A\cos(\omega x) + B\sin(\omega x) \\
&\quad + \left(-\frac{1}{\omega}\int_0^x f(s)\sin(\omega s)\,ds\right)\cos(\omega x) + \left(\frac{1}{\omega}\int_0^x f(s)\cos(\omega s)\,ds\right)\sin(\omega x) \\
&= A\cos(\omega x) + B\sin(\omega x) + \frac{1}{\omega}\int_0^x f(s)\Big(\cos(\omega s)\sin(\omega x) - \sin(\omega s)\cos(\omega x)\Big)\,ds \\
&= A\cos(\omega x) + B\sin(\omega x) + \frac{1}{\omega}\int_0^x f(s)\sin\Big(\omega(x - s)\Big)\,ds
\end{aligned}
$$

The last simplification arises from the use of the standard sin addition and subtraction of angles formulae.

Next, apply the boundary conditions, $u'(0) = 0$ and $u'(L) = 0$. We find first, using Leibnitz's rule for derivatives of functions defined by integrals, that

$$
\begin{aligned}
u'(x) &= -\omega A\sin(\omega x) + \omega B\cos(\omega x) \\
&\quad + \frac{1}{\omega}f(x)\sin\Big(\omega(x - x)\Big) + \frac{1}{\omega}\int_0^x f(s)\omega\,\cos\Big(\omega(x - s)\Big)\,ds \\
&= -\omega A\sin(\omega x) + \omega\cos(\omega x) + \int_0^x f(s)\,\cos\Big(\omega(x - s)\Big)\,ds.
\end{aligned}
$$

Hence,

$$
\begin{aligned}
u'(0) &= 0 = \omega B \\
u'(L) &= 0 = -\omega A\sin(L\omega) + B\omega\cos(L\omega) + \int_0^L f(s)\,\cos\Big(\omega(L - s)\Big)\,ds.
\end{aligned}
$$

It immediately follows that $B = 0$ and to satisfy our boundary conditions, we must have

$$
0 = -\omega A\sin(L\omega) + \int_0^L f(s)\,\cos\Big(\omega(L - s)\Big)\,ds.
$$

We now see an interesting result. We can determine a unique value of A only if $\sin(L\omega) \neq 0$. Otherwise, since we assume $\omega \neq 0$, if $\omega L = n\pi$ for any integer $n \neq 0$, we find the value of A cannot be determined as we have the equation

$$
0 = -\omega A \times 0 + \int_0^L f(s)\,\cos\left(\frac{n\pi}{L}(L - s)\right)\,ds.
$$

In the case that $\omega L = n\pi$, we see we must also insist that f satisfy

$$
\int_0^L f(s)\,\cos\left(\frac{n\pi}{L}(L - s)\right)\,ds = 0.
$$

Otherwise, if $\omega L \neq n\pi$, we can solve for A to obtain

$$
A = \frac{1}{\omega\sin(L\omega)}\int_0^L f(s)\,\cos\Big(\omega(L - s)\Big)\,ds.
$$

This leads to the solution

$$u(x) = \frac{\cos(\omega x)}{\omega \sin(L\omega)} \int_0^L f(s) \cos\left(\omega(L-s)\right) ds + \frac{1}{\omega} \int_0^x f(s) \sin\left(\omega(x-s)\right) ds$$

28.2.3.1   The Kernel Function

We can manipulate this solution quite a bit. The following derivation is a bit tedious, so grab a cup of your favorite brew while you work through the details!

$$\begin{aligned}
u(x) &= \frac{1}{\omega \sin(L\omega)} \int_0^L f(s) \cos(\omega x) \cos\left(\omega(L-s)\right) ds + \frac{1}{\omega} \int_0^x f(s) \sin\left(\omega(x-s)\right) ds \\
&= \frac{1}{\omega \sin(L\omega)} \int_0^x f(s) \left(\cos(\omega x) \cos\left(\omega(L-s)\right) + \sin(L\omega) \sin\left(\omega(x-s)\right) \right) ds \\
&\quad + \frac{1}{\omega \sin(L\omega)} \int_x^L f(s) \cos(\omega x) \cos\left(\omega(L-s)\right) ds
\end{aligned}$$

Now we can use trigonometric identities to simplify these expressions.

$$\begin{aligned}
\cos(\omega x)\cos(\omega(L-s)) &+ \sin(L\omega)\sin(\omega(x-s)) = \cos(\omega x)\Big(\cos(\omega L)\cos(\omega s) + \sin(\omega L)\sin(\omega s)\Big) \\
&\quad + \sin(L\omega)\Big(\sin(\omega x)\cos(\omega s) - \sin(\omega s)\cos(\omega x)\Big) \\
&= \cos(\omega x)\cos(\omega L)\cos(\omega s) + \cos(\omega x)\sin(\omega L)\sin(\omega s) \\
&\quad + \sin(L\omega)\sin(\omega x)\cos(\omega s) - \sin(\omega L)\sin(\omega s)\cos(\omega x) \\
&= \cos(\omega x)\cos(\omega L)\cos(\omega s) + \sin(L\omega)\sin(\omega x)\cos(\omega s) \\
&= \cos(\omega s)\Big(\cos(\omega x)\cos(\omega L) + \sin(L\omega)\sin(\omega x)\Big) \\
&= \cos(\omega s)\cos\left(\omega(L-x)\right).
\end{aligned}$$

Using this rewrite of the first integral's term, we find

$$\begin{aligned}
u(x) &= \frac{1}{\omega \sin(L\omega)} \int_0^x f(s) \cos(\omega s) \cos\left(\omega(L-x)\right) ds \\
&\quad + \frac{1}{\omega \sin(L\omega)} \int_x^L f(s) \cos(\omega x) \cos\left(\omega(L-s)\right) ds
\end{aligned}$$

We can then define the *kernel* function k_ω by

$$k_\omega(x, s) = \frac{1}{\omega \sin(L\omega)} \begin{cases} \cos(\omega s) \cos(\omega(L-x)) & 0 \le s \le x \\ \cos(\omega x) \cos(\omega(L-s)) & x < s \le L \end{cases}$$

Note that k_ω is continuous on the square $[0, L] \times [0, L]$ and that it is also a symmetric functions as $k_\omega(x, s) = k_\omega(s, x)$. We can thus say that for any $\omega L \ne n\pi$, we find the solution to our nonhomogeneous boundary value problem can be written as

$$u(x) = \int_0^L f(s) k_\omega(x, s) ds$$

28.2.3.2   Conclusions

We can say all of this more strongly. If $\omega L \neq n\pi$ for any integer $n \neq 0$, we have

$$\begin{pmatrix} u'' + \omega^2 u & = f, & 0 \leq x \leq L \\ u'(0) & = 0 \\ u'(L) & = 0 \end{pmatrix} \iff u(x) = \int_0^L f(s)\, k_\omega(x,s)\, ds.$$

We also know that for $f = 0$, the homogeneous problem has nonzero solutions

$$u_n(x) = \cos(\omega_n(x))$$

where $\omega_n L = n\pi$ for any integer $n \neq 0$. Hence, the nonzero functions u_n satisfy the boundary value problem

$$\begin{aligned} u'' + \omega_n^2 u &= 0,\ 0 \leq x \leq L \\ u'(0) &= 0 \\ u'(L) &= 0 \end{aligned}$$

Note we can rewrite this as

$$\begin{aligned} u_n'' &= -\omega_n^2 u_n,\ 0 \leq x \leq L \\ u_n'(0) &= 0 \\ u_n'(L) &= 0 \end{aligned}$$

The model above can be rewritten in a more abstract form by defining the operator $\mathscr{L}$ acting on a suitable domain of functions. Let $\mathscr{D}(\mathscr{L})$ denote the domain of $\mathscr{L}$ which for us will be

$$\mathscr{D}(L) = C^2([0,L]) \cap \{x \in C^2([0,L]) \text{ with } x'(0) = 0; x'(L) = 0\}$$

where $C^2([0,L])$ is the set of functions that are twice differentiable on $[0,L]$. We only assume the second derivative exists, but of course, if f is continuous and u is also continuous, since $u'' = f - \Theta u$, we actually know we want the second derivative to be continuous also. However, we can relax the *smoothness* of f and get situations where we lose the continuity of the second derivative at various points and even its existence itself. But that is a more advanced topic for later. For convenience, we will denote this collection of functions by $C^2([0,L]) \cap \{BC\}$. The operator $\mathscr{L}$ is then defined on this domain to be

$$\mathscr{L}(u) = u''.$$

So we can rewrite our model again on $C^2([0,L]) \cap \{BC\}$ as $\mathscr{L}(u_n) = -\omega_n^2 u$ which looks exactly like the standard eigenvalue equation for a square matrix: $Ax = cx$. So, the *operator* $\mathscr{L}$ must play the role of the matrix A, the scalar $-\omega_n^2$ is the eigenvalue with u_n, the eigenvector. Since u_n is a function here, we typically call it an eigenfunction instead. As usual, since eigenvectors must be nonzero, we want our eigenfunctions to be nonzero functions as well. You can see our eigenfunctions here are indeed nonzero. Our work thus tells us the eigenvalues of $\mathscr{L}$ are $-\omega_n^2 = -\frac{n^2\pi^2}{L^2}$ for any nonzero integer n with associated eigenfunctions $u_n(x) = \cos(\omega_n x)$. Furthermore, there are infinitely many of them! There can never be more than a finite number of eigenvalues in the $Ax = cx$ eigenvalue problem, so this is a very new turn of events. Note we can also think of the solution $u(x) = \int_0^L f(s)\, k_\omega(x,s)\, ds$ as a way of defining the inverse of the operator $\mathscr{L} + \omega^2$, $(\mathscr{L} + \omega^2)^{-1}$. Finally, this inverse operator is very interesting. We can define the new operator $\mathscr{J}_\omega$ on the domain

$C([0, L])$, the set of all functions continuous on the interval $[0, L]$, by

$$\mathscr{I}_\omega(u) \;=\; \int_0^L f(s)\, k_\omega(x, s)\, ds.$$

Hence $(\mathscr{L} + \omega^2)^{-1} = \mathscr{I}_\omega(u)$ for $\omega \neq \omega_n^2$. This will always transform the external function f into a continuous function u which solves the original model. However, it is *much* more general. As long as there is some notion of the integrability of f, we can do this transformation and that means our way of solving the model using a kernel function leads us to more general ways of looking at solutions. This actually brings us lots of new insights, but that discussion will come later. Finally note $(\mathscr{L} + \omega^2)^{-1}$ for ω an eigenvalue simply does not exist.

28.3   Linear Partial Differential Equations

Now let's look at a specific partial differential equation model (PDE) near and dear to our heart: the general cable model which is a modified diffusion PDE. The ideas we present here are pretty standard for any linear PDE and it is a lot better really to go through the details on a specific one. Pay attention to the flow of the ideas and you will be able to apply them to other situations with a bit of thought. The general cable model is

$$\begin{aligned}
\beta^2 \frac{\partial^2 \Phi}{\partial x^2} - \Phi - \alpha \frac{\partial \Phi}{\partial t} &= 0, \text{ for } 0 \leq x \leq L, \; t \geq 0, \\
\frac{\partial \Phi}{\partial x}(0, t) &= 0, \\
\frac{\partial \Phi}{\partial x}(L, t) &= 0, \\
\Phi(x, 0) &= f(x).
\end{aligned}$$

for positive constants α and β. The domain is the usual half infinite $[0, L] \times [0, \infty)$ where the spatial part of the domain corresponds to the length of the dendritic cable in a neuron model. The conditions $u(0, t) = 0$ and $u(L, t) = 0$ are known as *Dirichlet Boundary conditions*. The solution to a model such as this is a function $\Phi(x, t)$ which is sufficiently smooth to have partial derivatives with respect to the needed variables continuous for all the orders required. For these problems, the highest order we need is the second order partials.

One way to find the solution is to assume we can separate the variables so that we can write $\Phi(x, t) = u(x)w(t)$. If we make this separation assumption, we will find solutions that must be written as what are called infinite series and to solve the boundary conditions, we will have to be able to express boundary functions as series expansions. We assume a solution of the form $\Phi(x, t) = u(x)\, w(t)$ and compute the needed partials. This leads to a the new equation

$$\beta^2 \frac{d^2 u}{dx^2}\, w(t) - u(x)w(t) - \alpha u(x)\frac{dw}{dt} \;=\; 0.$$

Rewriting, we find for all x and t, we must have

$$w(t)\left(\beta^2 \frac{d^2 u}{dx^2} - u(x)\right) \;=\; \alpha u(x)\frac{dw}{dt}.$$

Rewriting, we have

$$\frac{\beta^2 \frac{d^2 u}{dx^2} - u(x)}{u(x)} = \frac{\alpha \frac{dw}{dt}}{w(t)}, \ 0 \le x \le L, \ t > 0.$$

The only way this can be true is if both the left and right hand side are equal to the separation constant Θ. This leads to the decoupled equations Equation 28.1 and Equation 28.2.

$$\alpha \frac{dw}{dt} = \Theta \, w(t), \ t > 0, \tag{28.1}$$

$$\beta^2 \frac{d^2 u}{dx^2} = (1 + \Theta) \, u(x), \ 0 \le x \le L, \tag{28.2}$$

We also have boundary conditions. Our assumption leads to the following boundary conditions in x:

$$\frac{du}{dx}(0) \, w(t) = 0, \ t > 0,$$

$$\frac{du}{dx}(L) \, w(t) = 0, \ t > 0.$$

Since these equations must hold for all t, this forces

$$\frac{du}{dx}(0) = 0, \tag{28.3}$$

$$\frac{du}{dx}(L) = 0. \tag{28.4}$$

Equations 28.1 - 28.4 give us the boundary value problem in $u(x)$ we need to solve. Then, we can find w.

28.3.1 Determining the Separation Constant

The model is then

$$u'' - \frac{1 + \Theta}{\beta^2} u = 0$$

$$\frac{du}{dx}(0) = 0,$$

$$\frac{du}{dx}(L) = 0.$$

We are looking for nonzero solutions, so any choice of separation constant Θ that leads to a zero solution will be rejected.

28.3.1.1 Case I: Separation Constant is Positive

The model to solve is

$$u'' - \frac{\omega^2}{\beta^2} u = 0$$

$$u'(0) = 0,$$

$$u'(L) = 0.$$

with characteristic equation $r^2 - \frac{\omega^2}{\beta^2} = 0$ with the real roots $\pm\frac{\omega}{\beta}$. The general solution of this second order model is given by

$$u(x) \;=\; A\cosh\!\left(\frac{\omega}{\beta}x\right) + B\sinh\!\left(\frac{\omega}{\beta}x\right)$$

which tells us

$$u'(x) \;=\; A\frac{\omega}{\beta}\sinh\!\left(\frac{\omega}{\beta}x\right) + B\frac{\omega}{\beta}\cosh\!\left(\frac{\omega}{\beta}x\right)$$

Next, apply the boundary conditions, $u'(0) = 0$ and $u'(L) = 0$. Hence,

$$u'(0) \;=\; 0 = B$$
$$u'(L) \;=\; 0 = A\sinh\!\left(L\frac{\omega}{\beta}\right)$$

Hence, $B = 0$ and $A\sinh\!\left(L\frac{\omega}{\beta}\right) = 0$. Since sinh is never zero when ω is not zero, we see $A = 0$ also. Hence, the only u solution is the trivial one and we can reject this case.

28.3.1.2 Case II: Separation Constant is Zero

The model to solve is now

$$u'' \;=\; 0$$
$$u'(0) \;=\; 0,$$
$$u'(L) \;=\; 0.$$

with characteristic equation $r^2 = 0$ with the double root $r = 0$. Hence, the general solution is now

$$u(x) \;=\; A + Bx$$

Applying the boundary conditions, $u(0) = 0$ and $u(L) = 0$. Hence, since $u'(x) = B$, we have

$$u'(0) \;=\; 0 = B$$
$$u'(L) \;=\; 0 = BL$$

Hence, $B = 0$ but the value of A can't be determined. Hence, any arbitrary constant which is not zero is a valid nonzero solution. Choosing $A = 1$, let $u_0(x) = 1$ be our chosen nonzero solution for this case. We now need to solve for w in this case. Since $\Theta = -1$, the model to solve is

$$\frac{dw}{dt} \;=\; -\frac{1}{\alpha}w(t),\; 0 < t$$

The general solution is $w(t) = Ce^{-\frac{1}{\alpha}t}$ for any value of C. Choose $C = 1$ and we set

$$w_0(t) \;=\; e^{-\frac{1}{\alpha}t}.$$

Hence, the product $\phi_0(x,t) = u_0(x)\,w_0(t)$ solves the boundary conditions. That is

$$\phi_0(x,t) \;=\; e^{-\frac{1}{\alpha}t}.$$

is a solution.

28.3.1.3 Case III: Separation Constant is Negative

$$u'' + \frac{\omega^2}{\beta^2} u = 0$$
$$u'(0) = 0,$$
$$u'(L) = 0.$$

The general solution is given by

$$u(x) = A \cos\left(\frac{\omega}{\beta} x\right) + B \sin\left(\frac{\omega}{\beta} x\right)$$

and hence

$$u'(x) = -A\frac{\omega}{\beta} \sin\left(\frac{\omega}{\beta} x\right) + B\frac{\omega}{\beta} \cos\left(\frac{\omega}{\beta} x\right)$$

Next, apply the boundary conditions to find

$$u'(0) = 0 = B$$
$$u'(L) = 0 = A \sin\left(L\frac{\omega}{\beta}\right)$$

Hence, $B = 0$ and $A \sin\left(L\frac{\omega}{\beta}\right) = 0$. Thus, we can determine a unique value of A only if $\sin\left(L\frac{\omega}{\beta}\right) \neq 0$. If $\omega \neq \frac{n\pi\beta}{L}$, we can solve for A and find $A = 0$, but otherwise, A can't be determined. So the only solutions are the trivial or zero solutions unless $\omega L = n\pi\beta$. Letting $\omega_n = \frac{n\pi\beta}{L}$, we find a a nonzero solution for each nonzero value of A of the form

$$u_n(x) = A \cos\left(\frac{\omega_n}{\beta} x\right) = A \cos\left(\frac{n\pi}{L} x\right).$$

For convenience, let's choose all the constants $A = 1$. Then we have an infinite family of nonzero solutions $u_n(x) = \cos\left(\frac{n\pi}{L} x\right)$ and an infinite family of separation constants $\Theta_n = -1 - \omega_n^2 = -1 - \frac{n^2\pi^2\beta^2}{L^2}$. We can then solve the w equation. We must solve

$$\frac{dw}{dt} = -\frac{(1 + \omega_n^2)}{\alpha} w(t), \ t \geq 0.$$

The general solution is

$$w(t) = B_n e^{-\frac{1+\omega_n^2}{\alpha} t} = B_n e^{-\left(\frac{1}{\alpha} + \frac{n^2\pi^2\beta^2}{\alpha L^2}\right) t}$$

Choosing the constants $B_n = 1$, we obtain the w_n functions

$$w_n(t) = e^{-\left(\frac{1}{\alpha} + \frac{n^2\pi^2\beta^2}{\alpha L^2}\right) t}$$

Hence, any product

$$\phi_n(x,t) \;=\; u_n(x)\,w_n(t)$$

will solve the model with the x boundary conditions and any finite sum of the form, for arbitrary constants A_n

$$\Psi_N(x,t) \;=\; \sum_{n=1}^{N} A_n \phi_n(x,t) \;=\; \sum_{n=1}^{N} A_n u_n(x)\,w_n(t)$$

$$=\; \sum_{n=1}^{N} A_n \cos\left(\frac{n\pi}{L}x\right) e^{-\frac{L^2+n^2\pi^2\beta^2}{\alpha L^2}t}$$

Adding in the $1 + \Theta = 0$ case, we find the most general finite term solution has the form

$$\Phi_N(x,t) \;=\; A_0\phi_0(x,t) + \sum_{n=1}^{N} A_n\phi_n(x,t) \;=\; A_0 u_0(x)w_0(t) + \sum_{n=1}^{N} A_n u_n(x)\,w_n(t)$$

$$=\; A_0 e^{-\frac{1}{\alpha}t} + \sum_{n=1}^{N} A_n \cos\left(\frac{n\pi}{L}x\right) e^{-\frac{L^2+n^2\pi^2\beta^2}{\alpha L^2}t}.$$

Now these finite term solutions do solve the boundary conditions $\frac{\partial \Phi}{\partial x}(0,t) = 0$ and $\frac{\partial \Phi}{\partial x}(L,t) = 0$, but how do we solve the remaining condition $\Phi(x,0) = f(x)$? To do this, we note since we can assemble the finite term solutions for any value of N, no matter how large, it is clear we should let $N \to \infty$ and express the solution as

$$\Phi(x,t) \;=\; A_0\phi_0(x,t) + \sum_{n=1}^{\infty} A_n\phi_n(x,t) \;=\; A_0 u_0(x)w_0(t) + \sum_{n=1}^{\infty} A_n u_n(x)\,w_n(t)$$

$$=\; A_0 e^{-\frac{1}{\alpha}t} + \sum_{n=1}^{\infty} A_n \cos\left(\frac{n\pi}{L}x\right) e^{-\frac{L^2+n^2\pi^2\beta^2}{\alpha L^2}t}.$$

This is the form that will let us solve the remaining boundary condition. We can see series of the form

$$A_0 e^{-\frac{1}{\alpha}t} + \sum_{n=1}^{\infty} A_n \cos\left(\frac{n\pi}{L}x\right) e^{-\frac{L^2+n^2\pi^2\beta^2}{\alpha L^2}t}.$$

are like Fourier series although in terms of two variables. We will show these series converge pointwise for x in $[0, L]$ and all t for the right choice of coefficients (A_n). Further, we will show that we can take the partial derivative of this series solution with respect to x term by term to obtain

$$\sum_{n=1}^{\infty} -A_n \frac{n\pi}{L} \sin\left(\frac{n\pi}{L}x\right) e^{-\frac{L^2+n^2\pi^2\beta^2}{\alpha L^2}t}.$$

This series evaluated at $x = 0$ and $x = L$ gives 0 and hence the derivative boundary conditions are satisfied. Indeed we will show the solution $\Phi(x,t)$ given by

$$\Phi(x,t) \;=\; A_0 e^{-\frac{1}{\alpha}t} + \sum_{n=1}^{\infty} A_n \cos\left(\frac{n\pi}{L}x\right) e^{-\frac{L^2+n^2\pi^2\beta^2}{\alpha L^2}t}.$$

for appropriate constants (A_n) is a well-behaved solution on our domain. The right choice of constants (A_n) comes from the remaining boundary condition

$$\Phi(x,0) \;=\; f(x), \text{ for } 0 \le x \le L.$$

We know

$$\Phi(x,0) \;=\; A_0 + \sum_{n=1}^{\infty} A_n \cos\left(\frac{n\pi}{L}x\right)$$

and so rewriting in terms of the series solution, for $0 \le x \le L$, we find

$$A_0 + \sum_{n=1}^{\infty} A_n \cos\left(\frac{n\pi}{L}x\right) \;=\; f(x)$$

The Fourier series for appropriate f is given by

$$f(x) \;=\; B_0 + \sum_{n=1}^{\infty} B_n \cos\left(\frac{n\pi}{L}x\right)$$

with

$$B_0 \;=\; \frac{1}{L}\int_0^L f(x),$$

$$B_n \;=\; \frac{2}{L}\int_0^L f(x)\cos\left(\frac{n\pi}{L}x\right)dx.$$

Then, setting these series equal, we find that the solution is given by $A_n = B_n$ for all $n \ge 0$.

Homework

Exercise 28.3.1 *Do the cable problem analysis for*

$$4\frac{\partial^2\Phi}{\partial x^2} - \Phi - 0.7\frac{\partial\Phi}{\partial t} \;=\; 0, \textit{ for } 0 \le x \le 6, \; t \ge 0,$$

$$\frac{\partial\Phi}{\partial x}(0,t) \;=\; 0, \quad \frac{\partial\Phi}{\partial x}(6,t) = 0,$$

$$\Phi(x,0) \;=\; f(x).$$

for $f(x)$ is a pulse centered at $x = 3$ of height 10 and width 2. Find the first four terms of the solution Φ.

Exercise 28.3.2 *Do the cable problem analysis for*

$$9\frac{\partial^2\Phi}{\partial x^2} - \Phi - 2.6\frac{\partial\Phi}{\partial t} \;=\; 0, \textit{ for } 0 \le x \le 12, \; t \ge 0,$$

$$\Phi(0,t) \;=\; 0, \quad \Phi(6,t) = 0,$$

$$\Phi(x,0) \;=\; f(x).$$

for $f(x)$ is a C^∞ bump function centered at $x = 1$ with height 10 and width 1. Find the first four terms of the solution Φ.

Exercise 28.3.3 *Do the cable problem analysis for*

$$4\frac{\partial^2 \Phi}{\partial x^2} - \Phi - 0.7\frac{\partial \Phi}{\partial t} = 0, \textit{ for } 0 \le x \le 6, \ t \ge 0,$$

$$\frac{\partial \Phi}{\partial x}(0,t) = 0, \quad \frac{\partial \Phi}{\partial x}(6,t) = 0,$$

$$\Phi(x,0) = f(x).$$

for $f(x) = 5x(6-x)$. Find the first four terms of the solution Φ.

Exercise 28.3.4 *Do the cable problem analysis for*

$$9\frac{\partial^2 \Phi}{\partial x^2} - \Phi - 2.6\frac{\partial \Phi}{\partial t} = 0, \textit{ for } 0 \le x \le 12, \ t \ge 0,$$

$$\Phi(0,t) = 0, \quad \Phi(12,t) = 0,$$

$$\Phi(x,0) = f(x).$$

for $f(x) = 10x(x-12)x^4$. Find the first four terms of the solution Φ.

28.3.2   Convergence Analysis for Fourier Series Revisited

Let's look at the Fourier Sine and Fourier Cosine series expansions again. You can see that applying the separation of variables technique, the boundary constraints for the partial differential equation model lead us to second order linear differential equations of the form

$$\mathscr{L}(u) + \Theta u = f$$

plus boundary conditions.

Note this technique always gives us the eigenvalue - eigenfunction problem $\mathscr{L}(u) = -\Theta u$ and depending on the boundary conditions, we have been finding the eigenfunctions are sin or cos orthogonal families when $\mathscr{L}(u) = u''$. This $\mathscr{L}$ is the simplest example of a **Stürm - Liouville** problem (SLP) whose eigenfunctions are the nice families we have seen. If $\mathscr{L}$ is a different SLP, the eigenfunctions could be Bessel functions, Legendre polynomials, Laguerre polynomials or others. The eigenfunctions for these models are also mutually orthogonal and if they are the family $\{w_n\}$ then we would look for expansions of the data function f in terms of this orthogonal family, $f = \sum_n A_n w_n$ where the coefficients A_n would be the generalized Fourier coefficients of f associated with the orthogonal family $\{w_n\}$ as defined in Definition 28.3.1. We write the summation as just over n because it might start at 0 or 1 and we don't want to clutter the notation.

Definition 28.3.1 Generalized Fourier Coefficients

> *Let $\{w_n\}$ be a mutually orthogonal family of continuous functions in $C([a,b])$. Note $\|w_n\|_2$ is the length of each function. The family $\{\hat{w}\}_n$ defined by $\hat{w}_n = (1/\|w_n\|_2)\, w_n$ is a mutually orthogonal sequence of functions of length one in $C([a,b])$. The generalized Fourier coefficient of f with respect to this family is*
>
> $$A_n = \frac{1}{\|w_n\|_2^2} < f, w_n > = \frac{1}{\|w_n\|_2^2}\int_a^b f(s)w_n(s)ds$$

Comment 28.3.1 *For our usual $u_n(x) = \sin(n\pi x/L)$ and $v_n(x) = \cos(n\pi x/L)$ families, on the interval $[0,L]$, note the families $U_n(x) = \sin(n\pi(x-a)/(b-a))$ and $V_n(x) = \cos(n\pi(x-a)/(b-$*

a)) *are mutually orthogonal on* $[a, b]$ *and so the Fourier coefficients of* f *on* $[a, b]$ *would be*

$$A_n = \frac{1}{||U_n||_2^2} \int_a^b f(s)U_n(s)ds = \frac{1}{||U_n||_2^2} \int_a^b f(s)\sin\left(\frac{n\pi}{L}\left(s - a\right)\right)ds$$

$$B_n = \frac{1}{||V_n||_2^2} \int_a^b f(s)V_n(s)ds = \frac{1}{||U_n||_2^2} \int_a^b f(s)\cos\left(\frac{n\pi}{L}\left(s - a\right)\right)ds$$

Given the data function f on $[0, L]$, if we assume f', f'' and f''' are Riemann integrable, we can prove estimates for the Fourier coefficients like we did in Section 27.7.5.

Theorem 28.3.1 Bound on Fourier Coefficients

For $j = 1, 2$ *or* 3, *if* $f^{(j)}$ *exists and is Riemann Integrable on* $[0, L]$, $f^{j-1}(0) = f^{j-1}(L)$, *then we have the estimates*

$$\sum_{i=1}^n \left(\frac{i^2\pi^2}{L^2}\right)^j A_i^2 \leq \frac{2}{L} ||f^{(j)}||_2^2, \quad \sum_{i=1}^n \left(\frac{i^2\pi^2}{L^2}\right)^j B_i^2 \leq \frac{2}{L} ||f^{(j)}||_2^2$$

and the series $\sum_{i=1}^\infty (\frac{i\pi}{L})^{j-1} A_i \sin(\frac{i\pi}{L}x)$ *and* $\sum_{i=1}^\infty B_i (\frac{i\pi}{L})^{j-1} \cos(\frac{i\pi}{L}x)$ *converge uniformly on* $[0, L]$ *to continuous functions.*

Proof 28.3.1

We let A_n^L and B_n^L be the usual Fourier sine and cosine coefficients on $[0, L]$.
Case 1 f' is integrable on $[0, L]$ with $f(0) = f(L)$:
Since $f(0) = f(L)$, the periodic extension $\hat{f}$ is continuous on $[0, 2L]$ and $\hat{f}(0) = \hat{f}(2L)$. So we can apply our previous results to $\hat{f}$: Let $A_n^{',L}$ and $B_n^{',L}$ be the Fourier Sine and Cosine coefficients for f' on $[0, L]$. We have

$$\sum_{i=1}^n \frac{L}{2}(A_i')^2 \leq ||f'||_2^2, \quad \sum_{i=1}^n \frac{L}{2}(B_i')^2 \leq ||f'||_2^2.$$

where $A_i^{',L} = -\frac{i\pi}{L}B_i^L$ and $B_i^{',L} = \frac{i\pi}{L}A_i^L$. From this we established the estimates

$$\sum_{i=1}^n \frac{i^2\pi^2}{2L}(B_i^L)^2 \leq ||f'||_2^2, \quad \sum_{i=1}^n \frac{i^2\pi^2}{2L}(A_i^L)^2 \leq ||f'||_2^2$$

Then, using the Cauchy - Schwartz Inequality, we found for the partial sums T_n^s of the Fourier sine series on $[0, L]$ that

$$T_m^s(x) - T_n^s(x) = \sum_{i=n+1}^m A_i^L \sin\left(\frac{i\pi}{L}x\right) = \sum_{i=n+1}^m \left(\frac{i\pi}{L}A_i^L\right)\left(\frac{L}{i\pi}\sin\left(\frac{i\pi}{L}x\right)\right)$$

$$\leq \sqrt{\sum_{i=n+1}^m \frac{i^2\pi^2}{L^2}|A_i^L|^2} \sqrt{\sum_{i=n+1}^m \frac{L^2}{i^2\pi^2}\left|\sin\left(\frac{i\pi}{L}x\right)\right|^2}$$

$$\leq \sqrt{\sum_{i=n+1}^m \frac{i^2\pi^2}{L^2}|A_i^L|^2} \sqrt{\sum_{i=n+1}^m \frac{L^2}{i^2\pi^2}}$$

Letting the partial sums of the Fourier cosine series be T_n^c, we find

$$T_m^c(x) - T_n^c(x) \leq \sqrt{\sum_{i=n+1}^{m} \frac{i^2\pi^2}{L^2}|B_i^L|^2} \sqrt{\sum_{i=n+1}^{m} \frac{L^2}{i^2\pi^2}}$$

Since the Fourier sine and Cosine Series derivative arguments gave the estimates

$$\sum_{i=1}^{n} \frac{i^2\pi^2}{2L}(A_i^L)^2 \leq ||f'||_2^2, \quad \sum_{i=1}^{n} \frac{i^2\pi^2}{2L}(B_i^L)^2 \leq ||f'||_2^2$$

we have

$$\left| T_m^s(x) - T_n^s(x) \right| \leq ||f'||_2 \sqrt{\frac{L}{2}} \sqrt{\sum_{i=n+1}^{m} \frac{L^2}{i^2\pi^2}}$$

$$\left| T_m^c(x) - T_n^c(x) \right| \leq ||f'||_2 \sqrt{\frac{L}{2}} \sqrt{\sum_{i=n+1}^{m} \frac{L^2}{i^2\pi^2}}$$

Since the series $\sum_{i=1}^{\infty} \frac{L^2}{i^2\pi^2}$ converges, this says the sequence of partial sums (T_n^s) and (T_n^c) satisfies the UCC for series. So the sequence of partial sums converges uniformly to continuous functions T^s and T^c on $[0, L]$ which by uniqueness of limits must be f in each case.

Case 2 *f'' is integrable on $[0, L]$ and $f'(0) = f'(L)$. In this case the extension $\hat{f}'$ satisfies $\hat{f}'(0) = \hat{f}'(0)$ and is continuous. For this case, we need to find the Fourier coefficients of f'', A_i'' and B_i''. The usual arguments find*

$$\sum_{i=1}^{n} \frac{L}{2}(A_i'')^2 \leq ||f''||_2^2, \quad \sum_{i=1}^{n} \frac{L}{2}(B_i'')^2 \leq ||f''||_2^2.$$

The arguments presented in Case (1) can then be followed almost exactly. Assuming f'' exists on $[0, L]$, consider the Fourier Sine series for f'' on $[0, L]$.

Recall, for the even and odd extensions of f', f_o' and f_e', we know the Fourier sine and cosine coefficients of f_o', $A_{i,o}'$, and f_e', $B_{i,o}'$, satisfy

$$A_{i,o}'^{,2L} = \frac{1}{L} \int_0^{2L} f_o'(s) \sin\left(\frac{i\pi}{L}s\right) ds = A_i'^{,L} = \frac{2}{L} \int_0^L f'(s) \sin\left(\frac{i\pi}{L}s\right) ds, \ i \geq 1$$

$$B_{0,o}'^{,2L} = \frac{1}{2L} \int_0^{2L} f_o'(s) ds = B_0'^{,L} = \frac{1}{L} \int_0^L f'(s) ds, \ i = 0$$

$$B_{i,o}'^{,2L} = \frac{1}{L} \int_0^L f_o'(s) \cos\left(\frac{i\pi}{L}s\right) = B_i'^{,L} = \frac{2}{L} \int_0^L f'(s) \cos\left(\frac{i\pi}{L}s\right) ds, \ i \geq 1$$

For $i > 0$

$$\frac{1}{L} \left\langle f_o'', \sin\left(\frac{i\pi}{L}x\right) \right\rangle = \frac{1}{L} \int_0^{2L} f_o''(x) \sin(\frac{i\pi}{L}x) dx$$

$$= \frac{1}{L} \left(f_o'(x) \sin(i\frac{i\pi}{L}x) \Big|_0^{2L} - \int_0^{2L} f_o'(x)\frac{i\pi}{L} \cos(i\frac{i\pi}{L}x) dx \right)$$

$$= -\frac{1}{L} \int_0^{2L} f_o'(x) \frac{i\pi}{L} \cos(i\frac{i\pi}{L}x)\, dx = -\frac{i\pi}{L^2} \left\langle f_o'(x), \cos\left(\frac{i\pi}{L}x\right) \right\rangle$$

$$= -\frac{i\pi}{L} B_i'^{,L} = -\frac{i^2\pi^2}{L^2} A_i^L$$

If we assume f'' is integrable, this leads to coefficient bounds. We look at

$$\left\langle f_o'' - \left(-\sum_{i=1}^n \frac{i^2\pi^2}{L^2} A_i^L \sin\left(\frac{i\pi}{L}x\right) \right), f_o'' - \left(-\sum_{j=1}^n \frac{j^2\pi^2}{L^2} A_j^L \sin\left(\frac{j\pi}{L}x\right) \right) \right\rangle >$$

$$= \; < f_o'', f_o'' > + 2 \sum_{i=1}^n \frac{i^2\pi^2}{L^2} A_i^L \left\langle f_o'', \sin\left(\frac{i\pi}{L}x\right) \right\rangle$$

$$+ \sum_{i=1}^n \sum_{j=1}^n \frac{i^2\pi^2}{L^2} \frac{j^2\pi^2}{L^2} A_i^L A_j^L \left\langle \sin\left(\frac{i\pi}{L}x\right), \sin\left(\frac{j\pi}{L}x\right) \right\rangle$$

$$= < f_o'', f_o'' > + 2 \sum_{i=1}^n \frac{i^2\pi^2}{L^2} A_i^L \left\langle f_o'', \sin\left(\frac{i\pi}{L}x\right) \right\rangle$$

$$+ \sum_{i=1}^n \frac{i^4\pi^4}{L^4} (A_i^L)^2 \left\langle \sin\left(\frac{i\pi}{L}x\right), \sin\left(\frac{i\pi}{L}x\right) \right\rangle$$

Substituting in for $< f_o'', \sin(\frac{i\pi}{L}x) >$, we have

$$\left\langle f_o'' - \left(-\sum_{i=1}^n \frac{i^2\pi^2}{L^2} A_i^L \sin\left(\frac{i\pi}{L}x\right) \right), f_o'' - \left(-\sum_{j=1}^n \frac{j^2\pi^2}{L^2} A_j^L \sin\left(\frac{j\pi}{L}x\right) \right) \right\rangle >$$

$$= < f_o'', f_o'' > - \sum_{i=1}^n L \frac{i^4\pi^4}{L^4} (A_i^L)^2$$

Hence,

$$\sum_{i=1}^n L \frac{i^4\pi^4}{L^4} (A_i^L)^2 \;\leq\; \|f_o''\|_2^2 = 2\|f''\|_2^2$$

implying

$$\sum_{i=1}^n \frac{i^4\pi^4}{L^4} A_i^2 \;\leq\; \frac{2}{L} \|f''\|_2^2$$

A similar analysis for the cosine components leads to

$$\sum_{i=1}^n \frac{i^4\pi^4}{L^4} B_i^2 \;\leq\; \frac{2}{L} \|f''\|_2^2$$

Let T_n denote the nth partial sum of $\sum_{i=1}^n \frac{i\pi}{L} A_i \sin(\frac{i\pi}{L}x)$ on $[0, L]$. Then, the difference of the nth and m^{th} partial sum for $m > n$ gives

$$T_m(x) - T_n(x) \;=\; \sum_{i=n+1}^m \frac{i\pi}{L} A_i^L \sin\left(\frac{i\pi}{L}x\right)$$

$$= \sum_{i=n+1}^{m} \left(\frac{i^2 \pi^2}{L^2} A_i^L \right) \left(\frac{L}{i\pi} \sin\left(\frac{i\pi}{L} x \right) \right)$$

Now apply our analogue of the Cauchy - Schwartz inequality for series.

$$T_m(x) - T_n(x) \leq \sum_{i=n+1}^{m} \left(\frac{i^2 \pi^2}{L^2} |A_i^L| \right) \frac{L}{i\pi} \left| \sin\left(\frac{i\pi}{L} x \right) \right|$$

$$\leq \sqrt{\sum_{i=n+1}^{m} \frac{i^4 \pi^4}{L^4} |A_i^L|^2} \sqrt{\sum_{i=n+1}^{m} \frac{L^2}{i^2 \pi^2} \left| \sin\left(\frac{i\pi}{L} x \right) \right|^2}$$

$$\leq \sqrt{\sum_{i=n+1}^{m} \frac{i^4 \pi^4}{L^4} |A_i^L|^2} \sqrt{\sum_{i=n+1}^{m} \frac{L^2}{i^2 \pi^2}}$$

Since

$$\sum_{i=1}^{n} \frac{i^4 \pi^4}{L^4} (A_i^L)^2 \leq \frac{L}{2} \|f'\|_2^2.$$

We have

$$\left| T_m(x) - T_n(x) \right| \leq \|f'\|_2 \sqrt{\frac{L}{2}} \sqrt{\sum_{i=n+1}^{m} \frac{L^2}{i^2 \pi^2}}.$$

Since the series $\sum_{i=1}^{\infty} \frac{L^2}{i^2 \pi^2}$ converges, this says the sequence of partial sums (T_n) satisfies the UCC for series and so the sequence of partial sums converges uniformly to a function T. The same sort of argument works for $\sum_{i=1}^{n} \frac{i\pi}{L} A_i \cos(\frac{i\pi}{L} x)$ on $[0, L]$. So there is a function Y which this series converges to uniformly on $[0, L]$.

Case 3 f''' *is integrable on* $[0, L]$:
We have $\hat{f}''$ is continuous and $\hat{f}''(0) = \hat{f}'' L$. So we can apply our previous work to f'''. The calculations are quite similar and are left to you. ■

28.3.3   Fourier Series Convergence Analysis

Let's put all that we know about the convergence of Fourier Series together now.

- Assume f has continuous derivatives f' and f'' on $[0, L]$ and assume f and f' are zero at both 0 and L. The Fourier Sine and Cosine series for f then converge uniformly on $[0, L]$ to continuous functions S and C. We also know the Fourier Sine and Fourier Cosine series converge to f on $[0, L]$. Further, we know the Derived Series of the Fourier Sine and Fourier Cosine series converge uniformly on $[0, L]$ due to Theorem 28.3.1.

  Let's check the conditions of the derivative interchange theorem, Theorem 24.4.2 applied to the sequence of partial sums (T_n^s) and (T_n^c) for the Fourier Sine and Cosine Series.

  1. T_n^s and T_n^c are differentiable on $[0, L]$: **True**.
  2. $(T_n^s)'$ and $(T_n^c)'$ are Riemann Integrable on $[0, L]$: **True as each is a polynomial of sines and cosines**.
  3. There is at least one point $t_0 \in [0, L]$ such that the sequence $(T_n^s(t_0))$ and $(T_n^c(t_0))$ converges. **True as these series converge on** $[0, L]$.

4. $(T_n^s)' \xrightarrow{\text{unif}} W^s$ and $(T_n^c)' \xrightarrow{\text{unif}} W^c$ on $[0, L]$ where both limit functions W^s and W^c are continuous. **True because of Theorem 28.3.1**.

The conditions of Theorem 24.4.2 are thus satisfied and we can say there are functions U^s and U^c on $[0, L]$ so that $T_n^s \xrightarrow{\text{unif}} U^s$ on $[0, L]$ with $(U^s)' = W^s$ and $T_n^c \xrightarrow{\text{unif}} U^c$ on $[0, L]$ with $(U^c)' = W^c$. Since limits are unique, we then have $U^s = f$ with $f' = W^s$. and $U^c = f$ with $f' = W^c$. This is the statement that we can take the derivative of the Fourier Sine and Fourier Cosine Series termwise. That is

$$f'(t) = \left(\sum_{n=1}^{\infty} A_n \sin(\frac{n\pi}{L}x) \right)' = \sum_{n=1}^{\infty} \frac{n\pi}{L} A_n \cos(\frac{n\pi}{L}x)$$

$$f'(t) = \left(\sum_{n=1}^{\infty} B_n \cos(\frac{n\pi}{L}x) \right)' = \sum_{n=1}^{\infty} -\frac{n\pi}{L} B_n \sin(\frac{n\pi}{L}x)$$

- Now assume f also has a continuous third derivative on $[0, L]$. The arguments we just presented can be used with some relatively obvious modifications to show

$$f''(t) = \left(\sum_{n=1}^{\infty} \frac{n\pi}{L} A_n \cos(\frac{n\pi}{L}x) \right)' = \sum_{n=1}^{\infty} -\frac{n^2\pi^2}{L^2} A_n \sin(\frac{n\pi}{L}x)$$

$$f''(t) = \left(\sum_{n=1}^{\infty} -\frac{n\pi}{L} B_n \sin(\frac{n\pi}{L}x) \right)' = \sum_{n=1}^{\infty} -\frac{n^2\pi^2}{L^2} B_n \cos(\frac{n\pi}{L}x)$$

- If we assume the continuity of higher order derivatives, we can argue in a similar fashion to show that term by term higher order differentiation is possible.

It is possible to do this sort of analysis for more general functions f but to do so requires we move to a more general notion of integration called Lebesgue integration. This is for future discussions to be done with better tools!

28.3.4 Homework

Exercise 28.3.5 *If f and g are two nonzero continuous functions on $[a, b]$ which are orthogonal, prove f and g are linearly independent.*

Exercise 28.3.6 *If $\{f_1, \ldots, f_n\}$ are nonzero continuous functions on $[a, b]$ which are mutually orthogonal, prove it is a linearly independent set.*

Exercise 28.3.7 *If $(w_n)_{n \geq 1}$ is an orthogonal family of continuous functions on $[a, b]$ prove $(w_n)_{n \geq 1}$ is a linearly independent set.*

Exercise 28.3.8 *If f is continuous on $[0, L]$ and $(w_n)_{n \geq 1}$ is an orthonormal family of continuous functions on $[0, L]$ prove $\sum_{n=1}^{\infty} (< f, w_n >)^2 \leq \|f\|_2^2$.*

Exercise 28.3.9 *Is $\sqrt{t}$ in the span of $\{1, t, t^2, \ldots, t^n, \ldots\}$ for $t \geq 0$?*

28.3.5 Convergence of the Cable Solution: Simplified Analysis

The formal series solution is

$$\Phi(x, t) = B_0 e^{-\frac{1}{\alpha}t} + \sum_{n=1}^{\infty} B_n \cos\left(\frac{n\pi}{L}x\right) e^{-\frac{L^2 + n^2\pi^2\beta^2}{\alpha L^2}t}.$$

We can show how we establish that Φ is indeed the solution to the cable model by setting all the parameters here to the value of 1 to make the analysis more transparent. Hence, we set $L = 1, \alpha = 1$ and $\pi\,\beta = 1$. The solution is then

$$\Phi(x,t) \;=\; B_0 e^{-t} + \sum_{n=1}^{\infty} B_n \cos(n\pi x)\, e^{-(1+n^2)t}.$$

Now, we can estimate the B_n coefficients as

$$|B_0| \;\leq\; \frac{1}{L}\int_0^L |f(x)|dx \leq \frac{1}{L}\int_0^L |f(x)|dx,$$

$$|B_n| \;\leq\; \frac{2}{L}\int_0^L |f(x)|\left|\cos\left(\frac{n\pi}{L}x\right)\right|dx \leq \frac{2}{L}\int_0^L |f(x)|dx.$$

Letting $C = \frac{2}{L}\int_0^L |f(x)|dx$, we see $|B_n| \leq C$ for all $n \geq 0$. Thus, for any fixed $t > 0$, we have the estimate

$$\sum_{n=1}^{\infty} |B_n||\cos(n\pi x)|\, e^{-(1+n^2)t} \;\leq\; e^{-t}\sum_{n=1}^{\infty} C\, e^{-n^2 t}.$$

and by the Weierstrass Theorem for Uniform Convergence of Series, we see the series on the left hand side converges uniformly to $\Phi(x,t)$ on the interval $[0,1]$ as long as $t > 0$. Since each of the individual functions in the series on the left hand side is continuous and convergence is uniform, we know the limit function for the left hand side series is continuous on $[0,1]$ for each $t > 0$. Hence, the function $\Phi(x,t)$ is continuous on the domain $[0,1] \times [t,\infty)$ for any positive t.

At $t = 0$, the series for $\Phi(x,t)$ becomes

$$\Phi(x,0) \;=\; B_0 + \sum_{n=1}^{\infty} B_n \cos(n\pi x).$$

which is the Fourier cosine series for f on $[0,1]$. Now from our previous work with the Fourier cosine series, we know this series converges uniformly to $f(x)$ as long as f is continuous on $[0,1]$ with a derivative. So if we assume f' is continuous on $[0,1]$ and $f^{(1)}(0) = f^{(1)}(L)$, we know $\Phi(x,0)$ converges uniformly to a continuous function of x which matches the data f. Indeed, we see

$$\lim_{t\to 0^+} |\Phi(x,t)| \;=\; \lim_{t\to 0^+}\left(B_0 e^{-t} + \sum_{n=1}^{\infty} B_n \cos(n\pi x)\, e^{-(1+n^2)t}\right)$$

$$=\; f(x) = \Phi(x,0).$$

Hence, we know $\Phi(x,t)$ is continuous on $[0,1] \times [0,\infty)$. Now let's look at the partial derivatives. If we take the partial derivative with respect to t of the series for $\Phi(x,t)$ term by term, we find the function $D_t(x,t)$ given by

$$D_t(x,t) \;=\; -B_0 e^{-t} - \sum_{n=1}^{\infty} B_n(1+n^2)\cos(n\pi x)\, e^{-(1+n^2)t}$$

The series portion for $D_t(x,t)$ satisfies the estimate

$$\sum_{n=1}^{\infty} |B_n|(1+n^2)|\cos(n\pi x)|\, e^{-(1+n^2)t} \;\leq\; \sum_{n=1}^{\infty} C(1+n^2)\, e^{-n^2 t}.$$

For $t > 0$, applying the Weierstrass Uniform Convergence Test, the series for $D_t(x, t)$ converges uniformly. Since it is built from continuous functions, we know limit function is continuous for $t > 0$ because the convergence is uniform. Finally, applying the Derivative Interchange Theorem, we see the series for $\frac{\partial \Phi}{\partial t}$ is the same as the series $D_t(x, t)$ we found by differentiating term by term.

At $t = 0$, we have the series

$$D_t(x, 0) = -B_0 - \sum_{n=1}^{\infty} B_n (1 + n^2) \cos(n\pi x)$$

From our earlier discussions, if we know f''' is continuous and the second derivative matches at 0 and L, the series $\sum_{n=1}^{\infty} n^2 B_n \cos(n\pi x)$ converges uniformly and differentiation term by term is permitted. Also, the series $\sum_{n=1}^{\infty} B_n \cos(n\pi x)$ converges uniformly since we know f' is continuous. We conclude

$$\Phi_t = \partial_t \left(B_0 e^{-t} + \sum_{n=1}^{\infty} B_n \cos(n\pi x) \, e^{-(1+n^2)t} \right)$$

The formal partial derivatives of the cable series solution with respect to x are then

$$D_x(x, t) = -\sum_{n=1}^{\infty} B_n n\pi \sin(n\pi x) \, e^{-(1+n^2)t}$$

$$D_{xx}(x, t) = -\sum_{n=1}^{\infty} B_n n^2 \pi^2 \cos(n\pi x) \, e^{-(1+n^2)t}$$

Now we can estimate the B_n coefficients as usual with the constant C as before. Hence, $|B_n| \leq C$ for all $n \geq 0$. The series for the two partial derivatives with respect to x satisfies

$$\sum_{n=1}^{\infty} |B_n| n\pi |\sin(n\pi x)| \, e^{-(1+n^2)t} \leq e^{-t} \sum_{n=1}^{\infty} Cn\pi \, e^{-n^2 t}$$

and

$$\sum_{n=1}^{\infty} |B_n| n^2 \pi^2 |\sin(n\pi x)| \, e^{-(1+n^2)t} \leq e^{-t} \sum_{n=1}^{\infty} Cn^2 \pi^2 \, e^{-n^2 t}$$

Again, these estimates allow us to use the Weierstrass Theorem for Uniform Convergence to conclude the series for $D_x(x, t)$ and $D_{xx}(x, t)$ converge uniformly. Since these series are built from continuous functions, we then know the limit function is continuous for $t > 0$ since the convergence is uniform. Thus, the series for $\frac{\partial \Phi}{\partial x}$ is the same as the series $D_x(x, t)$ we find by differentiating term by term. Further, the series for $\frac{\partial^2 \Phi}{\partial x^2}$ is the same as the series $D_{xx}(x, t)$ we also find by differentiating term by term.

At $t = 0$, we have

$$D_x(x, 0) = -\sum_{n=1}^{\infty} B_n n\pi \sin(n\pi x)$$

$$D_{xx}(x, 0) = -\sum_{n=1}^{\infty} B_n n^2 \pi^2 \cos(n\pi x)$$

which converge uniformly if f''' is continuous and the Derivative Interchange Theorem tells us that differentiation is justified term by term.

Hence, on $[0, 1] \times (0, \infty)$ we can compute

$$
\begin{aligned}
\frac{\partial^2 \Phi}{\partial x^2} - \Phi - \frac{\partial \Phi}{\partial t} &= -\sum_{n=1}^{\infty} B_n n^2 \pi^2 \cos(n\pi x)\, e^{-(1+n^2)t} - B_0 e^{-t} - \sum_{n=1}^{\infty} B_n \cos(n\pi x)\, e^{-(1+n^2)t} \\
&\quad + B_0 e^{-t} + \sum_{n=1}^{\infty} B_n (1 + n^2 \pi^2) \cos(n\pi x)\, e^{-(1+n^2)t} \\
&= 0.
\end{aligned}
$$

So that we see our series solution satisfies the partial differential equation. To check the boundary conditions, because of continuity, we can see

$$
\frac{\partial \Phi}{\partial x}(0, t) = \left(-\sum_{n=1}^{\infty} B_n n\pi \sin(n\pi x)\, e^{-(1+n^2)t} \right)\Bigg|_{x=0} = 0
$$

$$
\frac{\partial \Phi}{\partial x}(1, t) = \left(-\sum_{n=1}^{\infty} B_n n\pi \sin(n\pi x)\, e^{-(1+n^2)t} \right)\Bigg|_{x=1} = 0.
$$

and finally, the data boundary condition gives

$$
\Phi(x, 0) = f(x).
$$

Note, we can show that higher order partial derivatives of Φ are also continuous even though we have only limited smoothness in the data function f. Hence, the solutions of the cable equation smooth irregularities in the data. However, we cannot see this from our analysis because we must assume continuity in the third derivative of the data to insure this. But there are other techniques we can use to show these solutions are very smooth which we cover in more advanced classes. The analysis for the actual cable model solution is handled similarly, of course. The arguments we presented in this chapter are quite similar to the ones we would use to analyze the smoothness qualities of the solutions to other linear partial differential equations with boundary conditions. We didn't want to go through all of them in this text as we don't think there is a lot of value in repeating these arguments over and over. Just remember that to get solutions with this kind of smoothness requires that the series we formally compute converge uniformly. If we did not have that information, our series solutions would not have this amount of smoothness. However, there are other tools we can bring to bear but that is another story!

Also, if the data f was a C^∞ bump function on $[0, L]$ centered at $L/2$, we would have continuity in all orders of the data f and hence all order partials of the solution Φ would be smooth.

28.3.6 Homework

Let $\pi_n = \{t_i\}_{i=0}^{n}$ be the uniform partition of the interval $[0, L]$ with $\|\pi_n\| = L/n$. Let ϕ_i be the C^∞ bump function defined on $[t_i, t_{i+1}]$ centered at the midpoint $(t_i + t_{i+1})/2$ of height 1. Let f be any continuous function on $[0, L]$.

Exercise 28.3.10 *Prove the functions ϕ_i form an orthonormal set.*

Exercise 28.3.11 *Prove* $< f, \phi_i > \le (L/n) f_{av,i}$, *where* $f_{av,i}$ *is the average value of* f *on the subinterval* $[t_i, t_{i+1}]$.

Exercise 28.3.12 *Show that there is some constant* $A > 0$ *so that* $\|\phi_i\|_2 = A$ *for all* i.

Exercise 28.3.13 *Find the best approximation* y *of* f *to the subspace of* $C([0, L])$ *spanned by* $\{\phi_i\}_{i=1}^n$ *and prove* $\|y\|_\infty$ *is bounded above by* $L f_{av}/A^2$ *where* f_{av} *is the average value of* f *on* $[0.L]$.

28.4 Power Series for ODE

You all probably know how to solve ordinary differential equations like $y'' + 3y' + 4y = 0$ which is an example of a linear second order differential equation with constant coefficients. To find a solution we find twice differentiable function $y(x)$ which satisfies the given dynamics. But what about a model like $x\,y'' + 4x^2 y' + 5y = 0$ or $y'' + 3xy' + 6x^3 y = 0$? The *coefficients* here are not constants; they are polynomial functions of the independent variable x. You don't usually see how to solve such equations in your first course in ordinary differential equations. In this course, we have learned more of the requisite analysis that enables us to understand what to do. However, the full proofs of some of our assertions will elude us as there are still theorems in this area that we do not have the background to prove. These details can be found in older books such as (E. Ince (3) 1956) which is a reprint of the original 1926 text.

We are going to solve such models such as

$$p(x)\ y'' + q(x)y' + r(x)y = 0$$
$$y(x_0) = y_0, \quad y'(x_0) = y_1$$

using a power series solution given by $y(x) = \sum_{n=0}^{\infty} a_n (x - x_0)^n$. We must use a powerful theorem from (E. Ince (3) 1956).

Theorem 28.4.1 Power Series Solutions for ODEs

> Let $p(x)\ y'' + q(x)y' + r(x)y = 0$ *be a given differential equation where* $p(x)$, $q(x)$ *and* $r(x)$ *are polynomials in* x. *We say* a *is an* **ordinary point** *of this equation if* $p(a) \ne 0$ *and otherwise, we say* a *is a* **singular point**. *The solution to this equation is* $y(x) = \sum_{n=0}^{\infty} a_n (x - a)^n$ *at* **any** *ordinary point* a *and the radius of convergence* R *of this solution satisfies* $R = d$ *where* d *is the distance from* a *to the nearest root of the coefficient polynomial functions* p.

Comment 28.4.1 *In general, d is the distance to the nearest singularity of the coefficient functions whose definition is more technical than being a root of a polynomial if the coefficient functions were not polynomials.*

Proof 28.4.1
Chapter 5 and Chapter 15 in (E. Ince (3) 1956) discusses the necessary background to study n^{th} *order ordinary differential equations of the form*

$$w^{(n)} + p_1(z)\, w^{(n-1)} + \ldots p_{n-1}(z)w^{(1)} + p_n(z)\, w = 0$$
$$w^{(i)}(z_0) = z_i, \quad 0 \le i \le n - 1$$

where $w(z)$ *is a function of the complex variable* $z = x + iy$ *in the complex plane* $\mathbb{C}$. *To understand this sort of equation, you must study what is called* functions of a complex variable. *Each complex* *coefficient function is assumed to be analytic at z_0 which means we can write*

$$p_1(z) = \sum_{k=0}^{\infty} a_k^1 (z - z_0)^k$$

$$p_2(z) = \sum_{k=0}^{\infty} a_k^2 (z - z_0)^k$$

$$\vdots = \vdots$$

$$p_i(z) = \sum_{k=0}^{\infty} a_k^i (z - z_0)^k$$

$$\vdots = \vdots$$

$$p_n(z) = \sum_{k=0}^{\infty} a_k^n (z - z_0)^k$$

where each of these power series converges in the ball in the complex plane centered at z_0 of radius r_i. Hence, there is a smallest radius $r = \min\{r_1, \ldots, r_n\}$ for which they all converge. In the past, we have studied convergence of power series in a real variable $x - x_0$ and the radius of convergence of such a series gives an interval of convergence $(r - x_0, r + x_0)$. Here, the interval of convergence becomes a circle in the complex plane

$$B_r(z_0) = \{z = x + iy : \sqrt{(x - x_0)^2 + (y - y_0)^2} < r\}$$

$z_0 = x_0 + iy_0$. *However, the coefficient functions need not be so nice. They need not be analytic at the point z_0. A more general class of coefficient function are those whose power series expansions have a more general form at a given point ζ.*

*A point ζ is called a **regular singular point** of this equation if each of the coefficient function p_i can be written as $p_i(z) = (z - \zeta)^{-i} P_i(z)$ where P is analytic at ζ. Thus, if $\zeta = 0$ was a **regular singular point** we could write $P_i(z) = \sum_{k=0}^{\infty} b_i z^k$ and have*

$$p_1(z) = \frac{1}{z} P_1(z) = \frac{1}{z} \sum_{k=0}^{\infty} b_k^1 z^k$$

$$p_2(z) = \frac{1}{z^2} P_2(z) = \frac{1}{z^2} \sum_{k=0}^{\infty} b_k^2 z^k$$

$$\vdots = \vdots$$

$$p_i(z) = \frac{1}{z^i} P_i(z) = \frac{1}{z^i} \sum_{k=0}^{\infty} b_k^i z^k$$

$$\vdots = \vdots$$

$$p_n(z) = \frac{1}{z^n} P_n(z) = \frac{1}{z^i} \sum_{k=0}^{\infty} b_k^n z^k$$

Now our model converted to the complex plane is $p(z)\ y'' + q(z)y' + r(z)y = 0$ which can be rewritten as

$$y'' + \frac{q(z)}{p(z)} y' + \frac{r(z)}{p(z)} y = 0.$$

*Thus $p_1(z) = \frac{q(z)}{p(z)}$ and $p_2(z) = \frac{r(z)}{p(z)}$. Let's consider an example: say $p(z) = z$, $q(z) = 2z^2 + 4z - 3$ and $r(z) = -z + 2$. Then $p_1(z) = \frac{2z^2 + 4z - 3}{z}$ and $p_2(z) = \frac{-z + 2}{z} = \frac{-z^2 + 2z}{z^2}$ and so $z = 0$ is a regular singular point. Note if $p(z) = z^2$, the root of p occurs with multiplicity two and the requirements of a regular singular point are **not** met. Note the zero of $p(z)$ here is just $z = 0$ and so we can solve this equation at a point like $z = 1$ as a power series expansion $y(z) = \sum_{k=0}^{\infty} a_k(z - 1)^k$ and the radius of convergence $R = 1$ as that is the distance to the zero $z = 0$ of $p(z)$.*

The proof that the radius of convergence is the R mentioned is a bit complicated and is done in Chapter 16 of (E. Ince (3) 1956). Once you have taken a nice course in complex variable theory it should be accessible. So when you get a chance, look it up! ∎

It is best to do examples!

28.4.1 Constant Coefficient ODEs

Let's start with constant coefficient problems.

Example 28.4.1 *Solve the model $y'' + y = 0$ using power series methods.*

Solution *The coefficient functions here are constants, so the power series solution can be computed at any point a and the radius of convergence will be $R = \infty$. Let's find a solution at $a = 0$. We assume $y(x) = \sum_{n=0}^{\infty} a_n x^n$. From our study of power series, we know the first and second derived series have the same radius of convergence and differentiation can be done term by term. Thus,*

$$y' = \sum_{n=1}^{\infty} n a_n x^{n-1}, \quad y'' = \sum_{n=2}^{\infty} n(n-1) a_n x^{n-2}$$

The model we need to solve then becomes

$$\sum_{n=2}^{\infty} n(n-1) a_n x^{n-2} + \sum_{n=0}^{\infty} a_n x^n = 0$$

The powers of x in the series have different indices. Change summation variables in the first one to get

$$k = n - 2 \implies \sum_{n=2}^{\infty} n(n-1) a_n x^{n-2} = \sum_{k=0}^{\infty} (k+2)(k+1) a_{k+2} x^k$$

as $k = n - 2$ tells us $n = k + 2$ and $n - 1 = k + 1$. Since the choice of summation variable does not matter, relabel the k above back to an n to get

$$\sum_{k=0}^{\infty} (k+2)(k+1) a_{k+2} x^k = \sum_{n=0}^{\infty} (n+2)(n+1) a_{n+2} x^n$$

The series problem to solve is then

$$\sum_{n=0}^{\infty} (n+2)(n+1) a_{n+2} x^n + \sum_{n=0}^{\infty} a_n x^n = 0$$

Hence, we have, for all x

$$\sum_{n=0}^{\infty} \{(n+2)(n+1) a_{n+2} + a_n\} x^n = 0$$

The only way this can be true is if each of the coefficients of x^n vanish. This gives us what is called a **recursion relation** *the coefficients must satisfy. For all $n \geq 0$, we have*

$$(n+2)(n+1)a_{n+2} + a_n = 0 \implies a_{n+2} = -\frac{a_n}{(n+2)(n+1)} \qquad (28.5)$$

By direct calculation, we find

$$
\begin{aligned}
a_2 &= -\frac{a_0}{1 \cdot 2} \\
a_3 &= -\frac{a_1}{2 \cdot 3} = -\frac{a_1}{3!} \\
a_4 &= -\frac{a_2}{3 \cdot 4} = 1\left(-\frac{a_0}{1 \cdot 2}\right)\frac{1}{3 \cdot 4} = \frac{a_2}{1 \cdot 2 \cdot 3 \cdot 4} = \frac{a_2}{4!} \\
a_5 &= -\frac{a_3}{4 \cdot 5} = \frac{a_1}{5!} \\
&\vdots
\end{aligned}
$$

In general, it is impossible to detect a pattern in these coefficients, but here we can find a pattern easily. We see

$$
\begin{aligned}
a_{2k} &= (-1)^k \frac{a_0}{(2k)!} \\
a_{2k+1} &= (-1)^k \frac{a_1}{(2k+1)!}
\end{aligned}
$$

We have found the solution is

$$y(x) = a_0 + a_1 x + a_0 (-1)^2 \frac{x^2}{2!} + a_1(-1)^3 \frac{x^3}{3!} + a_0(-1)^4 \frac{x^4}{4!} + a_1(-1)^5 \frac{x^5}{5!} + \cdots$$

Define two new series by

$$
\begin{aligned}
y_0(x) &= 1 + (-1)^2 \frac{x^2}{2!} + \cdots + (-1)^k \frac{x^{2k}}{(2k)!} + \cdots \\
y_1(x) &= x + (-1)^3 \frac{x^3}{3!} + \cdots + (-1)^k \frac{x^{2k+1}}{(2k+1)!} + \cdots
\end{aligned}
$$

It is easy to prove that for any convergent series, $\alpha \sum_{n=0}^{\infty} a_n x^n = \sum_{n=0}^{\infty} \alpha a_n x^n$. The series $\sum_{n=0}^{\infty} (-1)^{2n} \frac{x^{2n}}{(2n)!}$ and $\sum_{n=0}^{\infty} (-1)^{2n+1} \frac{x^{2n+1}}{(2n+1)!}$ converge for all x by the ratio test, so we can write the solution $y(x)$ as

$$
\begin{aligned}
y(x) &= a_0 \sum_{n=0}^{\infty} (-1)^{2n} \frac{x^{2n}}{(2n)!} + a_1 \sum_{n=0}^{\infty} (-1)^{2n+1} \frac{x^{2n+1}}{(2n+1)!} \\
&= a_0\, y_0(x) + a_1\, y_1(x)
\end{aligned}
$$

Since the power series for y_0 and y_1 converge on $\Re$, we also know how to compute their derived series to get

$$y_0'(x) = \sum_{n=1}^{\infty} (-1)^{2n} (2n) \frac{x^{2n-1}}{(2n)!}$$

$$y_1'(x) = \sum_{n=0}^{\infty} (-1)^{2n+1}(2n+1)\frac{x^{2n}}{(2n+1)!}$$

Next, note $y_0(x)$ and $y_1(x)$ are linearly independent on $\Re$ as if

$$c_0\, y_0(x) + c_1\, y_1(x) = 0, \quad \forall\, x$$

then we also have

$$c_0\, y_0'(x) + c_1\, y_1'(x) = 0, \quad \forall\, x$$

Since these equations must hold for all x, in particular they hold for $x = 0$ giving

$$\begin{aligned} c_0\, y_0(0) + c_1\, y_1(0) &= c_0(1) + c_1(0) = 0 \\ c_0\, y_0'(0) + c_1\, y_1'(0) &= c_0\,(0) + c_1\,(1) = 0 \end{aligned}$$

In matrix - vector form this becomes

$$\begin{bmatrix} 1 & 0 \\ 0 & 1 \end{bmatrix} \begin{bmatrix} c_0 \\ c_1 \end{bmatrix} = \begin{bmatrix} 0 \\ 0 \end{bmatrix}$$

The determinant of the coefficient matrix of this linear system is positive and so the only solution is $c_0 = c_1 = 0$ implying these functions are linearly independent on $\Re$. Note y_0 is the solution to the problem

$$y'' + y = 0, \quad y(0) = 1,\ y'(0) = 0$$

and y_1 solves

$$y'' + y = 0, \quad y(0) = 0,\ y'(0) = 1$$

and, as usual the solution to

$$y'' + y = 0, \quad y(0) = \alpha,\ y'(0) = \beta$$

is the linear combination of the two linearly independent solutions y_0 and y_1 giving

$$y(x) = \alpha y_0(x) + \beta y_1(x)$$

Of course, from earlier courses and our understanding of the Taylor Series expansions of $\cos(x)$ and $\sin(x)$, we also know $y_0(x) = \cos(x)$ and $y_1(x) = \sin(x)$. The point of this example, is that we can do a similar analysis for the more general problem with polynomial coefficients.

Example 28.4.2 *Solve the model $y'' + 6y' + 9y = 0$ using power series methods.*

Solution *Again, the coefficient functions here are constants, so the power series solution can be computed at any point a and the radius of convergence will be $R = \infty$. Let's find a solution at $a = 0$. We assume $y(x) = \sum_{n=0}^{\infty} a_n x^n$. Thus,*

$$y' = \sum_{n=1}^{\infty} n a_n x^{n-1}$$

$$y'' = \sum_{n=2}^{\infty} n(n-1) a_n x^{n-2}$$

The model we need to solve then becomes

$$\sum_{n=2}^{\infty} n(n-1)a_n x^{n-2} + 6\sum_{n=1}^{\infty} na_n x^{n-1} + 9\sum_{n=0}^{\infty} a_n x^n \;=\; 0$$

The powers of x in the series have different indices. We change the summation variables in both derivative terms to get

$$k = n-1 \;\Longrightarrow\; \sum_{n=1}^{\infty} na_n x^{n-1} = \sum_{k=0}^{\infty} (k+1)a_{k+1} x^k$$

$$k = n-2 \;\Longrightarrow\; \sum_{n=2}^{\infty} n(n-1)a_n x^{n-2} = \sum_{k=0}^{\infty} (k+2)(k+1)a_{k+2} x^k$$

Since the choice of summation variable does not matter, relabel the k above back to an n to get

$$\sum_{k=0}^{\infty} (k+1)a_{k+1} x^k \;=\; \sum_{n=0}^{\infty} (n+1)a_{n+1} x^n$$

$$\sum_{k=0}^{\infty} (k+2)(k+1)a_{k+2} x^k \;=\; \sum_{n=0}^{\infty} (n+2)(n+1)a_{n+2} x^n$$

The series problem to solve is then

$$\sum_{n=0}^{\infty} (n+2)(n+1)a_{n+2} x^n + 6\sum_{n=0}^{\infty} (n+1)a_{n+1} x^n + 9\sum_{n=0}^{\infty} a_n x^n \;=\; 0$$

Hence, we have, for all x

$$\sum_{n=0}^{\infty} \{(n+2)(n+1)a_{n+2} + 6(n+1)a_{n+1} + 9\,a_n\}x^n \;=\; 0$$

The only way this can be true is if each of the coefficients of x^n vanishes. This gives us what is called a **recursion relation** *the coefficients must satisfy. For all $n \geq 0$, we have*

$$(n+2)(n+1)a_{n+2} + 6(n+1)a_{n+1} + 9\,a_n \;=\; 0 \;\Longrightarrow\; a_{n+2} = -\frac{6(n+1)a_{n+1} + 9a_n}{(n+2)(n+1)}$$

By direct calculation, we find

$$a_2 \;=\; -\frac{6a_1 + 9a_0}{1\cdot 2} = -3a_1 + \frac{9}{2}a_0$$

$$a_3 \;=\; -\frac{6(2)a_2 + 9a_1}{2\cdot 3} = -2a_2 - \frac{3}{2}a_1$$

$$=\; -2\left(-3a_1 + \frac{9}{2}a_0\right) - \frac{3}{2}a_1 = 6a_1 - 9a_0 - \frac{3}{2}a_1 = -9a_0 + \frac{9}{2}a_1$$

$$a_4 \;=\; -\frac{6(3)a_3 + 9a_2}{3\cdot 4} = -\frac{3}{2}a_3 - \frac{3}{4}a_2$$

$$=\; -\frac{3}{2}\left(-9a_0 + \frac{9}{2}a_1\right) - \frac{3}{4}\left(-3a_1 + \frac{9}{2}a_1\right)$$

$$=\; -\frac{27}{4}a_1 + \frac{27}{2}a_0 + \frac{9}{4}a_1 - \frac{27}{8}a_0 = -\frac{9}{2}a_1 + \frac{81}{8}a_0$$

$$\vdots$$

We have found the solution is

$$y(x) = a_0 + a_1 x + \left(-3a_1 + \frac{9}{2}a_0\right)x^2 + \left(-9a_0 + \frac{9}{2}a_1\right)x^3$$
$$+ \left(-\frac{9}{2}a_1 + \frac{81}{8}a_0\right)x^4 + \cdots$$

Define two new series by

$$y_0(x) = 1 + \frac{9}{2}x^2 - 9x^3 + \frac{81}{8}x^4 + \cdots$$
$$y_1(x) = x - 3x^2 + \frac{9}{2}x^3 - \frac{9}{2}x^4 + \cdots$$

Hence, we can write the solution as

$$y(x) = a_0 \left\{1 + \frac{9}{2}x^2 - 9x^3 + \frac{81}{8}x^4 + \cdots\right\}$$
$$+ a_1 \left\{x - 3x^2 + \frac{9}{2}x^3 - \frac{9}{2}x^4 + \cdots\right\}$$
$$= a_0\, y_0(x) + a_1\, y_1(x)$$

Our theorem guarantees that the solution $y(x)$ converges in $\Re$, so the series y_0 we get setting $a_0 = 1$ and $a_1 = 0$ converges too. Setting $a_0 = 0$ and $a_1 = 1$ then shows y_1 converges. Since the power series for y_0 and y_1 converge on $\Re$, we also know how to compute their derived series to get

$$y_0'(x) = 9x - 27x^2 + \frac{81}{2}x^3 + \cdots$$
$$y_1'(x) = 1 - 6x + \frac{27}{2}x^2 - 18x^3 + \cdots$$

Next, note $y_0(x)$ and $y_1(x)$ are linearly independent on $\Re$ as if

$$c_0\, y_0(x) + c_1\, y_1(x) = 0, \quad \forall\, x$$

then we also have

$$c_0\, y_0'(x) + c_1\, y_1'(x) = 0, \quad \forall\, x$$

Since these equations must hold for all x, in particular they hold for $x = 0$ giving

$$c_0\, y_0(0) + c_1\, y_1(0) = c_0(1) + c_1(0) = 0$$
$$c_0\, y_0'(0) + c_1\, y_1'(0) = c_0(0) + c_1(1) = 0$$

which tells us $c_0 = c_1 = 0$ implying these functions are linearly independent on $\Re$. Note y_0 is the solution to the problem

$$y'' + 6y' + 9y = 0, \quad y(0) = 1, \ y'(0) = 0$$

The general solution is $y(x) = Ae^{-3x} + Bxe^{-3x}$ and for these initial conditions, we find $A = 1$ and $B = 3$ giving $y_0(x) = (1 + 3x)e^{-3x}$. Using the Taylor Series expansion of e^{-3x}, we find

$$
\begin{aligned}
y^{-3x} + 3x\, e^{-3x} &= \left(1 - 3x + \frac{9}{2}x^2 - \frac{27}{6}x^3 + \cdots\right) \\
&\quad + 3x\left(1 - 3x + \frac{9}{2}x^2 - \frac{27}{6}x^3 + \cdots\right) \\
&= 1 + (3x - 3x)x + \left(\frac{9}{2} - 9\right)x^2 + \left(-\frac{27}{6} + \frac{27}{2}\right)x^3 + \cdots \\
&= 1 - \frac{9}{2}x^2 - 9x^3 + \cdots
\end{aligned}
$$

which is the series we found using the power series method! Further, y_1 solves

$$y'' + 6y' + 9y = 0, \quad y(0) = 0,\ y'(0) = 1$$

Then y_1 solves

$$y'' + 6y' + 9y = 0, \quad y(0) = 0,\ y'(0) = 1$$

The general solution is $y(x) = Ae^{-3x} + Bxe^{-3x}$ and for these initial conditions, we find $A = 0$ and $B = 1$ giving $y_1(x) = x\, e^{-3x}$. Using the Taylor Series expansion of e^{-3x}, we find

$$
\begin{aligned}
x\, e^{-3x} &= x\left(1 - 3x + \frac{9}{2}x^2 - \frac{27}{6}x^3 + \cdots\right) \\
&= x - 3x^2 + \frac{9}{2}x^3 - \frac{9}{2}x^4 \cdots
\end{aligned}
$$

which is the series we found using the power series method! As usual the solution to

$$y'' + 6y' + 9y = 0, \quad y(0) = \alpha,\ y'(0) = \beta$$

is the linear combination of the two linearly independent solutions y_0 and y_1 giving

$$y(x) = \alpha y_0(x) + \beta y_1(x)$$

28.4.2   Homework

Exercise 28.4.1 *Solve*

$$y'' + 4y = 0$$

using the Power Series method.

1. *Use our theorems to find the radius of convergence R.*

2. *Find the two solutions y_0 and y_1.*

3. *Show y_0 and y_1 are linearly independent.*

4. *Write down the Initial Value Problem y_0 and y_1 satisfy.*

5. *Find the solution to this model with $y(0) = 3$ and $y'(0) = 2$.*

6. *Express y_0 and y_1 in terms of traditional functions.*

Exercise 28.4.2 *Solve*

$$y'' + y' - 6y = 0$$

using the Power Series method.

1. *Use our theorems to find the radius of convergence R.*

2. *Find the two solutions y_0 and y_1.*

3. *Show y_0 and y_1 are linearly independent.*

4. *Write down the Initial Value Problem y_0 and y_1 satisfy.*

5. *Find the solution to this model with $y(0) = -1$ and $y'(0) = 4$.*

6. *Express y_0 and y_1 in terms of traditional functions.*

28.4.3   Polynomial Coefficient ODEs

Now let's look at problems with polynomial coefficients.

Example 28.4.3 *Solve the model $y'' + x^2\, y = 0$ using power series methods.*

Solution *The coefficient functions here are $p(x) = 1$, $q(x) = 0$ and $r(x) = x^2$. Since $p(x) = 1$ has no roots, so the power series solution can be computed at any point a and the radius of convergence is $R = \infty$. Let's find a solution at $a = 0$. We assume $y(x) = \sum_{n=0}^{\infty} a_n x^n$. Thus,*

$$y' = \sum_{n=1}^{\infty} n a_n x^{n-1}$$

$$y'' = \sum_{n=2}^{\infty} n(n-1) a_n x^{n-2}$$

The model we need to solve then becomes

$$\sum_{n=2}^{\infty} n(n-1) a_n x^{n-2} + x^2 \sum_{n=0}^{\infty} a_n x^n = 0$$

$$\sum_{n=2}^{\infty} n(n-1) a_n x^{n-2} + \sum_{n=0}^{\infty} a_n x^{n+2} = 0$$

The powers of x in the series have different indices. Change summation variables as follows:

$$k = n - 2 \implies \sum_{n=2}^{\infty} n(n-1) a_n x^{n-2} = \sum_{k=0}^{\infty} (k+2)(k+1) a_{k+2} x^k$$

$$k = n + 2 \implies \sum_{n=0}^{\infty} a_n x^{n+2} = \sum_{k=2}^{\infty} a_{k-2} x^k$$

Since the choice of summation variable does not matter, relabel the k above back to an n to get

$$\sum_{k=0}^{\infty} (k+2)(k+1) a_{k+2} x^k = \sum_{n=0}^{\infty} (n+2)(n+1) a_{n+2} x^n$$

$$\sum_{k=2}^{\infty} a_{k-2}x^k \;=\; \sum_{n=2}^{\infty} a_{n-2}x^n$$

The series problem to solve is then

$$\sum_{n=0}^{\infty} (n+2)(n+1)a_{n+2}x^n + \sum_{n=2}^{\infty} a_{n-2}x^n \;=\; 0$$

Hence, we have, for all x, several pieces which must add up to zero: the $n=0$ and $n=1$ case which are isolated from the sum $\sum_{n=2}^{\infty}$ and the coefficients belonging to the full summation $\sum_{n=2}^{\infty}$. We have

$$2\cdot 1\, a_2\, x^0 + 3\cdot 2\, a_3\, x^1 + \sum_{n=2}^{\infty} \{(n+2)(n+1)a_{n+2} + a_{n-2}\}x^n \;=\; 0$$

The only way this can be true is if each of the coefficients of x^n vanishes. This gives us two isolated conditions and the **recursion relation** *the coefficients must satisfy.*

$$2a_2 \;=\; 0 \Longrightarrow a_2 = 0 \tag{28.6}$$

$$6a_3 \;=\; 0 \Longrightarrow a_3 = 0 \tag{28.7}$$

$$(n+2)(n+1)a_{n+2} + a_{n-2} \;=\; 0 \Longrightarrow a_{n+2} = -\frac{a_{n-2}}{(n+1)(n+2)}, \;\; \forall n \geq 2 \tag{28.8}$$

By direct calculation, we find

$$a_4 \;=\; -\frac{a_0}{3\cdot 4}$$

$$a_5 \;=\; -\frac{a_1}{4\cdot 5}$$

$$a_6 \;=\; -\frac{a_2}{5\cdot 6} = 0$$

$$a_7 \;=\; -\frac{a_3}{6\cdot 7} = 0$$

$$a_8 \;=\; -\frac{a_4}{7\cdot 8} = \frac{a_0}{3\cdot 4\cdot 7\cdot 8}$$

$$a_9 \;=\; -\frac{a_5}{8\cdot 9} = \frac{a_1}{4\cdot 5\cdot 8\cdot 9}$$

$$a_{10} \;=\; -\frac{a_6}{9\cdot 10} = 0$$

$$a_{11} \;=\; -\frac{a_7}{10\cdot 11} = 0$$

$$a_{12} \;=\; -\frac{a_8}{11\cdot 12} = -\frac{a_0}{3\cdot 4\cdot 7\cdot 8\cdot 11\cdot 12}$$

$$\vdots$$

In general, it is impossible to detect a pattern in these coefficients, but we can determine this:

$$a_2 = a_6 = a_{10} = 0 \;\;\Longrightarrow\;\; a_{4n+2} = 0,\; n \geq 0$$

$$a_3 = a_7 = a_{11} = 0 \;\;\Longrightarrow\;\; a_{4n+3} = 0,\; n \geq 0$$

$$a_4 \;=\; (-1)^1 \frac{a_0}{3\cdot 4}, \; a_8 = (-1)^2 \frac{a_0}{3\cdot 4\cdot 7\cdot 8}, \; a_{12} = (-1)^3 \frac{a_0}{3\cdot 4\cdot 7\cdot 8\cdot 11\cdot 12}$$

$$a_5 \;=\; (-1)^1 \frac{a_1}{4\cdot 5}, \; a_9 = (-1)^2 \frac{a_1}{4\cdot 5\cdot 8\cdot 9}, \; a_{13} = (-1)^3 \frac{a_1}{4\cdot 5\cdot 8\cdot 9\cdot 12\cdot 13}$$

We have found the solution is

$$
\begin{aligned}
y(x) &= a_0 + a_1 x + a_4 x^4 + a_5 x^5 + a_8 x^8 + a_9 x^9 + a_{12} x^{12} + a_{13} x^{13} + \cdots \\
&= a_0 + a_1 x + (-1)^1 \frac{a_0}{3 \cdot 4} x^4 + (-1)^1 \frac{a_1}{4 \cdot 5} x^5 \\
&\quad + (-1)^2 \frac{a_0}{3 \cdot 4 \cdot 7 \cdot 8} x^8 + (-1)^2 \frac{a_1}{4 \cdot 5 \cdot 8 \cdot 9} x^9 + \cdots \\
&= a_0 \left(1 + (-1)^1 \frac{1}{3 \cdot 4} x^4 + (-1)^2 \frac{1}{3 \cdot 4 \cdot 7 \cdot 8} x^8 + \cdots \right) \\
&\quad + a_1 \left(x + (-1)^1 \frac{1}{4 \cdot 5} x^5 + +(-1)^2 \frac{1}{4 \cdot 5 \cdot 8 \cdot 9} x^9 + \cdots \right)
\end{aligned}
$$

Define two new series by

$$
\begin{aligned}
y_0(x) &= 1 + (-1)^1 \frac{1}{3 \cdot 4} x^4 + (-1)^2 \frac{1}{3 \cdot 4 \cdot 7 \cdot 8} x^8 + \cdots \\
y_1(x) &= x + (-1)^1 \frac{1}{4 \cdot 5} x^5 + +(-1)^2 \frac{1}{4 \cdot 5 \cdot 8 \cdot 9} x^9 + \cdots
\end{aligned}
$$

We can thus write the solution $y(x)$ as

$$
y(x) = a_0 \, y_0(x) + a_1 \, y_1(x)
$$

Since the power series for y_0 and y_1 converge on $\Re$, we also know how to compute their derived series to get

$$
y_0'(x) = -\frac{4}{3 \cdot 4} x^3 + \cdots, \quad y_1'(x) = 1 - \frac{5}{4 \cdot 5} x^4
$$

Next, note $y_0(x)$ and $y_1(x)$ are linearly independent on $\Re$ as if

$$
c_0 \, y_0(x) + c_1 \, y_1(x) = 0, \quad \forall x
$$

then we also have

$$
c_0 \, y_0'(x) + c_1 \, y_1'(x) = 0, \quad \forall x
$$

Since these equations must hold for all x, in particular they hold for $x = 0$ giving

$$
\begin{aligned}
c_0 \, y_0(0) + c_1 \, y_1(0) &= c_0(1) + c_1(0) = 0 \\
c_0 \, y_0'(0) + c_1 \, y_1'(0) &= c_0(0) + c_1(1) = 0
\end{aligned}
$$

So the only solution is $c_0 = c_1 = 0$ implying these functions are linearly independent on $\Re$. Note y_0 is the solution to the problem

$$
y'' + x^2 y = 0, \quad y(0) = 1, \, y'(0) = 0
$$

and y_1 solves

$$
y'' + x^2 y = 0, \quad y(0) = 0, \, y'(0) = 1
$$

and, as usual the solution to

$$
y'' + y = 0, \quad y(0) = \alpha, \, y'(0) = \beta
$$

is the linear combination of the two linearly independent solutions y_0 and y_1 giving

$$y(x) \quad = \quad \alpha y_0(x) + \beta y_1(x)$$

Example 28.4.4 Solve the model $y'' + 6xy' + (9x^2 + 2)y = 0$ using power series methods.

Solution Again, the coefficient functions here are constants, so the power series solution can be computed at any point a and the radius of convergence will be $R = \infty$. Let's find a solution at $a = 0$. We assume $y(x) = \sum_{n=0}^{\infty} a_n x^n$. Thus,

$$y' \quad = \quad \sum_{n=1}^{\infty} n a_n x^{n-1}$$

$$y'' \quad = \quad \sum_{n=2}^{\infty} n(n-1) a_n x^{n-2}$$

The model we need to solve then becomes

$$\sum_{n=2}^{\infty} n(n-1) a_n x^{n-2} + 6x \sum_{n=1}^{\infty} n a_n x^{n-1} + (9x^2 + 2) \sum_{n=0}^{\infty} a_n x^n \quad = \quad 0$$

This can be rewritten as

$$\sum_{n=2}^{\infty} n(n-1) a_n x^{n-2} + 6 \sum_{n=1}^{\infty} n a_n x^n + 9 \sum_{n=0}^{\infty} a_n x^{n+2} + 2 \sum_{n=0}^{\infty} a_n x^n \quad = \quad 0$$

We change the summation variables in these series to get the powers of x to match in all the summations. This gives

$$k = n - 2 \quad \Longrightarrow \quad \sum_{n=2}^{\infty} n(n-1) a_n x^{n-2} = \sum_{k=0}^{\infty} (k+2)(k+1) a_{k+2} x^k$$

$$k = n + 2 \quad \Longrightarrow \quad \sum_{n=0}^{\infty} a_n x^{n-2} = \sum_{k=2}^{\infty} a_{k-2} x^k$$

Since the choice of summation variable does not matter, relabel the k above back to an n to get

$$\sum_{k=0}^{\infty} (k+2)(k+1) a_{k+2} x^k \quad = \quad \sum_{n=0}^{\infty} (n+2)(n+1) a_{n+2} x^n$$

$$\sum_{k=2}^{\infty} a_{k-2} x^k \quad = \quad \sum_{n=2}^{\infty} a_{n-2} x^n$$

The series problem to solve is then

$$\sum_{n=0}^{\infty} (n+2)(n+1) a_{n+2} x^n + 6 \sum_{n=1}^{\infty} n a_n x^n + 9 \sum_{n=2}^{\infty} a_{n-2} x^n + 2 \sum_{n=0}^{\infty} a_n x^n \quad = \quad 0$$

Hence, we have, for all x

$$(2a_2 + 2a_0) x^0 + (6a_3 + 6a_1 + 2a_1) x^1 +$$

$$\sum_{n=2}^{\infty} \{(n+2)(n+1)a_{n+2} + 6na_n + 9\,a_{n-2} + 2a_n\}x^n = 0,$$

The only way this can be true is if each of the coefficients of x^n vanishes. This gives us what is called a **recursion relation** *the coefficients must satisfy. For all $n \geq 0$, we have*

$$2a_2 + 2a_0 = 0 \Longrightarrow a_2 = -a_0$$
$$6a_3 + 8a_1 = 0 \Longrightarrow a_3 = -\frac{4}{3}a_1$$
$$(n+2)(n+1)a_{n+2} + 6na_n + 9\,a_{n-2} + 2a_n = 0 \Longrightarrow$$
$$a_{n+2} = -\frac{(6n+2)a_n + 9a_{n-2}}{(n+2)(n+1)}, \ \forall n \geq 2$$

We can thus compute the coefficients using this recurrence relationship and generate a solution we can write as $y(a_0, a_1, x)$. As usual, we define the two linearly independent solutions by $y_0(x) = y(1, 0, x)$ and $y_1(x) = y(0, 1, x)$. Hence, we can write the solution as

$$y(a_0, a_1, x) = a_0 + a_1 x + a_2 x^2 + \cdots = a_0\,y_0(x) + a_1\,y_1(x)$$

Our theorem guarantees that the solution $y(x)$ converges in $\Re$, so the series y_0 we get setting $a_0 = 1$ and $a_1 = 0$ converges too. Setting $a_0 = 0$ and $a_1 = 1$ then shows y_1 converges. We can easily show the power series for y_0 and y_1 converge on $\Re$, we also know how to compute their derived series to get

$$y'(a_0, a_1, x) = a_1 + 2a_2 x + \cdots$$

and so $y_0'(0) = y'(1, 0, 0) = 0$, $y_1'(0) = y'(0, 1, 0) = 1$. Next, note $y_0(x)$ and $y_1(x)$ are linearly independent on $\Re$ as if

$$c_0\,y_0(x) + c_1\,y_1(x) = 0, \quad \forall x \Longrightarrow c_0\,y_0'(x) + c_1\,y_1'(x) = 0, \quad \forall x$$

and since these equations must hold for all x, in particular they hold for $x = 0$ giving

$$c_0\,y_0(0) + c_1\,y_1(0) = c_0(1) + c_1(0) = 0$$
$$c_0\,y_0'(0) + c_1\,y_1'(0) = c_0\,(0) + c_1\,(1) = 0$$

which tells us $c_0 = c_1 = 0$ implying these functions are linearly independent on $\Re$. So you can see this argument is pretty standard. As usual y_0 is the solution to the problem

$$y'' + 6xy' + (9x^2 + 2)y = 0, \quad y(0) = 1, \ y'(0) = 0$$

Further, y_1 solves

$$y'' + 6xy' + (9x62 + 2)y = 0, \quad y(0) = 0, \ y'(0) = 1$$

The solution to

$$y'' + 6y' + 9y = 0, \quad y(0) = \alpha, \ y'(0) = \beta$$

is the linear combination of the two linearly independent solutions y_0 and y_1 giving

$$y(x) = \alpha y_0(x) + \beta y_1(x)$$

To finish up here, let's show you how to use MATLAB to compute these coefficients and to plot *the approximate solutions. First, plotting series solutions always means we plot partial sums and so we need efficient ways to compute polynomial evaluations. We will use Horner's method. A short example should show you what is going on. To evaluate $h(x) = 1 + 2x + 3x^2 + 4x^3$ the straightforward way requires 4 additions and 2 multiplies for the $2x$ term, 3 multiplies for the $3x^2$ term as we have to square x and finally 4 multiplies for the $4x^3$ term. This adds up to $4+2+3+4 = 13$ operations. If we use Horner's method, we organize the multiplies like this:*

$$
\begin{aligned}
a &= 3 + 4x \\
b &= 2 + x\,a = 2 + 3x + 4x^2 \\
c &= 1 + x\,b = 1 + 2x + 3x^2 + 4x^3
\end{aligned}
$$

Now let the coefficients of h be $A = [1; 2; 3; 4]'$. Then these computations can be organized as a loop.

Listing 28.4: **Sample Horner's Method Loop**

```
  p = A(n);
2 for k=n-1:-1:1
    p = p*x + A(k)
```

Note, unrolling this loop, this gives for n=4

Listing 28.5: **Sample Horner's Method Loop Details**

```
  p = A(4) = 4;
  % k = 3
  p = A(4)*x + A(3) = 4x + 3; one multiply, one add
  % k = 2
5 p = (4x+3)*x + A(2) = 4x^2 + 3x + 2; one multiple, one add
  %k = 1
  p = (4x^2+3x+2)*x + 1 = 4x^3 + 3x^2 +2x + 1
  for k=n-1:-1:1
    p = p*x + A(k)
```

So we can implement Horner's method using a loop. Note Horner's method here takes one multiply and one add for each of the three steps for a total of 6 operations. The savings is considerable when there are high powers of x in the polynomial. Our partial sums could easily go to powers such as x^{10} so using Horner's method is smart. The code is straightforward.

Listing 28.6: **Horner's Method**

```
1 function pval = HornerV(a,z)
  %
  % a     column vector of polynomial coefficients
  % z     the vector of points to evaluate polynomial at
  % pval  vector to hold polynomial values for each z value
6 %
  n = length(a);
  m = length(z);
  pval = a(n)*ones(m,1);
```

```
   for  k=n-1:-1:1
11    pval = z'.*pval + a(k);
   end
   end
```

For the problem at hand, we can write code to find the coefficients. This code will have to be edited for each problem of course. For this one, we use the function `PowerSeriesODE` to calculate the coefficients.

Listing 28.7: **Finding the Power Series Coefficients for the ODE**

```
   function A = PowerSeriesODE(a0,a1,n)
   %
   % a0, a1 = arbitrary constants
   % n = number of terms to find in PS solution
 5 %
   % Recurrence Relation to use
   % Coefficients are off by one
   % because of MATLAB's indexing
   A = zeros(n+1,1);
10 A(1) = a0;
   A(2) = a1;
   A(3) = - A(1);
   A(4) = -(4/3)*A(2);
   for i = 3:n+1
15    A(i+2) = -((6*i+2)*A(i)+9*A(i-2))/((i+1)*(i+2));
   end
```

We can then find y_0, y_1 and plot them as follows:

Listing 28.8: **Finding the Solutions to a Power Series ODE**

```
   X = linspace(0,1.7,31);
   % Get y_0 for 15 coefficients
   Y0 = PowerSeriesODE(1,0,15);
 4 % Get y_1 for 15 coefficients
   Y1 = PowerSeriesODE(0,1,15);
   %Evaluate both solutions on [0,1.7]
   p0 = HornerV(Y0,X);
   p1 = HornerV(Y1,X);
 9 % plot Y0 and Y1
   plot(X,p0,'-',X,p1,'+');
   xlabel('x');
   ylabel('Solution');
   title('Linearly Independent Solutions to y'''' + 6xy'' + (9x^2 + 2) y
       = 0 on [0,1.7]');
14 legend('Y0','Y1');
   % solve problem with ICs y(0) = -1, y'(0) = 2
   Y = PowerSeriesODE(-1,2,15);
   %Evaluate solution on [0,1.7]
   p = HornerV(Y,X);
19 % plot solution to IVP
```

```
plot(X,p);
xlabel('x');
ylabel('Solution');
title('Linearly Independent Solutions to y'''' + 6xy'' + (9x^2 + 2) y
    = 0 on [0,1.7], y(0) = -1, y''(0) = 2');
```

We see the two solutions y_0 and y_1 in Figure 28.3. The solution to the IVP is shown in Figure 28.4.

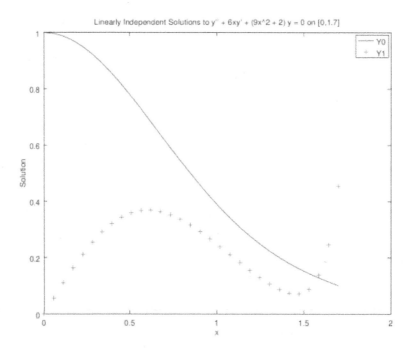

Figure 28.3: The solutions y_0 and y_1 to $y'' + 6xy' + (9x^2 + 2)y = 0$ on $[0, 1.7]$.

28.4.4 Homework

Exercise 28.4.3 *Solve*

$$y'' + 3xy' + 5y = 0$$

using the Power Series method.

1. *Use our theorems to find the radius of convergence R.*

2. *Find the recursion relationship.*

3. *Modify the PowerSeriesODE function to find the coefficients.*

4. *Find the two solutions y_0 and y_1 numerically.*

5. *Show y_0 and y_1 are linearly independent.*

6. *Write down the Initial Value Problem y_0 and y_1 satisfy.*

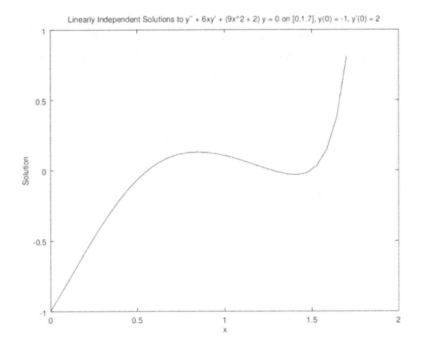

Figure 28.4: The solution to $y'' + 6xy' + (9x^2 + 2)y = 0$, $y(0) = -1, y'(0) - 2$ on $[0, 1.7]$.

7. *Provide plots for the solutions y_0 and y_1.*

8. *Find the solution to this model with $y(0) = -3$ and $y'(0) = 2$. numerically and provide plots.*

Exercise 28.4.4 *Solve*

$$y'' + (3x_1^2)y' + 5xy = 0$$

using the Power Series method.

1. *Use our theorems to find the radius of convergence R.*

2. *Find the recursion relationship.*

3. *Modify the PowerSeriesODE function to find the coefficients.*

4. *Find the two solutions y_0 and y_1 numerically.*

5. *Show y_0 and y_1 are linearly independent.*

6. *Write down the Initial Value Problem y_0 and y_1 satisfy.*

7. *Provide plots for the solutions y_0 and y_1.*

8. *Find the solution to this model with $y(0) = 1$ and $y'(0) = -2$. numerically and provide plots.*

Part IV

Summing It All Up

Chapter 29

Summary

We have now come to the end of these notes which cover a one year senior level course in basic analysis. There is much more to teach you and the next step is the material in (Peterson (15) 2020) which continues our coverage of $\Re^n$ in more detail. We have not covered all of the things we wanted to but we view that as a plus: there is more to look forward to! In particular, we hope we have encouraged your interest in these extensions of the ideas from the first volume on analysis to a bit more of calculus on $\Re^n$. Our aims here were to very carefully introduce you to a careful way of thinking where we pay close attention to the consequences of assumptions. This is a hard road and most of you are also taking a course or two in abstract algebra which asks you to think carefully also but in a very different way. You need both as the more tools at your disposal and the more ways you have of thinking about how to build a model or to phrase and prove a conjecture, the better off you are. Full understanding of the material in these notes comes slowly but if you are interested in using ideas like this in your future work, it is worth the intellectual effort. There are many books that discuss these ideas although fewer that use a tool such as MATLAB/ Octave as part of the discussion and you need to find the ones that fit your style of self learning. Old books are often useful as they are cheap and so you don't have to invest a lot. One thing you should remember is that it is a great idea to have 5 - 10 books in each subject on your bookshelf as each author will approach coverage of an idea differently and there is a lot to say for listening to the different voices. We have picked out a few oldies but goodies:

- **Introduction to Mathematical Analysis** (Douglas (2) 1996) is a classic analysis text about 20 years old which covers all the basics and has many more problems you can sink your teeth into.

- **Advanced Calculus: An Introduction to Analysis** (Fulks (6) 1978) is my favorite of the old books. It has great coverage, lots of problems and has a good balance between abstraction and problem solving. We highly recommend finding one used and reading through it carefully.

- **Introduction to Classical Real Analysis** (Stromberg (19) 1981) is a very hard book and the most advanced one we know. We used this to develop projects for this book. A typical problem here is one with 10 parts so it makes a great project if it is annotated carefully. Of course, Stromberg does not do that so that is why it is a hard read. It is extremely abstract but chock full of interesting things that are hardly covered anymore. You should invest in this one.

- **Schaum's Outline of Advanced Calculus** (R. Wrede and M. Spiegel (16) 2010) is a great book on how to use advanced calculus. Lots of worked out problems on multiple integration, line integrals, coordinate systems and transformations that are hard to find now. In our day, we took courses in mathematical physics that covered a lot of this. Also, there used to be courses called advanced engineering mathematics which were a lot like a course in mathematical physics but the course we call that now is not the same. You should get one of these used to have on your shelf.

- **Drawing Physics: 2,600 Years of Discovery from Thales to Higgs** (D. Lemons (1) 2018) is a fun read and shows you how drawing pictures to help capture abstraction is a time honored way to help us understand complicated things. This is highly recommended.

- **Things to Make and Do in the Fourth Dimension: A Mathematician's Journey Through Narcissistic Numbers, Optimal Dating Algorithms, at Least Two Kinds of Infinity and More** (M. Parker (8) 2015) is a joy. All the stuff we talk about in such careful ways is laid bare in interesting and amusing prose! Good for party conversation!

- **What Can Be Computed? A Practical Guide to the Theory of Computation** (J. MacCormick (7) 2018) talks about what kinds of computation we can do and how much time they will take. The level of abstraction is well within your reach at this point and since we believe solving interesting problems uses these ideas as well as deep mathematics and science, you should start working your way through a book like this. Do it side by side with code. You have read through the code we use in this text and it should have made you think a bit more about how to write efficient code. Here, we just use code to illuminate something and we rarely think about such a thing as efficiency. We occasionally talk about data structures but you should start to learn more.

Let's summarize what we have done in the first volume. The first volume (Peterson (13) 2020) is a critical course in the use of abstraction and is a primary text for a typical junior - senior year course in basic analysis. It can also be used as a supplementary text for anyone whose work requires that they begin to assimilate more abstract mathematical concepts as part of their professional growth after graduation or as a supplement to deficiencies in undergraduate preparation that leaves them unprepared for the jump to the first graduate level analysis course. Students in other disciplines, such as biology and physics, also need a more theoretical discussion of these ideas and the writing is designed to help such students use this book for self study as their individual degree programs do not have enough leeway in them to allow them to take this course. This text is for a two semester sequence.

First Semester: Sequences, Continuity and Differentiation: This semester is designed to cover through the consequences of differentiation for functions of one variable and to cover the basics leading up to extremal theory for functions of two variables. Along the way, a lot of attention is paid to developing the theory underlying these ideas properly. The usual introductory chapters on the real line, basic logic and set theory are not covered as students have seen that before and frankly are bored to see it again. The style here is to explain very carefully with many examples and to always leave pointers to higher level concepts that would be covered in the other texts. The study of convex functions and lower semicontinuity are also introduced so that students can see there are alternate types of smoothness that are useful.

The number e is developed and all of its properties from the sequence definition $e = \lim_n (1+1/n)^n$. Since ideas from sequences of functions and series are not yet known, all of the properties of the exponential and logarithm function must be explained using limit approaches which helps tie together all of the topics that have been discussed. The pointer to the future here is that the exponential and logarithm function are developed also in the second semester by defining the logarithm as a Riemann integral and all the properties are proven using other sorts of tools. It is good for the students to see alternate pathways.

Note that this text eschews the development of these ideas in a metric space setting although we do talk about metrics and norms as appropriate. It is very important to develop the derivative and its consequences on the real line and while there is a simplicity and economy of expression if convergence and so on is handled using a general metric, the proper study of differentiation in $\Re^n$ is not as amenable to the metric space choice of exposition, That sort of discussion is done in a later text. Chapters 1 - 14 are the first semester of undergraduate analysis.

Second Semester: Riemann Integration, Sequences and Series of Functions The second half of this text is about developing the theory of Riemann Integration and sequences and series of functions carefully. Also, student horizons are expanded a bit by showing them some basic topology in $\Re^2$ and $\Re^3$ and revisiting sequential and topological compactness in these higher dimensions. This ties in well with the last chapter of semester one. These ideas are used at the end to prove the pointwise convergence of the Fourier Series of a function which is a great application of these ideas. In the text, Chapters 15 - 28 are for the second semester.

Once the first year of training is finished, there is a lot more to do. The first book on basic analysis essentially discusses the abstract concepts underlying the study of calculus on the real line. A few higher dimensional concepts are touched on such as the development of rudimentary topology in $\Re^2$ and $\Re^3$, compactness and the tests for extrema for functions of two variables, but that is not a proper study of calculus concepts in two or more variables. A full discussion of the $\Re^n$ based calculus is quite complex and we do some of that in the second volume (Peterson (15) 2020) where the focus is on differentiation in $\Re^n$ and important concepts about mappings from $\Re^n$ to $\Re^m$ such as the inverse and implicit function theorem and change of variable formulae for multidimensional integration. These topics alone require much discussion and setup. These topics intersect nicely with many other important applied and theoretical areas which are no longer covered in mathematical science curricula. The knowledge here allows a quantitatively inclined person to more properly develop multivariable nonlinear ODE models for themselves and to learn the proper background to study differential geometry and manifolds among many other applications.

We think of the pathway to analysis competence as achievable using a sequence of courses. Hence, we have selected a set of core ideas to develop in five books all written in one voice so there is continuity in presentation and style. These core ideas are as follows:

- **Basic Analysis One**: This covers the fundamental ideas of calculus on the real line. Hence, sequences, function limits, continuity and differentiation, compactness and all the usual consequences as well as Riemann integration, sequences and series of functions and so forth. It is important to add pointers to extensions of these ideas to more general things regularly even though they are not gone over in detail. The problem with analysis on the real line is that the real line is everything: it is a metric space, a vector space, a normed linear space and an inner product space and so on. So many ideas that are actually separate things are conflated because of the special nature of $\Re$. **Basic Analysis One** is a two semester sequence which is different from the other basic analysis texts to be discussed.

- **Basic Analysis Two**: A proper study of calculus in $\Re^n$ is no longer covered in most undergraduate and graduate curricula. Most students are learning this on their own or not learning it properly at all even though they get a Master's or Ph.D. in mathematics. This course covers differentiation in $\Re^n$, integration in $\Re^2$ and $\Re^3$, the inverse and implicit function theorem and connections to these things to optimization. It is not possible to cover many things:

  – Vector calculus such as Stokes Theorem and surface integrals should be covered as part of a study of manifolds within a course on differential geometry. Courses in differential geometry are hard to find in most universities. Note these ideas are really useful in physics yet most mathematics departments no longer serve the needs of their physics colleagues in this matter. Further, more and more physics courses, chemistry courses and so forth are delayed and not taken until later than the sophomore year. Hence, the reinforcement of ideas from calculus by using them in science is being lost. For example, in a Physics 3 course on electromagnetics, Gauss's law which connects the surface integrals, line integrals and Stokes theorem into Maxwell's equations and circuits used to be a great way for Calculus 3 students to see how the ideas in their mathematics course were useful. However, the loss of coverage of these ideas in Calculus 3 has impacted how Physics 3 is taught which a loss of mathematical rigor there as well.

– Since $\Re^n$ integration and differentiation are not covered, if you teach optimization such as Kuhn Tucker theory, you do not have access to previous exposure to the inverse and implicit function theorem which means the full level of rigor can not be applied.

– Complex analysis requires a good understanding of complex line integrals which requires a good understanding of Riemann integration. Most students do not take real undergraduate analysis prior to taking complex analysis and so again the right level of rigor is not attained. Also, it is possible to develop much of this material using a power series approach which is done in the second semester of undergraduate real analysis which again most students do not have.

– The course called advanced engineering mathematics generally covers Stürm Liouville equation and the solution of linear Partial Differential equations (PDEs) using separation of variables. This requires the use of Fourier series and to truly understand why these series solutions solve the PDEs needs a careful discussion of the interchange of partial differentiation and series which is not covered at all in the first year of undergraduate real analysis.

Hence, even with the above partial list, one can see it is not possible to cover everything a student needs in a **linear progression of courses**. So what must be done is train students to think and read on their own. We believe training students to be able to do **self study** is a key requirement here.

- **Basic Analysis Three**: This covers the basics of three new kinds of spaces: metric spaces, normed linear spaces and inner product spaces. Since the students do not know measure theory at this point, many of the examples come from the sequence spaces ℓ^p. However, one can introduce many of the needed ideas on operators, compactness and completeness even with that restriction. This includes what is called **Linear Functional Analysis**. In general, the standards are discussed: the Hahn - Banach Theorem, the open and closed mapping theorems and some spectral theory for linear operators of various kinds.

- **Basic Analysis Four**: This covers the extension of what is meant by the length of an interval to more general ideas on the *length* or *measure* of a set. This is done abstractly first and then specialized to what is called **Lebesgue Measure**. In addition, more general notions are usually covered along with many ideas on the convergence of functions with respect to various notions.

- **Basic Analysis Five**: Here, we connect topology and analysis more clearly. There is a full discussion of topological and linear topological spaces, differential geometry, some degree theory and distribution theory.

So here's to the future! So enjoy your journey and get started!

You can see this analysis sequence laid out in Figure 29.1. There you see how all the material from this text and the others fits together. The arrangement of this figure should make it clear you to you both how much you now know and how much you can still learn! Since what we discuss in the first four volumes is still essentially what is a *primer* for the start of learning even more analysis and mathematics at this level, in Figure 29.1 we have explicitly referred to our texts using that label. The first volume is the one in the figure we call **A Primer On Analysis**, the second is **Primer Two: Escaping The Real Line**, the third is **Primer Three: Basic Abstract Spaces**, the fourth (this text) is **Primer Four: Measure Theory** and the fifth is *Primer Five: Functional Analysis.* and from that

Figure 29.1: The general structure of the five core analysis courses.

you can plan out your studies.

Part V

References

References

[1] D. Lemons. *Drawing Physics: 2,600 Years of Discovery from Thales to Higgs*. MIT Press, 2018.

[2] S. Douglas. *Introduction to Mathematical Analysis*. Addison-Wesley Publishing Company, 1996.

[3] E. Ince. *Ordinary Differential Equations*. Dover Books On Mathematics, 1956.

[4] John W. Eaton, David Bateman, Søren Hauberg, and Rik Wehbring. *GNU Octave version 5.2.0 manual: a high-level interactive language for numerical computations*, 2020. URL https://www.gnu.org/software/octave/doc/v5.2.0/.

[5] Free Software Foundation. *GNU General Public License Version 3*, 2020. URL http://www.gnu.org/licenses/gpl.html.

[6] W. Fulks. *Advanced Calculus: An Introduction to Analysis*. John Wiley & Sons, third edition, 1978.

[7] J. MacCormick. *What Can Be Computed? A Practical Guide to the Theory of Computation*. Princeton University Press, 2018.

[8] M. Parker. *Things to Make and Do in the Fourth Dimension: A Mathematician's Journey Through Narcissistic Numbers, Optimal Dating Algorithms, at Least Two Kinds of Infinity and More*. Farrar, Straus and Giroux, 2015.

[9] MATLAB. *Version Various (R2010a) - (R2019b)*, 2018 - 2020. URL https://www.mathworks.com/products/matlab.html.

[10] R. McElreath and R. Boyd. *Mathematical Models of Social Evolution: A Guide for the Perplexed*. University of Chicago Press, 2007.

[11] J. Peterson. *Basic Analysis V: Functional Analysis and Topology*. CRC Press, Boca Raton, Florida 33487, 2020.

[12] J. Peterson. *Basic Analysis IV: Measure Theory and Integration*. CRC Press, Boca Raton, Florida 33487, 2020.

[13] J. Peterson. *Basic Analysis I: Functions of a Real Variable*. CRC Press, Boca Raton, Florida 33487, 2020.

[14] J. Peterson. *Basic Analysis III: Mappings on Infinite Dimensional Spaces*. CRC Press, Boca Raton, Florida 33487, 2020.

[15] J. Peterson. *Basic Analysis II: A Modern Calculus in Many Variables*. CRC Press, Boca Raton, Florida 33487, 2020.

[16] R. Wrede and M. Spiegel. *Schaum's Outline of Advanced Calculus*. McGraw Hill, 2010.

[17] H. Sagan. *Advanced Calculus of Real Valued Functions of a Real Variable and Vector - Valued Functions of a Vector Variable*. Houghton Mifflin Company, 1974.

[18] G. Simmons. *Introduction to Topology and Modern Analysis*. McGraw-Hill Book Company, 1963.

[19] K. Stromberg. *Introduction to Classical Real Analysis*. Wadsworth International Group and Prindle, Weber and Schmidt, 1981.

Part VI

Detailed Index

Index